Food Fortification

In a world that is constantly evolving, our understanding of nutrition and its impact on human health has grown exponentially. Food, once merely a source of sustenance, is now recognized as a powerful tool for improving public health and well-being. Organized into four sections, ***Food Fortification: Trends and Technologies*** presents a comprehensive exploration of food fortification—from its historical roots to its modern applications.

Part I introduces the concept of food fortification as a potential strategy for the control of micronutrient malnutrition and the role of micronutrients in human health, recommended dietary allowance, and source. It also details the deficiency, prevalence, populations under risk, and factors contributing to micronutrient deficiency. Part II summarizes the prevalence, causes, and consequences of vitamin deficiencies. It lays a framework for national and international fortification programs. In addition, it provides information about case studies, the impact of fortification on food textural and sensory properties, as well as challenges with currently used fortification methods.

Part III provides technical information on various minerals that can be used to fortify foods, including their chemistry, absorption, metabolism, and biological role. It also reviews their applications in specific food vehicles. Part IV describes the key steps involved in food bioactive fortification. This section also deals with the fortification of multigrain flour and challenges associated with PUFA fortification. It also highlights the important roles of encapsulation on bioavailability, with examples of fortification in dairy, egg, bakery, confectionery, and other products.

This book delves into the critical realm of fortifying our food supply to address the complex nutritional challenge and is a tribute to the progress that has been made in food fortification over the past few decades, as well as a call to action for the work that still lies ahead.

Food Fortification

Trends and Technologies

Edited by
Khalid Bashir, Kulsum Jan,
Vaibhav Kumar Maurya, and Amita Shakya

CRC Press
Taylor & Francis Group
Boca Raton London New York

CRC Press is an imprint of the
Taylor & Francis Group, an **informa** business

First edition published 2024
by CRC Press
2385 NW Executive Center Drive, Suite 320, Boca Raton FL 33431

and by CRC Press
4 Park Square, Milton Park, Abingdon, Oxon, OX14 4RN

CRC Press is an imprint of Taylor & Francis Group, LLC

© 2024 selection and editorial matter, Khalid Bashir, Kulsum Jan, Vaibhav Kumar Maurya, and Amita Shakya, individual chapters, the contributors

ISBN: 978-0-367-72392-7 (hbk)
ISBN: 978-0-367-74987-3 (pbk)
ISBN: 978-1-003-16066-3 (ebk)

DOI: 10.1201/9781003160663

Typeset in Times
by Deanta Global Publishing Services, Chennai, India

Dedicated To Our Beloved Parents

*"My Lord! Bestow On Them Thy Mercy as They Had Mercy Upon Us,
When We Were Small"*

(Aameen)

Contents

PART I Fortification: Necessity and Strategy

PART II Vitamin Fortification

PART III Mineral Fortification

PART IV Other Food Bioactive Fortification

Foreword

It is my great pleasure to write the foreword for *Food Fortification: Trends and Technologies*. The aim of this book was to provide a single resource for academic, industry, and government scientists on the different approaches available to fortify foods with bioactive agents, such as vitamins, minerals, nutraceuticals, and probiotics. Food fortification is required to reduce micronutrient deficiencies that negatively impact human health and wellbeing, as well as to boost the beneficial health effects of foods. Micronutrient malnutrition is currently a major cause of chronic disease around the globe and could be addressed using well-designed foods. Fortification of foods with nutraceuticals and probiotics may be able to improve health by reducing the risk of chronic diseases (like obesity, diabetes, chronic heart disease, hypertension, cancer, or eye disease) or increasing human performance (like mood, focus, and energy levels). However, functional foods and beverages containing these bioactive agents must be carefully designed to ensure they are affordable, accessible, convenient, and desirable, as well as to ensure that the bioactive agents are bioavailable, safe, and efficacious. This book brings together a diverse range of researchers around the world who are experts in different aspects of the design, testing, and application of encapsulation technologies for fortification of foods. These authors focus on the fortification of foods with different kinds of vitamins, minerals, phytochemicals, polyunsaturated fatty acids, dietary fibers, bioactive peptides, and probiotics.

The book provides a comprehensive exploration of food fortification—from its historical roots to its modern applications. The authors have shown the importance of advanced food fortification strategies to tackle malnutrition and other diet-related diseases. The delicate balance between providing essential nutrients while avoiding overconsumption, the cultural nuances of dietary habits, and the global effort to address specific nutrient deficiencies are explored. This book contains a summary of the knowledge obtained by food scientists, nutritionists, and health professionals who have dedicated their time to advancing the science and practice of food fortification. The book will serve as a valuable resource for anyone seeking a deeper understanding of this important field, including students, academics, food manufacturers, healthcare professionals, and policymakers.

Micronutrient deficiency causes a wide range of non-specific physiological impairments in addition to the more evident clinical manifestations. These impairments include decreased resistance to infections, metabolic problems, and delayed or impaired physical and mental development. Micronutrient deficiency has enormous adverse effects on public health, and it is particularly important when developing plans for combating diseases like HIV/AIDS, malaria, tuberculosis, and other chronic conditions linked to diet. Contrary to common belief, micronutrient malnutrition is not just a problem for developing nations, with problems also being observed in certain populations in developed nations. This is especially true of iron and iodine deficiency, which is currently the most common micronutrient deficiency in the world. Additionally, increasing consumption of highly processed, energy-dense foods with low levels of micronutrients in industrialized nations (and increasingly in those undergoing social and economic change) is expected to have a negative impact on micronutrient status and intake. The simplest way to avoid micronutrient deficiency is to eat a balanced diet that provides adequate amounts of each nutrient. Unfortunately, this is far from being possible everywhere since it requires access to appropriate types and amounts of nutrient-rich foods, as well as suitable dietary habits. From this perspective, food fortification has the twin benefit of being able to provide micronutrients to significant populations without demanding significant adjustments to eating habits. In fact, fortification has been used in industrialized countries for more than 80 years as a way to replace micronutrients lost during food processing, particularly some of the B vitamins. This practice has played a significant role in the eradication of chronic diseases linked to deficiencies in these vitamins. A growing number of developing nations are now committed to, or are considering, fortification programs as a result of increased awareness of the widespread prevalence and detrimental effects of micronutrient malnutrition, as well as taking into account changes in food systems (such as a greater reliance on centrally processed foods). Indeed, highly successful fortification programs have already been established in some regions.

This book is organized into four sections. Part I introduces the concept of food fortification as a potential strategy for the control of micronutrient malnutrition and the role of micronutrients in human health, recommended dietary allowance, and sources. It also details the deficiency, prevalence, populations at risk, and factors contributing to micronutrient deficiency. Part II summarizes the prevalence, causes, and consequences of vitamin deficiencies. It lays a framework for national and international fortification programs. In addition, it provides information about case studies, the impact of fortification on food textural and sensory properties, as well as challenges with currently used fortification methods. This section also deals with the use of nanotechnologies for fortification and bioavailability. Part III provides technical information on various minerals that can be used to fortify foods, including their chemistry, absorption, metabolism, and biological role. It also reviews their applications in specific food vehicles. Part IV describes the key steps involved in food bioactive fortification. This section also deals with the fortification of multigrain flour and challenges associated with PUFA fortification. It highlights the important roles of encapsulation on bioavailability, with examples of fortification in dairy, egg, bakery, confectionery, and other products. Information about the implementation of monitoring and evaluating systems of micronutrient levels in national and international programs is also provided.

This book is a tribute to the progress that has been made in food fortification over the past few decades, as well as a call to action for the work that still lies ahead. Together, we can build a world where every person has access to the nutrients they need to thrive, and where food fortification is not just a concept but a cornerstone of a healthier, more equitable future.

David J. McClements
Distinguished Professor
University of Massachusetts

Editors

Dr. Khalid Bashir (Ph.D.) graduated from the Islamic University of Science and Technology (IUST), J&K, as founding batch and has a doctorate from the National Institute of Food Technology Entrepreneurship and Management (NIFTEM), Haryana, India. Dr. Khalid was the recipient of a Senior Research Fellowship from CSIR, Government of India, during his PhD. He works as an assistant professor in the Department of Food Technology, Jamia Hamdard, New Delhi, India. Dr. Khalid has published many research and review papers and books with national and international publishers and has delivered several invited, oral, and poster presentations. He is the recipient of UGC startup grant and serves as Academic Counselor for PG Diplomas (FSQM & DMT) offered by IGNOU. He is also program coordinator of the Diploma in Bakery and Confectionery Technology. He is a life member of the Association of Food Scientists and Technologists India (AFSTI); Nutrition Society of India (NSI); Indian Science Congress Association (ISCA); Probiotic Association of India; Association of Official Analytical Chemists (AOAC) India; and Indian Meat Science Association (IMSA). Dr. Khalid has edited books with Springer, Elsevier, CRC, AAP, and RSC.

Dr. Kulsum Jan (Ph.D.) is presently working as an assistant professor at the Department of Food Technology, Jamia Hamdard. She is the recipient of a Senior Research Fellowship from CSIR, Government of India, with Young Scientist, Best Paper, and Best Poster awards at various national and international conferences. Dr. Jan has been a part of ePG Pathshala, a program launched by MHRD, Government of India. She was also awarded a Post-Doctoral Fellowship from CSIR, Government of India. She has done exceptionally well in the department by receiving various awards for outstanding research. She was granted a UGC research project for the valorization of agricultural wastes. Dr. Jan has also served as an organizing committee member of various conferences organized by the Department of Food Technology, Jamia Hamdard. Dr. Jan acts as an Academic Councilor for the IGNOU course PG Diploma in FSQM, and is a life member of the Association of Food Scientists and Technologists India (AFSTI) and the Probiotic Association of India. Dr. Jan has published her research in several reputed and peer-reviewed journals along with several book chapters in edited books published by national and international publishers.

Dr. Vaibhav Kumar Maurya (Ph. D.) is presently associated as Field Application Specialist with the leading multi-national company PerkinElmer. He completed his Ph.D. at the National Institute of Food Technology Entrepreneurship and Management, Sonepat, Haryana, India, in vitamin research and food fortification. He is a knowledgeable person with hands-on expertise in various cutting edge techniques such as LC-MS/MS, GC-MS/MS, and HPLC for nutrient and pesticide residue analysis. He is the author of many high-impact scientific publications in peer-reviewed journals and book chapters. He is the editor of the book series *Micro and Nanoengineering in Food Science* (Springer).

Dr. Amita Shakya (Ph.D.) is presently serving as an Assistant Professor at Amity University Chhattisgarh. She completed her Ph.D. at the Dept. of Agriculture and Environmental Sciences of the National Institute of Food Technology Entrepreneurship and Management, Sonepat, Haryana, India. She is an interdisciplinary researcher with more than ten years of experience and is presently exploring the food–environment–health nexus. She is the author of many scientific publications and book chapters in peer-reviewed high-impact journals and international publishers.

Contributors

N. Arno
BET Bioscience Extraction Technologies Inc.,
Abbotsford, B.C. Canada

C. G. Awuchi
Department of Physical Sciences
Kampala International University
Kampala, Uganda

S. Ayushi
Department of Food Science and Technology
National Institute of Food Technology
    Entrepreneurship and Management
Kundli Sonepat, India

R.I. Barbhuiya
Department of Food Process Engineering
National Institute of Technology Rourkela
Rourkela, India

K. Bashir
Department of Food Technology, Jamia
    Hamdard
New Delhi, India

J. Bora
Department of Food Technology
Jamia Hamdard, India

Komal Chauhan
Department of Food Science and Technology
National Institute of Food Technology
    Entrepreneurship and Management
Haryana, India

R.S. Das
Section of Food and Nutrition
School of Agriculture and Food Science
University College Dublin
Belfield, Ireland
and
Teagasc, Food Research Centre
Ashtown, Ireland

M. Habib
Department of Food Technology
Jamia Hamdard, India

M. Hoque
School of Food and Nutritional Sciences
University College Cork
Cork, Ireland
and
Teagasc Food Research Centre
Ashtown, Ireland

K. Jan
Department of Food Technology
Jamia Hamdard, India

S. Jan
Department of Food Science and Technology
National Institute of Food Technology
    Entrepreneurship and Management
Kundli Sonepat, India

S. Joshi
Department of Food Technology
Jamia Hamdard, India

D. Karley
Amity Institute of Biotechnology
Amity University Chhattisgarh
Raipur, India

P. Karthik
Department of Food Technology
Faculty of Engineering
Karpagam Academy of Higher Education
Coimbatore, India

A. L. Khan
Chemical Engineering Division
Bhabha Atomic Research Centre
Mumbai, India

H. K. Khurana
Department of Food Technology
Jamia Hamdard, India

P. Kiran Kumar
Department of Chemistry
ATME College of Engineering
Mysuru, India

V. Kiran
Department of Food Science and Technology
National Institute of Food Technology,
    Entrepreneurship and Management
Kundli, Sonepat, India

V. Kolla
Amity Institute of Biotechnology
Amity University Chhattisgarh
Raipur, India

V. K. Maurya
Centre for Food Research and Analysis
National Institute of Food Technology
    Entrepreneurship and Management
Haryana, India
and
India Field Application Specialist
PerkinElmer, India

D. J. McClements
University of Massachusetts
Amherst, USA

A. Mobeen
Informatics and Big Data
CSIR – Institute of Genomics and Integrative
    Biology
New Delhi, India

S. Morya
Department of Food Technology & Nutrition
Lovely Professional University
Phagwara, India

D. Nath
Department of Food Process Engineering
National Institute of Technology Rourkela
Rourkela, India

K. K. Nayak
Amity Institute of Biotechnology
Amity University Chhattisgarh
Raipur, India

S. Nazir
Department of Food Technology
Jamia Hamdard
New Delhi, India

I. Qureshi
Department of Food Technology
Jamia Hamdard, India

S. M. Rahman
Department of Food Technology
Jamia Hamdard, India

D. Sandhu
Department of Food Technology and Nutrition
School of Agriculture
Lovely Professional University
Phagwara, India

A. Shakya
Amity Institute of Biotechnology
Amity University Chhattisgarh
Raipur, India

S. Singh
Department of Food Technology
Jamia Hamdard, India

D. Swarnima
Amity Institute of Food Technology
Amity University
Noida, India

N. Tabassum
Department of Post-Harvest Engineering and
    Technology
Faculty of Agricultural Sciences
Aligarh Muslim University
Aligarh, India

M. Vijaykrishnaraj
School of Food Science and Biotechnology
Zhejiang Gongshang University
Hangzhou, China

K. Yogesh
Department of Food Technology
Faculty of Science and Humanities
SRM University
Delhi-NCR Sonipat, India

Preface

In a world that is constantly evolving, our understanding of nutrition and its impact on human health has grown exponentially. Food, once merely a source of sustenance, is now recognized as a powerful tool for improving public health and well-being. This book, *Food Fortification: Trends and Technologies*, delves into the critical realm of fortifying our food supply to address the complex nutritional challenges facing our global population. Today, a dietary vitamin and mineral deficiency contributes significantly to the micronutrient deficiencies that affect more than 2 billion people worldwide. The severity of these deficiencies and their health effects, particularly in young children and pregnant women as they impair fetus and child growth, cognitive development, and infection resistance, are what make them important for public health. Although people from all demographic groups and geographical areas may be impacted, the cause of the most prevalent and serious issues are usually found amongst poor, food-insecure, and vulnerable households in developing nations. The prevalence of infectious diseases, lack of access to a range of foods, ignorance of proper dietary practises, and poverty are the major contributing causes. Thus, micronutrient deficiency is a significant barrier to socioeconomic growth, perpetuating a cycle of underdevelopment and to the detriment of already underprivileged groups. Its long-lasting consequences on learning capacity, productivity, and health are accompanied by substantial social and public expenditures that limit labour capacity as a result of high rates of illness and disability.

The motivation to compile this book arises from a profound awareness of the significant role that food fortification plays in shaping the health and well-being of populations, especially those in vulnerable and resource-limited settings. While remarkable progress has been made in this arena, there remain considerable challenges and opportunities to optimize the fortification of foods in a manner that benefits all segments of society. This book delves into the economic, regulatory, and ethical considerations that shape food fortification and policies at the global, national, and local levels. As readers turn the pages, a comprehensive strategy on the food fortification techniques, micronutrient deficiency, chemistry, absorption, metabolism of vitamins and minerals, along with the challenges that have arisen will be explored.

Our mission is clear: to shed light on the transformative power of food fortification, to inspire innovation, and to empower individuals and communities to make informed choices about their diets. Through this collective effort, we aspire to contribute to a healthier world where food fortification is recognized as a cornerstone of public health and a symbol of our commitment to nourishing the global population.

We extend our gratitude to the dedicated authors and contributors who have shared their expertise and experiences to make this book possible. We also thank the readers for their interest in this critical topic, and we hope that the knowledge imparted here will catalyse positive change in the way we think about and utilize food fortification.

Together, let us embark on a journey toward a brighter and healthier future for all through the science and practice of food fortification.

Editors

Acknowledgments

We would like to express our deepest gratitude to the authors who played a significant role in the creation and completion of this book. Their support, encouragement, and contributions have been invaluable.

First and foremost, we want to thank our families for their unwavering support throughout this journey. Their patience, understanding, and belief in our work have been our greatest source of strength.

We extend our appreciation to the talented individuals who have generously shared their insights and experiences during the writing process. Your expertise has added depth and authenticity to the pages of this book.

We are indebted to Department of Food Technology, Jamia Hamdard, Department of Biotechnology, Amity University, Chhattisgarh, and PerkinElmer, New Delhi, for providing us with the resources and environment necessary for this endeavor. Their support has been instrumental in allowing us to focus on our work.

We are immensely proud to acknowledge the extraordinary achievement of our scholar authors. Their dedication, creativity, and hard work have culminated in this literary masterpiece that we are honored to present to the world.

Our heartfelt thanks go out to our seniors, friends, and colleagues who offered their encouragement, feedback, and a listening ear throughout this journey. Their friendship has been a constant source of motivation.

Lastly, we want to express our gratitude to the readers of this book. Your interest in the subject matter and your engagement with the content are what make the effort of writing worthwhile.

Thank you all for being a part of this book's journey. Your contributions, whether big or small, have made a significant impact.

Editors

Part I

Fortification

Necessity and Strategy

1 Introduction to Food Fortification

Shumaila Jan, Mehvish Habib, Sakshi Singh,
Vaibhav K. Maurya, and Kulsum Jan

1.1 INTRODUCTION

One of the biggest issues throughout the globe is malnutrition, and it is mostly children and lactating women who are particularly at risk. Malnutrition has been closely linked to poor immunological response, poor muscular and respiratory function, overall increased problems, slow wound healing, protracted rehabilitation, longer duration of time in the hospital, and higher death rates. It may be described in terms of undernutrition, overnutrition, and micronutrient deficiencies (Table 1.1 and 1.2). It appears in a variety of ways, from hunger to obesity (Siddiqui et al., 2020, Bailey et al., 2015). One of the important Sustainable Development Goals (SDGs) of the United Nations is to eradicate hunger. The measurement of hunger is in itself a difficult problem. International organizations including the Food and Agriculture Organization (FAO), World Health Organization (WHO), and World Food Programme (WFP) describe hunger in different ways; these include chronic undernourishment, a shortage of food supply, food insecurity, reduced food intake accompanied by physical symptoms brought on by hunger, and continual anxiety about where and when one's next meal will be. Long-term hunger causes undernourishment and mortality. While undernutrition is a cause of hunger, undernutrition is not always present when there is no sign of hunger. The global hunger index is presented in Figure 1.1 and Table 1.3. Simply stated, high intakes, inadequate dietary intake, and improper nutrient utilization all contribute to malnutrition. It is likely that several micronutrient deficits are widespread in certain regions of the globe and in specific demographic groups on the basis of what is known about the occurrence of deficits in individual micronutrients (Grebmer et al., 2019).

People who eat meals with low nutritional quality or who have increased nutrient needs because of rapid development rates and/or the existence of bacterial illnesses or parasites are more prone to have coexisting micronutrient deficits. Low intakes of bioavailable zinc and iron, retinol (preformed vitamin A), calcium, vitamin B6, vitamin B2 (riboflavin), and vitamin B12 are particularly common in diets low in foods from animal sources. Fresh fruits and vegetables are often absent from meals of low quality, which suggests that vitamin (ascorbic acid), beta-carotene (provitamin A), and folate intake will also be insufficient. When grains are milled, a number of minerals are lost, including folate, iron, zinc, and different B vitamins (thiamine, riboflavin, and niacin). Thus, those who consume a lot of refined cereals are more likely to lack any one of these micronutrients. It is most probable that the levels of B vitamins, selenium, vitamin A (retinol), and iodine contents in the breast milk of underfed nursing mothers with a restricted diet and various micronutrient deficits would be low (Norman et al., 2021).

Vitamin A deficiency is a widespread nutritional disorder affecting millions of people, particularly in developing regions with limited access to diverse diets (Figures 1.2 and 1.3). This deficiency arises from an inadequate intake or poor absorption of vitamin A, a vital fat-soluble nutrient essential for maintaining proper vision, immune function, and cellular growth. Insufficient levels of vitamin A can lead to various health complications, including night blindness, dry and damaged

TABLE 1.1

Prevalence of Iron Deficiency (Our World in Data 2023)

African countries	% deficiency	European countries	% deficiency
Algeria	34.30%	Austria	14.60%
Botswana	43.30%	Bulgaria	24.10%
Burkina Faso	76.60%	Belaris	17.80%
Central African Republic	73.60%	Finland	11.80%
Chad	66.30%	France	14.70%
Democratic republic of Congo	64.90%	Germany	15.10%
Egypt	32.20%	Greece	16.70%
Ethiopia	52.10%	Hungary	18.10%
Kenya	42.80%	Kazakhstan	23%
Libya	26.60%	Norway	13.50%
Morocco	30.40%	Portugal	14.30%
Mali	79%	Romania	27.10%
Madagascar	49.50%	Russia	21.90%
Mozambique	68.20%	Sweden	14.60%
Mauritania	65.50%	Syria	32.90%
Niger	72%	Spain	14.60%
Nigeria	68.90%	Ukraine	25.60%
Namibia	46.10%		
Sudan	50.80%		
Somalia	51.80%		
South Africa	44.40%		
Zambia	55.10%		
Zimbabwe	37.80%		
Asian countries		**North America**	
Afghanistan	44.90%	Canada	13.20%
Armenia	19.60%	United state	6.10%
Bangladesh	43.10%	Mexico	21.70%
Bhutan	44.70%	**South America**	
Cambodia	49%	Argentina	19%
Georgia	26.60%	Bolivia	36.90%
India	53.40%	Colombia	22.20%
Kyrgyzstan	33.40%	Ecuador	23.50%
Magnolia	21.70%	Peru	29.60%
Myanmar	49.60%	Uruguay	25.10%
Nepal	44.60%	Venezuela	27.90%
Oman	24.30%		
Pakistan	53%		
Uzbekistan	21.90%		
Vietnam	22.90%		
Yemen	79.50%		

skin, impaired immunity, and an increased risk of infectious diseases, particularly in children and pregnant women. Addressing vitamin A deficiency requires a comprehensive approach, emphasizing dietary diversification, fortified food programs, and public health awareness campaigns to alleviate its significant impact on global health and wellbeing (Norman et al., 2021).

TABLE 1.2

Global Prevalence of Vitamin A (Our World in Data 2023)

African countries	% deficiency	European countries	% deficiency
Algeria	15.70%		
Botswana	26.10%	Bulgaria	18.30%
Burkina Faso	54.30%	Belarus	17.40%
Central African Republic	68.20%	Hungary	7.00%
Chad	50.10%	Kazakhstan	27.10%
Congo	61.10%	Romania	16.20%
Egypt	11.90%	Ukraine	23.80%
Ethiopia	46.10%		
Kenya	84.40%		
Libya	9.00%		
Morocco	40.40%	**North America**	
Mali	58.60%	Canada	-
Madagascar	42.10%	United state	-
Mozambique	68.80%	Mexico	26.80%
Mauritania	47.70%	**South America**	
Niger	67%	Argentina	14.30%
Nigeria	29.50%	Bolivia	21.80%
Namibia	17.50%	Colombia	5.90%
Sudan	27.80%	Ecuador	14.70%
Somalia	61.70%	Peru	14.90%
South Africa	16.90%	Uruguay	11.90%
Zambia	54.10%	Venezuela	9.40%
Zimbabwe	35.80%		
Asian countries			
Afghanistan	64.50%	Syria	12.10%
Armenia	0.60%	Myanmar	36.70%
Bangladesh	27.70%	Nepal	32.30%
Bhutan	22.00%	Oman	5.50%
Cambodia	22.30%	Pakistan	12.50%
Georgia	30.90%	Uzbekistan	53.10%
India	62.00%	Vietnam	12.00%
Kyrgyzstan	26.30%	Yemen	27.00%
Mognolia	19.80%	Russia	14.10%

Maternal and child undernutrition is a critical global health issue for pregnant and lactating women, as well as infants and young children. Insufficient access to nutritious food, inadequate healthcare, and poor sanitation and hygiene practices contribute to this problem. Maternal undernutrition can lead to complications during pregnancy, babies with low birth weight, and increased maternal mortality rates, while child undernutrition results in stunted growth, wasting, and deficiencies in essential nutrients. These conditions have severe and long-lasting consequences on physical and cognitive development, weaken the immune system, and increase the risk of infections and diseases. Combating maternal and child undernutrition requires a comprehensive approach involving nutrition education, improved healthcare access, and strategies to enhance food security and hygiene, ensuring a healthier future for both mothers and children (Heidkamp et al., 2021).

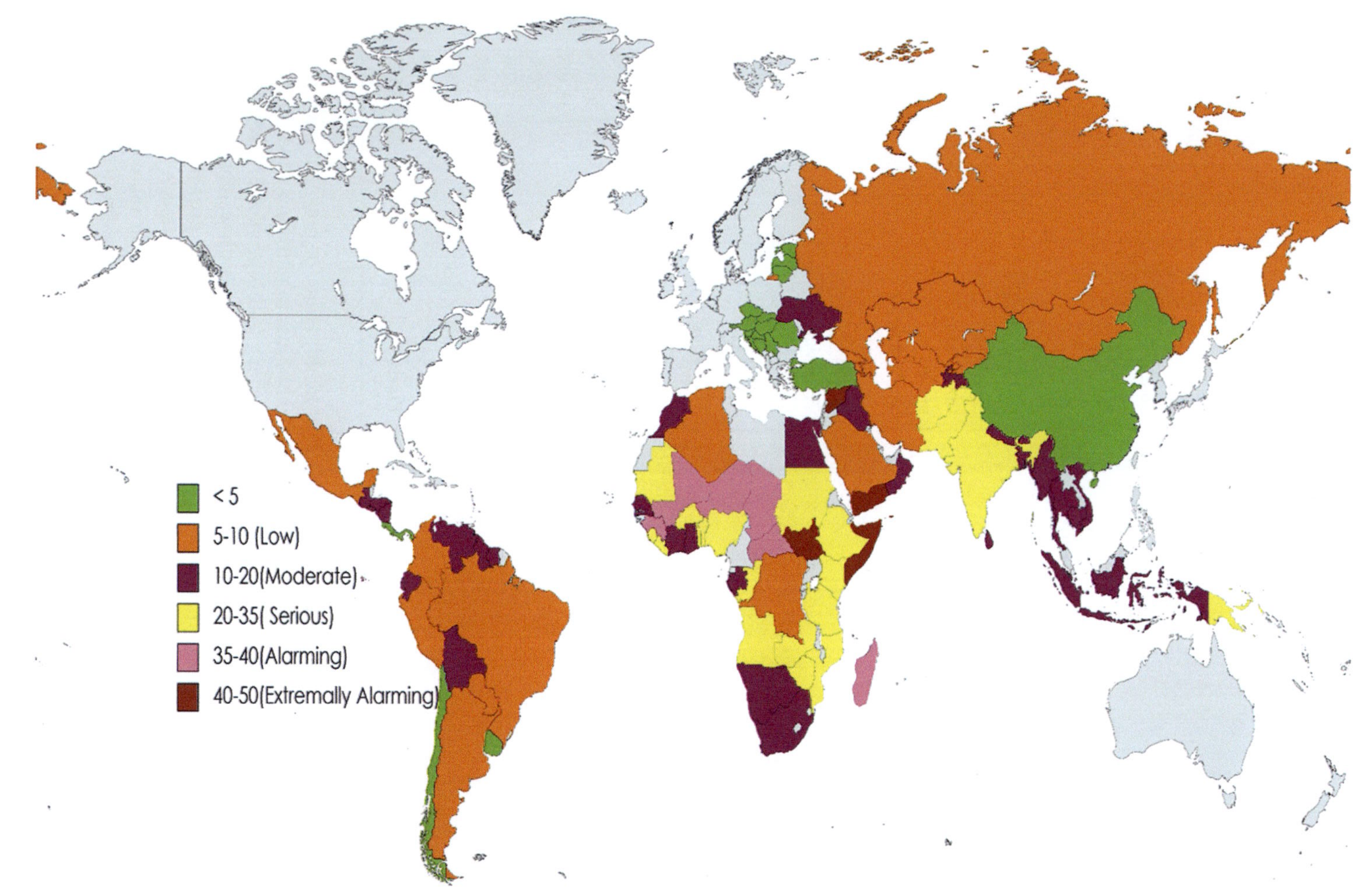

FIGURE 1.1   Global Hunger Index (Source: Our World in Data, 2023).

TABLE 1.3

Global Prevalence of Preschool Global Hunger index (Our World in Data 2023)

African countries	% deficiency	European countries	% deficiency
Algeria	15.70%	Austria	-
Botswana	26.10%	Bulgaria	18.30%
Burkina Faso	54.30%	Belarus	17.40%
Central African Republic	68.20%	Finland	-
Chad	50.10%	France	-
Democratic republic of Congo	61.10%	Germany	-
Egypt	11.90%	Greece	-
Ethiopia	46.10%	Hungary	7.00%
Kenya	84.40%	Kazakhstan	27.10%
Libya	9.00%	Norway	-
Morocco	40.40%	Portugal	-
Mali	58.60%	Romania	16.20%
Madagascar	42.10%	Russia	14.10%
Mozambique	68.80%	Sweden	-
Mauritania	47.70%	Syria	12.10%
Niger	67%	Spain	-
Nigeria	29.50%	Ukraine	23.80%
Namibia	17.50%		
Sudan	27.80%		
Somalia	61.70%		
South Africa	16.90%		
Zambia	54.10%		
Zimbabwe	35.80%		
Asian countries		**North America**	
Afghanistan	64.50%	Canada	-
Armenia	0.60%	United state	-
Bangladesh	27.70%	Mexico	26.80%
Bhutan	22.00%	**South America**	
Cambodia	22.30%	Argentina	14.30%
Georgia	30.90%	Bolivia	21.80%
India	62.00%	Colombia	5.90%
Kyrgyzstan	26.30%	Ecuador	14.70%
Mognolia	19.80%	Peru	14.90%
Myanmar	36.70%	Uruguay	11.90%
Nepal	32.30%	Venezuela	9.40%
Oman	5.50%		
Pakistan	12.50%		
Uzbekistan	53.10%		
Vietnam	12.00%		
Yemen	27.00%		

The deficiency of micronutrients leads to the development of new techniques like food fortification in order to increase the intake of micronutrients. One of the best ways to avoid nutritional deficiencies is via food fortification, which stands out among public health treatments. Nutritional deficiencies are a global issue that have negative effects on both health and the economy. The danger of Fe deficiency alone, together with the range of unfavorable functional implications it is linked

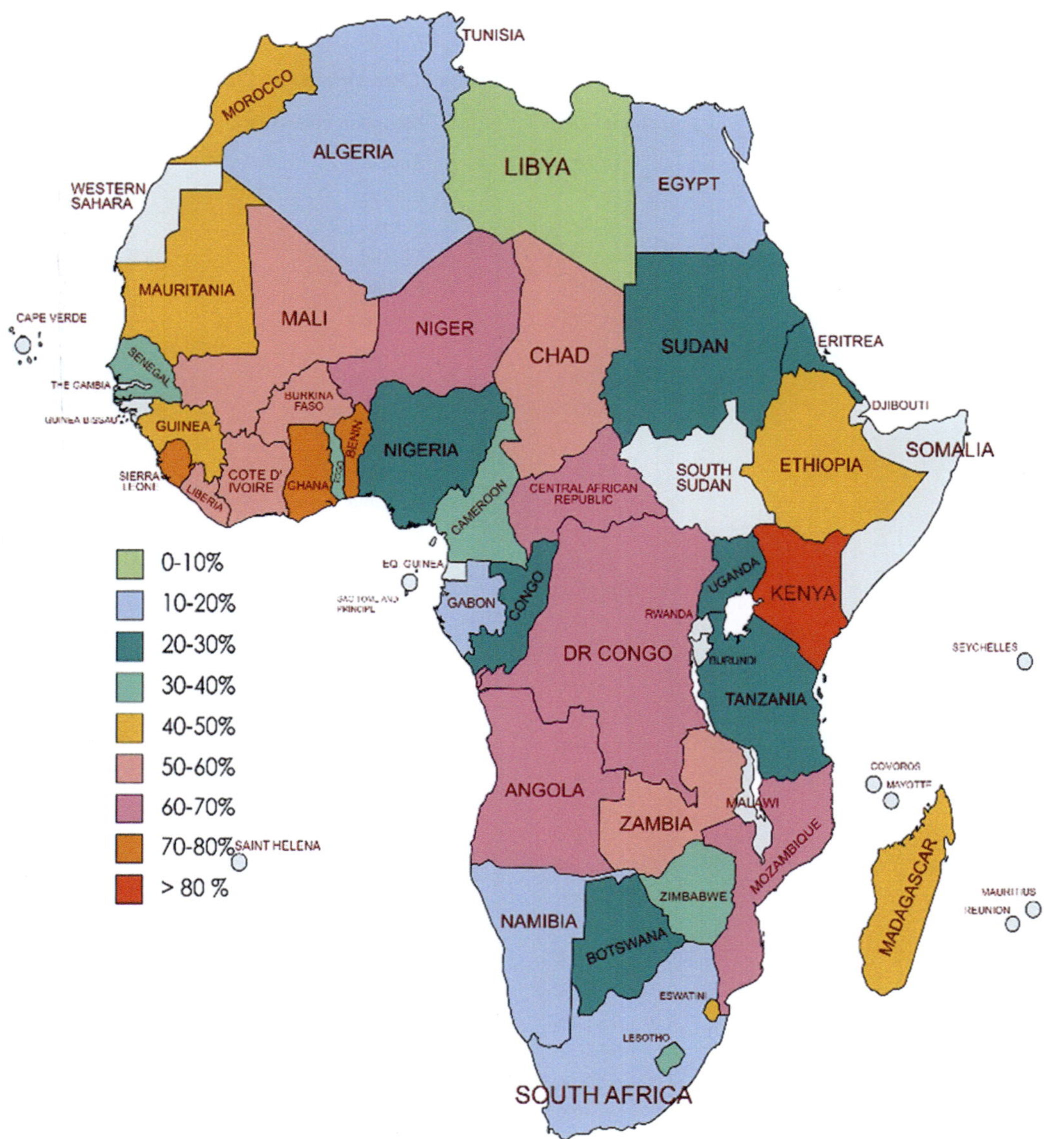

FIGURE 1.2 Geographical distribution of Vitamin A deficiency in children (Our World in Data, 2023).

with throughout the life cycle, affects over two billion people worldwide (ACC/SCN, 2000). The growing demand for foods containing biologically active ingredients is a result of consumers' greater knowledge of the potential health advantages of their diet. The current challenge for the food sector is to add functional ingredients, like antioxidants, to their goods so that they can be consumed by people (Manuela et al., 2020). Malnutrition causes a number of disorders, both infectious and chronic, which negatively affect the quality of life as well as epidemiological variables like mortality and morbidity (Verma, 2015). In order to reduce micronutrient deficiency, food fortification has been employed as a practical and affordable method (Method & Tulchinsky, 2015). Additionally, guidelines for food fortification programs have been published. Fortification remains one of the primary approaches used to address major worldwide concerns connected to malnutrition (Figure 1.4) (Reginald et al., 2004). A practical method of preventing micronutrient deficiency has been food fortification (Method & Tulchinsky, 2015). Utilizing items produced by the industry raises the

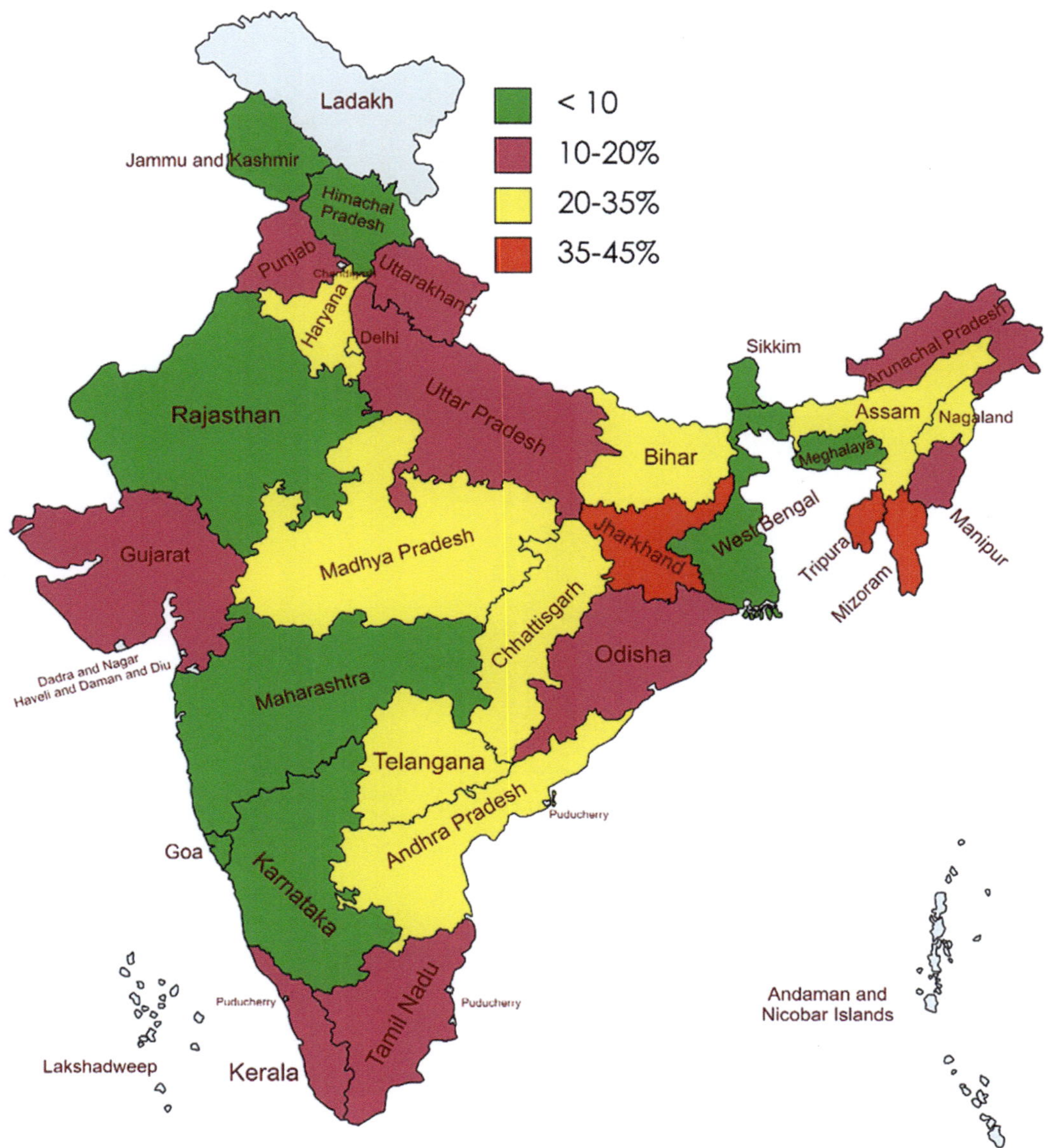

FIGURE 1.3   Geographical distribution of vitamin A deficiency in India (Our World in Data, 2023).

micronutrient content of the diet to decrease deficiency. Population coverage is influenced by the dietary source, but the effect relies on the supplementary micronutrient intakes, their bioavailability, and their bioconversion.

Food fortification is a well-known effective and extremely economical approach for delivering micronutrients to large populations and for lowering nutritional problems in children under the age of five. Because only synthetic nutrients are utilized and they are not always accessible, mass and targeted conventional food fortifications are not very sustainable and are difficult to execute in impoverished nations. Due to the availability of healthy and popular food components, food-to-food fortification is becoming more and more asserted in these nations. In Africa, there have been successful reports of using regional foods as food fortificants in food-to-food fortification (Bhagwat et al., 2014).

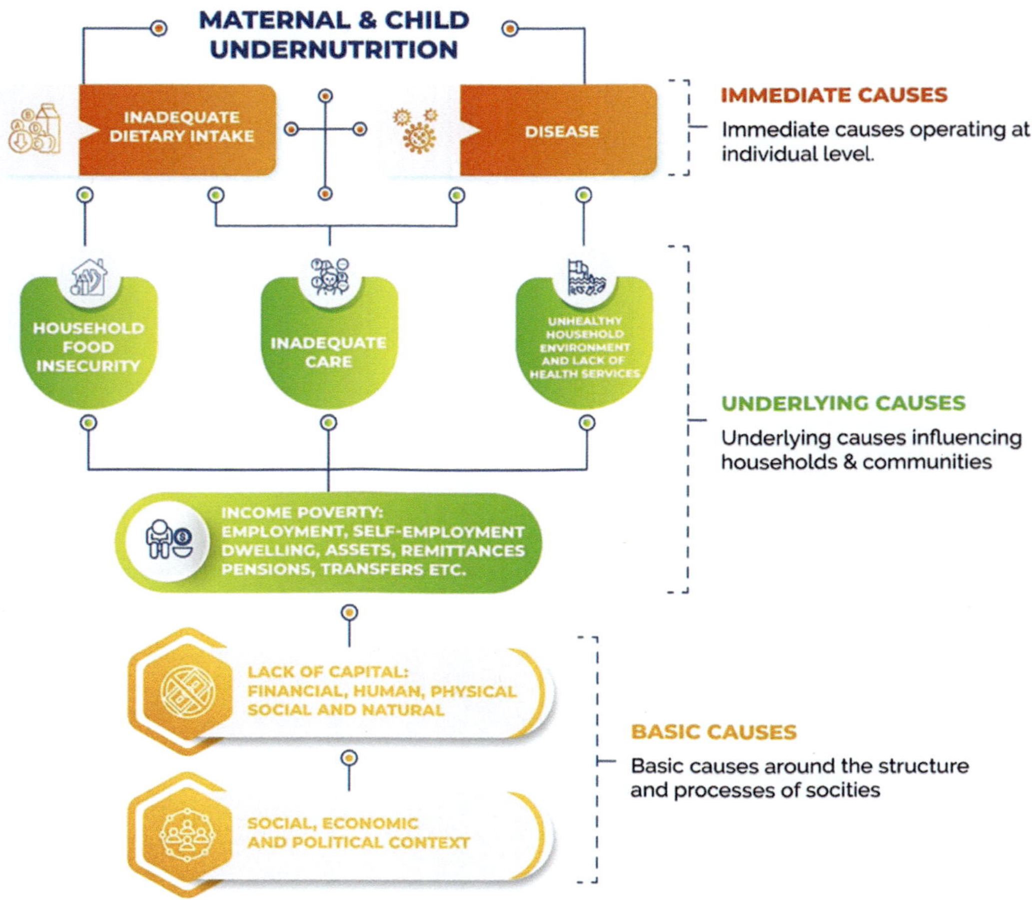

FIGURE 1.4   Factors responsible for child and maternal undernutrition.

Recent years have seen a substantial increase in food fortification methods, which have attracted a lot of criticism. In order to address micronutrient malnutrition, two-thirds of all nations reportedly need food fortification, however many do not always translate legislation into better nutrition. This is according to the *Global Fortification Data Exchange*. They assert that if nations bridge the gaps, food fortification might become the next big global health success (Global Fortification Data Exchange). Food fortification with bioactive compounds is a difficult problem to solve because suitable and food-compatible encapsulation systems must be available to maintain the chemical stability of the bioactives in processing as well as storage conditions prior to consumption, without impairing the sensory as well as physicochemical characteristics of the food. Consequently, each bioactive should have a specific use in mind (Adinepour et al., 2022).

The four primary measures of supplementation, fortification, nutrition education, and dietary diversification were suggested in Health Organization/Food Agriculture Organization recommendations from 2006. It has been shown that food fortification is the most effective fond cost-effective technique out of the four. However, the guidelines themselves identify two significant barriers to traditional fortification: consistent fortifier distribution in food vehicles, which mostly consist of basic foods, and producer assessment of internal and external fortification norms and standards compliance. In order to supplement traditional fortification, researchers imagined a new technique called food-to-food fortification. This approach entails nutrient-rich food-based fortifiers being

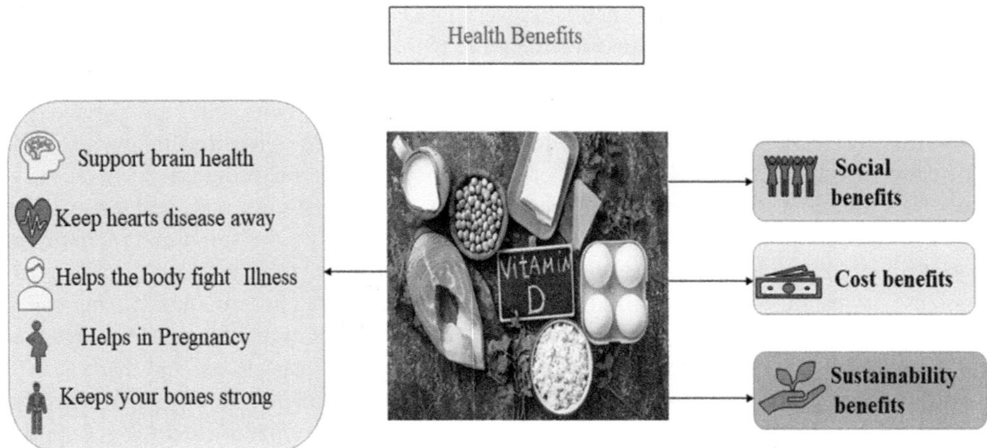

FIGURE 1.5   Major advantages of fortification.

added to food carriers. The main benefit of using food-based fortifiers is that they have the potential to increase the fortified food's bioavailability, provide extra nutrients, and lead to dietary diversification. Additionally, it makes it easier to use underutilized crops as food-based fortifiers. The nutritional, ecological, and economic advantages of underutilized crops have led to their recognition as possible useful food sources (Adinepour et al., 2022).

1.1.1 Advantages of fortification

Biofortification has an edge over many other approaches to improving a person's nutritional status since it addresses the whole population via everyday meals. Many processed and fortified foods are out of the price range of the poor, and introducing them into daily diets via alternative channels like free distribution includes several challenges including raising nutrition knowledge, introducing the produced product, and implementing it (which might be a difficult project if the community is uneducated as it is in most of such cases). Such goods are time-limited endeavors since continuing them demands a great deal of effort, money, and labor (Bouis et al., 1991) (Figure 1.5).

1.1.1.1 Cost-effective

By incorporating certain genotypes into their seed variety, biofortified seeds enhance the crop's variety. These cultivars can provide nutrient-dense staple foods and can withstand a variety of environmental stresses. According to Hoddinott et al. (2013), food fortification is a practical method for enhancing the population's nutrition status and has major economic advantages (Boy et al., 2009). Micronutrients have regularly been regarded as the most cost-efficient development that provides big benefits for a low cost, according to, for example, the Copenhagen Consensus (Spohrer et al., 2013). Iodized salt, for instance, may be purchased for as low as $0.05 USD yearly, and fortified maize and wheat with folic and iron acid, for as little as $0.12. These two fortified commodities have lifetime expenditures of less than USD 15 per person and may provide a return on investment of over USD 26 via enhanced productivity and cost savings in healthcare. Additionally, each dollar invested in fortification generates USD 9 in economic benefits (Hoddinott et al., 2012). In feasibility and cost-efficiency research carried out in the Philippines, fortified powdered milk was shown to be a cost-efficient strategy for treating iron insufficiency in children. The findings showed that food fortification was the most economical treatment for iron deficiency, with costs estimated at USD 66 for every disability adjusted life year (DALY), as opposed to USD 179 and USD 103 for supplementation and dietary diversification, correspondingly. For biofortification, the anticipated health benefit-to-cost ratio was USD 17 in benefits for every USD 1 invested (Spohrer et al., 2013). In spite of the great economic potential of food fortification, there are a number of obstacles that make it difficult to establish the right conditions for a scale-up of food fortification throughout the world. These restrictions include a lack of national laws on food enhancement and a low private–public collaboration. The difficulties of large-scale food fortification (LSFF) programs in low and middle-income countries (LMICs) as well as the efficiency of food fortification programs were recently the subject of a systematic study. The authors mentioned the absence of rules and food laws as further obstacles to the expansion of food fortification initiatives, as well as the small and medium enterprise (SME) sector's limited technical and economic capabilities (Hoddinott et al., 2013). In the Philippines, for instance, salt is required to be iodized, but owing to poor manufacturing capacity, more than two-thirds of the salt is imported and not fortified. Additionally, manufacturers in the country create salt without a license, bypassing any government inspections, and there aren't any high-quality labs, which makes it impossible to check and monitor the quality of fortified foods (Van et al., 2013). A thorough examination of the regulatory control of fortified foods is provided by Luthringer et al. (2015).

1.1.1.2 Social

The World Food Summit's statement from 1996 states the requirement for "Everyone to have access to safe and nutritious food, consistent with the right to adequate food and the fundamental right of everyone to be free from hunger" (FAO, 1996). The Universal Declaration of Human Rights, 1948 and the International Covenant on Economic, Social, and Cultural Rights, 1966 both mention the right to food. As a result, fighting hunger as well as malnutrition is not just a moral obligation but also a crucial component of national legislation that must be upheld in accordance with human rights obligations in many nations. Unfortunately, in many LMICs, ensuring the right to food and nourishment is a difficult and sometimes overwhelming issue. In particular, given the COVID-19 epidemic, imposed lockdowns and economic slowdowns have increased food insecurity as well as malnutrition. So, before, during, and after a pandemic, food fortification may be a vital strategy to reduce the malnutrition risk. In order to attain the SDGs and the 2030 Agenda for Sustainable Development (2030 Agenda), data and international collaboration will be used to create a world free of hunger as well as malnutrition in all of its aspects. In fact, SDG17 promotes "the global partnership for sustainable development, complemented by the use of multi-stakeholder partnerships" as a means of implementation of the 2030 Agenda (Sustainable Development Goals Knowledge Platform, 2018). The public-private cooperation must be enhanced in order to have positive multi-sectoral effects on society. Initiatives for food fortification include both governments and civil society, although the main actors are private sector partners from the food industry. Particularly civil society may contribute to increased responsibility, adherence to rules, etc. The majority/all of the initiatives that include food fortification involve public-private partnerships (PPPs), as well as participation with funders, civil society, consumers, and other stakeholders. Finally, food fortification has been successful in delivering fortified food to vulnerable people while also disseminating information on diets when combined with Social Safety Net Programs (SSNPs), like distributions to the poor or to vulnerable groups, school feeding programs, food aid during emergency situations, and food for work programs (World Food Programme and Sight and Life, 2017). The Public Distribution System (PDS), Integrated Child Development Scheme (ICDS), and the Mid-Day Meal (MDM) Program were three SSNPs in Gujarat, India, that the government employed as platforms to spread wheat flour fortification among recipients. Through this, they discovered that switching from enriched wheat flour to wheat grain significantly boosted the consumption of micronutrients among its SSNP participants and that it was a highly economical strategy (Fiedler et al.,2012). The long-term efficacy of SSNP initiatives like school meals is a problem. In South Africa, in spite of biscuits providing 10% of the daily RDA for vitamin A and 50% of the daily RDA for carotene, vitamin A deficiency has not been eliminated in almost four years after the program's implementation (Stuijvenberg et al.,2005). Another issue arose during the most recent COVID lockdowns when

SMEs had more trouble obtaining the mineral and vitamins premix needed for the fortification operations, especially from the foreign premix suppliers (Carducci et al., 2021).

1.1.1.3 Sustainable

In contrast to other methods of fortification or supplementation, the cycle will continue once the crop is given a new genome without any more expense. Once this happens, its products as well as seeds will also include the new genotype (Olson et al., 2021).

The following are a few other benefits of crop biofortification:

- Seed that has been fortified does not reduce yield. Additionally, it provides significant collateral advantages like plants with disease resistance, which boosts agricultural output.
- Addresses the issue of starvation in a bigger region.
- Nutritious value of regular diets increases.
- Increased germplasm diversity and an improvement in plant or crop quality.

1.1.1.4 Health

According to mortality statistics from the WHO, an estimated 0.8 million fatalities per year (1.5 % of the total) may be attributable to iron insufficiency, and a comparable amount can be ascribed to vitamin A deficiency (WHO, 2013). This represents a huge loss of life (WHO, 2006). LSFF may have an impact on public health in both LMICs and HICs, according to a significant body of research. Iodine, vitamin A, folic acid, and iron fortification have considerably decreased the incidence of severe illness, as per the recent study of 50 research from LMICs (Mkambula et al., 2020). The prevalence of NTDs, goitre, anemia, and other health outcomes was shown to be positively impacted by industrial food fortification in a 2019 analysis that sought to quantify the effect of fortification on health as well as nutrition outcomes in LMICs (Mkambula et al., 2020). With larger advantages seen by individuals who are most at risk of deficiency, population-wide fortification initiatives were linked to a 34% fall in anemia from better iron storage, a 74% fall in the probability of goiter, and a 41% fall in the probability of NTDs (Kraemer et al., 2019). The possible advantages of food fortification are shown in Figure 1.6 for all life stages.

Numerous studies examining the effects of food fortification on the availability of micronutrients on a national scale have shown highly encouraging results. , During 2011 to 2012 research conducted in two areas of West Java, Indonesia, assessed the effect of widespread vitamin fortification. Breastfeeding mothers' breast milk now includes more vitamin A, along with a significant decline in vitamin A deficiency and an improvement in everyone's vitamin A status. Anemia incidence among children and women of childbearing age between the ages of 1 and 7 significantly decreased nationwide after an analysis of the impact of iron fortification on anemia occurrence in Costa Rica. Children's anemia dropped from 1 to 4% and women's anemia from 18 to 10%, while iron deficiency in children dropped from 27 to 7% (Martorell et al., 2015). Other studies have shown the opposite results, namely that food fortification initiatives have minimal impact and merely give safe upper limits, in spite of the enormous increases in nutritional status that food fortification programs have brought about. For instance, research carried out among Brazilian children aged under 6 years found no link between the occurrence of anemia and iron-fortified wheat. The study used four population-based surveys performed over a four-year period to investigate food intake and hemoglobin levels. The outcomes showed a sharp rise in childhood anemia. The research observed that people consumed 100 g of fortified flour on a daily average, but the poor nutritional status of children with low iron bioavailability nullified the benefits of fortified wheat (Dary et al., 2008). Additionally, a systematic review conducted in both LIMCs and high-income countries (HICs) didn't discover any evidence of a link between zinc deficiency and the impact "of multiple micronutrient fortification on child growth outcomes like height/length-for-age z-score (HAZ/LAZ) (MD 0.09, 95% confidence interval [CI] 0.01 to 0.18; eight studies, 2889 participants) (RR 0.84, 95% CI 0.65 to 1.08; five

	PREGNANCY	LACTING MOTHER	6-23 MONTHS	2-5 YEARS	5-18 YEARS	WRA (15-49 YEARS)	ADULT MEN	ELDERLY
MICRO NUTRIENT	VERY HIGH	VERY HIGH	VERY HIGH	HIGH	MODERATE TO HIGH	MODERATE TO HIGH	LOW TO MODERATE	MODERATE TO HIGH
AMOUNT OF FOOD EATEN	MODERATE	MODERATE	LOW	LOW, INCREASING WITH AGE	INCREASING WITH AGE	MODERATE	HIGH	MODERATE
POTENTIAL TO BENEFIT	HIGH	HIGH	LOW	LOW, INCREASING WITH AGE	INCREASING WITH AGE	HIGH	HIGH	HIGH
POTENTIAL FULLY MEET NEED	LOW	LOW	NO	LOW, INCREASING WITH AGE	INCREASING WITH AGE	HIGH	HIGH	HIGH

FIGURE 1.6 Potential benefits of food fortification across the life cycle (Olson et al.,2021).

studies, 1490 participants; low-quality evidence) (Osendarp et al., 2018). In other situations, fortified supplemental meals were linked to increased diarrhea episodes while having no effect on development and just a little influence on children's anemia (Heidkamp et al., 2021). Additionally, debates concerning the efficiency and security of fortification efforts globally and in certain states are still going on (whether or not eating meals fortified with micronutrients might have a negative impact on one's health as a result of the body accumulating these nutrients). To properly fortify foods, the Tolerable Upper Intake Level (UL) for each micronutrient must be established. The Guidelines on Food Fortification with Micronutrients, which change based on the local environment, were developed by the WHO and include a technique for computing and establishing the acceptable upper limit. However, problems may occur if fortified foods are consumed in large quantities by the same population and numerous meal delivery systems are used (Assuncao et al., 2013). There are growing worries about the acceptable limits after research in Guatemala revealed that intake of wheat flour fortified with acid folic differed dramatically across persons from various socioeconomic positions (rich women consumed 15 times more than poor women) (Dary et al., 2008). Children in Cameroon who consume a variety of vitamin A-fortified meals, like wheat or sugar flour with edible oil, may have an excess of UL. This was the scenario as a consequence of the widespread eating of these meals in cities (Engle-stone et al.,2019). Numerous exposures to micronutrients, such as those in fortified meals, over a long period of time, may have negative consequences, but when fortification is adequately controlled, the danger of toxicity is limited (Bruins et al., 2015). For instance, the safety of vitamin A fortification is often noted as a problem in food fortification programs, but a recent assessment found that the danger of women and young children consuming too much vitamin A from fortified foods is probably small (Allen et al., 2002).

1.1.2 Strategies for malnutrition

According to Soeters et al. (2008), malnutrition is a condition that results in altered body composition (reduced fat-free mass, but precisely body cell mass), as well as diminished function. Both overeating and undereating fall under this category. Despite having a major impact on surgical outcomes, orthopedic surgeons have typically not focused on managing malnutrition (Bohl et al., 2016). Contrary to earlier research that has implemented different indicators such as transferrin or prealbumin, these investigations defined malnutrition as serum albumin levels below 3.5 or an absolute lymphocyte count below 1500. These numbers, however, are not absolute and have been called into doubt in a number of studies (Jensen et al., 2010). Chronic disease-related malnutrition, starvation-related malnutrition, and acute injury or disease-state-related malnutrition were categorized into three groups by a consensus reached through a series of meetings held at the European Society for Parenteral and Enteral Nutrition Congresses and the American Society for Enteral and Parenteral Nutrition (Norman et al., 2008). Malnutrition may be categorized according to its aetiology, which simplifies the problem and helps in choosing the best treatment methods.

According to McWhirter et al. (1994), malnutrition can affect as many as 20%–50% of patients in hospitals, whereas Pruzansky et al. (2014) indicated that it affects 9%–39% of orthopedic patients. Additionally, malnutrition was found in the 26% of patients undergoing arthroplasty at an urban academic centre (Raj et al., 2002), and other studies indicate that this prevalence may be as high as 50% (Jaberi et al., 2008).

To improve the consumption of micronutrients by individuals in developing nations, and more especially by the poor, effective techniques must be created. The fortification approach must be economical and long-lasting. The first approach can include expanding the variety of foods ingested via dietary diversification. Ideally, every person should have access to a varied diet that is nutritionally balanced. However, since so many people cannot afford such a diet, this will not be a workable option. As an alternate technique, the provision of dietary supplements via different initiatives has been implemented. This method has shown to be quite successful in the short term when programs that provide the supplements are appropriate. However, when funding for these programs is cut off, consumers are seldom interested in continuing their supplement usage due to the expense, problems with availability, and unfavorable side effects. In many developing nations, the practice of fortifying basic meals with vitamins and minerals is becoming more popular. Due to its numerous benefits over dietary supplements, food fortification has been a popular method in many industrialized nations.

1.1.2.1 Supplementation

The most often suggested strategy to alleviate vitamin deficiency is supplementation. Providing relatively large dosages of micronutrients in the form of syrups, tablets, or capsules is known as supplementation. Supplementation provides the ideal amounts of micronutrients in a highly absorbable form. However, putting a supplements program into place is not simple. First, the amount of the supplement must be determined so as not to exceed the needed iron intake. Second, a dependable distribution system has to be put up. Adherence to supplement usage by the target demographic is also crucial, especially when supplements are to be taken for an extended length of time (Galaniha et al., 2020).

In addition to being more suitable for senior people, oral protein as well as energy supplements are also often utilized because they may be safer and simpler to deliver than nasogastric enteral feeds. Oral supplements, however, may not be utilized as efficiently in older persons due to issues with their desire and capacity to take them. . The formulation and time of administration in relation to meals may be relevant in addition to flavor. Additionally, efforts must be made to provide senior citizens with the regular meals and snacks they need as well as any necessary feeding help (Galaniha et al., 2020).

Milne et al. (2009) studied providing extra energy and protein to senior citizens in danger of malnutrition. From the 62 trials, a total of 10,187 randomized participants have been included. The intervention duration was up to 18 months. The reviewers claim that using supplements seems to result in a slow but steady weight increase. This updated review found no evidence of a protective effect on overall mortality, however, there might be a protective effect on death in malnourished individuals. The severity of complications may be decreased by supplementation. There were differences in the reported acceptance of supplements among trials. There have been some recorded negative effects, like diarrhea or nausea. However, there were problems with the design and grade of the study. In order to verify the positive influence on the severity of complications, to determine whether there is an advantageous impact on mortality for senior citizens who are undernourished, and to show whether these effects can be seen in elderly people who are malnourished, additional study is required to establish if protein and energy supplements could reduce morbidity and enhance functional status in older adults who have a fragile health status.

Corriea et al. (2019) presented findings on the effects of prolonged malnutrition and vitamin A supplementation on child development. The development of 200 million children in preschool is inadequate. There are many causes of developmental delays in children. This study's main objective is to determine whether vitamin A supplementation is linked to better development as well as how the effect of nutritional status may moderate this association. In a population-based study, 8000 families with 1232 infants and toddlers aged 0 to 35 months in Ceará, Brazil, were chosen as a representative sample. Developmental state of the child, nutritional determinants, and confounding variables were among the variables studied. Cox regression models were used to assess the main effects and interactions. Supplementing with vitamin A has been shown to be protective against delays in cognitive as well as motor development that are influenced by interactions with dietary status. In contrast to stunted children, who did not benefit from supplementation (adjusted PRR = 097 [0.39–2.40]), well-nourished supplemented children showed a 67% lower risk of cognitive delay (adjusted PRR = 0·33 [0·21–0·53]). Supplementing with vitamin A has a protective impact on child growth, but not in children who are stunted. Given the significance of this mediator, it appears that supplementation is useful in fostering child development, particularly when combined with an effort to enhance children's nutritional status.

Rusell (2007) examined the effect of dietary deficiency on medical expenses and the cost-effectiveness of using oral nutritional supplements. Costs associated with disease-related malnutrition and the use of oral nutritional supplements in hospital and community settings were studied in depth by the UK's National Institute for Health and Clinical Excellence. Prices for disease-related malnutrition were calculated for 2003 based on reported rates of malnutrition in different types of care settings, rates of artificial nutrition assistance provided in the home, lengths of hospital stays, national reference pricing for healthcare services, and the costs of nutritional items. Studies examining the utilization of nutritional supplements taken orally were found to yield outcomes that could be evaluated economically. Factors included the quantity of the product consumed, the duration of the hospital stay, and any issues. It was predicted that £7.3 billion (€10.5 billion) was spent last year on the care of individuals at medium or high risk of disease-related malnutrition. Spending on people over 65 and hospital care accounted for more than half of total expenditures. The usage of supplements resulted in an average cost reduction of £700 (€1000) per patient for abdominal surgery. Additionally, by supplementing orthopedic surgery, aged care, and procedures before elective surgery, there was a net cost saving.

Persson et al. (2007) conducted research on nutritional supplementation as well as dietary guidance in elderly individuals at malnutrition risk. After being released from a geriatric service, individuals at risk of protein-energy malnutrition (PEM) received combination nutritional therapy. As per the brief form of the mini nutritional assessment, patients (n = 108, age 85±6 years) at risk of malnutrition were randomly assigned to intervention (I, n=51) and control (C, n=57) groups for dietary counseling that included liquid and multivitamin supplements. At the beginning of the trial and four months later, participants' body weight, biochemical indicators (such as insulin-like growth factor I (IGF-I)), the Katz Activities of Daily Living (ADL) index, the Mini Mental Status Examination (MMSE), and their quality of life (QoL) as measured by the SF-36 were evaluated. On the basis of "treated-as-protocol" and "intention-to-treat," respectively, statistical analyses were conducted. Fifty-four patients completed the trial in accordance with the guidelines: 25 in the C-group (85+7 years, 72% females) and 29 in the I-group (86+7 years, 29 patients, 66% females). The combined nutritional intervention had a significantly good impact on weight maintenance, according to both research methods. The Katz ADL index improved in the I-group, according to treated-as-protocol studies ($p<0.001$; $p<0.05$ across the groups). In the I-group, serum IGF-I levels raised ($p <0.001$), whereas they remained unaltered in the C-group ($p = 0.07$ between the groups). It was determined that QoL was poor and had remained unchanged during nutritional therapy. ADL functioning was enhanced and weight loss was stopped in older patients who were in danger of malnutrition and had been discharged.

1.1.2.2 Biofortification

By breeding nutrients into food crops, a process known as biofortification, more micronutrients may be provided over time and at a relatively cheap cost. Even though they cannot deliver the same amount of vitamins and minerals as supplements or industrially fortified meals can, biofortified staple foods may increase daily adequacy of micronutrient intakes over the lifetime (Bouis, 2011). Note that not all population groups are anticipated to benefit from biofortification in treating or eliminating micronutrient deficiencies. The issue of micronutrient malnutrition cannot be resolved by a single intervention, but biofortification supports current interventions by providing micronutrients to the most vulnerable populations in a cost- and sustainably-effective way (Qaim et al., 2007; McClafferty and Pfeiffer, 2007; Nestel et al., 2006; Bouis, 1999).

Rural inhabitants who are malnourished and maybe with restricted access to various diets, vitamins, and commercially fortified foods may benefit from biofortification. The goal of the biofortification strategy is to enhance cultivars that already have beneficial agronomic and nutritional traits, including high yield, by introducing the nutrient-dense micronutrient feature. These crops' marketed surpluses may find their way into retail businesses, where they will eventually reach clients in rural and then urban areas. Fortification and supplementation are examples of complementary programs that start in urban areas, in contrast (Nestel et al., 2006).

In contrast to the recurring expenses needed for supplemental as well as commercial fortification programs, a one-time investment in plant breeding may generate planting materials that are micronutrient-rich for farmers to cultivate for years to come. The benefits of the initial investment can be increased by evaluating the performance of varieties grown for one nation in other geographies and adapting them to other conditions. The cost of developing nutritionally improved crops and institutionalizing nutrient content as a valid breeding objective for the crop development pipelines of national and international research centres is low compared to the cost of ongoing monitoring and maintenance of these traits in crops (Saltzman et al., 2017).

Agronomic, transgenic, and conventional biofortification are now the three most used methods. Through fertilizers, agronomic biofortification can temporarily enhance micronutrient levels. For instance, foliar application of zinc fertilizer in Pakistan and India may increase grain zinc content in wheat by up to 20 parts per million (ppm) but only during the application season (Zou, 2012). Nearly the full goal increment was reached, as sought by plant breeders and nutritionists. Unlike maize, where foliar spray had a negligible effect (Kalayci et al., 2011; Phattarakul et al., 2012), rice can achieve around half of the required increment.

Plants with the required nutritional and agronomic traits may be produced via classic plant breeding, which crosses parent lines with high mineral or vitamin levels over many generations. Transgenic approaches may be helpful when a crop naturally lacks the nutrient (like rice lacks provitamin A) or when sizable amounts of readily available micronutrients cannot be properly absorbed into the crop. To guarantee that the transgenes are stably transmitted and to incorporate the transgenic line into types that farmers want, several years of traditional breeding are needed once a transgenic line has been produced. Despite the fact that transgenic breeding may sometimes provide micronutrient enhancements that are superior to those offered by traditional breeders, many nations lack the legal infrastructure required to allow for the release and marketing of these cultivars (Bouis, 2011).

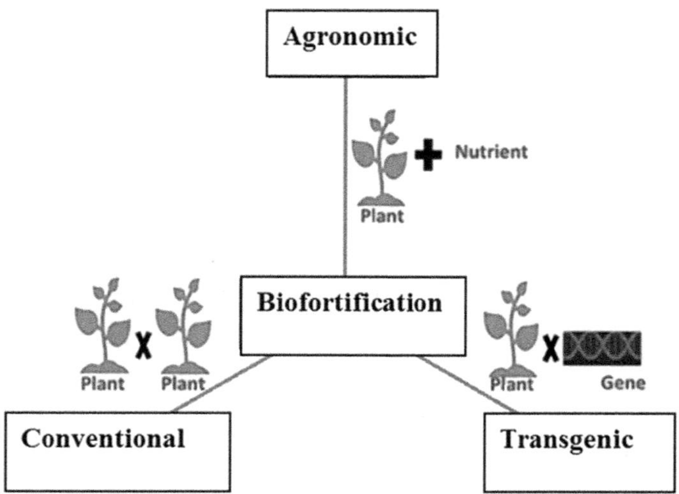

FIGURE 1.7 Different approaches to achieve biofortification.

1.1.2.2.1 Biofortification through the Mode of Mineral Fertilization

By raising the concentration of micronutrients in a target crop's edible parts, biofortification tries to improve the nutritional quality of the crop without degrading other agronomic characteristics like insect resistance, drought tolerance, or yield (Klikocka et al., 2018). Conventional biofortification, agronomic biofortification, and transgenic biofortification are the three main methods to achieve nutritional security via biofortification (Figure 1.7).

The conventional approach entails selecting high-yielding crop varieties that already naturally contain higher amounts of required nutrients and cross-breeding them with other varieties with conventional techniques to make staple crops with the necessary nutritional and agronomic attributes. The content of micronutrients in sections of edible plants may be increased by precision genetic modification. Non-genetic techniques, such as agronomic biofortification, may be more efficient in boosting the micronutrient content in food plants. In fact, this method enhances grain nutritional value as well as yields (Cakmak and Kutman,2017). According to Kumar et al. (2018), N, Zn, and Fe fertilization is a sustainable and cost-efficient way to enhance plant growth, yield-contributing traits, and Fe, Zn content in rice crops.

Agronomic biofortification is the application of micronutrient fertilizers to the soil and/or plant to increase the number of micronutrients in the field crop's edible portion, as shown in Figure 1.8 (Cakmak and Kutman, 2017). Various biofortified crops, such as pulses, cereals, fodder crops, oilseeds, and so on, have been developed to this point around the world. In this procedure, the right nutrient is externally supplied either before or during the development of plants. While soil application provides an acceptable quantity of nutrients for root absorption, foliar treatment increases the content of the nutrient in the leaf to facilitate its transfer to other plant parts. The availability of micronutrients and, thus, the success rate of biofortification are influenced by soil characteristics like pH, its calcareous nature, etc. Thus, a crop's mineral enrichment is influenced by the technique of fertilization application, the characteristics of the soil, and amendments (Cakmak and Kutman, 2017).

1.1.2.2.2 Biofortification through Agronomic Management Practices

The primary aim of management practices is to improve soil quality, which will then increase the availability of micronutrients for plant uptake and, ultimately, their content in food. The efficiency of a plant's use of nutrients is impacted by various factors, like the physical, chemical, and biological features of the soil. Integrated soil fertility management (IFSM), water management, and tillage

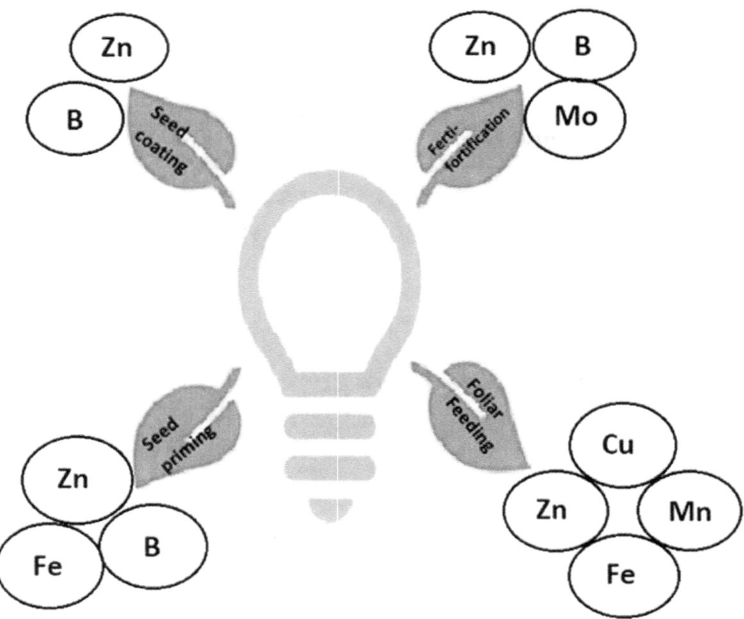

FIGURE 1.8 Different modes of mineral fertilization.

techniques are common management techniques. The usage of mineral fertiliser, organic inputs, and enhanced germplasm is all part of the integrated soil fertility management method (Selim et al., 2020). Additionally, promoting the sustainability of soil organic matter, organic compounds support a number of advantages for soil health, including cation exchange capability, water holding capacity, and soil structure. However, because they only release nutrients at predetermined rates, crops rarely receive the nutrients they need for the best yield. Therefore, using only organic inputs and mineral fertilisers is insufficient to fill the gap between the needs of grain crops and mineral shortage. As a result, the two therapies work well together because they have beneficial interactions and complementary roles (Padbhushan et al., 2021).

Due to the rarity of perfect soil for plant growth, substantial attention has been placed on improving soil status, particularly in terms of its chemical composition. Numerous methods for soil amendments have been effectively utilized as agricultural activities have increased (Figure 1.9). The pH of the soil is one of the chemical soil characteristics that have a significant influence on plant growth, nutrient availability, and the elimination of dangerous hazardous substances. Therefore, achieving both quantitative and qualitative crop output requires optimizing soil pH. It either entails lowering the pH of alkaline/calcareous soils or raising the pH of acidic soils.

1.1.2.2.3 Biofortification through Green Technology

By making nutrients more accessible to crops, microorganisms are utilized in green technology in improving the soil's nutritional status. Overuse of chemical fertilizers has a negative impact on both human health and environmental ecology. In order to improve soil fertility and plant development, microorganisms that harbor biofertilizers may prove to be a successful and environmentally acceptable alternative to chemical fertilizers (Kumar et al., 2018).

The synthesis of several plant growth hormones, siderophores, hydrolytic enzymes, HCN, as well as the solubilization of Zn, K, and P are only a few of the direct and indirect ways that plant growth-promoting microorganisms enhance plant health (Kaur et al., 2020). Intrinsic plant-based processes including organic acid production, chelator secretion, and phyto-siderophores are inadequate on their own to transform the nutrients into their utilizable form. Zn is bound to and released

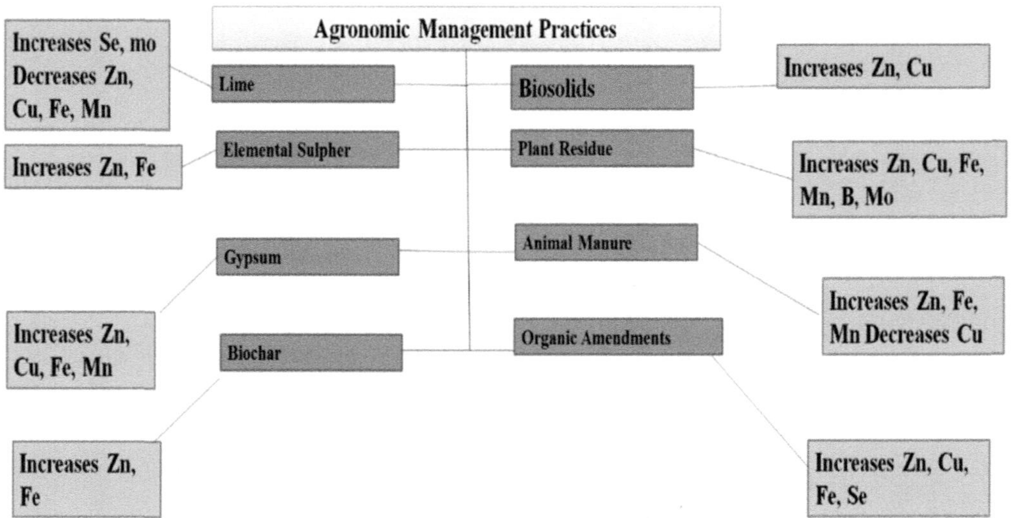

FIGURE 1.9 Various amendments as part of agronomic management techniques to improve the mineral content of plants (Padbushan et al., 2021).

at the root surface by chelating compounds generated by microorganisms, increasing Zn availability and eventually resulting in Zn biofortification in plants. Numerous studies have been done on the advantages of microorganisms for crop biofortification. Biological fertilizers are mostly derived from bacteria and fungi (Table 1.4).

1.1.2.3 Microencapsulation

Small particles known as microcapsules are created by microencapsulating solids, liquids, and gases with thin polymeric coverings (Gharsallaovi et al., 2007). The polymer isolates the heart and keeps it from being overexposed by acting as a protective layer. This membrane disintegrates in reaction to a specific stimulus, enabling the release of the core at the precise time or place (Suave, 2006).

Traditional techniques of vitamin incorporation often affect the physical, chemical, and functional aspects of foods that have been fortified. Microencapsulation is a very useful technique for food fortification since it provides staple sources of micronutrients in a bioavailable form. Mineral reactivity and vitamin volatility in processed meals are addressed by this method. It keeps the active ingredients apart from other food components in a safe setting, avoiding unwanted changes in fortified foods. This method could be the best approach to simultaneously deliver two or more micronutrients in a secure and bioavailable form. Microencapsulation may improve the sensory qualities of fortified meals by hiding unwanted colors and tastes from the fortificants and avoiding interactions between the fortificants and the food carrier. Due to microencapsulation-based technology, the fortificants may be administered in acceptable forms that closely resemble the physical traits of the chosen food vehicles in terms of size, shape, colour, and appearance. Using a microencapsulation-based technique, a broad variety of staple foods with particle sizes ranging from a few hundred microns to several millimetres may be fortified with many micronutrients. To avoid particle segregation, which might result in under- or overdosing, micronutrients should be added in a method that either binds to the food particles or creates agglomerated premixes that suit the meal's particle size and, if feasible, density. The addition of micronutrients must be evenly distributed throughout the meal and be undetectable to the customer, according to recent advancements in harvest and post-harvest technology in fisheries. In order to avoid changing the flavours of the meal, the whole distribution system must match the dish's colour and appearance (Guidelines for Micronutrient Fortification by the FSSAI, 2021, Cakmak and Kutman., 2017).

TABLE 1.4

Biofortification through Green Technology

Microorganisms	Micronutrient Affected	Crop	References
Arthrobacter sp. DS-179and *Arthrobacter sulfonivorans* (DS-68)	Fe and Zn	Wheat	Singh et al. (2017)
Rhizophagus irregularis	Zn		Watts-Williams et al. (2018)
PGPB and AM fungi	Zn and Fe	Wheat	Yadav et al. (2020)
Endophytes	Fe and Zn	Wheat	Singh et al. (2018)"
Bacillus subtilis DS-178 and *Arthrobacter* sp. DS-179, *Enterococcus hirae* DS-163 and *Arthrobacter sulfonivorans* DS-68	Fe and Zn	Wheat	Singh et al. (2018)

1.1.2.4 Micronization

Food particles are mechanically reduced to a micron size by the process of micronization, which utilizes intense shearing. In the past, micronization methods had only been used in the pharmaceutical sector, but in the last few years, a large number of studies have shown their immense potential in the food sector. When the size is lowered to microns, physicochemical and functional qualities like swelling capacity, water holding capacity, solubility, as well as antioxidant capacity increase. Additionally, it promotes food stability by exposing antimicrobial substances to the surface, neutralizing germs, and inhibiting enzymes. Micronization slowed retrogradation, induced gelatinization, reduced syneresis, sped up fermentation, created more stable products, and increased the dissolution rate. These one-of-a-kind improvements were discovered as an outcome of a decrease in particle size and an increase in surface area. To provide readers and researchers a brief review of recent advancements and innovation, a thorough study has been done on solvent-free micronization by supercritical fluid, solvent-free micronization by grinding, homogenizing, electrospraying, and other methods of micronization (FSSAI Guidelines on Micronutrient Fortification, 2021, Kumar et al., 2018).

1.1.2.5 Fortification

When speaking about food, fortification, "enhancement," or "enrichment" refers to a strategy designed to raise the quantity of a certain nutrient in a food product in order to improve its nutritional value(s). The purposeful addition of one or more microelements to food items is known as fortification, according to the WHO, and is done to enhance their content, rectify particular deficiencies, or improve health (Allen et al., 2006). Additionally, it is feasible to add active ingredients to a product that aren't necessarily present in a particular type of food (National Academy of Sciences, 2003; WHO, 2015). Functional foods should be classified as fortified goods. A scientifically confirmed health impact may only be induced or triggered if the added substance is present in sufficient quantities (such as lowering the risk of disease). As a result, not all fortified goods comply with this criterion, such as juices enhanced with ascorbic acid meant to halt oxidation processes (Siró et al., 2008). However, the substance used to fortify meals shouldn't have an adverse impact on how other nutrients are absorbed and metabolized or be poisonous at the dose employed (Dwyer et al., 2015).

The twentieth century saw the emergence of the present focus on enhancing food meant for human consumption. The first recorded example of such practices, however, dates back to 400 BCE, and Doctor Melanpus is a well-known example. He added iron filings to wine to increase the endurance of Persian warriors (Wesley and Bulusu, 2011). Jean Baptiste Boussingault, a French scientist, noticed that inhabitants in Colombia's Antioquia area who ate salt from an abandoned mine did not get goiter in the 1820s. The salt was contaminated with iodine, as he subsequently demonstrated. In order to prevent thyroid problems, he suggested that salt be widely enriched with iodine (Preedy et al., 2009; Venkatesh Mannar, 2011). The importance of enrichment's positive effects on health led to a number of nations beginning to systematically iodize salt, although this advice wasn't followed through with until 1920.

Margarine was first fortified with vitamins A and D, milk with vitamin D, wheat and bread with vitamins B1, B2, and B3, and iron, and margarine with vitamins B1, B2, and B3. This practice started during World War II (Nathoo et al., 2005; Bulusu and Wesley, 2011). The market for products of this kind is expanding in both developed and developing nations as food fortification continues to gain appeal (Martorell et al., 2015; Velu et al., 2014; Das et al., 2013; Black et al., 2012).

The traditional approach to food fortification involves incorporating one or more nutrients into the finished food product while it is being produced. When a substantial loss occurs when preparing food, as it does with the amounts of riboflavin, thiamine, and niacin in refined grains, the added components may include those already present in processed food but in decreased quantities. They might also contain substances that aren't found in the product naturally like calcium added to orange juice. Both natural and artificial sources for the added compounds are possible (Liyanage and Hettiarachchi, 2011).

Iodine, vitamin A and calcium, iron, B-group vitamins, and vitamin D (especially B12) are the most commonly added as supplements to food products since their deficiency is most frequently seen (Bailey, 2015). It should be noted that the primary alternative for utilizing food supplements is industrial fortification. However, their effectiveness in reducing nutritional deficiencies is challenged by a variety of concerns, such as inadequate absorption of micronutrients from these sources, cross-interaction between specific elements of dietary supplements, and contamination in low-quality goods. The use of the food supplements has increased in popularity in emerging nations despite the fact that their efficacy in reducing nutrient deficiencies is constrained by a number of problems, including low bioavailability of micronutrients from certain sources, cross-interaction between contaminants in low-quality products, specific food supplement ingredients, and discrepancies between declared and determined nutrient contents (Rzymski, 2018; Poniedziałek, 2018; Yetley, 2007).

The most common kind of industrial fortification used worldwide is the iodization of table salt. Over 70% of the world's population now utilizes iodized salt, which was first made mandatory in Switzerland (1922), USA (1924), and New Zealand. Currently, 130 nations throughout the globe have adopted this practice. Iodination levels range from 8mg/kg of salt to 110 mg per kg and vary from country to country It is important to keep in mind that iodine fortification must be implemented with regular monitoring to reduce the dangers of both insufficient and excessive iodine intake (Skeaff and Charlton, 2011).

Due to the severity of this element's deficit, food products marketed to children, women of reproducing age, as well as pregnant women are the principal recipients of industrial iron fortification. The increased risk of perinatal mortality in both mothers and their unborn children as well as child mental impairment are the most severe consequences of iron deficiency anemia (IDA). In children of various age groups, IDA also leads to subpar psychomotor development and lower outcomes on cognitive function tests (Baltussen et al., 2004). Due to their widespread use, relatively simple production, and excellent product durability, breakfast cereals are a suitable and frequently chosen carrier for iron (IFT, 2007). Additionally, it is feasible to utilize an at-home method of iron fortification, in which a carer for the child or infant at home adds micronutrient powders (MNPs) containing iron to the food that the child or infant consumes. Nearly 40 countries now have home food fortification programs in place or are planning to do so (Paganini, 2016; Paganini and Zimmermann, 2017).

When it comes to females that are fertile, dietary fortification with folic acid appears especially reasonable. The fetus's development depends on folic acid. Its absence may cause neural tube abnormalities, which can result in birth problems such as anencephaly or spina bifida. Since this vitamin is sensitive to sunlight and high temperatures and is mostly found in green vegetables, it may be difficult to meet daily dietary requirements.

Folic acid enrichment in flour has positive effects on lowering folic acid deficits in women of reproductive age and the prevalence of neural tube abnormalities in their progeny. It should be noted that the introduction of fortified grain products, as opposed to folic acid supplementation alone, seems to be more effective in decreasing the frequency of neural tube abnormalities in some places (Persad et al., 2002; Cortés and Hertrampf, 2004; De Wals et al., 2007; Abdollahi et al., 2011).

The primary purpose of fortification is to prevent nutritional deficits and their accompanying illnesses. However, industrial fortification can also be directed towards particular occupational groups that, as a result of challenging living circumstances and restricted food access, need specialized, adapted products that fit their unique needs. Soldiers are one of these groups. To alleviate nutrition shortages during WWII, chocolate bars were given to US soldiers (Henry and Chappell, 2000). As of right now, soldiers are given food items like candies, chocolate bars, and chewing gum that have been supplemented with vitamins, probiotics, and minerals. These foods not only help soldiers in combating nutritional deficiencies but can also help them avoid certain diseases like allergies (quercetin), diarrhea (probiotics), or hypotension (caffeine). Iron-fortified dietary items were also produced for female soldiers due to the decreased levels of iron after rigorous physical exercise, such as battle training (McClung et al., 2009; Karl et al., 2010). The personnel of space stations are another set of specialists who have a particular need for reinforced goods. The biggest vitamin D shortages are seen in cases where astronauts are not subjected to sunlight because spacewalks are conducted while wearing spacesuits, and because spacecraft windows are shielded from damaging UV rays even though they also inhibit vitamin D production on the skin. Members of the spaceship crew meet their vitamin D needs by consuming milk, yogurt, cereals, and orange juices that have been fortified with vitamin D. Additionally, diets for astronauts must be fortified with calcium, iron, vitamin C, and vitamin K (Smith et al., 2012).

1.1.2.5.1 *Benefits of Industrial Fortification*

a) Improved micronutrient status
b) Decreased infant and pregnant woman mortality
c) Decreased likelihood of micronutrient deficiencies' negative health effects, such as osteoporosis, blindness, and anemia
d) High availability of the additional ingredient
e) Low chance of negative consequences because of regulatory requirements and a number of safety tests that have been conducted
f) No need to alter dietary routine
g) Supporting the agriculture industry via increased financial investments and farmer education
h) The potential to make up for product nutritional value deficits
i) High degree of acceptability because problematic technologies are not used

1.1.2.5.2 *Types of Fortification*

(a) **Mandatory Fortification**

Governments may impose mandatory fortification requirements on food producers, requiring them to add certain micronutrients to specific foods or categories of foods. Mandatory fortification gives a high level of confidence that the selected food(s) will be properly fortified and accessible at all times, particularly when accompanied by a well-resourced compliance and information dissemination system. When deciding on the precise structure of a required fortification rule, governments must make sure that the combination of the food vehicle and the fortificants is safe for both the target group and non-target groups while also being dependable and effective for the target group. Basic items such as different varieties of wheat, salt, and sugar are available for purchase on the retail market for use by customers, in addition to ingredients in processed foods and processed foods that are fortified at the time of production. Due to their frequent usage and availability, simple products are most suited for mass fortification (namely, activities aimed at the whole population), while certain refined designed meals are often the better medium for targeted fortification projects (i.e. those aimed at specific population groups). Food fortification with micronutrients including folic acid, iodine, iron, and vitamin A is legally required everywhere in the globe. The most popular form of compulsory mass fortification is probably the iodization of salt. For instance, the Philippine Act Promoting Salt Iodization Nationwide needs that all food-grade salt intended for human consumption to have a minimum amount of iodine fortification. This kind of legal precedent is used in many other nations. Adding vitamin A to sugar and margarine, fortifying wheat with iron (often in combination with the restoration of vitamins B1, B2, and niacin), and, more recently, adding vitamin B12 and folic acid are more instances of obligatory mass fortification. The sorts of food delivery systems that must be fortified are often identified by a specific function, a physical or inherent quality, or both. A necessary flour, for instance, can be specified as white or wholemeal, milled from a particular grain, or used for bread making. Alternatively, the fortification criterion may only apply to items that have particular markings and labels. For example, only flour and other grain products labelled enriched are permitted by law to include extra folic acid in the United States (and some other essential micronutrients) Similarly, only salt that is designated as "iodized salt" must be iodized in Australia and New Zealand. Mass fortification can still be accomplished in these circumstances, even if the potential public health effect is less predictable, especially if the marked fortified goods represent a sizable and steady percentage of the market for that food class as a whole. (FSSAI Guidelines on Micronutrient Fortification, 2021, Dary et al., 2008, Das et al., 2013).

(b) **Voluntary Fortification**

When a food producer voluntarily decides to fortify specific foods in response to permission granted by a food law, or when the government encourages the food producer to do so under special circumstances, fortification is defined as voluntary. Industry and consumers are typically the driving forces behind voluntary fortification, as they aim to gain potential health benefits from increased micronutrient intakes. The state, on the other hand, may also be the guiding force. It is hardly unexpected that voluntary fortification may have small to significant impacts on public health given the vast range of factors that might cause it. The nutritional value of a person's fundamental diet will determine whether or not someone who regularly consumes fortified foods will experience any obvious advantages. However, governments must continue to exercise reasonable oversight over voluntary fortification via food laws or other cooperative agreements, such as industry standards of practice. The degree of regulation needs to at the very least be proportionate to the underlying danger. Such regulatory restrictions may guarantee that fortified foods are safe for all consumers while also enabling businesses to produce fortified foods that provide customers with nutritional and/or other health advantages. Scientific data that is generally accepted may indicate that the prospective advantages are plausible or likely without being demonstrable. Policymakers must guarantee that voluntary fortification agreements do not mislead or deceive consumers. They also must make sure that the market marketing of fortified foods does not conflict with or undermine any national food and nutrition policies that promote healthy eating. This might be done by passing laws defining the kinds of foods that can be fortified voluntarily as well as the acceptable pairings of micronutrients and foods. Voluntary fortification is currently permitted in many nations; however, the permitted food categories differ significantly from nation to nation. The United States permits the fortification of a considerably larger range of products than Scandinavian nations that only permit it for a small subset of foods. Similar to this, the fortifiers that have been authorized vary from a small number to almost all important micronutrients. Market factors have a significant impact on the industry's adoption of fortification practices. For instance, the majority of packaged breakfast cereals in many developed nations are moderately fortified with different combinations of micronutrients, which are frequently differentiated based on the target market; the remainder are either heavily or extensively fortified (as permitted) and/or unfortified. Other food categories, such as dairy products and fruit juices, have greater degrees of fortification variability, which is influenced by brand identity and customer differentiation. Industry interest in fortification for certain authorized categories may not exist (FSSAI Guidelines on Micronutrient Fortification, 2021, Bouis, 1999, Black et al., 2012).

1.1.2.6 Selection of Fortificants

In fact, a variety of technical and statutory elements influence the choice of a food vehicle-fortificant combination. Mandatory mass fortification is especially well suited for foods like oils, cereals, dairy products, drinks, and different condiments like sugar, sauces (like soy sauce), and salt. These foods have some, all, or none of the following traits in common:

- A significant majority of the population consumes them, including (or particularly) the demographic segments that are most at risk of insufficiency.
- They are regularly ingested in sufficient and mostly constant proportions.
- They may be handled centrally (central processing is preferred for a variety of reasons but principally because it is simpler to establish quality control measures when there are fewer sites where fortificants are applied. Monitoring and enforcement methods are therefore probably more efficient).
- They allow for the relatively simple and inexpensive addition of a nutrition premix while ensuring a uniform distribution of the ingredient across product batches. They are put to use shortly after being produced and purchased. Foods that are bought and consumed shortly after being processed often retain their vitamins better and undergo fewer sensory alterations.
- The selection of a fortificant molecule involves striking a balance between acceptable cost, dietary bioavailability, and the acceptance of any sensory modifications. The following are the important factors to take into account and issues to be aware of when choosing the best chemical form of a particular micronutrient:
- **Sensory problems**. Fortificants must be stable within predetermined parameters and must not separate from the food matrix or produce intolerable sensory issues (like flavour, colour, texture, or odour) at the amount of fortification planned. If extra packaging is required to increase the fortificant's stability, it would be beneficial that doing so would not dramatically increase the product's price and render it expensive for the client.
- **Interactions.** Prior to the implementation of a fortification program, it is critical to evaluate and investigate the likelihood or potential for interactions between the added micronutrient and the food vehicle as well as with other nutrients (either added or naturally present), particularly any interactions that may impair the fortificant's ability to be utilized metabolically.
- **Cost**. The price of fortification must not make the product more expensive or less competitive than similarly-priced unfortified foods.
- **Bioavailability**. The fortificant, which must be capable to be adequately well absorbed from the food vehicle, must be able to enhance the micronutrient status of the target population.
- The safety issue is equally crucial. For fortification to be efficient, the amount of intake needed must be suitable for a balanced diet (Hurrell 2022).

1.1.2.7 National and International Laws and Programs Regarding Fortification

National and international laws regarding food fortification vary by country and region. However, there are some common principles and guidelines that many countries follow. At the national level, many countries have established laws to mandate the fortification of certain staple foods with specific nutrients. These laws aim to address nutrient deficiencies prevalent in their population. For example, in many countries, wheat flour is fortified with iron and folic acid to combat anemia, while iodine is added to salt to prevent iodine deficiency disorders. Other commonly fortified foods include rice, maize, cooking oil, and sugar. National laws regarding food fortification often specify the types and levels of nutrients to be added, the specific food products to be fortified, and the labeling requirements for fortification. These laws may also establish monitoring and enforcement mechanisms to ensure compliance. At the international level, various organizations and agencies provide guidelines and recommendations for food fortification. One of the key international bodies in this field is the World Health Organization (WHO), which sets global standards and provides technical guidance on food fortification. The WHO's guidelines help countries establish their own fortification regulations and ensure the safety and effectiveness of fortification programs (Keats et al., 2019).

In addition, the FAO and the Codex Alimentarius Commission also provide international guidelines on food fortification. The Codex Alimentarius is a collection of internationally recognized food standards, codes of practice, and guidelines adopted by the Codex member countries. These standards often include provisions related to food fortification (Keats et al., 2019).

Overall, national and international laws regarding food fortification aim to improve public health by addressing nutrient deficiencies through the addition of specific nutrients to staple foods. These laws are essential for ensuring the safety, quality, and effectiveness of fortified foods and helping to prevent nutrient deficiencies on a large scale (Tam et al.,2020)

Global and national experiences demonstrate that partnerships with organizations/parties that can contribute in the following crucial areas—management, advocacy, capacity development, regulatory monitoring, and implementation—are most effective in achieving food fortification goals. A worldwide humanitarian group called Sight and Life seeks to eradicate all types of malnutrition. The initiatives, which include many projects centred on food fortification, are supported by data and create strategic, long-lasting alliances to put sustainable solutions into practise and enhance the lives of people who are most in need. Key stakeholders are brought together by Sight and Life as a catalyst to address the transdisciplinary issues surrounding food fortification solutions (Olson et al., 2021).

(a) **India**

In the Indian state of Andhra Pradesh, micronutrient deficiencies are often seen in both mothers and children. In response to this issue, the PM's office announced the National Nutrition Mission in the early months of 2018 with the goal of reducing vitamin and mineral deficiencies via the cost-effective fortification of basic foods. Rice is the most efficient food source for the poorest people among the many Andhra Pradesh staples. It is also one of just two staples that, when properly fortified, can transport a variety of vitamins and minerals. Through the government's three primary food supplementation programs—the MDM, ICDS, and PDS—it is distributed to the most nutritionally vulnerable people (Tam et al., 2020).

Following expert discussions, the Food Safety and Standards Authority of India (FSSAI) created requirements for fortified rice based on solid clinical data from eminent Indian academic institutions. Fortification of rice has now gathered pace, and 15 states have currently developed plans to begin implementing it via ICDS, MDM, and PDS. In order to improve these initiatives, Tata Trust, Sight and Life, and the local government have used continuous blending, a promising and efficient blending method. This sort of strategy is being used for the first time in India for fortified rice under extensive government initiatives (Olson et al., 2020).

(b) **Rwanda**

Five staple foods, including flour of wheat and maize (fortified with iron, vitamin A, folic acid, niacin, zinc, vitamin B1, and vitamin B12), sugar (fortified with vitamin A), edible oils (fortified with vitamin A), and salt (fortified with vitamin A), began to be fortified by the Rwandan government's Ministry of Health in October 2019 (fortified with iodine). Rwanda has created obligatory food fortification standards and supported the adoption of new food fortification legislation that makes it necessary for all locally produced and imported food goods to be one of the five essential foods. Partners like Sight and Life have helped with this endeavor. Currently, fortified flours and certain imported eatable oils from adjacent nations are available on the market due to food manufacturers like MINIMEX, SOSOMA, and African Improved Foods.

2019 saw the establishment of Rwanda's own FDA. Together with Sight and Life, a forum for food fortification was also created where all stakeholders can gather to address issues relating to the application of the food fortification rule. Members of the Rwanda National Fortification Alliance, for instance, identified VAT taxation as one of their greatest challenges, and important stakeholder groups like the private sector, CSOs, and NGOs supported discussions with the Ministry of Finance to reduce or waive VAT taxes on fortified flour as a crucial fortification incentive (Olson et al., 2020).

(c) **Ghana**

Women and children in Ghana continue to have high rates of micronutrient deficiencies. A recent micronutrient survey by the GHS (Ghana Health Service) revealed deficiencies in important micronutrients like iron and folate and vitamin A, particularly in pregnant women. Additionally, 50 and 66% of preschool-aged children and non-pregnant women, correspondingly, suffer from anemia, and 30% of preschool children suffer from iron and vitamin A deficiencies (UNICEF, 2020).

In 2013, a partnership was formed between Sight and Life, DSM, the German Federal Ministry for Economic Cooperation and Development, and the Children's Investment Fund Foundation. (Olson et al., 2020).

1.1.2.8 Efficacy Trials

Efficacy trials in food fortification refer to studies that evaluate the effectiveness of adding specific nutrients to food to enhance the nutritional status of populations or specific target groups. These trials aim to assess the impact of fortified foods on nutrient intake, body nutrient stores, and health outcomes (Waller et al., 2020).

The efficacy trials in food fortification typically involve the following steps:

i). **Formulating the fortified food:** The nutrient(s) to be added are selected based on the targeted population's nutritional needs and the feasibility of fortification. The fortification process and the desired nutrient levels are determined.

ii). **Designing the trial**: The study design should be robust and should include an appropriate control group or comparison group to assess the efficacy of the fortification intervention.

iii). **Participant recruitment**: Individuals or communities eligible for the trial are recruited. Sample size is determined based on statistical power calculations, taking into account the expected effect size of the fortification intervention.

iv). **Randomization**: To the control group or the fortified meal group, participants are randomly allocated. The randomization procedure increases the likelihood that any differences between the two groups are due to the intervention and not another factor.

v). **Intervention implementation**: Participants in the fortified food group receive the fortified food product, while the control group may receive a non-fortified food product or a placebo. The intervention is typically provided for a specific duration, often weeks or months.

vi). **Data collection:** Various measures are taken during the trial, including dietary intake assessments, blood and urine samples to measure nutrient levels, anthropometric measurements, and health outcomes.

vii). **Statistical analysis**: The collected data are evaluated using appropriate statistical methods to determine the effect of fortified food on nutrient intake, nutrient levels, and health outcomes. The outcomes are compared between the fortified food group and the control group.

viii). **Reporting and dissemination**: The findings of the efficacy trial are compiled in a report or manuscript and are disseminated through scientific journals, conferences, or other channels to inform policymakers, nutritionists, and relevant stakeholders (Waller et al., 2020).

Efficacy trials in food fortification play a crucial role in determining the impact and effectiveness of fortification interventions. These studies help inform the development of evidence-based fortification programs and policies aimed at enhancing the nutritional status of populations (Waller et al., 2020).

1.3　CONCLUSION

Food fortification is a powerful and cost-effective strategy to address and prevent nutritional deficiencies, especially in populations with limited access to diverse and nutritious food sources. Through the addition of essential minerals, vitamins, and other nutrients to commonly consumed foods, fortification aims to enhance public health and reduce the prevalence of various deficiencies. In conclusion, food fortification is a valuable tool in combating nutrient deficiencies and promoting public health. By working hand in hand with nutrition education and other health interventions, fortification can contribute significantly to improving the overall well-being and quality of life for populations worldwide.

REFERENCES

Abdollahi, Z., Elmadfa, I., Djazayery, A., et al. 2011. Efficacy of flour fortification with folic acid in women of childbearing age in Iran. *Annals of Nutrition and Metabolism*, 58, 188–196.

Adinepour, F., Pouramin, S., Rashidinejad, A., Jafari, S.M. 2022. Fortification/enrichment of milk and dairy products by encapsulated bioactive ingredients. *Food Research International*, 157, 111212.

Allen, L., de Benoist, B., Dary, O., Hurrell, R. 2006. *Guidelines on Food Fortification with Micronutrients*. Paris: World Health Organization, Food and Agricultural Organization of the United Nations.

Allen, L.H., Haskell, M. 2002. Estimating the potential for vitamin A toxicity in women and young children. *Journal of Nutrition*, 132, 2907S–2919S.

Assunção, M., Santos, I., Barros, A., Gigante, D., Victora, C.G. 2013. Flour fortification with iron has no impact on anaemia in urban Brazilian children—Corrigendum. *Public Health Nutrition*, 16, 188.

Bailey, R.L., West, K.P. Jr., Black, R.E. 2015. The epidemiology of global micronutrient deficiencies. *Annals of Nutrition and Metabolism*, 66, 22–33.

Baltussen, R., Knai, C., Sharan, M. 2004. Iron fortification and iron supplementation are cost-effective interventions to reduce iron deficiency in four subregions of the world. *Journal of Nutrition*, 134, 2678–2674.

Bhagwat, S., Gulati, D., Sachdeva, R., Sankar, R. 2014. Food fortification as a complementary strategy for the elimination of micronutrient deficiencies: case studies of large scale food fortification in two Indian states. *Asia Pacific Journal of Clinical Nutrition*, 23. s29-s37

Black, L.J., Seamans, K.M., Cashman, K.D., Kiely, M. 2012. An updated systematic review and meta-analysis of the efficacy of vitamin D food fortification. *Journal of Nutrition*, 142, 1102–1108.

Bohl, D.D., Shen, M.R., Kayupov, E., et al. 2016. Hypoalbuminemia independently predicts surgical site infection, pneumonia, length of stay, and readmission after total joint arthroplasty. *Journal of Arthroplasty*, 31(1), 15.

Bouis, H.E. 1999. Economics of enhanced micronutrient density in food staples. *Field Crops Research*, 60, 165–173.

Bouis, H.E., Hotz, C., McClafferty, B., Meenakshi, J.V., Pfeiffer, W.H. 2011. Biofortification: A new tool to reduce micronutrient malnutrition. *Food and Nutrition Bulletin*, 32 (Suppl. 1), 31S–40S.

Boy, E., Mannar, V., Pandav, C., de Benoist, B., Viteri, F., Fontaine, O., Hotz, C. 2009. Achievements, challenges, and promising new approaches in vitamin and mineral deficiency control. *Nutrition Reviews*, 67, S24–S30.

Bruins, M.J., Mugambi, G., Verkaik-Kloosterman, J., Hoekstra, J., Kraemer, K., Osendarp, S., Melse-Boonstra, A., Gallagher, A.M., Verhagen, H. 2015. Addressing the risk of inadequate and excessive micronutrient intakes: Traditional versus new approaches to setting adequate and safe micronutrient levels in foods. *Food and Nutrition Research*, 59, 26020.

Bulusu, S., Wesley, A.S. 2011. Addressing micronutrient malnutrition through food fortification. In *Public Health Nutrition in Developing Countries* (ed. S. Chander-Vir), pp. 795–843. Woodhead Publishing India.

Cakmak, I., Kutman, U.B. 2017. Agronomic biofortification of cereals with zinc: A review. *European Journal of Soil Science*, 69, 172–180.

Carducci, B., Keats, E.C., Ruel, M., Haddad, L., Osendarp, S.J.M., Bhutta, Z. 2021. A Food systems, diets and nutrition in the wake of COVID-19. *Nature Food*, 2, 68–70.

Charlton, K., Skeaff, S. 2011. Iodine fortification: Why, when, what, how, and who? *Current Opinion in Clinical Nutrition and Metabolic Care*, 14, 618–624.

Dary, O. 2008. Establishing safe and potentially efficacious fortification contents for folic acid and vitamin B12. *Food and Nutrition Bulletin*, 29(2) (Suppl. 2), S214–S224.

Das, J.K., Salam, R.A., Kumar, R., Bhutta, Z.A. 2013. Micronutrient fortification of food and its impact on woman and child health: A systematic review. *Systematic Reviews*, 2, 67.

De Wals, P., Tairou, F., Van Allen, M.I., et al. 2007. Reduction in neural-tube defects after folic acid fortification in Canada. *New England Journal of Medicine*, 357, 135–142.

Dwyer, J., Wiemer, K., Dary, O., et al. 2015. Fortification and health: Challenges and opportunities. *Advances in Nutrition*, 6, 124–131.

Engle-Stone, R., Vosti, S.A., Luo, H., Kagin, J., Tarini, A., Adams, K.P., French, C., Brown, K.H. 2019. Weighing the risks of high intakes of selected micronutrients compared with the risks of deficiencies. *Annals of the New York Academy of Sciences*, 1446, 81–101.

FAO. 1996. *Rome Declaration on World Food Security*. Rome: FAO.

Fiedler, J., Babu, S., Smitz, M.F., Lividini, K., Bermudez, O. 2012. Indian social safety net programs as platforms for introducing wheat flour fortification: A case study of Gujarat, India. *Food and Nutrition Bulletin*, 33, 11–30.

FSSAI Guidelines on Micronutrient Fortification. 2021. https://fssai.gov.in/upload/uploadfiles/files/Compendium_Food_Fortification_Regulations_04_03_2021.pdf

Gharsallaoui, A., Roudaut, G., Chambin, O., Voilley, A., Saurel, R. 2007. Applications of spray-drying in microencapsulation of food ingredients: An overview. *Food Research International*, 40(9), 1107–1121.

Galaniha, L.T., McClements, D.J., Nolden, A. 2020. Opportunities to improve oral nutritional supplements for managing malnutrition in cancer patients: A food design approach. *Trends in Food Science and Technology*, 102, 254–260.

Heidkamp, R.A., Piwoz, E., Gillespie, S., Keats, E.C., D'Alimonte, M.R., Menon, P., … Bhutta, Z.A. 2021. Mobilising evidence, data, and resources to achieve global maternal and child undernutrition targets and the Sustainable Development Goals: An agenda for action. *The Lancet*, 397(10282), 1400–1418.

Henry, M.R., Chappell, M. 2000. *The US Army in World War II*. Osprey Publishing. UK.

Hertrampf, E., Cortés, F. 2004. Folic acid fortification of wheat flour: Chile. *Nutrition Reviews*, 62, S44–S48.

Hoddinott, J., Rosegrant, M., Torero, M. 2013. Hunger and malnutrition. In *Global Problems, Smart Solutions: Costs and Benefits* (ed. B. Lomborg), pp. 332–367. New York: Cambridge University Press and Copenhagen Consensus Center.

Hoddinott, J., Rosegrant, M., Torero, M. 2012. *Investments to Reduce Hunger and Undernutrition; Copenhagen Consensus Center Working Paper March*. Copenhagen: Copenhagen Consensus Center.

Hurrell, R. F. 2022. Ensuring the efficacious iron fortification of foods: a tale of two barriers. *Nutrients*, 14(8), 1609.

Jaberi, F.M., Parvizi, J., Haytmanek, C.T., et al. 2008. Procrastination of wound drainage and malnutrition affect the outcome of joint arthroplasty. *Clinical Orthopaedics and Related Research*, 466(6), 1368.

Jensen, G.L., Mirtallo, J., Compher, C., International Consensus Guideline Committee et al. 2010. Adult starvation and disease-related malnutrition: A proposal for etiology-based diagnosis in the clinical practice setting from the International Consensus Guideline Committee. *JPEN. Journal of Parenteral and Enteral Nutrition*, 34(2), 156.

Kalayci, M., Arisoy, Z., Ceikic, C., Kaya, Y., Savasli, E., Tezel, M., Onder, O., et al. 2011. The effects of soil and foliar applications of zinc on grain zinc concentrations of wheat and maize. Presentation Presented at the 3rd International Zinc Symposium, Hyderabad, India, October 10–14.

Karl, J.P., Lieberman, H.R., Cable, S.J., et al. 2010. Randomized, double-blind, placebocontrolled trial of an iron-fortified food product in female soldiers during military training: Relations between iron status, serum hepcidin, and inflammation. *American Journal of Clinical Nutrition*, 92, 93–100.

Kaur, T., Rana, K.L., Kour, D., Sheikh, I., Yadav, N., Kumar, V., Yadav, A.N., Dhaliwal, H.S., Saxena, A.K. 2020. Microbe-mediated biofortification for micronutrients: Present status and future challenges. In Vijai Kumar Gupta (ed.), *New and Future Developments in Microbial Biotechnology and Bioengineering (Ed Rastegari et al.)*, pp. 1–17. Amsterdam: Elsevier.

Keats, E.C., Neufeld, L.M., Garrett, G.S., Mbuya, M.N., Bhutta, Z.A. 2019. Improved micronutrient status and health outcomes in low-and middle-income countries following large-scale fortification: Evidence from a systematic review and meta-analysis. *The American Journal of Clinical Nutrition*, 109(6), 1696–1708.

Klikocka, H., Marks, M. 2018. Sulphur and nitrogen fertilization as a potential means of agronomic biofortification to improve the content and uptake of microelements in spring wheat grain DM. *Journal of Chemistry*, 2018, 9326820.

Kraemer, K., van Zutphen, K.G. 2019. Translational and implementation research to bridge evidence and implementation. *Annals of Nutrition and Metabolism*, 75, 144–148.

Kumar, D., Dhaliwal, S.S., Naresh, R.K., Salaria, A. 2018. Agronomic biofortification of paddy through nitrogen, zinc and iron fertilization: A review. *International Journal of Current Microbiology and Applied Sciences*, 7, 2942–2953.

Liyanage, C., Hettiarachchi, M. 2011. Food fortification. *Ceylon Medical Journal*, 54, 124–127.

Luthringer, C., Rowe, L.A., Vossenaar, M., Garrett, G.S. 2015. Regulatory monitoring of fortified foods: Identifying barriers and good practices. *Global Health Science and Practice*, 3, 446–461.

Manuela, P., Drakula, S., Cravotto, G., et al. 2020. Biological activity and sensory evaluation of cocoa by-products NADES extracts used in food fortification. *Innovative Food Science and Emerging Technologies*, 66, 102514

Martorell, R., Ascencio, M., Tacsan, L., et al. 2015. Effectiveness evaluation of the food fortification program of Costa Rica: Impact on anemia prevalence and hemoglobin concentrations in women and children. *American Journal of Clinical Nutrition*, 101, 210–217.

Martorell, R., Ascencio, M., Tacsan, L., Alfaro, T., Young, M.F., Addo, O.Y., Dary, O., Flores-Ayala, R. 2015. Effectiveness evaluation of the food fortification program of Costa Rica: Impact on anemia prevalence and hemoglobin concentrations in women and children. *American Journal of Clinical Nutrition*, 101, 210–217.

McClung, J.P., Karl, J.P., Cable, S.J., et al. 2009. Longitudinal decrements in iron status during military training in female soldiers. *British Journal of Nutrition*, 102, 605–609.

McWhirter, J.P., Pennington, C.R. 1994. Incidence and recognition of malnutrition in hospital. *BMJ*, 308, 945.

Method, A., Tulchinsky, T.H. 2015. Food fortification: African countries can make more progress. *Advances in Nutrition and Food Technology: Open Access*, 2015, S22–S28.

Milne, A.C., Potter, J., Vivanti, A., Avenell, A. 2009. Protein and energy supplementation in elderly people at risk from malnutrition. *Cochrane Database of Systematic Reviews*, 2, 1–98.

Mkambula, P., Mbuya, M.N.N., Rowe, L.A., Sablah, M., Friesen, V.M., Chadha, M., Osei, A.K., Ringholz, C., Vasta, F.C., Gorstein, J. 2020. The unfinished agenda for food fortification in low- and middle-income countries: Quantifying progress, gaps and potential opportunities. *Nutrients*, 29, 354.

Nathoo, T., Holmes, C.P., Ostry, A. 2005. An analysis of the development of Canadian food fortification policies: The case of vitamin B. *Health Promotion International*, 20, 375–382.

National Academy of Sciences. 2003. *Dietary Reference Intakes: Guiding Principles for Nutrition Labeling and Fortification*. Washington, DC: National Academy of Sciences.

Nestel, P., Bouis, H.E., Meenakshi, J.V., Pfeiffer, W. 2006. Biofortification of staple food crops. *Journal of Nutrition*, 136, 1064–1067.

Norman, K., Haß, U., Pirlich, M. 2021. Malnutrition in older adults—Recent advances and remaining challenges. *Nutrients*, 13(8), 2764.

Norman, K., Pichard, C., Lochs, H., Pirlich, M. 2008. Prognostic impact of disease-related malnutrition. *Clinical Nutrition*, 27(1), 5.

Olson, R., Gavin-Smith, B., Ferraboschi, C., Kraemer, K. 2021. Food fortification: The advantages, disadvantages and lessons from sight and life programs. *Nutrients*, 13(4), 1118.

Osendarp, S.J.M., Martínez, H., Garrett, G.S., Neufeld, L., de Regil, L., Vossenaar, M., Darnton-Hill, I. 2018. Large-scale food fortification and biofortification in low- and middle-income countries: A review of programs, trends, challenges, and evidence gaps. *Food and Nutrition Bulletin*, 39(2), 315–331.

Padbhushan, R., Sharma, S., Kumar, U., Rana, D.S., Kohli, A., Kaviraj, M., Parmar, B., Kumar, R., Annapurna, K., Sinha, A.K., et al. 2021. Meta-analysis approach to measure the effect of integrated nutrient management on crop performance, microbial activity, and carbon stocks in Indian soils. *Frontiers in Environmental Science*, 9, 724702.

Paganini, D., Zimmermann, M.B. 2017. The effects of iron fortification and supplementation on the gut microbiome and diarrhea in infants and children: A review. *American Journal of Clinical Nutrition*, 106, 1688S–1693S.

Persad, V.L., Van den Hof, M.C., Dubé, J.M., Zimmer, P. 2002. Incidence of open neural tube defects in Nova Scotia after folic acid fortification. *Canadian Medical Association Journal*, 167, 241–245.

Persson, M., Hytter-Landahl, Å., Brismar, K., & Cederholm, T. (2007). Nutritional supplementation and dietary advice in geriatric patients at risk of malnutrition. *Clinical Nutrition*, 26(2), 216–224.

Pfeiffer, W.H., McClafferty, B. 2007. HarvestPlus: Breeding crops for better nutrition. *Crop Science*, 47, S88–S105.

Phattarakul, N., Rerkasem, B., Li, L.J., Wu, H., Zou, C.Q., Ram, H., Sohu, V.S., et al. 2012. Biofortification of rice grain with zinc through zinc fertilization in different countries. *Plant and Soil*, 361, 131–141.

Poniedziałek, B., Niedzielski, P., Kozak, L., et al. 2018. Monitoring of essential and toxic elements in multi-ingredient food supplements produced in European Union. *Journal of Consumer Protection and Food Safety*, 13, 41–48.

Preedy, V., Burrow, G., Watson, R. 2009. *Comprehensive Handbook of Iodine: Nutritional, Biochemical, Pathological and Therapeutic Aspects*. Cambridge, MA: Academic Press.

Pruzansky, J.S., Bronson, M.J., Grelsamer, R.P. 2014. Prevalence of modifiable surgical site infection risk factors in hip and knee joint arthroplasty patients at an urban academic hospital. *Journal of Arthroplasty*, 29(2), 272.

Qaim, M., Stein, A.J., Meenakshi, J.V. 2007. Economics of biofortification. *Agricultural Economy*, 37(s1), 119–133.

Rai, J., Gill, S.S., Kumar, B.R. 2002. The influence of preoperative nutritional status in wound healing after replacement arthroplasty. *Orthopedics*, 25(4), 417.

Russell, C.A. 2007. The impact of malnutrition on healthcare costs and economic considerations for the use of oral nutritional supplements. *Clinical Nutrition Supplements*, 2(1), 25–32.

Rzymski, P., Klimaszyk, P. 2018. Is the yellow knight's mushroom edible or not? A systematic review and critical viewpoints on the toxicity of Tricholoma equestre. *Comprehensive Reviews in Food Science and Food Safety*, 17, 1309–1324.

Saltzman, A., Birol, E., Oparinde, A., Andersson, M.S., Asare-Marfo, D., Diressie, M.T., … Zeller, M. 2017. Availability, production, and consumption of crops biofortified by plant breeding: Current evidence and future potential. *Annals of the New York Academy of Sciences*, 1390(1), 104–114.

Selim, M.M. 2020. Introduction to the integrated nutrient management strategies and their contribution to yield and soil properties. *International Journal of Agronomy*, 2, 2821678.

Siddiqui, F., Salam, R.A., Lassi, Z.S., Das, J.K. 2020. The intertwined relationship between malnutrition and poverty. *Frontiers in Public Health*, 8, 453.

Singh, B.R., Timsina, Y.N., Lind, O.C., Cagno, S., Janssens, K. 2018. Zinc and iron concentration as affected by nitrogen fertilization and their localization in wheat grain. *Frontiers in Plant Science*, 9, 307.

Singh, D., Rajawat, M.V.S., Kaushik, R. 2017. Beneficial role of endophytes in biofortification of Zn in wheat genotypes varying in nutrient use efficiency grown in soils sufficient and deficient in Zn. *Plant and Soil*, 416(1–2), 107–116.

Siró, I., Kápolna, E., Kápolna, B., Lugasi, A. 2008. Functional food: Product development, marketing and consumer acceptance–a review. *Appetite*, 51(3), 456–456.

Smith, S.M., Davis-Street, J., Neasbitt, L., Zwart, S.R. 2012. *Space Nutrition*. NASA, USA.

Soeters, P.B., Reijven, P.L., Schols, J.M., Halfens, R.J., Meijers, J.M., van Gemert, W.G.2008. A rational approach to nutritional assessment. *Clinical Nutrition*, 43(4), 706-716.

Spohrer, R., Larson, M., Maurin, C., Laillou, A., Capanzana, M., Garrett, G.S. 2013. The growing importance of staple foods andcondiments used as ingredients in the food industry and implications for large-scale food fortification programs in Southeast Asia. *Food and Nutrition Bulletin*, 34(2) (Suppl. 2), S50–S61.

Suave, J., Dall'Agnol, E.C., Pezzin, A.P.T., Silva, D.A.K., Meier, M.M., Soldi, V. 2006. Microencapsulação: Inovação em diferentes áreas. *Revista Saúde e Ambiente/Health and Environment Journal*, 7(2), 12–20.

Sustainable Development Goals Knowledge Platform. 2018. Sustainable development goal strengthen the means of implementation and revitalize the global partnership for sustainable development. Available online: https://sustainabledevelopment.un. org/sdg17 (accessed on 6 September 2020).

Tam, E., Keats, E.C., Rind, F., Das, J.K., Bhutta, Z.A. 2020. Micronutrient supplementation and fortification interventions on health and development outcomes among children under-five in low-and middle-income countries: A systematic review and meta-analysis. *Nutrients*, 12(2), 289.

Van den Wijngaart, A., Bégin, F., Codling, K., Randall, P., Johnson, Q.W. 2013. Regulatory monitoring systems of fortified salt and wheat flour in selected ASEAN countries. *Food and Nutrition Bulletin*, 34, S102–S111.

Van Stuijvenberg, M.E. 2005. Using the school feeding system as a vehicle for micronutrient fortification: Experience from South Africa. *Food and Nutrition Bulletin*, 26, S213–S219. Nutrients 2021, 13, 1118 12 of 12.

Velu, G., Ortiz-Monasterio, I., Cakmak, I., et al. 2014. Biofortification strategies to increase grain zinc and iron concentrations in wheat. *Journal of Cereal Science*, 9(3), 365–372.

Venkatesh Mannar, M.G. 2011. Universal salt iodization (USI). In *Public Health Nutrition in Developing Countries* (ed. S. Chander-Vir), pp. 562–574. Woodhead Publishing India.

Verma, A. 2015. Food fortification: A complementary strategy for improving micronutrient malnutrition (MNM) status. *Food Science Research Journal*, 6(2), 381–389.

von Grebmer, K., Bernstein, J., Mukerji, R., Patterson, F., Wiemers, M., Chéilleachair, R.N., Fritschel, H. 2019. Global hunger index by severity, map in 2019 global hunger index: The challenge of hunger and climate change.

Waller, A.W., Andrade, J.E., Mejia, L.A. 2020. Performance factors influencing efficacy and effectiveness of iron fortification programs of condiments for improving anemia prevalence and iron status in populations: A systematic review. *Nutrients*, 12(2), 275.

Watts-Williams, S.J., Cavagnaro, T.R. 2018. Arbuscular mycorrhizal fungi increase grain zinc concentration and modify the expression of root ZIP transporter genes in a modern barley (Hordeum vulgare) cultivar. *Plant Science: An International Journal of Experimental Plant Biology*, 274, 163–170.

WHO. 2015. *General Principles for the Addition of Essential Nutrients to Foods*. CAC/GL 9-1987. Geneva: World Health Organization.

WHO. 2013. World Health Report. Research for Universal Health Coverage. Geneva: World Health Organization.

WHO, F.A.O. 2006. *Guidelines on Food Fortification with Micronutrients*. Rome: World Health Organization: Rome: FAO.

World Food Programme and Sight and Life. Scaling up Rice Fortification in Asia. 2017. Available online: https://sightandlife. org/wp-content/uploads/2017/02/SAL_WFP_Suppl.pdf (accessed on 10 October 2020).

Yetley, E.A. 2007. Multivitamin and multimineral food supplements: Definitions, characterization, bioavailability, and drug interactions. *American Journal of Clinical Nutrition*, 85, 269–276.

2 Micronutrients

*Sweta Joshi, Nazia Tabassum,
Ahmed Mobeen, and Jinku Bora*

2.1 INTRODUCTION

Micronutrients are essential nutrients that our body needs in small amounts to function properly. These nutrients include vitamins and minerals, which are required for various physiological processes such as metabolism, growth, and development. Despite being required in small amounts, micronutrients play a critical role in maintaining good health and preventing chronic diseases. The fundamental role of micronutrients in human metabolism and physiology is in the preservation, improvement, and prevention of health and disease. The body's balance, physiological functionality, and a child's optimal growth and development all depend on adequate intakes. More than 2 billion people, according to the World Health Organization (WHO), are low in critical micronutrients (Shergill-Bonner, 2017; Chilimba et al., 2012; Longchamp et al., 2013, 2015).

Minerals are required in small amounts, however they are essential for various physiological functions such as building strong bones, transmitting nerve impulses, and maintaining a healthy immune system. There are two types of minerals: macrominerals and trace minerals. Macrominerals are required in larger amounts, while trace minerals are required in smaller amounts. Examples of trace minerals, commonly termed as micronutrients, include iron, zinc, selenium, iodine, etc. (Silva et al., 2019; Detzel and Wieser, 2015; Martinez-Navarrete et al., 2002).

Micronutrient deficiencies can lead to various health problems, including anemia, goiter, osteoporosis, and rickets (Table 2.1 and Table 2.2). Micronutrient deficiencies are common in developing countries, where access to a diverse range of foods is limited. In developed countries, micronutrient deficiencies are less common, but they still occur, especially among certain populations such as pregnant women and the elderly. The best way to ensure adequate intake of micronutrients is to consume a balanced and varied diet that includes a wide range of fruits, vegetables, whole grains, lean proteins, and low-fat dairy products. In some cases, dietary supplements may be necessary to ensure adequate intake of certain micronutrients, but their excessive intake may sometimes prove to be harmful (Al-Fartusie and Mohssan, 2017; Pérez-Granados and Vaquero, 2002; Jugdaohsingh, 2007; Chilimba et al., 2012; Longchamp et al., 2013, 2015). The chapter summarizes in detail the sources, bioavailability, role in health, and deficiency diseases of different micronutrients.

2.1.1 IRON

2.1.1.1 Sources and Bioavailability

It is one of the elements found abundantly on earth. Even having geological abundance, iron is not readily available for uptake due to its tendency to form highly insoluble oxides in the presence of oxygen (Wood et al., 2005; Quintero-Gutiérrez et al., 2008). To enhance its bioavailability, various cellular mechanisms have advanced. Iron in the human body exists as heme enzymes, heme compounds: hemoglobin or myoglobin, nonheme compounds i.e., flavin-iron enzymes, transferring, ferritin and in complex forms bounded to the protein i.e., hemoprotein (McDowell, 1992). The hemoglobin present in circulating erythrocytes and myoglobin accounts for about two-thirds and 15% of the body's iron respectively, whereas 25% is present in a readily mobilizable iron store involved in

TABLE 2.1

Source, Functions, and Deficiencies of Various Micronutrients

Micronutrient	Food Sources	RDA/AIs (Adults>19 Years)	Functions	Deficiency
Iron	Meat and meat products, eggs, oysters, spinach, lentils, beans,	8–18 mg	Hemoglobin formation, oxygen transport	Anemia
Manganese	Pineapple, pecans, peanuts, oatmeal, legumes, spinach	4.0 mg/day	Carbohydrate, amino acid & cholesterol metabolism, calcium absorption, protection against free radical damage	Bone demineralization & Poor growth in children, skin rash
Copper	Liver, crabs, fish, nuts, whole grain	1.7–2.0 mg/day	Required for connective tissue formation, as well as normal brain and nervous system function	Reduced absorption of iron, anemia
Zinc	Oysters, crab, meat, poultry, mushroom, legumes	8–11 mg	Necessary for normal growth, immune function, and wound healing	Growth impairment, sexual dysfunction, inflammation, gastrointestinal symptoms, or cutaneous involvement
Iodine	Seaweed, fish, shellfish, iodized salt, yogurt, eggs, milk, cheese	150 µg/day	Assists in thyroid regulation and normal metabolism of cells	Iodine deficiency disorders (IDDs), include endemic goiter, hypothyroidism, cretinism, decreased fertility rate, increased infant mortality, and mental retardation
Fluorine	Fruit juice, drinking water, crab	3–4 mg	Necessary for the development of bones and teeth	Dental caries, osteoporosis
Selenium	Brazil nuts, pork, fish, shellfish	40 µg/day	Important for thyroid health, reproduction and defense against oxidative damage	Autoimmune thyroid conditions such as Hashimoto's disease and Graves' disease
Chromium	Meat, shellfish, fruits, vegetables, nuts, brewer's yeast, wine	50 µg/day	Breakdown of fats and carbohydrates, maintain blood sugar level	Increased blood sugar, cholesterol, diabetes, heart disease
Cobalt	Fish, nuts, broccoli, spinach, oats		Component of vitamin B12, involved in production of RBCs	Anemia
Vanadium	Mushroom, shellfish, grains, beverages, parsley		Treats diabetes, low blood sugar, high cholesterol, heart disease, cancer, tuberculosis, syphilis, water retention (edema) Improves athletic performance in weight training	Growth retardation, bone deformities, infertility (animals) Not essential (human)

(*Continued*)

TABLE 2.1 (CONTINUED)

Source, Functions, and Deficiencies of Various Micronutrients

Micronutrient	Food Sources	RDA/AIs (Adults>19 Years)	Functions	Deficiency
Silicon	Whole grain and their products, green beans		Essential for structural integrity & development of connective tissue	Abnormal tissue growth
Nickel	Soy products, tofu, cocoa, cashew, figs		Hormone regulation, lipid metabolism	Disrupts metabolism
Molybdenum	Legumes, whole grain, dairy products, meat	45 µg/day	Aids metabolism, prevents toxin build up in body	Rare (brain dysfunction)

TABLE 2.2

Recommended Dietary Allowance And Estimated Average Requirements for Indians, 2020

Groups	Iron (mg/day)	Zinc (mg/day)	Iodine (µg/day)
Infants (0-12 months)	46 µg/day–5 mg/day	----	100–130
Children (1-9 years)	9–16	5–8	90–120
Adolescence (10-17 years)	21–32	9–12	150
Men	17–19	12–17	150
Women	21–29	10–13.2	150
Pregnant	35–40	14.5	250
Lactating women	21–23	14	280

many cellular functions (Takeda, 2003; Tuschl et al., 2012; Chilimba et al., 2012; Longchamp et al., 2013, 2015). The iron from dietary sources occurs in two forms, heme and nonheme. The principal source of heme iron (hemoglobin and myoglobin) is meat, fish, and poultry, and sources for non-heme iron are legumes, pulses, fruits, cereals, and vegetables. The availability of non-heme iron in diets is significantly higher than heme iron, whereas the bioavailability of heme iron (15%–35%) is higher than non-heme iron (2%–20%) and is also susceptible to influence from other dietary components such as calcium, vitamin C, phytates, polyphenols, etc. (Hurrell, 2002). Therefore, consumption of mainly plant-based diets often results in reduced dietary iron bioavailability (Zimmermann and Hurrell, 2007) and increases the chances of iron deficiency (Haider et al., 2018).

The amount of iron being absorbed from the food being ingested is very low due to many factors. The enterocytes, members of the solute carrier group of membrane transport proteins, are involved in the absorption of iron in the duodenum and upper jejunum parts of the body (16). Normally, at physiological pH, ferrous iron (Fe+2) gets oxidized to the insoluble ferric (Fe+3) form. The gastric juice in the duodenum lowers the pH, resulting in the reduction of the ferric (Fe+3) form by ferric reductases in the intestinal lumen. This causes the transport of ferrous iron across the apical membrane of enterocytes. If the production of gastric pH is reduced, the absorption of iron is significantly reduced. The iron after being transferred from the duodenal mucosa into the blood, is transported by transferring to the cells or the bone marrow for the process of erythropoiesis, i.e. Production of red blood cells (RBCs) (Hurrell, 1997; Frazer and Anderson, 2005; Nadadur et al., 2008).

The metabolism of dietary heme is not dependent on pH of the duodenum, making its absorption more effective than that of inorganic iron. The absorption of dietary heme is not affected by phytate and polyphenols. Owing to their higher concentration of hemoglobin, red meats are a potent source of iron. Iron is present in a bound form and is transported in the body by transferrin and stored in ferritin molecules. There is no physiological mechanism for the excretion of excess iron once absorbed from the body except through blood loss from bleeding, menstruation, and pregnancy (Abbaspour et al., 2014; Vasiluk et al., 2019).

2.1.1.2   Role of Iron in Human Health

Iron plays a crucial role in various metabolic processes like oxygen transport, electron transport, and DNA synthesis. Recently, there have been studies on the effect of iron overload in the pathogenesis and manifestations of COVID-19 disease, including disease hypercoagulation, hyperferritinemia, inflammation, and immune dysfunction. Even though iron is essential for all, due to iron dysregulation and overload, the produced free unbound iron is capable of generating highly reactive oxygen species (ROS). ROS damages nucleic acids, proteins, and lipids in the cells. The iron-catalyzed lipid damage causes nonapoptotic cell death known as ferroptosis, which results in a series of inflammatory reactions (Habib et al., 2021). Being a pro-oxidant, high iron levels in the body is associated with the risk of the development of type 2 diabetes, gestational diabetes, and polycystic ovarian syndrome (Rajpathak et al., 2009). The dysregulation in iron homeostasis at different levels, including absorption, systemic transportation, and cellular uptake and storage, leads to the growth of cancer cells, which need more iron for their growth and metabolism (Salnikow, 2021).

2.1.1.3   Deficiency

Deficiency of iron affects billions of people worldwide and is found to be more prevalent in developing nations, particularly affecting children, pregnant and premenopausal women. It arises due to a physiological increase in iron requirement in children, adolescents, and pregnant women (Camaschella, 2019) which is not met by regular iron intake due to excessive blood loss or reduced iron absorption from the diet. Symptoms of iron deficiency include fatigue, lethargy, dizziness, pallor, headache, and low concentration; however, it may also be asymptomatic. Dietary iron in meat (around 30–70% heme iron) has high absorption efficiency (15–35%) while plant (non-heme) has low (10%) and is also susceptible to influence from other dietary components such as calcium, vitamin C, phytates, polyphenols, etc. (Hurrell, 2002). Therefore, consumption of mainly plant-based diets often results in reduced dietary iron bioavailability (Zimmermann and Hurrell, 2007), and increases the chances of iron deficiency (Haider et al., 2018).

Iron deficiency results first in the reduction and then depletion of stored iron, particularly macrophage and hepatocyte iron stores, thereby causing a reduction in hemoglobin concentration. This results in anemia, an iron deficiency disorder that causes clinical and functional impairments and disorders, in severe cases resulting in death. However, anemia may also result due to other nutritional deficiencies and disorders such as malaria, HIV, tuberculosis, etc. (Lynch, 2005). Anemia due to malnutrition or insufficient dietary intake is termed as nutritional anemia while that caused by excessive bleeding resulting in low iron availability is referred to as hypochromic microcytic anemia, i.e., low concentration of heme in the blood (Marks, 2019; Silva et al., 2019; Zimmermann, 2008; Shubham et al., 2020).

According to 2016, WHO estimates, 41.7% of children (< 5 years), 40.1% of pregnant women, and 32.5% of non-pregnant women were anemic worldwide (Pasricha et al., 2021) resulting in a higher child and maternal mortality rates, especially among low-income groups. In developing countries, such as India, this incidence is very high with >50% women of reproductive age and around 39% adolescent girls suffering from anemia (Premkumar et al., 2018; Shubham et al., 2020).

Iron requirement increases significantly in pregnant women to meet the demands of the fetus and placenta during the massive expansion of red blood cell production. Non-fulfillment of this demand often results in premature birth, low birth weight, or even mortality (infant and mother). Blood contains 0.4–0.5 mg/ml of iron, and heavy menstrual blood loss in premenopausal women affects nearly 20% of women (Samani et al., 2018). On the other hand, in men and postmenopausal women, the main cause is gastrointestinal blood loss due to bleeding lesions, colorectal cancer, inflammatory bowel disease, or ulcers resulting from various drug consumption or disorders such as celiac or Crohn's disease (Pasricha et al., 2021). Heart patients also suffer from iron deficiency (~60%), and the risk of heart failure increases in such cases. Even blood donation may increase the risk of iron deficiency as each whole blood donation costs the donor around 250 mg of iron. This could be prevented by reducing the frequency of donation and also by providing the donor with iron supplementation (Kiss et al., 2015; Coppinger and Diamond, 2001; Pennington, 1991; Powell et al., 2005).

2.1.2 FLUORINE

2.1.2.1 Sources and Bioavailability

The negative ion of the fluorine element, fluoride (F), is ubiquitous in the environment and is an inevitable part of the biosphere and human health. It is present in air, food, and water. The major amount of F in the environment comes from the weathering process of parent rocks, volcanicity, wind-blown dust, and application of phosphate fertilizer, which contributes F to soils and some to groundwater. The mineralization of the element takes place in the plants grown in F-rich soils. It is estimated that from the manufacturing of bricks, about 1.8 Mt a^{-1} of F is drained into the environment, making it the largest source of F release. The absorbed protein gets stored in bones and teeth, and the unabsorbed fluoride is excreted through urine. The most effective method of making fluoride accessible to the general public is through its administration into the community drinking water. Among foods, tea is rich in fluoride content, and excess drinking of tea can lead to dental and skeletal fluorosis (Fuge, 2019). The ingested fluoride gets absorbed 77% in the duodenum and around 25% in the stomach (Buzalaf, 2011; Al-Fartusie and Mohssan, 2017; Fraczek and Pasternak, 2013).

2.1.2.2 Role in Human Health

Fluoride is present in toothpaste and mouthwashes for maintenance of teeth health. The fluorine in low concentration is known to be beneficial for humans. It promotes dental health, strong bones, and helps in the prevention of osteoporosis. If taken in higher concentrations, fluoride causes dental and skeletal fluorosis (Fuge, 2019; Bertinato and L'Abbe, 2004, Harris, 2001; Fraczek and Pasternak, 2013).

2.1.2.3 Deficiency

Fluorine deficiency is a rare condition as fluorine (fluoride) is practically found ubiquitously. It occurs when there is a lack of fluoride in the diet. Fluoride helps in the strengthening of teeth and bones and is commonly found in water, toothpaste, and certain foods. The population at risk of developing fluorine deficiency are those residing in areas with water supply low in fluoride or where the water is non-fluorinated. The symptoms include tooth decay, brittle bones, and an increased risk of fractures. A higher intake of fluoride ($\geq$ 2 mg per liter of water) may produce mottling of teeth in children and an even higher intake ($\geq$ 4 mg per liter of water) may provide protection against osteoporosis in adults. Huge quantities of fluoride (>20 mg per day) when consumed everyday over a long period of time cause chronic fluoride toxicity, skeletal or dental fluorosis resulting in bone and joint disease, brown stains, and pitting of teeth. Therefore, a high-dose exposure to fluoride, especially through inhalation, may result in soft tissue damage and even death (Fordyce, 2011; Chilimba et al., 2012; Longchamp et al., 2013, 2015). Occupational exposure to fluoride during welding or mining operations may cause severe tissue damage, respiratory effects, cardiac arrests, and death. The latter two resulting from hypocalcemia and hyperkalemia, respectively. However, such incidences are rare and since the workers in such settings are also exposed to a variety of other substances, hence the role of fluoride in the health deterioration cannot be conclusively established (Fordyce, 2011; Pérez-Granados and Vaquero, 2002; Jugdaohsingh, 2007; Coppinger and Diamond, 2001).

2.1.3   Zinc

2.1.3.1   Source and Bioavailability

The dietary sources rich in zinc include red meat, liver, poultry, dairy products, eggs, fish, shellfish (Hotz et al., 2005; Pérez-Granados and Vaquero, 2002; Jugdaohsingh, 2007; Bertinato and L'Abbe, 2004, Harris, 2001), legumes, cereals, soy-based products, seeds, and nuts. The absorption of zinc is reduced in the presence of phytates, calcium, copper, magnesium, phosphorus, oxalate, and excess of iron concentration. Whereas, the absorption of zinc is increased with the intake of amino acids like glutamate, glycine, histidine, methionine, and citrate, ascorbate, picolinate, proprionate like low molecular weight ligands in comparison to freely available zinc (Rink, 2000; Cummings and Kovacic, 2009).

2.1.3.2   Role in Human Health

Zinc plays a crucial role in regulating hormone release, immunological response, reproduction, aging, cell growth, and nucleic acid metabolism. In cases of maintaining the function of the reproductive system of female's zinc is needed for ovulation, fertilization, normal pregnancy, fetal development, and parturition. For males, zinc helps in the development of spermatozoa, maintaining testosterone homeostasis, sperm count, density and motility and semen volume (Nasiadek et al., 2020). Zinc is an important element for the synthesis of protein in preterm born babies (Terrin, 2020; Takeda, 2003; Tuschl et al., 2012) and developmental neurogenesis during the prenatal period. The health of females before and during pregnancy influences the risk of diseases in the babies in their adulthood. The dysregulation in zinc homeostasis results in impaired immune function (Terrin, 2015; Mariani, 2006).

2.1.3.3   Deficiency

The role zinc plays for human health was realized quite late; only by the mid 20th century its significance in the growth and development of mammals reported (Turek and Fazel, 2009). Consequently, prevalence of zinc deficiency was also recognized and is reported to affect around two billion people in developing nations (Cavdar et al., 1983). Zinc deficiency occurs either due to inadequate intake or absorption of zinc from the diet or when there is increased loss of zinc from the body. It may also result from a lack of intake animal food and high dietary intake of phytate resulting in reduced bioavailability of zinc (Fischer Walker et al., 2009). Absorption of dietary zinc is around 40%; however, it is affected adversely by the presence of phytate and fibers. Symptoms of zinc deficiency in humans include loss of appetite, delayed growth and development in children, skin rashes, frequent infections, hair loss, diarrhea, impaired regeneration of wounds or poor wound healing, and decreased taste and smell activity.

Severe zinc deficiency affects several human organs, such as the gastrointestinal, skeletal, reproductive, immune, and central nervous systems. In such cases, zinc deficiency can lead to a condition called acrodermatitis enteropathica, characterized by skin lesions, diarrhea, and hair loss. It has been reported to be an inherited disease occurring due to malabsorption of zinc caused by a mutation in ZIP4 (intestinal zinc transporter) and can easily be treated by oral administration of zinc (Wang et al., 2002). Acrodermatitis enteropathica is a lethal, autosomal, and recessive trait that is usually found in infants of Italian, Armenian, and Iranian lineage. However, it can be completely treated with zinc supplementation. In the case of acquired zinc deficiency, patients receiving total parenteral nutrition without zinc supplementation are at risk. Moderate zinc deficiency results in growth retardation and hypogonadism in male adolescents. It results in rough skin, poor appetite, mental lethargy, delayed wound healing, dysfunction in cell-mediated immunity, and abnormal neurosensory changes. On the other hand, mild deficiency of zinc results in impaired growth in children, decreased serum testosterone level and oligospermia in men, impaired immune system, hyperammonemia, hypogeusia, reduced dark adaptation, and lean body mass (Turek and Fazel, 2009). Other medical and non-medical conditions may also result in zinc deficiency, such as sickle cell disease, chronic liver and renal disease, gastrointestinal disorders (such as malabsorption syndrome, Crohn's disease, regional ileitis, steatorrhea). Non-medical conditions refer to excessive sweating, trauma, blood loss, severe burn, and infection, etc.

2.1.4 Iodine

2.1.4.1 Source and Bioavailability

Iodine, one of the essential elements, is maximum absorbed in the thyroid gland and in the upper segment of the small intestine. The urine excretes out the unabsorbed iodine through the kidney. Iodized salt (IS) is one of the common effective methods for preventing iodine deficiency. Iodized oil pills and iodine-containing multi-vitamin supplements are also recommended by health clinicians for the human population especially for pregnant women (Sun et al., 2022; Chilimba et al., 2012; Longchamp et al., 2013, 2015; Detzel and Wieser, 2015; Martinez-Navarrete et al., 2002)

2.1.4.2 Role in Human Health

The iodine is important for the proper functioning of the thyroid gland. It is required to produce thyroid hormones, thyroxine, and triiodothyronine (Zimmermann, 2009).

2.1.4.3 Deficiency

Iodine is an essential mineral, required for the production of thyroid hormones, that is necessary for the regulation of growth and metabolism. Iodine deficiency leads to the enlargement of the thyroid gland (goitre) to compensate for the insufficient iodine levels. The low intake causes increased secretion of thyroid-stimulating hormone (TSH) which, in turn, stimulates thyroid hypertrophy and hyperplasia (). Goiter is basically swelling of the neck that causes discomfort, difficulty in swallowing, and changes in voice. This happens as the thyroid gland loses its ability to produce enough thyroid hormones, leading to a condition referred to as hypothyroidism. When iodine intake of populations or groups suffering from iodine deficiency increases, there is typically an increase in the occurrence of hyperthyroidism. Excessive fortification of iodine tends to increase the severity of the condition, particularly if the initial iodine deficiency was severe (Zimmermann and Boelaert, 2015; Chilimba et al., 2012; Longchamp et al., 2013, 2015).

Symptoms of hypothyroidism include fatigue, weight gain, dry skin, hair loss, cold intolerance, and cognitive impairment. Iodine deficiency often becomes a serious issue for pregnant women as iodine is essential for brain development of the fetus and a deficiency may lead to reduced intelligence, learning disability, and cognitive impairment in the child. Severe cases result in stillbirths, abortions, and cretinism, characterized by severe mental and physical disability (Cobra et al., 1997; Dillon and Milliez, 2000). Hence, pregnant and lactating women require an additional intake of iodine of around 25–30 μg to the recommended daily intake. The most efficient way to combat iodine deficiency is to add iodine to salt, which works as the simplest and most cost-effective method to ensure sufficient iodine intake. Regular consumption of iodine-rich foods such as seafood, fish, seaweeds, dairy products, and eggs also contributes to iodine uptake.

2.1.5 Copper

2.1.5.1 Source and Bioavailability

Copper is a naturally occurring metal that exists as two stable isotopes (63Cu and 65Cu), and as two radioactive isotopes (54Cu and 67Cu) with very short half-lives. Cu-rich foods include sesame seeds, roasted cashews, sunflower seeds, cocoa powder, lobster flesh, lamb liver, and drinking water (Ceko et al., 2014). The transport of copper across the cell membrane and into the cell is carried out by specialized pumps known as transporters. The copper is distributed throughout the body via the bloodstream by ceruloplasmin (95% Cu bound), albumin, and other proteins with a dihistidine binding site. Intracellular copper is directed to the Cu-required enzyme synthesis site, organelles, by special proteins called metallochaperones (Harris, 2000, 2001) and into subcellular compartments by another set of transporters (Bertinato and L'Abbe, 2004; Harris, 2001). The excess copper is transported to the liver for storage or biliary excretion (Harris, 2000).

The Cu concentration inside the body is tightly regulated for maintaining the metal homeostasis. The excess and accumulation of Cu initiates oxidative stress inducing Cu-dependent programmed cell death called cuproptosis (Chen et al., 2022) and contributes to the pathology of diseases like neurodegeneration and atherosclerosis (Bremner 1998).

2.1.5.2   Role in Human Health

The essential micronutrient Copper (Cu) acts as a key catalytic cofactor in various enzyme-catalyzed biological processes such as mitochondrial respiration, allosteric enzyme component, as a potential antioxidant having a critical role in the defense system, and bio compound synthesis. Some of the copper-containing enzymes are Haphaestin, which helps in transportation of iron intestinal mucosa into portal circulation, Superoxide dismutase involved in defense mechanism against reactive oxygen species, Dopamine involved in catecholamine metabolism, β-hydroxylase catalyzes conversion of dopamine to norepinephrine, Lysyl oxidase cross-linking of collagen and elastin, and many more (Stern, 2010; Takeda, 2003; Tuschl et al., 2012).

2.1.5.3   Deficiency

Copper, an essential trace mineral, plays a crucial role in various physiological processes in the human body, such as the formation of red blood cells, maintenance of healthy bones, connective tissues, and the function of the immune and nervous systems. Copper deficiency occurs in the case of inadequate consumption or absorption of copper and also during excessive loss of copper from the body. Deficiency due to inadequate intake of copper is rare in developed countries as copper can be found in a variety of foods (like meat, seafood, nuts, seeds, whole grains, legumes, etc.). However, malabsorption may result due to certain gastrointestinal conditions, such as celiac disease, Crohn's disease, or surgeries. It may also result due to increased demand for copper in conditions of pregnancy, lactation, wound healing, or due to hereditary diseases (Altarelli et al., 2019). Excessive losses of copper may occur due to major burns (>20% of body surface area burnt) as trace elements are lost through wound exudates (Jafari et al., 2018) and patients with major burn trauma may experience abnormally high urinary copper excretion (Voruganti et al., 2005). Even excessive presence of zinc may interfere with copper absorption and utilization in the body. In the past few years, cases reporting copper deficiency in the area of cardiovascular diseases and fat metabolism have increased significantly (DiNicolantonio et al., 2018; DiNicolantonio et al., 2018; Alvarez, 2021). Symptoms usually vary depending on the severity and duration of the deficiency, which includes anemia, fatigue, bone abnormalities, defects in pigmentation, changes in skin and hair, impaired immune function, neurological disturbances such as numbness, tingling, muscle weakness, and difficulty in walking. In severe cases, copper deficiency can lead to a condition called Menkes disease, which is a rare genetic disorder that can cause developmental delays, seizures, and other neurological problems (Danks, 1988; Sun et al., 2022; Bertinato and L'Abbe, 2004, Harris, 2001).

Copper deficiency can occur in individuals who consume diets that are very low in copper, such as those that rely heavily on processed foods, or in people who have certain medical conditions that interfere with copper absorption or utilization, such as celiac disease or Wilson's disease. Wilson's disease is another genetic disorder characterized by accumulation of copper in various organs (Bandmann et al., 2015), henceforth toxicity potential of copper cannot be ignored. Treatment typically involves consuming more copper-rich foods or taking copper supplements, and addressing any underlying medical conditions that may be contributing to the deficiency (Pérez-Granados and Vaquero, 2002; Jugdaohsingh, 2007; Pennington, 1991; Powell et al., 2005; Detzel and Wieser, 2015; Martinez-Navarrete et al., 2002).

2.1.6 Chromium

2.1.6.1 Source and Bioavailability

An element that is a member of the heavy metals family is chromium (Cr). Its oxidation states range from –2 to +6, but trivalent chromium (CrIII) and hexavalent chromium (CrVI) are the more stable forms that are found in nature [1]. Although its mechanisms are not entirely known, CrIII has a significant nutritional role. Salmon, eggs, broccoli, whole grains, and various crustaceans are good sources of CrIII, a trace element. Chromic chemicals can be absorbed by a variety of channels, including the oral cavity, the skin, and inhalation. When taken orally, CrIII has a very low absorption rate (1%), whereas CrVI is primarily absorbed in the duodenum with the aid of intestinal flora. Gastric fluids effectively convert ingested CrVI to CrIII (Alvarez, 2021; Coppinger and Diamond, 2001; Pennington, 1991; Powell et al., 2005).

2.1.6.2 Role in Human Health

As a heavy metal, chromium can accumulate in the body over time and cause bioaccumulation. This can lead to toxicity and a variety of pathophysiological defects, including allergic reactions, anemia, burns, and sores, particularly in the stomach and small intestine (Hossini, 2022; Takeda, 2003; Tuschl et al., 2012; Bertinato and L'Abbe, 2004, Harris, 2001).

2.1.6.3 Deficiency

Chromium levels in human tissues decline with age and result in disturbed glucose metabolism. It plays an important role in the metabolism of carbohydrates, fats, and proteins by enhancing the action of insulin, a hormone that regulates blood sugar levels (Kobla and Volpe, 2000). Chromium deficiency is rare in humans, but it can occur in people who consume a diet that is low in chromium or who have certain medical conditions that affect chromium absorption or utilization. Factors contributing to chromium deficiency may include inadequate dietary intake, impaired absorption or increased excretion and losses through urine, resulting in low levels in the body. Small amounts of chromium are even lost in hair, sweat, and bile (Kobla and Volpe, 2000). Diabetic patients are at a higher risk of chromium deficiency because they require more chromium to help regulate blood sugar levels. People suffering from certain digestive disorders, such as Crohn's disease or ulcerative colitis, may have difficulty absorbing chromium from their food. Pregnant women may also require more chromium to support the growth and development of the fetus. Strenuous exercise or athletes with high energy intake may be suffering from chromium deficiency due to increased urination and inadequate consumption of chromium. Symptoms of chromium deficiency are not well-defined as they are not so common and hence research on the subject is limited. Though they may include weight loss, fatigue, impaired glucose tolerance, and an increased risk of developing diabetes. However, these symptoms are not specific to chromium deficiency and can be caused by other factors as well. Some studies suggest that inadequate chromium levels may contribute to insulin resistance and impaired glucose tolerance, which are associated with conditions like type II diabetes mellitus (Kobla and Volpe, 2000; Racek, 2003; Takeda, 2003; Tuschl et al., 2012; Coppinger and Diamond, 2001). Chromium deficiency may be diagnosed by evaluation of dietary habits, assessment of symptoms, and blood tests to measure chromium levels in the body. In case of deficiency, chromium supplements may be recommended under medical supervision.

2.1.7 Cobalt

2.1.7.1 Source and Bioavailability

Iron (Fe) and nickel (Ni) are chemically very similar to cobalt (Co), a hard, silvery gray, and ductile metal element. Typically, cobalt compounds exist in two valence states: cobaltous (Co2+) and cobaltic (Co3+), with the former being the most readily available commercially and environmentally. Humans are frequently exposed to numerous Co compounds in daily life due to their widespread presence. The general public is mostly exposed via breathing in ambient air and consuming food and water that contain Co compounds (Leyssens et al., 2017). The fish, nuts, green leafy vegetables such as broccoli and spinach, and fresh cereals are significant sources of cobalt for human (Gál, 2008).

2.1.7.2 Role in Human Health

Cobalt (Co) is an essential element that is frequently consumed frequently in food and may be incrementally exposed through work, activity, and medical equipment. Cobalt's primary involvement in human health is based on its contribution to cobalamin (Cbl, vitamin B12). In humans, Cbl serves as a cofactor for the enzymes methylmalonyl-CoA mutase and methionine synthase. For good health, both enzymes are necessary (Yamada, 2013). Peak blood Co concentrations have been linked to adverse health effects, including cardiomyopathy and vision or hearing damage (Paustenbach et al., 2013).

2.1.7.3 Deficiency

In humans, cobalt deficiency is unheard of, however, in ruminants (like sheep, goats, and cattle), it results in anemia, muscular atrophy, and eventual death. Cobalt in the body serves as a part of vitamin B12, which is essential for the formation of red blood cells, proper nerve function, and DNA synthesis (Yamada, 2013). The disease is rare in humans, as it is found in many foods, including shellfish, organ meats, and some vegetables. However, certain health conditions can lead to cobalt deficiency, such as malabsorption disorders, chronic alcoholism, and certain types of anemia. The deficiency majorly arises due to inadequate dietary intake or impaired absorption of cobalamin. However, it can occur in individuals following a strict vegetarian diet that lacks animal-based products, as cobalamin is predominantly found in animal-derived foods. Symptoms of cobalt deficiency are often similar to those of vitamin B12 deficiency. It may include megaloblastic anemia with symptoms such as fatigue, weakness, shortness of breath, or neurological disorders, and developmental delays in infants. In severe cases, cobalt deficiency can lead to heart failure and death (Takeda, 2003; Tuschl et al., 2012; Panchal et al., 2017; Sun et al., 2022; Dillon and Milliez, 2000; Racek, 2003).

2.1.8 MANGANESE

2.1.8.1 Source and Bioavailability

The twelfth most prevalent element on earth, manganese (Mn), is a vital metal for human health. A trace elemental metal called manganese (Mn) is vital for preserving human health and function. It is present in a wide range of foods, including seafood, seeds, chocolates, teas, and whole grains including beans, rice, nuts, and whole grains (Peres et al., 2016; DiNicolantonio et al., 2018; Paustenbach et al., 2013).

2.1.8.2 Role in Human Health

The low doses of Mn are necessary to support a number of physiological processes in the body, including healthy growth, digestion, immunological response, and protection against oxidative stress via Mn superoxide dismutase (MnSOD) in mitochondria. If not enough Mn is consumed, one could develop Mn deficiency and show symptoms such as skeletal deficiency, decreased levels of fertility, impaired glucose tolerance, and impaired growth (Aschner and Aschner, 2005; Coppinger and Diamond, 2001; Fraczek and Pasternak, 2013).

2.1.8.3 Deficiency

Manganese deficiency is a rare condition that may occur in the case of insufficient amount of manganese in the body. It is an essential mineral required in small amounts for various physiological functions by humans. It plays a role in enzyme activation, metabolism, bone formation, and production of connective tissues and hormones. Symptoms of manganese deficiency in animals may include impaired growth and skeletal abnormalities (Strause et al., 1986), altered glucose and lipid metabolism (Baly et al., 1990; Taylor et al., 1996), reproductive and neurological issues. Inadequate manganese levels may affect bone development, thereby leading to skeletal abnormalities in children, such as short stature, bone deformities, and abnormal cartilage formation. Manganese is involved in carbohydrate and lipid metabolism, and hence its deficiency may impair glucose tolerance and insulin resistance, and lead to abnormalities in lipid profiles and fat metabolism, respectively. It may also reduce fertility and cause abnormal development of reproductive organs. Other symptoms may include fatigue, loss of appetite, nausea, vomiting, skin rashes, changes in hair color, etc. Some studies suggest that manganese may be associated with neurological symptoms, including impaired motor function, muscle rigidity, and tremors. However, more research is needed to establish a clear link between manganese deficiency and neurological issues. Manganese deficiency is rare in healthy individuals, as the body requires only small amounts of the mineral. However, individuals at risk of manganese deficiency include those with digestive disorders, such as Crohn's disease, and people who have undergone bariatric surgery, which can interfere with manganese absorption. It can also result due to a diet lacking in manganese-rich foods, such as whole grains, nuts, seeds, legumes, and green leafy vegetables. Treatment for manganese deficiency typically involves increasing the intake of manganese-rich foods. In some cases, supplements may be recommended under the guidance of a healthcare professional. However, excessive manganese intake can be toxic, especially during occupational exposure to manganese-laden dust by miners (Finley and Davis, 1999). Excessive manganese may accumulate in the liver, pancreas, bone, kidney, and brain, the latter of which is majorly targeted owing to its presence in the basal ganglia, thereby resulting in neurological disorders similar to Parkinson's disease (Takeda, 2003; Tuschl et al., 2012). Environmental concerns have also been raised due to its neurotoxicity, such as oxidative stress, mitochondrial dysfunction, protein misfolding, endoplasmic reticulum stress, autophagy dysregulation, apoptosis, and disruption of other metal homeostasis. People with iron deficiency also tend to be at a higher risk of manganese poisoning due to its increased absorption, especially in vegetarians (Bertinato and L'Abbe, 2004, Harris, 2001; Detzel and Wieser, 2015; Martinez-Navarrete et al., 2002; Chilimba et al., 2012; Longchamp et al., 2013, 2015).

2.1.9 Selenium

2.1.9.1 Source and Bioavailability

Selenium (Se) is an essential micronutrient vital for all age groups. The bioavailability of Selenium (Se) is enhanced in the presence of vitamins A, D, and E and protein rich foods. The foods rich in Se include cereals, meat and dairy products, fish, seafood, milk, nuts, bread, mushrooms, garlic, asparagus, kohlrabi (Fraczek and Pasternak, 2013). Due to low protein content of vegetables and fruits, they possess low selenium content (dos Santos et al., 2017). Astragalus bisulcatus and other varieties of Brassicaceae are known to accumulate high amounts of selenium.

2.1.9.2 Role in Human Health

Selenium is incorporated in known 25 selenoproteins that can mediate the biological effects of the element in supporting antioxidant defense systems, maintaining thyroid functioning, and anti-inflammatory responses. In insufficient amounts, Se contributes to the pathophysiology of diseases like neurodegenerative diseases, diabetes, cancer, and cardiovascular disorders (Thomson, 2004; Ibrahim, 2019). Its deficiency during pregnancy can result in miscarriages, pre-term deliveries, intrauterine growth retardation, preeclampsia, thyroid dysfunctions, gestational diabetes, and cholestasis (Zachara, 2018).

2.1.9.3 Deficiency

Selenium is an essential micronutrient that is important for various bodily functions, including thyroid function, immune system function, and antioxidant defense. Human selenium intake largely depends on the concentration of selenium in the soils and from them into the harvested edible plants (dos Reis et al., 2017). A selenium deficiency or toxicity can lead to a range of health problems, such as thyroid dysfunction, weakened immune system, and infertility issues. Kabata-Pendias and Mukherjee (2007) and Fordyce (2012) found the range of selenium toxicity and dietary deficiency to be >400 µg/day and <40 µg/day respectively. The general symptoms of selenium deficiency in humans include muscle weakness and inflammation, fragile RBCs, abnormal skin coloration, heart muscle dysfunction, susceptibility to cancer, Keshan and Kashin-Beck diseases, while its toxicity includes liver and kidney damage, blood clotting, necrosis of the heart and liver, hair and nail loss, nausea, and vomiting (Kabata-Pendias and Mukherjee, 2007). Keshan disease is a heart condition that may cause heart failure due to cardiomyopathy, i.e., enlargement of the heart. It is a rare disease but may occur due to severe and prolonged selenium deficiency, and its occurrence is usually reported in areas with extremely low selenium levels in the soil. On the other hand, Kashin-Beck disease is an endemic osteoarthropathy, i.e., stunting feet and hands. It results in joint deformity and impairment of movement in extremities, resulting in shortened fingers and toes, and dwarfism in extreme cases (Fordyce, 2012).

Selenium is essential for the conversion of inactive thyroid hormone (T4) to its active form (T3) and hence a deficiency disrupts thyroid hormone metabolism, quite often leading to hypothyroidism, symptoms of which include fatigue, weight gain, and cold intolerance (Coppinger and Diamond, 2001). Selenium is important for the proper functioning of the immune system function and a deficiency may lead to an increased risk of infections and illnesses. It is also involved in energy metabolism and the production of ATP (adenosine triphosphate), therefore a lack of selenium results in fatigue and weakness. Selenium deficiency has also been linked to hair (brittle, thin, and discolored) and nail (weak and easily breakable) abnormalities. Selenium is thought to have anti-cancer properties, and a deficiency may increase the incidence of certain cancers (such as prostate cancer). Deficiency of selenium may also result in male infertility, leading to reduced sperm motility and other fertility problems in females as well, such as increased risk of miscarriage. To address selenium deficiency, a balanced diet containing selenium-rich foods is essential. The recommended daily intake of selenium for adults is around 55 micrograms per day, and it can be obtained from sources such as Brazil nuts, seafood, meat, poultry, eggs, whole grains, and dairy products (Detzel and Wieser, 2015; Martinez-Navarrete et al., 2002; Pennington, 1991; Powell et al., 2005; Coppinger and Diamond, 2001).

2.1.10 VANADIUM

2.1.10.1 Source and Bioavailability

Vanadium is a natural omnipresent element, and acts as a pollutant if present in toxic concentrations in the environment.

2.1.10.2 Role in Human Health

The property of Vanadium and phosphate being homologues to each other is crucial in the physiological role of vanadium in human health. Vanadium in proper amounts exerts a positive effect on the treatment of type 2 diabetes, myocardial infarctions, viral, bacterial and parasitic infections, and cancer (Rehder, 2018).

2.1.10.3 Deficiency

Vanadium is a trace element that is found in small amounts in the human body. It is believed to play a role in several biological processes, including glucose and lipid metabolism, bone formation, and immune function. Humans are assumed to consume around 60 µg/day of vanadium daily via foods such as mushrooms, parsley, black pepper, etc (Gruzewska et al., 2014). However, vanadium deficiency is extremely rare in humans, as the body requires only small amounts of this element. On the other hand, animal studies have suggested that vanadium deficiency may lead to decreased insulin sensitivity and impaired glucose metabolism, as well as alterations in bone mineralization (Panchal et al., 2017). It is important to note that excessive vanadium intake can also be harmful, and can lead to toxicity. Symptoms of vanadium toxicity may include abdominal pain, vomiting, diarrhea, and changes in blood pressure and heart rate. Overall, while vanadium is an important trace element, deficiency is rare and excess can be harmful. Most people obtain sufficient amounts of vanadium through a balanced diet.

2.1.11 SILICON

2.1.11.1 Source and Bioavailability

Silicon (Si) is the second most abundant element in the Earth's crust, with an atomic weight of 28 (Exley, 1998). Due to its high affinity for oxygen, it occurs in the form of silica and silicates and is rarely found in its elemental form. The absorption mechanism of silicon is not well understood. The gastrointestinal tract is the entry point for silicon, where it gets absorbed in the human body and is then excreted through urine (Jugdaohsingh et al., 2002). The dietary sources with high concentrations of the element include whole grains such as barley, oats, rice bran, wheat bran, bread, biscuit, rice, pasta, cake and pastry items, beer, vegetables, namely beans (green, Kenyan, French), spinach, root vegetables, some herbs, seafood (mussels), and fruits like banana and dried fruits (Pennington, 1991; Powell et al., 2005; Schwarz, 1977). The dietary sources with low concentrations of silicon include meat and dairy products. The concentration of silicon in drinking water is due to the weathering of rocks and soil minerals at different rates in the water (Taylor et al., 1995).

2.1.11.2 Role in Human Health

Though the direct relation of silicon intake on bone health is not clear, studies shows its impact on collagen synthesis, its stabilization, and matrix mineralization. The consumption of silicon is crucial for bone and connective tissue health. It affects the bone mineral density, and lower concentration of silicon in the body can contribute to the occurrence of osteoporosis (Jugdaohsingh, 2007).

2.1.11.3 Deficiency

Silicon deficiency is not a well-established medical condition. Silicon is a natural element that is found in many foods, including whole grains, fruits, vegetables, and nuts. It is also present in some drinking water sources. Factors such as aging, reduction of estrogen levels, and exercise have been reported to decrease silicon absorption capacity. Hence, athletes are recommended to take an increased dosage of silicon (30–35 mg/day). Although silicon is not considered an essential nutrient, it plays a role in the formation and maintenance of connective tissues, bones, and cartilage (Pérez-Granados and Vaquero, 2002; Jugdaohsingh, 2007). It is also involved in the formation of collagen, a key component of skin, hair, nails, and other connective tissues. Therefore, some studies have suggested that a diet low in silicon may be associated with an increased risk of osteoporosis and other bone-related conditions. Still, not much evidence advocates silicon deficiency and is unlikely to occur unless there are underlying medical conditions or severely restricted diets (Pennington, 1991; Powell et al., 2005; Panchal et al., 2017).

2.1.12 NICKEL

2.1.12.1 Source and Bioavailability

Nickle, an element of group 10 in the periodic table, is the 24th most abundant element in the Earth's crust. Nickle has atomic number of 28, atomic weight of 58.6934, density of 8908 kg/m3, boiling point of 2913 °C, and melting point of 1455 °C.

The naturally occurring sources of nickel involve weathered rocks and soils, volcano activities, forest fire, windblown dust, combustion of coal, diesel oil, fuel oil, and waste and sewage (Von Burg, 1995). The dietary sources for nickel include vegetables like carrots, broccoli, spinach, asparagus, green beans, tomato, and cocoa, chocolate, and nuts (Carocci et al., 2016; Lavinia et al., 2018). The element also becomes part of the diet through different food processing operations carried out using stainless steel equipment and hand to mouth contact (Vasiluk et al., 2019).

2.1.12.2 Role in Human Health

Nickel in optimum quantities is important for plant growth and helps in various morphological and physiological functions. At higher concentrations, it inhibits photosynthetic electron transport and chlorophyll biosynthesis (Sreekanth et al., 2013). Owing to its resistance to oxidation property and price, nickel finds its utility in keys, paper clips, clothing, jewelry, zippers, snap buttons and belt buckles, stainless steel household utensils, coins, and electrical equipment (Henderson et al., 2012). Nickle is known to cause mitochondrial dysfunctions and oxidative stress resulting in the development of its toxicity. It can induce even detrimental health effects like lung fibrosis, allergic reactions, cardiovascular and kidney diseases, and cancers of nasal areas and lungs (Genchi et al., 2020).

2.1.12.3 Deficiency

Nickel deficiency is a rare condition that is characterized by insufficient levels of nickel in the body. Nickel is an essential trace element that plays a vital role in various biological processes, including the production of certain enzymes and the metabolism of fats and carbohydrates. In plants, nickel deficiency may result in dwarfism and pigmentation in leaves (Shukla et al., 2021). In humans, it is needed for heart muscle, liver, and kidneys. As such, no RDA has been established for nickel; however, the daily intake from food and water sources may range from 80–130 µg/day (Al-Fartusie and Mohssan, 2017). Symptoms of nickel deficiency can include skin rashes, skeletal abnormalities, digestive problems, and changes in hair and nail growth. However, because nickel is required in such small amounts, deficiency is uncommon and usually only occurs in people with certain medical conditions such as kidney disease or those who have undergone gastrointestinal surgeries that affect absorption of nutrients. Inadequate levels of nickel can potentially lead to heart-related complications. It may also cause biochemical changes, like a reduction in iron absorption which might further lead to anemia, interfering with zinc and carbohydrate metabolism (Al-Fartusie and Mohssan, 2017). Treatment for nickel deficiency typically involves increasing the intake of foods that are rich in nickel, such as whole grains, legumes, nuts, and leafy green vegetables. In some cases, a nickel supplement may be recommended by a healthcare professional. However, it is important to note that excessive intake of nickel can be toxic, and its exposure via air, food, or metals might result in nickel allergy, commonly also termed dermatitis. Nickel allergy is quite often observed in sensitive individuals wearing jewelry and is characterized by itching to severe burns (Thyssen et al., 2007). Serious health issues may also arise due to long-term exposure to nickel, such as cancer, kidney, respiratory, and cardiovascular diseases.

2.1.13 Molybdenum

2.1.13.1 Source and Bioavailability

Molybdenum was discovered in 1778 by Karl Scheele, a Swedish chemist. Molybdenum is easily found in different foods owing to its presence in the soil. Foods like nuts, legumes, vegetables, chicken, beef, pork, milk, eggs, and rice contain molybdenum in different amounts. Legumes are the richest source of the element. The bioavailability of molybdenum from foods is lower in comparison to molybdenum supplements. Studies show that the bioavailability of molybdenum from cress mixed with the element is higher (70–90%) than cress incorporated with the element directly into the plant tissues (50-80%). Increased molybdenum bioavailability can be achieved through the consumption of a mixed diet comprising different food categories (Novotny, 2011; Pennington, 1991; Powell et al., 2005; Takeda, 2003; Tuschl et al., 2012).

2.1.13.2 Role in Human Health

Molybdenum acts as a cofactor for sulfite oxidase, xanthine oxidase, aldehyde oxidase, and as a mitochondrial amidoxime-reducing component, each serving different functions in the human body. The enzyme Xanthine oxidase is involved in the oxidizing purines to uric acid (Harrison, 2002), and aldehyde oxidase ortholog contributes to the metabolism of different endogenous and exogenous N-heterocyclic compounds (TERAO et al., 1998). The sulfite oxidase catalyzes degradation reactions for conversion of sulfur amino acids into cysteine and methionine (Garrett et al.,1998), whereas mitochondrial amidoxime reducing component (mARC) becomes the catalytic segment of a three-component enzyme system with heme/cytochrome b5 and reduced nicotinamide adenine dinucleotide/flavin adenine dinucleotide–dependent cytochrome b5 reductase. The mARC, abundantly present in the liver and kidney, is important for detoxifying N-hydroxylated substrates (Wahl et al., 2010; Al-Fartusie and Mohssan, 2017).

2.1.13.3 Deficiency

Molybdenum is an essential trace mineral required by plants, animals, and humans for various biological processes. Its deficiency occurs due to a lack of adequate intake and is rarely found in humans. In plants, it plays a crucial role in converting inorganic nitrogen into organic form through an enzyme called nitrogenase. Hence, its deficiency in plants may lead to stunted growth, yellow leaves, and reduced flower formation and nitrogen metabolism (Shukla et al., 2021). In animals, molybdenum is required for the functioning of several enzymes involved in metabolic processes, and in case of deficiency, it may lead to decreased appetite, weight loss, impaired reproduction, skeletal abnormalities, and increased susceptibility to certain diseases. In humans, molybdenum plays a role in fighting cancer-causing nitrosamines, preventing dental caries, mouth and gum disorders, esophageal cancer, and sexual impotence in older people (Shukla et al., 2021). Molybdenum deficiency is rare in humans as the daily requirement for molybdenum is relatively low, and it is commonly found commonly in various foods. Moreover, the human body also has efficient mechanisms for molybdenum absorption and recycling. However, it may occur in individuals with specific health conditions, such as those with genetic disorders affecting molybdenum metabolism. Some studies have also reported low levels of molybdenum to be associated with impaired sulfur metabolism and increased sensitivity to sulfite-containing food and drugs. Excessive intake of molybdenum may cause toxicity, thereby interfering with copper absorption and metabolism and leading to copper deficiency (Takeda, 2003; Tuschl et al., 2012; Fraczek and Pasternak, 2013; DiNicolantonio et al., 2018).

2.1.14 Micronutrient Fortification/Supplementation

Fortification is a widely used method that primarily aims to increase the content of essential micronutrients in food, thereby enhancing their nutritional value while promoting public health with minimal potential drawbacks. Enhancement of the nutritional composition of food can be achieved by fortification or supplementation depending on the specific target population (Shubham et al., 2020; Dillon and Milliez, 2000; Paustenbach et al., 2013; Sun et al., 2022). Targeted micronutrient fortification and/or supplementation in foods can go a long way in countering and controlling malnutrition and micronutrient-related deficiencies among populations, especially rural communities, thereby eliminating the need for their reliance on pharmaceutical supplements. Specific laboratory techniques and measures must be undertaken to face and solve the technical challenges presented during the food fortification of specific micronutrients. For instance, double and quadruple fortification is being pursued that allows for the delivery of multiple micronutrients. Steps for food fortification include assessing the need among targeted groups and ways to deliver, such as commonly consumed foods among various social groups like rice, wheat, oil, salt, and tea are often chosen as vehicles for fortification. The selected food carrier and the targeted micronutrient must be synergistic, as they have implications for the bioavailability of nutrients. From the earlier parts of the chapter, we can summarize the main causes of iron deficiency as excessive blood loss, insufficient dietary intake of iron, or failed absorption from the diet to meet the physiological requirements of the human body. Also, since iron deficiency is the major cause of anemia, efforts must be undertaken to increase iron intake through food fortification, supplementation, and dietary diversification. Iron supplementation is basically the oral administration of pharmaceuticals containing iron and is employed in case of emergencies where there is an immediate need to increase the iron level in the body (Shubham et al., 2020; Pennington, 1991; Powell et al., 2005). On the other hand, iron fortification is increasing iron levels through food and mainly aims to target a specific group such as infants and premenopausal and pregnant women. Infants should be fed iron-fortified formula or weaning foods to ensure the required supply of iron as the incidence of iron deficiency is more in breastfed infants or cows' milk-fed infants. According to WHO recommendations, 2mg/kg/day of iron must be supplemented in the diet of children (6–23 months) of developing nations or of those living in areas with a high prevalence of anemia (40%). The prevalence of anemia in adolescent girls is also quite high, which may present working and reproductive limitations in them. The cause is mainly nutritional deficiency that can be addressed through iron-folic acid supplements, medications, and other dietary modifications (Wahl et al., 2010; Shubham et al., 2020). Pregnant women are especially recommended to start iron supplementation with low doses and get regularly screened for iron deficiency anemia. Since heme iron has higher absorption efficiency (15–35%) than non-heme iron (10%), a plant-based diet consisting mainly of non-heme iron often results in iron deficiency. This can be curbed by improving the iron bioavailability through fortification of iron in cereals and other plant-based products. It is advised to consume a mixed diet consisting of green leafy vegetables, cereals, pulses (germinated), eggs, roots, etc., as good non-heme iron sources and meat as a good source of heme iron. They should also include ascorbic acid, folic acid, B vitamins, etc., in their diet to increase the iron bioavailability in the human body. Food fortification must involve food that is widely available and is consumed by the mass (Hurrell and Egli, 2019), and also does not have any adverse chemical interaction with iron or iron-based compounds. A few examples include salt, sugar, cereal-based, and milk-based products (Detzel and Wieser, 2015; Martinez-Navarrete et al., 2002). More recently, bakery products such as biscuits, cookies, and bread have been fortified with iron as they are consumed by the masses. In India, pulses and rice are specifically aimed for fortification as they act as a cheap vehicle for iron delivery in low socio-economic households (Silva et al., 2019; Shubham et al., 2020; Al-Fartusie and Mohssan, 2017).

Excessive intake of fluorine may result in a condition called fluorosis, and hence it is important to consult with a healthcare professional before supplementing with fluoride. The most effective way of providing fluoride in most countries is via community drinking water (around 0.7–1.2 ppm), but when this level is low (< 0.3 ppm) supplementation is recommended (American Academy of Pediatric Dentistry, 2022). Fluoridation of water involves the use of sodium or potassium fluoride or hexafluorosilicic acid (H_2SiF_6) and its sodium salt (Na_2SiF_6) (Ponikvar, 2008). Fluoridation of salt is also sometimes recommended for communities with low fluorine content in natural water and inability to fluoridate the water (Pollick, 2013; Silva et al., 2019; Coppinger and Diamond, 2001).

Zinc deficiency can be caused by a variety of factors, including inadequate dietary intake, malabsorption syndromes, chronic diarrhea, liver disease, and alcoholism. Treatment would involve increased zinc intake through dietary changes or supplementation. This helps in reducing the incidence of infection, also cellular damage due to high oxidative stress, protects from liver damage and acute diarrhea (Tuerk and Fazel, 2009). Zinc supplementation thus helps in reducing morbidity and mortality from acute infectious diseases. Hence, zinc is considered a micronutrient of exceptional importance and mostly so in prenatal and postnatal development. Therefore, it is advised to increase the intake of zinc during pregnancy and lactation by 5 mg and 10 mg per day respectively. In infants and preschool children, zinc supplementation helps reduce the incidence of acute lower respiratory infections while in the elderly, it helps in wound healing and provides resistance to infections (Al-Fartusie and Mohssan, 2017; Shubham et al., 2020; Tuerk and Fazel, 2009). Increasing consumption of dietary supplements containing copper may cause health issues due to excessive copper uptake and exposure, such as free radical formation that may cause lipid peroxidation and interfere with bone metabolism, resulting in a reduction in bone cortex and bone strength (Qu et al., 2018). Excessive copper has also been linked with certain neurological diseases such as Parkinson,

Huntington, and Alzheimer (Silva et al., 2019). The most cost-effective and widespread method globally adopted for preventing iodine deficiency is salt iodization (Zimmermann, 2008). Iodine supplements may also be provided to susceptible groups, and it can be safely said that the small risks of iodine toxicity or excess are easily outweighed by the substantial risks of iodine deficiency. Iodization is also possible by using bread as a vehicle, wherein the baker's salt used had previously been enriched with iodine. The pharmaceutical industry may provide iodine in the form of tablets or drops of potassium iodine or iodate (Fraczek and Pasternak, 2013; Panchal et al., 2017).

Chromium supplementation helps in improving glucose tolerance and lipid metabolism, especially in type 2 diabetes mellitus patients. The varying kinds of chromium supplements used are chromium chloride, picolinate, and nicotinate. Chromium chloride is an inorganic compound, while the latter two are organic compounds and are considered to be more biologically available (Clarkson, 1997). Addressing the deficiency due to cobalt involves replenishing the cobalamin levels. If the deficiency is due to dietary factors, consumption of cobalamin-rich foods (such as meat, fish, eggs, and dairy products) must be increased. Additionally, vitamin B12 supplements or injections may also be required to restore cobalamin levels. However, it is essential to ensure that excessive intake is avoided as it has adverse effects on health. Manganese is mostly lost via excretion, and only a small percentage is absorbed. Hence, the risk of manganese toxicity is usually high when bile excretion is low, like in the case of neonate or liver disease (Trumbo et al., 2001; Aschner and Aschner, 2005). Manganese toxicity leads to the accumulation of the mineral in the liver, pancreas, bone, kidney, and brain, and diseases such as cirrhosis, dystonia, and Parkinson's are reported to occur (Chen, et al., 2018). Hence, high levels of manganese, particularly from supplements, must be prevented as it may lead to neurological symptoms, such as tremors, muscle spasms, and cognitive impairments. Therefore, it is crucial to follow the recommended dosage and avoid any risk of manganese toxicity.

Selenium biofortification has been reported by several authors for rice (Boldrin et al., 2013; Pandey and Gupta, 2015), maize (Chilimba et al., 2012; Longchamp et al., 2013, 2015) and wheat (Acuña et al., 2013; Galinha et al., 2015). Supplementation could be achieved by means of fertilizers, foliar spraying directly on plants or soil fortification using selenium enriched plant (dos Reis et al., 2017). Vanadium could be supplemented in the form of vanadyl sulfate in order to improve performance in weight training athletes at doses up to 60 mg/day. Since silicon prevents gastrointestinal aluminum absorption and reduces aluminum concentrations in various tissues, hence silicon supplementation plays a huge role in preventing their accumulation (Pérez-Granados and Vaquero, 2002). A significant portion of the silicon found in typical diets exists in the form of aluminosilicate and silica, compounds that make silicon less accessible. On the other hand, silicic acid, that is present in certain foods and beverages, is readily absorbed by the body. Since humans are prone to exposure from nickel during breathing air, eating, smoking or through skin exposure to metals, hence sensitive people must avoid exposure to nickel. Though small quantities of nickel are essential for our body, but its toxicity may cause skin irritation and other health problems (Bertinato and L'Abbe, 2004, Harris, 2001; Fraczek and Pasternak, 2013; Pennington, 1991; Powell et al., 2005).

Source: ICMR (Nutrient Requirements & RDA) for Indians. A report of the Expert group of the ICMR, 2010.

2.1.15 CONCLUSION

Micronutrients are essential for different metabolic activities and the preservation of tissue function in required amounts. The human body possesses a highly integrated system for regulating the flux of these nutrients throughout disease. Therefore, maintaining metabolism and tissue function requires an adequate intake, but giving excess tsupplements to people who don't need them could be hazardous.

REFERENCES

Abbaspour, N., Hurrell, R. and Kelishadi, R. (2014). Review on iron and its importance for human health. *Journal of Research in Medical Sciences: The Official Journal of Isfahan University of Medical Sciences*, 19(2), p. 164.

Acuña, J.J., Jorquera, M.A., Barra, P.J., Crowley, D.E. and de la Luz Mora, M. (2013). Selenobacteria selected from the rhizosphere as a potential tool for Se biofortification of wheat crops. *Biology and Fertility of Soils*, 49(2), pp. 175–185.

Al-Fartusie, F.S. and Mohssan, S.N. (2017). Essential trace elements and their vital roles in human body. *Indian Journal of Advances in Chemical Science*, 5(3), pp. 127–136.

Altarelli, M., Ben-Hamouda, N., Schneider, A. and Berger, M.M. (2019). Copper deficiency: Causes, manifestations, and treatment. *Nutrition in Clinical Practice*, 34(4), pp. 504–513.

Alvarez, C.C., Gómez, M.E.B. and Zavala, A.H. (2021). Hexavalent chromium: Regulation and health effects. *Journal of Trace Elements in Medicine and Biology*, 65, p. 126729.

American Academy of Pediatric Dentistry. (2022). *Fluoride therapy: The reference manual of pediatric dentistry*. Chicago, IL: American Academy of Pediatric Dentistry, pp. 317–320.

Aschner, J.L. and Aschner, M. (2005). Nutritional aspects of manganese homeostasis. *Molecular Aspects of Medicine*, 26(4–5), pp. 353–362.

Baly, D.L., Schneiderman, J.S. and Garcia-Welsh, A.L. (1990). Effect of manganese deficiency on insulin binding, glucose transport and metabolism in rat adipocytes. *The Journal of Nutrition*, 120(9), pp. 1075–1079.

Bandmann, O., Weiss, K.H. and Kaler, S.G. (2015). Wilson's disease and other neurological copper disorders. *The Lancet Neurology*, 14(1), pp. 103–113.

Bertinato, J. and L'Abbé, M.R. (2004). Maintaining copper homeostasis: Regulation of copper-trafficking proteins in response to copper deficiency or overload. *The Journal of Nutritional Biochemistry*, 15(6), pp. 316–322.

Boldrin, P.F., Faquin, V., Ramos, S.J., Boldrin, K.V.F., Ávila, F.W. and Guilherme, L.R.G. (2013). Soil and foliar application of selenium in rice biofortification. *Journal of Food Composition and Analysis*, 31(2), pp. 238–244.

Bremner, I. (1998). Manifestations of copper excess. *The American Journal of Clinical Nutrition*, 67(5), 1069S–1073S.

Buzalaf, M.A.R. (Ed.). (2011). *Fluoride and the oral environment* (Vol. 22). Karger, Switzerland: Karger Medical and Scientific Publishers.

Camaschella, C. (2019). Iron deficiency: Blood. *The Journal of the American Society of Hematology*, 133(1), pp. 30–39.

Carocci, A., Catalano, A., Lauria, G., Sinicropi, M.S. and Genchi, G. (2016). 16 A review on mercury toxicity in food. *Food Toxicology*, 315.

Çavdar, A.O., Arcasoy, A., Cin, S., Babacan, E. and Gözdasoğlu, S. (1983). Geophagia in Turkey: Iron and zinc deficiency, iron and zinc absorption studies and response to treatment with zinc in geophagia cases. *Progress in Clinical and Biological Research*, 129, pp. 71–97.

Ceko, M.J., Aitken, J.B. and Harris, H.H. (2014). Speciation of copper in a range of food types by X-ray absorption spectroscopy. *Food Chemistry*, 164, pp. 50–54.

Chen, L., Min, J. and Wang, F. (2022). Copper homeostasis and cuproptosis in health and disease. *Signal Transduction and Targeted Therapy*, 7(1), p. 378.

Chen, P., Bornhorst, J. and Aschner, M.A. (2018). Manganese metabolism in humans. *Front. Biosci. (Landmark Ed)* 23, pp. 1655–1679. PMID: 29293455.

Chilimba, A.D., Young, S.D., Black, C.R., Meacham, M.C., Lammel, J. and Broadley, M.R. (2012). Agronomic biofortification of maize with selenium (Se) in Malawi. *Field Crops Research*, 125, pp. 118–128.

Clarkson, P.M. (1997). Effects of exercise on chromium levels: Is supplementation required? *Sports Medicine*, 23(6), pp. 341–349.

Cobra, C., Muhilal, Rusmil, K., Rustama, D., Djatnika, Suwardi, S.S., Permaesih, D., Muherdiyantiningsih, Martuti, S. and Semba, R.D. (1997). Infant survival is improved by oral iodine supplementation. *The Journal of Nutrition*, 127(4), pp. 574–578.

Coppinger, R.J. and Diamond, A.M. (2001). Selenium deficiency and human disease. In *Selenium: Its molecular biology and role in human health*, pp. 219–233. Boston, MA: Springer US.

Cummings, J.E. and Kovacic, J.P. (2009). The ubiquitous role of zinc in health and disease. *Journal of Veterinary Emergency and Critical Care*, 19(3), pp. 215–240.

Danks, D.M. (1988). Copper deficiency in humans. *Annual Review of Nutrition*, 8(1), pp. 235–257.

Dary, O. & Hurrell, R. (2006). *Guidelines on food fortification with micronutrients*. World Health Organization, Food and Agricultural Organization of the United Nations: Geneva, Switzerland, 2006, 1–376.

Detzel, P. and Wieser, S. (2015). Food fortification for addressing iron deficiency in Filipino children: Benefits and cost-effectiveness. *Annals of Nutrition and Metabolism*, 66(suppl. 2), pp. 35–42.

Dillon, J.C. and Milliez, J. (2000). Reproductive failure in women living in iodine deficient areas of West Africa. *BJOG: An International Journal of Obstetrics and Gynaecology*, 107(5), pp. 631–636.

DiNicolantonio, J.J., Mangan, D. and O'Keefe, J.H. (2018). Copper deficiency may be a leading cause of ischaemic heart disease. *Open Heart*, 5(2), p. e000784.

dos Reis, A.R., El-Ramady, H., Santos, E.F., Gratão, P.L. and Schomburg, L. (2017). Overview of selenium deficiency and toxicity worldwide: Affected areas, selenium-related health issues, and case studies. In: Elizabeth A.H. Pilon-Smits, Lenny H.E. Winkel and Zhi-Qing Lin (eds.), *Selenium in plants: Molecular, physiological, ecological and evolutionary aspects*, pp. 209–230.

Exley, C. (1998). Silicon in life: A bioinorganic solution to bioorganic essentiality. *Journal of Inorganic Biochemistry*, 69(3), pp. 139–144.

Finley, J.W. and Davis, C.D. (1999). Manganese deficiency and toxicity: Are high or low dietary amounts of manganese cause for concern? *BioFactors*, 10(1), pp. 15–24.

Fischer Walker, C.L., Ezzati, M. and Black, R.E. (2009). Global and regional child mortality and burden of disease attributable to zinc deficiency. *European Journal of Clinical Nutrition*, 63(5), pp. 591–597.

Fordyce, F.M. (2011). Fluorine: human health risks. In: Nriagu J O (ed.) *Encyclopedia of Environmental Health*, 2, pp. 776–785. Burlington: Elsevier.

Fordyce, F.M. (2012). Selenium deficiency and toxicity in the environment. In *Essentials of medical geology*, Revised edition, 375–416. Dordrecht: Springer Netherlands.

Fraczek, A. and Pasternak, K. (2013). Selenium in medicine and treatment. *Journal of Elementology*, 18(1).

Frazer, D.M. and Anderson, G.J. (2005). Iron imports. I. Intestinal iron absorption and its regulation. *American Journal of Physiology-Gastrointestinal and Liver Physiology*, 289(4), pp. G631–G635.

Fuge, R. (2019). Fluorine in the environment, a review of its sources and geochemistry. *Applied Geochemistry*, 100, pp. 393–406.

Gál, J., Hursthouse, A., Tatner, P., Stewart, F. and Welton, R. (2008). Cobalt and secondary poisoning in the terrestrial food chain: Data review and research gaps to support risk assessment. *Environment International*, 34(6), pp. 821–838.

Galinha, C., Sánchez-Martínez, M., Pacheco, A.M., Freitas, M.D.C., Coutinho, J., Maçãs, B., Almeida, A.S., Pérez-Corona, M.T., Madrid, Y. and Wolterbeek, H.T. (2015). Characterization of selenium-enriched wheat by agronomic biofortification. *Journal of Food Science and Technology*, 52(7), pp. 4236–4245.

Garrett, R.M., Johnson, J.L., Graf, T.N., Feigenbaum, A. and Rajagopalan, K.V. (1998). Human sulfite oxidase R160Q: Identification of the mutation in a sulfite oxidase-deficient patient and expression and characterization of the mutant enzyme. *Proceedings of the National Academy of Sciences*, 95(11), pp. 6394–6398.

Genchi, G., Carocci, A., Lauria, G., Sinicropi, M.S. and Catalano, A. (2020). Nickel: Human health and environmental toxicology. *International Journal of Environmental Research and Public Health*, 17(3), p. 679.

Gruzewska, K., Michno, A., Pawelczyk, T. and Bielarczyk, H. (2014). Essentiality and toxicity of vanadium supplements in health and pathology. *Journal of Physiology and Pharmacology: An Official Journal of the Polish Physiological Society*, 65(5), pp. 603–611.

Habib, H.M., Ibrahim, S., Zaim, A. and Ibrahim, W.H. (2021). The role of iron in the pathogenesis of COVID-19 and possible treatment with lactoferrin and other iron chelators. *Biomedicine and Pharmacotherapy*, 136, p. 111228.

Haider, L.M., Schwingshackl, L., Hoffmann, G. and Ekmekcioglu, C. (2018). The effect of vegetarian diets on iron status in adults: A systematic review and meta-analysis. *Critical Reviews in Food Science and Nutrition*, 58(8), pp. 1359–1374.

Harris, E.D. (2000). Cellular copper transport and metabolism. *Annual Review of Nutrition*, 20(1), pp. 291–310.

Harris, E.D. (2001). Copper homeostasis: The role of cellular transporters. *Nutrition Reviews*, 59(9), pp. 281–285.

Harrison, R. (2002). Structure and function of xanthine oxidoreductase: Where are we now? *Free Radical Biology and Medicine*, 33(6), pp. 774–797.

Henderson, R.G., Durando, J., Oller, A.R., Merkel, D.J., Marone, P.A. and Bates, H.K. (2012). Acute oral toxicity of nickel compounds. *Regulatory Toxicology and Pharmacology*, 62(3), pp. 425–432.

Hossini, H., Shafie, B., Niri, A.D., Nazari, M., Esfahlan, A.J., Ahmadpour, M., ...Hoseinzadeh, E. (2022). A comprehensive review on human health effects of chromium: Insights on induced toxicity. *Environmental Science and Pollution Research*, 29(47), pp. 70686–70705.

Hotz, C., Lowe, N.M., Araya, M. and Brown, K.H. (2003). Assessment of the trace element status of individuals and populations: The example of zinc and copper. *The Journal of Nutrition*, 133(5), pp. 1563S–1568S.

Hotz, C., DeHaene, J., Woodhouse, L. R., Villalpando, S., Rivera, J. A., & King, J. C. (2005). Zinc absorption from zinc oxide, zinc sulfate, zinc oxide+ EDTA, or sodium-zinc EDTA does not differ when added as fortificants to maize tortillas. *The Journal of Nutrition*, 135(5), 1102–1105.

Hurrell, R. and Egli, I. (2019). Fortification chapter. In Robert T. Means Jr, (ed.), *Nutritional anemia: Scientific principles, clinical practice, and public health*, p. 16.

Hurrell, R.F. (1997). Bioavailability of iron. *European Journal of Clinical Nutrition*, 51, pp. S4–S8.

Hurrell, R. (2002). How to ensure adequate iron absorption from iron-fortified food. *Nutrition Reviews*, 60, pp. S7–S15.

Ibrahim, S.A., Kerkadi, A. and Agouni, A. (2019). Selenium and health: An update on the situation in the Middle East and North Africa. *Nutrients*, 11(7), p. 1457.

Jafari, P., Thomas, A., Haselbach, D., Watfa, W., Pantet, O., Michetti, M., Raffoul, W., Applegate, L.A., Augsburger, M. and Berger, M.M. (2018). Trace element intakes should be revisited in burn nutrition protocols: A cohort study. *Clinical Nutrition*, 37(3), pp. 958–964.

Jugdaohsingh, R. (2007). Silicon and bone health. *The Journal of Nutrition, Health and Aging*, 11(2), p. 99.

Jugdaohsingh, R., Anderson, S.H., Tucker, K.L., Elliott, H., Kiel, D.P., Thompson, R.P. and Powell, J.J. (2002). Dietary silicon intake and absorption. *The American Journal of Clinical Nutrition*, 75(5), pp. 887–893.

Kabata-Pendias, A. and Mukherjee, A.B. (2007). *Trace elements from soil to human*. Springer Science & Business Media.

Kiss, J.E., Brambilla, D., Glynn, S.A., Mast, A.E., Spencer, B.R., Stone, M., Kleinman, S.H., Cable, R.G. and National Heart, Lung, and Blood Institute. (2015). Oral iron supplementation after blood donation: A randomized clinical trial. *JAMA*, 313(6), pp. 575–583.

Kobla, H.V. and Volpe, S.L. (2000). Chromium, exercise, and body composition. *Critical Reviews in Food Science and Nutrition*, 40(4), pp. 291–308.

Lavinia, B., Florina, R. and Augustin, C. (2018). Is it possible a nickel-free diet? *Acta Medica Marisiensis*, 64.

Leyssens, L., Vinck, B., Van Der Straeten, C., Wuyts, F. and Maes, L. (2017). Cobalt toxicity in humans—A review of the potential sources and systemic health effects. *Toxicology*, 387, pp. 43–56.

Longchamp, M., Angeli, N. and Castrec-Rouelle, M. (2013). Selenium uptake in Zea mays supplied with selenate or selenite under hydroponic conditions. *Plant and Soil*, 362(1–2), pp. 107–117.

Longchamp, M., Castrec-Rouelle, M., Biron, P. and Bariac, T. (2015). Variations in the accumulation, localization and rate of metabolization of selenium in mature Zea mays plants supplied with selenite or selenate. *Food Chemistry*, 182, pp. 128–135.

Lynch, S.R. (2005). The impact of iron fortification on nutritional anaemia. *Best Practice and Research: Clinical Haematology*, 18(2), pp. 333–346.

Mariani, E., Cornacchiola, V., Polidori, M.C., Mangialasche, F., Malavolta, M., Cecchetti, R., ... Mecocci, P. (2006). Antioxidant enzyme activities in healthy old subjects: Influence of age, gender and zinc status: Results from the Zincage Project. *Biogerontology*, 7(5–6), pp. 391–398.

Marks, P.W. (2019). *Anemia: Clinical Approach. Concise Guide to Hematology*. In Hillard M. Lazarus and Alvin H. Schmaier (eds.), pp. 21–27.

Martınez-Navarrete, N., Camacho, M.M., Martınez-Lahuerta, J., Martınez-Monzó, J. and Fito, P.J.F.R.I. (2002). Iron deficiency and iron fortified foods—A review. *Food Research International*, 35(2–3), pp. 225–231.

McDowell, L.R. (1992). *Minerals in animal and human nutrition*. Academic Press Inchttps://doi.org/10.1016/B978-0-444-51367-0.X5001-6

Nadadur, S.S., Srirama, K. and Mudipalli, A. (2008). Iron transport & homeostasis mechanisms: Their role in health & disease. *Indian Journal of Medical Research*, 128(4), pp. 533–544.

Nasiadek, M., Stragierowicz, J., Klimczak, M. and Kilanowicz, A. (2020). The role of zinc in selected female reproductive system disorders. *Nutrients*, 12(8), p. 2464.

Novotny, J.A. (2011). Molybdenum nutriture in humans. *Journal of Evidence-Based Complementary and Alternative Medicine*, 16(3), pp. 164–168.

Panchal, S.K., Wanyonyi, S. and Brown, L. (2017). Selenium, vanadium, and chromium as micronutrients to improve metabolic syndrome. *Current Hypertension Reports*, 19(3), pp. 1–11.

Pandey, C. and Gupta, M. (2015). Selenium and auxin mitigates arsenic stress in rice (*Oryza sativa L.*) by combining the role of stress indicators, modulators and genotoxicity assay. *Journal of Hazardous Materials*, 287, pp. 384–391.

Pasricha, S.R., Tye-Din, J., Muckenthaler, M.U. and Swinkels, D.W. (2021). Iron deficiency. *The Lancet*, 397(10270), pp. 233–248.

Paustenbach, D.J., Tvermoes, B.E., Unice, K.M., Finley, B.L. and Kerger, B.D. (2013). A review of the health hazards posed by cobalt. *Critical Reviews in Toxicology*, 43(4), pp. 316–362.

Pennington, J.A.T. (1991). Silicon in foods and diets. *Food Additives and Contaminants*, 8(1), pp. 97–118.

Peres, T.V., Schettinger, M.R.C., Chen, P., Carvalho, F., Avila, D.S., Bowman, A.B. and Aschner, M. (2016). Manganese-induced neurotoxicity: A review of its behavioral consequences and neuroprotective strategies. *BMC Pharmacology and Toxicology*, 17, pp. 1–20.

Pérez-Granados, A.M. and Vaquero, M.P. (2002). Silicon, aluminium, arsenic and lithium: Essentiality and human health implications. *Journal of Nutrition, Health and Aging*, 6(2), pp. 154–162.

Pollick, H.F. (2013). Salt fluoridation: A review. *Journal of the California Dental Association*, 41(6), pp. 395–404.

Ponikvar, M. (2008). Exposure of humans to fluorine and its assessment. In *Fluorine and health*, Tressaud, A., & Haufe, G. (Eds.), pp. 487–549. Elsevier.

Powell, J.J., McNaughton, S.A., Jugdaohsingh, R., Anderson, S.H.C., Dear, J., Khot, F., ...Hodson, M.J. (2005). A provisional database for the silicon content of foods in the United Kingdom. *British Journal of Nutrition*, 94(5), pp. 804–812.

Premkumar, S., Ramanan, P.V. and Thanka, J. (2018). Anaemia in school children-looking beyond iron deficiency. *Journal of Evolution of Medical and Dental Sciences-JEMDS*, 7(45), pp. 4884–4887.

Qu, X., He, Z., Qiao, H., Zhai, Z., Mao, Z., Yu, Z. and Dai, K. (2018). Serum copper levels are associated with bone mineral density and total fracture. *Journal of Orthopaedic Translation*, 14, pp. 34–44.

Quintero-Gutiérrez, A.G., González-Rosendo, G., Sánchez-Muñoz, J., Polo-Pozo, J. and Rodríguez-Jerez, J.J. (2008). Bioavailability of heme iron in biscuit filling using piglets as an animal model for humans. *International Journal of Biological Sciences*, 4(1), p. 58.

Racek, J. (2003). Chromium as an essential element. *Casopis Lekaru Ceskych*, 142(6), pp. 335–339.

Rajpathak, S.N., Crandall, J.P., Wylie-Rosett, J., Kabat, G.C., Rohan, T.E. and Hu, F.B. (2009). The role of iron in type 2 diabetes in humans. *Biochimica et Biophysica Acta (BBA)-General Subjects*, 1790(7), pp. 671–681.

Rehder, D. (2018). Vanadium in health issues. *ChemTexts*, 4, pp. 1–7.

Rink, L. (2000). Zinc and the immune system. *Proceedings of the Nutrition Society*, 59(4), pp. 541–552.

Salnikow, K. (2021, November). Role of iron in cancer. In *Seminars in cancer biology* (Vol. 76, pp. 189–194). Academic Press.

Samani, R.O., Hashiani, A.A., Razavi, M., Vesali, S., Rezaeinejad, M., Maroufizadeh, S. and Sepidarkish, M. (2018). The prevalence of menstrual disorders in Iran: A systematic review and meta-analysis. *International Journal of Reproductive Biomedicine*, 16(11), p. 665.

Schwarz, K. (1977). Silicon, fibre, and atherosclerosis. *The Lancet*, 309(8009), pp. 454–457.

Shergill-Bonner, R. (2017). Micronutrients. *Paediatrics and Child Health*, 27(8), pp. 357–362.

Shubham, K., Anukiruthika, T., Dutta, S., Kashyap, A.V., Moses, J.A. and Anandharamakrishnan, C. (2020). Iron deficiency anemia: A comprehensive review on iron absorption, bioavailability and emerging food fortification approaches. *Trends in Food Science and Technology*, 99, pp. 58–75.

Shukla, K., Jagtap, A. D., Blackshire, J. L., Sparkman, D., and Karniadakis, G. E. (2021). A physics-informed neural network for quantifying the microstructural properties of polycrystalline nickel using ultrasound data: A promising approach for solving inverse problems. *IEEE Signal Processing Magazine*, 39(1), 68–77.

Silva, C.S., Moutinho, C., Ferreira da Vinha, A. and Matos, C. (2019). Trace minerals in human health: Iron, zinc, copper, manganese and fluorine. *International Journal of Science and Research Methodology*, 13(3), pp. 57–80.

Sreekanth, T.V.M., Nagajyothi, P.C., Lee, K.D. and Prasad, T.N.V.K.V. (2013). Occurrence, physiological responses and toxicity of nickel in plants. *International Journal of Environmental Science and Technology*, 10(5), pp. 1129–1140.

Stern, B.R. (2010). Essentiality and toxicity in copper health risk assessment: Overview, update and regulatory considerations. *Journal of Toxicology and Environmental Health. Part A*, 73(2–3), pp. 114–127.

Strause, L.G., Hegenauer, J., Saltman, P., Cone, R. and Resnick, D. (1986). Effects of long-term dietary manganese and copper deficiency on rat skeleton. *The Journal of Nutrition*, 116(1), pp. 135–141.

Sun, R., Fan, L., Du, Y., Liu, L., Qian, T., Zhao, M., ... Sun, D. (2022). The relationship between different iodine sources and nutrition in pregnant women and adults. *Frontiers in Endocrinology*, 13, p. 924990.

Takeda, A. (2003). Manganese action in brain function. *Brain Research Reviews*, 41(1), pp. 79–87.

Taylor, G.A., Newens, A.J., Edwardson, J.A., Kay, D.W. and Forster, D.P. (1995). Alzheimer's disease and the relationship between silicon and aluminium in water supplies in northern England. *Journal of Epidemiology and Community Health*, 49(3), p. 323.

Taylor, P.N., Patterson, H.H. and Klimis-Tavantzis, D.J. (1996). Manganese deficiency alters high-density lipoprotein subclass structure in the Sprague-Dawley rat. *The Journal of Nutritional Biochemistry*, 7(7), pp. 392–396.

Terao, M., Kurosaki, M., Demontis, S., Zanotta, S. and Garattini, E. (1998). Isolation and characterization of the human aldehyde oxidase gene: Conservation of intron/exon boundaries with the xanthine oxidoreductase gene indicates a common origin. *Biochemical Journal*, 332(2), pp. 383–393.

Terrin, G., Berni Canani, R., Di Chiara, M., Pietravalle, A., Aleandri, V., Conte, F. and De Curtis, M. (2015). Zinc in early life: A key element in the fetus and preterm neonate. *Nutrients*, 7(12), pp. 10427–10446.

Terrin, G., Boscarino, G., Di Chiara, M., Iacobelli, S., Faccioli, F., Greco, C., ... De Curtis, M. (2020). Nutritional intake influences zinc levels in preterm newborns: An observational study. *Nutrients*, 12(2), p. 529.

Thomson, C.D. (2004). Assessment of requirements for selenium and adequacy of selenium status: A review. *European Journal of Clinical Nutrition*, 58(3), pp. 391–402.

Thyssen, J.P., Linneberg, A., Menné, T. and Johansen, J.D. (2007). The epidemiology of contact allergy in the general population–prevalence and main findings. *Contact Dermatitis*, 57(5), pp. 287–299.

Trumbo, P., Yates, A.A., Schlicker, S. and Poos, M. (2001). Dietary reference intakes. *Journal of the American Dietetic Association*, 101(3), pp. 294–294.

Tuerk, M.J. and Fazel, N. (2009). Zinc deficiency. *Current Opinion in Gastroenterology*, 25(2), pp. 136–143.

Tuschl, K., Clayton, P.T., Gospe, S.M., Gulab, S., Ibrahim, S., Singhi, P., Aulakh, R., Ribeiro, R.T., Barsottini, O.G., Zaki, M.S., Del Rosario, M.L. (2012). Syndrome of hepatic cirrhosis, dystonia, polycythemia, and hypermanganesemia caused by mutations in SLC30A10, a manganese transporter in man. *The American Journal of Human Genetics*, 90(3), pp. 457–466.

Vasiluk, L., Sowa, J., Sanborn, P., Ford, F., Dutton, M.D. and Hale, B. (2019). Bioaccessibility estimates by gastric SBRC method to determine relationships to bioavailability of nickel in ultramafic soils. *Science of the Total Environment*, 673, pp. 685–693.

Von Burg, R. (1995). Toxicology update. *Journal of Applied Toxicology*, 15(6), pp. 495–499.

Voruganti, V.S., Klein, G.L., Lu, H.X., Thomas, S., Freeland-Graves, J.H. and Herndon, D.N. (2005). Impaired zinc and copper status in children with burn injuries: Need to reassess nutritional requirements. *Burns*, 31(6), pp. 711–716.

Wahl, B., Reichmann, D., Niks, D., Krompholz, N., Havemeyer, A., Clement, B., ... Bittner, F. (2010). Biochemical and spectroscopic characterization of the human mitochondrial amidoxime reducing components hmARC-1 and hmARC-2 suggests the existence of a new molybdenum enzyme family in eukaryotes. *Journal of Biological Chemistry*, 285(48), pp. 37847–37859.

Wang, K., Zhou, B., Kuo, Y.M., Zemansky, J. and Gitschier, J. (2002). A novel member of a zinc transporter family is defective in acrodermatitis enteropathica. *The American Journal of Human Genetics*, 71(1), pp. 66–73.

Wood, R.J., Ronnenberg, A.G., King, J.C., Cousins, R.J., Dunns, J.T., Burk, R.F. and Levander, O.A. (2005). In Catherine Ross, A., Caballero, B., Cousins, R.J., and Tucker, K.L. (eds.), *Modern nutrition in health and disease*. Lippincott's Illustrated Reviews Biochemistry. Lippincott Williams & Wilkins, pp. 248–270.

Yamada, K. (2013). Cobalt: Its role in health and disease. *Interrelations Between Essential Metal Ions and Human Diseases*, pp. 295–320.

Zachara, B.A. (2018). Selenium in complicated pregnancy. A review. *Advances in Clinical Chemistry*, 86, pp. 157–178.

Zimmermann, M.B. (2009). Iodine deficiency. *Endocrine Reviews*, 30(4), pp. 376–408.

Zimmermann, M.B. and Boelaert, K. (2015). Iodine deficiency and thyroid disorders. *The Lancet Diabetes and Endocrinology*, 3(4), pp. 286–295.

Zimmermann, M.B. and Hurrell, R.F. (2007). Nutritional iron deficiency. *The Lancet*, 370(9586), pp. 511–520.

Zimmermann, M.B. (2008). Iodine requirements and the risks and benefits of correcting iodine deficiency in populations. *Journal of Trace Elements in Medicine and Biology*, 22(2), pp. 81–92.

Part II

Vitamin Fortification

3 Vitamin D Fortification

Ayushi Saini, Mehvish Habib, Shumaila Jan,
Ab Lateef Khan, Kulsum Jan, and Komal Chauhan

3.1 INTRODUCTION

Vitamin D is a fat-soluble vitamin that is present in plants and animals in two different forms. The form of vitamin D exist in plants is named as ergocalciferol or vitamin D_2. Colecalciferol or vitamin D_3 derive from animal sources. The structures of three forms of vitamin D are given in Figure 3.1. Fish liver oil and oily fish are the major natural sources of vitamin D_3 (Holick, 2013). In the presence of ultraviolet light from the sunlight, precursors aid in the non-enzymatic production of vitamin D_3 (Norman & Powell; 2014; Haussler et al., 2013). Because of this characteristic, vitamin D it is also known as the "sunshine vitamin" (Nair & Maseeh, 2012). The two forms are different from each other as the vitamin D_2 has methyl group at 24th C atom and poses double bonds between the 22nd and 23rd C atoms (Saponaro et al., 2020, Haussler et al., 2013). Mushrooms exposed to UV light from artificial sources and sunshine yield vitamin D2 in its natural form, which was first produced in the 1930s. Vitamin D2 is commercially produced from yeast exposed to ultraviolet radiation, while vitamin D3 may be obtained from fish oils or synthesized from lanolin (Holick, 2013). Lack of vitamin D can lead to adult osteoporosis and rickets in youngsters. Researchers found out that exposure to the sunlight could decrease the prevalence of rickets in children, and hence vitamin D was considered as a major vitamin for healthy bones. Therefore, in the 1930s, the fortification of vitamin D started in various food like bread, soda, custard etc. (Holick, 2012). In order to achieve their vitamin D requirements, the populations of the US and Canada significantly rely on fortified foods and dietary supplements due to the low intake of naturally vitamin D-rich foods in these

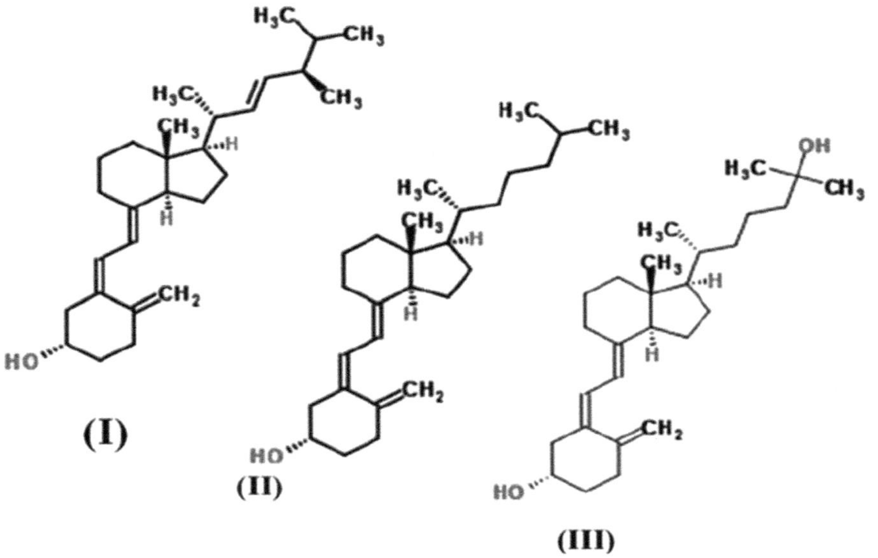

FIGURE 3.1 Chemical composition of three naturally occurring dietary forms of vitamin D (I) ergocalciferol (II) cholecalciferol, and (III) 25(OH) cholecalciferol (Borel et al., 2015).

DOI: 10.1201/9781003160663-5

countries. (Calvo et al., 2004). However, as a result of the outbreak of vitamin D intoxication in Europe in the 1950s, the US Food and Drug Administration limited fortification to only milk and cereals in the 1950s. However, fortified milk does not appear to be very helpful at preventing vitamin D deficiency in the general population due to the high frequency of lactose intolerance among Asians and Native Americans as well as due to milk allergies (Tangpricha et al., 2003). In that scenario, vitamin D fortification of fruit juice might also be an effective way to avoid vitamin D insufficiency (Tangpricha et al., 2003).

There are numerous ways to increase vitamin D content in dietary products. The first is the direct addition of vitamin D to food, although vitamin D's homogeneity, stability, and bioavailability in the matrix play key roles in the effectiveness of vitamin D fortification. Effective methods for delivering vitamin D in food matrix include emulsification, microencapsulation, and nanoencapsulation (Vieira & Souza, 2022; Maurya & Aggarwal, 2017a; Maurya et al., 2023).

3.1.1 METABOLISM OF VITAMIN D

Vitamin D_3 synthesized in the skin from 7-dehydrocholestrol when UV light from the sun with the spectrum 280–320 UVB break the benzene ring producing pre-vitamin D_3 which also forms D_3 when thermally induced (Bikle, 2014). The functional form of vitamin D known as 1, 25-dehydroxyvitamin D. Vitamin D_2 and vitamin D_3 both are inactive forms and are activated by enzymes inside our body (Chang & Lee, 2019).

Vitamin D is first delivered to the liver when it is ingested through diet or generated in the skin. (Holick, 2013). The primary circulating form of vitamin D, 25(OH)D (calcidiol), first undergoes 25-hydroxylation in the liver (Maurya & Aggarwal, 2017b). Then, under the influence of parathyroid hormone (PTH) and other mediatory compounds along with hypophosphatemia and growth hormone, it is converted into the active form, 1,25(OH)2D (calcitriol) in kidney via 1-α hydroxylation (Chang & Lee, 2019). This form of vitamin D is activated by the enzyme 25-hydroxyvitamin D-1α-hydroxylase (CYP27B1) binds to the nuclear vitamin D receptor (VDR) and acts as a steroid hormone and regulates gene expression (Hewison, M.; 2012). VDR's main biological function is to encourage intestinal calcium absorption and enterocyte differentiation, which enables the body to absorb calcium (Chang & Lee, 2019).

3.1.2 SOURCES OF VITAMIN D

The skin's exposure to UVB radiation of sun is a prime source of vitamin D for the majority of individuals (Nair & Maseeh, 2012). There are three possible sources of vitamin D: food sources, endogenous production triggered by UVB exposure, and dietary supplementation. In humans, vitamin D is particularly synthesized inside the pores and skin after being subjected to UVB while some of it is derived from nutritional sources (Prietl et al., 2013; Hossein-nezhad & Holick, 2013). Only some food sources clearly comprise vitamin D. These include fatty fishes inclusive of salmon, mackerel, and herring; cod liver oil; and mushrooms subjected to ultraviolet radiations. Fortified foods like milk, various yoghurts and cheeses, some breads, and some orange juices are among the most common vitamin D-fortified foods and supplements (Holick, 2012; Hossein-nezhad & Holick, 2013). Sources of vitamin D with the quantity of vitamin D present in the food are shown in Table 3.1. The transition of 7-dehydrocholesterol to vitamin D_3 has been observed to be inhibited by sunscreen and clothes (Christakos et al., 2012). As man evolved closer to the equator, they developed a dark hue called melanin to protect themselves from the harmful effects of sunlight such as skin cancer. Although melanin content in the skin also lowers the vitamin D production, it is not so concentrated in skin to inhibit all of the UV rays, therefore vitamin D can still be produced (Wacker & Holick, 2013, Holick, 2012).

TABLE 3.1
Food Sources of Vitamin D

Food Items	Quantity of Vitamin D in IU	Reference
Cod liver oil	1360 IU/ tsp	Nih.gov
Salmon	447/ 3 ounce	Cuomo et al. (2017)
Mackerel	250 IU/ 3.5 oz	Holick (2012)
Swordfish	566.6 / 3 ounce	Cuomo et al. (2017)
Raw fresh Mushrooms	100 IU/3.5 oz	Holick (2013).
Sundried Mushrooms	1600 IU/3.5 oz	Holick (2012)
Milk	115–124 IU/ 1 cup	Cuomo et al. (2017)
Sardines canned in oil	46 IU/ 2 sardines	Nih.gov
Tuna	154/ 3 ounce	Cuomo et al. (2017)
Egg yolk	20 IU	Holick (2013).
Fortified milk	100 IU / 8 oz	Holick (2012)

3.1.3 Importance and role of vitamin D

The fundamental duty of vitamin D is to keep the body's calcium and phosphorus levels at a healthy level so that metabolic functions could run smoothly (Wimalawansa, 2012). When vitamin D is transformed to its most active form in the kidney it adheres to the VDR and acts as a steroid hormone and circulated to the all-other tissues of the body (Christakos et al., 2012). Calcium receptors in the parathyroid gland notice a decrease in the plasma level of ionized calcium when there is hypocalcemia. The parathyroid gland secretes PTH, which stimulates 1-alpha-hydroxylationin kidneys to make create vitamin D, which in turn elevates calcium transport within the intestinal tract, bones, and kidneys as well as regulating osteoblast and osteoclast activity. As plasma calcium returns to normal, PTH secretion decreases (Chang & Lee, 2019). Vitamin D deficiency can also decrease muscle strength, as the regulation of calcium homeostasis is critical for muscular contraction and relaxation (Battault et al., 2013; Saponaro et al., 2020).

Clinical applications have shown that vitamin D_3 analogs could be used the treatment of various hyper-proliferative disorders in the skin (Bikle, 2014). Additionally, studies have indicated that vitamin D can help prevent cancer, particularly colon and breast cancer (Holick, 2013, Bikle, 2014). Vitamin D sufficiency has been shown to reduce cardiovascular disease in several studies (Norman & Powell, 2014; Battault et al., 2013). Studies have also shown that vitamin D could lower the chances of diabetes, as it is correlated with insulin resistance (Battault et al., 2013, Bikle, 2014).

VDRs are found throughout the body, including antigen-presenting cells, and have been linked to innate and adaptive immunity (Chang & Lee, 2019; Saponaro et al., 2020). There are numerous lung disorders, such as pneumonia, community-acquired pneumonia, ARDS, sepsis, heart failure, and deadly pulmonary infections, that are negatively associated with vitamin D concentration. Vitamin D also protects against acute respiratory illnesses and lung infections, as well as acting as an anti-inflammatory. Because of these reasons, vitamin D has recently been hypothesized to be a potential nutrient that could be used to treat or prevent COVID-19 (Saponaro et al., 2020). A cohort study was done in 2020 which showed that vitamin D supplementation prevented the risk of COVID-19. Several other studies concluded that Vitamin D supplementation may help to minimize the occurrence, severity, and risk of death associated with COVID-19 (Demir et al., 2021).

Supplementation with 1,25(OH)2D$_3$ has been demonstrated to be beneficial in the treatment of autoimmune illnesses such as experimental autoimmune encephalomyelitis (EAE) and collagen-induced arthritis in animals (CIA) (Sassi et al., 2018).

3.1.4 RDA AND HEALTH STATUS OF VITAMIN D

From the initial years of life to the age of 50, adequate vitamin D intake (AI) is estimated to be between 200 and 600 IU per day for both male and female. For persons 51 to 70 years old, the daily requirement rises to 600 IU/day, and for people above 70 years old, it climbs to 800 IU/ day (Harinarayan & Akhila, 2019; Cuomo et al., 2017; Pludowski et al., 2018). According to the ICMR (2010) RDA for vitamin D should be 400 IU/ day (10 µg/ day) for Indians (FSSAI, 2020). Recommended dietary allowance (RDA) of vitamin D is given in Table 1.2. The committee of the Institute of Medicine (IOM) concluded in 2012 that a blood 25(OH) D level of 50 nmol/L (20 ng/ ml) is essential for normal bone health (Manzoor et al., 2018; Pludowski et al., 2018). Moreover, the majority of studies that utilized 25(OH)D concentrations to investigate relationships between health and disease risk suggested higher 25(OH)D concentrations, such as those in the range of 30–50 ng/ml or 40–60 ng/ml, rather than 20 ng/ml as the required minimum concentration for human well-being. A 25(OH)D level of >30 ng/mL (75 nmol/L) is more appropriate for preventing preclinical osteomalacia (Pludowski et al., 2018). As per the Endocrine Society Practice Guidelines (ESPG); vitamin D sufficiency level is in the range of 30–100 ng/ ml and deficiency level is < 20 ng/ ml. Vitamin D status according to ESPG and IOM is shown in Table 3.3 (Lhamo et al., 2016; Ramasamy, 2020).

Vitamin D fortification depends on consumption dose of fortified food. In Canada, fluid milk is marketed as containing 44% of the daily recommended consumption in each serving of 250 ml. In Canada, all margarines are fortified with 530 IU/100 g of vitamin D. Depending on the product's energy content and intended use, different amounts of vitamin D may be added; for example, for carefully prepared liquid diets, the amount may be no less than 100 IU and no more than 400 IU per 1000 kcal as long as the desired total energy intake is 2500 kcal. Most ready-to-eat cereals typically have a fortification level of 10–35% daily value, or 40–140 IU (Calvo et al., 2004). Vanaspati is fortified in India with 200 IU of Vitamin D per 100 g (Kuchay & Mithal, 2018).

3.1.5 DEFICIENCY OF VITAMIN D

Vitamin D health status is estimated or computed via 25(OH)D serum concentration (Amrein et al., 2020; Chang & Lee, 2019; Gupta, 2014). 25(OH)D serum level less than 20 ng/mL is considered vitamin D deficiency, while 21–29 ng/mL is considered vitamin D insufficiency as shown in Table 1.3 (Holick, M. F., 2010). The definition of severe vitamin D insufficiency uses a threshold of less than 10 or 12 ng/mL since it significantly increases the risk of rickets and osteomalacia. Worldwide, vitamin D insufficiency is extremely prevalent., but still it is not diagnosed properly and mostly remain untreated (Gupta, 2014). Prevalence of vitamin D deficiency in general population worldwide is shown in **Figure 3.2**

TABLE 3.2

Recommended Dietary Allowance (RDA) of Vitamin D

Age	USA (IOM) RDA for vitamin D	ICMR RDA for Indians
0–12 y	400 IU	400 IU
1–13 y	600 IU	400 IU
14–18 y	600 IU	400 IU
19–50 y	600 IU	400 IU
51–70 y	600 IU	400 IU
>70 y	800 IU	400 IU
Pregnancy and lactation	600 IU	400 IU

Source: RDA according to IOM USA [19-21, 25] and ICMR India.

TABLE 3.3
Vitamin D Health Status

Vitamin D Health Status	ESPG	IOM
Vitamin D deficiency – associated with osteomalacia in adults as well as rickets in infants and toddlers.	< 50 nmol/L (20ng/ml)	<30 nmol/ L (<20 ng/ ml)
Insufficiency of vitamin D – in healthy people, commonly considered to be insufficient for bone and overall health.	52.5–72.5 nmol/L (21–29ng/ml)	30–50 nmol/L (12–20 ng/ ml)
Sufficient – in healthy people, this is generally regarded appropriate for bone and overall health.	72–250 nmol/L (30–100 ng/ml)	50 nmol/L (20ng/ ml)
Toxicity – associated with the possibility of negative consequences.	>150 ng/ml	> 50 ng/ ml

Source: Vitamin D health status according to ESPG and IOM

A lack of vitamin D affects the metabolism of calcium, phosphorus, and bones. Vitamin D deficiency (VDD) decreases dietary calcium and phosphorus absorption, which raises PTH levels and leads to the development of secondary hyperparathyroidism (Nair & Maseeh, 2012, Wimalawansa, 2012). Vitamin D deficiency has skeletal repercussions such as osteoporosis in adults and rickets in children, as well as a greater chance of fractures (Holick, 2010). VDD also results in muscle weakness, making it more difficult for young children to stand and walk and for older people to balance (Nair & Maseeh, 2012; Holick, 2010). VDD has an impact not only on skeletal health but also on non-skeletal health (Gupta, 2014). Non-skeletal diseases include the higher risk of cancers especially breast and colon cancers. Chronic diseases including type 2 diabetes, (Holick, 2010), several cardiovascular diseases, (Norman & Powell, 2014, Holick, 2012), autoimmune disorders, and acute respiratory infections (Holick, 2010) have an increased risk with the deficiency of vitamin D.

Therefore, adequate intake of vitamin D via sources like mushrooms, cod liver oil, and oily fish fortified food, and sun exposure to skin is advisable for maintaining good health. In light-skinned people, 10 to 15 minutes of sun exposure of skin from 10am to 3pm during spring, summer, and autumn could supply enough vitamin D. Darker skinned people, on the other hand, have more epithelial melanin, which suggests that skin vitamin D production requires more sunlight exposure (Chang & Lee, 2019). Still, there are several factors that could create the higher chances of VDD, which is discussed in next section.

3.1.5.1 Factors Contributing to the Chances of Vitamin D Deficiency

Even though exposure to sunshine is the best source of vitamin D, there are a number of factors that could influence how the vitamin is generated in the skin. These include:

i). **Skin color**: The melanin content in our skin could inhibit the UVB radiation and therefore vitamin D production. African population has enough melanin into their epidermis that it could lessen vitamin D production up to 90% (Holick, 2012).

ii). **Surface area exposed**: Vitamin D production is influenced by the area of skin exposed and the length of time spent in the sun. The amount of clothing worn owing to cultural or religious reasons, as well as the use of sunscreen, can obstruct optimal epidermal biosynthesis (Siddique et al., 2021).

iii). **Environmental factor**: Zenith angle of sun is a major factor concerning vitamin D synthesis. When the sunlight falls directly like in summer the biosynthesis of vitamin D is more but during winters the sunrays are in oblique angle therefore reducing the vitamin D production. Population living towards 33° N could not synthesize enough quantity of vitamin D under their skin (Chang & Lee, 2019; Holick, 2012). Furthermore, increasing ozone and

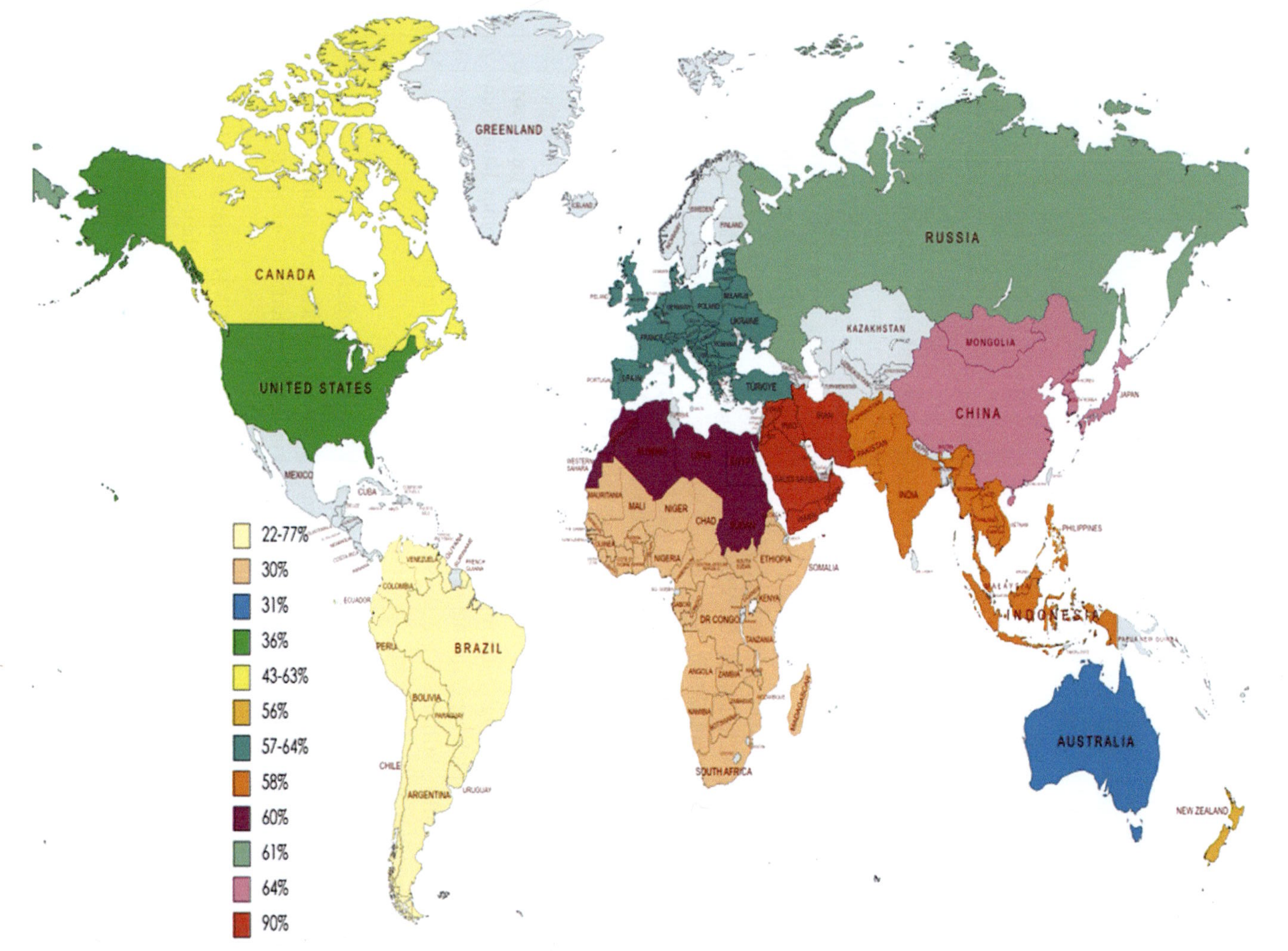

FIGURE 3.2 Prevalence of vitamin D deficiency in general population worldwide (Our World in Data, 2023).

nitrogen dioxide levels in the air, clouds; and fog have been shown to interfere with vitamin D production (Siddique et al., 2021; Hossein & Holick, 2013).

iv). **Age**: Human vitamin D status is also affected by aging, as people age, their capacity to synthesize vitamin D decreases, even when UVB radiation is abundant. On the other hand,young generations could efficiently produce vitamin D through sun exposure (Siddique et al., 2021; Maurya & Aggarwal, 2017b). Because of that reason, elderly people need to take vitamin D from dietary sources.

In India, a considerable percentage of the population is socioeconomically disadvantaged. The poorest people in society cannot get enough nutrition because it is expensive for them and therefore could face scarcity of vitamin D. Also, vegetarianism, which undoubtedly reduces vitamin D-rich dietary options, is practiced by the majority of Indians (Gupta, 2014).

The possible solutions to decrease the prevalence of vitamin D deficiency could be dietary sources, but they are limited as well as costly, and there are not many sources for the vegetarian population except milk and milk products. Another important source for vitamin D is sun exposure, but due to lifestyle patterns like being indoors, clothing choices due to religious beliefs etc., it is not enough for the production of significant amounts of vitamin D. Therefore, strategies such as supplementation and fortification are being adopted these days, which we will discuss in the next section.

3.1.5.2 Strategies to Fight against Vitamin D Deficiency

(a) Supplementation: Dietary supplements are substances having one or more concentrated nutrients that are meant to be added to a person's daily diet when it is unbalanced or deficient in certain nutrients (Hassan et al., 2020). Vitamins, minerals, and herbs, as well as bio-chemicals, amino acids, enzymes, and animal extracts, are all included. Tablets, soft gels, and gel caps are all common kinds of dietary supplements (Ghosh et al., 2018). Nutritional supplements have a number of advantages, including high nutrient content in tiny amounts, unique nutrient compositions, absence of unwanted companion components like lipids, cholesterol, and purines, and comprehensive covering of specialized sporting needs (Hassan et al., 2020).

To maintain physiologic serum vitamin D levels and health, patients at risk of VDD might require vitamin D supplements in high dosage for e.g., 2000 IU/day or 50,000 IU every 2 to 4 weeks, and persons who do not maintain therapy regularly or malabsorb the vitamin need doses in higher amounts such as 50,000 IU (Wimalawansa, 2012). Cholecalciferol and alfacalcidol are some of the most commonly available supplements in India. The most popular D_3 supplement is 60,000 IU, which is available in dry powder in pouches or as oil-based capsules (Gupta, 2014).

The main disadvantage of supplements is they can cause toxicity because users may take higher amount of doses more frequently. Vitamin D toxicity causes hypercalcemia and hyperphosphatemia, which leads to mineralization of numerous soft tissues, such as the kidneys, heart, and epidermal tissue (Zittermann et al., 2013). Vitamin D overdose symptoms include poor appetite, weight loss, abdominal cramps, nausea, frequent urination, and polydipsia in children as well as life-threatening dehydration in extreme cases (Vogiatzi et al., 2014). Toxicity is relatively uncommon with doses less than 5000 IU/day (Wimalawansa, 2012). Another disadvantage of dietary supplements is that they are comparatively expensive.

In order to overcome the disadvantage of the supplements as mentioned above, food fortification might be one of the best strategies to meet vitamin D sufficiency.

3.1.6 Food Fortification

Micronutrient deficiencies are generally caused by insufficient consumption of foods high in nutrients and nutrient losses as a result of unhealthy diets, illnesses, and other factors (Olson et al., 2021). Malnutrition, a serious public health problem, particularly in underdeveloped nations, is commonly caused by micronutrient deficiencies. Malnutrition of this type causes mortality and morbidity, disability, and lowers productivity (Chadare et al., 2019). The most successful preventive measure against malnutrition brought on by a lack of micronutrients is thought to be food fortification (Bhagwat et al., 2014). Fortification can also be used to balance the total nutritional content of a diet, restore nutrients destroyed during food processing, or create products more attractive to consumers (Dwyer et al., 2015). Among the available health strategies, food fortification has been identified as the most cost-effective strategy to combat nutritional insufficiency disorders (Handu et al., 2021). Fortification is the inclusion of one or more micronutrients, for example vitamins or minerals, to the food products, even if they're not naturally present in the food, in order to prevent micronutrient deficiency (Lourdes et al., 2012).

The type of fortification that will be most appropriate and effective in a particular country depends on a number of factors, including the frequency of particular micronutrient deficiencies, the population most affected, dietary compositions, available infrastructure, and capacities for food processing and production systems (Olson et al., 2021).

Fortification is classified into three categories:

i). Mandatory or mass fortification: this is done for staple foods like rice, flour, and milk; it is mandatory and supervised by the government.
ii). Targeted fortification: this is done for food that is produced for special purposes such as infant formula.
iii). Voluntary fortification: this is done by food manufacturers with an intention to add value to the packaged food, hence it can help increase sales (Gupta, 2014).

The foods to be fortified and the micronutrient concentrations must be carefully chosen in order to pinpoint the best delivery systems for food fortification and to focus on the population at risk of deficiency without overdosing on other population subgroups (Dwyer et al., 2015). In order to prevent overconsumption, certain nutrients, like folic acid and vitamin D, are particularly constrained by laws governing which foods can be fortified and at what doses. Contrarily, there are no restrictions on the addition of vitamin A to food other than those set by good manufacturing practices (Dwyer et al., 2014).

3.1.6.1 Vitamin D Fortification

In order to consume the recommended daily allowance (RDA) and maintain plasma 25(OH)D levels above 30 ng/ml, certain population subsets, such as newborns, older people (not adequately exposed to sun radiation), and patients (affected by fat metabolism and genetic variations in protein associated to vitamin D uptake in the intestine), must rely on vitamin D from supplements and fortified food (Wimalawansa, 2012). Food fortification in milk fortified with vitamin D has historically been used as a means to combat nutritional-deficiency rickets (Lourdes et al., 2012). In many developing nations, the most regularly employed fortification carriers are some of the most commonly consumed foods, such as milk, oils and fats, sugar, cereal flour, and salt (Chadare et al., 2019). Fortification of fresh fruits and vegetables, fresh animal products like meat, fish, sweets, or snacks such as candies or carbonated beverages, and the indiscriminate adding of nutrients to foods are all judged unsuitable and inappropriate (Dwyer et al., 2014). In recent years, the food industry has fortified calcium and vitamin D-fortified drinks, omega-3 fatty acid-fortified breads, and vegetable oil spreads with plant sterols available to consumers looking for products with added health advantages (Lourdes et al., 2012).

Food fortification with vitamin D can be a successful strategy, and many nations have adopted either mandated or voluntary food fortification in order to address the widespread vitamin D deficiency (Zahedirad et al., 2019). In the United States, food fortification with vitamin D gained popularity in the 1930s and 1940s. At the time, vitamin D fortified milk was quite prominent (Pilz et al., 2019). Nowadays, milk and milk products, fruit beverages such as orange juice and vegetable oils are the most common foods fortified with vitamin D. The Canadian Food and Drug Regulations have made it compulsory to fortify milk and margarine (Handu et al., 2021). According to India's national food standard, milk can be fortified with vitamin D at an amount of 0.5–0.75 g/100 ml (Cashman & O'Dea, 2019). Milk fortification in dairy industries is done with the addition of premix in milk. The two distinct types of vitamin premix are oil-based and water-dispersible formulations. These can be used with different dairy processing methods. After the cream separator, milk must typically have oil-based vitamin premix added to it. Before the separator, at any point in the milk flow, or both, the water-dispersible vitamin premix can be utilized (Yeh et al., 2017).

Many countries have adopted food and drinks, such as milk and milk products, cereal based products, bread, plant-based drinks, margarine, vegetable oils, fruit juices etc. for the fortification of vitamin D (Zahedirad et al., 2019). According to the literature reported, consuming breads fortified with vitamin D causes a rise in 25(OH)D serum concentrations, which in turn lowers PTH levels and improves bone health (Souza et al., 2022). Cake baked with wheat flour enriched with vitamin D demonstrated an 84% retention of the vitamin (Bajaj & Singhal, 2021). Additionally, it has been observed that drinking milk that has been fortified with vitamin D increases the 25 (OH)D blood concentration (Itkonen et al., 2018). A meta-analysis on randomized controlled trials (RCTs) conducted worldwide to assess the effectiveness of fortification discovered that meals fortified with vitamin D raised circulating 25(OH)D concentrations in a dose-dependent manner (Spiro & Buttriss, 2014; Black et. al., 2012). According to Leskauskaite et al., (2016), milk products that were fortified with emulsified vitamin D showed improved processing stability and shelf life. Additionally, since vitamin D is stable in storage, pasteurized process cheeses have a promising future as a source of vitamin D3 throughout their shelf life (Upreti et al., 2002). Furthermore, no appreciable loss in vitamin D was observed during the preparation of vitamin D fortified cheese from fortified milk. A study was conducted in which starch-based nanocarriers of vitamin D3 were created, and sensory testing on enriched milk with these nanocarriers revealed that the nanocarriers did not significantly differ from the control milk sample and could effectively hide the aftertaste and improve vitamin D3's low solubility (Hasanvand et al., 2015). In another study, a thin-film hydration-sonication process was used to create vitamin D_3 nanoliposomes, and the study's findings suggested that liposomal nanoparticles might be used as a viable carrier for adding vitamin D_3 to beverages (Mohammadi et al., 2014).

The proportion of ingested vitamin D that eventually ends up in systemic blood circulations and hence imparts relevant physiological activities is known as bioavailability of vitamin D to the human body (Maurya et al., 2020). Vitamin D bioavailability is influenced by several factors. Out of those factors, bioaccessibility is one of the important factors. Studies conducted on bioaccessibility have shown that infant formula, partially defatted milk and whole milk, had bioaccessibility of 70, 53, and 54% respectively. However, UV-treated yeast white bread bioaccessibility was found to be only 15%, and it did not increase the blood serum level of 25(OH) D. The reason was that the yeast cells captured vitamin D_2 and prevented it from being incorporated into the mixed micelles (Lavelli et al., 2021). Therefore, the selection of food matrix and correct technology for fortification is necessary for enhancing the bioaccessibility of vitamin D.

3.1.7 Several Technologies Used for Food Fortification

Fortifying vitamin D has proven difficult for the food industry due to its volatility and diverse distribution in food matrix. Emulsification, microencapsulation, and nanoencapsulation are a few of the methods used to fortify vitamin D. Milk and milk-based products are frequently supplemented with vitamin D. Direct addition homogenization follows the dissolution of vitamin D in an organic solvent, such as ethanol and butter oil, and can also be done to fortify foods with vitamin D (Handu et al., 2021).

(a) Emulsification

TABLE 3.4

Techniques and Methods Used for the Encapsulation of Vitamin D

Techniques	Matrix	Method	Results	Reference
Nanoencapsulation	Nanoliposomes	Thin film sonication	Encapsulated vitamin D3 and reported high stability against degradation.	(Mohammadi et al., 2014)
Emulsions	Nanoemulsions	Phase inversion method	Encapsulated vitamin D3 stable nanoparticles could be used to fortify milk-based drinks.	(Maurya & Aggarwal, 2019a)
Emulsion	Oil in water emulsion	Microchannel emulsification	Formulated stable emulsions and encapsulated vitamin D2.	(Khalid et al., 2015)
Microencapsulation	Polymer microparticles of gum Arabic- chitosan	Spray drying	Fortified vitamin D3 encapsulated particles in pear juice.	(Dima et al., 2020)
Homogenization	Vitamin D3 encapsulated in reassembled casein complex	Ultra high-pressure homogenization	Resistance to thermal degradation and enhanced bioavailability	(Haham et al., 2012)

Emulsions are employed to enclose, safeguard, and transport lipophilic substances. They are colloidal dispersions of two immiscible phases. Emulsions are divided into two groups: water-in-oil (w/o) and oil-in-water (o/w), depending on whether the dispersed phase is made up of oil or water (Vieira & Souza, 2022). Casein, whey protein isolate (WPI), medium-chain triglycerides (MCT), carboxymethyl chitosan, zein, and Tween 20 are examples of food-grade ingredients that have been used in various emulsion techniques in several studies to create vitamin D nanomaterials (Handu et al., 2021). Additionally, it has been found that phase inversion-derived nanoparticles are better suited for adding nutrients to drinks like buttermilk and lassi (Maurya & Aggarwal, 2017c, 2019a, 2019b).

(b) Microencapsulation

Environmental elements such as temperature, oxygen, light, and humidity have a high vulnerability to vitamin D, causing losses during manufacturing and storage. As a result, encapsulated vitamin D appears to be a viable strategy that could improve not just bioavailability but also stability in the face of environmental and gastrointestinal tract circumstances (Santos et al., 2021). Microencapsulation is the process of insulating a bioactive core material with secondary wall materials that shield it from the outside environment (Maurya et al., 2020). The diameter of microcapsules ranges from 5 to 300 micrometers (Esmaeili et al., 2021). The two most common ways for microencapsulating vitamin D in beverage products are spray-drying and coacervation (Vieira & Souza, 2022).

1. **Spray drying**: One of the earliest methods for encapsulating bioactive chemicals is spray drying. In a dispersion comprising wall components, vitamin D needs to be homogenized. The homogenized dispersion must next be introduced into the spray dryer, where hot air atomization causes the evaporation of water, culminating in the development of nanomaterials. To investigate the impact of temperature on the physicochemical features of spray-dried whey nanoparticles containing vitamin D, various combinations of maltodextrin,

gum arabic, modified starch, and whey protein concentrate were used to encapsulate vitamin D. Spray drying was used to encapsulate vitamin D2 in casein micelles, increasing its stability and bioavailability (Maurya et al., 2020).

2. **Coacervation**: Under ideal circumstances, two or more electrolytes that are oppositely charged are combined to produce two liquid phases: a diluted polymer phase that serves as a continuous medium and a polymeric solution incorporated into the encapsulating system. Other microencapsulation methods with significant potential for the creation of vitamin D-fortified beverages include spray drying, nanostructured lipid carriers, liposomes, micro- or nanoemulsions, nanoemulsions, and polymer complexation (Vieira & Souza, 2022).

(c) Nanoencapsulation

A growing body of research has established a link between dietary vitamin D consumption and a reduced risk of developing a number of chronic diseases (Calvo et al., 2013; Green and Miller 2022; Hohman et al., 2011; Keegan et al., 2013; Urbain et al., 2011). However, vitamin D has a low absorption rate, which dramatically lowers its effectiveness as a disease-fighting substance. In order to develop unique materials and structures for a variety of uses within the food industry, nanotechnology has become an essential tool. In order to protect bioactive compounds from harmful environments and to boost their solubility in aqueous systems, nanoencapsulation technology makes it easier to shield bioactive ingredients into nanoparticles. Additionally, encapsulation methods enable controlled release and the administration of appropriate amounts, both of which aid in preventing the hypervitaminosis syndrome and its likely negative effects. In order to increase the health advantages of vitamin D by encapsulation, protection, and/or controlled/sustained release, several encapsulated nanoparticles (Ens) have been constructed and studied for their potential usage as delivery systems for the vitamin (Gonnet et al., 2010). An interesting strategy for increasing vitamin D's effectiveness in humans is to increase its bioavailability. Significant progress has recently been made in developing ENs to increase vitamin D bioavailability (Guttoff et al., 2015). The size of nanocapsules ranges from 50 to 1000 nanometers (Esmaeili et al., 2021). This method boosts bioactive solubility while also improving release behavior, cellular absorption, and bioavailability (Handu et al., 2021). Controlled release of nanoencapsulated bioactive chemicals under gastrointestinal circumstances improves their stability and allows them to be used in human food matrices (Melo et al., 2021). Nanosuspensions, nanoliposomes, nano emulsions, and cyclodextrin carriers are some of the technologies employed in food fortification. Starches, sugars, lipids, gums, chitosan, gelatin, and maltodextrin are some of the coating materials used to encapsulate fortificants (Handu et al., 2021). The breakdown of vitamin D in simulated intestinal fluids was reduced by encapsulating it in micro liposomes fabricated from food-grade materials. Recently, lipid-based ENs have been employed to increase the bioaccessibility of lipophilic vitamins, including nano emulsions, liposomes, micelles, and solid lipid nanoparticles (Müllertz et al., 2010; Santos and Meireles, 2010). Encapsulating lipophilic vitamin D in lipid-derived encapsulated nanoparticles (nanoemulsions) has been utilized frequently to increase their absorption. The lipophilic vitamin D is transported across the aqueous mucous layer by mixed micelles that are created as a result of the digestion of nano emulsions, and these micelles then make the vitamin accessible to brush-bordered enterocytes for absorption. Encapsulated nanoparticles boost the gastrointestinal tract's (GIT) ability to absorb lipophilic compounds, increase their solubility in intestinal juice, and lessen the loss of first-pass metabolism in the gut and liver (Maurya & Aggarwal, 2017a). A novel encapsulation technique that uses a phase inversion-based procedure could be used to create a relatively stable nanodelivery system for vitamin D encapsulation, protection, and high bioavailability (Maurya & Aggarwal, 2019).

3.1.8 FUTURE PROSPECTUS

A major public health issue is vitamin D deficiency. However, few nations have a dedicated strategy to prevent population deficiencies. Food fortification is an effective strategy to deal with vitamin D deficiency. However, there is a need to fortify vitamin D in various kinds of staple food categories, such as flours and other cereal- and millet-based products, so that it can reach to the wide range of population socioeconomically disadvantaged sectors as well as vegan and vegetarian populations in India. The guidelines regarding the fortification of vitamin D in cereal and millet- based products needed to be set up by competent authority.

3.1.9 CONCLUSION

In addition to being essential for maintaining bone health, vitamin D also protects against a number of ailments, including cancer and chronic illnesses. Despite the fact that endogenous synthesis through sunlight is the main source of vitamin D, the majority of the population suffers from vitamin D deficiency due to factors such as lifestyle changes, religious beliefs, clothing, and melanin content of the skin. One of the greatest approaches to combat vitamin D deficiency is fortification. Milk and milk products, cereal-based products, edible oils, and infant formula are some of the vehicles for the fortification of vitamin D. However, interactions with food components are unavoidable, leading to the loss of vitamin D during preparation and storage. The bioavailability, uniform distribution in food, and stability of the fortified food are major concern in food industries. Microencapsulation and nanoencapsulation are some of the efficient methods to enhance the stability and bioavailability and for the controlled release of vitamin D from fortified foods in the gastrointestinal tract. The right choice of the strategy for fortification of vitamin D could help eradicate vitamin D deficiency from the world in a cost-effective manner.

REFERENCES

Amrein, K., Scherkl, M., Hoffmann, M., Neuwersch-Sommeregger, S., Köstenberger, M., Berisha, A. T., … Malle, O. (2020). Vitamin D deficiency 2.0: An update on the current status worldwide. *European Journal of Clinical Nutrition*, 74(11), 1498–1513.

Bajaj, S. R., & Singhal, R. S. (2021). Fortification of wheat flour and oil with vitamins B12 and D3: Effect of processing and storage. *Journal of Food Composition and Analysis*, 96, 103703.

Battault, S., Whiting, S. J., Peltier, S. L., Sadrin, S., Gerber, G., & Maixent, J. M. (2013). Vitamin D metabolism, functions and needs: From science to health claims. *European Journal of Nutrition*, 52(2), 429–441.

Bhagwat, S., Gulati, D., Sachdeva, R., & Sankar, R. (2014). Food fortification as a complementary strategy for the elimination of micronutrient deficiencies: Case studies of large-scale food fortification in two Indian States. *Asia Pacific Journal of Clinical Nutrition*, 23, Suppl 1:S4–S11.

Bikle, D. D. (2014). Vitamin D metabolism, mechanism of action, and clinical applications. *Chemistry and Biology*, 21(3), 319–329.

Black, L. J., Seamans, K. M., Cashman, K. D., & Kiely, M. (2012). An updated systematic review and meta-analysis of the efficacy of vitamin D food fortification. *The Journal of Nutrition*, 142(6), 1102–1108.

Borel, P., Caillaud, D., & Cano, N. J. (2015). Vitamin D bioavailability: State of the art. *Critical Reviews in Food Science and Nutrition*, 55(9), 1193–1205.

Calvo, M. S., Whiting, S. J., & Barton, C. N. (2004). Vitamin D fortification in the United States and Canada: Current status and data needs 1–4. *American Journal of Clinical Nutrition*, 80. https://academic.oup .com/ajcn/article/80/6/1710S/4690516.

Calvo, M. S., Babu, U. S., Garthoff, L. H., Woods, T. O., Dreher, M., Hill, G., & Nagaraja, S. (2013). Vitamin D2 from light-exposed edible mushrooms is safe, bioavailable and effectively supports bone growth in rats. *Osteoporosis International*, 24(1), 197–207. https://doi.org/10.1007/s00198-012-1934-9.

Cashman, K. D., & O'Dea, R. (2019). Exploration of strategic food vehicles for vitamin D fortification in low/lower-middle income countries. *The Journal of Steroid Biochemistry and Molecular Biology*, 195, 105479.

Chadare, F. J., Idohou, R., Nago, E., Affonfere, M., Agossadou, J., Fassinou, T. K., … Hounhouigan, D. J. (2019). Conventional and food-to-food fortification: An appraisal of past practices and lessons learned. *Food Science and Nutrition*, 7(9), 2781–2795.

Chang, S. W., & Lee, H. C. (2019). Vitamin D and health-The missing vitamin in humans. *Pediatrics and Neonatology*, 60(3), 237–244.

Christakos, S., Ajibade, D. V., Dhawan, P., Fechner, A. J., & Mady, L. J. (2012). Vitamin D: Metabolism. *Rheumatic Disease Clinics*, 38(1), 1–11.

Cuomo, A., Giordano, N., Goracci, A., & Fagiolini, A. (2017). Depression and vitamin D deficiency: Causality, assessment, and clinical practice implications. *Neuropsychiatry*, 7(5), 606–614.

de Lourdes Samaniego-Vaesken, M., Alonso-Aperte, E., & Varela-Moreiras, G. (2012). Vitamin food fortification today. *Food and Nutrition Research*, 56(1), 5459.

de Melo, A. P. Z., da Rosa, C. G., Noronha, C. M., Machado, M. H., Sganzerla, W. G., da Cunha Bellinati, N. V., … Barreto, P. L. M. (2021). Nanoencapsulation of vitamin D3 and fortification in an experimental jelly model of Acca sellowiana: Bioaccessibility in a simulated gastrointestinal system. *LWT*, 145, 111287.

Demir, M., Demir, F., & Aygun, H. (2021). Vitamin D deficiency is associated with COVID-19 positivity and severity of the disease. *Journal of Medical Virology*, 93(5), 2992–2999.

Dima, C., Milea, A. S., Constantin, O. E., Stoica, M., Ivan, A. S., Alexe, P., & Stanciuc, N. (2020). Fortification of pear juice with vitamin D3 encapsulated in polymer microparticles: Physico-chemical and microbiolgical characterization. *Journal of Agroalimentary Processes and Technologies*, 26, 140–148.

Dwyer, J. T., Wiemer, K. L., Dary, O., Keen, C. L., King, J. C., Miller, K. B., … Bailey, R. L. (2015). Fortification and health: Challenges and opportunities. *Advances in Nutrition*, 6(1), 124–131.

Dwyer, J. T., Woteki, C., Bailey, R., Britten, P., Carriquiry, A., Gaine, P. C., … Smith Edge, M. (2014). Fortification: New findings and implications. *Nutrition Reviews*, 72(2), 127–141.

Esmaeili, M., Yekta, R., Abedi, A. S., Ghanati, K., Derav, R. Z., Houshyarrad, A., … Mahmoudzadeh, M. (2021). Encapsulating vitamin D: A feasible and promising approach to combat its deficiency. *Pharmaceutical Sciences*, 28(2), 194–207.

F Holick, M. (2013). Vitamin D, sunlight and cancer connection. *Anti-Cancer Agents in Medicinal Chemistry (Formerly Current Medicinal Chemistry-Anti-Cancer Agents)*, 13(1), 70–82.

Ghosh, S., Pahari, S., & Roy, T. (2018). An updated overview on food supplement. *Asian Journal of Research in Chemistry*, 11(3), 691–697.

Gonnet, M., Lethuaut, L., & Boury, F. (2010). New trends in encapsulation of liposoluble vitamins. *Journal of Controlled Release*, 146(3), 276–290.

Green, R., & Miller, J. W. (2022). Vitamin B12 deficiency. InVitamins and Hormones (January 1, Vol. 119, pp. 405–439). Academic Press.

Gupta, A. (2014). Vitamin D deficiency in India: Prevalence, causalities and interventions. *Nutrients*, 6(2), 729–775.

Guttoff, M., Saberi, A. H., & McClements, D. J. (2015). Formation of vitamin D nanoemulsionbased delivery systems by spontaneous emulsification: Factors affecting particle size and stability. *Food Chemistry*, 171, 117–122.

Haham, M., Ish-Shalom, S., Nodelman, M., Duek, I., Segal, E., Kustanovich, M., & Livney, Y. D. (2012). Stability and bioavailability of vitamin D nanoencapsulated in casein micelles. *Food and Function*, 3(7), 737–744.

Handu, S., Jan, S., Chauhan, K., & Saxena, D. C. (2021). Vitamin D fortification: A perspective to improve immunity for COVID-19 infection. *Functional Food Science*, 1(10), 50–66.

Harinarayan, C. V., & Akhila, H. (2019). Modern India and the tale of twin nutrient deficiency–calcium and vitamin D–nutrition trend data 50 years-retrospect, introspect, and prospect. *Frontiers in Endocrinology*, 10, 493.

Hasanvand, E., Fathi, M., Bassiri, A., Javanmard, M., & Abbaszadeh, R. (2015). Novel starch based nanocarrier for vitamin D fortification of milk: Production and characterization. *Food and Bioproducts Processing*, 96, 264–277. https://doi.org/10.1016/j.fbp.2015.09.007.

Hassan, S., Egbuna, C., Tijjani, H., Ifemeje, J. C., Olisah, M. C., Patrick-Iwuanyanwu, K. C., … Ephraim-Emmanuel, B. C. (2020). Dietary supplements: Types, health benefits, industry and regulation. In Chukwuebuka Egbuna and Genevieve Dable Tupas (eds.), *Functional Foods and Nutraceuticals* (pp. 23–38). Springer, Cham.

Haussler, M. R., Whitfield, G. K., Kaneko, I., Haussler, C. A., Hsieh, D., Hsieh, J. C., & Jurutka, P. W. (2013). Molecular mechanisms of vitamin D action. *Calcified Tissue International*, 92(2), 77–98.

Hewison, M. (2012). An update on vitamin D and human immunity. *Clinical Endocrinology*, 76(3), 315–325.

Hohman, E. E., Martin, B. R., Lachcik, P. J., Gordon, D. T., Fleet, J. C., & Weaver, C. M. (2011). Bioavailability and efficacy of vitamin D2 from UVirradiated yeast in growing, vitamin D-deficient rats. *Journal of Agriculture and Food Chemistry*, 59(6), 2341–2346.

Holick, M. F. (2010). The vitamin D deficiency pandemic: A forgotten hormone important for health. *Public Health Reviews*, 32(1), 267–283.

Holick, M. F. (2012). The D-lightful vitamin D for child health. *Journal of Parenteral and Enteral Nutrition*, 36, 9S–19S.

Hossein-nezhad, A., & Holick, M. F. (2013, July). Vitamin D for health: A global perspective. In *Mayo Clinic Proceedings* (Vol. 88, No. 7, pp. 720–755). Elsevier.

https://ods.od.nih.gov/factsheets/VitaminD-HealthProfessional/.

https://www.fssai.gov.in/upload/advisories/2020/01/5e159e0a809bbLetter_RDA_08_01_2020.pdf.

ICMR (2010). Nutrient requirements and recommended dietary allowances for Indians. A report of the expert group of the Indian Council of Medical Research, New Delhi.

Itkonen, S. T., Erkkola, M., & Lamberg-Allardt, C. J. (2018). Vitamin D fortification of fluid milk products and their contribution to vitamin D intake and vitamin D status in observational studies—A review. *Nutrients*, 10(8), 1054.

Keegan, R.-J. H., Lu, Z., Bogusz, J. M., Williams, J. E., & Holick, M. F. (2013). Photobiology of vitamin D in mushrooms and its bioavailability in humans. *Dermato-Endocrinology*, 5(1), 165–176. https://doi.org /10.4161/ derm.23321.

Khalid, N., Kobayashi, I., Wang, Z., Neves, M. A., Uemura, K., Nakajima, M., & Nabetani, H. (2015). Formulation characteristics of triacylglycerol oil-in-water emulsions loaded with ergocalciferol using microchannel emulsification. *RSC Advances*, 5(118), 97151–97162.

Kuchay, M. S., & Mithal, A. (2018). Vitamin D deficiency in India. *Journal of the Indian Medical Association*, 116(10), pp. 41–44+55). Evangel Publishing.

Lavelli, V., D'Incecco, P., & Pellegrino, L. (2021). Vitamin D incorporation in foods: Formulation strategies, stability, and bioaccessibility as affected by the food matrix. *Foods*, 10(9), 1989.

Leskauskaite, D., Jasutiene, I., Malinauskyte, E., Kersiene, M., & Matusevicius, P. (2016). Fortification of dairy products with vitamin D3. *International Journal of Dairy Technology*, 69(2), 177–183.

Lhamo, Y., Chugh, P. K., & Tripathi, C. D. (2016). Vitamin D supplements in the Indian market. *Indian Journal of Pharmaceutical Sciences*, 78(1), 41.

Manzoor, M. S., Musafa, U., Naseem, N., Gilani, S., & Tahir, H. (2018). Development of Vitamin D Status in School Going Children by School Meal Planning in Selected School of District Pakpattan. *Indian Journal of Nutrition*, 5(2).

Maurya, V. K., & Aggarwal, M. (2017a). Enhancing bio-availability of vitamin D by nano-engineered based delivery systems-An overview. *International Journal of Current Microbiology and Applied Sciences*, 6(7), 340–353.

Maurya, V. K., & Aggarwal, M. (2017b). Factors influencing the absorption of vitamin D in GIT: An overview. *Journal of Food Science and Technology*, 54(12), 3753–3765.

Maurya, V. K., & Aggarwal, M. (2017c). Impact of Aqueous/ethanolic goji Berry (Lycium barbarum) fruit extract supplementation on vitamin D stability in yoghurt. *International Journal of Current Microbiology and Applied Sciences*, 6(8), 2016–2029.

Maurya, V. K., & Aggarwal, M. (2019a). Fabrication of nano-structured lipid carrier for encapsulation of vitamin D3 for fortification of 'Lassi'; A milk based beverage. *The Journal of Steroid Biochemistry and Molecular Biology*, 193, 105429.

Maurya, V. K., & Aggarwal, M. (2019b). A phase inversion based nanoemulsion fabrication process to encapsulate vitamin D3 for food applications. *The Journal of Steroid Biochemistry and Molecular Biology*, 190, 88–98.

Maurya, V. K., Bashir, K., & Aggarwal, M. (2020). Vitamin D microencapsulation and fortification: Trends and technologies. *The Journal of Steroid Biochemistry and Molecular Biology*, 196, 105489.

Maurya, V. K., Shakya, A., Bashir, K., Jan, K., & McClements, D. J. (2023). Fortification by design: A rational approach to designing vitamin D delivery systems for foods and beverages. *Comprehensive Reviews in Food Science and Food Safety*, 22(1), 135–186.

Mohammadi, M., Ghanbarzadeh, B., & Hamishehkar, H. (2014). Formulation of nanoliposomal vitamin D3 for potential application in beverage fortification. *Advanced Pharmaceutical Bulletin*, 4, 569–575.

Müllertz, A., Ogbonna, A., Ren, S., & Rades, T. (2010). New perspectives on lipid and surfactant-based drug delivery systems for oral delivery of poorly soluble drugs. *Journal of Pharmacy and Pharmacology*, 62(11), 1622–1636.

Nair, R., & Maseeh, A. (2012). Vitamin D: The "sunshine" vitamin. *Journal of Pharmacology and Pharmacotherapeutics*, 3(2), 118.

Norman, P. E., & Powell, J. T. (2014). Vitamin D and cardiovascular disease. *Circulation Research*, 114(2), 379–393.

Olson, R., Gavin-Smith, B., Ferraboschi, C., & Kraemer, K. (2021). Food fortification: The advantages, disadvantages and lessons from sight and life programs. *Nutrients*, 13(4), 1118.

Pilz, S., Zittermann, A., Trummer, C., Theiler-Schwetz, V., Lerchbaum, E., Keppel, M. H.Grübler, M.R., März, W., & Pandis, M. (2019). Vitamin D testing and treatment: A narrative review of current evidence. *Endocrine Connections*, 8(2), R27-43.

Pludowski, P., Holick, M. F., Grant, W. B., Konstantynowicz, J., Mascarenhas, M. R., Haq, A., … Wimalawansa, S. J. (2018). Vitamin D supplementation guidelines. *The Journal of Steroid Biochemistry and Molecular Biology*, 175, 125–135.

Prietl, B., Treiber, G., Pieber, T. R., & Amrein, K. (2013). Vitamin D and immune function. *Nutrients*, 5(7), 2502–2521.

Ramasamy, I. (2020). Vitamin D metabolism and guidelines for vitamin D supplementation. *The Clinical Biochemist Reviews*, 41(3), 103.

Santos, D. T., & Meireles, M. A. (2010). Carotenoid pigments encapsulation: Fundamentals, techniques and recent trends. *Open Chemical Engineering Journal*, 4, 42–50.

Santos, M. B., de Carvalho, C. W. P., & Garcia-Rojas, E. E. (2021). Microencapsulation of vitamin D3 by complex coacervation using carboxymethyl tara gum (Caesalpinia spinosa) and gelatin A. Food Chemistry, 343, 128529.

Saponaro, F., Saba, A., & Zucchi, R. (2020). An update on vitamin D metabolism. *International Journal of Molecular Sciences*, 21(18), 6573.

Sassi, F., Tamone, C., & D'Amelio, P. (2018). Vitamin D: Nutrient, hormone, and immunomodulator. *Nutrients*, 10(11), 1656.

Siddique, M. T., Hafeez, A., Mearaj, S., & Abbas, S. (2021). Overview of vitamin D, its status and consequences: Challenges and prospects for Pakistani population: A review. *Biomedical Letters*, 7(1), 1–11.

Souza, S. V., Borges, N., & Vieira, E. F. (2022). Vitamin D-fortified bread: Systematic review of fortification approaches and clinical studies. *Food Chemistry*, 372, 131325.

Spiro, A., & Buttriss, J. L. (2014). Vitamin D: An overview of vitamin D status and intake in Europe. *Nutrition Bulletin*, 39(4), 322–350. https://doi.org/10.1111/nbu.12108.

Tangpricha, V., Koutkia, P., Rieke, S. M., Chen, T. C., Perez, A. A., & Holick, M. F. (2003). Fortification of orange juice with vitamin D: A novel approach for enhancing vitamin D nutritional health 1–3. *American Journal of Clinical Nutrition*, 77. www.stat.ucla.edu.

Upreti, P., Mistry, V. V., & Warthesen, J. J. (2002). Estimation and fortification of vitamin D3 in pasteurized process cheese. *Journal of Dairy Science*, 85(12), 3173–3181.

Urbain, P., Singler, F., Ihorst, G., Biesalski, H.-K., & Bertz, H. (2011). Bioavailability of vitamin D2 from UV-B-irradiated button mushrooms in healthy adults deficient in serum 25-hydroxyvitamin D: A randomized controlled trial. *European Journal of Clinical Nutrition*, 65(8), 965–971.

Vieira, E. F., & Souza, S. (2022). Formulation strategies for improving the stability and bioavailability of vitamin D-fortified beverages: A review. *Foods*, 11(6), 847.

Vogiatzi, M. G., Jacobson-Dickman, E., DeBoer, M. D., & Drugs, and Therapeutics Committee of The Pediatric Endocrine Society. (2014). Vitamin D supplementation and risk of toxicity in pediatrics: A review of current literature. *The Journal of Clinical Endocrinology and Metabolism*, 99(4), 1132–1141.

Wacker, M., & Holick, M. F. (2013). Sunlight and vitamin D: A global perspective for health. *Dermato-Endocrinology*, 5(1), 51–108.

Wimalawansa, S. J. (2012). Vitamin D in the new millennium. *Current Osteoporosis Reports*, 10(1), 4–15.

Yeh, E. B., Barbano, D. M., & Drake, M. A. (2017). Vitamin fortification of fluid milk. *Journal of Food Science*, 82(4), 856–864. https://doi.org/10.1111/1750-3841.13648.

Zahedirad, M., Asadzadeh, S., Nikooyeh, B., Neyestani, T. R., Khorshidian, N., Yousefi, M., & Mortazavian, A. M. (2019). Fortification aspects of vitamin D in dairy products: A review study. *International Dairy Journal*, 94, 53–64.

Zittermann, A., Prokop, S., F Gummert, J., & Borgermann, J. (2013). Safety issues of vitamin D supplementation. *Anti-Cancer Agents in Medicinal Chemistry (Formerly Current Medicinal Chemistry-Anti-Cancer Agents)*, 13(1), 4–10.

4 Vitamin A Fortification and Encapsulation
Trends and Technology

Vaibhav Kumar Maurya, Ranjan Singh,
Amita Shakya, and Shumaila Jan

4.1 INTRODUCTION

Since the first description for vitamin A dates back to as early as 1500 BC in Egyptian literature where Ebers Papyrus mentions liver juice or ox liver extract in curing semi-darkness conditions (night blindness) (Olson, 1990). During the course of time accumulating scientific discoveries have revealed the multifaceted role of vitamin A in human health. Besides its traditional role in vision, vitamin A is implicated in several disorders including, but not limited to, anemia, immune dysfunction, increased susceptibility to infection, poor growth, cardiovascular diseases, cancer cell growth, mortality, and mortality neurodegenerative diseases (Blomhoff, 1994). Due to the lack of vitamin A synthesis pathway in humans, they are compelled to acquire it through dietary sources. Carotenoids, naturally occurring pigments produced by algae, bacteria, and plants, serve as precursor molecules for vitamin A (Maurya et al., 2021). β-carotene, α-carotene, and β-cryptoxanthin are some of the most common carotenoids that demonstrate high vitamin A activity and exist in different physicochemical forms (Maurya, Aggarwal, Ranjan, & Gothandam, 2020; Maurya, Singh, et al., 2020). However, the carotenoids in many fruits and vegetables often have low oral bioavailability, which ultimately limits their bioactivity and efficiency in combating diseases (Maurya, Singh, et al., 2020).

WHO data clearly indicate the spreading magnitude of vitamin A deficiency across the globe, irrespective of biological, social, and geographical differences. Furthermore, its poor water solubility, susceptibility to environmental factors (pH, temperature, ionic strength, humidity), and photodegradation pose several challenges when addressing its deficiency through oral delivery. Diet diversification, biofortification, supplementation, and food fortification seem to be the most feasible strategies to address its deficiency (Aghajanzadeh, Kashaninejad, & Ziaiifar, 2016; Maurya, Shakya, Bashir, Kushwaha, & McClements, 2022). Moreover, loss during processing and storage, and interactions with other components of the matrix in which vitamin A is incorporated, result in reduction in its bioavailability.

To overcome these challenges, researchers are coming up with new fortification methods. Nevertheless, these fortification methods should be able to overcome low water dispersibility, chemical instability, and poor bioavailability of vitamin A (Maurya et al., 2022). In this chapter article, we start by discussing molecular structure, physicochemical properties, and bioavailability. Then we proceed to a discourse on emerging techniques focused on novel encapsulation techniques.

4.2 VITAMIN A: DISCOVERY

The first inscription of Vitamin A dates back to 1500 BC in Egyptian literature when liver juice or ox liver extract was utilized to cure night blindness. However, the first scientific reports for vitamin

DOI: 10.1201/9781003160663-6

A as a nutrient come from McCollum and Davis's study where rat growth was reinstated after feeding food comprising pure casein, carbohydrates, and an extract of butter or egg (McCollum & Davis, 1915). The active factor present in the extract is called as "fat soluble A," later known as "vitamin A" (Drummond & Coward, 1920). Contemporary researchers developed a technique to synthesize retinol (1947) and β-carotene (1950) in laboratory (Isler, Klaui, & Solms, 1947). The linkage between vision and vitamin A was established by George Wald (Wald, 1933).

Later, it was observed that several compounds have vitamin A activity, and researchers grouped all these compounds as "retinoids" and demarcated them as "any compound, other than carotenoids, exhibiting qualitatively the biological activity of retinol" (Sporn, Dunlop, Newton, & Smith, 1976). The retinoid group is comprised of retinylesters, retinol, and its metabolites: retinaldehyde and all the retinol isomers. However, the International Union of Pure and Applied Chemistry-International Union of Biochemistry (IUPAC-IUB) defined retinoids as

a class of compounds consisting of four isoprenoid units joined in a head-to-tail manner, that is a class of compounds derived from a monocyclic parent compound containing five carbon–carbon double bonds and a functional group at the end of the acyclic portion.

(IUPAC-IUB, 1983)

4.3 PHYSICOCHEMICAL PROPERTIES

All retinol and its analogues possess an electron-rich section that attracts electron-deficient ligands such as free radicals (Figure 4.1). Consequently, retinoid compounds become susceptible to chemical conversion in the presence of oxidants, transition metals, and free radicals as a result of isomerization (at the 9, 11, and 13 positions) and/or oxidative degradation (Carlotti, Rossatto, & Gallarate,

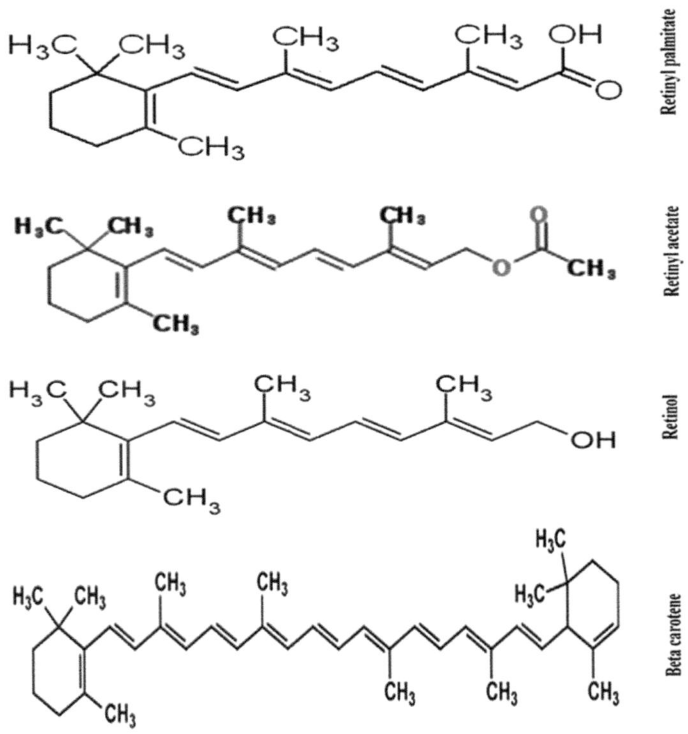

FIGURE 4.1 Chemical structure of vitamin A and its analogues (Source: Maurya et al., 2020a).

TABLE 4.1

Physicochemical Attributes of Vitamin A Analogues

Property	Molar mass (g/mol)	Formula	LogP	Water solubility (mg/cm3)	Density (g/cm3)	Refractive index	Melting point ($^\circ$C)
Retinol	286.5	$C_{20}H_{30}O$	6.2	0	1	1.55	63
Retinyl acetate	328.5	$C_{22}H_{32}O_2$	6.9	0	1	1.53	57
Retinyl palmitate	524.9	$C_{36}H_{60}O_2$	14.8	0	0.9	1.51	28
Retinyl propionate	342.5	$C_{23}H_{34}O_2$	7.1	0	0.94	1.53	<20
β-Carotene	536.87	$C_{40}H_{56}$	14.76	0	0.94	1.56	180

2002). Additionally, retinoids are susceptible to thermal degradation and are attributed to heat-induced isomerization reactions or formation of 13-cis isomers (Panfili, Manzi, & Pizzoferrato, 1998). The bioactivity of retinoids varies with the chemical transformations; hence it is rational to understand key reactions that control the degradation/transformation when vitamin A is incorporated in supplements or foods. The chemical susceptibility can be minimized with an esterified form, such as retinyl [acetate, palmitate, or propionate which is less prone to degradation (Table 4.1)]. All these factors ultimately result in reducing vitamin A bioavailability.

4.4 BIOAVAILABILITY, SOURCES, AND BIOMARKERS FOR VITAMIN A

Bioavailability of vitamin A can be defined as the fraction of ingested vitamin A which is ultimately absorbed and utilized by the human body. An array of factors affects the bioavailability of vitamin A, including but not limited to the complexity of the food matrix, location of the vitamin in food, particle size of the food, and degree food processing conditions (Maurya, Aggarwal, et al., 2020; Maurya, Singh, et al., 2020). Human meet their obligatory need for vitamin A through diet, which could be either plant-derived or animal-based. Leafy vegetables, carrots, broccoli, fruits, liver, chicken and tuna/salmon fish are the major sources of vitamin A. The availability of vitamin A-rich food items in the food basket, religious constraints, and socioeconomics are major factors that govern the vitamin A status in a given population. There are several tests that help in deciding deficiency or sufficiency status of vitamin A in an individual. The loss of vision or ocular examination is the best test for vitamin A evaluation. However, serum vitamin A is the most accepted biomarker for evaluating its deficiency evaluation. Vitamin A serum vitamin A level below 0.70 μmol/L is recognized as a deficient state (WHO, 2021c).

4.5 VITAMIN A DEFICIENCY STATUS ACROSS THE GLOBE

A WHO report estimated 2 billion people across the globe living with one or multiple micronutrient deficiency (WHO, 2021b). Figure 4.2 describes the vitamin A status across the globe. Approximately 254 million pre-school children are suffering from vitamin deficiencies. Nearly 19.8 million pregnant women are living with low vitamin A serum levels, and it is projected that two-thirds of this population is living in South and Southeast Asia (Wirth et al., 2017; Wiseman, Bar-El Dadon, & Reifen, 2017; Xu et al., 2021).

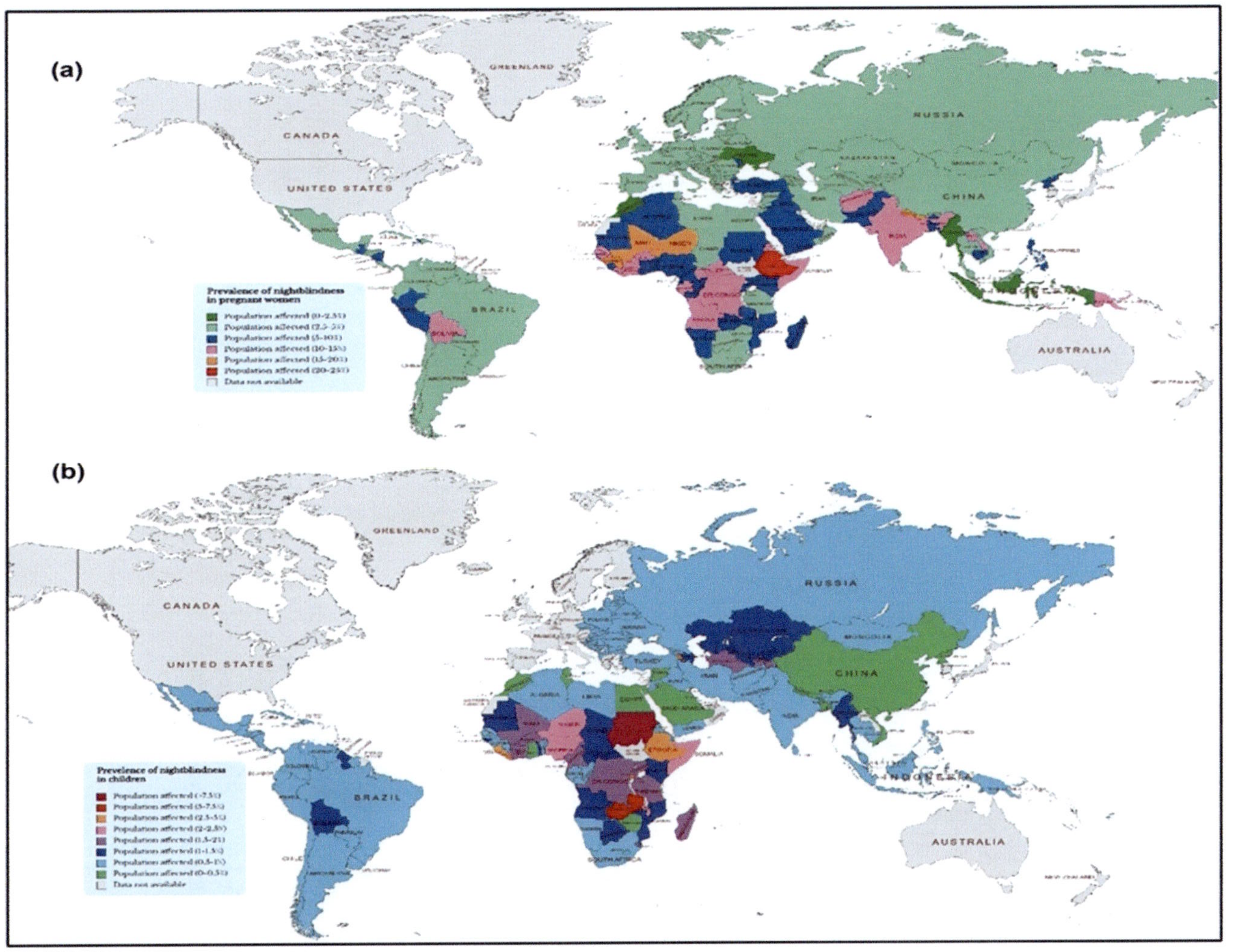

FIGURE 4.2 Geographical distribution of vitamin A deficiency; (a) in children and (b) pregnant women from 1995 to 2005 reported by WHO (2021). Countries with a gross domestic product (GDP) ≥ US$1500 are excluded, as they are expected to be free from vitamin A deficiency. Note: Vitamin A serum level ≤0.7μmol/L is considered vitamin A deficient. Source: Our World in Data (2021) with permission.

4.5.1 Vitamin A: Daily Need and RDA

Literature clearly demonstrates the gap between intake requirement and current supply (Table 4.2). The daily requirement of vitamin A depends on the age, gender, and physiological status of an individual. Therefore, WHO has set different recommended dietary allowances (RDAs) based on population ages: 400–600 µg/day for 1–8 years, 600–800 µg/day for above 8 years, 900–1000 µg/day for men, and 800–900 µg/day for women (WHO, 2021c).

4.5.2 Vitamin A Deficiency: Impact on Human Health and Limiting Factors

Vitamin A deficiency can be simply reversed with consumption of vitamin A in sufficient quantity. Suboptimal vitamin A serum level results in corneal necrosis, corneal xerosis, corneal ulceration, temporary or permanent loss of vision or xerophthalmia (McConnell, 2000; Semba, 2007). Vitamin A deficiency is found to be linked with several biochemical functions which play a key role in cell growth, embryonic development, epithelium repair, spermatogenesis, reproduction, and immunological functions (Blomhoff, 1994).

Number of factors donate to vitamin A deficiency including exclusion of vitamin A-rich food in the diet, religious constraints, cultural practices, and socioeconomics (Maurya, Aggarwal, et al., 2020; Maurya et al., 2022). It is observed that the vitamin A deficiency becomes more prevailing in underdeveloped countries, which could be attributed to the inclusion of cereal-based food matrix (deficiency in vitamin A) as a major proportion in the diet. Further, cooking practices and storage conditions equally contribute to vitamin A degradation before consumption. Additionally, low vitamin A bioavailability also contributes to its deficiency (Borel & Desmarchelier, 2017; Lin, Liang, Williams, & Zhong, 2018), Metabolism, age, health status, body weight, and genetics also affect the absorption of vitamin A (Castenmiller, West, Linssen, van het Hof, & Voragen, 1999; Rock et al., 1998) and are comprehensively reviewed in our previous articles (Maurya, Aggarwal, et al., 2020; Maurya, Singh, et al., 2020).

4.6 APPROACHES TO ADDRESSING VITAMIN A DEFICIENCY

Literature reports mainly four approaches to curtail vitamin A deficiency, which are diet diversification, biofortification, dietary supplements, and food fortification.

4.6.1 Vitamin A: Diet Diversification

Dietary diversification strategy relies on inclusion of vitamin A-rich food in daily diets in adequate amount. Effectiveness of this approach depends on multiple factors including socioeconomic status, vitamin A-rich food accessibility, religious or cultural practices, and bioavailability of vitamin A in food included in the diet. Awareness about the importance of diet diversity seems to be key to diet diversification (Maurya et al., 2022).

TABLE 4.2
Vitamin A Supply Across the World

Source/Region	Region	Africa	Americas	Asia	Europe	Western Pacific
Plant source	µg RE/day	654	519	378	467	781
Animal source	µg RE/day	122	295	53	271	216
Total	µg RE/day	775	814	431	738	997

4.6.2 Vitamin A: Biofortification

Human rely on plants for food supply and cereals like wheat, rice, maize, and cassava are a major source of diet. Though, cereals are deficient in vitamin A leading to nutritional deficiency. Improving the vitamin A content of staple food like seems to be one of important approaches to fight against vitamin A deficiency. As per the WHO definition "biofortification is a process by which the nutritional quality of food crops is improved through agronomical practices, conventional plant breeding, or modern biotechnology" (WHO, 2021a). Vitamin A biofortification of crops can be carried out by supplying specific micronutrients through fertilizer, which can help improve the vitamin A content, or by soil irrigation management. Plant breeding is one potential strategy that helps in improving vitamin A in crops. For instance, Indian farmers developed a vitamin A-rich carrot variety called "Madhuban Gajar" through plant breeding. Additionally, genetic engineering approaches have been equally effective in developing different crops rich in vitamin and listed in Table 4.3.

4.6.3 Vitamin A: Supplementation and Food Fortification

Supplements are a vitamin A rich formulation which is used to obtain vitamin A. These supplements are formulated with different kinds of polymers whose impact on humans remains unknown.

Food fortification is recognized as the most effective technique to address any micronutrient deficiency. Food fortification is the practice of intentionally adding a micronutrient in a target food matrix. Food fortification could be mandatory or voluntary depending on the rules and legislation of the country. According to the Fortification Data Exchange report, 153 countries follow fortification practices.

4.7 PREREQUISITES FOR FOOD FORTIFICATION

Before opting for food fortification, information on the following points may offer better understanding: target population, current deficiency level in the target population, RDA of the population, difference between current intake and RDA, potency of vitamin derivative, target food matrix and its fortification dose, and physicochemical challenges posed by matrix and food processing processes.

TABLE 4.3

Vitamin A Biofortified Crops Develop Using Genetic Engineering

Target gene	Desired nutrient	Biofortified crops
Phytoenesynthase gene	Provitamin A	Cassava
Maize PSY and Crtl	Total carotenoids	Maize
Zmpsy, Pacrtl, Gllycb, Glbch, ParacrtW	β-Carotene	
Crtl, LCYe, CHYe, CrtB, Crtl, CrtY	Total carotenoids	Potato
Cauliflower or organegene	β-carotene	Tomato
Cycopeneb-cyclasegene	Precursor β-carotene	Wheat
Maize PSY and Crtl, CrtB, Crtlgene)	Total carotenoids	Mustard
Bacterial phytoenesynthase (CrtB	β-carotene	Rice
Phytoene synthase(psy) and lycopene β-cyclase (β-lcy)	β-carotene	
PSY-1, CRT-I, PMI	Carotenoids	Sorghum
crt B and *crtl*	Vitamin A	Canola

4.8 LIMITING FACTOR IN VITAMIN A FORTIFICATION

The challenges in vitamin A fortification vary with its derivative and target food matrix. For example, an oil-based food matrix can be a suitable vehicle and can be easily fortified by solubilizing vitamin A in it through either homogenization or emulsification. The presence of unsaturated bonds makes vitamin A prone to chemical degradation, particularly against acids, heat, or radiation. Several studies have been conducted to evaluate the vitamin A degradation pattern when it is fortified in different food matrices (Table 1.4). It is observed that the deposition of vitamin A on the inner surfaces of packaging materials, heterogeneous distribution, poor water solubility, and low bioavailability are major challenges in vitamin A fortification (Maurya et al., 2022). There has been continuous research exploration to develop novel technology that can able to overcome these challenges.

4.9 ENCAPSULATION: TOOL TO IMPROVE VITAMIN A STABILITY AND BIOAVAILABILITY

Nanotechnology offers a plethora of microencapsulation techniques for vitamins that have facilitated researchers to carve various nanodelivery systems with desired attributes, including uniform distribution within food matrices, high protection against environmental factors, and high stability against food processing. Microencapsulation relies on nanodelivery systems which basically consist of vitamin A encapsulated by secondary wall materials that not only protect the vitamin from external environment but also offer desired functionalities, including improved bioavailability through targeted and controlled release and optimized dose delivery, thus avoiding the overdosing (Maurya et al., 2022).

Vitamin A is fortified in different food matrices that differ in their complexity, micronutrients, and supporting structure despite having low water solubility, instability against environmental conditions, high melting point, photo-chemical susceptibility, prone to oxidation, and heterogenous distribution in food matrices, which finally lessens its bioavailability. For that reason, researchers are coming up with an array of delivery systems tuned with desired functionalities that are not limited to improved stability, better homogeneity, high dispersibility, enhanced solubility, better targeted and controlled release, and ultimately enhanced bioavailability (Gonçalves, Estevinho, & Rocha, 2016).

Delivery systems are defined as carrier systems in which bioactive core is encompassed within a nano-/microstructure to not only protect the bioactive core from unfavorable environmental conditions (oxidation, pH, temperature, ionic strength, humidity, and pressure) but also release the bioactive core at a defined rate at absorption site in the GIT. A detailed and comprehensive literature review on carotenoid delivery systems has already been discussed in our previous article (Maurya, Aggarwal, et al., 2020; Maurya et al., 2022) hence, here only those delivery systems are included which are used to encapsulate retinol or retinol derivatives only. However, the delivery systems developed for vitamin A delivery can primarily be classified into two classes: polymer-based delivery systems (PBDSs) and lipid-based delivery systems (LBDSs). The chronological development in nanodelivery systems dedicated to vitamin A is depicted in Figure 4.3

4.9.1 POLYMER-BASED DELIVERY SYSTEMS (PBDSs)

This class of delivery systems utilizes the indigenous diversity of natural or synthetic polymers to encapsulate a bioactive core. To minimize the long-term health risk associated with synthetic polymers, PBDSs that are intended for food applications are carved with naturally occurring biopolymers including polysaccharides, proteins, and lipids. However, these naturally occurring polymers inherit challenges like specific and particular thermal treatment requirements and poor control on preparation methods, resulting in porous micro-/nanoparticle formation and fading the aim of

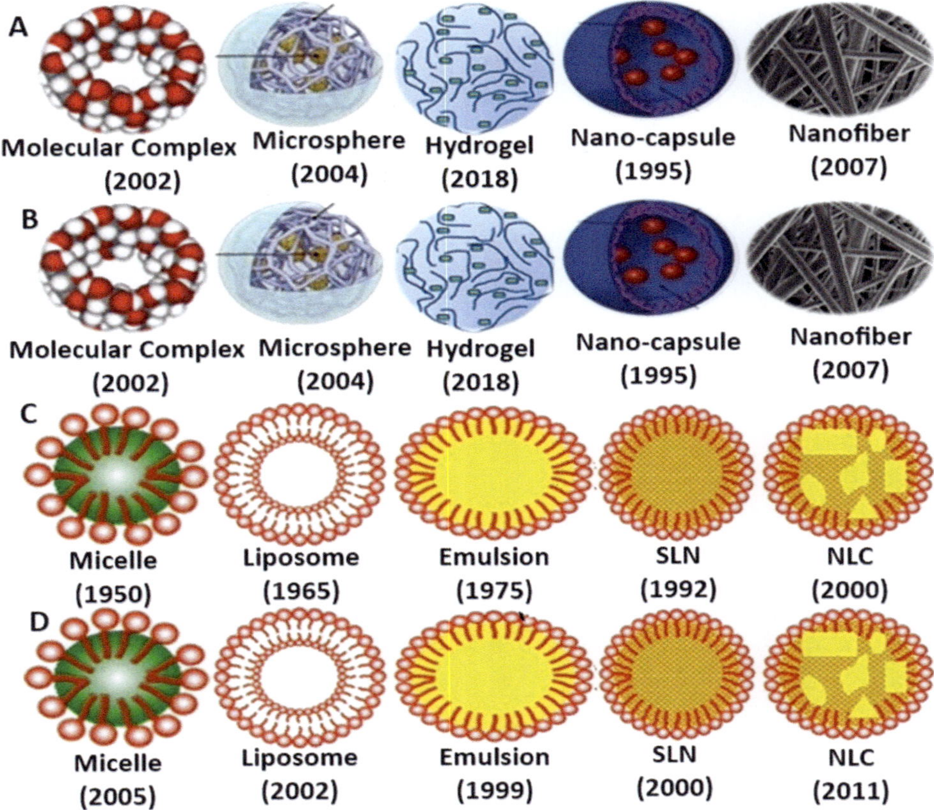

FIGURE 4.3 Chronological development of delivery system for vitamin A encapsulation: (A) historical events in the evolution of polymer-based delivery systems; (B) historical events in the application of polymer-based delivery systems for encapsulating vitamin A; (C) historical events in the evolution of lipid-based delivery systems; (D) historical events for applying lipid-based delivery systems for encapsulating vitamin A.

encapsulation. An array of PBDSs has been tested to check their potential in vitamin A delivery; however, emphasis has been given only to only those PBDSs made from natural biopolymers and polymers that generally recognized as safe (GRS) (Table 4.4). The GRS class of delivery systems involves nano-/microcapsules, nano-/microspheres, colloidal nano-/microemulsions, hydrogel micelles, nanofibers, and molecular inclusion (Figure 4.3 A and B).

4.9.1.1 Molecular Complexes

Molecular complexes have been well adopted to encapsulate, protect, and deliver a variety of food bioactives. The complex formation between the host molecule and bioactive compounds happens only when water molecules are available during the process. Several polymers have been utilized to develop micro-/nanoparticles by molecular complexation, but cyclodextrins are the most exploited molecules. Cyclodextrins are macrocyclic oligosaccharides containing α(1,4)-linked glucopyranose subunits, which offer a distinctive hydrophobic central cavity and hydrophilic outer surface (Kumar & Singhal, 2021). These molecules have a cage-like supramolecular structure where they can accommodate several hydrophobic bioactives like vitamin A (Fourmentin, Crini, & Lichtfouse, 2018). Literature on animal model and human consumption studies clearly indicates the ability of cyclodextrin to improve the bioavailability of fat-soluble compounds, including vitamin A, when they are fortified in the food matrix(Astray, Mejuto, & Simal-Gandara, 2020; Loftsson, 2004; Zaibunnisa, Aini Marhanna, & Ainun Atirah, 2011).

TABLE 4.4

Encapsulation Techniques Adopted for Vitamin A Encapsulation

Class of delivery system	Name of delivery system	Technique/fabrication methods	Key components	References
Polymer-based delivery systems (PBDSs)	Molecular complexes	Solvent evaporation	Cyclodextrins	(Semenova et al., 2002a)
		Spray drying	β-lactoglobulin	
		Stirring	13-cis-RA with -cyclodextrin, hydroxypropyl-cyclodextrin (HP-CD)	
		Hydration	WPI	
		Hot-extrusion	Modified starch	(Pinkaew et al., 2012)
			Sucrose, butylated hydroxytoluene (BHT), and fractionated	
			Coconut oil	
		Homogenization	Succinated chitosan	(Huang et al., 2013)
		Co-solvent dissolution	Pectin	
		Freeze-drying	β-cyclodextrins	
		Precipitation	Cyclodextrin	
		Stirring	Sodium caseinate,	
			Whey protein concentrate (WPC),	
			Milk protein concentrate (MPC)	
		Air bath oscillation	B-Lactoglobulins	(Tang et al., 2017)
		Heating	Ovalbumin	
		Heating	Native ovalbumin,	
			high methoxyl pectin	
		Stirring	Modified sodium caseinate	(Gupta et al., 2018)
	Microsphere	Solvent evaporation technique	Poly(methylmethacrylate-co-methacrylicacid[poly-(MMA-co-MAA)],	(Lee et al., 2004)
			Polyethylenimine, ,3-dicyclohexyl-Carbodimide (DCC), N-hydroxy succinimide (NHS), Tween 20	
		Ultrasonication	Chitosan	(Kim et al., 2006)
		Solvent evaporation		
		Solvent displacement	Ethyl cellulose,	(Arayachukeat et al., 2011)

(Continued)

TABLE 4.4 (CONTINUED)

Encapsulation Techniques Adopted for Vitamin A Encapsulation

Class of delivery system	Name of delivery system	Technique/fabrication methods	Key components	References
			Poly(ethylene glycol)-4-methoxycinnamoylphth aloylchitosan (PCPLC).	
		Supercritical emulsion extraction	Poly(lactic-*co*-glycolic)acid (PLGA)	(Della Porta et al., 2011)
		Dehydration/rehydration	Poly(ethylene adipate), PEA; poly(butylene adipate),	(Cho, 2012)
			PBA; poly(hexamethylene adipate), PHMA; and three polycaprolactones, PCL	
		High pressure homogenization	Lecithin	(Ghouchi-Eskandar et al., 2012)
			Miglyol®812	
	Nanosphere	Ultrasonication and solvent evaporation	Chitosan	(Syed et al., 2013)
		Stirring	Poly(ethylene glycol), CaCl$_2$,	(Müller et al., 2015)
		Co-solvent evaporation	Bacterial cellulose	(Numata et al., 2015)
			Poly(ethylene oxide)-b-poly(caprolactone) (PEO-b-PCL)	
		Impregnation	Montmorillonite K-10	(Calabrese et al., 2016)
			Sepiolite (SEP)	
	Microgel	Precipitation polymerisation	Poly(*N*-vinylcaprolactam) (PVCL)	(Schroeder, 2018)
	Microcapsule	High-pressure homogenizer	WPI	(Beaulieu et al., 2002)
		Emulsification	Cremophor	(Taha et al., 2004)
			EL	
			Soybean oil	
			Hydroxyl propyl methyl cellulose	
		Spray-cooling	Hydrogenated oil	(Zimmermann et al., 2004)
		High pressure homogenization	Silica nanoparticles,	(Eskandar et al., 2009a)
			Caprylic/capric triglyceride (Miglyol®812),	

(Continued)

TABLE 4.4 (CONTINUED)

Encapsulation Techniques Adopted for Vitamin A Encapsulation

Class of delivery system	Name of delivery system	Technique/fabrication methods	Key components	References
		Spray-cooling	Palm hydrogenated oil	(Pinkaew et al., 2012)
		Sonication	Alginic acid Sodium salt	(Kislitza et al., 2009)
		Super critical fluid	Poly(l-lactic acid), Pluronic F68 (F68) and Pluronic F127	(Sane & Limtrakul, 2009)
		Ionic cross-linking	Chitosan, Polyoxyethylene 20 sorbitan monolaurate (Tween 20), soya lecithin	(Albertini et al., 2010)
		Spray drying	Arabic gum	(Gonçalves et al., 2017)
		Microfluization	Pectin	(Noh et al., 2019)
			Tween 80	
		Molecular complexation	β-Lactoglobulin	(탕가문, 2015b)
	Nanocapsules	Stirring	Silicon glycol copolymer	(Oliveira et al., 2014)
		Stirring	Chitosan	(Park et al., 2015)
			Zein	
		Single emulsion technique	PLA	(Puntel et al., 2015)
		Homogenization	Silibinin,	(Pan et al., 2016)
			Squalene	
			deoxycholic acid	
			Hexadecyl palmitate	
			Soybean phosphatidylcholine	
		Emsulsification	Chitosan	(Pisetpackdeekul et al., 2016)
		Nanoprecipitation	PLA	(Yildirim et al., 2017)
	nanofiber	Electrospinning	Cellulose acetate	(Taepaiboon et al., 2007)
Lipid based delivery systems (LBDSs)	Micelle	Solvent evaporation	Distearoylphosphatidylcholine (DSPC)/cholesterol	(Kawakami et al., 2005)
		Solvent evaporation Method	PEG-poly(aspartic acid), PEG-poly(benzyl L-aspartate)	(Satoh et al., 2009)

(Continued)

TABLE 4.4 (CONTINUED)

Encapsulation Techniques Adopted for Vitamin A Encapsulation

Class of delivery system	Name of delivery system	Technique/fabrication methods	Key components	References
		Emulsification	Sodium caseinate polysorbate 80	(Loewen, 2014)
		High Pressure Homogenization	Casein	(Mohan, 2014)
		Stirring	Casein	(Loewen et al., 2018)
	Emulsion	Emulsification	Polyoxyethylene hydrogenated castor oil, Polyoxyethylenediisostearate, Polyoxyethylenedioleate	(Yoshida et al., 1999)
		High pressure homogenization	Caprylic/capric triglyceride (Miglyol®812), Soybean Lecithin	(Eskandar et al., 2009b)
		Dehydration–rehydration method	Egg phosphatidylcholine (EggPC) and 1,2-distearoyl-snglycero-3-phosphoethanolamine-N-polyethylene glycol 2000 (DSPE-PEG2000)	(Chansri et al., 2006)
		Ultrasonication	Tween20 Poly(ethylene oxide)-block-poly(caprolactone) blockpoly(ethylene oxide) (PEO-PCL-PEO)	(Cho et al., 2012)
		Dehydration–rehydration method	Egg phosphatidylcholine (EggPC) and 1,2-distearoyl-snglycero-3-phosphoethanolamine-N-polyethylene glycol 2000 (DSPE-PEG2000)	(Chansri et al., 2006)
		Heating and stirring	Silicone fluid Anionic self-emulsifying wax	(Moyano & Segall, 2011)
		Emulsification, Sol–gel method	Tetraethyl orthosilicate, Hydroxypropyl cellulose, Poly(ethylene lycol)-block-poly(propylene glycol)-block-poly(ethylene glycol) Sorbitanmonoleate,	(Hwang et al., 2005)

(Continued)

TABLE 4.4 (CONTINUED)

Encapsulation Techniques Adopted for Vitamin A Encapsulation

Class of delivery system	Name of delivery system	Technique/fabrication methods	Key components	References
		High pressure homogenizer	Polyethylene glycol N-trimethyl chitosan (TMC), Lecithin Cholesterol	(He et al., 2013)
		Homogenization	Tween-20 solution or canola oil	(Chaudhari & Nitin, 2015)
		Homogenization	Canola oil	(Pan et al., 2015)
	Liposome	Solvent evaporation and rehydration method	PC	(Hwang & Ludescher, 2002)
		Dehydration/rehydration	Dimyristoyl PC Dipalmitoyl PC	(Lee et al., 2003)
		dehydration–rehydration method	soybean phosphatidylcholine (PC) and cholesterol (CH)	(Lee et al., 2005)
		Dehydration/rehydration	Tween 20, Span 60 and cholesterol, Sephadex G-75, Phospholipon 100 and Phospholipon 100H	(Carafa et al., 2008)
		Extrusion	krill phospholipid extract (KPE), soybean phosphatidylcholine (SPC), and dipalmitoyl phosphatidylcholine (DPPC).	(Monroig et al., 2007)
		film dispersion method	N-trimethyl chitosan	(Wen et al., 2010)
		Dehydration/rehydration	Distearoylphosphatidylcholine (DSPC/cholesterol)	(Uzzaman & Grace, 2013)
		Solvent evaporation/rehydration	Distearoyl-L-phosphatidylcholine (DSPC),	(Uzzaman & Grace, 2013)

(Continued)

TABLE 4.4 (CONTINUED)

Encapsulation Techniques Adopted for Vitamin A Encapsulation

Class of delivery system	Name of delivery system	Technique/fabrication methods	Key components	References
		film hy- dration–sonication method	Cholesterol Lecithin	(Pezeshky et al., 2016)
		dehydration-rehydration	Cholesterol L-_-phosphatidylcholine (PC)	(Ko & Lee, 2010)
		thin-film hydration method	phospholipids	(Clares et al., 2014)
		Film hydration and sonication	Distearoyl-L-phosphatidylcholine (DSPC),	(Siddikuzzaman & Grace, 2014)
		Film hydration	Cholesterol Distearoyl-Lphosphatidylcholine (DSPC) and cholesterol	(Berlin Grace & Rimashree, 2015)
		Extrusion	Dipalmitoyl-sn-glycero-3-phosphatidylcholine monohydrate(DPPC), N-(fluorescein-5-tiocarbamoyl)-1,2,dihexadeca noyl-sn-glycero-3-phosphoethanolaminetri ethylammonium salt (fluorescein-DHPE), N-(carbonyl-methoxipolyethyleneglycol-20 00)-1,2-distearoyl-sn-glycero-3-phosphoethano lamine (DSPE-mPEG2000)	(Cristiano et al., 2017)
	SLN	Hot homogenization	Compritol 888 ATO (glyceryl behenate, tribehenin), Miglyol 812 (caprylic/capric triglycerides)	(Jenning et al., 2000)
		Hot homogenization	· Compritol 888 ATO (glyceryl behenate, tri- behenin) Miglyol 812 (caprylic / capric triglycerides) cream cooled down to ambient temperature. · glycerides)	(Jenning et al., 2000)

(Continued)

TABLE 4.4 (CONTINUED)

Encapsulation Techniques Adopted for Vitamin A Encapsulation

Class of delivery system	Name of delivery system	Technique/fabrication methods	Key components	References
		Hot homogenization	Compritol 888 ATO (glyceryl behenate, tribehenin)	(Jenning et al., 2000b)
		Hot homogenization	Cetyl palmitate, glyceryl behenate, and palmitic acid	(Carlotti et al., 2005)
		Hot homogenization	Precirol ATO5, Pluronic F68	(Carafa et al., 2008)
		hot homogenization technique	Witepsol; Precirol ATO	(Cerreto et al., 2013)
		Hot homogenization	Gelucire 50/13®, Precirol ATO5®. Dicetyl phosphate (DCP)	(Jeon et al., 2005)
		Hot homogenization	Cetyl palmitate, Caprylic/capric triglycerides, Polyglyceryl-3 methylglucosedistearate	(Jung et al., 2013)
		Sonication	Labrafac1 lipophile, Labrasol1:Plurol1oleique	(Clares et al., 2014)
		High pressure homogenization	Gelucire 44/14® (G), Tween 80®	(Argimón et al., 2017)
	NLC	High Pressure Homogenization	Caprol PGE-860), Soybean Lecithin	(Xia & Kong, 2011)
		Hot homogenization method	Precirol ATO 5, Miglyol 812 (caprylic/capric triglycerides)	(Pezeshki et al., 2014)
		High pressure homogenization	Soybean Lecithin	(Kong et al., 2011)

The literature dictates a range of fabrication methods for molecular complex preparation, particularly isoelectric precipitation, chemical modification, and solvent evaporation(Adeoye & Cabral-Marques, 2017; Arima, Motoyama, & Higashi, 2017; Chilajwar, Pednekar, Jadhav, Gupta, & Kadam, 2014; Shafaei et al., 2017). To align with the specificity of this article, emphasis has been given only to only those methodologies that are utilized to form molecular complexes with retinol or its derivatives (Table 1.4).

Study on the application of molecular complex for vitamin A binding was initiated by Manaco in 2000 by studying the ability of trans thyretin to form a complex with retinol(Monaco, 2000). Further, an aqueous gel containing vitamin A/cyclodextrin complex was developed for topical use (Sandrine Weisse et al., 2009; S. Weisse, Perly, Creminon, Ouvrard-Baraton, & Djedai'Ni-Pilard, 2004)(Zhang et al., 2023). From then onward, several researchers have utilized cyclodextrins to deliver vitamin A. β-cyclodextrin has been recognized as the most widely adopted supramolecule for vitamin A molecular complex formation. The first food application for a molecular complex came into the picture when β-lactoglobulin/vitamin A molecular complex was prepared using spray drying for milk fortification purpose(Reynolds, 2006). Similarly, modified starch/vitamin A molecular complex was used to fortify vitamin using extrusion (Pinkaew, Wegmuller, & Hurrell, 2012).

Despite providing high stability to encapsulated food bioactives, molecular complexes fail to address several challenges like premature release of entrapped food bioactives, poor loading/carrying capacity, high cost, and low regulatory compliance. Also, cyclodextrins are legally prohibited for food applications in many countries. To overcome this, other biopolymers with fair binding capacity for vitamin A, such as β-lactoglobulin, whey protein isolate (WPI), ovalbumin, pectin, chitosan, and casein were used, which show unique binding ability to host lipophilic food bioactives in their lipophilic patches. Furthermore, these biopolymers promise improved stability against environmental factors through several mechanisms including (a) designed molecular complexes based on micro-/nanoparticles offer high solubility of lipophilic food bioactives, resulting in homogeneous distribution within the target food matrix, (b) the lipophilic patches/helical cavity as well as the "slim" and hydrophobic alkyl chain of these biopolymers offer enhanced binding affinity for lipophilic bioactives like vitamin A; (c) facilitate improved linkage for biopolymer-surfactant-food bioactives (vitamin A) in the ternary system (surfactant-oil-water) (Maurya & Aggarwal, 2017; Maurya, Aggarwal, et al., 2020; Maurya et al., 2021; Maurya et al., 2022).

4.9.1.2 Micro-/Nanospheres

Micro-/nanospheres are fabricated with synthetic/natural polymers with a particle size range of 1–1000 nm (nanospheres) or 1–1000 μm (microspheres). These structures are carved by dispersing a hydrophilic polymer or blend of polymers in an organic phase in the presence of cross-linking agents such as hexamethylenetetramine, triazines formaldehyde, melamine formaldehyde, glutaraldehyde, etc. Lipophilic bioactives like vitamin A get encapsulated either into the inner core of nano-microspheres or incorporated within the solid polymeric matrix of nano-/microspheres.

Several preparation methods for nano-/microspheres are reported including single coacervation phase separation, polymerization, single emulsion, double emulsion (Arayachukeat, Wanichwecharungruang, & Tree-Udom, 2011; Cho, 2012). These structures are recognized for their easy optimization process, flexibility in tuning their functionalities as per requirement, such as site directed and control of release of encapsulated bioactives, efficacy in bioactive delivery, enhanced biocompatibility, and *in vivo* stability.

Despite having a great reputation in the pharmaceutical field for drug delivery, the potential of micro-/nanospheres is not fully explored for vitamin A encapsulation, to be more precise, in food application. The first depiction of vitamin A encapsulation in micro-/nanospheres was initiated with Torrado's study, where researchers have evaluated *in vitro* and *in vivo* release for vitamin A encapsulated within albumin-based microspheres (Torrado, Torrado, & Cadórniga, 1992).

From then onward, several researchers have designed micro-/nanospheres differing in their composition to assess their ability to be an alternative delivery system for vitamin A (Calabrese et al., 2016;

Müller et al., 2015). However, the literature lacks the reports where nano-/microspheres have been used for food fortification of vitamin A. Nevertheless, nano-/microspheres do not involve sophisticated instruments and are easy to scale, but they do pose several challenges such as poor drug carrying capacity, being prone to enzymatic degradation, and premature release of encapsulated bioactive.

4.9.1.3   Nano-/Microhydrogels

These delivery systems are three-dimensional soft structures prepared with a water-soluble cross-linking agent, embracing a wide range of physicochemical properties. McConnell (2000) prepared Retin-A microgel, a retinoic acid, for topical application; it was the first report on the use of hydrogels for delivering vitamin A (McConnell, 2000). Since then, the literature has been updated with excellent articles demonstrating several preparation methods for nano-/microhydrogels, including cross-linking, inverse-suspension polymerization, and sonication methods (Huynh, 2012; Klein & Poverenov, 2020; Nascimento, Casanova, Silva, de Carvalho Teixeira, & de Carvalho, 2020). These delivery systems promise sustained-/targeted drug delivery under defined stimuli (pH, thermal, mechanical change, or enzymatic action), low systemic side effects, minimal loss during digestion, unavoidable interaction with other components of the target food matrix, reduced systemic effects, and hence ensure high bioavailability.

Despite these attractive attributes, there is a scarcity of data on its use for food applications. Though these microstructures have been used for the encapsulation of β-carotene (Perrechil, Maximo, & Cunha, 2020; Silva, Feltre, Hubinger, & Sato, 2020) but to the best of our knowledge, only single report has been published where nano-/microgels were fabricated to encapsulate retinol or its derivatives for food fortification (Schroeder, 2018). Poor drug loading capacity (Muhamad, Fen, Hui, & Mustapha, 2011; Qu & Luo, 2020), susceptibility against oxidation, and premature release of encapsulated bioactive are matters of prime concern which limits its adaptation in the food industry (Muhamad et al., 2011; Zhang et al., 2016).

4.9.1.4   Micro-/Nanocapsules

These delivery systems come under vesicular delivery systems where bioactive resides within the cavity embracing an inner liquid core surrounded by a polymeric membrane and significantly differ in their particle size like nanospheres (1-1000 nm) and microspheres (1-1000 μm). Several fabrication methods are reported but spray-drying and solvent displacement are the most adopted techniques for nano-/microcapsules preparation (Belhadj Slimen et al., 2014; Dima, Assadpour, Dima, & Jafari, 2020; Rehman et al., 2020; Sharma & Borah, 2021). Nano-/microcapsules are anticipated as an alternative to liposomes because of their triggered release under defined stimulus and its low-cost (Dhakal & He, 2020; Maurya, Aggarwal, et al., 2020; Maurya et al., 2021; Maurya et al., 2022; Sharma & Borah, 2021). Though, nano-/microcapsules are well recognized for encapsulating β-carotene (Maurya et al., 2021) but data is limited on their application for vitamin A encapsulation. The use of nano-/microcapsules in food fortification initiated with salt fortification with microcapsule designed to evaluate vitamin A stability in fortified salt (Zimmermann et al., 2004). Improved retinyl palmitate stability was attained by encapsulating it into microcapsules along with iron and iodine (Wegmüller, Zimmermann, Bühr, Windhab, & Hurrell, 2006). Further, researchers have found β-lactoglobulin-based vitamin A microcapsule suitable for food fortification (Tang, Cho, Kim, Wang, & Hwang, 2017).Vitamin A food fortification using nano-/microcapsules could be limited due to challenges, including complexity in the preparation process, the requisite of synthetic polymers, and instability of vitamin A adsorbed on the surface of nano-/microcapsules' polymeric outer crust, inability to address the stability issues like sedimentation, fusion, and leakage (Maurya, Aggarwal, et al., 2020; Maurya et al., 2021; Maurya, Singh, et al., 2020).

4.9.1.5   Nanofibers

Nanofibers are recognized as one of the most suitable delivery systems for encapsulation of heat-labile bioactives due to their nanoscale dimensions, quick wetting properties, temperature independence, and rapid operation (Zhao et al., 2020). Freeze drying, centrifugal spinning, and electrospinning are the most accepted fabrication techniques for nanofibers (Kumar et al., 2019; Nagarajan et al., 2021; Rajaram, Angaiah, & Lee, 2023). A number of natural or synthetic polymers are used to develop nanofibers; natural polymers include chitosan, cellulose, cyclodextrins, starch, pullulan, gelatin, zein protein, egg albumin, whey protein, soy protein while synthetic polymers involve hydroxypropyl, polyvinyl alcohol, cellulose acetate, ethyl cellulose, hydroxypropyl methyl cellulose, and methyl cellulose (Rezaei, Nasirpour, & Fathi, 2015; Zhang et al., 2023). Despite having most of the desired functionalities, nanofibers remain underutilized for vitamin A encapsulation (Taepaiboon, Rungsardthong, & Supaphol, 2007) and data remain scarcer when it comes to its use in food fortification. Porous nature for the resultant product, susceptibility to oxidative degradation of encapsulated vitamin A could be the most prominent limiting factors to the application of nanofibers for vitamin A encapsulation (Zhang et al., 2023).

4.9.2 Lipid-based Delivery Systems (LBDSs)

This class of delivery systems is generally derived from lipids or their derivatives, including oil, phospholipids, and surfactant. Lipid-based delivery systems are well accepted in the food industry due to their high drug loading capacity, better control on drug release at the target site, ability to co-encapsulate hydrophilic as well as lipophilic bioactives, enhanced stability, high compatibility with the GIT, avoidance of toxic solvents and excipients, cost-effectiveness, and easy scale-up (Gonçalves et al., 2016; Loveday & Singh, 2008; Sauvant, Cansell, Sassi, & Atgié, 2012). With technological evolution, an array of lipid-based delivery systems has come into the picture, including liposomes, noisomes, solid lipid carriers, bilosomes, nano-/microemulsions, cubosomes, nanostructured lipid carriers, and many others (Gonçalves et al., 2016; Loveday & Singh, 2008; Mohammadi et al., 2023) (Figure 4.3C and D). The studies carried out with LBDSs that have been adopted for vitamin A encapsulation have been summarized in Table 4.4.

4.9.2.1 Micelles and Microemulsions

Micelles are colloidal dispersions with a particle size of 5 to 100 nm and belong to a vesicle delivery system comprising particulate matter (known as the dispersed phase) dispersed in a continuous phase (Sun & Bandara, 2019). Interestingly, the hydrophilic bioactive resides on the micelles' outer crust while the hydrophobic bioactive (like vitamin A) remains in its core (Kiss, 2020). Excellent reviews describing fabrication methods, structural characterization, and the technological evolution of micelle design for food bioactives delivery are already present in the research database (Gothwal, Khan, & Gupta, 2016; Khan, Hemar, Li, Yang, & De Leon-Rodriguez, 2023; Kiryukhin, Lim, & Chia, 2023). Micelles offer several advantages over polymeric delivery systems, including better penetration across physiological barriers, enhanced solubility for lipophilic bioactives (Ranadheera, Liyanaarachchi, Chandrapala, Dissanayake, & Vasiljevic, 2016), improved penetration across physiological barriers (Kim, 2016), minimal toxicity and other adverse effects, and efficient drug distribution from the systemic circulation by body organs (Gonçalves et al., 2016; McClements & Xiao, 2017). These qualities attracted researchers to test micelles' ability to encapsulate, protect, and deliver vitamin A. The first evidence of vitamin A incorporation in micelles was reported in 1970, where researcher investigated vitamin absorption from a micellar solution in the rat intestine (Kakemi, Sezaki, Muranishi, & Yano, 1970). However, it took more than four decades to use micelles for food application; in 2014 micelles' containing vitamin A were designed to fortify milk by researchers(Loewen, 2014). Poor drug carrying capacity, low stability of encapsulated bioactives, and premature release could be the challenges and reasons that limited micelles' application for food fortification (Gonçalves et al., 2016; McClements & Xiao, 2017).

4.9.2.2 Nanoemulsions

Nano-/microemulsions are colloidal dispersion delivery systems differing in their particle size, i.e., microemulsions (diameter $\leq$ 100 nm) and nanoemulsions (diameter $\leq$ 50 nm) (McClements, 2012).

A range of fabrication methods for emulsions are reported, including high-pressure homogenization, microfluidization, supercritical fluid method, spontaneous emulsification, phase inversion emulsion phase, inversion, and phase-inversion temperature method (Islam et al., 2023; Pires et al., 2023; Rostamabadi et al., 2021; Trombino, Curcio, & Roberta, 2021).

Nano-/microemulsions are one of most widely used delivery systems for encapsulation of bioactive compounds like vitamin A due to their attractive attributes such as high drug loading capacity, high biocompatibility with the GIT, cost effectiveness, and hassle free scale up (Gonçalves et al., 2016). Despite their success in the pharmaceutical and cosmetics sectors, their food application is still limited. This could be due to some practical and technical hurdles including limited availability of food-grade surfactants (Nishitani Yukuyama, Tomiko Myiake Kato, Lobenberg, & Araci Bou-Chacra, 2017), difficulty in the preparation process (as most of them require organic solvents), poor drug carrying capacity, and premature release during storage(Choi & McClements, 2020; Maurya et al., 2022)

4.9.2.3 Liposomes

Liposomes, spherical liquid structures, belong to vesicular delivery system where an aqueous core is surrounded by a single lipid layer (unilamellar) or multiple lipid bilayers (multilamellar liposomes). Liposomes are the most widely utilized delivery system for bioactive delivery as they offer high biocompatibility with animal tissues due to their similarity with the cell membrane, ability to encapsulate lipophilic and lipophobic bioactives, and better control over their composition, texture, particle size, and release (Ajeeshkumar et al., 2021). Emulsification, sonication, lipid film hydration, solvent dispersion method, membrane extrusion, supercritical fluid, and detergent removal methods are reportedly used to fabricate liposomes(Ajeeshkumar et al., 2021; Chaves, Ferreira, Baldino, Pinho, & Reverchon, 2023; D'Souza & Zhang, 2023; Mohammadi et al., 2023). These microstructures have been widely used for β-carotene encapsulation (Maurya et al., 2021; Perrechil et al., 2020; Silva et al., 2020)but reports on their use in vitamin A fortification are limited. This could be due to the associated challenges including difficulties in scale-up, aggregation, sedimentation, fusion, and instability against environmental factors that might lead to degradation and premature release of encapsulated vitamin A (Dhakal & He, 2020; Maurya et al., 2021; Maurya et al., 2022).

4.9.2.4 Solid Lipid Nanoparticles

To address the challenges associated with liposomes, researchers have come up with an advanced delivery system referred to as solid lipid nanoparticles (SLNs), where the liquid oil phase is replaced with a solid lipid (Maurya et al., 2022; Nahum & Domb, 2021; Yaghmur & Mu, 2021). These delivery systems offer desired attributes such as enhanced stability for encapsulants, targeted and controlled release for bioactives, large surface volume ratio, flexible particle size distribution, and ease in scaling up (Duan et al., 2020; Maurya et al., 2022; Nahum & Domb, 2021; Qushawy & Nasr, 2020). High-pressure homogenization under cold and hot conditions (Amoabediny et al., 2018), evaporation or diffusion (Amoabediny et al., 2018; Hwang & Ludescher, 2002), phase inversion methods(Aditya & Ko, 2015), supercritical fluid (supercritical fluid extraction of emulsions) (Trucillo & Campardelli, 2019), film dispersion method (Wen, Al Gailani, Yin, & Rashidinejad, 2018), homogenization or high-speed assisted with ultrasonication (Zardini, Mohebbi, Farhoosh, & Bolurian, 2018), solvent emulsification (Trotta, Debernardi, & Caputo, 2003), extrusion (Mujica-Álvarez et al., 2020) and spray drying (Trotta et al., 2003) are the fabrication techniques used for SLNs. However, several excellent in-depth reviews focused on designing and fabrication techniques for SLNs are already available (Duan et al., 2020; Nahum & Domb, 2021; Qushawy & Nasr, 2020).

The encapsulation of vitamin A in SLN is still in its infancy, particularly when it comes to food application. Table 4.4 summarizes particular studies where SLNs have been used for vitamin A encapsulation. Unexpected particle size growth, drug expulsion during polymeric transition during storage, random gelation tendency, burst releases, poor drug carrying capacity, and unpredictable dynamics of polymeric transitions are some of the major concern while using SLN for vitamin A fortification (da Silva Santos, Ribeiro, & Santana, 2019; Paliwal, Paliwal, Kenwat, Kurmi, & Sahu, 2020).

4.9.2.5 Nanostructured Lipid Carriers (NLC)

Researchers have come up with nanostructured lipid carriers; next-generation delivery systems, to address the challenges associated with SLNs where the solid oil phase is substituted with an unstructured solid-lipid core matrix (Gomaa, Fathi, Eissa, & Elsabahy, 2021; Maurya et al., 2022). Several preparation methods for NLCs are explored, including high-pressure homogenization under cold and hot conditions (Sun et al., 2016), supercritical fluid (supercritical fluid extraction of emulsion) (Liu et al., 2016), solvent emulsification evaporation (Palaria et al., 2023), solvent displacement (Gonçalves, Lellis-Santos, Curi, Lajolo, & Genovese, 2014), solvent diffusion (Hejri, Khosravi, Gharanjig, & Hejazi, 2013), phase inversion (Gomes et al., 2019), melt emulsification (Hentschel, Gramdorf, Müller, & Kurz, 2008; Oliveira, Michelon, de Figueiredo Furtado, Sinigaglia-Coimbra, & Cunha, 2016) Hentschel, sonication (Tamjidi, Shahedi, Varshosaz, & Nasirpour, 2013), spray dry (Hentschel et al., 2008), and solvent evaporation.

The unstructured solid matrix of NLCs not only offers high stability for encapsulated bioactive but also promises high drug load and better control on the release (Gasa-Falcon, Odriozola-Serrano, Oms-Oliu, & Martín-Belloso, 2020). Despite having these desirable attributes, data is scarce on applications for vitamin A encapsulation in food applications. So far there is not even a single report generated on the application of NLCs for vitamin A food fortification (Maurya, Aggarwal, et al., 2020; Maurya et al., 2021).

4.9.3 Technological Evolution in the Vitamin A Delivery System

Based on the high dispensability of vitamin A pharmaceutical formulation in aqueous media, it was assumed that encapsulation can increase the solubility of vitamin A in food matrices. This hypothesis was verified by various devoted studies, such as 100 fold increase in its solubility when encapsulated with β-cyclodextrin (Qi & Shieh, 2002)while it was 10000 folds in case of hydroxypropyl β-cyclodextrin (Jarho, Urtti, Järvinen, Pate, & Järvinen, 1996). Though these cyclic molecules do increase vitamin A solubility, their drug loading capacity remains very low. To address this challenge, it was hypothesized that an emulsion system can ensure high drug encapsulation efficiency (EE) as well as high stability. This hypothesis was supported by devoted studies with 85% (EE) (Yoshida, Sekine, Matsuzaki, Yanaki, & Yamaguchi, 1999) and 86% (EE) (Ghouchi-Eskandar, Simovic, & Prestidge, 2012). Further, it was assumed that the carrier oil used in the emulsion system may affect the EE of vitamin A. This assumption was verified by another research team (Hwang & Ludescher, 2002); where the EE of vitamin A varied from 28-53% when the proportion of carrier oil in the delivery system varied significantly (Hwang & Ludescher, 2002). This vitamin A loading capacity was further improved in various nanomaterials, such 97% (silica nanoparticles) (Ghouchi Eskandar, Simovic, & Prestidge, 2009), 95% (liposome) (Mj & Mm, 2005), 93% in oil/cyclodextrin bead (Trichard, Fattal, Besnard, & Bochot, 2007) and 74–100% (solid lipid carrier) (Jee, Lim, Park, & Kim, 2006). After achieving high loading efficiency, the retention of encapsulated vitamin A still remained an evident problem that needs to be addressed. However, the use of hard fat instead of carrier oils might be a possible way out and is supported by various studies where high retention of vitamin A was achieved when carrier oils were replaced by hard fat (Argimón et al., 2017; Clares et al., 2014; Jenning, Schäfer-Korting, & Gohla, 2000). Along with it, the use of various kinds of nanostructures could also be a game-changing approach to address the vitamin retention issue. However, the stability of vitamin A in the emulsion system was found to be dependent on the type emulsion system, i.e. O (oil)/W (water)/O>W/O>O/W (Yoshida et al., 1999). Further, it was also assumed that the retention of encapsulated vitamin A can also be improved by coating in microcapsules. To achieve higher retention of vitamin A, encapsulated nanosystems were coated with microcapsules (Albertini, Di Sabatino, Calogerà, Passerini, & Rodriguez, 2010; Taha, Al-Saidan, Samy, &

Khan, 2004; Zimmermann et al., 2004), nanospheres (Cho, 2012; Della Porta, Campardelli, Falco, & Reverchon, 2011) and nanoglobules (Ghouchi Eskandar et al., 2009). Further, vitamin A stability can also be improved by cross-linking (Ezpeleta et al., 1996) and varying the surfactant (Taha et al., 2004). Conversely it was also observed that the use of semi-solid incipient mixture as a substitute for corn oil enhanced vitamin A stability against oxidative degradation (Halbaut, Barbé, Aroztegui, & De La Torre, 1997). Stability of encapsulated vitamin A can be improved by covering the capsule system with a multiple-layer coating (Albertini et al., 2010). The authors demonstrated improved vitamin A stability with high drug loading capacity (42%) and high encapsulation efficiency of 94% during a 1-year storage period (Albertini et al., 2010).

The degradation of trans double bonds in the isoprenoid side chain of vitamin A may result in loss of vitamin A activity, so the chemical stability of encapsulated vitamin A is another issue to be addressed for successful encapsulation. It was observed that the microemulsion containing vitamin A displayed delayed degradation as compared to vitamin A dissolved in methanol (Hwang, Lim, Park, & Kim, 2004). Likewise, slower degradation rate was observed for vitamin A encapsulated in liposomes derived with phosphatidylcholine than that of free retinol (Arsić & Vuleta, 1999; Y.-I. Hwang & Ludescher, 2002). However, contradictory observations were reported for vitamin A encapsulated in liposomes, i.e. higher photodegradation in liposomes as compared to free retinol (Young & Gregoriadis, 1996). Chemical stability of vitamin A can also be improved by complex formation with cyclodextrin, by preventing isomerization as well as photodegradation (Liu, 2003; Munoz-Botella, Martın, Del Castillo, Lerner, & Menendez, 2002; Yap, Liu, Thenmozhiyal, & Ho, 2005). Further, vitamin A encapsulated in solid lipid nanoparticles also exhibited decelerated degradation than vitamin A solubilized in methanol (Jee et al., 2006). In conclusion, the chemical stability of vitamin A in the encapsulated system is attributed to physiochemical properties of the developed nanostructure. Furthermore, these physiochemical properties of nanostructures are influenced by a broad spectrum of variables such as quantity and variety of the emulsifier, physical state (liquid/solid), interfacial layer thickness, quantity and type of carrier oils (degree of saturation and chain length for fatty acid), presence/absence of oxidants, and method of preparation.

4.10 VITAMIN A NANOPARTICLES AND THEIR APPLICATION

Prevailing vitamin A deficiency and customer awareness are the major driving factors for vitamin A fortification intervention where foods are used as a platform for vitamin A delivery. However, it brings various challenges such as solubilization of vitamin A, susceptibility against physicochemical stress (pH, oxygen, temperature, UV, pressure), undesirable interaction with food matrices, homogeneity, and customer acceptability. Most of the vitamin A fortification intervention adopts the direct addition of vitamin A in food matrices followed by homogenization. This brings various challenges such as loss of activity, degradation, heterogeneous distribution, undesirable interaction, alteration in food appearance, and taste, hence affecting the customer acceptability. In order to address these issues, various encapsulation techniques are adopted for the fortification of target food. Despite having an array of benefits, encapsulation techniques still remain untapped for commercial and industrial food applications. Vitamin A encapsulated in lactoglobulin and the application of this encapsulated system for milk fortification is the first evidence of the use of encapsulation technique for food application (Liu, 2003). Further, the spray drying technique was applied for fortifying salt with a vitamin A capsule system along with iodine and iron (Zimmermann et al., 2004), Similarly, vitamin A encapsulated in hydrogenated palm fat and lecithin along with iodine and iron has great potential for food fortification (Wegmüller et al., 2006). Further, liposome incorporating vitamin A was applied for an indirect fortification method to enrich *Artemia nauplii* (fish larvae) (Monroig, Navarro, Amat, & Hontoria, 2007). Conversely, the high chemical stability of vitamin A in chitosan-derived microcapsule systems makes it very suitable capsule system for food fortification (Albertini et al., 2010). Likewise, vitamin A premix was developed and applied for the production of vitamin A-enriched artificial rice fortification (Pinkaew et al., 2012). Similarly, re-assembled casein micelles encapsulating vitamin A and D displayed great stability during the storage period which has great potential in milk fortification (Loewen, 2014).The potential of micellar system for the encapsulation of vitamin A was further recognized for milk fortification (Mohan, 2014). Further, β-Lactoglobulin-Vitamin A molecular complexes have shown their potential in food application (Tang et al., 2017). N-Vinylcaprolactam, ethylene glycol diacrylate, and 2,2'-Azobis[2-methylpropionamidine] dihydrochloride-based microgel encapsulating vitamin A displayed great potential for the development of space food (Schroeder, 2018).

4.11   CONCLUSIONS

The encapsulation seems to be an indispensable means to improve the bioavailability of vitamin A. Accumulating reports have highlighted that nanoparticles can be applied to improve their potential health benefits in humans to combat the associated disorders. It is noticeable that vitamin A displays greater bioavailability when incorporated in lipid-derived nanoparticles as compared to polymer-based nanoparticles. Additionally, it was also observed that the quantity, degree of saturation of carrier oil, nature of surfactant, and adopted encapsulation techniques are the key to defining the bioavailability of encapsulated vitamin A. A more systematic mechanistic approach is needed to carve the correlation between the nanoparticle's attributes and their effects on the biological fate of incorporated lipophilic vitamin A. An update on this for vitamin A may offer solid scientific information for the rational designing of novel nanoparticles incorporating to improve the bioavailability of vitamin A, which could be applicable to other vitamins as well as lipophilic bioactive compounds.

REFERENCES

Adeoye, O., & Cabral-Marques, H. (2017). Cyclodextrin nanosystems in oral drug delivery: A mini review. *International Journal of Pharmaceutics*, 531(2), 521–531.

Aditya, N. P., & Ko, S. (2015). Solid lipid nanoparticles (SLNs): Delivery vehicles for food bioactives. *RSC Advances*, 5(39), 30902–30911.

Aghajanzadeh, S., Kashaninejad, M., & Ziaiifar, A. M. (2016). Effect of infrared heating on degradation kinetics of key lime juice physicochemical properties. *Innovative Food Science and Emerging Technologies*, 38, 139–148.

Ajeeshkumar, K. K., Aneesh, P. A., Raju, N., Suseela, M., Ravishankar, C. N., & Benjakul, S. (2021). Advancements in liposome technology: Preparation techniques and applications in food, functional foods, and bioactive delivery: A review. *Comprehensive Reviews in Food Science and Food Safety*, 20(2), 1280–1306.

Albertini, B., Di Sabatino, M., Calogerà, G., Passerini, N., & Rodriguez, L. (2010). Encapsulation of vitamin A palmitate for animal supplementation: Formulation, manufacturing and stability implications. *Journal of Microencapsulation*, 27(2), 150–161.

Amoabediny, G., Haghiralsadat, F., Naderinezhad, S., Helder, M. N., Akhoundi Kharanaghi, E., Mohammadnejad Arough, J., & Zandieh-Doulabi, B. (2018). Overview of preparation methods of polymeric and lipid-based (niosome, solid lipid, liposome) nanoparticles: A comprehensive review. *International Journal of Polymeric Materials and Polymeric Biomaterials*, 67(6), 383–400.

Arayachukeat, S., Wanichwecharungruang, S. P., & Tree-Udom, T. (2011). Retinyl acetate-loaded nanoparticles: Dermal penetration and release of the retinyl acetate. *International Journal of Pharmaceutics*, 404(1–2), 281–288.

Argimón, M., Romero, M., Miranda, P., Mombrú, Á. W., Miraballes, I., Zimet, P., & Pardo, H. (2017). Development and characterization of vitamin A-loaded solid lipid nanoparticles for topical application. *Journal of the Brazilian Chemical Society*, 28, 1177–1184.

Arima, H., Motoyama, K., & Higashi, T. (2017). Potential use of cyclodextrins as drug carriers and active pharmaceutical ingredients. *Chemical and Pharmaceutical Bulletin*, 65(4), 341–348.

Arsić, I., & Vuleta, G. (1999). Influence of liposomes on the stability of vitamin A incorporated in polyacrylate hydrogel. *International Journal of Cosmetic Science*, 21(4), 219–225.

Astray, G., Mejuto, J. C., & Simal-Gandara, J. (2020). Latest developments in the application of cyclodextrin host-guest complexes in beverage technology processes. *Food Hydrocolloids*, 106, 105882.

Beaulieu, L., Savoie, L., Paquin, P., & Subirade, M. (2002). Elaboration and characterization of whey protein beads by an emulsification/cold gelation process: Application for the protection of retinol. *Biomacromolecules*. 3(2), 239–248.

Belhadj Slimen, I., Najar, T., Ghram, A., Dabbebi, H., Ben Mrad, M., & Abdrabbah, M. (2014). Reactive oxygen species, heat stress and oxidative-induced mitochondrial damage. A review. *International Journal of Hyperthermia*, 30(7), 513–523.

Blomhoff, R. (1994). *Vitamin A in health and disease*. Boca Raton: CRC Press.

Borel, P., & Desmarchelier, C. (2017). Genetic variations associated with vitamin A status and vitamin A bioavailability. *Nutrients*, 9(3), 246.

Calabrese, I., Liveri, M. L. T., Ferreira, M. J., Bento, A., Vaz, P. D., Calhorda, M. J., & Nunes, C. D. (2016). Porous materials as delivery and protective agents for vitamin A. *RSC Advances*, 6(71), 66495–66504.

Carafa, M., Marianecci, C., Salvatorelli, M., Di Marzio, L., Cerreto, F., Lucania, G., & Santucci, E. (2008). Formulations of retinyl palmitate included in solid lipid nanoparticles: Characterization and influence on light-induced vitamin degradation. *Journal of Drug Delivery Science and Technology*. 18(2), 119–124.

Carlotti, M. E., Rossatto, V., & Gallarate, M. (2002). Vitamin A and vitamin A palmitate stability over time and under UVA and UVB radiation. *International Journal of Pharmaceutics*, 240(1–2), 85–94. doi:10.1016/S0378-5173(02)00128-X.

Carlotti, M. E., Sapino, S., Trotta, M., Battaglia, L., Vione, D., & Pelizzetti, E. (2005). Photostability and stability over time of retinyl palmitate in an O/W emulsion and in SLN introduced in the emulsion. *Journal of Dispersion Science and Technology*. 26(2), 125–138.

Castenmiller, J. J. M., West, C. E., Linssen, J. P. H., van het Hof, K. H., & Voragen, A. G. J. (1999). The food matrix of spinach is a limiting factor in determining the bioavailability of β-carotene and to a lesser extent of lutein in humans. *The Journal of Nutrition*, 129(2), 349–355.

Cerreto, F., Paolicelli, P., Cesa, S., Amara, H. M. A., D'Auria, F. D., Simonetti, G., & Casadei, M. A. (2013). Solid lipid nanoparticles as effective reservoir systems for long-term preservation of multidose formulations. *Aaps Pharmscitech*. 14, 847–853.

Chansri, N., Kawakami, S., Yamashita, F., & Hashida, M. (2006). Inhibition of liver metastasis by all-trans retinoic acid incorporated into O/W emulsions in mice. *International Journal of Pharmaceutics*. 321(1–2), 42–49.

Chaudhari, A., & Nitin, N. (2015). Role of oxygen scavengers in limiting oxygen permeation into emulsions and improving stability of encapsulated retinol. *Journal of Food Engineering*. 157, 7–13.

Chaves, M. A., Ferreira, L. S., Baldino, L., Pinho, S. C., & Reverchon, E. (2023). Current applications of liposomes for the delivery of vitamins: A systematic review. *Nanomaterials*, 13(9), 1557.

Chilajwar, S. V., Pednekar, P. P., Jadhav, K. R., Gupta, G. J. C., & Kadam, V. J. (2014). Cyclodextrin-based nanosponges: A propitious platform for enhancing drug delivery. *Expert Opinion on Drug Delivery*, 11(1), 111–120.

Cho, E.-C. (2012). Effect of polymer characteristics on the thermal stability of retinol encapsulated in aliphatic polyester nanoparticles. *Bulletin of the Korean Chemical Society*, 33(8), 2560–2566.

Choi, S. J., & McClements, D. J. (2020). Nanoemulsions as delivery systems for lipophilic nutraceuticals: Strategies for improving their formulation, stability, functionality and bioavailability. *Food Science and Biotechnology*, 29(2), 149–168.

Clares, B., Calpena, A. C., Parra, A., Abrego, G., Alvarado, H., Fangueiro, J. F., & Souto, E. B. (2014). Nanoemulsions (NEs), liposomes (LPs) and solid lipid nanoparticles (SLNs) for retinyl palmitate: Effect on skin permeation. *International Journal of Pharmaceutics*, 473(1–2), 591–598.

D'Souza, G. G. M., & Zhang, H. (2023). *Liposomes: Methods and protocols*, Vol. 2622. Springer Nature. Humana New York, NY.

da Silva Santos, V., Ribeiro, A. P. B., & Santana, M. H. A. (2019). Solid lipid nanoparticles as carriers for lipophilic compounds for applications in foods. *Food Research International*, 122, 610–626.

Della Porta, G., Campardelli, R., Falco, N., & Reverchon, E. (2011). PLGA microdevices for retinoids sustained release produced by supercritical emulsion extraction: Continuous versus batch operation layouts. *Journal of Pharmaceutical Sciences*, 100(10), 4357–4367.

Dhakal, S. P., & He, J. (2020). Microencapsulation of vitamins in food applications to prevent losses in processing and storage: A review. *Food Research International*, 109326.

Dima, C., Assadpour, E., Dima, S., & Jafari, S. M. (2020). Bioavailability of nutraceuticals: Role of the food matrix, processing conditions, the gastrointestinal tract, and nanodelivery systems. *Comprehensive Reviews in Food Science and Food Safety*, 19(3), 954–994.

Drummond, J. C., & Coward, K. H. (1920). Researches on the fat-soluble accessory factor (vitamin A). VI: Effect of heat and oxygen on the nutritive value of butter. *Biochemical Journal*, 14(6), 734.

Duan, Y., Dhar, A., Patel, C., Khimani, M., Neogi, S., Sharma, P., … Vekariya, R. L. (2020). A brief review on solid lipid nanoparticles: Part and parcel of contemporary drug delivery systems. *RSC Advances*, 10(45), 26777–26791.

Eskandar, N. G., Simovic, S., & Prestidge, C. A. (2009a). Chemical stability and phase distribution of all-trans-retinol in nanoparticle-coated emulsions. *International Journal of Pharmaceutics*. 376: 186–194.

Eskandar, N. G., Simovic, S., & Prestidge, C. A. (2009b). Nanoparticle coated emulsions as novel dermal delivery vehicles. *Current Drug Delivery*. 6(4), 367–373.

Ezpeleta, I., Irache, J. M., Stainmesse, S., Chabenat, C., Gueguen, J., Popineau, Y., & Orecchioni, A.-M. (1996). Gliadin nanoparticles for the controlled release of all-trans-retinoic acid. *International Journal of Pharmaceutics*, 131(2), 191–200.

Fourmentin, S., Crini, G., & Lichtfouse, E. (2018). *Cyclodextrin fundamentals, reactivity and analysis*, Vol. 16, pp. 1–55. Heidelberg: Springer. doi:10.1007/978-3-319-76159-6_1.

Gasa-Falcon, A., Odriozola-Serrano, I., Oms-Oliu, G., & Martín-Belloso, O. (2020). Nanostructured lipid-based delivery systems as a strategy to increase functionality of bioactive compounds. *Foods*, 9(3), 325.

Ghouchi Eskandar, N., Simovic, S., & Prestidge, C. A. (2009). Nanoparticle coated submicron emulsions: Sustained in-vitro release and improved dermal delivery of all-trans-retinol. *Pharmaceutical Research*, 26(7), 1764–1775.

Ghouchi-Eskandar, N., Simovic, S., & Prestidge, C. A. (2012). Solid-state nanoparticle coated emulsions for encapsulation and improving the chemical stability of all-trans-retinol. *International Journal of Pharmaceutics*, 423(2), 384–391.

Gomaa, E., Fathi, H. A., Eissa, N. G., & Elsabahy, M. (2021). Methods for preparation of nanostructured lipid carriers. *Methods*. doi:10.1016/j.ymeth.2021.05.003.

Gomes, G. V. L., Sola, M. R., Rochetti, A. L., Fukumasu, H., Vicente, A. A., & Pinho, S. C. (2019). β-carotene and α-tocopherol coencapsulated in nanostructured lipid carriers of murumuru (Astrocaryum murumuru) butter produced by phase inversion temperature method: Characterisation, dynamic in vitro digestion and cell viability study. *Journal of Microencapsulation*, 36(1), 43–52.

Gonçalves, A., Estevinho, B. N., & Rocha, F. (2016). Microencapsulation of vitamin A: A review. *Trends in Food Science and Technology*, 51, 76–87.

Gonçalves, A., Estevinho, B. N., & Rocha, F. (2017). Design and characterization of controlled-release vitamin A microparticles prepared by a spray-drying process. *Powder Technology*. 305, 411–417.

Gonçalves, A. E. d. S. S., Lellis-Santos, C., Curi, R., Lajolo, F. M., & Genovese, M. I. (2014). Frozen pulp extracts of camu-camu (Myrciaria dubia McVaugh) attenuate the hyperlipidemia and lipid peroxidation of Type 1 diabetic rats. *Food Research International*, 64, 1–8.

Gothwal, A., Khan, I., & Gupta, U. (2016). Polymeric micelles: Recent advancements in the delivery of anti-cancer drugs. *Pharmaceutical Research*, 33(1), 18–39.

Gupta, C., Arora, S., Syama, M. A., & Sharma, A. (2017). Preparation of milk protein-vitamin A complexes and their evaluation for vitamin A binding ability. *Food Chemistry*. 237, 141–149.

Gupta, C., Arora, S., Syama, M. A., & Sharma, A. (2018). Physicochemical characterization of native and modified sodium caseinate-Vitamin A complexes. *Food Research International*, 106, 964–973.

Halbaut, L., Barbé, C., Aroztegui, M., & De La Torre, C. (1997). Oxidative stability of semi-solid excipient mixtures with corn oil and its implication in the degradation of vitamin A. *International Journal of Pharmaceutics*, 147(1), 31–40.

Hejri, A., Khosravi, A., Gharanjig, K., & Hejazi, M. (2013). Optimisation of the formulation of β-carotene loaded nanostructured lipid carriers prepared by solvent diffusion method. *Food Chemistry*, 141(1), 117–123.

Hentschel, A., Gramdorf, S., Müller, R. H., & Kurz, T. (2008). β-carotene-loaded nanostructured lipid carriers. *Journal of Food Science*, 73(2), N1–N6.

Herath, T. (2007). Effect of whey protein isolate on the oxidative stability of Vitamin A: A thesis presented in partial fulfilment of the requirements for the degree of Masters of Science in Nutrition Science. Institute of Food, Nutrition and Human Health, Massey University, Palmerston North (Doctoral dissertation, Massey University).

Huynh, P. T. (2012). *Solvent-free beta-carotene nanoparticles for food fortification*. New Brunswick: Rutgers The State University of New Jersey.

Hwang, Y.-I., & Ludescher, R. D. (2002). Stabilization of retinol through incorporation into liposomes. *Journal of Biochemistry and Molecular Biology*, 35(4), 358–363.

Hwang, S. R., Lim, S.-J., Park, J.-S., & Kim, C.-K. (2004). Phospholipid-based microemulsion formulation of all-trans-retinoic acid for parenteral administration. *International Journal of Pharmaceutics*, 276(1–2), 175–183.

Hwang, S. M., Lee, S. M., Choi, J. H., Park, K., Joo, J, Lim, J. H., & Kim, H. (2012). Microstructure and dielectric characteristics of high-k tetragonal ZrO2 films with various thicknesses processed by sol–gel method. *Journal of Nanoscience and Nanotechnology*. 12(4), 3350–3354.

Islam, F., Saeed, F., Afzaal, M., Hussain, M., Ikram, A., & Khalid, M. A. (2023). Food grade nanoemulsions: Promising delivery systems for functional ingredients. *Journal of Food Science and Technology*, 60(5), 1461–1471.

Isler, O., Klaui, H., & Solms, U. (1947). Vitamins A and carotene-3. Industrial preparation and production. *The Vitamins*, 1.

IUPAC-IUB, Joint Commission on Biochemical Nomenclature. (1983). Nomenclature of retinoids: Recommendations 1981. *Archives of Biochemistry and Biophysics*, 224(2), 728–731.

Jarho, P., Urtti, A., Järvinen, K., Pate, D. W., & Järvinen, T. (1996). Hydroxypropyl-β-cyclodextrin increases aqueous solubility and stability of anandamide. *Life Sciences*, 58(10), 181–185.

Jee, J.-P., Lim, S.-J., Park, J.-S., & Kim, C.-K. (2006). Stabilization of all-trans retinol by loading lipophilic antioxidants in solid lipid nanoparticles. *European Journal of Pharmaceutics and Biopharmaceutics*, 63(2), 134–139.

Jenning, V., Thünemann, A. F., & Gohla, S. H. (2000). Characterisation of a novel solid lipid nanoparticle carrier system based on binary mixtures of liquid and solid lipids. *International Journal of Pharmaceutics*. 199(2), 167–177.

Jeon, B.J., Kim, N.C., Han, E.M., and Kwak, H.S. 2005. Application of microencapsulated isoflavone into milk. *Archives of Pharmacology Research*. 7, 859–865.

Kakemi, K., Sezaki, H., Muranishi, S., & Yano, A. (1970). Mechanism of drug absorption from micellar solution. I. Absorption of solubilized vitamin A from the rat intestine. *Chemical and Pharmaceutical Bulletin*, 18(8), 1563–1568.

Kawakami, Y., Tsurugasaki, W., Nakamura, S., & Osada, K. (2005). Comparison of regulative functions between dietary soy isoflavones aglycone and glucoside on lipid metabolism in rats fed cholesterol. *The Journal of Nutritional Biochemistry*. 16(4), 205–212.

Khan, M. A., Hemar, Y., Li, J., Yang, Z., & De Leon-Rodriguez, L. M. (2023). Fabrication, characterization, and potential applications of re-assembled casein micelles. *Critical Reviews in Food Science and Nutrition*, 1–25.

Kim, J.-S. (2016). Liposomal drug delivery system. *Journal of Pharmaceutical Investigation*, 46(4), 387–392.

Kiryukhin, M. V., Lim, S. H., & Chia, C. Y. (2023). Design and use of microcarriers for the delivery of nutraceuticals. Materials Science and Engineering in Food Product. *Development*, 93–116.

Kislitza, O., Manaenkov, O., & Savin, A. (2009). Microencapsulation of retinol-acetate in alginate microspheres. In XVIIth International Conference on Bioencapsulation, Groningen, Netherlands.

Kiss, É. (2020). Nanotechnology in food systems: A review. *Acta Alimentaria*, 49(4), 460–474.

Klein, M., & Poverenov, E. (2020). Natural biopolymer-based hydrogels for use in food and agriculture. *Journal of the Science of Food and Agriculture*, 100(6), 2337–2347.

Kong, M., Chen, X. G., Kweon, D. K., & Park, H. J. (2011). Investigations on skin permeation of hyaluronic acid based nanoemulsion as transdermal carrier. *Carbohydrate Polymers*. 86(2), 837–843.

Kumar, T. S. M., Kumar, K. S., Rajini, N., Siengchin, S., Ayrilmis, N., & Rajulu, A. V. (2019). A comprehensive review of electrospun nanofibers: Food and packaging perspective. *Composites Part B: Engineering*, 175, 107074.

Kumar, Y., & Singhal, S. (2021). Food applications of cyclodextrins. *Sustainable Agriculture Reviews 55: Micro and Nano Engineering in Food Science*, 1, 201–238.

Lee, S. C., Lee, K. E., Kim, J. J., & Lim, S. H. (2005). The effect of cholesterol in the liposome bilayer on the stabilization of incorporated retinol. *Journal of Liposome Research*. 15(3–4), 157–166.

Lee, K. J., & Dabrowski, K. (2004). Long-term effects and interactions of dietary vitamins C and E on growth and reproduction of yellow perch, Perca flavescens. *Aquaculture*, 230(1-4), 377–389.

Lin, Q., Liang, R., Williams, P. A., & Zhong, F. (2018). Factors affecting the bioaccessibility of β-carotene in lipid-based microcapsules: Digestive conditions, the composition, structure and physical state of microcapsules. *Food Hydrocolloids*, 77, 187–203.

Liu, Y. (2003). *Beta-lactoglobulin complexed vitamins A and D in skim milk: Shelf life and bioavailability.* North Carolina State University.

Liu, R., Wang, S., Sun, L., Fang, S., Wang, J., Huang, X., … Liu, C. (2016). A novel cationic nanostructured lipid carrier for improvement of ocular bioavailability: Design, optimization, in vitro and in vivo evaluation. *Journal of Drug Delivery Science and Technology*, 33, 28–36.

Loewen, A. J. (2014). *Optimizing the loading of vitamin A and vitamin D into re-assembled casein micelles and investigating the effect of micellar complexation on vitamin D stability.* (Doctoral dissertation, University of British Columbia).

Loewen, A., Chan, B., & Li-Chan, E. C. (2018). Optimization of vitamins A and D3 loading in re-assembled casein micelles and effect of loading on stability of vitamin D3 during storage. *Food Chemistry.* 240, 472–481.

Loftsson, T. (2004). BrewsterM. E. Masson M.-Role of cyclodextrins in improving oral drug delivery. *American Journal of Drug Delivery*, 2(4), 261–275.

Loveday, S. M., & Singh, H. (2008). Recent advances in technologies for vitamin A protection in foods. *Trends in Food Science and Technology*, 19(12), 657–668. doi:10.1016/j.tifs.2008.08.002.

Maurya, V. K., & Aggarwal, M. (2017). Enhancing bio-availability of vitamin D by nano-engineered based delivery systems-An overview. *International Journal of Current Microbiology and Applied Sciences*, 6(7), 340–353.

Maurya, V. K., Aggarwal, M., Ranjan, V., & Gothandam, K. M. (2020). Improving bioavailability of vitamin A in food by encapsulation: An update. *Nanoscience in Medicine*, 1, 117–145.

Maurya, V. K., Singh, J., Ranjan, V., Gothandam, K. M., Bohn, T., & Pareek, S. (2020). Factors affecting the fate of β-carotene in the human gastrointestinal tract: A narrative review. *International Journal for Vitamin and Nutrition Research*, 92(5–6), 385–405.

Maurya, V. K., Shakya, A., Aggarwal, M., Gothandam, K. M., Bohn, T., & Pareek, S. (2021). Fate of β-carotene within loaded delivery systems in food: State of knowledge. *Antioxidants*, 10(3), 426.

Maurya, V. K., Shakya, A., Bashir, K., Kushwaha, S. C., & McClements, D. J. (2022). Vitamin A fortification: Recent advances in encapsulation technologies. *Comprehensive Reviews in Food Science and Food Safety*, 21(3), 2772–2819.

McClements, D. J. (2012). Nanoemulsions versus microemulsions: Terminology, differences, and similarities. *Soft Matter.* 8(6), 1719–1729.

McClements, D. J., & Xiao, H. (2017). Is Nano safe in foods? Establishing the factors impacting the gastrointestinal fate and toxicity of organic and inorganic food-grade nanoparticles. *npj Science of Food*, 1(1), 1–13.

McCollum, E. V., & Davis, M. (1915). The essential factors in the diet during growth. *Journal of Biological Chemistry*, 23(1), 231–246.

McConnell, C. F. (2000). Retinoid therapy for cutaneous disease: Current concepts. *Current Problems in Dermatology*, 12(5), 246–248. doi:10.1016/S1040-0486(00)80010-1.

Mj, F. C., & Mm, J. S. (2005). In vitro percutaneous absorption of all-trans retinoic acid applied in free form or encapsulated in stratum corneum lipid liposomes. *International Journal of Pharmaceutics*, 297(1–2), 134–145.

Mohammadi, M. A., Farshi, P., Ahmadi, P., Ahmadi, A., Yousefi, M., Ghorbani, M., & Hosseini, S. M. (2023). Encapsulation of vitamins using nanoliposome: Recent advances and perspectives. *Advanced Pharmaceutical Bulletin*, 13(1), 48.

Mohan, M. S. (2014). *Casein micelles and their properties: Polydispersity, association with vitamin A and effect of ultra-high pressure homogenization.* PhD diss., University of Tennessee, https://trace.tennessee.edu/utk_graddiss/2844

Monaco, H. L. (2000). The transthyretin-retinol-binding protein complex. *Biochimica et Biophysica Acta (BBA)-Protein Structure and Molecular Enzymology*, 1482(1–2), 65–72.

Monroig, Ó., Navarro, J. C., Amat, F., & Hontoria, F. (2007). Enrichment of artemia nauplii in vitamin A, vitamin C and methionine using liposomes. *Aquaculture*, 269(1–4), 504–513.

Muhamad, I. I., Fen, L. S., Hui, N. H., & Mustapha, N. A. (2011). Genipin-cross-linked kappa-carrageenan/carboxymethyl cellulose beads and effects on beta-carotene release. *Carbohydrate Polymers*, 83(3), 1207–1212.

Mujica-Álvarez, J., Gil-Castell, O., Barra, P. A., Ribes-Greus, A., Bustos, R., Faccini, M., & Matiacevich, S. (2020). Encapsulation of vitamins A and E as spray-dried additives for the feed industry. *Molecules*, 25(6), 1357.

Müller, W. E. G., Tolba, E., Dorweiler, B., Schröder, H. C., Diehl-Seifert, B., & Wang, X. (2015). Electrospun bioactive mats enriched with Ca-polyphosphate/retinol nanospheres as potential wound dressing. *Biochemistry and Biophysics Reports*, 3, 150–160.

Munoz-Botella, S., Martın, M., Del Castillo, B., Lerner, D., & Menendez, J. (2002). Differentiating geometrical isomers of retinoids and controlling their photo-isomerization by complexation with cyclodextrins. *Analytica Chimica Acta*, 468(1), 161–170.

Nagarajan, K. J., Ramanujam, N. R., Sanjay, M. R., Siengchin, S., Surya Rajan, B., Sathick Basha, K., … Raghav, G. R. (2021). A comprehensive review on cellulose nanocrystals and cellulose nanofibers: Pretreatment, preparation, and characterization. *Polymer Composites*, 42(4), 1588–1630.

Nahum, V., & Domb, A. J. (2021). Recent developments in solid lipid microparticles for food ingredients delivery. *Foods*, 10(2), 400.

Nascimento, L. G. L., Casanova, F., Silva, N. F. N., de Carvalho Teixeira, A. V. N., & de Carvalho, A. F. (2020). Casein-based hydrogels: A mini-review. *Food Chemistry*, 314, 126063.

Nishitani Yukuyama, M., Tomiko Myiake Kato, E., Lobenberg, R., & Araci Bou-Chacra, N. (2017). Challenges and future prospects of nanoemulsion as a drug delivery system. *Current Pharmaceutical Design*, 23(3), 495–508.

Noh, M. F. M., & Mustar, R. D. N. G. (2019). Vitamin A in health and disease. Vitamin A. IntechOpen. doi:10.5772/intechopen.84460.

Numata, S., Kinoshita, M., Tajima, A., Nishi, A., Imoto, I., & Ohmori, T. (2015). Evaluation of an association between plasma total homocysteine and schizophrenia by a Mendelian randomization analysis. *BMC Medical Genetics*. 16(1), 1–6.

Oliveira, D. R. B., Michelon, M., de Figueiredo Furtado, G., Sinigaglia-Coimbra, R., & Cunha, R. L. (2016). β-carotene-loaded nanostructured lipid carriers produced by solvent displacement method. *Food Research International*, 90, 139–146.

Olson, J. A. (1990). Vitamin A. Handbook of vitamins, 3, 1–50.

Palaria, B., Tiwari, V., Tiwari, A., Aslam, R., Kumar, A., Sahoo, B. M., … Kumar, S. (2023). Nanostructured lipid carriers: A promising carrier in targeted drug delivery system. *Current Nanomaterials*, 8(1), 23–43.

Paliwal, R., Paliwal, S. R., Kenwat, R., Kurmi, B. D., & Sahu, M. K. (2020). Solid lipid nanoparticles: A review on recent perspectives and patents. *Expert Opinion on Therapeutic Patents*, 30(3), 179–194.

Pan, C. W., Dirani, M., Cheng, C. Y., Wong, T. Y., & Saw, S. M. (2015). The age-specific prevalence of myopia in Asia: A meta-analysis. *Optometry and Vision Science*. 92(3), 258–266.

Panfili, G., Manzi, P., & Pizzoferrato, L. (1998). Influence of thermal and other manufacturing stresses on retinol isomerization in milk and dairy products. *Journal of Dairy Research*, 65(2), 253–260.

Park, S., Ham, J. O., & Lee, B. K. (2015). Effects of total vitamin A, vitamin C, and fruit intake on risk for metabolic syndrome in Korean women and men. *Nutrition*. 31(1), 111–118.

Perrechil, F. A., Maximo, G. J., & Cunha, R. L. (2020). Microbeads of sodium caseinate and κ-carrageenan as a β-carotene carrier in aqueous systems. *Food and Bioprocess Technology*, 13(4), 661–669.

Pezeshky, A., Ghanbarzadeh, B., Hamishehkar, H., Moghadam, M., & Babazadeh, A. (2016). Vitamin A palmitate-bearing nanoliposomes: Preparation and characterization. *Food Bioscience*. 13, 49–55.

Pezeshki, A., Ghanbarzadeh, B., Mohammadi, M., Fathollahi, I., & Hamishehkar, H. (2014). Encapsulation of vitamin A palmitate in nanostructured lipid carrier (NLC)-effect of surfactant concentration on the formulation properties. *Advanced Pharmaceutical Bulletin*. 4, 563.

Pinkaew, S., Wegmuller, R., & Hurrell, R. (2012). Vitamin A stability in triple fortified extruded, artificial rice grains containing iron, zinc and vitamin A. *International Journal of Food Science and Technology*, 47(10), 2212–2220.

Pires, P. C., Fernandes, M., Nina, F., Gama, F., Gomes, M. F., Rodrigues, L. E., … Santos, A. O. (2023). Innovative aqueous nanoemulsion prepared by phase inversion emulsification with exceptional homogeneity. *Pharmaceutics*, 15(7), 1878.

Pisetpackdeekul, P., Supmuang, P., Pan-In, P., Banlunara, W., Limcharoen, B., Kokpol, C., & Wanichwecharungruang, S. (2016). Proretinal nanoparticles: Stability, release, efficacy, and irritation. *International Journal of Nanomedicine*. 11, 3277.

Puntel, A., Maeda, A., Golczak, M., Gao, S. Q., Yu, G., Palczewski, K., & Lu, Z. R. (2015). Prolonged prevention of retinal degeneration with retinylamine loaded nanoparticles. *Biomaterials*. 44, 103–110.

Qi, Z. H., & Shieh, W. J. (2002). Aqueous media for effective delivery of tretinoin. *Journal of Inclusion Phenomena and Macrocyclic Chemistry*, 44(1/4), 133–136.

Qu, B., & Luo, Y. (2020). Chitosan-based hydrogel beads: Preparations, modifications and applications in food and agriculture sectors–A review. *International Journal of Biological Macromolecules*, 152, 437–448.

Qushawy, M., & Nasr, A. L. I. (2020). Solid lipid nanoparticles (SLNs) as nano drug delivery carriers: Preparation, characterization and application. *International Journal of Applied Pharmaceutics*, 12(1), 1–9.

Rajaram, R., Angaiah, S., & Lee, Y. R. (2023). Polymer supported electrospun nanofibers with supramolecular materials for biological applications–a review. *International Journal of Polymeric Materials and Polymeric Biomaterials*, 72(13), 1042–1058.

Ranadheera, C. S., Liyanaarachchi, W. S., Chandrapala, J., Dissanayake, M., & Vasiljevic, T. (2016). Utilizing unique properties of caseins and the casein micelle for delivery of sensitive food ingredients and bioactives. *Trends in Food Science and Technology*, 57, 178–187.

Rehman, A., Tong, Q., Jafari, S. M., Assadpour, E., Shehzad, Q., Aadil, R. M., … Ashraf, W. (2020). Carotenoid-loaded nanocarriers: A comprehensive review. *Advances in Colloid and Interface Science*, 275, 102048.

Reynolds, R. (2006). *Spray drying of beta-lactoglobulin-vitamin A and beta-lactoglobulin-vitamin D complexes.*

Rezaei, A., Nasirpour, A., & Fathi, M. (2015). Application of cellulosic nanofibers in food science using electrospinning and its potential risk. *Comprehensive Reviews in Food Science and Food Safety*, 14(3), 269–284.

Rock, C. L., Lovalvo, J. L., Emenhiser, C., Ruffin, M. T., Flatt, S. W., & Schwartz, S. J. (1998). Bioavailability of β-carotene is lower in raw than in processed carrots and spinach in women. *The Journal of Nutrition*, 128(5), 913–916.

Rostamabadi, H., Falsafi, S. R., Boostani, S., Katouzian, I., Rezaei, A., Assadpour, E., & Jafari, S. M. (2021). Design and formulation of Nano/micro-encapsulated natural bioactive compounds for food applications. In *Application of nano/microencapsulated ingredients in food products* (pp. 1–41). Elsevier.

Sane, A., & Limtrakul, J. (2009). Formation of retinyl palmitate-loaded poly (l-lactide) nanoparticles using rapid expansion of supercritical solutions into liquid solvents (RESOLV). *The Journal of Supercritical Fluids.* 51(2), 230–237.

Satoh, K., Makimura, K., Hasumi, Y., Nishiyama, Y., Uchida, K., & Yamaguchi, H. (2009). Candida auris sp. nov., a novel ascomycetous yeast isolated from the external ear canal of an inpatient in a Japanese hospital. *Microbiology and Immunology.* 53(1), 41–44.

Sauvant, P., Cansell, M., Sassi, A. H., & Atgié, C. (2012). Vitamin A enrichment: Caution with encapsulation strategies used for food applications. *Food Research International*, 46(2), 469–479.

Schroeder, R. (2018). Microgels for long-term storage of vitamins for extended spaceflight. *Life Sciences in Space Research*, 16, 26–37.

Semba, R. D. (2007). Nutritional blindness (vitamin A deficiency disorders). In *Handbook of nutrition and ophthalmology* (pp. 1–119). Springer Science & Business Media, Humana Press.

Semenova, E. M., Cooper, A., Wilson, C. G., & Converse, C. A. (2002). Stabilization of all-trans-retinol by cyclodextrins: A comparative study using HPLC and fluorescence spectroscopy. *Journal of Inclusion Phenomena and Macrocyclic Chemistry.* 44(1), 155–158.

Shafaei, Z., Ghalandari, B., Vaseghi, A., Divsalar, A., Haertlé, T., Saboury, A. A., & Sawyer, L. (2017). β-Lactoglobulin: An efficient nanocarrier for advanced delivery systems. *Nanomedicine: Nanotechnology, Biology and Medicine*, 13(5), 1685–1692.

Sharma, R., Borah, A., Sharma, R., & Borah, A. (2021). Prospect of microcapsules as a delivery system in food technology: A review. *The Pharma Innovation Journal.* 10(5), 182–191.

Silva, K. C. G., Feltre, G., Hubinger, M. D., & Sato, A. C. K. (2020). Protection and targeted delivery of β-carotene by starch-alginate-gelatin emulsion-filled hydrogels. *Journal of Food Engineering*, 290, 110205.

Sporn, M. B., Dunlop, N. M., Newton, D. L., & Smith, J. M. (1976, May). Prevention of chemical carcinogenesis by vitamin A and its synthetic analogs (retinoids). In *Federation Proceedings* (Vol. 35, No. 6, pp. 1332–1338).

Sun, X., & Bandara, N. (2019). Applications of reverse micelles technique in food science: A comprehensive review. *Trends in Food Science and Technology*, 91, 106–115.

Sun, C., Dai, L., Liu, F., & Gao, Y. (2016). Simultaneous treatment of heat and high-pressure homogenization of zein in ethanol–water solution: Physical, structural, thermal and morphological characteristics. *Innovative Food Science & Emerging Technologies.* 34, 161–170.

Syed, E. U., Wasay, M., & Awan, S. (2013). Vitamin B12 supplementation in treating major depressive disorder: A randomized controlled trial. *The Open Neurology Journal.* 7, 44.

Taepaiboon, P., Rungsardthong, U., & Supaphol, P. (2007). Vitamin-loaded electrospun cellulose acetate nanofiber mats as transdermal and dermal therapeutic agents of vitamin A acid and vitamin E. *European Journal of Pharmaceutics and Biopharmaceutics*, 67(2), 387–397.

Taha, E. I., Al-Saidan, S., Samy, A. M., & Khan, M. A. (2004). Preparation and in vitro characterization of self-nanoemulsified drug delivery system (SNEDDS) of all-trans-retinol acetate. *International Journal of Pharmaceutics*, 285(1–2), 109–119.

Tamjidi, F., Shahedi, M., Varshosaz, J., & Nasirpour, A. (2013). Nanostructured lipid carriers (NLC): A potential delivery system for bioactive food molecules. *Innovative Food Science and Emerging Technologies*, 19, 29–43.

Tang, J. W., Cho, H., Kim, J., Wang, Z. G., & Hwang, K. T. (2017). Optimization of microencapsulation of β -lactoglobulin–vitamin A using response surface methodology. *Journal of Food Processing and Preservation*, 41(1), e12747.

Torrado, S., Torrado, J. J., & Cadórniga, R. (1992). Topical application of albumin microspheres containing vitamin A drug release and availability. *International Journal of Pharmaceutics*, 86(2), 147–152. doi:10.1016/0378-5173(92)90191-4.

Trichard, L., Fattal, E., Besnard, M., & Bochot, A. (2007). α-cyclodextrin/oil beads as a new carrier for improving the oral bioavailability of lipophilic drugs. *Journal of Controlled Release*, 122(1), 47–53.

Trombino, S., Curcio, F., & Cassano, R. (2021). Nano-and micro-technologies applied to food nutritional ingredients. *Current Drug Delivery*. 18(6), 670–678.

Trotta, M., Debernardi, F., & Caputo, O. (2003). Preparation of solid lipid nanoparticles by a solvent emulsification–diffusion technique. *International Journal of Pharmaceutics*, 257(1–2), 153–160.

Trucillo, P., & Campardelli, R. (2019). Production of solid lipid nanoparticles with a supercritical fluid assisted process. *The Journal of Supercritical Fluids*, 143, 16–23.

Uzzaman, S., & Grace, V. (2013). Antioxidant potential of all-trans retinoic acid (ATRA) and enhanced activity of liposome encapsulated ATRA against inflammation and tumor-directed angiogenesis. *Immunopharmacology and Immunotoxicology*. 35(1–6), 165–174.

Vilanova, N., & Solans, C. (2015). Vitamin A Palmitate–β-cyclodextrin inclusion complexes: Characterization, protection and emulsification properties. *Food Chemistry*. 175, 529–535.

Visentini, F. F., Sponton, O. E., Perez, A. A. et al. (2017b). Biopolymer nanoparticles for vehiculization and photochemical stability preservation of retinol. *Food Hydrocolloids*. 70, 363–370.

Wald, G. (1933). Vitamin A in the retina. *Nature*, 132(3330), 316–317.

Wegmüller, R., Zimmermann, M. B., Bühr, V. G., Windhab, E. J., & Hurrell, R. F. (2006). Development, stability, and sensory testing of microcapsules containing iron, iodine, and vitamin A for use in food fortification. *Journal of Food Science*, 71(2), S181–S187.

Weisse, S., Kalimouttou, S., Lahiani-Skiba, M., Djedaini-Pilard, F., Perly, B., & Skiba, M. (2009). Investigations on topically applied vitamin A loaded amphiphilic cyclodextrin nanocapsules. *Journal of Nanoscience and Nanotechnology*, 9(1), 640–645.

Weisse, S., Perly, B., Creminon, C., Ouvrard-Baraton, F., & Djedai'Ni-Pilard, F. (2004). Enhancement of vitamin A skin absorption by cyclodextrins. *Journal of Drug Delivery Science and Technology*, 14(1), 77–86. doi:10.1016/S1773-2247(04)50009-6.

Wen, J., Al Gailani, M., Yin, N., & Rashidinejad, A. (2018). Liposomes and niosomes. In Shahin Roohinejad, Ralf Greiner, Indrawati Oey and Jingyuan Wen (eds.), *Emulsion-Based Systems for Delivery of Food Active Compounds: Formation, Application. Health and Safety*, Shahin Roohinejad, Ralf Greiner, Indrawati Oey and, Jingyuan Wen (eds.), Vol. 263.

Wen, Z., Liu, B., Zheng, Z., You, X., Pu, Y., & Li, Q. (2010). Preparation of liposomes entrapping essential oil from Atractylodes macrocephala Koidz by modified RESS technique. *Chemical Engineering Research and Design*. 88(8), 1102–1107.

World Health Organization. (2009). Global prevalence of vitamin A deficiency in populations at risk 1995–2005: WHO global database on vitamin A deficiency.

WHO. (2021a). Biofortification of staple crops. Retrieved from https://www.who.int/elena/titles/biofortification/en/.

WHO. (2021b). Malnutrition. Retrieved 17 September 2021, from https://www.who.int/news-room/fact-sheets/detail/malnutrition.

WHO. (2021c). Vitamin A deficiency. https://www.who.int/publications/i/item/WHO-NUT-95.3

Wirth, J. P., Petry, N., Tanumihardjo, S. A., Rogers, L. M., McLean, E., Greig, A., … Rohner, F. (2017). Vitamin A supplementation programs and country-level evidence of vitamin A deficiency. *Nutrients*, 9(3), 190. doi:10.3390/nu9030190.

Wiseman, E. M., Bar-El Dadon, S., & Reifen, R. (2017). The vicious cycle of vitamin A deficiency: A review. *Critical Reviews in Food Science and Nutrition*, 57(17), 3703–3714. doi:10.1080/10408398.2016.1160362.

Xu, Y., Shan, Y., Lin, X., Miao, Q., Lou, L., Wang, Y., & Ye, J. (2021). Global patterns in vision loss burden due to vitamin A deficiency from 1990 to 2017. *Public Health Nutrition*, 24(17), 5786–5794. doi:10.1017/S1368980021001324.

Yaghmur, A., & Mu, H. (2021). Recent advances in drug delivery applications of cubosomes, hexosomes, and solid lipid nanoparticles. *Acta Pharmaceutica Sinica B*. 11(4), 871–885.

Yap, K. L., Liu, X., Thenmozhiyal, J. C., & Ho, P. C. (2005). Characterization of the 13-cis-retinoic acid/cyclodextrin inclusion complexes by phase solubility, photostability, physicochemical and computational analysis. *European Journal of Pharmaceutical Sciences*, 25(1), 49–56.

Yoshida, K., Sekine, T., Matsuzaki, F., Yanaki, T., & Yamaguchi, M. (1999). Stability of vitamin A in oil-in-water-in-oil-type multiple emulsions. *Journal of the American Oil Chemists' Society*, 76(2), 1–6.

Young, A. M., & Gregoriadis, G. (1996). Photolysis of retinol in liposomes and its protection with tocopherol and oxybenzone. *Photochemistry and Photobiology*, 63(3), 344–352.

Zaibunnisa, A. H., Aini Marhanna, M. N. A., & Ainun Atirah, M. (2011). Characterisation and solubility study of γ-cyclodextrin and B-carotene complex. *International Food Research Journal*. 18(3).

Zardini, A. A., Mohebbi, M., Farhoosh, R., & Bolurian, S. (2018). Production and characterization of nanostructured lipid carriers and solid lipid nanoparticles containing lycopene for food fortification. *Journal of Food Science and Technology*, 55(1), 287–298.

Zhang, R., Zhang, Z., Kumosani, T., Khoja, S., Abualnaja, K. O., & McClements, D. J. (2016). Encapsulation of β-carotene in nanoemulsion-based delivery systems formed by spontaneous emulsification: Influence of lipid composition on stability and bioaccessibility. *Food Biophysics*, 11(2), 154–164.

Zhang, Y., Chen, J., Zhang, Z., Zhu, H., Ma, W., Zhao, X., … Naeem, A. (2023). Solvent-free loading of vitamin A palmitate into β-cyclodextrin metal-organic frameworks for stability enhancement. *AAPS PharmSciTech*, 24(5), 136.

Zhao, L., Duan, G., Zhang, G., Yang, H., He, S., & Jiang, S. (2020). Electrospun functional materials toward food packaging applications: A review. *Nanomaterials*, 10(1), 150.

Zimmermann, M. B., Wegmueller, R., Zeder, C., Chaouki, N., Biebinger, R., Hurrell, R. F., & Windhab, E. (2004). Triple fortification of salt with microcapsules of iodine, iron, and vitamin A. *The American Journal of Clinical Nutrition*, 80(5), 1283–1290.

5 Vitamin C Encapsulation and Fortification

Amita Shakya, Vaibhav Kumar Maurya, Khalid Bashir,
Kulsum Jan, and David Julian McClements

5.1 INTRODUCTION

Vitamin C or ascorbic acid, a six-carbon structure, is historically a well-recognized hydrophilic bioactive, demonstrating antioxidant activity (Johnston et al., 2007). This molecule has always been a matter of research in the scientific community due to its association with an array of physiological functions. As humans lack an essential enzyme in the biosynthetic pathway (L-gulono-1,4 lactone oxidase), they depend on dietary sources to meet the daily vitamin C requirement (30–40 mg per day for adults) (Combs & McClung, 2016; Davies et al., 1991). It is a well-established fact that its inadequate consumption may undermine its disease-combating potential and also may raise several health concerns (Carr & Rowe, 2020; Packer, 1997). Besides its traditional role in scurvy rectification, its pleiotropic role penetrates several essential biological functions including homeostasis, collagen synthesis, neurotransmitter production, osteogenesis, amino acid synthesis, reduction in hypertension, atherogenesis and prevention of infection, cold, and carcinogenesis (Granger & Eck, 2018; Iqbal et al., 2004; Packer, 1997). An increase in health awareness further creates a demand for vitamin C-rich functional foods. The World Health Organization (WHO) has also recognized food fortification as the most effective, safe, and cheapest means to address any micronutrient deficiency (Orriss, 1998). The incorporation of vitamin C in the food matrix poses several challenges to food technologists due to its susceptibility to environmental factors (pH, temperature, humidity, radiation, and ionic strength) and food processing conditions, hence diminishing its bioavailability in fortified food (Abbas et al., 2012; Caritá et al., 2020; Comunian et al., 2020; Lešková, 2006). Researchers have attempted to address the abovementioned challenges that have occurred in vitamin fortification (Maurya & Aggarwal, 2017; Maurya et al., 2020a; Maurya et al., 2020b; Maurya et al., 2021; Maurya et al., 2022), but the literature available so far, especially on vitamin C, is either partial or incomplete in order that a clear understanding of the challenges and limitations associated with vitamin C fortification can be obtained (Abbas et al., 2012; Combs & McClung, 2016; Comunian et al., 2020). So, here we review vitamin C regarding its molecular structure, chemistry, physicochemical attributes, sources, recommended dietary allowance, strategies adopted to address its deficiency, challenges in its food fortification, and encapsulation techniques adopted to address these challenges.

5.2 DISCOVERY, CHEMISTRY, AND NOMENCLATURE

The linkage of vitamin C deficiency with scurvy dates back to around 1700 BC when Ebers Papurus described the distinctive features of scurvy. Some pieces of information were also created by the great Indian surgeon Susrutra (400 BC), Hippocrates, and Chang Chi (200 AD) in this regard. The first evidence-based literature traced back between 1500 to 1800 AD, when at least 2 million sea sailors died due to scurvy (McCord, 1959). James Lind created the first evidence-based written document "A treatise of scurvy" in 1753 where he highlighted the significance of oranges, lemons,

DOI: 10.1201/9781003160663-7

and fresh green vegetables in the cure and prevention of scurvy (Lind, 1757). Since then, it took 175 years for Albert Szent-Gyorgyi to extract vitamin C (unknowingly mentioned sugar-like crystal) from the ox adrenal gland and publish his observation in the Biochemical Journal as "Observation on the function of peroxidase systems and the chemistry of the adrenal cortex: description of a new carbohydrates derivates" (Szent-Györgyi, 1928). Fortunately, W M Haworth elucidated the structure and named it hexuronic acid. Later, King and Waugh (1932) extracted it from lemon juice and referred to it as ascorbic acid and published their observation with the title "The chemical nature of vitamin C" (King & Waugh, 1932). Finally, in 1937, Albert Szent-Gyorgyi was awarded the Nobel prize in Physiology or Medicine for his work on the linkage of biological combustion processes with vitamin C and fumaric acid catalysis (Carpenter, 1988).

Vitamin C or ascorbic acid is a low molecular weight carbohydrate having an ene-diol structure which makes it a natural, universal and essential electron donor. The present ene-diol structure poses degradation against environmental factors (Rucker et al., 2007). Recent advancements in chemical engineering and drug discovery enable researchers to perform structural modifications which resulted in several vitamin C derivatives (Figure 5.1) differing in their hydrophilic nature

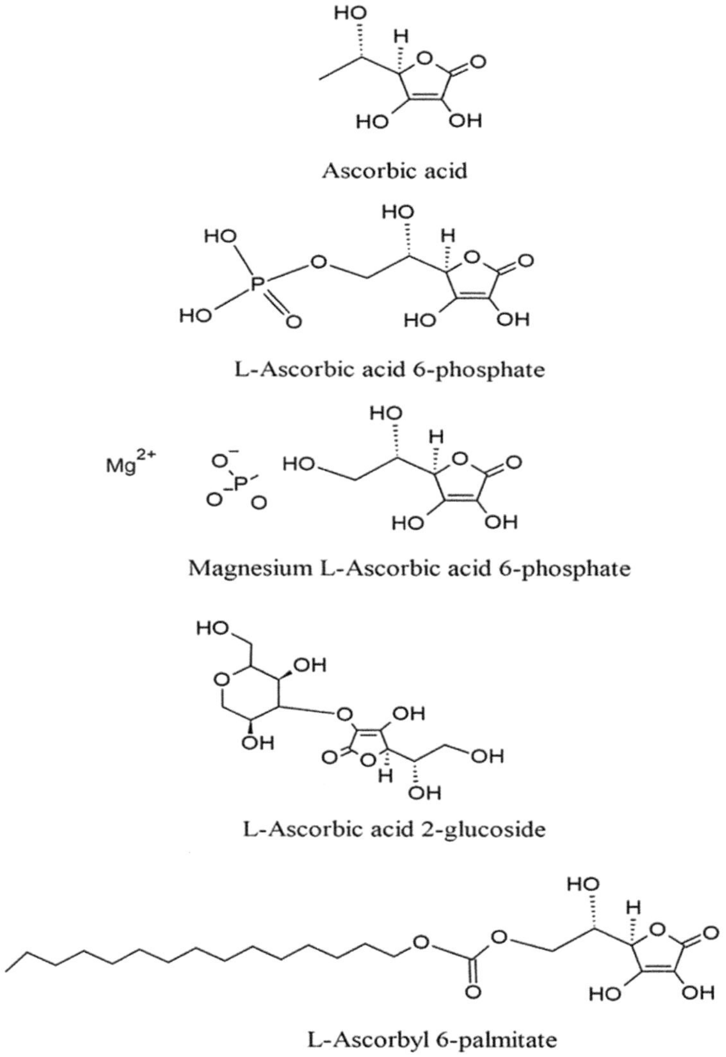

FIGURE 5.1 Vitamin C derivatives.

including (i) hydrophilic ascorbic acid: L-Ascorbic acid 6-phosphate, Magnesium L-Ascorbic acid 6-phosphate and L-Ascorbic acid 2-glucoside (ii) hydrophilic ascorbic acid: L-Ascorbyl 6- palmitate and tetra-isopalmitoyl ascorbic acid (Caritá et al., 2020). These derivatives also differ in their biopotency and can be classified into two classes depending on their potency (Caritá et al., 2020): (i) strongly biopotent: Ascorbic acid 2-O-α-glucoside, 6-Bromo-6-deoxy-l-ascorbic acid, L-ascorbate 2-phosphate and L-ascorbate 2-triphosphate; and (ii) weakly bipotent: L-ascorbyl palmitate, L-ascorbyl-2-sulfate and L-ascorbate-O-methyl ether.

5.3 BIOSYNTHESIS

Glucuronic acid pathway is responsible for ascorbic acid biosynthesis, which is present in probably all green plants as well as higher animal species (Chatterjee, 2009). The essential enzymes required for this pathway reside in the kidney of amphibians, reptiles, and primitive birds (Chatterjee, 1973). Unfortunately, humans lack an essential enzyme (L-gulono-1,4 lactone oxidase) of the ascorbic acid biosynthetic pathway due to evolution; hence, they rely on dietary sources (Chatterjee, 1973; Drouin et al., 2011; Jukes & King, 1975).

The uptake of dietary vitamin C is absorbed through passive diffusion at a higher dose while at a lower dose, it takes place through carrier-mediated active transport (Lykkesfeldt & Tveden-Nyborg, 2019). The efficiency of vitamin C absorption in human adults reaches 80-90% (at < 180 mg/day) while it declines when the dose reaches approximately 1 g/day (Lykkesfeldt & Tveden-Nyborg, 2019).

5.4 SOURCE

Human depends on dietary sources to meet their vitamin C daily requirement. Vegetables and fruits are major contributors (approximately 90%) of vitamin C in the human diet, while rest of the proportion comes from animal-derived sources (Carr & Rowe, 2020). Table 5.1 describes the potential sources of vitamin C.

5.4.1 STABILITY IN FOODS

Susceptibility against environmental factors, the content of vitamin C in most of the food matrix dramatically reduces during its processing as well as storage (Lešková, 2006). Degradation of biologically active vitamin C is attributed to the hydrolytic opening of the lactone ring, hence yielding biologically inactive 2,3- deketogluconic acid (Lešková, 2006). This degradation occurs in the presence of oxygen as well as trace metals and is further stimulated by temperature, heat, and alkaline pH (Table 5.2). This can be correlated with a significant loss in vitamin C observed during storage if food items undergo cooking or any other food processing conditions. For instance, a potato loses its vitamin C by 50% in 5 months and 65% in 8 months of storage. A similar reduction pattern in vitamin C is recorded in the case of apple (50%) and cabbage (40%) during winter storage (Wang et al., 1992). The degree of degradation significantly varies with differences in food processing methods and water activity. For instance, a potato loses its vitamin C when it goes through boiling (Combs & McClung, 2016). In contrast, food processing techniques (especially quick heating methods) may protect vitamin C content as they inactive oxidase enzymes responsible for vitamin C oxidation (Combs & McClung, 2016).

5.4.2 VITAMIN C BIOAVAILABILITY

The bioactivity of any bioactive compound relies on the proportion absorbed rather than the ingested amount. In general, a fraction of the total ingested vitamin is uptaken in its biologically active state, which is referred to as a bioavailable vitamin (Combs & McClung, 2016). The bioavailability

TABLE 5.1
Vitamin C Source and Its Content

Food categories	Food commodity	Vitamin C content (mg/kg)	Reference
Food with animal origin	Milk (cow)	0–10	(Kon & Watson, 1937; Matsui, 2012)
	Milk (human)	50	(Romeu-Nadal et al., 2006)
Vegetables	Apple	50	(Klimczak & Gliszczyńska-Świgło, 2015)
	Banana	90	(Hernández et al., 2006)
	Cherry	70–100	(Pfendt et al., 2003)
	Grapes	418	(Chebrolu et al., 2012)
	Guava	2280	(Bashir & Abu-Goukh, 2003)
	Lemon	530	(Kumar et al., 2013)
	Melon	80–370	(Laur & Tian, 2011)
	Orange	590	(Pfendt et al., 2003)
	Peach	70	(Liu et al., 2015)
	Raspberry	260	(Klesk et al., 2004)
	Rose hips	4260	(Leahu et al., 2014)
	Strawberry	590	(Koyuncu & Dilmaçünal, 2010)
	Tangerine	270	(Izuagie & Izuagie, 2007)
Fruits	Asparagus	60	(Padayatty et al., 2003)
	Broccoli	890	(Somsub et al., 2008)
	Cabbage	370	(Park et al., 2014)
	Carrot	30	(Matějková & Petříková, 2010)
	Celery	30	(Sowbhagya, 2014)
	Collards	350	(Phillips et al., 2010)
	Kale	120	(Padayatty et al., 2003)
	Onion	70	(Colina-Coca et al., 2014)
	Pepper	800–1280	(Padayatty et al., 2003)

of vitamin C is governed by several factors, including quantity ingested, the complexity of the food matrix, degree of processing, and host-related factors. Theoretically, the biological activity of vitamin C in food differing in their matrix remains equivalent to that of purified ascorbic acid if its content remains within the nutritional range (15-200 mg) (Combs & McClung, 2016). High quantity of vitamin C consumption (>1000 mg) results in a decline in its bioavailability (up to 50% efficiency) (Combs & McClung, 2016). Since the biologically inactive (dehydroascorbic acid) can also be reduced during digestion to regain its bioactivity. In contrast to naturally occurring vitamin C, several chemically modified vitamin C derivatives (such as ascorbate 2-monophosphate, ascorbate 2-sulfate, and ascorbate 2-triphosphate; mixtures of these are referred to as ascorbate polyphosphate) offer better bioavailability by providing greater physicochemical stability (Combs & McClung, 2016).

5.5 VITAMIN C DEFICIENCY

5.5.1 INDICATORS FOR VITAMIN C

In general, vitamin C intake is carried out either by dietary assessment or plasma/serum vitamin C levels. Dietary assessment involves Food Frequency Questionnaires and Diet History Questionnaires. Plasma/serum vitamin C level is accepted as a universal biomarker for vitamin C status. Deficiency,

TABLE 5.2
Vitamin C Retention in Fortified Foods during Their Storage

Food categories	Food commodity	Storage conditions	Storage time (days)	Vitamin C retention (%)	References
Milk	Fortified milk	25 °C/ 3-layered packaging material	30	1	(Lešková, 2006)
		25 °C/ 6-layered packaging material	30	49	
		25 °C/ 6-layered packaging material	120	25	
	Fortified milk	4 °C	5	90.6	(Biringen Löker et al., 2003)
	Evaporated milk	23 °C	365	75	(De Ritter, 1976)
Cereal based food	Bread	25°C, polyethylene bags	7	15	(Park et al., 1997b)
	Fibre fortified bread	25 °C/ moisture 45%	7	3	(Park et al., 1997a)
	Bread without fiber	25 °/ moisture 37%	7	14	
	Bread fortified with L-ascorbate 2-monophosphate and reduced Iron	25 °C	6	52	(Wang et al., 1995a)
	Bread fortified with ascorbic acid and reduced Iron	25 °C	6	18	
	Ready-to-eat cereals	23 °C	365	71	(De Ritter, 1976)
	Ready-to-eat cereals	Room temperature	360	60	(Steele, 1976)
	Cereals	40 °C	90		(Anderson et al., 1976)
		22 °C	180		
	Bran flakes	25 °C/7% moisture	30	95	(Wang et al., 1995b)
		40 °C/ 11% moisture	30	20	
Fruit beverages	Strawberry drink	4–6 °C	90	67.7	(Murtaza et al., 2004)
	Yellow passion fruit juice	37 °C	14	0	(Talcott et al., 2003)
	Blood orange juice	4.5 °C	49	25.1	(Choi et al., 2002)
	Powder fruit drinks	21 °C	1	84	(Mehansho et al., 2003)
	Dry fruit drink mix	23 °C	365	94	(De Ritter, 1976)
	Apple juice	23 °C	365	68	(De Ritter, 1976)
	Cranberry juice	23 °C	365	81	
	Grapefruit juice	23 °C	365	81	
	Pineapple juice	23 °C	365	78	
	Grape drink	23 °C	365	76	
	Orange drink	23 °C	365	80	

(Continued)

TABLE 5.2 (CONTINUED)

Vitamin C Retention in Fortified Foods during Their Storage

Food categories	Food commodity	Storage conditions	Storage time (days)	Vitamin C retention (%)	References
Vegetable beverages	Tomato juice	23 °C	365	80	(De Ritter, 1976)
	Vegetable juice 68 0.44	23 °C	365	68	(De Ritter, 1976)
Carbonated drinks	Carbonated beverages	23 °C	365	60	(De Ritter, 1976)
Cola beverages	Cola drinks fortified with ascorbic acid	15 °C	365	83.1	(De Ritter, 1976)
	Cola drinks fortified with L-ascorbate 2-monophosphate	15 °C	365	97	
	Cola drinks fortified with L-ascorbate 2-polyphosphate	15 °C	365	97.7	
	Cola drinks fortified with ascorbic acid	25 °	365	70	
	Cola drinks fortified with L-ascorbate 2-monophosphate	25 °	365	90	
	Cola drinks fortified with L-ascorbate 2-polyphosphate	25 °	365	95.4	
	Cola drinks fortified with ascorbic acid	35 °C	365	63.8	
	Cola drinks fortified with L-ascorbate 2-monophosphate	35 °C	365	68.4	
	Cola drinks fortified with L-ascorbate 2-polyphosphate	35 °C	365	93.8	
Coffee product	Cocoa powder	23 °C	365	97	(De Ritter, 1976)
Fruit/ vegetable flakes	Dried apple chips	7 °C, RH 45%	270	80.4	(Konopacka & Markowski, 2004)
		18 °C, RH 90%	270	63.1	
	Potato flakes fortified with ascorbic acid	25 °C	129	18	(Wang et al., 1992)
	Potato flakes fortified with L-ascorbate 2-monophosphate,	25 °C	129	88	
	Potato flakes fortified with L-ascorbate 2-polyphosphate	25 °C	129	84	

suboptimal, and sufficiency of vitamin C are defined by its serum level as <11 μmol/L (deficient), ≥11–28 μmol/L (suboptimal), >28 μmol/L (sufficient) (Carr & Rowe, 2020).

5.5.2 VITAMIN C DEFICIENCY ACROSS THE GLOBE

Despite increasing concerns over micronutrient deficiency across the globe., irrespective of geography, ethnicity, lifestyle, and economic status, vitamin deficiency status has never been fully reported (Fain, 2004; Nakade et al., 2020; Rowe & Carr, 2020). Several researchers have attempted to produce an important piece of information by targeting a specific population. For instance, explore the vitamin C deficiency status (Figure 5.2). For instance, the EPIC-Norfolk study conducted a study with more than 22,400 participants (with an age range of 40–79 years) indicates a higher prevalence of vitamin D deficiency in male participants over female participants (Canoy et al., 2005; McCall et al., 2019). Similarly, the National Diet and Nutrition Survey (UK 1994–1995) (Bates et al., 1999) and the MONICA study (Glasgow, Scotland, 1992) demonstrated higher vitamin C deficiency, being 14% (British Population) and 20% (Scottish population) (Wrieden et al., 2000). The high prevalence of vitamin C deficiency in the European population and the USA is reported in several studies with smaller population size(Wrieden et al., 2000). Vitamin C deficiency status becomes more severe in low middle-income countries. For example, the Quinto, Ecuador female population was found to be 60% deficient while Brazil and African populations were also observed to have low vitamin C (Charlton et al., 2005; de Oliveira et al., 2008; García et al., 2012; Halestrap & Scheenstra, 2018; Hamer et al., 2008; Kiondo et al., 2012; Nwagha et al., 2012; Ugwa et al., 2016; Villalpando et al., 2003). A significant proportion of the Indian and Chinese populations also demonstrated hypovitaminosis C (Frankenfeld et al., 2012; Lam et al., 2013; Ravindran et al., 2011).

5.5.3 VITAMIN C DEFICIENCY AND HEALTH CONCERNS

An array of factors govern the vitamin C status and requirements, which are but not limited to host-related factors, environmental factors, food consumption factors, and economic status (Carr & Rowe, 2020; Maxfield & Crane, 2018). Due to its essential multifaceted role in human physiology, its suboptimum level invites both various communicable and non-communicable diseases. Several researchers have demonstrated its essential role in the prevention of diabetes, cardiometabolic disorders, and cancer (Gordon et al., 2020; Nowak, 2021). Literature also advocates a strong correlation with immunological function, hormonal regulation, connective tissue development, neurotransmitter production, and several other functions yet to be explored (Carr & Lykkesfeldt, 2021; Chisnall & Macknight, 2017; Combs & McClung, 2016; Granger & Eck, 2018).

5.5.4 RECOMMENDED DIETARY ALLOWANCE

Though the daily vitamin C requirement varies with geographical locations, economic status, and other host-related factors, the Institute of Medicine has defined 90 mg/day (adult men) and 75 mg/day (women) as recommended dietary allowance (RDA) (Brauchla et al., 2021; Carr & Lykkesfeldt, 2021; Frei & Traber, 2001; Levine et al., 1999; Levine et al., 2001). These doses are established based on the vitamin C intake to retain maximal neutrophil concentration with minimal urinary excretion. Smoking causes depletion of vitamin C in the body, so cigarette lovers are recommended to intake 35 mg of additional vitamin C to maintain the optimal vitamin C serum level (Combs & McClung, 2016; Smith & Hodges, 1987).

5.6 STRATEGY ADOPTED FOR VITAMIN C DEFICIENCY

Researchers and policymakers have developed several strategies to address vitamin deficiency: (i) diet diversification; (ii) biofortification; (iii) food fortification; and (iv) dietary supplements (Maurya et al., 2020b; Maurya et al., 2021; Maurya et al., 2022).

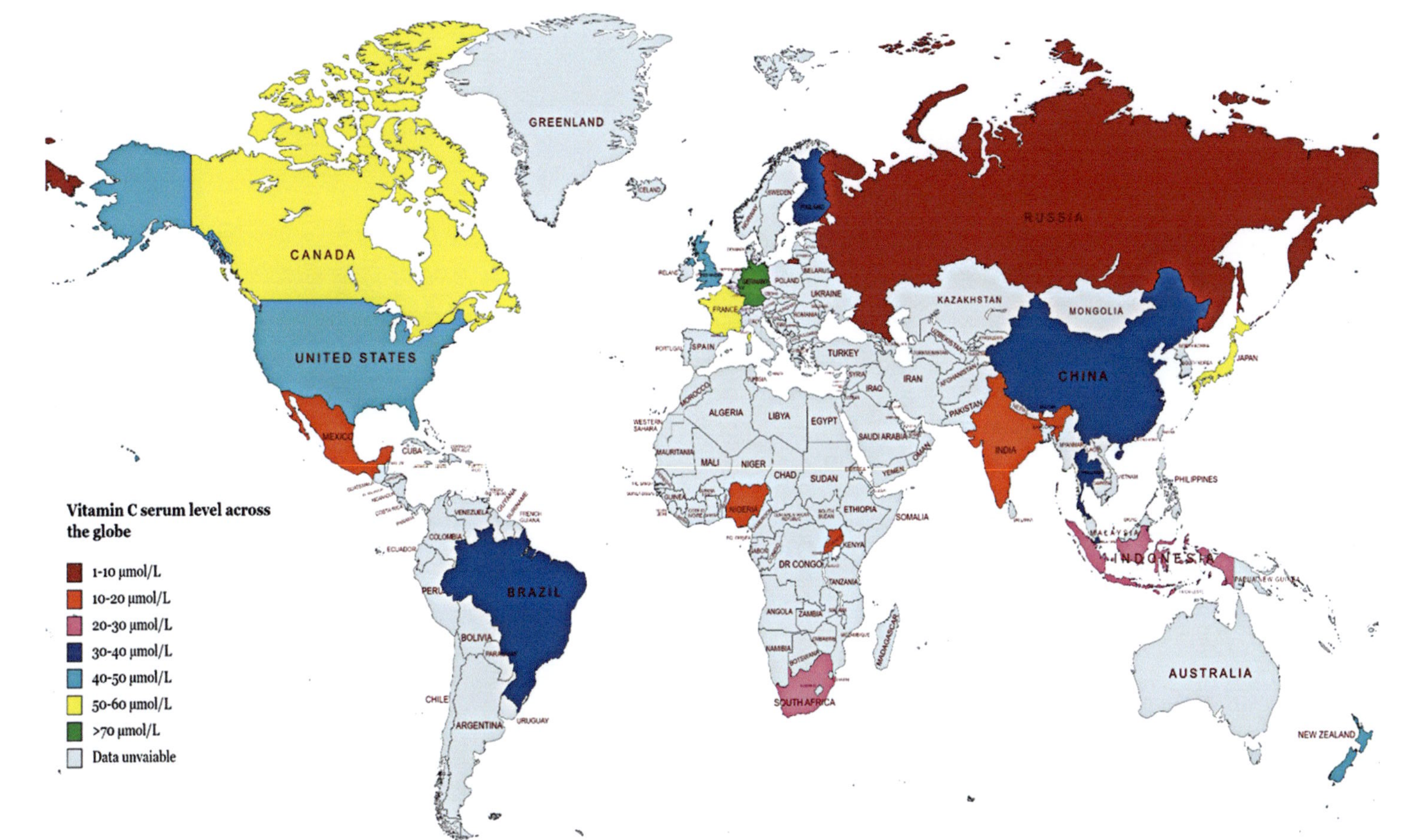

FIGURE 5.2   Vitamin C serum level across the globe.

5.6.1 Diet Diversification

Diet diversification method.s recommend the inclusion of vitamin C-rich food items (lemon, oranges, kiwi, and other fruits and vegetables) in the daily diet (Marik & Liggett, 2019; Nair et al., 2016; Ojiewo et al., 2013). Though this method does not involve additional efforts and special formulation, its success relies on an array of factors including regional food consumption patterns, food accessibility, lifestyle, socioeconomic status, and religious boundaries (Caritá et al., 2020; Maurya et al., 2020b; Maurya et al., 2022). In addition, consumption of vitamin C-rich food items exhibiting higher bioavailability is key to its success (Caritá et al., 2020). Spreading awareness to the general population regarding the significance of diet diversity on nutrition may help in uprooting the vitamin C deficiency.

5.6.2 Supplementation

Today, changes in lifestyle have compelled the population to incline toward supplements as an alternative to food due to time constraints (Lykkesfeldt & Poulsen, 2010; Spoelstra-de Man et al., 2018). Supplements are pharmaceutical formulations that contain a high dose of vitamin C. These supplements involve several non-food grade excipients and high concentrations of vitamin C or its derivatives, the long-term impacts of which on human health are still a matter of concern. While designing vitamin C supplements, the primary emphasis is given to its pharmacology, pharmacokinetics, and bioavailability regardless of several factors cum food matrix.

5.6.3 Food fortification: Need and Opportunity

Food fortification practice is recognized as the most effective, safe, and cheapest method to address nutritional deficiency. This method becomes even more effective if a food distribution network already exists (Maurya et al., 2020b; Maurya et al., 2022). Incorporation of vitamin C into the food matrix poses several challenges to food technologists as well as the industry, which could be attributed to a lack of awareness about the role of nutrients in human health, insufficient availability of data on the existing food basket and consumption patterns, and finally lack of technology to design and produce reasonably fortified food without affecting its organoleptic attributes (Dwyer et al., 2015; Korzeniowska et al., 2017).

The increasing prevalence of vitamin C across the globe. and the significance of functional foods created a demand for the development of vitamin C fortified foods. Before opting for vitamin C fortification, food processors need to have a better understanding of given points: target population (for mass population or specific groups like children/pregnant women), vitamin C serum level of the target population, RDA of the targeted population, the gap between current vitamin C intake and RDA, potency of available vitamin C derivatives, screening of the most suitable food matrix, degradation pattern of vitamin C derivatives in a target food matrix, compatibility of check between selected vitamin C derivative and target food matrix, and currently available fortification method.s.

5.7 CHALLENGES AND OPPORTUNITIES IN VITAMIN C FORTIFICATION

Several practices have been adopted by food technologists to fortify micronutrients in food. Some of the most adopted methods are mixing, homogenization, dissolution, and extrusion. Several food matrices have been fortified with vitamin C, which is reviewed in Table 5.2. A significant loss in vitamin C content has been witnessed in fortified foods, which have been compiled in Table 5.2. Further, environmental factors, trace metals, and food processing also stimulate its degradation (Abe-Matsumoto et al., 2020; Burch, 2011; Crotti et al., 2008; Cvetković & Jokanović, 2009; Dennison & Kirk, 1982; Dennison & Kirk, 1978; Godoy et al., 2021; Ilic & Ashoor, 1988; Kirk et al., 1977; Lipasek et al., 2011; Majid et al., 2019; Masamba & Mndalira, 2013; Ottaway, 1993; Peleg,

2017; Polydera et al., 2003; Rahman et al., 2015; Remini et al., 2015; Tikekar et al., 2011; Zhang et al., 2015). These factors pose negative impacts on vitamin C bioavailability in fortified foods, hence limiting its health-promoting activity. Fortunately, all these limitations faced during vitamin C fortification can be easily rectified using encapsulation techniques, which offer several delivery systems with desired functionalities.

5.8 ENCAPSULATION TECHNOLOGIES

Today, a range of food items is being fortified with vitamin C, but its direct incorporation into the food matrix may cause inevitable interactions that may result in compromising its bioavailability, taste, quality, overall acceptability, and marketability (Dwyer et al., 2015; Maurya et al., 2020b; Maurya et al., 2022). Experience gained in the preparation of pharmaceutical formulations, food processors are now adopting several encapsulation techniques to encapsulate, protect, and deliver vitamin C with given objectives: (i) avoid the inevitable interaction between vitamin C and components present in the food matrix; (ii) protect vitamin C against environmental factors (pH, temperature, moisture, and oxidation) and food processing conditions; (iii) ensure enhanced bioavailability by offering the controlled and site-specific release mechanism.

There are an array of encapsulation techniques utilized to encapsulate and stabilize vitamin C (Abbas et al., 2012; Comunian et al., 2020). Regrettably, there is no universal stabilization method available that can perform universally in all types of food matrices differing in their complexity, composition, and internal environments. Further, data pertaining to vitamin C encapsulation is either partial or discrete to make a clear understanding of vitamin C encapsulation. Furthermore, the use of encapsulated Vitamin C for food fortification purposes is even rarer. So, in this section, we have attempted to put all discrete/partial information available on vitamin C encapsulation on one common platform. The encapsulation technique. adopted for vitamin C encapsulation so far can be grouped into two classes: (i) Microencapsulation techniques and (ii) Nanoencapsulation techniques.

5.8.1 Microencapsulation techniques

The selection of microencapsulation technique. solely depends on several factors: physiological properties of bioactive, encapsulation condition, target food, and release mechanism for release of the bioactive core. Encapsulation techniques reported so far in the literature include spray drying, fluidized bed coating, spray chilling/cooling, extrusion, and liposome are more likely to be selected for vitamin C encapsulation for fortification purposes (Table 5.3).

5.8.1.1 Spray Drying

Spray drying is one of the oldest known encapsulation techniques explored to encapsulate bioactive compounds. To encapsulate vitamin C, it obligates the solubilization of bioactive compounds in the dispersion encompassing wall material (Coimbra et al., 2021). Then this homogenized dispersion system needs to be fed to the spray dryer and nebulized by hot air resulting in the formation of microparticles due to water evaporation (Figure 5.3). In spray drying, the encapsulation process is governed by several factors, including consistency of the dispersion system, the rheology of the dispersion system, quantity and quality of emulsifier applied, inlet and outlet temperature, feed rate, and flow rate of hot air supplied (Rezvankhah et al., 2020). Several researchers have described the mechanism and its potential to encapsulate food bioactive in their review (Furuta & Neoh, 2021; Jafari et al., 2021; Mudalip et al., 2021; Piñón-Balderrama et al., 2020; Samborska et al., 2022). Despite having enormous potential, spray drying also inherits several challenges, for instance, poor encapsulation efficiency, porous end products, the inclusion of high temperature, vitamin C sensitivity, and premature release of encapsulated Vitamin C.

TABLE 5.3
Microencapsulation Technique Adopted for Vitamin C Encapsulation

Encapsulation techniques	Wall material	Particle characterization				Potential application	References
		Particle size	Encapsulation efficiency	Release behaviours	Stability/Morphology		
Spray drying	Casein	5.8–14.8		Fast release	Stable at low pH. The resulting product is irregular and porous	Food and infant formula	(Yan et al., 2021)
	Sodium alginate and chitosan, modified chitosan	3 μm	41.8 to 55.6%	Sustained release	Rough surface: Microparticle derived with chitosan	Pharmaceutical and food	(Estevinho et al., 2016)
	Chitosan and tripolyphosphate	4.1–7.3	58.3–68.7	Fast release	Spherical smooth surface microparticle	Pharmaceutical and food	(Desai & Park, 2006)
	Chitosan	6.1–9.0	45.5–58.30	Fast release	Spherical smooth surface microparticle	Pharmaceutical	(Desai & Park, 2005)
	Pea protein isolates, cowpea protein isolates	1.23–8.37			High vitamin C retention (65 - 69.30%)/ Irregular shape	Food application	(Pereira et al., 2009)
	Gum Arabic and modified starch	1087 μm–1245 μm		Shown controlled release of AA during *in vitro* digestion	Offered high vitamin C retention during the storage period (9 weeks)	Pharmaceutical and food	(Leyva-López et al., 2019)
	Taro starch	14.5 μm–18.7 μm	20.9 ± 0.30%.		High retention 80% after 6 weeks storage	Nutraceutical supplements	(Hoyos-Leyva et al., 2018)
	Arabic gum	9.3	>97		Microparticle offered 17% higher retention than that of vitamin C compared to free vitamin C	Encapsulation of bioactive for bakery products	(Alvim et al., 2016)
	Eudragit® RL, L and RS.		>95	Slow release		Pharmaceutical	(Esposito et al., 2002)
	Maltodextrin and starch	4.75–7.6	100		High vitamin C retention (81-85%) after 60 days at room temperate Irregular and porous	Pharmaceuticals	(Finotelli & Rocha-Leão, 2005)
	Maltodextrin and gum Arabic		>95		High retention after 300 days	Encapsulation of bioactive for bakery products	(Righetto & Netto, 2006)
	Pea protein and sodium-carboxymethylcellulose	1.83–8.21	>84	Fast release	Pea protein microparticle: Quite irregular, shriveled, and rough sodium-carboxymethylcellulose homogeneous and smooth	Food and Pharmaceuticals	(Pierucci et al., 2006)
	Starch, gum arabic, and gelatin	8.0–20.5	10,30		High vitamin C stability at ambient condition/Polyhedric microcapsules		(Trindade & Grosso, 2000)
	Sodium Alginate and Gum Arabic	2.88–14.09 μm	> 90%		Spherical regular shape/ Stable at higher temperature (188 °C)	Nutraceutical supplements and food fortification	(Barra et al., 2019)

(Continued)

TABLE 5.3 (CONTINUED)
Microencapsulation Technique Adopted for Vitamin C Encapsulation

Encapsulation techniques	Wall material	Particle characterization				Potential application	References
		Particle size	Encapsulation efficiency	Release behaviours	Stability/Morphology		
Spray chilling/spray cooling	Hydrogenated vegetable fat and stearic acid	31.2 μm	97.8		Microparticle have shown 13% higher retention than that of vitamin C compared to free vitamin C	Suitable for bakery products	(Osman et al., 2016)
	Oleic acid (OA) and lauric acid (LA)	18–67 μm	89–98	Slow release in aqueous medium	Microparticle were present in agglomerates	Nutraceutical supplements and food fortification	(Sartori et al., 2015)
	Palm oil and hydrogenated palm oil	98–181 μm	80.22–93.51	slow controlled release behavior	Crystalline microparticle	Nutraceutical supplements and food fortification	(Carvalho et al., 2019)
	Palm oil and hydrogenated palm oil	84.63 ± 1.20 μm			74.25%–83.07% vitamin C retention after 45 days	Nutraceutical supplements and food fortification	(Carvalho et al., 2021b)
Complex coacervation	Gum Arabic and gelatin	52–84	98	Controlled release under defined conditions	32 to 44% vitamin C retention after 34 days storage	Nutraceutical supplements and food fortification	(Comunian et al., 2013)
Supercritical fluid (SC-CO$_2$) assisted encapsulation	Vitamin E and liposomes	0.911	32.97		High emulsion stability under cold storage for 20 days	Nutraceutical supplements and food fortification	(Tsai & Rizvi, 2017)
Microchannel emulsification	Soybean oil	15–18		High bioavailability	Narrow size distribution	Nutraceutical supplements and food fortification	(Khalid et al., 2014)
Microfluidic technique	Chitosan and Na$_2$CO$_3$/palm fat	195–343	73.4–96.6		High Vitamin C retention (56–99%) at 4°C High Vitamin C retention (46–98%) at 20°C	fortified food products	(Comunian et al., 2014)
Fluidized bed coating	Ethylcellulose/Polymethacrylate/ waxy coating material	>315		Microparticle having Al-stearate showed the best release profile	Agglomeration of microparticle	Pharmaceutical	(Knezevic et al., 1998)
Liposome	Cholesterol,DL-α-tocopherol and phosphatidylcholine		53–55	Controlled release behaviour	Multilamellar microparticles	Infant food formulations	(Kirby et al., 1991)
	Milk-based phospholipids	1.0	10		High retention under cold conditions/ resistant to pH variation/ unilamellar microparticle	Food applications	(Farhang et al., 2012)
	DL-α-tocopherol, egg phosphatidylcholine and cholesterol	0.2–1.0	59		Stable against pasteurization	Milk fortification	(Sharma & Lal, 2005)
Melt extrusion	Maltodextrin	500–1000		Sustained release	High vitamin C retention (70%)/ Crystalline	Bakery products	(Bouquerand, 2007)
	Maltodextrin	500–1000	96	Sustained release	High retention/Large particle size	Food fortification	(Chang et al., 2010)
	Fructo-oligosaccharide	300–1000		Sustained release	Provide high stability to encapsulated Vitamin C/crystalline	Fortificaiotn of low moisture containing foods	(Comunian et al., 2020)
Melt dispersion	Carnauba wax	~50	<100		Small size capsules/Porous microparticles	Food fortification	(Uddin et al., 2001)
Emulsion solvent evaporation	Ethylcellulose				ough and flexible microparticles/less porous microparticles	Food fortification	(Uddin et al., 2001)
	Arabic gum and maltodextrin	55–107		Controlled release	Crystalline	Food formulations	(Özdemir & Gökmen, 2015)

(Continued)

TABLE 5.3 (CONTINUED)
Microencapsulation Technique Adopted for Vitamin C Encapsulation

		Particle characterization					
Encapsulation techniques	Wall material	Particle size	Encapsulation efficiency	Release behaviours	Stability/Morphology	Potential application	References
Pickering emulsions	Modified cellulose and chitosan	620 nm	90.3	Controlled release	Susceptible to degradation and are used in Pickering emulsion	Pharmaceuticals	(Baek et al., 2021)
Emulsions and coacervation	Gelatin and sodium caseinate system		65–97	Controlled release	Irregular and porous microparticles	Pharmaceutical/ nutraceutical	(Fraj et al., 2021)
Spray coating	Polyacylglycerol monostearate		80.7–94.2	Slow release (9.2% after 12 d) in beverages	Improve retention oxidation and moisture	Beverage/ Milk fortification	(Lee et al., 2004)
	Medium-chain triacylglycerol	2–5	88.9–95.0	Higher degree of release	High protection against oxidation	Beverage/ Milk fortification	(Lee et al., 2003)
Co-crystallization	Lactose and sucrose	2–30	>90		Low drug loading capacity with over 90% stability. This method offers low drug loading capacity, but high stability/ crystalline	Food fortification	(Kim et al., 2001)
Immobilization/dispersion	Sodium alginate and hydrated zinc oxide	359	Sustained release (~90% release after 6h)		Enhanced stability/gel like structure	Food fortification	(Desai et al., 2005)
Cross-Linking and Coacervation	Chitosan and Alginate	2.6	Controlled release in GIT		Stable against pH/acidic condition	Pharmaceuticals	(Han et al., 2008)

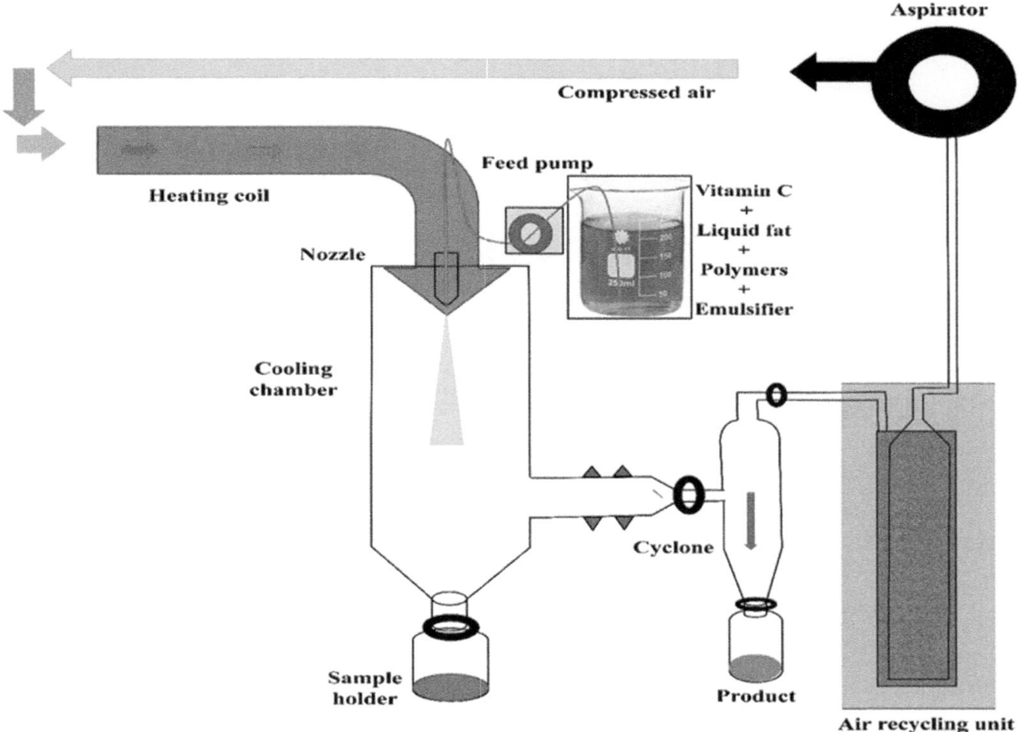

FIGURE 5.3 Schematic diagram of spray drying encapsulation method).

5.8.1.2 Spray Chilling/Cooling

Spraying chilling/cooling has great similarity to spray drying, the only difference is that it mandates lipid in the molten phase. Nevertheless, blended dispersion containing vitamin C is nebulized through heated nozzles using cooled air (Figure 5.4). Spray chilling/cooling is more compatible for encapsulating heat-labile bioactives. Additionally, the resultant solid lipid microparticles demonstrate high reproducibility, improved bioavailability, stability to bioactive and great potential for scaling up (Jaskulski et al., 2017; Okuro et al., 2013; Oxley, 2012). Here are a few reports available on the use of the spray chilling technique for vitamin C encapsulation. For instance, Alvin et al. (2016) attempted vitamin C encapsulation by spray chilling method using stearic acid hydrogenated vegetable fat and studied the stability of vitamin stability in biscuits fortified with developed microparticles. Unlikely, the resultant microparticles of spray chilling remain mostly non-porous as well as resistant to agitation, hence offering an effective barrier against moisture and oxygen.

5.8.1.3 Fluidized Bed Coating

Fluidized bed coating encapsulation method involves dispersing bioactive compounds in hot molten polymeric wall materials like wax or fat along with hot airflow, leading to the development of a coating around a bioactive core (Coronel-Aguilera & San Martín-González, 2015; Dewettinck & Huyghebaert, 1999; Dewettinck & Huyghebaert, 1998; Guignon et al., 2002). The involvement of hot air does make this technique susceptible to degradation.

5.8.1.4 Melt Extrusion

This encapsulation technique. involves the dispersion of bioactive compounds into a molten mass of carbohydrate-based wall material followed by extruding the resultant dispersion through an

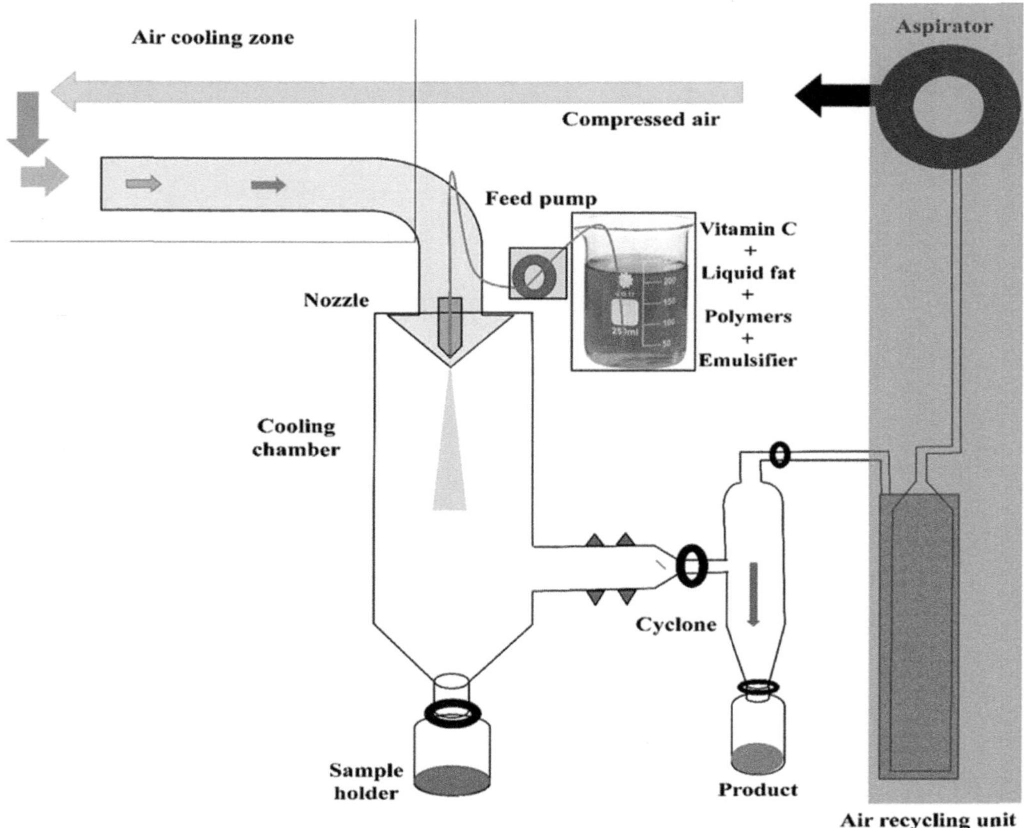

FIGURE 5.4   Schematic diagram of spray chilling or spray cooling encapsulation technique.

extruder barrel under high temperature and pressure (Chokshi & Zia, 2004; Repka et al., 2012). The last stage involves cooling of the resultant microparticles and particle size reduction.

5.8.1.5   Melt Dispersion

Melt dispersion encapsulation method relies on the emulsification of molten polymeric wall material into an aqueous phase under continuous stirring, followed by solidifying the microparticle in an ice bath or cold water. The exclusion of organic solvent for encapsulation makes the melt dispersion method highly suitable for vitamin C encapsulation (Carvalho et al., 2021a; İvedi et al., 2022). However, only single study has been carried out to evaluate the efficiency of the melt dispersion method to encapsulate vitamin C and observed that the obtained microparticles had high encapsulation efficiency, sustained-release kinetics, and non-porous in nature. It is also noted that the melt dispersion method is an easy technique compared to spray drying but remains costly and does not provide better control over particle structure and morphology.

5.8.1.6   Emulsion Solvent Evaporation

This encapsulation technique. involves the formation of a homogeneous mixture of ascorbic acid and polymeric wall materials (like ethylcellulose) dissolved in an organic solvent. The obtained homogeneous mixture is then fed into a continuous phase (aqueous/organic phase) dropwise, followed by the evaporation of the organic solvent (Kim et al., 2002; O'Donnell & McGinity, 1997; Rosca et al., 2004). The resultant microparticle is then separated from the continuous phase, followed by drying.

The inclusion of the organic phase makes the emulsion solvent evaporation technique unsuitable for food fortification purposes, yet some researchers have developed vitamin C loaded nanoparticles using emulsion solvent evaporation techniques.

5.8.1.7 Co-Crystallization

Co-crystallization encapsulation techniques rely on the preparation of a supersaturated mixture of sucrose syrup and bioactive compounds at elevated temperature to prevent crystallization, where the bioactive core acts as a seed for triggering nucleation when the mixture is allowed to cool down (Chezanoglou & Goula, 2021; Quast et al., 2020). Despite being easy-to-use, the inclusion of sucrose makes it unpopular in the food industry due to sugar-linked health concerns.

5.8.2 Nanoencapsulation Techniques for Vitamin C

The notorious role of nanoparticles makes nanotechnology the most widely adopted multidisciplinary approach in different fields such as agriculture, pharmaceuticals, cosmetics, and food as well. Nanotechnology has allowed researchers to design and carve nanoparticles with improved functionalities, including enhanced stability against environmental factors (pH, oxidation, thermal, and enzymatic degradation) and food processing conditions, improved bioavailability, and prevention of inevitable interaction with food components (Caritá et al., 2020; Comunian et al., 2020; Walia et al., 2019). Literature reports an array of nanoencapsulation techniques that allow the researcher to carve a range of nanodelivery systems that differ in their functionalities (Table 5.4). Ideally, a nanodelivery system should encapsulate, protect against all kinds of processing conditions and environmental factors, deliver bioactive compounds with desirable bioavailability without affecting the organoleptic properties of the target food matrix, and finally should be fabricated with food-grade materials.

5.8.2.1 Lipid-based Nanodelivery System

This class includes a range of nanodelivery systems, which include nanoemulsion, solid lipid carrier, nanoliposome, and nanostructured lipid carriers. These lipid carriers are generally exploited to enhance vitamin C lipophilicity within the target food matrix.

5.8.2.2 Liposome/Nanoliposome

These nanoparticles are designed in such a way that they can encapsulate both hydrophilic (in the inner cavity) as well as lipophilic (in the lipid bilayer) bioactives (Figure 5.5). The literature reports a range of fabrication methods for liposomes (Ajeeshkumar et al., 2021; Bozzuto & Molinari, 2015; Daeihamed et al., 2017; Esposto et al., 2021; Maherani et al., 2011). In general, these fluid spherical nanostructures either encompass an aqueous core by a single bilayer (unilamellar liposomes) or multiple lipid bilayers (multilamellar liposomes). A liposome having a smaller particle size (≤ 200 nm) is also referred to as nanoliposome. The ability to accommodate both lipophilic as well as hydrophilic bioactives makes liposomes the most exploited delivery system for bioactive encapsulation. In addition to this, it remains the most biocompatible delivery system as it resembles the natural plasma membrane. Several researchers have encapsulated Vitamin C into liposomes in order to evaluate its potential to encapsulate, protect, and deliver vitamin C (Kirby et al., 1991; Sharma & Lal, 2005) and have also used it for milk fortification purposes (Farhang et al., 2012). Vitamin C fortification using liposomes becomes more effective when it is intended for a food matrix high in its moisture content. The literature also reports the use of nanoliposomes for vitamin C encapsulation. However, the encapsulation of vitamin C in liposomes/nanoliposomes confirmed high physicochemical stability, but its utilization for food fortification remained intact. This could be attributed to its inclusion of soya-based phospholipids, which inherit intense flavour.

TABLE 5.4

Nanodelivery Systems Adopted for Vitamin C Encapsulation

Class of nanodelivery system	Nanodelivery system	Fabrication process	Wall materials	Particle characterization		Key outcomes	Reference
				Particles size	Encapsulation efficiency		
Lipid derived nanodelivery system	Nanoliposome	Film evaporation and micro fluidization	Soy phosphatidylcholine	70–130	48–50	Enhanced vitamin C stability Reduced lipid oxidation, agglomeration, and premature release of encapsulated Vitamin C Improve physicochemical stability	(Zhou et al., 2014)
		Extrusion	Soy phospholipid Krill		<100	Multilamellar liposomes demonstrated higher stability than unilamellar	(Monroig et al., 2007)
		Dehydration–rehydration	Soy phosphatidylcholine	140–220	38	High potential for food fortification	(Marsanasco et al., 2011)
		Micro fluidization	Soy phosphatidylcholine	~100	~62	Vitamin C stability can be enhanced by the addition of sucrose and applying freeze drying	(Yang et al., 2013)
		Film hydration-ultrasonication	Lecithin	373	42	Highly stable nanoparticle	(Yang et al., 2010)
		Hydration with extrusion	Hydrogenated soy phosphatidylcholine	<120	~100	Boosted antitumor activity	(Lipka et al., 2013)
	Nanostructured lipid carrier	High pressure homogenization	Witepsol®, Miglyol 812®tegocare450®Carbopol 940®	221	71.1	High stability under cold condition	(Üner et al., 2005)
		High pressure homogenization	Labrasol, Tristearin Phospholipid-90NG	268	87	Offer great drug target delivery	(Jain et al., 2016)
	Solid lipid carrier	High pressure homogenization		228	67.6	High stability under cold condition	(Üner et al., 2005)

(Continued)

TABLE 5.4 (CONTINUED)

Nanodelivery Systems Adopted for Vitamin C Encapsulation

Class of nanodelivery system	Nanodelivery system	Fabrication process	Wall materials	Particle characterization		Key outcomes	Reference
				Particles size	Encapsulation efficiency		
Polymer derived nanodelivery system	Chitosan nanoparticles	Ionic gelation	Chitosan Sodium tripolyphosphate	90-350	~60	Ph controlled release Fast release in phosphate buffer Slow release in hcl	(Alishahi et al., 2011)
		Oil-in-water emulsions and ionic gelation	Chitosan Sodium tripolyphosphate	30–100	39–77	Better encapsulation efficiency	(Yoksan et al., 2010)
		Ionic gelation	Chitosan Sodium tripolyphosphate	186–201	10–12	Improved stability against heat processing	(Jang & Lee, 2008)
		Chitosan	Ionic gelation	375–503	83–89	High vitamin C encapsulation Enhanced shelf life	(Aresta et al., 2013)
		N,N,N-trimethyl chitosan	Ionic gelation	~530	N/A	Enhanced vitamin C stability	(de Britto et al., 2012)
		Chitosan Sodium tripolyphosphate	Ionic gelation	185	~50	Controlled release	(Alishahi et al., 2011)
	Starch nanoparticles	Potato starch	Ultrasonication	N/A	42–80	High stability against heat processing	(Shabana et al., 2019)
	Nanofiber	Polyvinyl alcohol	Electrospinning process	50	NA	Porous in nature Fast release of encapsulated Vitamin C	(Tehrani & Amiri, 2022)

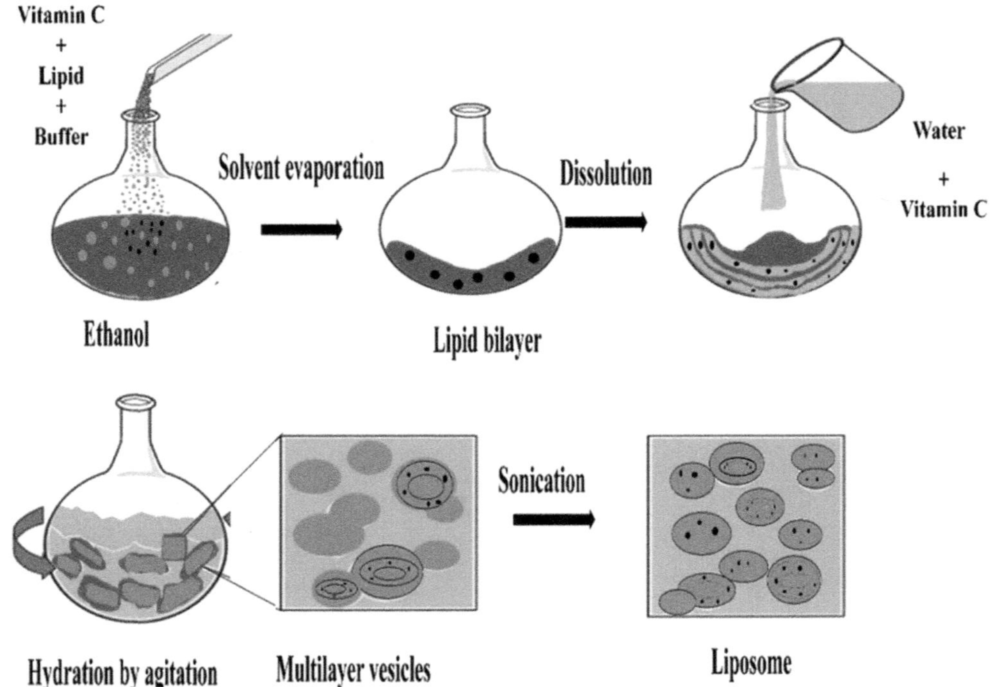

FIGURE 5.5   Schematic diagram for liposome encapsulation methods.

5.8.2.3 Nanoemulsion

Emulsification involves the blending of at least two immiscible phases (lipid and water) where one phase is distributed as small spherical droplets in another phase (Figure 5.6). Based on the spatial arrangement of these two phases, the emulsion system can be classified as oil in water (O/W) or water in oil (W/O). The stabilization of the above-mentioned immiscible phases is generally facilitated by the use of surfactants and emulsifiers (Loveday & Singh, 2008). The literature describes the emulsion system as one of the most widely utilized encapsulation techniques for vitamins. Researchers have created several nanoemulsions that differ in their composition, fabrication process, and potency to encapsulate and deliver vitamin C. The choice of nanoemulsion for vitamin C encapsulation is primarily governed by the quality and quantity of the carrier oil and surfactant.

5.8.2.4 Solid-Lipid Nanoparticles

This delivery system seems to be the most appropriate means for vitamin encapsulation as it is the hybrid structure of a liposome and emulsion system hence offering an array of benefits such as greater encapsulation efficiency, a high drug loading capacity, and greater physicochemical stability against food processing conditions as well as environmental factors encountered during processing and storage. The literature describes a range of fabrication methods for solid lipid nanoparticles (Basha et al., 2021; Cruz et al., 2015; Hou et al., 2003; Mehnert & Mäder, 2012). Researchers have tried to evaluate its potential to encapsulate, protect, and deliver vitamin C (Table 5.4). Exclusion of encapsulated bioactive during rearrangement of the crystalline lattice of solid fat used and poor drug loading capacity made it the least popular delivery system.

5.8.2.5 Nanostructured Lipid Carriers

This nano delivery system resolves the limitation associated with solid lipid carriers by replacing solid lipid material with a blend of solid and liquid lipid materials. Literature dictates several

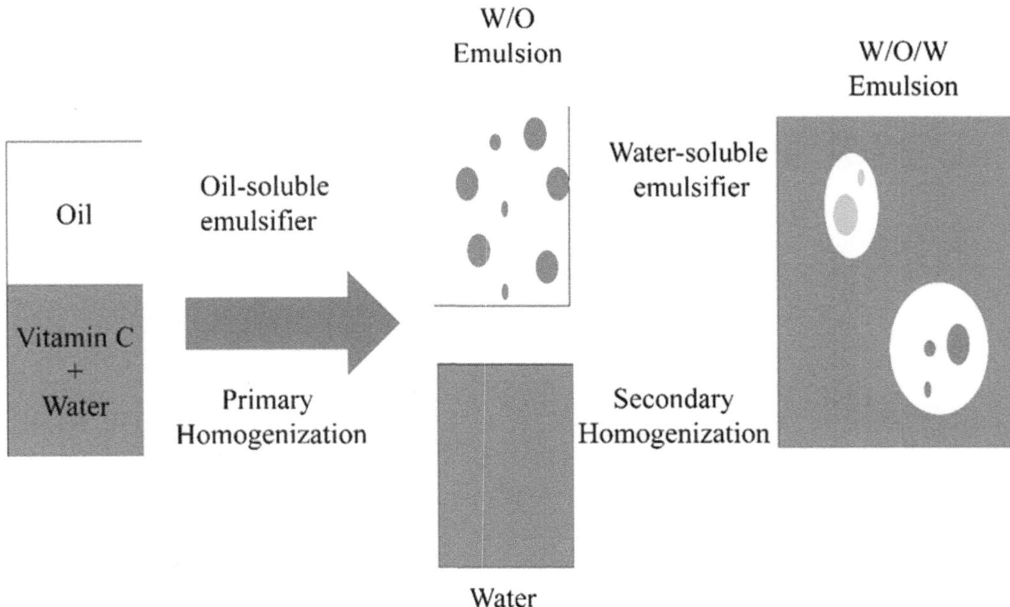

FIGURE 5.6 Schematic diagram for emulsification method.

fabrication methods for nanostructured lipid carriers, primarily being hot/cold homogenization, microemulsification, supercritical fluid method, solvent displacement injection, solvent emulsification-diffusion method, solvent emulsification-evaporation, melt emulsification, and phase inversion methods (Beloqui et al., 2016; Cruz et al., 2015; Iqbal et al., 2012; Obeidat et al., 2010). Few researchers have encapsulated Vitamin C into a nanostructured lipid carrier (Table 5.4) but its full potential still remain untapped.

5.8.2.6 Polymer-derived Nano Delivery Systems

In the pharmaceutical field, researchers have exploited the inherent diversity of biopolymers to design an array of nano delivery systems to encapsulate drugs differing in their polarity, hydrophobicity, suitability, and physicochemical stability. These are the most common delivery systems of this class. The literature describes several fabrication methods for these polymeric nanoparticles (Akbari-Alavijeh et al., 2020; Garavand et al., 2022; Sundar et al., 2010). Researchers have attempted vitamin C encapsulation in nanospheres, nanocapsules, nanohydrogels, and nanofibers. The porous nature of the end product, involvement of organic solvent, inclusion of non-food grade ingredients, and poor control on its reproducibility make polymeric nanoparticles unpopular in vitamin C encapsulation, especially when it comes to food fortification (Maurya et al., 2020a; Maurya et al., 2022).

5.8.2.7 Nature-Inspired Nano Delivery System

These nano delivery systems are referred to as those nanoparticles that have the ability to accommodate bioactive compounds in their naturally existing cavity or hydrophilic/hydrophobic patches (Kumar & Singhal, 2021). Literature dictates several nature inspired nano delivery systems, primarily cyclodextrin, caseins, and amylose nanoparticles (Kumar & Singhal, 2021; Narayanan et al., 2022). Casein has the ability to form micelles that can accommodate vitamin C in their internal cavity. Similarly, cyclodextrin also possesses an internal void where it can encapsulate bioactive compounds like vitamin C (Table 5.4). The porous structure resultant product makes encapsulated Vitamin C susceptible to environmental factors as well as food processing conditions encountered during its preparation, food fortification, and storage.

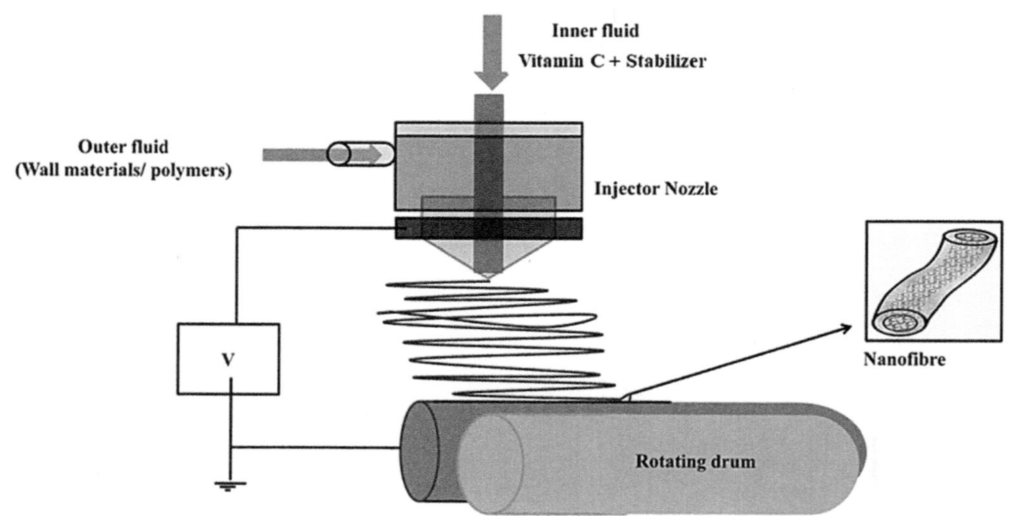

FIGURE 5.7　Schematic diagram for electrospinning encapsulation technique.

5.8.2.8　Special Equipment-based Nano Delivery Systems

Literature dictates novel nanoparticles which are fabricated with special types of equipment such as spraying, nano spray dryer, and electrospinning. In contrast to spray drying techniques, the nano spray drying technique provides nanoparticles with a particle size in the nano range and a narrow dispersity range. The unique properties can be harnessed to design and develop novel nanoparticles that can offer better encapsulation, higher protection, and enhanced bioavailability for vitamin C. Similarly, the electrospinning technique is an extremely adaptable technique that exploits electrostatic force, spraying, and spinning to develop novel nanoparticles with tuned functionalities (Figure 5.7). Melted polymeric material containing bioactive compounds is fed to the spinning disc, which acts as one of the electrodes hence applying a high voltage on droplets leaving the spinning disc (Wen et al., 2017). Though these specially equipped based encapsulation techniques seem to be an ideal option for vitamin C encapsulation due to the exclusion of organic solvents in the process and high physicochemical for encapsulated bioactive, but low production rate and lack of reproducibility make them unpopular in food industries.

5.9　EXPERIENCE IN FOOD FORTIFICATION WITH ENCAPSULATED VITAMIN C

Increasing awareness about the significance of micronutrient deficiency has been a major driving force for designing and development of fortified foods. Several researchers have attempted vitamin encapsulation to overcome the limitations associated with food fortification. The first evidence-based documents on the use of vitamin C encapsulation come from a study where researchers have developed a complex coacervation method to design vitamin C loaded microparticles to fortify chicken frankfurters (He et al., 2017). Another team also developed vitamin C loaded microparticles using the spray chilling technique to fortify chicken frankfurter sausage. Similarly, vitamin C encapsulation was carried out by pseudo-boiling layer using maltodextrin to fortify noncarbonated soft drinks(Diachkova et al., 2019). Another team also formulated vitamin C loaded microparticles using spray drying and spray chilling techniques for developing vitamin C fortified bakery products (Alvim et al., 2016).

5.10 CONCLUSION

Vitamin C has always been a matter of discussion due to its inherent pleiotropic role in human health. However, this versatile compound is one of the most utilized compounds not only in the food sector but also in pharmaceuticals, cosmetics, nutraceuticals, and several others. The physiological role of vitamin C encompasses several biological activities, including prevention of diseases and hypovitaminosis C, skin diseases, tumors, cancer, diabetes, and synthesis of connective tissues, immunoglobulins, osteogenesis, neurotransmitters, and several more. Awareness about micronutrient deficiency, changes in lifestyle and food consumption patterns has created huge demand for functional food with desirable micronutrient bioavailability. The development of functional food fortified with vitamin C faces several limitations due to its instability and susceptibility to environmental factors encountered during processing and storage. Encapsulation techniques seem the most suitable means to address these challenges as they promise encapsulation, better protection, and enhanced bioavailability of vitamin C even after incorporation within the target food matrix. Therefore, it can be concluded that the development of functional foods fortified with vitamin using the encapsulation technique. not only increases the diversity in the present food basket but also helps minimize the magnitude of vitamin C deficiency.

REFERENCES

Abbas, S., Da Wei, C., Hayat, K., Xiaoming, Z. 2012. Ascorbic acid: Microencapsulation techniques and trends—A review. *Food Reviews International*, 28(4), 343–374.

Abe-Matsumoto, L.T., de Araújo, Y.A., Medeiros, M.L. 2020. Stability of vitamin C in enriched jelly. *Brazilian Journal of Analytical Chemistry*, 7(27), 14–19.

Ajeeshkumar, K.K., Aneesh, P.A., Raju, N., Suseela, M., Ravishankar, C.N., Benjakul, S. 2021. Advancements in liposome technology: Preparation techniques and applications in food, functional foods, and bioactive delivery: A review. *Comprehensive Reviews in Food Science and Food Safety*, 20(2), 1280–1306.

Akbari-Alavijeh, S., Shaddel, R., Jafari, S.M. 2020. Encapsulation of food bioactives and nutraceuticals by various chitosan-based nanocarriers. *Food Hydrocolloids*, 105, 105774.

Alishahi, A., Mirvaghefi, A., Tehrani, M.R., Farahmand, H., Koshio, S., Dorkoosh, F.A., Elsabee, M.Z. 2011. Chitosan nanoparticle to carry vitamin C through the gastrointestinal tract and induce the non-specific immunity system of rainbow trout (Oncorhynchus mykiss). *Carbohydrate Polymers*, 86(1), 142–146.

Alvim, I.D., Stein, M.A., Koury, I.P., Dantas, F.B.H., Cruz, C.L.d.C.V. 2016. Comparison between the spray drying and spray chilling microparticles contain ascorbic acid in a baked product application. *LWT - Food Science and Technology*, 65, 689–694.

Anderson, R.H., Maxwell, D.L., Mulley, A.E., Fritsch, C.W. 1976. Effects of processing and storage on micronutrients in breakfast cereals. *Food Technology*.

Aresta, A., Calvano, C.D., Trapani, A., Cellamare, S., Zambonin, C.G., De Giglio, E. 2013. Development and analytical characterization of vitamin(s)-loaded chitosan nanoparticles for potential food packaging applications. *Journal of Nanoparticle Research*, 15(4), 1592.

Baek, J., Ramasamy, M., Willis, N.C., Kim, D.S., Anderson, W.A., Tam, K.C. 2021. Encapsulation and controlled release of vitamin C in modified cellulose nanocrystal/chitosan nanocapsules. *Current Research in Food Science*, 4, 215–223.

Barra, P.A., Márquez, K., Gil-Castell, O., Mujica, J., Ribes-Greus, A., Faccini, M. 2019. Spray-drying performance and thermal stability of L-ascorbic acid microencapsulated with sodium alginate and gum Arabic. *Molecules*, 24(16), 2872.

Basha, S.K., Dhandayuthabani, R., Muzammil, M.S., Kumari, V.S. 2021. Solid lipid nanoparticles for oral drug delivery. *Materials Today: Proceedings*, 36, 313–324.

Bashir, H.A., Abu-Goukh, A.-B.A. 2003. Compositional changes during guava fruit ripening. *Food Chemistry*, 80(4), 557–563.

Bates, C.J., Prentice, A., Cole, T.J., van der Pols, J.C., Doyle, W., Finch, S., Smithers, G., Clarke, P.C. 1999. Micronutrients: Highlights and research challenges from the 1994–5 national diet and nutrition survey of people aged 65 years and over. *British Journal of Nutrition*, 82(1), 7–15.

Beloqui, A., Solinís, M.Á., Rodríguez-Gascón, A., Almeida, A.J., Préat, V. 2016. Nanostructured lipid carriers: Promising drug delivery systems for future clinics. *Nanomedicine: Nanotechnology, Biology and Medicine*, 12(1), 143–161.

Biringen Löker, G., Uğur, M., Yıldız, M. 2003. A parrtial supplementation of pasteurized milk with vitamin C, iron and zinc. *Food/Nahrung*, 47(1), 17–20.

Bouquerand, P. 2007. Extruded glassy vitamin C particles. European Patent, 1836902, 26.

Bozzuto, G., Molinari, A. 2015. Liposomes as nanomedical devices. *International Journal of Nanomedicine*, 10, 975–999.

Brauchla, M., Dekker, M.J., Rehm, C.D. 2021. Trends in vitamin C consumption in the United States: 1999–2018. *Nutrients*, 13(2), 420.

Burch, R. 2011. The stability and shelf life of vitamin-fortified foods. In: D. Kilcast, P. Subramaniam (Eds.), *Food and beverage stability and shelf life*. Woodhead Publishing, pp. 743–754.

Canoy, D., Wareham, N., Welch, A., Bingham, S., Luben, R., Day, N., Khaw, K.-T. 2005. Plasma ascorbic acid concentrations and fat distribution in 19 068 British men and women in the European prospective investigation into cancer and nutrition Norfolk cohort study. *The American Journal of Clinical Nutrition*, 82(6), 1203–1209.

Caritá, A.C., Fonseca-Santos, B., Shultz, J.D., Michniak-Kohn, B., Chorilli, M., Leonardi, G.R. 2020. Vitamin C: One compound, several uses: Advances for delivery, efficiency and stability. *Nanomedicine: Nanotechnology, Biology and Medicine*, 24, 102117.

Carpenter, K.J. 1988. *The history of scurvy and vitamin C*. Cambridge University Press.

Carr, A.C., Lykkesfeldt, J. 2021. Discrepancies in global vitamin C recommendations: A review of RDA criteria and underlying health perspectives. *Critical Reviews in Food Science and Nutrition*, 61(5), 742–755.

Carr, A.C., Rowe, S. 2020. Factors affecting vitamin C status and prevalence of deficiency: A global health perspective. *Nutrients*, 12(7), 1963.

Carvalho, A.S.d., Rezende, S.C.d., Caleja, C., Pereira, E., Barros, L., Fernandes, I., Manrique, Y.A., Gonçalves, O.H., Ferreira, I.C.F.R., Barreiro, M.F. 2021a. β-carotene colouring systems based on solid lipid particles produced by hot melt dispersion. *Food Control*, 129, 108262.

Carvalho, J.D.d.S., Oriani, V.B., de Oliveira, G.M., Hubinger, M.D. 2019. Characterization of ascorbic acid microencapsulated by the spray chilling technique using palm oil and fully hydrogenated palm oil. *LWT*, 101, 306–314.

Carvalho, J.D.d.S., Oriani, V.B., de Oliveira, G.M., Hubinger, M.D. 2021b. Solid lipid microparticles loaded with ascorbic acid: Release kinetic profile during thermal stability. *Journal of Food Processing and Preservation*, 45(6), e15557.

Chang, D., Abbas, S., Hayat, K., Xia, S., Zhang, X., Xie, M., Kim, J.M. 2010. Encapsulation of ascorbic acid in amorphous maltodextrin employing extrusion as affected by matrix/core ratio and water content. *International Journal of Food Science and Technology*, 45(9), 1895–1901.

Charlton, K.E., Kolbe-Alexander, T.L., Nel, J.H. 2005. Micronutrient dilution associated with added sugar intake in elderly black South African women. *European Journal of Clinical Nutrition*, 59(9), 1030–1042.

Chatterjee, I. 1973. Evolution and the biosynthesis of ascorbic acid. *Science*, 182(4118), 1271–1272.

Chatterjee, I. 2009. The history of vitamin C research in India. *Journal of Biosciences*, 34(2), 185.

Chebrolu, K.K., Jayaprakasha, G.K., Jifon, J., Patil, B.S. 2012. Production system and storage temperature influence grapefruit vitamin C, limonoids, and carotenoids. *Journal of Agricultural and Food Chemistry*, 60(29), 7096–7103.

Chezanoglou, E., Goula, A.M. 2021. Co-crystallization in sucrose: A promising method for encapsulation of food bioactive components. *Trends in Food Science and Technology*, 114, 262–274.

Chisnall, M., Macknight, R. 2017. Importance of vitamin C in human health and disease. In: *Ascorbic acid in plant growth, development and stress tolerance (Ed Hossain)*. Springer, pp. 491–501.

Choi, M., Kim, G., Lee, H. 2002. Effects of ascorbic acid retention on juice color and pigment stability in blood orange (Citrus sinensis) juice during refrigerated storage. *Food Research International*, 35(8), 753–759.

Chokshi, R., Zia, H. 2004. *Hot-melt extrusion technique: A review. Iranian Journal of Pharmaceutical Research*, 3(1), 3-16.

Coimbra, P.P.S., Cardoso, F.d.S.N., Goncalves, E.C.B.d.A. 2021. Spray-drying wall materials: Relationship with bioactive compounds. *Critical Reviews in Food Science and Nutrition*, 61(17), 2809–2826.

Colina-Coca, C., de Ancos, B., Sánchez-Moreno, C. 2014. Nutritional composition of processed onion: S-Alk (en) yl-L-cysteine sulfoxides, organic acids, sugars, minerals, and vitamin C. *Food and Bioprocess Technology*, 7(1), 289–298.

Combs Jr, G.F., McClung, J.P. 2016. *The vitamins: Fundamental aspects in nutrition and health*. Academic press.

Comunian, T., Babazadeh, A., Rehman, A., Shaddel, R., Akbari-Alavijeh, S., Boostani, S., Jafari, S.M. 2020. Protection and controlled release of vitamin C by different micro/nanocarriers. *Critical Reviews in Food Science and Nutrition*, 1–22.

Comunian, T.A., Abbaspourrad, A., Favaro-Trindade, C.S., Weitz, D.A. 2014. Fabrication of solid lipid microcapsules containing ascorbic acid using a microfluidic technique. *Food Chemistry*, 152, 271–275.

Comunian, T.A., Thomazini, M., Alves, A.J.G., de Matos Junior, F.E., de Carvalho Balieiro, J.C., Favaro-Trindade, C.S. 2013. Microencapsulation of ascorbic acid by complex coacervation: Protection and controlled release. *Food Research International*, 52(1), 373–379.

Coronel-Aguilera, C.P., San Martín-González, M.F. 2015. Encapsulation of spray dried β-carotene emulsion by fluidized bed coating technology. *LWT – Food Science and Technology*, 62(1), 187–193.

Crotti, L., Dresch, G., Ramallo, L.A. 2008. Stability of vitamin C-fortified yerba mate and sensory evaluation of the drink. *Journal of Food Processing and Preservation*, 32(2), 306–318.

Cruz, Z., García-Estrada, C., Olabarrieta, I., Rainieri, S. 2015. Chapter 16. Lipid nanoparticles: Delivery system for bioactive food compounds. In: L.M.C. Sagis (Ed.), *Microencapsulation and microspheres for food applications*. Academic Press, pp. 313–331.

Cvetković, B.R., Jokanović, M.R. 2009. Effect of preservation method and storage condition on ascorbic acid loss in beverages. *Acta Periodica Technologica*, (40), 1–7.

Daeihamed, M., Dadashzadeh, S., Haeri, A., Faghih Akhlaghi, M. 2017. Potential of liposomes for enhancement of oral drug absorption. *Current Drug Delivery*, 14(2), 289–303.

Davies, M.B., Austin, J., Partridge, D.A. 1991. *Vitamin C: Its chemistry and biochemistry*. Royal Society of Chemistry.

de Britto, D., de Moura, M.R., Aouada, F.A., Mattoso, L.H.C., Assis, O.B.G. 2012. N,N,N-trimethyl chitosan nanoparticles as a vitamin carrier system. *Food Hydrocolloids*, 27(2), 487–493.

de Oliveira, A.M., Rondó, P.H.C., Mastroeni, S.S., Oliveira, J.M. 2008. Plasma concentrations of ascorbic acid in parturients from a hospital in Southeast Brazil. *Clinical Nutrition*, 27(2), 228–232.

De Ritter, E. 1976. Stability characteristics of vitamins in processed foods. *Food Technology*.

Dennison, D.B., Kirk, J.R. 1982. Effect of trace mineral fortification on the storage stability of ascorbic acid in a dehydrated model food system. *Journal of Food Science*, 47(4), 1198–1200.

Dennison, D.B., Kirk, J.R. 1978. Oxygen effect on the degradation of ascorbic acid in a dehydrated food system. *Journal of Food Science*, 43(2), 609–618.

Desai, K.G., Park, H.J. 2006. Effect of manufacturing parameters on the characteristics of vitamin C encapsulated tripolyphosphate-chitosan microspheres prepared by spray-drying. *Journal of Microencapsulation*, 23(1), 91–103.

Desai, K.G.H., Liu, C., Park, H.J. 2005. Characteristics of vitamin C immobilized particles and sodium alginate beads containing immobilized particles. *Journal of Microencapsulation*, 22(4), 363–376.

Desai, K.G.H., Park, H.J. 2005. Encapsulation of vitamin C in tripolyphosphate cross-linked chitosan microspheres by spray drying. *Journal of Microencapsulation*, 22(2), 179–192.

Dewettinck, K., Huyghebaert, A. 1999. Fluidized bed coating in food technology. *Trends in Food Science and Technology*, 10(4–5), 163–168.

Dewettinck, K., Huyghebaert, A. 1998. Top-spray fluidized bed coating: Effect of process variables on coating efficiency. *LWT – Food Science and Technology*, 31(6), 568–575.

Diachkova, A.V., Tikhonov, S.L., Tikhonova, N.V., Tolmachev, V.O. 2019. Production of vitaminized food products as a new development vector for the agro-industrial complex. *IOP Conference Series: Earth and Environmental Science*, 315(2), 022043.

Drouin, G., Godin, J.-R., Pagé, B. 2011. The genetics of vitamin C loss in vertebrates. *Current Genomics*, 12(5), 371–378.

Dwyer, J.T., Wiemer, K.L., Dary, O., Keen, C.L., King, J.C., Miller, K.B., Philbert, M.A., Tarasuk, V., Taylor, C.L., Gaine, P.C. 2015. Fortification and health: Challenges and opportunities. *Advances in Nutrition*, 6(1), 124–131.

Esposito, E., Cervellati, F., Menegatti, E., Nastruzzi, C., Cortesi, R. 2002. Spray dried Eudragit microparticles as encapsulation devices for vitamin C. *International Journal of Pharmaceutics*, 242(1–2), 329–334.

Esposto, B.S., Jauregi, P., Tapia-Blácido, D.R., Martelli-Tosi, M. 2021. Liposomes vs. chitosomes: Encapsulating food bioactives. *Trends in Food Science and Technology*, 108, 40–48.

Estevinho, B.N., Carlan, I., Blaga, A., Rocha, F. 2016. Soluble vitamins (vitamin B12 and vitamin C) microencapsulated with different biopolymers by a spray drying process. *Powder Technology*, 289, 71–78.

Fain, O. 2004. Vitamin C deficiency. *La Revue de medecine interne*, 25(12), 872–880.

Farhang, B., Kakuda, Y., Corredig, M. 2012. Encapsulation of ascorbic acid in liposomes prepared with milk fat globule membrane-derived phospholipids. *Dairy Science and Technology*, 92(4), 353–366.

Finotelli, P.V., Rocha-Leão, M.H. 2005. Microencapsulation of ascorbic acid in maltodextrin and capsul using spray-drying. In *Anais do 4° Mercosur Congress on Process Systems Engineering*.

Fraj, J., Petrović, L., Đekić, L., Budinčić, J.M., Bučko, S., Katona, J. 2021. Encapsulation and release of vitamin C in double W/O/W emulsions followed by complex coacervation in gelatin-sodium caseinate system. *Journal of Food Engineering*, 292, 110353.

Frankenfeld, C.L., Lampe, J.W., Shannon, J., Gao, D.L., Li, W., Ray, R.M., Chen, C., King, I.B., Thomas, D.B. 2012. Fruit and vegetable intakes in relation to plasma nutrient concentrations in women in Shanghai, China. *Public Health Nutrition*, 15(1), 167–175.

Frei, B., Traber, M.G. 2001. The new US dietary reference intakes for vitamins C and E. *Redox Report*, 6(1), 5–9.

Furuta, T., Neoh, T.L. 2021. Microencapsulation of food bioactive components by spray drying: A review. *Drying Technology*, 39(12), 1800–1831.

Garavand, F., Cacciotti, I., Vahedikia, N., Rehman, A., Tarhan, Ö., Akbari-Alavijeh, S., Shaddel, R., Rashidinejad, A., Nejatian, M., Jafarzadeh, S., Azizi-Lalabadi, M., Khoshnoudi-Nia, S., Jafari, S.M. 2022. A comprehensive review on the nanocomposites loaded with chitosan nanoparticles for food packaging. *Critical Reviews in Food Science and Nutrition*, 62(5), 1383–1416.

García, O.P., Ronquillo, D., Caamaño, M.d.C., Camacho, M., Long, K.Z., Rosado, J.L. 2012. Zinc, vitamin A, and vitamin C status are associated with leptin concentrations and obesity in Mexican women: Results from a cross-sectional study. *Nutrition and Metabolism*, 9(1), 59.

Godoy, H.T., Amaya-Farfan, J., Rodriguez-Amaya, D.B. 2021. Degradation of vitamins. In: *Chemical changes during processing and storage of foods*. Elsevier, pp. 329–383.

Gordon, D.S., Rudinsky, A.J., Guillaumin, J., Parker, V.J., Creighton, K.J. 2020. Vitamin C in health and disease: A companion animal focus. *Topics in Companion Animal Medicine*, 39, 100432.

Granger, M., Eck, P. 2018. Dietary vitamin C in human health. *Advances in Food and Nutrition Research*, 83, 281–310.

Guignon, B., Duquenoy, A., Dumoulin, E.D. 2002. Fluid bed encapsulation of particles: Principles and practice. *Drying Technology*, 20(2), 419–447.

Halestrap, P., Scheenstra, S. 2018. Outbreak of scurvy in Tana River County, Kenya: A case report. *African Journal of Primary Health Care and Family Medicine*, 10(1), 1–3.

Hamer, D.H., Sempértegui, F., Estrella, B., Tucker, K.L., Rodríguez, A., Egas, J., Dallal, G.E., Selhub, J., Griffiths, J.K., Meydani, S.N. 2008. Micronutrient deficiencies are associated with impaired immune response and higher burden of respiratory infections in elderly Ecuadorians. *The Journal of Nutrition*, 139(1), 113–119.

Han, J., Guenier, A.-S., Salmieri, S., Lacroix, M. 2008. Alginate and chitosan functionalization for micronutrient encapsulation. *Journal of Agricultural and Food Chemistry*, 56(7), 2528–2535.

He, J., Zhu, Q., Dong, X., Pan, H., Chen, J., Zheng, Z.-P. 2017. Oxyresveratrol and ascorbic acid O/W microemulsion: Preparation, characterization, anti-isomerization and potential application as antibrowning agent on fresh-cut lotus root slices. *Food Chemistry*, 214, 269–276.

Hernández, Y., Lobo, M.G., González, M. 2006. Determination of vitamin C in tropical fruits: A comparative evaluation of methods. *Food Chemistry*, 96(4), 654–664.

Hou, D., Xie, C., Huang, K., Zhu, C. 2003. The production and characteristics of solid lipid nanoparticles (SLNs). *Biomaterials*, 24(10), 1781–1785.

Hoyos-Leyva, J.D., Chavez-Salazar, A., Castellanos-Galeano, F., Bello-Perez, L.A., Alvarez-Ramirez, J. 2018. Physical and chemical stability of l-ascorbic acid microencapsulated into taro starch spherical aggregates by spray drying. *Food Hydrocolloids*, 83, 143–152.

Hussein, H., Awad, S., El-Sayed, I., & Ibrahim, A. (2020). Impact of chickpea as prebiotic, antioxidant and thickener agent of stirred bio-yoghurt. *Annals of Agricultural Sciences*, 65(1), 49–58.

Ilic, D.B., Ashoor, S.H. 1988. Stability of vitamins A and C in fortified yogurt. *Journal of Dairy Science*, 71(6), 1492–1498.

Iqbal, K., Khan, A., Khattak, M. 2004. Biological significance of ascorbic acid (vitamin C) in human health-a review. *Pakistan Journal of Nutrition*, 3(1), 5–13.

Iqbal, M.A., Md, S., Sahni, J.K., Baboota, S., Dang, S., Ali, J. 2012. Nanostructured lipid carriers system: Recent advances in drug delivery. *Journal of Drug Targeting*, 20(10), 813–830.

İvedi, İ., Güneşoğlu, B., Karavana, S.Y., Kartal, G.E., Erkan, G., Sarışık, A.M. 2022. Using spraying as an alternative method for transferring capsules containing shea butter to denim and non-denim fabrics : Preparation of microcapsules for delivery of active ingredients. *Johnson Matthey Technology Review*, 66(1), 90–102.

Izuagie, A., Izuagie, F. 2007. Iodimetric determination of ascorbic acid (vitamin C) in citrus fruits. *Research Journal of Agriculture and Biological Sciences*, 3(5), 367–369.

Jafari, S.M., Arpagaus, C., Cerqueira, M.A., Samborska, K. 2021. Nano spray drying of food ingredients; materials, processing and applications. *Trends in Food Science and Technology*, 109, 632–646.

Jain, A., Garg, N.K., Jain, A., Kesharwani, P., Jain, A.K., Nirbhavane, P., Tyagi, R.K. 2016. A synergistic approach of adapalene-loaded nanostructured lipid carriers, and vitamin C co-administration for treating acne. *Drug Development and Industrial Pharmacy*, 42(6), 897–905.

Jang, K.-I., Lee, H.G. 2008. Stability of chitosan nanoparticles for l-ascorbic acid during heat treatment in aqueous solution. *Journal of Agricultural and Food Chemistry*, 56(6), 1936–1941.

Jaskulski, M., Kharaghani, A., Tsotsas, E., Krokida, M.K. 2017. Encapsulation methods: Spray drying, spray chilling and spray cooling. In: *Thermal and nonthermal encapsulation methods*. CRC Press, pp. 67–114.

Johnston, C.S., Steinberg, F.M., Rucker, R.B. 2007. Ascorbic acid. *Handbook of Vitamins*, 4, 489–520.

Jukes, T.H., King, J.L. 1975. Evolutionary loss of ascorbic acid synthesizing ability. *Journal of Human Evolution*, 4(2), 85–88.

Khalid, N., Kobayashi, I., Neves, M.A., Uemura, K., Nakajima, M., Nabetani, H. 2014. Formulation of mono-disperse water-in-oil emulsions encapsulating calcium ascorbate and ascorbic acid 2-glucoside by microchannel emulsification. *Colloids and Surfaces A: Physicochemical and Engineering Aspects*, 459, 247–253.

Kim, B.K., Hwang, S.J., Park, J.B., Park, H.J. 2002. Preparation and characterization of drug-loaded polymethacrylate microspheres by an emulsion solvent evaporation method. *Journal of Microencapsulation*, 19(6), 811–822.

Kim, S.-S., Han, Y.-J., Hwang, T.-J., Roh, H.-J., Hahm, T.-S., Chung, M.-S., Shin, S.-G. 2001. Microencapsulation of ascorbic acid in sucrose and lactose by cocrystallization. *Food Science and Biotechnology*, 10(2), 101–107.

King, C.G., Waugh, W.A. 1932. The chemical nature of vitamin C. *Science*, 75(1944), 357–358.

Kiondo, P., Tumwesigye, N.M., Wandabwa, J., Wamuyu-Maina, G., Bimenya, G.S., Okong, P. 2012. Plasma vitamin C assay in women of reproductive age in Kampala, Uganda, using a colorimetric method. *Tropical Medicine and International Health*, 17(2), 191–196.

Kirby, C., Whittle, C., Rigby, N., Coxon, D., Law, B. 1991. Stabilization of ascorbic acid by microencapsulation in liposomes. *International Journal of Food Science and Technology*, **26**(5), 437–449.

Kirk, J., Dennison, D., Kokoczka, P., Heldman, D. 1977. Degradation of ascorbic acid in a dehydrated food system. *Journal of Food Science*, 42(5), 1274–1279.

Klesk, K., Qian, M., Martin, R.R. 2004. Aroma extract dilution analysis of cv. meeker (Rubus idaeus L.) red raspberries from Oregon and Washington. *Journal of Agricultural and Food Chemistry*, 52(16), 5155–5161.

Klimczak, I., Gliszczyńska-Świgło, A. 2015. Comparison of UPLC and HPLC methods for determination of vitamin C. *Food Chemistry*, 175, 100–105.

Knezevic, Z., Gosak, D., Hraste, M., Jalsenjako, I. 1998. Fluid-bed microencapsulation of ascorbic acid. *Journal of Microencapsulation*, 15(2), 237–252.

Kon, S.K., Watson, M.B. 1937. The vitamin C content of cow's milk. *Biochemical Journal*, 31(2), 223.

Konopacka, D., Markowski, J. 2004. Retention of ascorbic acid during apple chips production and storage. *Polish Journal of Food and Nutrition Sciences*, 13(3), 237–242.

Korzeniowska, M., Wojdylo, A., Barrachina, A.A.C. 2017. Advances in food fortification with vitamins and co-vitamins. In: Agnieszka Saeid (Ed.), *Food biofortification technologies*. CRC Press, pp. 61–96.

Koyuncu, M.A., Dilmaçünal, T. 2010. Determination of vitamin C and organic acid changes in strawberry by HPLC during cold storage. *Notulae Botanicae Horti Agrobotanici Cluj-Napoca*, 38(3), 95–98.

Kumar, G.V., Kumar, A., Raghu, K., Patel, G.R., Manjappa, S. 2013. Determination of vitamin C in some fruits and vegetables in Davanagere city,(Karanataka)–India. *International Journal of Pharmacy & Life Sciences*, 4(3), 2489–2491.

Kumar, Y., Singhal, S. 2021. Food applications of cyclodextrins. In: V.K Maurya, K.M Gothandam, S. Ranjan, N. Dasgupta, E. Lichtfouse (Eds.), *Sustainable agriculture reviews 55: Micro and nano engineering in food science*, Vol. 1. Springer International Publishing, pp. 201–238.

Lam, T.K., Freedman, N.D., Fan, J.-H., Qiao, Y.-L., Dawsey, S.M., Taylor, P.R., Abnet, C.C. 2013. Prediagnostic plasma vitamin C and risk of gastric adenocarcinoma and esophageal squamous cell carcinoma in a Chinese population. *The American Journal of Clinical Nutrition*, 98(5), 1289–1297.

Laur, L.M., Tian, L. 2011. Provitamin A and vitamin C contents in selected California-grown cantaloupe and honeydew melons and imported melons. *Journal of Food Composition and Analysis*, 24(2), 194–201.

Leahu, A., Damian, C., Oroian, M., Ropciuc, S., Rotaru, R. 2014. Influence of processing on vitamin C content of rosehip fruits. *Scientific Papers Animal Science and Biotechnologies*, 47, 116–120.

Lee, J.-B., Ahn, J., Lee, J., Kwak, H.-S. 2004. L-ascorbic acid microencapsulated with polyacylglycerol monostearate for milk fortification. *Bioscience, Biotechnology, and Biochemistry*, 68(3), 495–500.

Lee, J.B., Ahn, J., Kwak, H.S. 2003. Microencapsulated ascorbic acid for milk fortification. *Archives of Pharmacal Research*, 26(7), 575–580.

Lešková, A.S.-M.M.-E. 2006. Vitamin C degradation during storage of fortified foods. *Journal of Food and Nutrition Research*, 45(2), 55–61.

Levine, M., Rumsey, S.C., Daruwala, R., Park, J.B., Wang, Y. 1999. Criteria and recommendations for vitamin C intake. *JAMA*, 281(15), 1415–1423.

Levine, M., Wang, Y., Padayatty, S.J., Morrow, J. 2001. A new recommended dietary allowance of vitamin C for healthy young women. *Proceedings of the National Academy of Sciences*, 98(17), 9842–9846.

Leyva-López, R., Palma-Rodríguez, H.M., López-Torres, A., Capataz-Tafur, J., Bello-Pérez, L.A., Vargas-Torres, A. 2019. Use of enzymatically modified starch in the microencapsulation of ascorbic acid: Microcapsule characterization, release behavior and in vitro digestion. *Food Hydrocolloids*, 96, 259–266.

Lind, J. 1757. *A treatise on the scurvy*. A. Millar.

Lipasek, R.A., Taylor, L.S., Mauer, L.J. 2011. Effects of anticaking agents and relative humidity on the physical and chemical stability of powdered vitamin C. *Journal of Food Science*, 76(7), C1062–C1074.

Lipka, D., Gubernator, J., Filipczak, N., Barnert, S., Süss, R., Legut, M., Kozubek, A. 2013. Vitamin C-driven epirubicin loading into liposomes. *International Journal of Nanomedicine*, 8, 3573–3585.

Liu, H., Cao, J., Jiang, W. 2015. Evaluation and comparison of vitamin C, phenolic compounds, antioxidant properties and metal chelating activity of pulp and peel from selected peach cultivars. *LWT – Food Science and Technology*, 63(2), 1042–1048.

Loveday, S.M., Singh, H. 2008. Recent advances in technologies for vitamin A protection in foods. *Trends in Food Science and Technology*, 19(12), 657–668.

Lykkesfeldt, J., Poulsen, H.E. 2010. Is vitamin C supplementation beneficial? Lessons learned from randomised controlled trials. *British Journal of Nutrition*, 103(9), 1251–1259.

Lykkesfeldt, J., Tveden-Nyborg, P. 2019. The pharmacokinetics of vitamin C. *Nutrients*, 11(10), 2412.

Maherani, B., Arab-Tehrany, E., Mozafari, R.M., Gaiani, C., Linder, M. 2011. Liposomes: A review of manufacturing techniques and targeting strategies. *Current Nanoscience*, 7(3), 436–452.

Majid, I., Hussain, S., Nanda, V. 2019. Impact of sprouting on the degradation kinetics of color and vitamin C of onion powder packaged in different packaging materials. *Journal of Food Processing and Preservation*, 43(1), e13849.

Marik, P.E., Liggett, A. 2019. Adding an orange to the banana bag: Vitamin C deficiency is common in alcohol use disorders. *Critical Care*, 23(1), 1–6.

Marsanasco, M., Márquez, A.L., Wagner, J.R., del V. Alonso, S., Chiaramoni, N.S. 2011. Liposomes as vehicles for vitamins E and C: An alternative to fortify orange juice and offer vitamin C protection after heat treatment. *Food Research International*, 44(9), 3039–3046.

Masamba, K.G., Mndalira, K. 2013. Vitamin C stability in pineapple, guava and baobab juices under different storage conditions using different levels of sodium benzoate and metabisulphite. *African Journal of Biotechnology*, 12(2).

Matějková, J., Petříková, K. 2010. Variation in content of carotenoids and vitamin C in carrots. *Notulae Scientia Biologicae*, 2(4), 88–91.

Matsui, T. 2012. Vitamin C nutrition in cattle. *Asian-Australasian Journal of Animal Sciences*, 25(5), 597.

Maurya, V.K., Aggarwal, M. 2017. Enhancing bio-availability of vitamin D by Nano-engineered based delivery systems-an overview. *International Journal of Current Microbiology and Applied Sciences*, 6(7), 340–353.

Maurya, V.K., Aggarwal, M., Ranjan, V., Gothandam, K.M. 2020a. Improving bioavailability of vitamin A in food by encapsulation: An update. In: *Nanoscience in medicine*, Vol. 1. Springer, pp. 117–145.

Maurya, V.K., Bashir, K., Aggarwal, M. 2020b. Vitamin D microencapsulation and fortification: Trends and technologies. *The Journal of Steroid Biochemistry and Molecular Biology*, 196, 105489.

Maurya, V.K., Shakya, A., Aggarwal, M., Gothandam, K.M., Bohn, T., Pareek, S. 2021. Fate of β-carotene within loaded delivery systems in food: State of knowledge. *Antioxidants*, 10(3), 426.

Maurya, V.K., Shakya, A., Bashir, K., Kushwaha, S.C., McClements, D.J. 2022. Vitamin A fortification: Recent advances in encapsulation technologies. *Comprehensive Reviews in Food Science and Food Safety.*

Maxfield, L., Crane, J.S. 2018. *Vitamin C deficiency.*

McCall, S.J., Clark, A.B., Luben, R.N., Wareham, N.J., Khaw, K.-T., Myint, P.K. 2019. Plasma vitamin C Levels: Risk factors for deficiency and association with self-reported functional health in the European prospective investigation into cancer-Norfolk. *Nutrients*, 11(7), 1552.

McCord, C.P. 1959. SCURVY as an occupational disease: The Sappington memorial lecture. *Journal of Occupational Medicine*, 1(6), 315–318.

Mehansho, H., Mellican, R.I., Hughes, D.L., Compton, D.B., Walter, T. 2003. Multiple-micronutrient fortification technology development and evaluation: from lab to market. *Food and Nutrition Bulletin*, 24(4_suppl_1), S111–S119.

Mehnert, W., Mäder, K. 2012. Solid lipid nanoparticles: Production, characterization and applications. *Advanced Drug Delivery Reviews*, 64, 83–101.

Monroig, Ó., Navarro, J.C., Amat, F., Hontoria, F. 2007. Enrichment of artemia nauplii in vitamin A, vitamin C and methionine using liposomes. *Aquaculture*, 269(1), 504–513.

Mudalip, S.K.A., Khatiman, M.N., Hashim, N.A., Man, R.C., Arshad, Z.I.M. 2021. A short review on encapsulation of bioactive compounds using different drying techniques. *Materials Today: Proceedings*, 42, 288–296.

Murtaza, M.A., Huma, N., Javaid, J., Shabbir, M.A., Din, G.M.U., Mahmood, S. 2004. Studies on stability of strawberry drink stored at different temperatures. *International Journal of Agriculture and Biology*, 6(1), 58–60.

Nair, M.K., Augustine, L.F., Konapur, A. 2016. Food-based interventions to modify diet quality and diversity to address multiple micronutrient deficiency. *Frontiers in Public Health*, 3, 277.

Nakade, M., Jungari, M.L., Ambad, R., Dhingra, G. 2020. Status of vitamins and minerals in pregnancy: Still A point of concern in central India. *International Journal of Current Research and Review*, 12(14).

Narayanan, G., Shen, J., Matai, I., Sachdev, A., Boy, R., Tonelli, A.E. 2022. Cyclodextrin-based nanostructures. *Progress in Materials Science*, 124, 100869.

Nowak, D. 2021. *Vitamin C in human health and disease*, Vol. 13. Multidisciplinary Digital Publishing Institute, pp. 1595.

Nwagha, U.I., Iyare, E.E., Ejezie, F.E., Ogbodo, S.O., Dim, C.C., Anyaehie, B.U. 2012. Parity related changes in obesity and some antioxidant vitamins in non-pregnant women of South-Eastern Nigeria. *Nigerian Journal of Clinical Practice*, 15(4), 380–384.

O'Donnell, P.B., McGinity, J.W. 1997. Preparation of microspheres by the solvent evaporation technique. *Advanced Drug Delivery Reviews*, 28(1), 25–42.

Obeidat, W.M., Schwabe, K., Müller, R.H., Keck, C.M. 2010. Preservation of nanostructured lipid carriers (NLC). *European Journal of Pharmaceutics and Biopharmaceutics*, 76(1), 56–67.

Ojiewo, C., Tenkouano, A., Hughes, J.D.A., Keatinge, J.D.H. 2013. Diversifying diets: Using indigenous vegetables to improve profitability, nutrition and health in Africa. In: *Diversifying Food and Diets: Using agricultural biodiversity to improve nutrition and health*, pp. 291–302.

Okuro, P.K., de Matos Junior, F.E., Favaro-Trindade, C.S. 2013. Technological challenges for spray chilling encapsulation of functional food ingredients. *Food Technology and Biotechnology*, 51(2), 171.

Orriss, G.D. 1998. Food fortification: Safety and legislation. *Food and Nutrition Bulletin*, 19(2), 109–116.

Osman, A., Goda, H.A., Abdel-Hamid, M., Badran, S.M., Otte, J. 2016. Antibacterial peptides generated by alcalase hydrolysis of goat whey. *LWT – Food Science and Technology*, 65, 480–486.

Ottaway, P.B. 1993. Stability of vitamins in food. In: *The technology of vitamins in food*. Springer, pp. 90–113.

Oxley, J.D. 2012. Spray cooling and spray chilling for food ingredient and nutraceutical encapsulation. In: *Encapsulation technologies and delivery systems for food ingredients and nutraceuticals*. Elsevier, pp. 110–130.

Özdemir, K.S., Gökmen, V. 2015. Effect of microencapsulation on the reactivity of ascorbic acid, sodium chloride and vanillin during heating. *Journal of Food Engineering*, 167, 204–209.

Packer, L. 1997. *Vitamin C in health and disease.* CRC Press.

Padayatty, S.J., Katz, A., Wang, Y., Eck, P., Kwon, O., Lee, J.-H., Chen, S., Corpe, C., Dutta, A., Dutta, S.K. 2003. Vitamin C as an antioxidant: Evaluation of its role in disease prevention. *Journal of the American College of Nutrition*, 22(1), 18–35.

Park, H., Seib, P.A., Chung, O.K. 1997a. Fortifying bread with a mixture of wheat fiber and psyllium husk fiber plus three antioxidants. *Cereal Chemistry*, 74(3), 207–211.

Park, H., Seib, P.A., Chung, O.K., Seitz, L.M. 1997b. Fortifying bread with each of three antioxidants. *Cereal Chemistry*, 74(3), 202–206.

Park, S., Arasu, M.V., Lee, M.-K., Chun, J.-H., Seo, J.M., Lee, S.-W., Al-Dhabi, N.A., Kim, S.-J. 2014. Quantification of glucosinolates, anthocyanins, free amino acids, and vitamin C in inbred lines of cabbage (Brassica oleracea L.). *Food Chemistry*, 145, 77–85.

Peleg, M. 2017. Theoretical study of aerobic vitamin C loss kinetics during commercial heat preservation and storage. *Food Research International*, 102, 246–255.

Pereira, H.V.R., Saraiva, K.P., Carvalho, L.M.J., Andrade, L.R., Pedrosa, C., Pierucci, A.P.T.R. 2009. Legumes seeds protein isolates in the production of ascorbic acid microparticles. *Food Research International*, 42(1), 115–121.

Pfendt, L.B., Vukašinović, V.L., Blagojević, N.Z., Radojević, M.P. 2003. Second order derivative spectrophotometric method for determination of vitamin C content in fruits, vegetables and fruit juices. *European Food Research and Technology*, 217(3), 269–272.

Phillips, K.M., Tarrago-Trani, M.T., Gebhardt, S.E., Exler, J., Patterson, K.Y., Haytowitz, D.B., Pehrsson, P.R., Holden, J.M. 2010. Stability of vitamin C in frozen raw fruit and vegetable homogenates. *Journal of Food Composition and Analysis*, 23(3), 253–259.

Pierucci, A.P.T.R., Andrade, L.R., Baptista, E.B., Volpato, N.M., Rocha-Leão, M.H.M. 2006. New microencapsulation system for ascorbic acid using pea protein concentrate as coat protector. *Journal of Microencapsulation*, 23(6), 654–662.

Piñón-Balderrama, C.I., Leyva-Porras, C., Terán-Figueroa, Y., Espinosa-Solís, V., Álvarez-Salas, C., Saavedra-Leos, M.Z. 2020. Encapsulation of active ingredients in food industry by spray-drying and nano spray-drying technologies. *Processes*, 8(8), 889.

Polydera, A.C., Stoforos, N.G., Taoukis, P.S. 2003. Comparative shelf life study and vitamin C loss kinetics in pasteurised and high pressure processed reconstituted orange juice. *Journal of Food Engineering*, 60(1), 21–29.

Quast, L.B., Farina, S.G., Quast, E., Vieira, M.A., Queiroz, M.B. 2020. Co-crystallized honey with sucrose: Evaluation of process and product characterization. *Journal of Food Processing and Preservation*, 44(11), e14876.

Rahman, M.S., Al-Rizeiqi, M.H., Guizani, N., Al-Ruzaiqi, M.S., Al-Aamri, A.H., Zainab, S. 2015. Stability of vitamin C in fresh and freeze-dried capsicum stored at different temperatures. *Journal of Food Science and Technology*, 52(3), 1691–1697.

Ravindran, R.D., Vashist, P., Gupta, K., S., Young, S., I., Maraini, G., Camparini, M., Jayanthi, R., John, N., Fitzpatrick, K.E., Chakravarthy, U. 2011. Prevalence and risk factors for vitamin C deficiency in north and south India: A two centre population based study in people aged 60 years and over. *PLoS One*, 6(12), e28588.

Remini, H., Mertz, C., Belbahi, A., Achir, N., Dornier, M., Madani, K. 2015. Degradation kinetic modelling of ascorbic acid and colour intensity in pasteurised blood orange juice during storage. *Food Chemistry*, 173, 665–673.

Repka, M.A., Shah, S., Lu, J., Maddineni, S., Morott, J., Patwardhan, K., Mohammed, N.N. 2012. Melt extrusion: Process to product. *Expert Opinion on Drug Delivery*, 9(1), 105–125.

Rezvankhah, A., Emam-Djomeh, Z., Askari, G. 2020. Encapsulation and delivery of bioactive compounds using spray and freeze-drying techniques: A review. *Drying Technology*, 38(1–2), 235–258.

Righetto, A.M., Netto, F.M. 2006. Vitamin C stability in encapsulated green West Indian cherry juice and in encapsulated synthetic ascorbic acid. *Journal of the Science of Food and Agriculture*, 86(8), 1202–1208.

Romeu-Nadal, M., Morera-Pons, S., Castellote, A., Lopez-Sabater, M. 2006. Rapid high-performance liquid chromatographic method for vitamin C determination in human milk versus an enzymatic method. *Journal of Chromatography B*, 830(1), 41–46.

Rosca, I.D., Watari, F., Uo, M. 2004. Microparticle formation and its mechanism in single and double emulsion solvent evaporation. *Journal of Controlled Release*, 99(2), 271–280.

Rowe, S., Carr, A.C. 2020. Global vitamin C status and prevalence of deficiency: A cause for concern? *Nutrients*, 12(7), 2008.

Rucker, R.B., Zempleni, J., Suttie, J.W., McCormick, D.B. 2007. *Handbook of vitamins*. CRC Press.

Samborska, K., Poozesh, S., Barańska, A., Sobulska, M., Jedlińska, A., Arpagaus, C., Malekjani, N., Jafari, S.M. 2022. Innovations in spray drying process for food and pharma industries. *Journal of Food Engineering*, 110960.

Sartori, T., Consoli, L., Hubinger, M.D., Menegalli, F.C. 2015. Ascorbic acid microencapsulation by spray chilling: Production and characterization. *LWT - Food Science and Technology*, 63(1), 353–360.

Shabana, S., Prasansha, R., Kalinina, I., Potoroko, I., Bagale, U., Shirish, S.H. 2019. Ultrasound assisted acid hydrolyzed structure modification and loading of antioxidants on potato starch nanoparticles. *Ultrasonics Sonochemistry*, 51, 444–450.

Sharma, R., Lal, D. 2005. Fortification of milk with microencapsulated vitamin C and its thermal stability.

Smith, J.L., Hodges, R.E. 1987. Serum levels of vitamin C in relation to dietary and supplemental intake of vitamin C in smokers and nonsmokers. *Annals of the New York Academy of Sciences*, 498, 144–152.

Somsub, W., Kongkachuichai, R., Sungpuag, P., Charoensiri, R. 2008. Effects of three conventional cooking methods on vitamin C, tannin, myo-inositol phosphates contents in selected Thai vegetables. *Journal of Food Composition and Analysis*, 21(2), 187–197.

Sowbhagya, H. 2014. Chemistry, technology, and nutraceutical functions of celery (Apium graveolens L.): An overview. *Critical Reviews in Food Science and Nutrition*, 54(3), 389–398.

Spoelstra-de Man, A.M.E., Elbers, P.W.G., Oudemans-Van Straaten, H.M. 2018. Vitamin C: Should we supplement? *Current Opinion in Critical Care*, 24(4), 248.

Steele, C.J. 1976. *Cereal fortification--Technological problems.* Cereal Foods World (USA).

Sundar, S., Kundu, J., Kundu, S.C. 2010. Biopolymeric nanoparticles. *Science and Technology of Advanced Materials*, 11(1), 014104.

Szent-Györgyi, A. 1928. Observations on the function of peroxidase systems and the chemistry of the adrenal cortex: Description of a new carbohydrate derivative. *Biochemical Journal*, 22(6), 1387–1409.

Talcott, S.T., Percival, S.S., Pittet-Moore, J., Celoria, C. 2003. Phytochemical composition and antioxidant stability of fortified yellow passion fruit (Passiflora edulis). *Journal of Agricultural and Food Chemistry*, 51(4), 935–941.

Tehrani, E., Amiri, S. 2022. Synthesis and characterization PVA electro-spun nanofibers containing encapsulated vitamin C in chitosan microspheres. *The Journal of the Textile Institute*, 113(2), 212–223.

Tikekar, R.V., Anantheswaran, R.C., LaBorde, L.F. 2011. Ascorbic acid degradation in a model apple juice system and in apple juice during ultraviolet processing and storage. *Journal of Food Science*, 76(2), H62–H71.

Trindade, M.A., Grosso, C.R.F. 2000. The stability of ascorbic acid microencapsulated in granules of rice starch and in gum arabic. *Journal of Microencapsulation*, 17(2), 169–176.

Tsai, W.-C., Rizvi, S.S.H. 2017. Simultaneous microencapsulation of hydrophilic and lipophilic bioactives in liposomes produced by an ecofriendly supercritical fluid process. *Food Research International*, 99(1), 256–262.

Uddin, M.S., Hawlader, M.N.A., Zhu, H.J. 2001. Microencapsulation of ascorbic acid: Effect of process variables on product characteristics. *Journal of Microencapsulation*, 18(2), 199–209.

Ugwa, E.A., Iwasam, E.A., Nwali, M.I. 2016. Low serum vitamin C status among pregnant women attending antenatal care at general hospital Dawakin kudu, Northwest Nigeria. *International Journal of Preventive Medicine*, 7.

Üner, M., Wissing, S.A., Yener, G., Müller, R.H. 2005. Solid lipid nanoparticles (SLN) and nanostructured lipid carriers (NLC) for application of ascorbyl palmitate. *Die Pharmazie - an International Journal of Pharmaceutical Sciences*, 60(8), 577–582.

Villalpando, S., Montalvo-Velarde, I., Zambrano, N., García-Guerra, A., Ramírez-Silva, C.I., Shamah-Levy, T., Rivera, J.A. 2003. Vitamins A, and C and folate status in Mexican children under 12 years and women 12–49 years: A probabilistic national survey. *Salud Publica de Mexico*, 45, 508–519.

Walia, N., Dasgupta, N., Ranjan, S., Ramalingam, C., Gandhi, M. 2019. Food-grade nanoencapsulation of vitamins. *Environmental Chemistry Letters*, 17(2), 991–1002.

Wang, X.Y., Kozempel, M.G., Hicks, K.B., Seib, P.A. 1992. Vitamin C stability during preparation and storage of potato flakes and reconstituted mashed potatoes. *Journal of Food Science*, 57(5), 1136–1139.

Wang, X., Seib, P., Ra, K. 1995a. L-ascorbic acid and its 2-phosphorylated derivatives in selected foods: Vitamin C fortification and antioxidant properties. *Journal of Food Science*, 60(6), 1295–1300.

Wen, P., Zong, M.-H., Linhardt, R.J., Feng, K., Wu, H. 2017. Electrospinning: A novel nano-encapsulation approach for bioactive compounds. *Trends in Food Science and Technology*, 70, 56–68.

Wrieden, W.L., Hannah, M.K., Bolton-Smith, C., Tavendale, R., Morrison, C., Tunstall-Pedoe, H. 2000. Plasma vitamin C and food choice in the third Glasgow Monica population survey. *Journal of Epidemiology and Community Health*, 54(5), 355–360.

Yan, B., Davachi, S.M., Ravanfar, R., Dadmohammadi, Y., Deisenroth, T.W., Pho, T.V., Odorisio, P.A., Darji, R.H., Abbaspourrad, A. 2021. Improvement of vitamin C stability in vitamin gummies by encapsulation in casein gel. *Food Hydrocolloids*, 113, 106414.

Yang, S., LIU, C.-M., LIU, W., LIU, W.-L., TONG, G.-H., ZHANG, Y., ZHENG, H.-J. 2010. Preparation and stability of vitamin C liposomes. *Food Science*, 20, 5–12.

Yang, S., Liu, C., Liu, W., Yu, H., Zheng, H., Zhou, W., Hu, Y. 2013. Preparation and characterization of nanoliposomes entrapping medium-chain fatty acids and vitamin C by lyophilization. *International Journal of Molecular Sciences*, 14(10), 19763–19773.

Yoksan, R., Jirawutthiwongchai, J., Arpo, K. 2010. Encapsulation of ascorbyl palmitate in chitosan nanoparticles by oil-in-water emulsion and ionic gelation processes. *Colloids and Surfaces B: Biointerfaces*, 76(1), 292–297.

Zhang, Z.-H., Zeng, X.-A., Brennan, C.S., Brennan, M., Han, Z., Xiong, X.-Y. 2015. Effects of pulsed electric fields (PEF) on vitamin C and its antioxidant properties. *International Journal of Molecular Sciences*, 16(10), 24159–24173.

Zhou, W., Liu, W., Zou, L., Liu, W., Liu, C., Liang, R., Chen, J. 2014. Storage stability and skin permeation of vitamin C liposomes improved by pectin coating. *Colloids and Surfaces B: Biointerfaces*, 117, 330–337.

6 B Vitamins Fortification
Need and Challenges

Vaibhav Kumar Maurya, Amita Shakya, and Varaprasad Kolla

6.1 INTRODUCTION

As per the World Health Organization, 2 billion people are observed to be suffering from undernourishment, which is directly linked to an insufficient intake of nutrients (Nantel and Tontisirin 2001; World Health 1999; WHO 2006). With awareness about the importance of nutrients in humans, the undernourished population has reduced from 23% to 13% (WHO 2006; Rasche 2020). However, the improvement in micronutrient status in the population is still not significant enough to make a difference. Since the world population relies on starchy staple foods for their diet, they lack the most of the micronutrients. Moreover, the availability of vitamins in these starchy staple food commodities is either null or very low. This puts major portions of the world population at risk of vitamin deficiency. Furthermore, insufficient micronutrient consumption has a more detrimental effect on infants, preschool children, women, and elderly people, resulting in impaired growth, health complications, underdevelopment of organs, and multiple birth defects (WHO 2006).

Accordingly, increased research activity has been observed in addressing vitamin deficiency. This chapter focuses on physicochemical attributes, food sources, recommended dietary allowance, and adopted fortification strategies for vitamin B. The chapter then goes on to highlight the different microencapsulation techniques adopted to encapsulate, protect, and deliver these vitamins. This knowledge will aid in better understanding and framing better strategies for fortifying foods with vitamin B complex.

6.2 VITAMIN B COMPLEX

Vitamin B complex is a class of water-soluble vitamins which includes eight vitamins: thiamine (vitamin B_1), riboflavin (vitamin B_2), niacin (vitamin B_3), pantothenic acid (vitamin B_5), pyridoxine (vitamin B_6), biotin (vitamin B_7), folate (vitamin B_9), and cobalamin (vitamin B_{12}) (Robinson 1951). These B vitamins play a vital role in several biochemical processes which are limiting factors in amino acids, carbohydrates, lipids, fatty acids, protein, neurotransmitters, cholesterol, S-adenosyl methionine, and nucleic acid metabolism (Maqbool et al. 2017). However, deficiency of B vitamins may directly result in anemia, skin disease, digestion problems, peripheral neuropathy, and psychiatric disorders (Sarwar, Sarwar, and Sarwar 2021). The deficiency of B vitamins varies across the globe and is briefly highlighted in **Table 6.1. Table 6.2 highlights the recommended dietary allowance for different B vitamins.**

Vitamin B_1 or thiamine is composed of a sulfur-containing thiazole moiety linked with a pyrimidine ring. Information on its discovery, role in human health, daily requirements, and physicochemical properties has been well-discussed in the literature (Polegato et al. 2019; Manzetti, Zhang, and van der Spoel 2014). Populations relying on stable foods poor in thiamine content like wheat flour, maize, cassava, rice, and sago are more susceptible to vitamin B_1 deficiency and prone to Beriberi disease (Titcomb and Tanumihardjo 2019). Its deficiency may cause several health complications like weight loss, anorexia, apathy, muscle weakness, short-term memory loss, irritability, confusion, and cardiovascular effects, and possibly congestive heart failure. Inclusion of fermented

TABLE 6.1

Summary of the Deficiency Prevalence of Vitamin B-Complex and Its Symptoms

Vitamin	Symptoms	Prevalence
Vitamin B_1	• Beriberi • Chronic Neurological • Cardiovascular Disease	Japan, Thailand
Vitamin B_2	• Wernicke-Korsakov syndrome • Dermatitis • Cheilosis • Glossitis	Developing countries
Vitamin B_3	• Anemia • Insomnia • Conjunctivitis	Developing countries, populations in famine conditions
Vitamin B_6	• Neuropathy • Seborrheic dermatitis • Glossitis • Cheilosis • Depression • Confusion • Seizures	Developing countries
Vitamin B_9	• Unexplained fatigue • Anemia • Muscle weakness	Insufficient data
Vitamin B_{12}	• Megaloblastic anemia • Atrophic gastritis • Poor balance • Memory trouble	Insufficient data

fish, shellfish, fern, eel, silkworm, and microorganisms may help in attaining vitamin B_1 sufficiency status.

Vitamin B_2 is composed of a flavin moiety linked with a ribitol side chain. There are excellent articles available on its discovery, importance in human health, daily requirements, and physicochemical attributes (Robinson 1951; Schellack, Harirari, and Schellack 2016). Vitamin B_2 deficiency condition is termed as ariboflavinosis. Swelling, redness of mouth and throat, sores on the outsides of the mouth, inflammation of the tongue, scaly dermatitis, and cheilitis are a few symptoms of Vitamin B_2 deficiency (Titcomb and Tanumihardjo 2019). Generally, strict vegetarian populations are more prone to Vitamin B_2 deficiency. The suboptimal consumption of vitamin B_2 results in serious health complications like Wernicke-Korsakov syndrome, dermatitis, cheilosis, glossitis, anemia, insomnia, and conjunctivitis (Mahabadi, Bhusal, and Banks 2022; Saedisomeolia and Ashoori 2018). Green leafy vegetables, milk and milk products, eggs, and meat are the primary sources of Vitamin B_2.

Vitamin B_3 is referred to as two compounds: niacinamide (also known as nicotinamide) and the synthetic form nicotinic acid. Several articles have been published on its discovery and physico-chemical attributes (Makarov, Trammell, and Migaud 2019; Lanska 2012). Populations relying on starchy diet are prone to niacin deficiency, and its deficiency may result in dermatitis, diarrhea, and dementia (Titcomb and Tanumihardjo 2019). The suboptimal status of vitamin B_3 can be reversed with the consumption of red meat: beef, beef liver, pork, poultry, fish, brown rice, nuts, legumes, and banana which are rich in niacin content.

TABLE 6.2

Recommended Dietary Allowance of Different B Vitamins

Age group	Thiamine (mg/day)	Riboflavin (mg/day)	Niacin (mg/day)	Vitamin B_6 (mg/day)	Folate (mg/day)	Vitamin B_{12} (mg/day)
1–3 years	0.4	0.4	6	0.5	150	0.9
4–8 years	0.5	0.5	8	0.6	200	1.2
9–13 years Male	0.7	0.8	12	1.0	300	1.8
14–18 years Male	1.0	1.3	16	1.3	400	2.4
19–50-year Male	1.0	1.3	16	1.3	400	2.4
51–70+ years Male	1.0	1.3	16	1.7	400	2.4
9–13 years Female	0.7	0.9	12	1.0	300	1.8
14–18 years Female	0.9	1.0	14	1.2	400	2.4
19–50-years Female	0.9	1.1	14	1.3	400	2.4
51–70+ years Female	0.9	1.1	14	1.5	600	2.4
19–50-years Pregnant women	1.2	1.4	18	1.9	600	2.8
19–50-years Lactation	1.2	1.6	17	1.9	700	2.8

Vitamin B_6 comprised of six vitamers including pyridoxine, pyridoxamine, pyridoxal and their phosphorylated forms. Several excellent articles on its history, importance in human health, and physico-chemical properties are already available in the existing literature (Hellmann and Mooney 2010; Parra, Stahl, and Hellmann 2018). Vitamin B_6 primarily acts as coenzyme and holds approximately 4% of all known enzymatic activity. It has been observed that populations living in low-income countries are more susceptible to vitamin B_6 deficiency and are found to be linked with neuropathy, seborrheic dermatitis, glossitis, cheilosis, depression, confusion, seizures, and several birth defects (Titcomb and Tanumihardjo 2019). The suboptimal status of vitamin B_6 can be minimized by including pork, poultry, peanuts, soybeans, oats, wheat germ, and banana in the diet.

Vitamin B_9 is composed of three moieties, a para-aminobenzoate, a pteridine ring, and glutamate tails with one or more L-glutamates. Information on its discovery, source, importance in human health, daily needs and physico-chemical properties is well reviewed in the literature (Naderi and House 2018; Guilland and Aimone-Gastin 2013). Different vitamers of folate differ in their chemical structure due to variation in oxidative state, length of glutamate tail, and substituent group linked on C-1. The World Health Organization reports approximately 1.6 billion people are affected by anemia across the globe, which is directly linked to either linked with iron or vitamin B_9 deficiency (Titcomb and Tanumihardjo 2019). Vitamin B_9 deficiency leads to several health complications, including but not limited to weakness, fatigue, shortness of breath, anencephaly, cleft palate, and spina bifida or neural tube defect.

Vitamin B_{12}, also known as cobalamin, is comprised of a nonorganic metal atom and is exclusively synthesized by archaea or bacteria. Due to the inability to synthesize vitamin B12, the population is generally deficient in vitamin B_{12} if they do not consume it from an external source (Rizzo and Laganà 2020). Populations staying in developing countries, vegans, and vegetarians are prone to its deficiency (Titcomb and Tanumihardjo 2019). Its deficiency is associated with several health complications, including poor intestinal microbial balance, atrophic gastritis, megaloblastic anemia, and memory trouble (Green et al. 2017; Brown 2005). The deficiency status of vitamin B_{12} can be minimized by including meat, liver, milk, seafood, and prebiotics microorganism in the diet.

6.3 STABILITY AND BIOAVAILABILITY

B vitamins are bioactive molecules and undergo significant degradation during food processing, storage, and distribution (Lešková 2006). The degree of degradation varies with the extent of processing, presence of oxidative agents, transition metals, heat, and pH. Loss of B vitamins against these environmental factors and process conditions is well-discussed in the literature (Ottaway 2010; Gregory Iii 1985; Godoy, Amaya-Farfan, and Rodriguez-Amaya 2021; Karmas and Harris 2012; Ryley and Kajda 1994; Herrera-Ardila et al. 2022). Oxidation, nixtamalization, thermal degradation, and the formation of inactive analogues are few major factors that govern the degradation of B vitamins (Ottaway 2010; Gregory Iii 1985; Godoy, Amaya-Farfan, and Rodriguez-Amaya 2021; Karmas and Harris 2012; Ryley and Kajda 1994; Herrera-Ardila et al. 2022).

The efficacy of B vitamins solely depends on the proportion of absorbed vitamins which are ultimately utilized by the human body rather than the quantity consumed. The ratio of B vitamins absorbed in their active form is known as their bioavailability (Combs and McClung 2016). The bioavailability of vitamins is influence by several factors, including but not limited to the quantity ingested, the complexity of the food matrix, the environmental factors to which they exposed, degree of processing, and storage and distribution conditions (Maurya, Bashir, and Aggarwal 2020; Maurya et al. 2023; Maurya et al. 2022; Maurya et al. 2020; Maurya and Aggarwal 2017; Maurya et al. 2021; Premjit, Pandey, and Mitra 2022; Yaman et al. 2021). Several researchers have evaluated the bioavailability of B vitamins in different food commodities and found it to be dependent on an array of factors such as the complexity of the food matrix, content, the location of vitamin in the tissue, presence/absence of bioavailability promoting/inhibiting factors (**Figure 6.1.**)

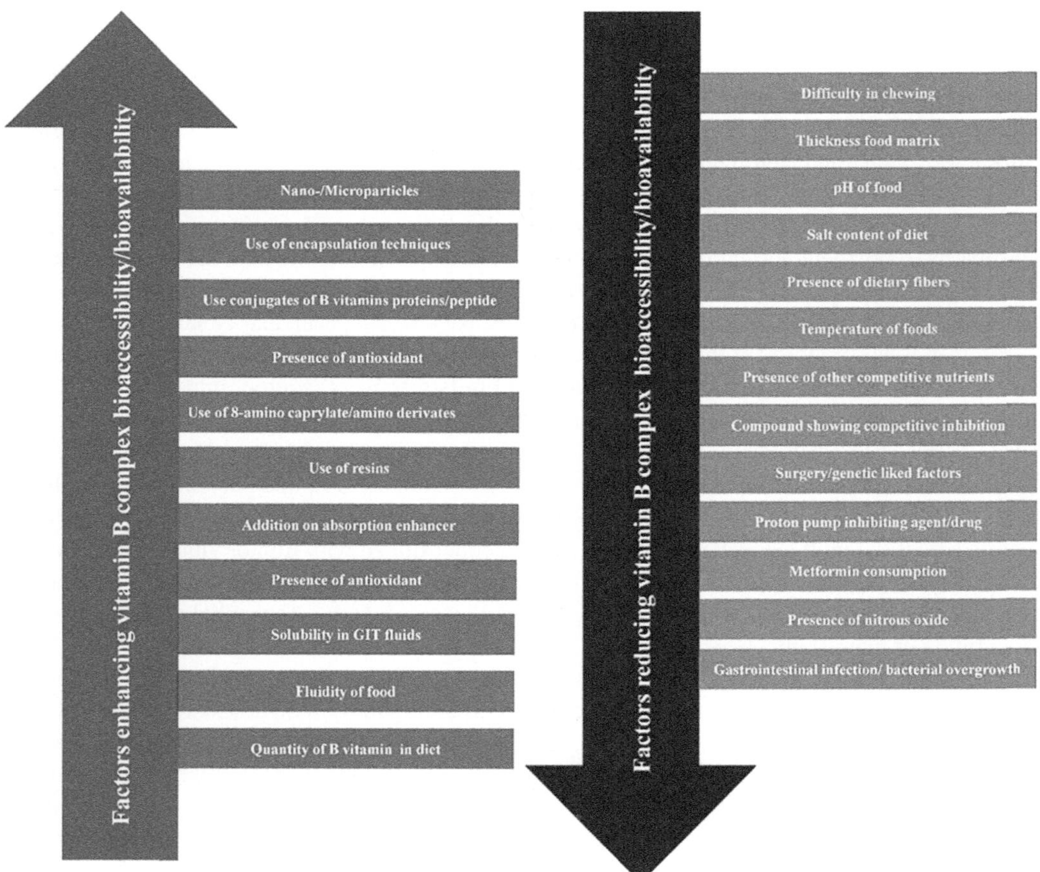

FIGURE 6.1 Factors affecting vitamin B complex bioavailability.

6.4 STRATEGIES ADOPTED TO ADDRESS B VITAMINS DEFICIENCY

Literature reports several methods to resolve vitamin deficiencies: (i) diet; (ii) biofortification; (iii) food fortification; and (iv) dietary supplements (Maurya et al. 2022, 2021; Maurya, Bashir, and Aggarwal 2020). In this section, we will be discussing these approaches with reference to B vitamins.

6.4.1 DIET DIVERSIFICATION

This approach relies on the inclusion of B vitamin-rich food in daily diet such as fish, milk, meat, pulses, and prebiotic organisms. However, diet diversification solely depends on the availability as well as affordability of vitamin-rich food items and consumption patterns of the populations (Maurya et al. 2022; Maurya, Bashir, and Aggarwal 2020; Caritá et al. 2020). The addition of B vitamin-rich food items, which are high bioavailability, is key to success of this approach (Caritá et al. 2020). Furthermore, increased awareness about health complications linked with B vitamins deficiency and alternative affordable food items is equally important.

6.4.2 SUPPLEMENTATION

Vitamin supplements are vitamin-containing pharmaceutical formulations available in the form of tablets, capsules, and powder alone or in combination with other combinations (Chugh and Lhamo 2012). Dietary supplements seem to be a successful tools to minimize the deficiency of B vitamins (Lykkesfeldt and Poulsen 2010; Spoelstra-de Man, Elbers, and Oudemans-Van Straaten 2018). These pharmaceutical supplements are designed to comprise B vitamins or their derivatives that can meet daily requirements (Schellack, Harirari, and Schellack 2016). These pharmaceutical formulations are fabricated to ensure the B vitamins' level during storage and after ingestion. The major challenges of B vitamin supplements are their affordability and desirability.

6.4.3 BIOFORTIFICATION

Biofortification strategies depend on improving the nutrient levels by means of crop management, selective breeding, and/or genetic engineering methods. Several excellent reports are available on the use genetic engineering for improving B vitamins (Chaturvedi, Chaudhary, and Tiwari 2021; Goyer 2017). Researchers have achieved a five-fold increase in thiamine in brown rice using overexpression regulatory genes like thi 1 and thi C (Dong et al. 2016). Lactobacillus plantarum strain CRL725 and Lactobacillus fermentum PBCC11 have been developed with improved riboflavin content (Russo et al. 2014). Transgenic cassava crops have been developed by inserting both AtPDX1.1 and AtPDX2 derived from *A. thaliana* to improve vitamin B_6 (Raschke et al. 2011). Researchers have used a biogenetic approach to overexpress GTPCHI and aminodeoxychorismate synthase to improve folate content in rice and potato (Blancquaert et al. 2015; Blancquaert et al. 2013). Researchers have also developed B12-producing Lactobacillus strains (Capozzi et al. 2012).

6.4.4 FOOD FORTIFICATION

Food fortification is a means of deliberately increasing the micronutrient content in food. WHO recognizes food fortification as the most effective, affordable, and safe method to resolve micronutrient deficiency in the mass population (Dary and Hurrell 2006). The effectiveness of any fortification program can be increased when a food supply network already exists (Maurya et al. 2022; Maurya, Bashir, and Aggarwal 2020). Though, it is vital to choose suitable food vehicles for fortification with B vitamins. Information on B vitamin requirements within the population, the current B vitamin serum status of the population, the consumption pattern of the population, and the current gap existing between the current B vitamin status and recommended allowance are key to designing a successful fortification program. Furthermore, an understanding of the physico-chemical attributes of B vitamins and their derivatives is also equally important. The literature contains several reports where fortification programs are evaluated for B vitamin fortification (Hirsch et al. 2002; Dwyer et al. 2015; Allen 2006). For instance, the Food and Drug Administration also mandates wheat flour fortification with 2.9 mg thiamin/pound. Currently, a total 66 countries are fortifying wheat flour, rice, or maize flour either on a mandatory or voluntary basis. Similarly, infants' food is fortified with vitamin B_2 (Liu et al. 1993; Bates et al. 1982). However, riboflavin-fortified salt is more common in China. The United States initiated a voluntary niacin fortification program in bread for regions affected by pellagra (du Plessis and Laubscher 1971). Today, 17 countries, particularly African countries, are fortifying breakfast cereals with vitamin B_6 (Tucker et al. 2004). A vitamin B_9 fortification program resulted in a 19% reduction in NTDs when the United States implemented it on mandatory basis (Honein et al. 2001). Some efforts are made to improve vitamin B_{12} in bread (Winkels et al. 2008). However, a significant loss is also reported in these B vitamin-fortified foods and is highlighted in several excellent articles (Godoy, Amaya-Farfan, and Rodriguez-Amaya 2021; Fuliaş et al. 2014; Okmen and Bayindirli 1999). The degree of B vitamin loss in fortified foods is governed by several factors including food processing methods, pH, temperature, oxidation, and the complexity of the matrix, therefore hampering its disease-combating potential (Godoy, Amaya-Farfan, and Rodriguez-Amaya 2021). Researchers are taking advantage of microencapsulation techniques to address the abovementioned challenges.

6.5 NANOTECHNOLOGY

Nanotechnology promises several benefits to researchers to manipulate physicochemical attributes of B vitamins, which are but not limited to enhancing stability, extending shelf life, controlled/sustained release, improving organoleptic attributes, and enhancing bioavailability. Literature reports discuss a range of nanomaterials adopted for food fortification purposes, which differs in their physicochemical nature, such as inorganic, organic, carbon, or composite (Maurya and Aggarwal 2017; Maurya et al., 2020; Maurya, Bashir, and Aggarwal 2020; Maurya et al. 2021; Maurya et al. 2023; Maurya et al. 2022). However, the organic nanomaterials (solid lipid carrier, nanostructured lipid carrier, liposome, nanoemulsion) and polymeric nanomaterials like carbohydrate or protein derived nanomaterials. In today's scenario, these nanomaterials seem to be indispensable means in today's scenario for fortifying food with B vitamins, providing protection, encapsulation, and deliver them at defined rate. Though literature reports exist on B Vitamin supplementation, biofortification, post-harvest fortification, and food fortification, data on nanofortification for B vitamins are still in its infancy stage.

6.5.1 NEED FOR B VITAMIN NANOFORTIFICATION

Researchers have started preferring nanofortification over traditional fortification techniques as it offers better stability, improved bioavailability, excellent consumer satisfaction, and sustained/controlled release. Nano fortification utilizes microencapsulation techniques to encapsulate micronutrients to improve polydisperse index, loading capacity, and encapsulation efficiency. Literature reports about four major fortification methods for B vitamins including microbial fortification, biofortification, home fortification, and commercial fortification. Nevertheless, none of these fortification methods is sufficient enough to address all the challenges associated with B vitamins fortification.

6.6 NANODELIVERY SYSTEMS USED FOR B VITAMIN NANO FORTIFICATION

The literature reports a range of nanodelivery systems which are used for nano fortification of B vitamins. The literature also reports a range of nanomaterials (**Figure 6.2.**) which are but not limited to micro/nano capsules, micro/nanospheres, hydrogel, nanofibers, nano/microemulsion,

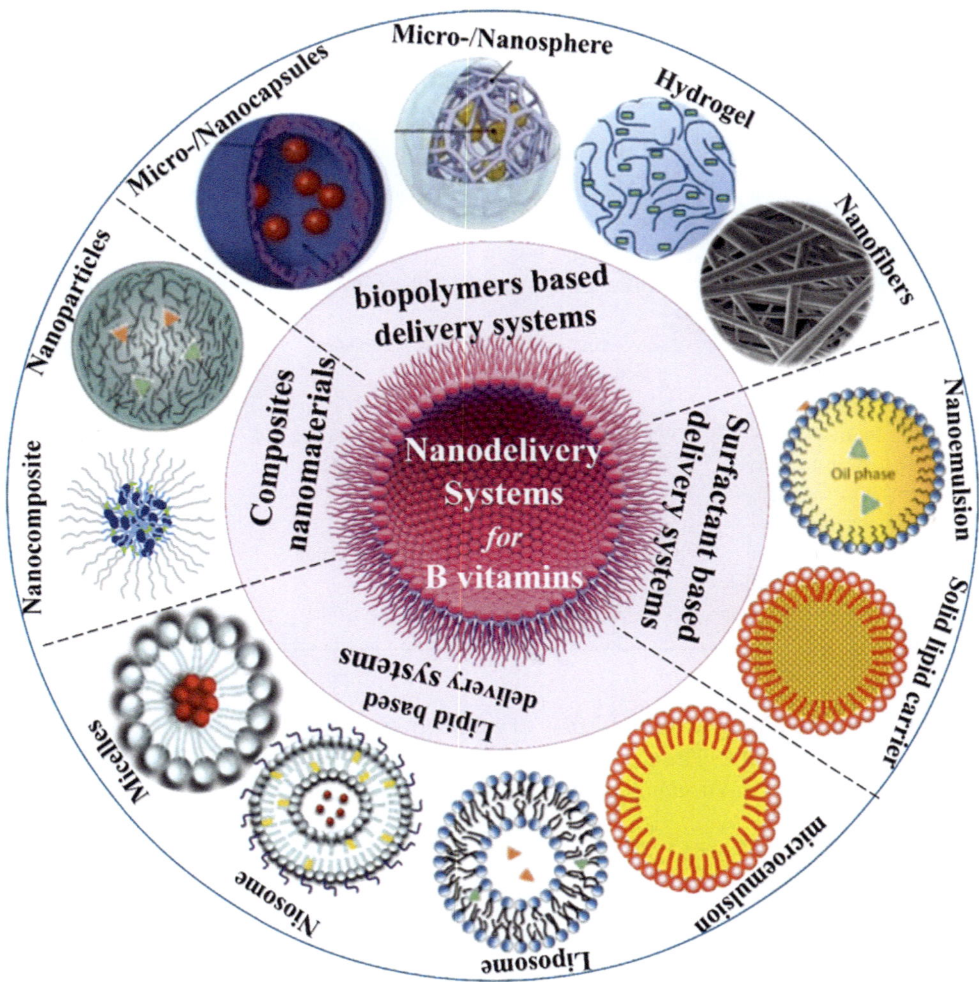

FIGURE 6.2 Nanoparticles adopted for encapsulation of bioactive compounds.

liposomes, noisomes, micelles, nanocomposites, and nanoparticle which are comprehensively discussed in our previous articles (Maurya and Aggarwal 2017; Maurya et al., 2020; Maurya, Bashir, and Aggarwal 2020; Maurya et al. 2021; Maurya et al. 2023; Maurya et al. 2022). For ease of understanding, these nanodelivery systems are classified into four major classes: organic nanomaterials, inorganic nanomaterials, carbon-based nanomaterials, and composite nanomaterials.

6.6.1 ORGANIC NANODELIVERY SYSTEMS

These delivery systems are the most commonly adopted for encapsulation and fortification of B vitamins due to their simple design, ease in fabrication, wide range in their biochemical attributes, high loading capacity, and improved bioavailability. Furthermore, these organic delivery systems are subcategorized depending upon their base materials like lipids, polysaccharides, proteins, and biopolymeric delivery systems.

6.6.1.1 Lipid-derived Nanodelivery Systems

These delivery systems are generally spherical microstructures that comprise one or multiple lipid bilayers to encompass an internal core containing water-soluble bioactives. These delivery systems

TABLE 6.3

Fabrication Method for Nanoparticle for B Vitamin Encapsulation

Vitamins	Purpose	Wall materials	Nanoencapsulation methods	References
Riboflavin	Functional food development	W1/O/W2 double emulsions with 4 different lipid sources	Nano emulsification	(Bou, Cofrades, and Jiménez-Colmenero 2014)
Riboflavin	Encapsulation & research purpose	Alginate and chitosan	Ionotropic gelation	(Azevedo et al. 2014)
Folic acid	Development of a natural delivery system	Lactoferrin, β-lactoglobulin	Coacervation	(Chapeau et al. 2016)
Folic acid	To study encapsulation efficiency and stability	Whey protein concentrate (WPC) and Commercial resistant starch.	Electrospraying and spray drying	(Pagano, Tiralti, and Perioli 2016)
Folic acid	Development of novel nanoparticles	Chitosan	Ultrasonication	(Bandara et al. 2018)
Cobalamin	Improve absorption	Soya bean protein	Cold gelation	(Zhang et al. 2015)

are generally prepared using extrusion, spray drying, self-assembly, coacervation, ion gelation, and electro-spraying (**Table 6.3**). The are found to offer enhanced stability, controlled/sustained release, better protection against environmental factors, and improved bioavailability for B vitamins.

6.6.1.2 Liposomes

Liposomes are considered as the most biocompatible, easy to use, stable, and suitable delivery systems for fortification. Researchers have also developed thiamine-loaded liposomes using phosphatidylcholine as the based materials. The literature reports a range of fabrication methods for liposome preparation, which are but not limited to homogenization, solvent evaporation and rehydration, emulsification, supercritical antisolvent based, film hydration-sonication techniques. Liposomes have been tested for encapsulation, protection, and delivery of several vitamins, including vitamin B_1 (Juhász et al. 2021), riboflavin (Ahmad et al. 2015; Ioniţă, Ion, and Carstocea 2003; Arsalan et al. 2020), niacin (He et al. 2018), vitamin B_6 (Abd-El-Azim et al. 2018), folic acid (Kumar, Huo, and Liu 2019; Lee and Low 1997) and vitamin B_{12} (Marchianò et al. 2022). However, these liposomes have demonstrated high encapsulation, protection, improved loading capacity, and improved bioavailability, but their full potential for food fortification still remains untapped.

6.6.1.3 Solid Lipid Carrier

These delivery systems are lipid spherical nanostructures and gaining attraction of research communities for their unparalleled properties like better encapsulation, solid outer interface, improved biocompatibility, and bioavailability. For example, riboflavin loaded lipid carrier was found to have better stability for encapsulated vitamin (Katouzian et al., 2017). Similarly vitamin B_{12} encapsulated in lipid carrier exhibited better stability against light and enhanced vitamin absorption efficiency (Couto, Alvarez, and Temelli 2017). The solid lipid carrier is adopted biomedical application but there are only few reports where these delivery systems are applied for food application.

6.6.1.4 Emulsion

Emulsions are small droplets dispersed in an aqueous phase. The literature reports a range of fabrication methods for emulsion preparation, including but not limited to homogenization, ultrasonication, solvent evaporation assisted with lyophilization, isoelectric precipitation, spontaneous emulsification, and microchannel emulsification (Maurya and Aggarwal 2017; Maurya et al., 2020; Maurya, Bashir, and Aggarwal 2020; Maurya et al. 2021; Maurya et al. 2023; Maurya et al., 2022). Researchers utilized different biopolymers to encapsulate folic acid using proteins as main wall materials (Assadpour, Jafari, and Maghsoudlou 2017). Similarly, researchers have optimized methods to encapsulate thiamine in chitosan-based nanoemulsions to improve vitamin stability (Tahir et al. 2023). Moreover, a few researchers developed fish oil-based riboflavin emulsions to study the impact of particle size and optical transparency on oxidation (Uluata, McClements, and Decker 2016). The literature reports several articles focusing on B encapsulation in emulsion systems for biomedical purposes, but its application for food applications is rare in existing literature.

6.6.1.5 Protein-based Delivery System

Proteins are one of the most suitable biopolymers for the fabrication of B vitamin-loaded nanoparticles due to their low cost, easy availability, high biocompatibility, and good digestibility. Proteins are one of the most explored polymeric materials for nanofortification purposes. For instance, researchers utilized soya protein to prepare vitamin B_{12} loaded nanoparticles (Zhang et al. 2015). Similarly, bovine serum albumin is used to encapsulate folic acid for the nanofortification of yoghurt (Zhang et al. 2015). It is also observed that casein nanoparticles have the ability to improve the oral bioavailability of folic acid by 50% (Penalva et al. 2015), and similarly, the potential of β-lactoglobulin to be a potential polymer for the preparation of suitable carriers for riboflavin delivery (Madalena et al. 2016). Several articles focusing on B encapsulation using protein-based nanoparticles can be seen with regard to its biomedical applications; however, their use in food fortification is still in its neonate stage.

6.6.1.6 Polysaccharide-based Delivery Systems

Wide physicochemical properties make polysaccharides one of the most suitable biopolymers to encapsulate B vitamins for nano fortification purposes. They are generally prepared using surface modification, homogenization, ultrasonication, and solvent evaporation. For instance, researchers improved B vitamin stability by encapsulating it in thymol-loaded chitosan nanoparticles (Ceylan et al., 2018). Similarly, alginate-pectin complex was made to improve folic acid bioavailability (Pamunuwa et al., 2020). In another study, researchers encapsulated folic acid and cobalamin in polylactic-co-glycolic acid nanoparticles to improve vitamins' bio accessibility (Ramalho, Loureiro, and Pereira 2021). Only few research articles are available on food-grade polysaccharide nanoparticles encapsulating vitamins (Acevedo-Fani, Soliva-Fortuny, and Martín-Belloso 2018) and need to be explored more for nanofortification purposes.

6.6.2 Inorganic Nanoparticles

Researchers are exploring the potential of inorganic nanoparticles for Nano fortification and have found more suitable for mineral Nano fortification. The literature reports a range of inorganic nanoparticles derived from silica, amino silicate, clays, calcium phosphate, and calcium carbonate (Jampilek, Kos, and Kralova 2019). Mesoporous silica is found to be a suitable nanoparticle for the encapsulation of folic acid and nano fortification of fruit-based beverages (Ruiz-Rico et al. 2017). In another study, researchers have also developed Simple powders and tablets of nanostructured hybrids to improve the encapsulation of folic acid (Pagano, Tiralti, and Perioli 2016). Similarly, cobalamin loaded nanoclays have also developed for nano fortification purposes and demonstrated improved absorption (Alavijeh, Sarvi, and Afarani 2017).

6.6.3 Carbon-based Nanoparticles

These nanoparticles involve carbon nanoparticles, carbon nanodots, and carbon nanotubes. They have immense potential for encapsulation of bioactive compounds but their utilization for nano fortification is still well-witnessed except a few published work (Zawari, Aghaei, and Monajjemi 2016).

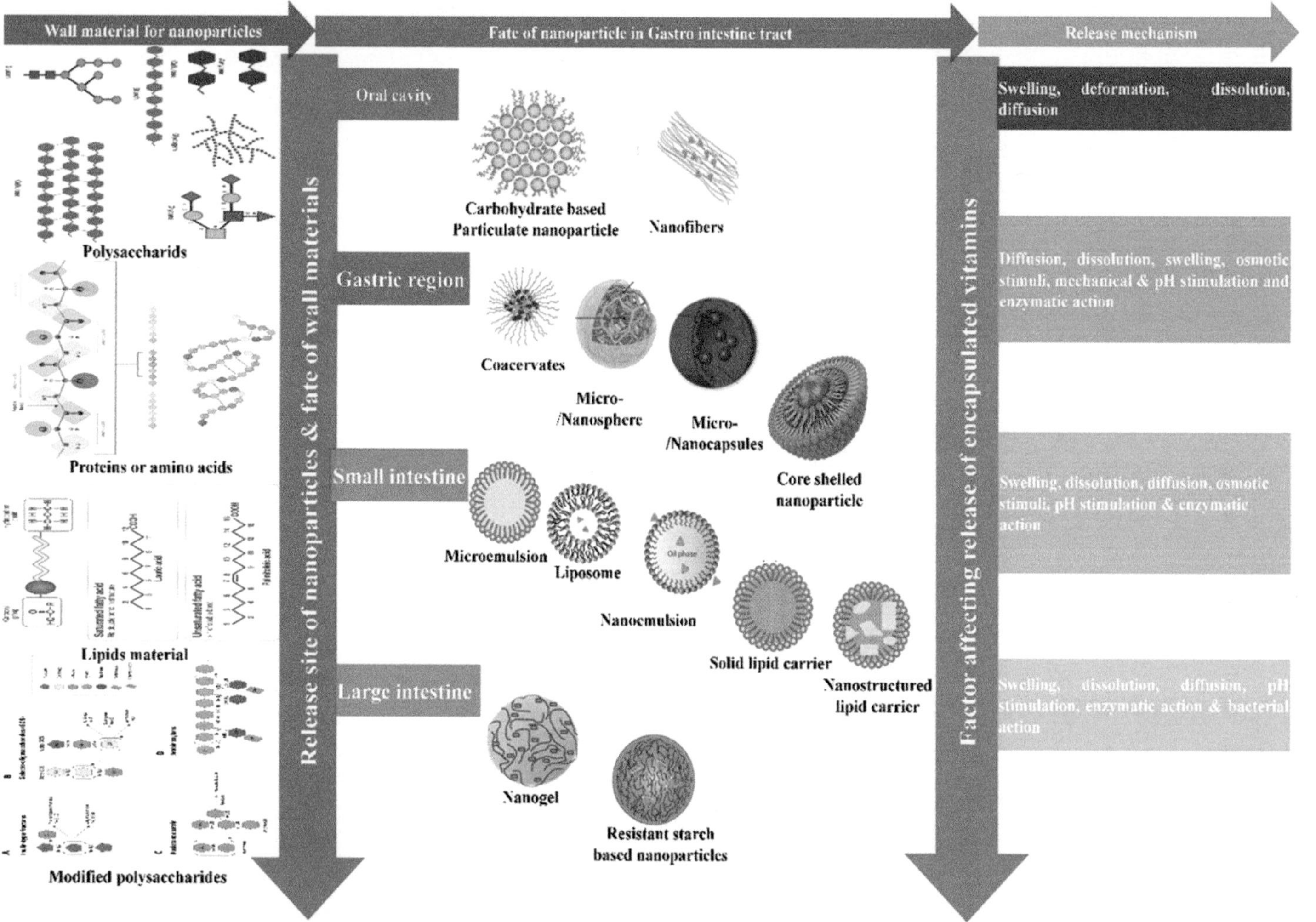

FIGURE 6.3　The digestion process of nanoparticles for releasing the encapsulated vitamins.

6.6.4 Composite Nanoparticles

These delivery systems are generally derived from proteins and lipid materials to improve loading capacity, enhance encapsulation efficiency, sustain/controlled release, and sited directed release with minimal cytotoxicity (Liu et al. 2018). Researchers have designed novel protein-lipid composition nanoparticles to improve the protection and encapsulation of bioactive compounds, but its use in nano fortification of B vitamins is still missing in the literature.

6.7 SAFETY, COMPLIANCE AND RISKS OF VITAMIN-LOADED NANOPARTICLES

It is essential that B vitamin nanoparticles should be safe for human consumption and should not pose health threats in long-term consumption (de Souza Simões et al. 2017; Dima et al. 2020). The nanoparticles should get digested and release the encapsulated vitamins to offer the health benefits. Figure 6.3 highlights the process through which these nanoparticle go. Table 6.2 highlights the recommended dietary allowance for different B vitamins) before getting excreted. Since these nanoparticles are fabricated with a range of surfactants and polymers that may have some adverse impact on human health, organic and biocompatible alternatives should be preferred (Chau, Wu, and Yen 2007). Likewise, application of alcohols, organic solvents, or synthetic chemicals in the fabrication process of nanoparticles should be avoided or completely removed prior to consumption (Chau, Wu, and Yen 2007). Generally, the effects of their short- and long-term impact on the human body should be evaluated (Chau, Wu, and Yen 2007). The Food and Drug Administration (FDA) in the United States has drafted a few recommendation for inclusion of nanoparticles in foods (Amenta et al. 2015). The European Food Safety Authority (EFSA) has also drafted few regulations on the application of nanoparticles in foods (Livney 2015). The procedure to carry out risk assessments of nanoparticles utilized in foods is well discussed in Hardy et al. (2018).

6.8 CONCLUSION

The obligatory role of B vitamins in human health is well recognized in the literature. With the inability to produce these vital compounds, humans rely on an external diet to meet their daily requirements. Several factors, including lack of vitamin B-rich foods in the food basket, ethnic/religious constraints, processing conditions, and exposure to several environmental factors, contribute to increased deficiency in the population, irrespective of age, gender, ethnicity, and geographical location. Though WHO recognizes food fortification as the most effective means to resolve B vitamin deficiency, incorporation of vitamins in food commodities inherits several challenges like loss during food processing, storage and distribution, cost, affordability, and consumer acceptability. Fortunately, these limitations can be easily overcome using microencapsulation techniques. A few attempts have been made to explore the potential of nanoparticles for nanofortification of B vitamins, which are briefly discussed in this article. However, extensive research exploration is still required to assess the potential of these nanoparticles for nanofortification purposes.

REFERENCES

Abd-El-Azim, Heba, Alyaa Ramadan, Noha Nafee, and Nawal Khalafallah. 2018. Entrapment efficiency of pyridoxine hydrochloride in unilamellar liposomes: Experimental versus model-generated data. *Journal of Liposome Research* 28(2):112–116.

Acevedo-Fani, Alejandra, Robert Soliva-Fortuny, and Olga Martín-Belloso. 2018. Photo-protection and controlled release of folic acid using edible alginate/chitosan nanolaminates. *Journal of Food Engineering* 229:72–82.

Ahmad, Iqbal, Arsalan Adeel, Syed Abid Ali, et al. 2015. Formulation and stabilization of riboflavin in liposomal preparations. *Journal of Photochemistry and Photobiology, B: Biology* 153:358–366.

Alavijeh, Mozhgan Akbari, Mehdi Nasiri Sarvi, and Zahra Ramazani Afarani. 2017. Properties of adsorption of vitamin B12 on nanoclay as a versatile carrier. *Food Chemistry* 219:207–214.

Allen, Lindsay H. 2006. New approaches for designing and evaluating food fortification programs. *The Journal of Nutrition* 136(4):1055–1058.

Amenta, Valeria, Karin Aschberger, Maria Arena, et al. 2015. Regulatory aspects of nanotechnology in the agri/feed/food sector in EU and non-EU countries. *Regulatory Toxicology and Pharmacology* 73(1):463–476.

Arsalan, Adeel, Iqbal Ahmad, Syed Abid Ali, et al. 2020. The kinetics of photostabilization of cyanocobalamin in liposomal preparations. *International Journal of Chemical Kinetics* 52(3):207–217.

Assadpour, Elham, Seid-Mahdi Jafari, and Yahya Maghsoudlou. 2017. Evaluation of folic acid release from spray dried powder particles of pectin-whey protein nano-capsules. *International Journal of Biological Macromolecules* 95:238–247.

Azevedo, Maria A., Ana I. Bourbon, António A. Vicente, and Miguel A. Cerqueira. 2014. Alginate/chitosan nanoparticles for encapsulation and controlled release of vitamin B2. *International Journal of Biological Macromolecules* 71:141–146.

Bandara, Subhani, Codi-anne Carnegie, Chevaun Johnson, et al. 2018. Synthesis and characterization of zinc/chitosan-folic acid complex. *Heliyon* 4(8).

Bates, C. J., A. M. Prentice, A. A. Paul, A. Prentice, B. A. Sutcliffe, and R. G. Whitehead. 1982. Riboflavin status in infants born in rural Gambia, and the effect of a weaning food supplement. *Transactions of the Royal Society of Tropical Medicine and Hygiene* 76(2):253–258.

Blancquaert, Dieter, Sergei Storozhenko, Jeroen Van Daele, et al. 2013. Enhancing pterin and para-aminobenzoate content is not sufficient to successfully biofortify potato tubers and Arabidopsis thaliana plants with folate. *Journal of Experimental Botany* 64(12):3899–3909.

Blancquaert, Dieter, Jeroen Van Daele, Simon Strobbe, et al. 2015. Improving folate (vitamin B9) stability in biofortified rice through metabolic engineering. *Nature Biotechnology* 33(10):1076–1078.

Bou, Ricard, Susana Cofrades, and Francisco Jiménez-Colmenero. 2014. Physicochemical properties and riboflavin encapsulation in double emulsions with different lipid sources. *LWT – Food Science and Technology* 59(2):621–628.

Brown, Kenneth L. 2005. Chemistry and enzymology of vitamin B12. *Chemical Reviews* 105(6):2075–2150.

Capozzi, Vittorio, Pasquale Russo, María Teresa Dueñas, Paloma López, and Giuseppe Spano. 2012. Lactic acid bacteria producing B-group vitamins: A great potential for functional cereals products. *Applied Microbiology and Biotechnology* 96(6):1383–1394.

Caritá, Amanda Costa, Bruno Fonseca-Santos, Jemima Daniela Shultz, Bozena Michniak-Kohn, Marlus Chorilli, and Gislaine Ricci Leonardi. 2020. Vitamin C: One compound, several uses. Advances for delivery, efficiency and stability. *Nanomedicine: Nanotechnology, Biology and Medicine* 24:102117.

Ceylan, Zafer, Mustafa Yaman, Osman Sağdıç, Ercan Karabulut, and Mustafa Tahsin Yilmaz. 2018. Effect of electrospun thymol-loaded nanofiber coating on vitamin B profile of gilthead sea bream fillets (Sparus aurata). *LWT* 98:162–169.

Chapeau, Anne-Laure, Guilherme M. Tavares, Pascaline Hamon, Thomas Croguennec, Denis Poncelet, and Saïd Bouhallab. 2016. Spontaneous co-assembly of lactoferrin and β-lactoglobulin as a promising biocarrier for vitamin B9. *Food Hydrocolloids* 57:280–290.

Chaturvedi, Siddhant, Roni Chaudhary, and Siddharth Tiwari. 2021. Contribution of crop biofortification in mitigating vitamin deficiency globally. *Genome Engineering for Crop Improvement*:112–130.

Chau, Chi-Fai, Shiuan-Huei Wu, and Gow-Chin Yen. 2007. The development of regulations for food nanotechnology. *Trends in Food Science and Technology* 18(5):269–280.

Chugh, Preeta K., and Y. Lhamo. 2012. An assessment of vitamin supplements in the Indian market. *Indian Journal of Pharmaceutical Sciences* 74(5):469.

Combs Jr, Gerald F., and James P. McClung. 2016. *The vitamins: Fundamental aspects in nutrition and health*. Academic Press.

Couto, Ricardo, Victor Alvarez, and Feral Temelli. 2017. Encapsulation of vitamin B2 in solid lipid nanoparticles using supercritical CO2. *The Journal of Supercritical Fluids* 120:432–442.

Dary, Omar, and Richard Hurrell. 2006. Guidelines on food fortification with micronutrients. In *World Health Organization, Food and Agricultural Organization of the United Nations*. Geneva:1–376.

de Souza Simões, Lívia, Daniel A. Madalena, Ana C. Pinheiro, Jose A. Teixeira, Antonio A. Vicente, and Oscar L. Ramos. 2017. Micro-and Nano bio-based delivery systems for food applications: In vitro behavior. *Advances in Colloid and Interface Science* 243:23–45.

Dima, Cristian, Elham Assadpour, Stefan Dima, and Seid Mahdi Jafari. 2020. Bioavailability of nutraceuticals: Role of the food matrix, processing conditions, the gastrointestinal tract, and nanodelivery systems. *Comprehensive Reviews in Food Science and Food Safety* 19(3):954–994.

Dong, Wei, Nicholas Thomas, Pamela C. Ronald, and Aymeric Goyer. 2016. Overexpression of thiamin biosynthesis genes in rice increases leaf and unpolished grain thiamin content but not resistance to Xanthomonas oryzae pv. oryzae. *Frontiers in Plant Science* 7:616.

du Plessis, J. P., W. Wittmann, M. E. J. Louw, A. Nel, P. van Twisk, and N. F. Laubscher. 1971. The clinical and biochemical effects of riboflavin and nicotinamide supplementation upon Ba. *South African Medical Journal* 45(19):530–537.

Dwyer, Johanna T., Kathryn L. Wiemer, Omar Dary, et al. 2015. Fortification and health: Challenges and opportunities. *Advances in Nutrition* 6(1):124–131.

Fuliaş, Adriana, Gabriela Vlase, Titus Vlase, Daniela Oneţiu, Nicolae Doca, and Ionuţ Ledeţi. 2014. Thermal degradation of B-group vitamins: B 1, B 2 and B 6: Kinetic study. *Journal of Thermal Analysis and Calorimetry* 118(2):1033–1038.

Godoy, Helena Teixeira, Jaime Amaya-Farfan, and Delia B. Rodriguez-Amaya. 2021. Degradation of vitamins. In Delia B. Rodriguez-Amaya and Jaime Amaya-Farfan (Eds), *Chemical changes during processing and storage of foods*. Elsevier.

Goyer, Aymeric. 2017. Thiamin biofortification of crops. *Current Opinion in Biotechnology* 44:1–7.

Green, Ralph, Lindsay H. Allen, Anne-Lise Bjørke-Monsen, et al. 2017. Vitamin B12 deficiency. *Nature Reviews Disease Primers* 3(1):1–20.

Gregory III, Jesse F. 1985. Chemical changes of vitamins during food processing. In *Chemical changes in food during processing*. Springer.

Guilland, Jean-Claude, and Isabelle Aimone-Gastin. 2013. Vitamin B9. *La Revue du Praticien* 63(8):1079–1081.

Hardy, A., D. Benford, T. Halldorsson, M. J. Jeger, H. K. Knutsen, and S. More. 2018. Guidance on risk assessment of the application of nanoscience and nanotechnologies in the food and feed chain: Part 1, human and animal health. *EFSA Journal* 16(7):5327.

He, Haisheng, Yi Lu, Jianping Qi, Weili Zhao, Xiaochun Dong, and Wei Wu. 2018. Biomimetic thiamine- and niacin-decorated liposomes for enhanced oral delivery of insulin. *Acta Pharmaceutica Sinica B* 8(1):97–105.

Hellmann, Hanjo, and Sutton Mooney. 2010. Vitamin B6: A molecule for human health? *Molecules* 15(1):442–459.

Herrera-Ardila, Yenny Mayerly, David Orrego, Andrés Felipe Bejarano-López, and Bernadette Klotz-Ceberio. 2022. Effect of heat treatment on vitamin content during the manufacture of food products at industrial scale. *Dyna* 89(223):127–132.

Hirsch, Sandra, Pía de la Maza, Gladys Barrera, Vivian Gattás, Margarita Petermann, and Daniel Bunout. 2002. The Chilean flour folic acid fortification program reduces serum homocysteine levels and masks vitamin B-12 deficiency in elderly people. *The Journal of Nutrition* 132(2):289–291.

Honein, Margaret A., Leonard J. Paulozzi, T. J. Mathews, J. David Erickson, and Lee-Yang C. Wong. 2001. Impact of folic acid fortification of the US food supply on the occurrence of neural tube defects. *JAMA* 285(23):2981–2986.

Ioniţă, M. A., R. M. Ion, and B. Carstocea. 2003. Photochemical and photodynamic properties of vitamin B2-- Riboflavin and liposomes. *Oftalmologia (Bucharest, Romania: 1990)* 58(3):29–34.

Jampilek, Josef, Jiri Kos, and Katarina Kralova. 2019. Potential of nanomaterial applications in dietary supplements and foods for special medical purposes. *Nanomaterials* 9(2):296.

Juhász, Ádám, Ditta Ungor, Egon Z. Várkonyi, Norbert Varga, and Edit Csapó. 2021. The pH-dependent controlled release of encapsulated vitamin B1 from liposomal nanocarrier. *International Journal of Molecular Sciences* 22(18):9851.

Karmas, Endel, and Robert S. Harris. 2012. *Nutritional evaluation of food processing*. Springer Science & Business Media.

Katouzian, Iman, Afshin Faridi Esfanjani, Seid Mahdi Jafari, and Sahar Akhavan. 2017. Formulation and application of a new generation of lipid nano-carriers for the food bioactive ingredients. *Trends in Food Science and Technology* 68:14–25.

Kumar, Parveen, Peipei Huo, and Bo Liu. 2019. Formulation strategies for folate-targeted liposomes and their biomedical applications. *Pharmaceutics* 11(8):381.

Lanska, Douglas J. 2012. The discovery of niacin, biotin, and pantothenic acid. *Annals of Nutrition and Metabolism* 61(3):246–253.

Lee, Robert J., and Philip S. Low. 1997. Folate-targeted liposomes for drug delivery. *Journal of Liposome Research* 7(4):455–466.

Lešková, Andrea Stešková-Monika Morochovičová-Emília. 2006. Vitamin C degradation during storage of fortified foods. *Journal of Food and Nutrition Research* 45(2):55–61.

Liu, Dong-sheng, Christopher J. Bates, T. A. Yin, X. B. Wang, and Cheng-qian Lu. 1993. Nutritional efficacy of a fortified weaning rusk in a rural area near Beijing. *The American Journal of Clinical Nutrition* 57(4):506–511.

Liu, Guangyu, Weijuan Huang, Oksana Babii, et al. 2018. Novel protein–lipid composite nanoparticles with an inner aqueous compartment as delivery systems of hydrophilic nutraceutical compounds. *Nanoscale* 10(22):10629–10640.

Livney, Yoav D. 2015. Nanostructured delivery systems in food: Latest developments and potential future directions. *Current Opinion in Food Science* 3:125–135.

Lykkesfeldt, Jens, and Henrik E. Poulsen. 2010. Is vitamin C supplementation beneficial? Lessons learned from randomised controlled trials. *British Journal of Nutrition* 103(9):1251–1259.

Madalena, Daniel A., Óscar L. Ramos, Ricardo N. Pereira, et al. 2016. In vitro digestion and stability assessment of β-lactoglobulin/riboflavin nanostructures. *Food Hydrocolloids* 58:89–97.

Mahabadi, Navid, Aakriti Bhusal, and Stephen W. Banks. 2022. Riboflavin deficiency. In StatPearls [Internet]. StatPearls Publishing.

Makarov, Mikhail V., Samuel A. J. Trammell, and Marie E. Migaud. 2019. The chemistry of the vitamin B3 metabolome. *Biochemical Society Transactions* 47(1):131–147.

Manzetti, Sergio, Jin Zhang, and David van der Spoel. 2014. Thiamin function, metabolism, uptake, and transport. *Biochemistry* 53(5):821–835.

Maqbool, Muhammad Amir, Muhammad Aslam, Waseem Akbar, and Zubair Iqbal. 2017. Biological importance of vitamins for human health: A review. *Journal of Agriculture: Basic Sciences* 2(3):50–58.

Marchianò, Verdiana, Maria Matos, Esther Serrano, et al. 2022. Lyophilised nanovesicles loaded with vitamin B12. *Journal of Molecular Liquids* 365:120129.

Maurya, Vaibhav Kumar, and Manjeet Aggarwal. 2017. Enhancing bio-availability of vitamin D by nano-engineered based delivery systems-An overview. *International Journal of Current Microbiology and Applied Sciences* 6(7):340–353.

Maurya, Vaibhav Kumar, Manjeet Aggarwal, Vijay Ranjan, and K. M. Gothandam. 2020. Improving bioavailability of vitamin A in food by encapsulation: An update *Nanoscience in Medicine* 1:117–145.

Maurya, Vaibhav Kumar, Khalid Bashir, and Manjeet Aggarwal. 2020. Vitamin D microencapsulation and fortification: Trends and technologies. *The Journal of Steroid Biochemistry and Molecular Biology* 196:105489.

Maurya, Vaibhav Kumar, Amita Shakya, Manjeet Aggarwal, Kodiveri Muthukaliannan Gothandam, Torsten Bohn, and Sunil Pareek. 2021. Fate of β-carotene within loaded delivery systems in food: State of knowledge. *Antioxidants* 10(3):426.

Maurya, Vaibhav Kumar, Amita Shakya, Khalid Bashir, Satish Chand Kushwaha, and David Julian McClements. 2022. Vitamin A fortification: Recent advances in encapsulation technologies. *Comprehensive Reviews in Food Science and Food Safety* 21(3):2772–2819.

Maurya, Vaibhav Kumar, Amita Shakya, Khalid Bashir, Kulsum Jan, and David Julian McClements. 2023. Fortification by design: A rational approach to designing vitamin D delivery systems for foods and beverages. *Comprehensive Reviews in Food Science and Food Safety* 22(1):135–186.

Naderi, Nassim, and James D. House. 2018. Recent developments in folate nutrition. *Advances in Food and Nutrition Research* 83:195–213.

Nantel, G., and K. Tontisirin. 2001. Human vitamin and mineral requirements. Report of a joint FAO/WHO expert consultation, Bangkok, Thailand. Food and nutrition division FAO, Rome, Italy. http://www.fao.org/3/a-y2809e.pdf.

Okmen, Zinet Aytanga, and A. Levent Bayindirli. 1999. Effect of microwave processing on water soluble vitamins: Kinetic parameters. *International Journal of Food Properties* 2(3):255–264.

Ottaway, P. Berry. 2010. Stability of vitamins during food processing and storage. In *Chemical deterioration and physical instability of food and beverages*. Elsevier.

Pagano, Cinzia, Maria Cristina Tiralti, and Luana Perioli. 2016. Nanostructured hybrids for the improvement of folic acid biopharmaceutical properties. *Journal of Pharmacy and Pharmacology* 68(11):1384–1395.

Pamunuwa, Geethi, Nipunika Anjalee, Diduli Kukulewa, Chapa Edirisinghe, Farrah Shakoor, and Desiree Nedra Karunaratne. 2020. Tailoring of release properties of folic acid encapsulated nanoparticles via changing alginate and pectin composition in the matrix. *Carbohydrate Polymer Technologies and Applications* 1:100008.

Parra, Marcelina, Seth Stahl, and Hanjo Hellmann. 2018. Vitamin B6 and its role in cell metabolism and physiology. *Cells* 7(7):84.

Penalva, Rebeca, Irene Esparza, Maite Agüeros, Carlos J. Gonzalez-Navarro, Carolina Gonzalez-Ferrero, and Juan M. Irache. 2015. Casein nanoparticles as carriers for the oral delivery of folic acid. *Food Hydrocolloids* 44:399–406.

Polegato, Bertha F., Amanda G. Pereira, Paula S. Azevedo, et al. 2019. Role of thiamin in health and disease. *Nutrition in Clinical Practice* 34(4):558–564.

Premjit, Yashaswini, Sachchidanand Pandey, and Jayeeta Mitra. 2022. Recent trends in folic acid (vitamin B9) encapsulation, controlled release, and mathematical modelling. *Food Reviews International* 39(8):1–35.

Ramalho, Maria Joao, Joana Angelica Loureiro, and Maria Carmo Pereira. 2021. Poly (lactic-co-glycolic acid) Nanoparticles for the Encapsulation and gastrointestinal Release of vitamin B9 and vitamin B12. *ACS Applied Nano Materials* 4(7):6881–6892.

Rasche, Andreas. 2020. 15The United Nations global compact and the sustainable development goals. In *Research handbook of responsible management*:228.

Raschke, Maja, Svetlana Boycheva, Michèle Crèvecoeur, et al. 2011. Enhanced levels of vitamin B6 increase aerial organ size and positively affect stress tolerance in Arabidopsis. *The Plant Journal* 66(3):414–432.

Rizzo, Gianluca, and Antonio Simone Laganà. 2020. A review of vitamin B12. *Molecular Nutrition*:105–129.

Robinson, Frank Alfred. 1951. *The vitamin B complex.* Рипол Классик.

Ruiz-Rico, María, Édgar Pérez-Esteve, María J. Lerma-García, María D. Marcos, Ramón Martínez-Máñez, and José M. Barat. 2017. Protection of folic acid through encapsulation in mesoporous silica particles included in fruit juices. *Food Chemistry* 218:471–478.

Russo, Pasquale, Vittorio Capozzi, Mattia Pia Arena, et al. 2014. Riboflavin-overproducing strains of Lactobacillus fermentum for riboflavin-enriched bread. *Applied Microbiology and Biotechnology* 98(8):3691–3700.

Ryley, Janice, and P. Kajda. 1994. Vitamins in thermal processing. *Food Chemistry* 49(2):119–129.

Saedisomeolia, Ahmad, and Marziyeh Ashoori. 2018. Riboflavin in human health: A review of current evidences. *Advances in Food and Nutrition Research* 83:57–81.

Sarwar, Muhammad Farhan, Muhammad Haroon Sarwar, and Muhammad Sarwar. 2021. Deficiency of vitamin B-Complex and its relation with body disorders. *B-Complex Vitamins-Sources, Intakes and Novel Applications*:79–100.

Schellack, Gustav, Pamela Harirari, and Natalie Schellack. 2016. B-complex vitamin deficiency and supplementation. *SA Pharmaceutical Journal* 83(4):14–19.

Spoelstra-de Man, Angélique M. E., Paul W. G. Elbers, and Heleen M. Oudemans-Van Straaten. 2018. Vitamin C: Should we supplement? *Current Opinion in Critical Care* 24(4):248–255.

Tahir, Iqmal, Justitia Millevania, Karna Wijaya, Roswanira Abdul Wahab, and Widi Kurniawati. 2023. Optimization of thiamine chitosan nanoemulsion production using sonication treatment. *Results in Engineering* 17:100919.

Titcomb, Tyler J., and Sherry A. Tanumihardjo. 2019. Global concerns with B vitamin statuses: Biofortification, fortification, hidden hunger, interactions, and toxicity. *Comprehensive Reviews in Food Science and Food Safety* 18(6):1968–1984.

Tucker, Katherine L., Beth Olson, Peter Bakun, Gerard E. Dallal, Jacob Selhub, and Irwin H. Rosenberg. 2004. Breakfast cereal fortified with folic acid, vitamin B-6, and vitamin B-12 increases vitamin concentrations and reduces homocysteine concentrations: A randomized trial. *The American Journal of Clinical Nutrition* 79(5):805–811.

Uluata, Sibel, D. Julian McClements, and Eric A. Decker. 2016. Riboflavin-induced oxidation in fish oil-in-water emulsions: Impact of particle size and optical transparency. *Food Chemistry* 213:457–461.

Winkels, Renate M., Ingeborg A. Brouwer, Robert Clarke, Martijn B. Katan, and Petra Verhoef. 2008. Bread cofortified with folic acid and vitamin B-12 improves the folate and vitamin B-12 status of healthy older people: A randomized controlled trial. *The American Journal of Clinical Nutrition* 88(2):348–355.

World Health Organization. 1999. *Thiamine deficiency and its prevention and control in major emergencies.* World Health Organization.

World Health Organization. 2006. *Guidelines on food fortification with micronutrients.* World Health Organization.

Yaman, Mustafa, Jale Çatak, Halime Uğur, et al. 2021. The bioaccessibility of water-soluble vitamins: A review. *Trends in Food Science and Technology* 109:552–563.

Zawari, M., H. Aghaei, and M. Monajjemi. 2016. Nano drug delivery of some vitamins such as B6 and C with carbon nanotube: A density functional theory study. *Quantum Matter* 5(1):147–154.

Zhang, Jing, Catherine J. Field, Donna Vine, and Lingyun Chen. 2015. Intestinal uptake and transport of vitamin B 12-loaded soy protein nanoparticles. *Pharmaceutical Research* 32(4):1288–1303.

7 Vitamin E and K Fortification

Yogesh Kumar, Swarnima Dey, and Kiran Verma

LIST OF ABBREVIATIONS

1. PUFA: Polyunsaturated fatty acids
2. MK: Menaquinone
3. Glu: Glutamate
4. ABCA1: ATP-binding cassette transporter A1
5. CEHC: Carboxy ethyl hydroxy chromanol
6. CYP: Cytochrome P450
7. LDL: Low-density lipoprotein
8. VLDL: Very low-density lipoprotein
9. HDL: High-density lipoprotein
10. α-TTP: α-Tocopherol transfer protein
11. SR-B1: Scavenger receptor of class B and type 1
12. PLTP: Phospholipid transfer protein
13. CETP: Cholesteryl ester transfer protein
14. LRP: LDL receptor protein
15. LPL: Lipoprotein lipase
16. KH: Vitamin K hydroquinone
17. QR: Quinone reductases
18. GGCX: Gamma-glutamyl carboxylase
19. RDA: Recommended Dietary Allowance
20. IOM: Institute of Medicine
21. EFSA: European Food Safety Authority
22. PTT: Partial thromboplastin time
23. SE: Spontaneous emulsification
24. EPI: Emulsion phase inversion
25. PIT: Phase inversion temperature
26. WOR: Water to oil phase ratio
27. HLB: Hydrophilic lipophilic balance
28. SOR: Surfactant to oil ratio
29. SOW: Surfactant oil water
30. GIT: Gastrointestinal tract
31. LCT: Long chain triglyceride
32. VKDB: Vitamin K deficiency bleeding

7.1 INTRODUCTION

Vitamin E is a widespread vitamin found in everyday foods. Vitamin E is also known as tocopherol, which means "to beget childbirth" and comes from the Greek words "tokos" and "pherein". The first explanation and significance of vitamin E was given by Evans and Bishop in 1922, classifying tocopherols (TOH) and tocotrienols (T3) as vitamins, including their α-, β-, γ-, and δ –types (Evans & Bishop, 1922). α-tocopherol predominates and is biologically active among the eight fat-soluble

DOI: 10.1201/9781003160663-9

derivatives, having a 6-hydroxyl group as the active site (Willson, 1983). Vitamin E, being a lipid soluble, is primarily present in oily plants (nuts and oilseeds). α-Tocopherol is abundant in sunflower oil, hazelnuts, almonds, and germ oil, while γ-Tocopherol is primarily found in walnuts, palm oil, and soybeans (Cardenas & Ghosh, 2013). T3 is available in various foods, including palm oil, rice bran oil, and cereals (Sen et al., 2007). T3s also found naturally in wheat germ, cocoa butter, barley, soybeans, and coconut oil. On the other hand, avocado, dried apricots, green olives, and some legumes contain no vitamin E, whereas vegetables and fruits contain low vitamin E (Wong & Radhakrishnan, 2012).

There are various factors that influence the bioavailability of vitamin E bioavailability: (1) the quantity of vitamin E present along with the consumption of interfering nutrients; (2) factors related to lifestyle; (3) proteins that help in the absorption of vitamin E and the variations in the efficiency of absorption of vitamin E, which is influenced by some diseases; (4) gender; (5) metabolism of vitamin E; and (6) genetic polymorphisms (Lisa Schmölz et al., 2016). When it comes to vitamin E, intestinal absorption is a key factor that limits its bioavailability. Vitamin E, as a lipid-soluble vitamin, is known to follow other lipophilic molecules and lipids through cellular uptake, hepatic metabolism, and intestinal absorption processes. As a result, the bioavailability of vitamin E depends on the availability of lipid-rich foods for intestinal absorption (Rigotti, 2007).

Vitamin E's main biological role is to shield polyunsaturated fatty acids (PUFA), low-density lipoprotein (LDL), and other components of cell membranes from free radical oxidation. The phospholipid bilayer of cell membranes is where vitamin E is found. It is particularly good at preventing lipid peroxidation, which is a set of chemical reactions that involve the oxidation of PUFAs. Lipid peroxidation products have been linked to a variety of diseases and health conditions (Duthie et al., 1992). While vitamin E is predominantly found in cell and organelle membranes, where it has the greatest protective effect, its concentration in these membranes can be as low as one molecule per 2000 phospholipid molecules. This means that it is quickly regenerated after reacting with free radicals, probably by other antioxidants (WHO, 2004).

7.2 VITAMIN K

Danish researcher Henrik Dam was the first to discover vitamin K in the 1930s. Vitamin K became known for its role in hemostasis and was given the name vitamin K for to represent "Koagulationsvitamin", or "Coagulation vitamin" (Dahlberg et al., 2021). The characteristic ring structure, i.e., naphthoquinone, is commonly shared by all the vitamin K family members, which differentiates it from other vitamins. Vitamin K_1 (phylloquinone) is the main supplement and dietary source of vitamin K and contains a phytol side chain (Dahlberg et al., 2017; Turck et al., 2017). Vitamin K_2 i.e., menaquinone (MK) contains isoprenoid residues that are again divided into MK-n, and these are indeed the naturally existing forms. Vitamin K_1 is mostly contained in dark green leafy and other green vegetables including spinach, broccoli, and brassica, whereas vitamin K_2 is mostly available in fermented foods including yogurt, curds, and natto (Turck et al., 2017). Also, vitamin K_3 (menadione) and K_5 are synthetic derivatives available (Kurosu & Begari, 2010) in animal feed and fungistatic agent, respectively (Merrifield & Yang, 1965). Vitamin K_2 (menaquinones) and MK-4 are produced by the bacterial fermentation in the colon and by phylloquinone catabolism in intestinal mucosa during absorption (Turck et al., 2017). Vitamin K acts as an important cofactor during the conversion of c-carboxylation of glutamate (Glu) to c-carboxyglutamate (Gla) that binds the protein with calcium ions and reacts with endothelial cells and phospholipids (Weston & Monahan, 2008). These proteins all together are known as Gla proteins, and their involvement in extrahepatic functions like carcinogenesis, bone metabolism, inflammation, and calcification has been linked to vitamin K dependent proteins, especially K_2 (Ozturk, 2017; Villa, 2017).

7.3 VITAMIN E CHEMISTRY, BIOLOGICAL ROLE, ABSORPTION, AND METABOLISM

Vitamin E absorption in the intestine is usually similar to dietary fat absorption. Vitamin E esters are hydrolyzed in the acidic environment of the stomach and then in the small intestine through enzymatic action of steryl-ester acylhydrolase, resulting in only the non-sterified vitamin E being absorbed (Nagy et al., 2013; Reboul, 2017). Bile acids are secreted by the liver that help with lipid digestion into the small intestine and the production of combined micelles where micelles and micellar lipids are transported into enterocytes through receptor-mediated transport or passive diffusion. There is no unique plasma transport protein for vitamin E. Vitamin E is introduced into chylomicrons and its secretion takes place in the lymphatic system by enterocytes with the participation of the ATP-binding cassette transporter A1 (ABCA1) in order to be transported in (Schmölz et al., 2016) the systemic circulation via the thoracic duct and further transferred to high-density lipoproteins (HDL) and then distributed to all tissues (Kayden & Traber, 1993). The absorption, metabolism, and transportation of vitamin E have been depicted in Figure 7.1.

The α-Tocopherol is absorbed by endocytosis, which gets accumulated in the endosome, after which it gets to the plasma membrane and its secretion takes place in systemic circulation along with lipoproteins. In the external endosomal membrane, α-tocopherol transferrin protein binds to α T and is carried to plasma membrane, where binding to phosphatidylinositol 4,5-bisphosphate causes a conformational shift that causes α T to be released and integrated into the membrane (Chung et al., 2016; Qian et al., 2005; Horiguchi et al., 2003). The side-chain fragmentation deprived of any changes in chromanol head, the portion of vitamin E constituents are not secreted into the bloodstream, gets degraded to carboxyethylhydroxychromanol (CEHC) metabolites (water-soluble) and excreted through bile and urine (Schmölz et al., 2016). The RRR—tocopherol is preferred in hepatic vitamin E secretion, most likely due to TTP's selectivity for the two R-congeners in combination with the non-T congeners' preferential metabolism (Flory et al., 2019). The substance with

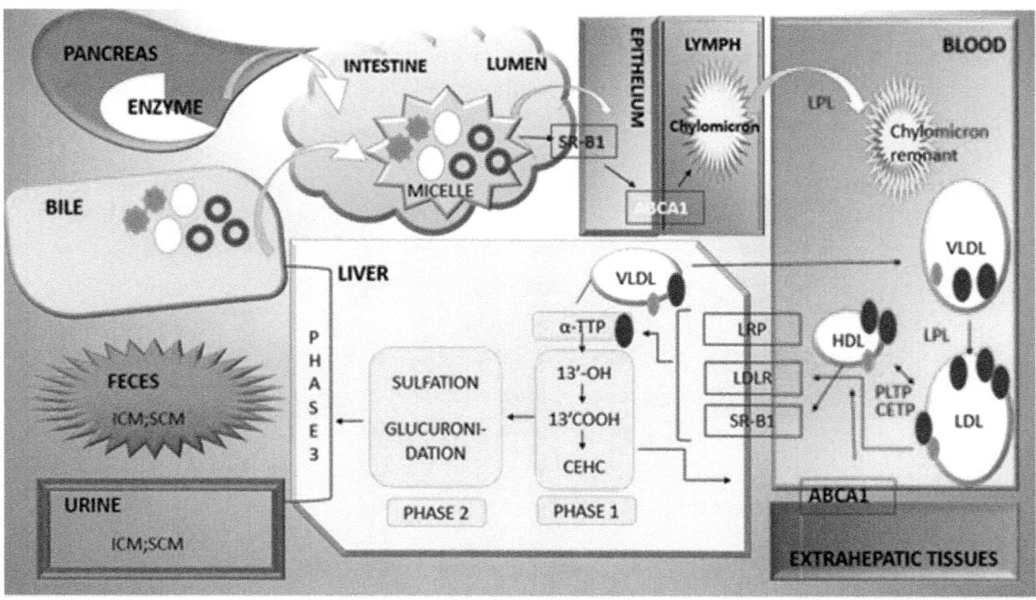

FIGURE 7.1 Depiction of vitamin E absorption and metabolism. Triglycerides; ICM: Intermediate chain metabolites; SCM: Short chain metabolites; (LDL: Low density lipoprotein; VLDL: Very low density lipoprotein; HDL: High density lipoprotein; α-TTP: α-Tocopherol transfer protein; 13'OH: 13'-hydroxychromanol; 13'-COOH: 13'-carboxychromanol; CEHC: Carboxy ethyl hydroxy chromanols; SR-B1: Scavenger receptor of class B and type 1; PLTP: Phospholipid transfer protein; CETP: Cholesteryl ester transfer protein; LRP: LDL receptor protein; LPL: Lipoprotein Lipase; ABCA1: ATP binding cassette transporter A1).

the highest vitamin E activity is known as α-tocopherol, which can be found both in its natural form as RRR-α-tocopherol extracted from plant sources and more frequently as synthetically made all-rac-α-tocopherol (Jensen & Lauridsen, 2007). There are four types of α-tocopherols like α-, β-, γ-, and δ-Tocopherol, which are produced and deposited in leaves and seeds of plants. The quantity and arrangement of the methyl groups on the chromanol ring determine their chemical differences. RRR stereochemistry is found in the natural form where the three asymmetrical carbon atoms of the phytol chain are connected to the chromanol ring (Azzi, 2018). RRR-α-tocopherol has more biological relevance when compared to all-rac-α-tocopherols as a result of these stereochemical distinctions (Kuchan et al., 2016).

Vitamin E metabolism is carried primarily in the liver and small intestine, involving a variety of enzymatic steps that are the same for all four tocopherol and tocotrienol congeners (Bardowell et al., 2012). The enzymatic process takes place in the endoplasmic reticulum, where the enzyme cytochrome P450 (CYP) reacts with terminal ω-hydroxylation of the side chain, resulting in the long-chain metabolite 13′-hydroxychromanol (13′-OH), which is the first and rate-limiting phase (Flory et al., 2019). In humans, the enzyme CYP4F2 involved in catalyzing the above reaction. On the other hand, 13′-carboxychromanol is obtained through the side chain oxidation of 13′-OH in peroxisomes. Subsequently, in mitochondria, β-oxidation results in the elimination of 2-carbon units in a stepwise manner, resulting in the formation of intermediate-chain metabolites and finally the water-soluble short-chain metabolite CEHC. The metabolites of vitamin E are excreted in the feces and urine. The lipid-soluble long-chain metabolites are found in feces, while the water-soluble short-chain metabolites are found in urine. The phase II enzymes UDP-glucuronosyltransferase and sulfotransferase conjugate about 90% of urinary metabolites, similar to xenobiotics. As a result, the metabolites of vitamin E in human and rat urine are glucuronidated and sulfate conjugates, respectively (Flory et al., 2019).

7.4 THE VITAMIN K CYCLE

Vitamin K cycle goes through a cyclic sequence of changes in sequence to participate in the transition from Glu to Gla (Figure 7.2). The cycle begins with the enzyme vitamin K reductase (VKR) which reduces vitamin K to vitamin K hydroquinone (KH2) and recovering of vitamin K_3 takes

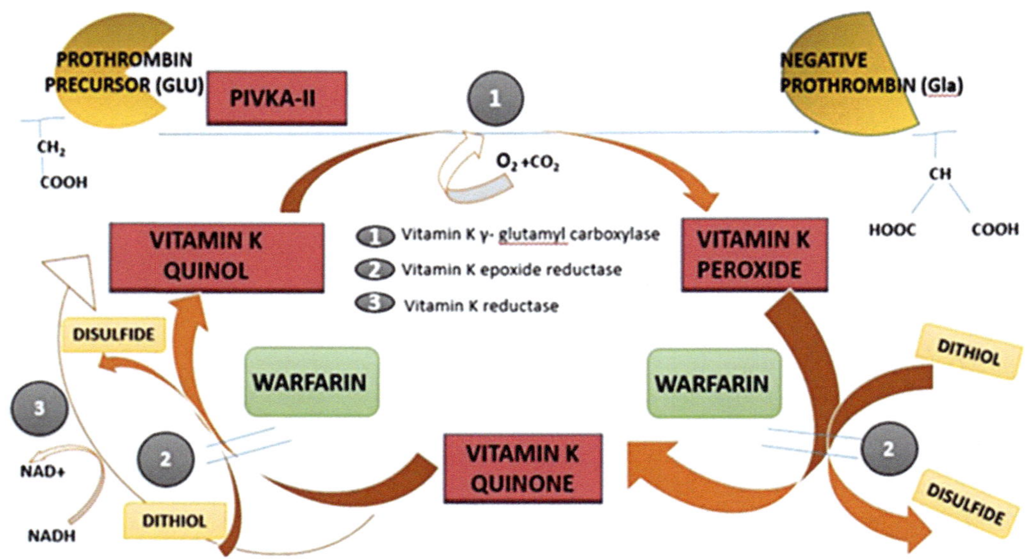

FIGURE 7.2 Schematic depiction of vitamin K cycle.

place in the presence of the enzymes quinone reductases 1 and 2 (QR1 and QR2). QR1 and QR2 promotes 2e⁻reduction, yielding KH2 and 1e⁻reduction yielding semiquinone, respectively (Gong et al., 2008) which is also known to produce reactive oxygen species (ROS) in live pancreatic acinar cells which are exposed to modified vitamin K_3 and dyed with fluorescent dye which are only subjected to QR2-mediated reduction (Criddle et al., 2006). The capability of vitamin K_3 redox cycling to proliferate reactive oxygen species (ROS) helps in cancer treatment. The carboxylation is further catalyzed by gamma-glutamyl carboxylase (GGCX), which uses KH2 along with oxygen and carbon dioxide. A sequence known as the propeptide allows the enzyme to identify its substrates. Since the vitamin K-dependent proteins are not entirely homogeneous, their affinity for GGCX is highly variable (Parker et al., 2014). During uremia, GGCX function in the liver and kidney is reduced, which could result in a higher danger of calcification and cardiovascular problems when dealing with chronic kidney illness (Dahlberg, 2017; Kaesler et al., 2014). During the carboxylation step, the hydroquinone is oxidized to vitamin K 2,3-epoxide (KO), which provides energy for the removal of proton from the gamma carbon of the glutamic acid (Glu) residue, yielding a carbanion, which further yields gamma-carboxyglutamic acid (Gla) through carboxylation. The gamma-carboxylation process is characterized by a cyclical transformation in which the driving factors are oxidized and reduced forms of vitamin K. Vitamin K antagonists, such as phenprocoumon and warfarin, inhibit these two enzymes, which has important medical implications in anticoagulation therapy. There are about 14 vitamin K-dependent proteins that have a wide range of functions in hemostasis, calcium metabolism, cell growth regulation, apoptosis, and signal transduction (Grober et al., 2015).

7.5 SOURCE AND INTAKE (RDA)

Vitamin E is a required micronutrient for all humans, and reaching its optimum level is thought to have health benefits. Vitamin E dietary intake recommendations have been recognized in several countries worldwide, and they mention a significant role as a free radical scavenger (antioxidant) in maintaining cell membrane integrity. Vitamin E intake guidelines currently range from 3 to 15 mg per day, depending on the country and the individual's age. In the United States, the Recommended Dietary Allowance (RDA) for vitamin E in adults is 15 mg of tocopherol in both men and women over the age of 14. A serum tocopherol concentration of less than 12 mol/L was identified as deficient by the Institute of Medicine (IOM) in the United States. The European Food Safety Authority (EFSA) recently concluded that the current RDA for vitamin E should be replaced by the following newly established Adequate Intake (AI): Men need 13 mg per day, women need 11 mg per day, and infants and children need 5–13 mg per day (depending on age) (Galli et al., 2017).

In India, RDA is given together by the Indian Council of Medical Research (ICMR) and the National Institute of Nutrition (NIN). The body requires a consumption of 0.8 mg/g of α-tocopherol from dietary essential fatty acids. Therefore, a recommended intake of approximately 7.5 - 10 mg tocopherol per day is equivalent to the FAO/WHO regulations, depending on the consumption of edible oil (RDA short report by NIN and ICMR, 2020) (Table 7.1).

Vitamin K deposits in the body are small, owing to the fact that they are quickly exhausted without daily dietary intake; the majority of consumed vitamin K is cleared within 24 hours. The K_2 activity varies with different types and distribution of intestinal bacteria, therefore its contribution to the regular vitamin K requirement remains unclear. In contrast, an insufficient vitamin K status seems to be caused by a decreased dietary intake of K_1. According to the National Academy of Sciences, 2 μg/day is the recommended dietary intake for newborns, 75 μg/day for adolescents, 120 μg/day for males, and 90 μg/day for females. However, the above quantities are normally insufficient to maintain the optimal vitamin K levels, which differ based on the ethnic origin and age of the subjects (Palermo et al., 2017). In India, the recommendation for vitamin K is 55 μg for adults as per the short report published by NIN and ICMR 2020, which is in tune with the recommendations of FAO/WHO.

TABLE 7.1

Vitamin E and K Sources, RDA, Functions, and Deficiency

Vitamins	Scientific Names	Sources	RDA	Functions	Diseases/Disorders	References
Vitamin E	Tocopherols	It is mostly found in oily plants like nuts, almonds, hazelnuts germ oil, and sunflower oil, palm oil and soyabeans.	According to ICMR and NIN, the requirement for alpha tocopherol is 0.8 mg/g of dietary essential fatty acids, which is 7.5-10 mg tocopherol per day for Indians.	Vitamin E protects membrane lipids from free radicals by acting as an antioxidant and reduces the incidence of chronic diseases such as heart disease and cancer. Apart from this, it plays the role of anti-inflammatory, hormone regulator, and cholesterol reducer.	It is very unlikely to have a deficiency but its deficiency leads to fragility of RBC/ hemolysis, lipid absorption and transport problems, neurological disorders, and muscle weakness.	Bjørneboe et al., 1990; Sen et al., 2007; Wong and Radhakrishnan, 2012; Cardenas and Ghosh, 2013; Schmölz et al., 2016; RDA short report by NIN and ICMR, 2020.
Vitamin K	Phylloquinone and Menaquinones	It is mostly contained in fermented foods like yogurt, curd and natto which is a Japanese food and also found in dark green leafy vegetables such as spinach, broccoli and brassica.	According to ICMR and NIN, the requirement for vitamin K is 55 µg for adults in India.	Vitamin K is a coenzyme that aids in the gamma carboxylation of glutamate residues in blood clotting factors, resulting in gamma carboxy glutamate, which chelates calcium ions during blood coagulation. It is needed to make a calcium-binding protein in the bones that is osteocalcin.	The deficiency leads to hemorrhagic disorders that are caused by abnormal blood coagulation (gastrointestinal bleeding, subcutaneous and intramuscular hemorrhage, postoperative bleeding).	Turck et al., 2017; RDA short report by NIN and ICMR, 2020.

7.6 DEFICIENCY AND TOXICITY

Under normal biological circumstances, vitamin E deficiency and toxicity are extremely unusual due to the fact that it is present in adipose tissues. As vitamin E is mobilized from adipose tissues for years, symptoms of mild deficiency appear after several years, typically decades. Severe deficiency of vitamin E causes acute symptoms like neuropathy and myopathy because it is needed for the proper development of the central nervous system. Ulatowski and Manor (2013) suggested two vitamin E deficiencies: (1) Primary deficiency, which arises from basic alteration of vitamin E status; (2) secondary deficiency, where low levels of vitamin E become secondary to global perturbations like metabolism and transport of lipoprotein and lipid malabsorption (Schmölz et al., 2016). Also, hereditary disorders along with dietary patterns have been linked to primary and secondary vitamin E deficiency or low vitamin E bioavailability (Schmölz et al., 2016).

Vitamin E deficiency reduces the erythrocytes, neurological dysfunction, and leads to myopathies in patients suffering from malabsorption and abetalipoproteinemia. In premature babies, vitamin E deficiency causes intraventricular hemorrhage in the brain, the development of bronchopulmonary dysplasia, anemia, and retrolental fibroplasia. Vitamin E deficiency is also responsible for the growth of some types of cancer, cardiovascular disease, and a weakened immune system (Bjørneboe et al., 1990).

α-Tocopherol concentrations less than 9 mmol/L for men and less than 12 mmol/L for women are considered deficient and negligible, respectively, in healthy adults. Dosage can be increased from 100 mg/d (150 IU/d) to 300 mg/d (450 IU/d) without any problems. There have been no reported side effects from consuming high quantities of vitamin E for a short-lived period of time. However, animal studies have shown that long-term high-dose supplementation interferes with blood clotting and is linked to an increased risk of haemorrhagic stroke. According to Miller et al. (2005), tocopherol, which was once considered safe as a food additive, stated an upsurge in mortality after consuming a high quantity of vitamin E for about a year. As a result, adverse effects were showcased with increased inclination for haemorrhagic conditions, and the upper intake level was put down to 1000 mg/d α-TOH for adults (Schmölz et al., 2016).

Vitamin E appears to be very safe, and supplements containing 100–200 mg of synthetic all-rac-α-tocopherol are commonly used. The feeding of supplements has been linked to pro-oxidant injury, but only at extremely high doses (e.g. >1000 mg/day) (Webb & Villamor, 2007). However, studies from the Netherlands found that giving 200 mg vitamin E per day for 15 months increased the severity of respiratory tract infections in people over the age of 60, which may be an indicator of a pro-oxidant effect (WHO, 2004).

7.7 VITAMIN K

Vitamin K deficiency in adults is characterized by a propensity to bleed in conjunction with low blood coagulation factor activity, which results in an elevation time of prothrombin (PT) or partial thromboplastin time (PTT). A low phylloquinone intake that is less than 10 µg/day, the symptoms for vitamin K deficiency and the weakening of regular haemostatic control in healthy people might take 2–3 weeks to develop. Human milk contains a limited quantity of vitamin K, which makes breastfed infants more prone to bleeding. For the prevention of hemorrhagic disease in infants, phylloquinone is given at a pharmacological dosage, either orally or through intramuscular injection. The Scientific Committee on Food, the Netherlands, has not defined a tolerable upper intake limit for vitamin K (Turck et al., 2017). For all age and sex demographic categories, an AI of 1 µg phylloquinone/kg body weight every day has been set. The recommended AIs for phylloquinone are 70 µg per day for all people, including pregnant and lactating women, 10 µg per day for babies 7 to 11 months old, 12 µg per day for children 1–3 years old, and 65 µg per day for children 15 to 17 years old (EFSA NDA Panel, 2017).

Natural K vitamins seem to have no toxic side effects when administered orally. The common clinical administration of phylloquinone at doses of 10–20 mg or higher attests to its apparent protection. Some patients have chronic fat malabsorption and consume doses of this size on a daily basis with no apparent side effects. Synthetic menadione or its salts, on the other hand, should be avoided for nutritional purposes, particularly for vitamin prophylaxis in neonates. In addition to lacking intrinsic biological activity, the unsubstituted 3-position's high reactivity has been linked to neonatal haemolysis and liver damage (WHO, 2004).

7.8   VITAMIN E AND K FORTIFICATION

As discussed earlier, vitamin E (lipid-soluble) categorizes into tocopherols and tocotrienols, where α-tocopherol and β-tocotrienol are bioactive molecules with elevated antioxidant activity and anti-carcinogenic properties. In the other hand, vitamin K takes part in calcium synthesis in the circulatory system and affects human bone health (Webb et al., 2011; Gonnet et al., 2010). Vitamin E and K are delicate lipid-soluble biologically active compounds with poor chemical stability and hydrophilicity, which affect their bioavailability rate. Since vitamin E and K are not synthesized in the human body, they must be obtained in sufficient quantities from food. As the intake of vitamin E and K food sources increases in our daily diet, all of us get enough benefit from these vitamins. As a result, the need for lipophilic vitamins to be added to the diet by food fortification has emerged. Government bodies are becoming more interested in foods fortified with vital micronutrients, vitamins that foster human well-being, and they are working to improve regulations and health issues on fortified foods with specific nutrients such as vitamins, minerals, and bioactive constituents. The Codex Alimentarius rules facilitate the establishment of the general criteria for fortified food commercialization among countries. Consumer-friendly fortified foods should be inexpensive, safe during fermentation, storage, and shipping, and contain no disproportionate or negligible amounts of the essential nutrient (Dey et al., 2022; Latham et al., 2001).

7.8.1   CHALLENGES IN VITAMIN E AND K FORTIFICATION

To avoid degradation, increase solubility, and thus oral bioavailability, efficient delivery mechanisms must be created (Grune et al., 2010; Teleki et al., 2013). Food fortification is a difficult problem to solve since it requires suitable and well-matched encapsulation systems (food-functional ingredients) that do not compromise the sensory and physicochemical properties of the food, as well as the chemical stability of bioactives during the manufacturing of fortified foods and storage until use (Augustin & Sangaunsri, 2015; Chen et al., 2013; Siegrist et al., 2008). Different kinds of techniques are already in use for encapsulation, which include spray drying, electrospinning, nanoprecipitation, air-bath oscillation, co-solvent desolation, molecular complexation, precipitation, hot homogenization, film dispersion, thin-film hydration, film hydration/sonication, dehydration/ rehydration, microfluidization, high-pressure homogenization, solvent evaporation, emulsification, spray-cooling/chilling, freeze-drying, fluidized-bed coating, extrusion, coacervation, and stirring (Maurya et al., 2020). The utilization of nanoemulsions is increasing day by day, such as in food, beverage, and pharmaceutical industries. As it is known, nanoemulsions act as a carrier system for delivering several kinds of functional lipophilic compounds (Dasgupta et al., 2015). Therefore, use of nanoemulsions for delivering fat-soluble vitamins E and K was discussed in this chapter. Nanoemulsions are dispersions that produce particle sizes in the diameters of less than 200 nm and are made up of two immiscible phases. Nanoemulsions are thermodynamically unstable but kinetically stable emulsions that vary in particle diameter from traditional emulsions (McClements, 2012). High-energy and low-energy approaches are utilized to make nanoemulsions, such as ultrasonication, microfluidization, and high-pressure processing-homogenizer. The difficulties in producing nanoemulsions using these methods are discussed.

7.8.1.1   Low Energy Methods

Spontaneous emulsification (SE), emulsion phase inversion (EPI), and phase inversion temperature (PIT) are the most common low-energy methods for creating food-grade nanoemulsions (Anton et al., 2009; Ostertag et al., 2012, Gulottta et al., 2014; Hategekimana et al., 2015). Out of the above three techniques, phase inversion is primarily used in the development of nanoemulsions by determining the basics of the system formed from a transient phase inversion or a catastrophic phase inversion. On the other hand, the phase inversion phenomenon is enhanced by changes in surfactant composition, electrolyte concentration, pH, or temperature, as well as the water-to-oil phase ratio of the system (Li et al., 2020). SE is accomplished by slowly stirring a previously formulated organic phase as a carrier like oil, lipid-soluble vitamins, and water-soluble surfactant into a liquid phase with specific speed and temperature (). Similarly, the EPI system is an isothermal low-energy process in which water is applied to the organic step under regulated conditions (water to oil phase ratio (WOR), temperature, and speed) (Mayer et al., 2013). In the literature, the effects of various parameters that are responsible for the stability of the encapsulation delivery system such as oil percentage (vitamin E and MCT), surfactant, phase temperature, phase inversion composition, ultrasonication, high-pressure homogenization, microfluidization, and mixing speed on particle diameter have been extensively analyzed (Laouini et al., 2012; Öztürk, 2017; Jiang et al., 2020). Small surfactants (hydrophilic) can vary in nature due to a unique molecular structure and should have a high hydrophilic-lipophilic balance (HLB) number. However, HLB specifies the preference of surfactants for the liquid and lipid phases, that are significant in phase inversion phenomena (i.e., water in oil to oil in water phases) in these methods. Surfactants with an HLB number greater than 10 contain a hydrophilic phase, and moderate HLB values lead to the production of an oil-in-water phase with smaller droplets in nanoemulsion (Guttoff et al., 2015). The EPI method is generally preferred to obtain nanoemulsions with a diameter less than 200 nm in which the optimal surfactant-to-oil (SOR) ratio was determined to be 1:1 (Mayer et al., 2013). As a result, the physicochemical pathways in the SE and EPI methods may be identical, despite the fact that the oil and water process addition phases differ.

The phase action of the surfactant-oil-water (SOW) combination to quickly move over a bicontinuous microemulsion state is critical for the creation of nanoemulsions using low-energy methods. The SOW state with an extremely viscous crystal-like fluid can form during high surfactant concentration. At low surfactant content, the SOW state is reduced during the formation of microemulsion, and thus larger droplets can be produced (). Propylene glycol, C_2H_5OH, and glycerol are used as solvents, which can alter the aqueous phase's bulk properties as well as the surfactant solutions, resulting in particle size changes in nanoemulsions (Öztürk, 2017; Saberi et al., 2013).

7.8.1.2 High Energy Methods

When using these methods to make nanoemulsion, a two-step protocol is used. In the first stage, the aqueous and oil phases are mixed to prepare coarse emulsions that are pressurized by the pneumatic force in the homogenizer into the narrow channels. In the contact field, emulsion droplets clash and meet, and the resulting disrupting forces like impact, cavitation, rupture, and shear forces make it easier to receive nanoparticles.

In high energy methods like ultrasonic probes in ultrasonic instruments, destructive forces are generated. The coarse emulsion size droplets can be disrupted to a target size by using power and product residence time of ultrasound within the disturbance region (Öztürk, 2017). In high energy methods, the factors affecting the droplet size of the formed nanoemulsion formed are the molecular geometry, emulsifier characteristics such as its composition, surface loading, interfacial stress, rate of adsorption, and adsorbed layer thickness. Vitamin E-nanoencapsulated delivery systems have been created using a variety of emulsifiers. Tween 80, quillaja saponin, and lecithin were found to be efficient at decreasing interfacial stress and creating nanoemulsions with a diameter less than 200 nm (Mao et al., 2010; Ozturk et al., 2015). Large molecular weight gum Arabic was unable to shape nanoemulsions because droplet breakup was difficult due to the emulsion aqueous phase's increased viscosity. Low molecular weight of size up to 12 kDa was able to shape small particles in the presence of enough emulsifier due to faster adsorption and a thin interfacial layer on the surface it produced. . With increased homogenization pressure and refined nanoemulsion composition, the particle size can be reduced by shear force and cavitation during the formation of vitamin E-enriched nanoemulsion (Mehmood et al., 2015).

When choosing a process for producing food-grade nanoemulsions, there are several factors to consider. First, there should be no apparent flocculation or coalescence, which make them physically stable. Along with this, nano-emulsions are prepared using surfactant concentrations less than 10%. These nanoemulsions efficiently deliver bioactive ingredients due to their large surface area, allowing rapid penetration of actives (Louini et al., 2012). Low-energy methods are more cost-effective compared to high-energy methods. However, in low-energy methods, the high concentrations of surfactant can cause toxicity issues, particularly in food products. To extract fine droplets using low-energy processes, only specific forms of oil and surfactants can be used, such as MCT and small size surfactants, respectively. On the other hand, high-energy methods enable food technologists to make food-grade nanoemulsions from a broad range of constituents, including food-grade carrier oil, natural biopolymers, and small-size surfactants.

Nanomaterials are presently used for a broad variety of uses such as nanosized powders to improve nutrient absorption; nutraceuticals nanoencapsulation for greater absorption, increased stability, or nano chelates; and targeted distribution to increase the efficiency of nutrient delivery without compromising the environmental food color or flavor. However, there is another side to the coin: nanomaterials can be toxic, and the following are some of the factors that make nanomaterials potentially toxic:

High aspect ratio: Despite their ability to encase long, thin nanostructures, macrophages are incapable of transporting to the proximal lymphatic node, where they can be cleared from the lungs by the mucociliary escalator (Dwyer et al., 2015). Size of the emulsion matters to transport the active ingredients to target site so it was found that the 50–100 nm is the optimum particle size for absorption (McClements et al., 2020).

Bio-persistence: Nanomaterials that travel through the flow of liquid or lipids may remain in the body indefinitely. It's impossible to say where the permanent material will take up biological space. The more permanent a nanomaterial is, the more likely it is to communicate with the body, whether positively or negatively. Also, nanoemulsions are digested by the upper gastrointestinal tract (lipases and proteinases) before reaching to the target site so the active compounds are not available (McClements et al., 2021).

Reactive surfaces: Some can produce reactive oxygen species, potentially increasing toxicity. It was found that the nanoemulsions responsible for the increasing the bioavailability of hydrophobic bioactive substances/undesirable toxic substances. Also, it was observed increases in β-carotene levels that may be responsible for lung cancer in smokers (McClements et al., 2021).

Composition and stability: Nanomaterials can provide a super physiologic concentration of materials to a very small room, bypassing the concept of volume distribution and thereby potentially increasing toxicity (Dwyer et al., 2015).

7.8.2 Types of Fortificant and Chemistry

Among the lipophilic vitamins, vitamins A, D, and E are the micronutrients which are used commonly for fortification of milk, bread, and cereal items (Öztürk, 2017). To fortify, nanotechnology applications have been utilized for the delivery system of lipophilic bioactive and one such emerging application is nanoemulsion. Nanoemulsions are of two types of emulsions: oil-in-water (o/w) and water-in-oil (w/o). Out of these, oil-in-water type expressed reliability in preventing, delivering, and stabilizing lipid-soluble bioactives such as vitamins (A, D, E and K) and omega-3, by encapsulating them in oil matrix (McClements, 2013). Emulsion formation necessitates the presence of a liquid phase, a liquid phase, an emulsifier, and either mechanical or physiochemical capacities (Tadros et al., 2004; McClements, 2011). An oil-in-water type contain three-component core-shell structure: a lipophilic core encasing a lipophilic bioactive agent, a hydrophilic shell (aqueous phase), and an amphiphilic interface containing the emulsifier/surfactant (McClements, 2011). Emulsifiers are surfactants that adsorb on the lipid-aqueous interface of newly formed droplets and reduce interfacial stress, resulting in smaller droplet scale emulsions. Based on the handling circumstances and emulsion structure, various types of food grade surfactants and emulsifiers are used to create nanoemulsions (Chanamai & McClements, 2002; Kralova & Sj€oblom, 2009). Due to the existence of hydrophilic groups and hydrophobic moieties in their structure, amphiphilic proteins and polysaccharides are biopolymers with large molecular sizes used as efficient emulsifier (Adjonu et al., 2014; Ozturk & McClements, 2016).

7.8.3 Novel and Emerging Techniques for Vitamin E and K Fortification

Nanotechnology in food science is gaining a lot more attention since it can be used to make delivery systems for biologically active compounds to improve their physico-chemical and free radical scavenging stability, as well as their bioaccessibility and bioavailability in the human body. Various types of nano-size particles (nanoemulsions, stable lipid nanoparticles, nanosuspensions) are efficient delivery systems for encapsulating and stabilizing lipophilic bioactives, according to the reports (Fathi et al., 2012; Silva et al., 2012). Lipid-based nanoemulsions delivery systems are excellent candidates for encapsulating and distributing lipophilic vitamins, and they can be inserted into drinks due to their translucent nature (Wang et al., 2021). Bioactive based nanoemulsions can also be encapsulated using spray dried, ultrasonication, or freeze dried to create a variety of solid foods (Öztürk, 2017; Wang et al., 2021). Table 7.2. summarized various novel technologies, including vacuum impregnation, nanoemulsion, micro-nano encapsulation, spray drying, inclusion complex, nanoliposomes, and electrospinning, for fortifying vitamin E and K in different food matrix.

7.9 VITAMIN E AND K BIOACCESSIBILITY AND BIOAVAILABILITY

Both vitamin E and K being lipophilic are reported to have chemical instability and poor water solubility, consequently leading to poor bioaccessibility and bioavailability. Lipid-soluble vitamins are highly prone to oxidative stress, and their efficiency may be lost during the manufacturing, storage, and usage of final products. Because of this, the utilization of these vitamins in commercial products is generally challenging (Yang & McClements, 2013a). Bioaccessibility of vitamins is determined using the in-vitro digestion method, wherein the food sample containing the vitamins is subjected to simulated gastrointestinal digestion. Simulated gastrointestinal digestion consists of an oral phase (imitating the oral conditions in the mouth), a gastric phase (imitating the stomach conditions), and an intestinal phase (imitating the small intestinal conditions). Bioavailability of vitamins is the amount absorbed in the blood after complete digestion of food. Bioaccessibility of vitamins is the amount of vitamin available in the intestine for absorption solubilized in digesta after complete digestion (Figure 7.3). When determining the bioaccessibility, the transformation of vitamins is important. The percentage of vitamin that is potentially in a bioactive form and available for absorption is considered as transformed (Li et al., 2018).

There are a number of physiological and physicochemical factors, such as chemical stability in acidic conditions and gastric enzymes, solubilization of hydrophobic vitamins in the micelle before absorption, and limited absorption in the body, which determine the bioaccessibility and bioavailability of ingested vitamins (Lee et al., 2020; Tan & McClements, 2021). After consumption, lipid digestion forms mixed micelles (micelles, vesicles, bilayers, and liquid crystals) that are helpful in solubilizing the oil-soluble vitamins released from the food matrix (Rigotti, 2007). These mixed micelles are colloidal types that have the capacity to solubilize the lipid-soluble constituents. Hence, the degree of solubilization of vitamins into the mixed micelles plays a critical part in bioaccessibility and bioavailability. There are numerous studies reporting the bioaccessibility and bioavailability of vitamin E and K, some of which are presented in Table 7.3. In the case of vitamin E, α-tocopherol, which is the most bioactive and abundant yet highly susceptible to oxidation, is chemically stable in

TABLE 7.2

Novel and Emerging Techniques for Vitamin E and K Fortification in Food Products

Technique	Preparation method	Matrix composition	Food products	References
Encapsulation	Chelation, reduction reaction	-	Fortified powder fruit drink (NutriStar)	Mehansho et al. (2003)
Nano-emulsion	High pressure homogenization, microfluidics, spray dry	Starch sodium octenyl succinate	Apple juice	Chen & Wagner, 2004
Mixing	Mechanical stirring	-	Skim milk (vitamin K)	Kruger et al. (2006)
Mixing	-	-	Yoghurt drink (vitamin K)	Knapen et al. (2015)
Inclusion complex	Electrospinning	β-cyclodextrin	Food, pharmaceutical	Celebioglu and uyar. (2017)
Encapsulation	Nanoliposomes	Gammaoryzanol, polyethylene glycol, lauric acid	Food	Amiri et al. (2018)
Encapsulation	Nanoemulsion	Lecithin	Food	Saratale et al. (2018)
Encapsulation	Spraydrying	Carboxy methyl starch, xanthum gum	Food	Jiang et al. (2019)
Emulsion system	Mechanical stirring, homogenization	Xanthum gum, guar gum	(Docosahexaenoic acid) DHA emulsion	Singh et al. (2019)
Emulsion system	Mechanical stirring	Sodium caseinate, xanthum gum	Cheese sauce	Patel et al. (2019)
Inclusion complex	Cross linking	β-cyclodextrin and soy soluble polysaccharide	Drug	Eid et al. (2020)
Encapsulation	Mechanical stirring	Chlorogenic acid, whey protein isolate and dextran	Food	Wang et al. (2020)
Nanocapsule	Mechanical stirring	Carboxy methyl cellulose	Packaging film (food)	Mirzaei-Mohkam et al. (2020)
Vacuum impregnation, microwave vacuum drying	Homogenization	Isotonic aqueous base (NaCl: 0.8%), Surfactants Tween 80. Span 60	Potato chips	Duarte-Correa et al. (2020)
Inclusion complex	emulsification	β-cyclodextrin	Drug	Ke et al. (2020)
Mixing	Mechanical stirring	-	Milk (vitamin K_2)	Koe, (2020)
Encapsulation	Spray drying (coacervation)	Chitosan and sodium lauryl ether sulfate	-	Budinčić et al. (2021)
Inclusion complex	Mechanical stirring	Cyclodextrin	-	Ogawa et al. (2021)
Emulsification	Solvent diffusion	Glycerol Monostearate (GMS) as solid lipids (GB to GMS ratio of 1:1), olive oil (solid to liquid weight ratios of 90:10 and 80:20) as a liquid lipid, soy lecithin as Emulsifier	Milk (vitamin E)	Kiani et al. (2021)

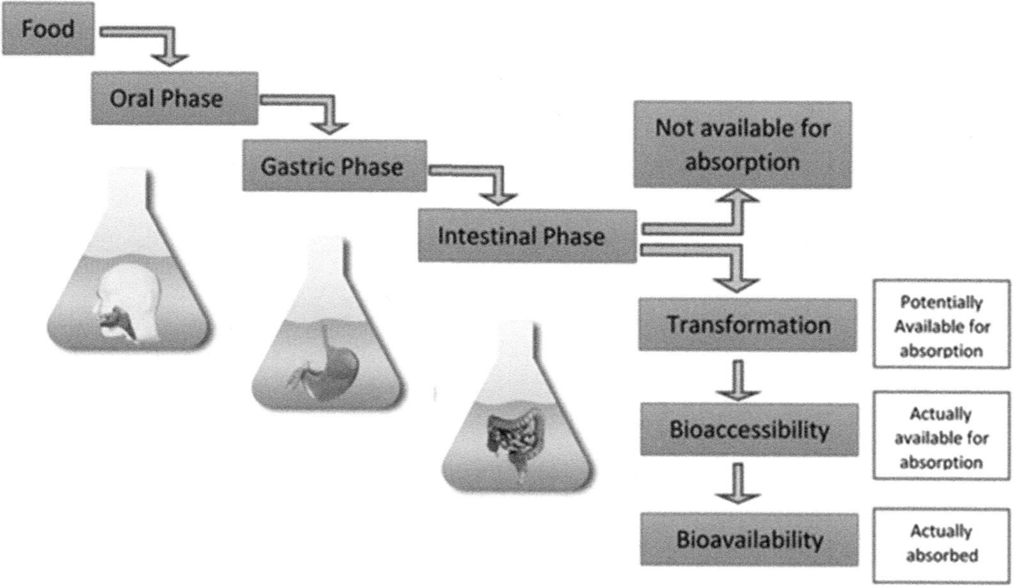

FIGURE 7.3   shows the process of determining the in-vitro bioaccessibility and bioavailability of vitamins.

the acetate form, i.e., α-tocopherol acetate, used for food fortification. After ingestion, α-tocopherol acetate is converted to vitamin E in the human system and absorbed by passive diffusion (Goncalves et al., 2015; Mayer et al., 2013). It has been reported that the chemical form of vitamin E plays a critical part in its bioaccessibility, whereas the a-tocopheryl acetate form of vitamin E is found to be less bioaccessible than the α-tocopherol form (Yang & McClements, 2013a).

Natural forms of vitamin K available are vitamin K_1 (phylloquinone) and vitamin K_2 (menaquinones, MK-n, where n represents the number of isoprene units) that are present in vegetable and animal products respectively. All dietary vitamin K homologues are utilized in biosynthesis of MK-4 in the body (Sakane et al., 2017; Sato et al., 2012). Menaquinones are not absorbed during digestion as they are strongly bound to insoluble material (bacteria, membrane remnants). Vitamin K is absorbed in the body by carrier-dependent proteins (Goncalves et al., 2015). To make the fat-soluble vitamins more bio-accessible, numerous attempts have been made which are discussed in detail in the following section.

7.9.1   Improving Bioaccessibility and Bioavailability in Fortified Foods

Vitamin E and vitamin K are extremely lipid soluble molecules with low hydrophilicity, and consequently the major focus has remained on increasing the water solubility by using suitable delivery systems for these lipophilic molecules so they can be incorporated into liquid-based products for varied food applications. Nanoemulsions are considered satisfactory for the encapsulation of lipid soluble constituents to deliver them into the human system. Nanoemulsions progress the stability of droplets to aggregation, gravitational separation, as well as enhance the shelf-life of commercial food products. The delivery system used for the encapsulation of vitamins in the form of emulsions should be able to protect the vitamins until they are released in the GIT after digestion. Such systems not only protect the vitamins under adverse conditions during processing, storage, transportation, and GIT but also help to enhance the bioaccessibility and bioavailability of the vitamins. To design suitable processes and delivery systems helpful in increasing the bioaccessibility of both the vitamins (E and K), diverse attempts have been made (Booth et al., 2000; Mayer et al., 2013; Sakane

TABLE 7.3

Previously Reported Bioaccessibility and Bioavailability of Vitamin E and Vitamin K

Vitamin	Method used	Bioaccessibility/bioaccessibility	Matrix/system	References
Vitamin E acetate (VE)	Microfluidization	Bioaccessibility >75%	VE, Medium chain triglycerides (MCT), Tween 80, VE, and citric buffer	(Mayer et al., 2013)
Vitamin E acetate	Emulsion phase inversion	Bioaccessibility >90%	VE, MCT; Tween 80; citric buffer, pH 3,	(Mayer et al., 2013)
vitamin E acetate	Oil-in-water emulsions	Bioaccessibility in corn oil, long chain triglyceride (LCT) > 35%	Vitamin E acetate, corn oil (LCT), surfactant (1 wt.% Q-NaturaleÒ), and 10 mM sodium phosphate buffer, pH 7.0	(Yang & McClements, 2013b)
vitamin E acetate	Oil-in-water emulsions	bioaccessibility in medium chain triglyceride (MCT) > 15%	Vitamin E acetate, medium chain triglyceride oil (MCT), surfactant (1 wt.% Q-NaturaleÒ), and 10 mM sodium phosphate buffer, pH 7.0	(Yang & McClements, 2013b)
Vitamin K_2	MK-4 or MK-7 single dose administration (420 µg)	Single intake of MK-7 and MK-7 reached maximum after 6 h and 48 h respectively	MK-4, MK-7	(Sato et al., 2012)
Vitamin K_2	Vitamin K_2 administered for 7 days (60 µg/day)	Baseline serum level of MK-4 was 2.2 ng/ml ± 0.38 and	MK-4, MK-7	(Sato et al., 2012)
phylloquinone (K_1), menaquinone-4 (MK-4)	K_1, MK-4	Bioavailability decreased from 30-50% to less than 0.2-2% after oral ingestion	K_1, MK-4, in buffer (0.15 M NaCl, 0.05 M Tris-HCl, pH 7.5)	(Groenen-Van Dooren, Ronden, Soute, & Vermeer, 1995)
Vitamin K	Polymeric Micelle encapsulation	Micelle, *in-vivo* absorption in plasma2304 ng/mL	Vitamin K, micelles of mPEG5000-b-p(HPMAm-lac2),	(van Hasselt et al., 2009)
Vitamin K	Encapsulation	Oil, *in-vivo* absorption in plasma 3928 ng/mL in oil		(van Hasselt et al., 2009)

et al., 2017). High lipophilicity is a major concern for the bioaccessibility and bioavailability, and in the case of vitamin K, the lipophilicity increases as the length of side' chain increases.

7.9.1.1 High Energy Processes

a) High-pressure homogenizers

High pressure homogenizers or high-pressure valve homogenizers are used in the food industry to fabricate conventional emulsions (Figure 7.4). To achieve emulsions of nano size, a coarse emulsion is generated and then directly added into the inlet of the high-pressure valve homogenizer (McClements & Rao, 2011). The coarse emulsion experiences a combination of disruptive forces as it passes through the valve, causing the larger droplets to be broken down into smaller ones. The size of emulsion droplets decreases with an increase in the number of passes it goes through the valve and with an increase in the processing pressure. Table 7.4 summarizes the advantages and disadvantages of high-energy and low energy methods.

b) Ultrasonicator

An ultrasonicator uses ultrasonic waves to create cavitation, which causes the disintegration of oil and water phases into smaller droplets. The sonicator probe homogenizes the liquid where the principle for ultrasonication is the same, generating forceful vibrations that produce cavitation that collapses the bubbles, generation, maturation, and this cycle continues in the liquid (Kumar et al., 2020). Collapsed bubbles cause the generation of intense disruptive forces in the immediate vicinity, further repeating the process of generation, maturation, and collapse of small bubbles (McClements & Rao, 2011).

c) Microfluidizers

Microfluidizer is a high-pressure homogenizer used traditionally in the pharmaceutical industry for emulsion-based applications. Microfluidizers have gained the interest of the food industry for similar applications like flavor emulsions, homogenized milk, and nutraceutical and many more such other applications. Microfluidizer involves the disruption of emulsion droplets when two streams

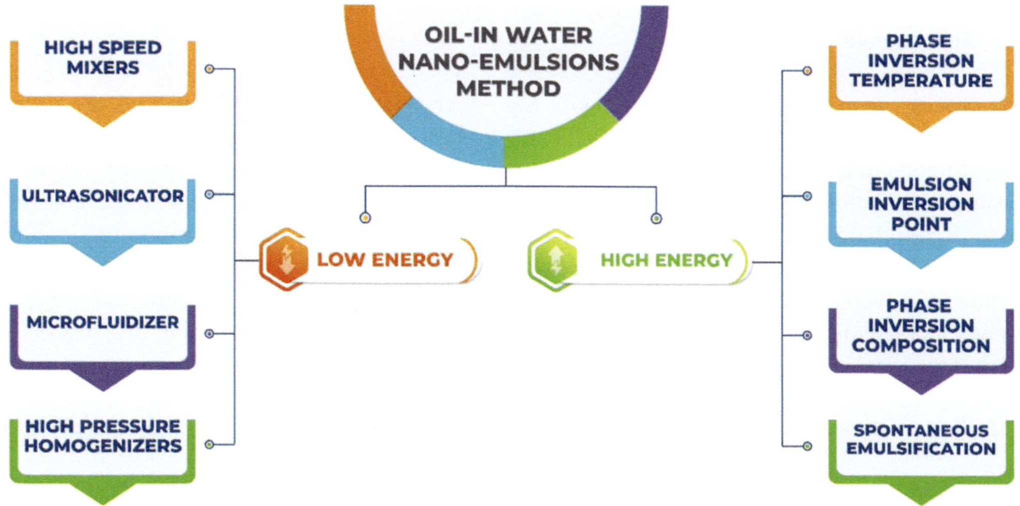

FIGURE 7.4 Approaches for the fabrication of oil-in-water nano-emulsions of vitamins

TABLE 7.4

Processes Used for Vitamin E and Vitamin K Emulsion Formulations

Processing Technique	Advantages	Disadvantages
High energy methods (High pressure valve homogenizers, High speed mixers, ultrasonicator, and Microfluidizers)	Allow good control of droplet size Nanoemulsions of extremely low particle size (up to 1 nm) Large selection options for integral components Highly efficient processing Faster processing is highly efficient. It is faster than low energy methods Equipment can be modified for continuous processing	High equipment and operating costs Low energy efficiency large amount of energy consumption Temperature increases during processing Equipment capacity and run time limit the processing Not applicable for thermolabile active ingredients such as retinoids, proteins, enzymes &nucleic acids)
Low energy methods (Spontaneous emulsification, Phase inversion temperature, Phase inversion composition, Emulsion inversion)	Reduced equipment and operating costs High energy efficiency Simplicity of implementation	Require higher amounts of surfactant than high energy methods Usually takes more time to complete the process Complexity of processing Requisites for precision Use of synthetic surfactants Necessity of thermal energy Requirement of the presence of phases Limited amount can be processed at a time

come in contact with each other coming from the opposite direction under high operation pressure in the interaction chamber. Nanoemulsions are generated with an increase in passes and operational pressure (Verma et al., 2021).

7.9.1.2 Low Energy Processes

a) Spontaneous emulsification

This process doesn't utilize the energy as in high energy processes. Nanoemulsions are formed spontaneously when liquids are mixed (McClements & Rao, 2011). The liquids used in this process are self-nano emulsifying drug delivery systems. During the formation of spontaneous nanoemulsion processes, conditions can vary depending upon the two phases (pH, temperature, and ionic strength) and their mixing (rate of addition, stirring speed, and order of addition). In this process, either both liquids are completely immiscible or one liquid is partially miscible. When both liquids are mixed, it increases the oil-water interfacial area, turbulence leading to the formation of spontaneous nanoemulsion. Spontaneous emulsification method can produce emulsion droplets of 140 nm in size.

b) Phase-inversion temperature process

Nanoemulsions are formed spontaneously using the phase inversion temperature process by altering the time-temperature profile of certain combinations of immiscible liquids in the presence of surfactants. Controlled emulsion transformation is used to change the emulsion type completely from one to another (oil-in-water to water-in-oil). Physicochemical properties of surfactant with temperature play a major role in this type of emulsion production (McClements & Rao, 2011). A nanoemulsion is designed from rapid cooling of an emulsion from a temperature slightly above the phase inversion temperature to well below the phase inversion temperature with constant stirring (McClements, 2012)

d) Phase inversion composition process

The phase inversion composition process is analogous to the phase inversion temperature process, but in this process, the preparation of the system is changed somewhat than altering the temperature of the system. With the addition of salt to an oil-in-water emulsion that is stabilized by an ionic surfactant, it can invert its phase to a water-in-oil emulsion (McClements & Rao, 2011).

e) Emulsion inversion point process

This process changes the type of emulsion through catastrophic-phase inversion, unlike phase inversion temperature and phase inversion composition, where altering the temperature and composition of the system is important (McClements & Rao, 2011).In this process, a catastrophic-phase inversion occurs where the ratio of the oil-to-water phases changes while the surfactant is kept constant. This inversion is produced by increasing or decreasing the volume of the dispersed phase above or below a critical level.

The processes explained above are to formulate nanoemulsions that have better qualities in terms of stability, storage, and bioaccessibility/bioavailability of vitamin E and vitamin K. In the following sections, the effects of micronutrients that are present in the food on bioaccessibility and bioavailability of these vitamins will be discussed.

7.9.2 SYNERGISTIC/ANTAGONISTIC EFFECT OF MICRONUTRIENTS ON BIOAVAILABILITY

To enhance the bioaccessibility and bioavailability of vitamin E and vitamin K, various mechanical and chemical processes are available and have been studied in detail (Lee et al., 2020; McClements, 2012). But are these processes enough to see the actual desirable effect on the absorption of these vitamins? As it is universally known, food is a complex matrix and has numerous compounds that may or may not be of our desire. As the mechanism of absorption of fat-soluble vitamins is a complex process as has been discussed earlier for both vitamin E and vitamin K. Researchers have reported that the absorption of tocopherol is not only passive absorption but also includes choles-terol transporters like CD36 (Cluster Determinant 36), SR-BI (Scavenger Receptor class B type I), and NPC1-L1 (Niemann–Pick C1-Like 1) (Reboul & Borel, 2011). There is a specific transporter for vitamin K (K_1) that apparently requires an energy-saturable transport mechanism for absorption, whereas menaquinones are absorbed by passive diffusion (Reboul & Borel, 2011).

The absorption of these vitamins may be affected in the presence of several chemical compounds. An inhibition in the uptake of α-tocopherol was detected in Caco-2 cells in the presence of various constituents, such as carotenoids, phenolic naringenin, and γ-tocopherol (Reboul et al., 2007). A review reporting an old study described that fat soluble vitamins may have antagonistic interactions that affect their intestinal absorption *in vivo*. The authors reported that in chick plasma, levels of α-tocopherol deflated when the chicks were fed with vitamin A rich diet as compared to those who had a control diet (Reboul & Borel, 2011). Researchers have reported that vitamins A and D decreased the uptake of vitamin E in a dose-dependent manner in Caco-2 cells, while vitamin K affected it negatively only at higher concentrations (Goncalves et al., 2015). But, in the case of vitamin K, all the fat-soluble vitamins significantly reduced the uptake of vitamin K (Goncalves et al., 2015). These two vitamins affect the absorption of each other at different concentration levels. Vitamin K uptake was altered by vitamin E and vice versa at high doses (Goncalves et al., 2015).

7.9.3 Challenges with Currently Adopted Methods

The challenges that are now of major concern for these vitamins remain with the absorption mechanism. The vitamins affect the uptake of other vitamins at different concentrations and combinations, showing that they cannot be used together for supplementation. In the case of vitamin E, there are options available for co-supplementation with other fat-soluble vitamins, which is not the case for vitamin K. There is very little data available on the absorption mechanism of vitamin K. Researchers have reported low absorption of vitamin K; 13% from spinach (Reboul & Borel, 2011). However, the absorption efficiency of vitamin K is presumed to be higher if it is taken as a supplement. Indeed, a study found that 80% of vitamin K was absorbed when it was consumed as a supplement (Reboul & Borel, 2011). The problem remains with the absorption from natural resources. There is a need to find a process that can help with the absorption of these vitamins from natural resources. There is no doubt that all the methods and processes that have been developed to isolate these vitamins from natural sources are available and will make these vitamins available in pharmaceutical and nutraceutical form. But, as the deficiencies related to these vitamins are exhaustive and universal, there is a need to focus on the methods that can make these vitamins accessible not only in the form of nutrition in the gut but also in natural form for the population who cannot afford pharmaceutical and nutraceutical forms of these vitamins.

7.10 CASE STUDIES

Inadequate consumption of vitamin E in the diet may lead to mild to moderate vitamin E deficiency. Its intestinal malabsorption is very common in growing children, the elderly population, and women of reproductive age. Human studies have shown that vitamin E can be useful in the treatment of anemia (Jilani & Iqbal, 2018). On the other hand, hereditary hemolytic anemias occur due to vitamin E deficiency. Fat-soluble vitamin E acts as an antioxidant that has been used in dermatology in the past 50 years. Due to free radical scavenging activity, it is used as vital constituent in several beautifying goods that protect the skin from solar radiation. Research on vitamin E has depicted that it has antitumorigenic and photoprotective activities (Keen & Hassan, 2016). In spite of the formulations of new cosmetics, first controlled clinical trials should be carried out that would be helpful in the optimization of clear dosages for oral and topical vitamin E consumption/application.

Hemorrhagic disease in the infant (HDN), now referred to as 'vitamin K deficiency bleeding' (VKDB), is a rare disorder and can lead to fatal bleeding. There are three types of VKDB: early, classical, and late. Within 24 hours, bleeding starts in early VKDB. Classical VKDB is the commonest variant and presents after 24 hours but within the first week of life, and late VKDB is uncommon and manifests between 2–12 weeks of age. The Child Health Division, Ministry of Health and Family Welfare, Government of India; has initiated a program to fight this fatal disease in infants (Government of India, 2014). Under this program, all the newborns delivered at public health facilities are to be administered with prophylactic vitamin K injections. A case has been reported where a newborn was diagnosed with late VKDB with multiple skin-colored to bluish, raised swellings on the trunk and upper extremities of 4 days duration. The newborn was given 7.5 mg vitamin K_1 intramuscularly once daily for 3 days and improvement in the conditions was substantial (Gahalaut & Chauhan, 2013). This life-threatening disease caused by deficiency of vitamin K can be cured easily.

7.11 FOOD APPLICATIONS

The increasing demand for nutrient-rich food motivated the food industry to develop functional food and nutraceuticals. Especially, nutrients like vitamins and minerals are the target bioactive compounds that are incorporated into the food matrix by various novel techniques. Functional foods are developed by the incorporation of vitamin E and K using novel techniques such as encapsulation (liposome, solid lipid nanoparticles, nanostructured lipid carriers, emulsion system, molecule complexes, electrospinning spray drying) (Maurya et al., 2020); mechanical techniques (spray drying, extrusion, spray-cooling) (Estevinho & Rocha, 2017); chemical techniques (coacervation and polymerization) (Riberio et al., 2021). Wall material like phosphatidylcholine, soybean lecithin, glycerol tri palmitate, whey protein isolate, sodium caseinate, cyclodextrin, and corn oil were applied. Nutraceuticals are used as dietary supplements, although they contain active compounds that can also be used for preparing pharmaceuticals. Vitamin E and K nutraceuticals are helpful in preventing cancer, cutaneous ulcers, yellow nail syndrome, claudication, vibration disease, collagen synthesis, wound healing, and epidermolysis bullosa (Reberio et al., 2021).

Vitamin E and K are supplemented in various food products, especially dairy, beverages, and meat products. Vitamin E supplementation in meat products (geese) increases the saturated fatty acids, mono-saturated fatty acids, and poly-unsaturated fatty acids (Nemati et al., 2020). Table 7.2 showed fortification of vitamin E and K in various food using novel techniques such as potato chips, milk, beverage, yogurt, packaging film, juices, and nutraceuticals, etc., and also used in the preparation of various drug. Maurya et al. (2021) described various delivery systems that help to release beta-carotene to the target site. Such systems may be helpful in releasing the fat-soluble vitamins E and K. These systems include inclusion complexes, micro-/nanospheres/nanohydrogels, micro-nanocapsules/nanofibers/micelles, micro-/nanoemulsions/liposomes, niosome solid-liquid nanoparticles, and nanostructured lipid carriers.

7.12 FUTURE PROSPECTS

In the context of the future, isomers of vitamin E and K could be studied for their molecular action and target system. Delivery systems have already well established for vitamin E; therefore, for vitamin K, various delivery systems could be developed, including encapsulation and inclusion complex systems. Additionally, the extraction of isomers of both vitamins should take place from natural sources and can be fortified with a novel delivery system in foods, which can be quantified using LC-MS/MS. The delivery system used in fortification should first standardize and optimized, and then *in vivo* trials should be conducted on the delivery system to check the bioavailability and toxicity of the vitamins. Therefore, further research is needed on the delivery system for vitamin E and K and its application in foods to know the processing techniques, interaction with human metabolism, and standardization.

7.13 CONCLUSION

Fat-soluble vitamins E and K are essential micronutrients that are beneficial to health. The amount of these vitamins is maintained by an essential daily diet as recommended. In the other hand, absence of vitamin E and K from the diet causes deficiency and toxicity that can be prevented by supplementation and fortification of these vitamins in daily food. Fortification and supplementation of vitamins with novel techniques should be introduced together with encapsulation and mechanical methods that are helpful to prevent degradation of these delicate compounds and their controlled delivery system. Vitamin E and K bioavailability in food could be enhanced by fortification or by the use of vitamins-nanomaterials in food. This chapter provides substantial information regarding the various deficiencies of vitamin E and K such as intraventricular hemorrhage in the brain, the development of bronchopulmonary dysplasia, anemia, and retrolental fibroplasia. Therefore, the study of metabolic and absorption activity of vitamin E and K and fortification of these vitamins using robust kinds of delivery systems helps to resolve the limitations on these vitamins.

ACKNOWLEDGMENTS

Kiran Verma would like to acknowledge Jayant Verma for artistic efforts for this chapter (Figure 7.3 and 7.4).

REFERENCES

Adjonu R, Doran G, Torley P, Agboola S (2014) Whey protein peptides as components of nanoemulsions: A review of emulsifying and biological functionalities. *J Food Eng* 122, 15–27.

Amiri S, Ghanbarzadeh B, Hamishehkar H, Hosein M, Babazadeh A, Adun P (2018) Vitamin E loaded nanoliposomes: Effects of gammaoryzanol, polyethylene glycol and lauric acid on physicochemical properties. *Colloids Interface Sci Commun* 26, 1–6.

Anton N, Vandamme TF (2009) The universality of lowenergy nano-emulsification. *Int J Pharm* 377(1–2), 142–147.

Augustin MA, Sanguansri L (2015) Challenges and solutionsto incorporation of nutraceuticals in foods. *Annu Rev Food Sci Technol* 6, 1.1–1.15.

Azzi A (2018) Many tocopherols, one vitamin E. *Mol. Asp. Med* 61, 92–103.

Bardowell SA, Ding X, Parker RS (2012) Disruption of P450-mediated vitamin E hydroxylase activities alters vitamin E status in tocopherol supplemented mice and reveals extra-hepatic vitamin E metabolism. *J Lipid Res* 53(12), 2667–2676.https://doi.org/10.1194/jlr.M030734.

Booth SL, McKeown NM, Lichtenstein AH, Morse MO, Davidson KW, Wood RJ,Gundberg C (2000) A hydrogenated form of vitamin K: Its relative bioavailability and presence in the food supply. *J Food Compos Anal* 13(4), 311–317.https://doi.org/10.1006/jfca.1999.0863.

Bjørneboe A, Bjørneboe GEA, Drevon CA (1990) Absorption, transport and distribution of vitamin E. *J Nutr* 120(3), 233–242.

Budinčić JM, Petrović L, Đekić L, Fraj J, Bučko S, Katona J, Spasojević L (2021) Study of vitamin E microencapsulation and controlled release from chitosan/sodium lauryl ether sulfate microcapsules. *Carbohydr Polym* 251, 116988.

Cardenas E, Ghosh R (2013) Vitamin E: A dark horse at the crossroad of cancer management. *Biochem Pharmacol* 86(7), 845–852. https://doi.org/10.1016/j.bcp.2013.07.018.

Celebioglu A, Uyar T (2017) Antioxidant vitamin E/cyclodextrin inclusion complex electrospun nanofibers: Enhanced water solubility, prolonged shelf life, and photostability of vitamin E. *J Agric Food Chem* 65(26), 5404–5412.

Chanamai R, McClements DJ (2002) Comparison of gum arabic, modified starch, and whey protein isolate as emulsifiers: Influence of pH, CaCl2 and temperature. *J Food Sci* 67(1), 120–125.

Chen B, McClements DJ, Decker EA (2013) Design of foods with bioactive lipids for improved health. *Annu Rev Food Sci Technol* 4, 35–56.

Chen CC, Wagner G (2004) Vitamin E nanoparticle for beverage applications. *Chem Eng Res Des* 82(11), 1432–1437.

Chung S, Ghelfi M, Atkinson J, Parker R, Qian J, Carlin C, Manor D (2016) Vitamin E and phosphoinositides regulate the intracellular localization of the hepatic α-tocopherol transfer protein. *J Biol Chem* 291(33), 17028–17039. https://doi.org/10.1074/jbc.M116.734210.

Criddle DN, Gillies S, Baumgartner-Wilson HK, Jaffar M, Chinje EC, Passmore S, Chvanov M, Barrow S, Gerasimenko OV, Tepikin AV, Sutton R, Petersen OH (2006) Menadione-induced reactive oxygen species generation via redox cycling promotes apoptosis of murine pancreatic acinar cells. *J Biol Chem* 281(52), 40485–40492.

Dahlberg S, Ede J, Schött U (2017) Vitamin K and cancer. *Scand J Clin Lab Inv* 77(8), 555–567.

Dahlberg S, Schött U, Kander T (2021) The effect of vitamin K on prothrombin time in critically ill patients: An observational registry study. *J Intensive Care* 9(1), 1–9.

Dasgupta S, Dutta J, Annamaneni S, Kudugunti N, Battini MR (2015) Association of vitamin D receptor gene polymorphisms with polycystic ovary syndrome among Indian women. *Indian J Med Res* September;142(3), 276.

Dasgupta N, Ranjan S, Mundra S, Ramalingam C, Kumar A (2016) Fabrication of food grade vitamin E nanoemulsion by low energy approach, characterization and its application. *Int J Food Prop* 19(3), 700–708.

Dey S, Saxena A, Kumar Y, Maity T, Trafdar A. (2022). Understanding the antinutritional factors and bioactive compounds of kodo millet (Paspalum scrobiculatum) and little millet (Panicum sumatrense). *Journal of Food Quality*, 2022, 1–19.

Duarte-Correa Y, Díaz-Osorio A, Osorio-Arias J, Sobral PJ, Vega-Castro O (2020) Development of fortified low-fat potato chips through vacuum impregnation and microwave vacuum drying. *Innov Food Sci Emerg Technol* 64, 102437.

Duthie GG, Wahle KW, Harris CI, Arthur JR, Morrice PC (1992) Lipid peroxidation, antioxidant concentrations, and fatty acid contents of muscle tissue from malignant hyperthermia-susceptible swine. *Arch. Biochem. Biophys.* August 1;296(2), 592–596.

Dwyer JT, Wiemer KL, Dary O, Keen CL, King JC, Miller KB, … Bailey RL (2015) Fortification and health: Challenges and opportunities. *Adv Nutr* 6(1), 124–131.

EFSA Panel on Dietetic Products, Nutrition and Allergies (NDA), Turck D, Bresson JL, Burlingame B, Dean T, Fairweather-Tait S, … Neuhäuser-Berthold M (2017) Dietary reference values for vitamin K. *EFSA J* 15(5), e04780.

Eid M, Sobhy R, Zhou P, Wei X, Wu D, Li B (2020) β-cyclodextrin-soy soluble polysaccharide based core-shell bionanocomposites hydrogel for vitamin E swelling controlled delivery. *Food Hydrocoll* July 1;104, 105751.

Estevinho BN, Rocha F (2017) A key for the future of the flavors in food industry. In *Nanotechnology Applications in Food1–19*. Elsevier. https://doi.org/10.1016/B978-0-12-811942-6.00001-7.

Evans HM, Bishop KS (1922) On the existence of a hitherto unrecognized dietary factor essential for reproduction. *Science* 56(1458), 650–651. https://doi.org/10.1126/science.56.1458.650.

Fathi M, Mozafari MR, Mohebbi M (2012) Nanoencapsulation of food ingredients using lipid based delivery systems. *Trends Food Sci Technol* 23(1), 13–27.

Flory S, Birringer M, Frank J (2019) Bioavailability and metabolism of vitamin E. In Peter Weber, Marc Birringer, Jeffrey B. Blumberg, Manfred Eggersdorfer, and Jan Frank (eds.), *Vitamin E in Human Health*, 31–41. Humana Press, Cham.

Gahalaut P, Chauhan S (2013) Vitamin K deficiency bleeding presenting as nodular purpura in infancy: A rare and life-threatening entity. *Indian J Dermatol* 58(5), 407. https://doi.org/10.4103/0019-5154.117334.

Galli F, Azzi A, Birringer M, Cook-Mills JM, Eggersdorfer M, Frank J, … Özer NK (2017) Vitamin E: Emerging aspects and new directions. *Free Radic Biol Med* 102, 16–36.

Goncalves A, Roi S, Nowicki M, Dhaussy A, Huertas A, Amiot MJ, Reboul E (2015) Fat-soluble vitamin intestinal absorption: Absorption sites in the intestine and interactions for absorption. *Food Chem* 172, 155–160. https://doi.org/10.1016/j.foodchem.2014.09.021.

Gong X, Gutala R, Jaiswal AK (2008) Quinone oxidoreductases and vitamin K metabolism. *Vitam Horm* 78, 85–101.

Gonnet M, Lethuaut L, Boury F (2010) New trends in encapsulation of liposoluble vitamins. *J Control Release* 146(3), 276–290.

Government of India. (2014) Operational guidelines: Injection vitamin K prophylaxis at birth. DOA 10.09.2022. http://nhm.gov.in/images/pdf/programmes/child-health/guidelines/Vitamin_K_Operational_Guidelines.pdf.

Gröber U, Schmidt J, Kisters K. (2015) Magnesium in prevention and therapy. *Nutrients*. September;7(9), 8199–8226.

Groenen-Van Dooren MMCL, Ronden JE, Soute BAM, Vermeer C (1995) Bioavailability of phylloquinone and menaquinones after oral and colorectal administration in vitamin K-deficient rats. *Biochem Pharmacol* 50(6), 797–801. https://doi.org/10.1016/0006-2952(95)00202-B.

Grune T, Lietz G, Palou A, Ross AC (2010) β- carotene is an important vitamin A source for humans. *J Nutr* 140(12), 2268S–2285S.

Guttoff M, Saberi AH, McClements DJ (2015) Formation of vitamin D nanoemulsion-based delivery systems by spontaneous emulsification: Factors affecting particle size and stability. *Food Chem* 171, 117–122.

Hategekimana J, Chamba MVM, Shoemaker CF, Majeeda H, Zhonga F (2015) Vitamin E nanoemulsions by emulsion phase inversion: Effect of environmental stress and long-term storage on stability and degradation in different carrier oil types. *Colloids Surf A* 483, 70–80.

Horiguchi M, Arita M, Kaempf-Rotzoll DE, Tsujimoto M, Inoue K, Arai H (2003) pH-dependent translocation of alpha-tocopherol transfer protein (alpha-TTP) between hepatic cytosol and late endosomes. *Genes Cells* 8(10), 789–800. https://doi.org/10.1046/j.1365-2443.2003.00676.x.

Jensen SK, Lauridsen C (2007) α -Tocopherol stereoisomers. *Vitam Horm* 76, 281–308.

Jiang M, Hong Y, Gu Z, Cheng L, Li Z, Li C (2019) Preparation of a starch-based carrier for oral delivery of vitamin E to the small intestine. *Food Hydrocoll* 91, 26–33.

Jiang T, Liao W, Charcosset C (2020) Recent advances in encapsulation of curcumin in nanoemulsions: A review of encapsulation technologies, bioaccessibility and applications. *Food Res Int* 132, 109035.

Jilani T, Iqbal MP (2018) Vitamin E deficiency in South Asian population and the therapeutic use of alpha-tocopherol (vitamin E) for correction of anemia. *Pak J Med Sci* 34(6), 1571–1575. https://doi.org/10.12669/pjms.346.15880.

Kaesler N, Magdeleyns E, Herfs M, Schettgen T, Brandenburg V, Fliser D, … Krüger T (2014) Impaired vitamin K recycling in uremia is rescued by vitamin K supplementation. *Kidney Int* 86(2), 286–293.

Kayden HJ, Traber MG (1993) Absorption, lipoprotein transport, and regulation of plasma concentrations of vitamin E in humans. *J Lipid Res* 34(3), 343–358.

Ke D, Chen W, Chen W, Yun YH, Zhong Q, Su X, Chen H (2020) Preparation and characterization of octenyl succinate β-cyclodextrin and vitamin E inclusion complex and its application in emulsion. *Molecules* 25(3), 654.

Keen M, Hassan I (2016) Vitamin E in dermatology. *Indian Dermatol Online J* 7(4), 311. https://doi.org/10.4103/2229-5178.185494.

Kiani A, Fathi M, Nasirpour A (2021) Production of novel vitamin E loaded nanostructure lipid carriers and lipid nanocapsules for milk fortification. *J Agr Sci Technol* 23(3), 545–558.

Knapen MH, Braam LA, Teunissen KJ, Zwijsen RM, Theuwissen E, Vermeer C (2015) Yogurt drink fortified with menaquinone-7 improves vitamin K status in a healthy population. *J Nutr Sci* 4, e35.

Koe T (2020) https://www.nutraingredients-asia.com/article/2020/03/24/vitamin-k2-milk-fortification-gains-steam-in-vietnam.

Kralova I, Sj€oblom J (2009) Surfactants used in food industry: A review. *J Disper Sci Technol* 30, 1363–1383.

Kruger MC, Booth CL, Coad J, Schollum LM, Kuhn-Sherlock B, Shearer MJ (2006) Effect of calcium fortified milk supplementation with or without vitamin K on biochemical markers of bone turnover in premenopausal women. *Nutrition* 22(11–12), 1120–1128.

Kuchan MJ, Jensen SK, Johnson EJ, Lieblein-Boff JC (2016) The naturally occurring α-tocopherol stereoisomer RRR-α-tocopherol is predominant in the human infant brain. *Br J Nutr* 116(1), 126–131.

Kumar Y, Singhal S, Tarafdar A, Pharande A, Ganesan M, Badgujar PC (2020) Ultrasound assisted extraction of selected edible macroalgae: Effect on antioxidant activity and quantitative assessment of polyphenols by liquid chromatography with tandem mass spectrometry (LC-MS/MS). *Algal Res* 52, 102114.

Kurosu M, Begari E (2010) Vitamin K2 in electron transport system: Are enzymes involved in vitamin K2 biosynthesis promising drug targets? *Molecules* 15(3), 1531–1553.

Laouini A, Fessi H, Charcosset C (2012) Membrane emulsification: A promising alternative for vitamin E encapsulation within nano-emulsion. *J Membr Sci* 423, 85–96.

Latham MC, Ash D, Ndossi G, Mehansho H, Tatala S (2001) Micronutrient dietary supplements-A new fourth approach. *Arch Latinoam Nutr* 51, 37–41.

Lee HJ, Shin C, Chun YS, Kim J, Jung H, Choung J, Shim SM (2020) Physicochemical properties and bio-availability of naturally formulated fat-soluble vitamins extracted from agricultural products for complementary use for natural vitamin supplements. *Food Sci Nutr* 8(10), 5660. https://doi.org/10.1002/FSN3.1804.

Li Z, Xu D, Yuan Y, Wu H, Hou J, Kang W, Bai B (2020) Advances of spontaneous emulsification and its important applications in enhanced oil recovery process. *Adv Colloid Interface Sci* 277, 102119.

Li ZL, Peng SF, Chen X, Zhu YQ, Zou LQ, Liu W, Liu CM. (2018) Pluronics modified liposomes for curcumin encapsulation: Sustained release, stability and bioaccessibility. *Food Res Int*, 246–253. https://doi.org/10.1016/j.foodres.2018.03.048.

Mao L, Yang J, Xu D, Yuan F, Gao Y (2010) Effects of homogenization models and emulsifiers on the physicochemical properties of b-carotene nanoemulsions. *J Disper Sci Technol* 31(7), 986–993.

Maurya VK, Bashir K, Aggarwal M (2020) Vitamin D microencapsulation and fortification: Trends and technologies. *J Steroid Biochem* 196, 105489.

Maurya VK, Aggarwal M, Ranjan V, Gothandam KM (2020) Improving bioavailability of vitamin A in food by encapsulation: An update. *Nanosci Med* 1, 117–145.

Maurya VK, Shakya A, Aggarwal M, Gothandam KM, Bohn T, Pareek S (2021) Fate of β-carotene within loaded delivery systems in food: State of knowledge. *Antioxidants* 10(3), 426.

Mayer S, Weiss J, McClements DJ (2013) Behavior of vitamin E acetate delivery systems under simulated gastrointestinal conditions: Lipid digestion and bioaccessibility of low-energy nanoemulsions. *J Colloid Interface Sci* 404, 215–222. https://doi.org/10.1016/j.jcis.2013.04.048.

McClements DJ (2012) Nanoemulsions versus microemulsions: Terminology, differences, and similarities. *Soft Matter* 8(6), 1719–1729. https://doi.org/10.1039/c2sm06903b.

McClements DJ (2013) Nano emulsion-based oral delivery systems for lipophilic bioactive components: Nutraceuticals and pharmaceuticals. *Ther Deliv* 4(7), 841–857.

McClements DJ (2020) Recent advances in the production and application of nano-enabled bioactive food ingredients. *Curr Opin Food Sci* 33, 85–90.

McClements DJ (2021) Advances in edible nanoemulsions: Digestion, bioavailability, and potential toxicity. *Prog Lipid Res* 81, 101081.

McClements DJ, Rao J (2011) Food-Grade nanoemulsions: Formulation, fabrication, properties, performance, biological fate, and Potential Toxicity. *Crit Rev Food Sci Nutr* 51(4), 285–330. https://doi.org/10.1080/10408398.2011.559558.

McClements DJ, Das AK, Dhar P, Nanda PK, Chatterjee N (2021) Nanoemulsion-based technologies for delivering natural plant-based antimicrobials in foods. *Front Sustain Food Syst* February 18;5, 643208.

Mehansho H, Mellican RI, Hughes DL, Compton DB, Walter T (2003) Multiple-micronutrient fortification technology development and evaluation: from lab to market. *Food Nutr Bull* 24(4_suppl_1), S111–S119.

Mehmood T (2015) Optimization of the canola oil based vitamin E nanoemulsions stabilized by food grade mixed surfactants using response surface methodology. *Food Chem* 183, 1–7.

Merrifield LS (1965) Yang HY. Vitamin K5 as a fungistatic agent. *Appl Microbiol* 13(5), 660–662.

Miller ER, Pastor-Barriuso R, Dalal D, Riemersma RA, Appel LJ, Guallar E (2005) Meta-analysis: High-dosage vitamin E supplementationmay increase all-cause mortality. *Ann Intern Med* 142, 37–46. PMID: 15537682. https://doi.org/10.7326/0003-4819-142-1-200501040-001.

Mirzaei-Mohkam A, Garavand F, Dehnad D, Keramat J, Nasirpour A (2020) Physical, mechanical, thermal and structural characteristics of nanoencapsulated vitamin E loaded carboxymethyl cellulose films. *Prog Org Coat* 138, 105383.

Nagy K, Ramos L, Courtet-Compondu M-C, Braga-Lagache S, Redeuil K, Lobo B, et al. (2013) Double-balloon jejunal perfusion to compare absorption of vitamin E and vitamin E acetate in healthy volunteers under maldigestion conditions. *Eur J Clin Nutr* 67(2), 202–206. https://doi.org/10.1038/ejcn.2012.183.

Nemati Z, Alirezalu K, Besharati M, Amirdahri S, Franco D, Lorenzo JM (2020) Improving the quality characteristics and shelf life of meat and growth performance in goose fed diets supplemented with vitamin E. *Foods* 9(6), 798.

Ogawa S, Shinkawa M, Hirase R, Tsubomura T, Iuchi K, Hara S (2021) Development of water-insoluble vehicle comprising natural cyclodextrin—Vitamin E complex. *Antioxidants* 10(3), 490.

Ostertag F, Weiss J, McClements DJ (2012) Low-energy formation of edible nanoemulsions: Factors influencing droplet size produced by emulsion phase inversion. *J Colloid Interface Sci* 388(1), 95–102.

Öztürk B (2017) Nanoemulsions for food fortification with lipophilic vitamins: Production challenges, stability, and bioavailability. *Eur J Lipid Sci Technol* 119(7), 1500539.

Ozturk B, Argin S, Ozilgen M, McClements DJ (2015) Formation and stabilization of nanoemulsion-based vitamin E delivery systems using natural biopolymers: Whey protein isolate and gum arabic. *Food Chem* 188, 256–263.

Ozturk B, McClements DJ (2016) Progress in natural emulsifiers for utilization in food emulsions. *Curr Opin Food Sci* 7, 1–6.

Palermo A, Tuccinardi D, D'Onofrio L, Watanabe M, Maggi D, Maurizi AR, … Manfrini S (2017) Vitamin K and osteoporosis: Myth or reality? *Metabolism* 70, 57–71.

Parker CH, Morgan CR, Rand KD, Engen JR, Jorgenson JW, Stafford DW (2014) A conformational investigation of propeptide binding to the integral membrane protein C-glutamyl carboxylase using nanodisc hydrogen exchange mass spectrometry. *Biochemistry* 53(9), 1511–1520.

Patel J, Al-Ghamdi S, Zhang H, Queiroz R, Tang J, Yang T, Sablani SS (2019) Determining shelf life of ready-to-eat macaroni and cheese in high barrier and oxygen scavenger packaging sterilized via microwave-assisted thermal sterilization. *Food Bioprocess Technol* 12(9), 1516–1526.

Qian J, Morley S, Wilson K, Nava P, Atkinson J, Manor D (2005) Intracellular trafficking of vitamin E in hepatocytes: The role of tocopherol transfer protein. *J Lipid Res* 46(10), 2072–2082. https://doi.org/10.1194/jlr.M500143-JLR200.

RDA short report by NIN and ICMR, 2020.

Reboul E (2017) Vitamin E bioavailability: Mechanisms of intestinal absorption in the spotlight. *Antioxidants (Basel)*. https://doi.org/10.3390/antiox6040095.

Reboul E, Borel P (2011) Proteins involved in uptake, intracellular transport and basolateral secretion of fat-soluble vitamins and carotenoids by mammalian enterocytes. *Prog Lipid Res* 50(4), 388–402. https://doi.org/10.1016/j.plipres.2011.07.001.

Reboul E, Thap S, Tourniaire F, André M, Juhel C, Morange S, … Borel P (2007) Differential effect of dietary antioxidant classes (carotenoids, polyphenols, vitamins C and E) on lutein absorption. *Br J Nutr* 97(3), 440–446. https://doi.org/10.1017/S0007114507352604.

Ribeiro AM, Estevinho BN, Rocha F (2021) One century of vitamin E: The progress and application of vitamin E encapsulation–A Review. *Food Hydrocoll*, 121, 106998.

Rigotti A (2007) Absorption, transport, and tissue delivery of vitamin E. *Mol Aspects Med* 28(5–6), 423–436. https://doi.org/10.1016/j.mam.2007.01.002.

Saberi AH, Fang Y, McClements DJ (2013) Effect of glycerol on formation, stability, and properties of vitamin-E enriched nanoemulsions produced using spontaneous emulsification. *J Colloid Interface Sci* 411, 105–113.

Sakane R, Kimura K, Hirota Y, Ishizawa M, Takagi Y, Wada A, … Suhara Y (2017) Synthesis of novel vitamin K derivatives with alkylated phenyl groups introduced at the ω-terminal side chain and evaluation of their neural differentiation activities. *Bioorg Med Chem Lett* 27(21), 4881–4884. https://doi.org/10.1016/j.bmcl.2017.09.038.

Saratale RG, Lee HS, Koo YE, Saratale GD, Kim YJ, Imm JY, Park Y (2018) Absorption kinetics of vitamin E nanoemulsion and green tea microstructures by intestinal in situ single perfusion technique in rats. *Food Res Int* 106, 149–155.

Sato T, Schurgers LJ, Uenishi K (2012) Comparison of menaquinone-4 and menaquinone-7 bioavailability in healthy women. *Nutr J* 11(1), 9–12. https://doi.org/10.1186/1475-2891-11-93.

Schmölz L, Birringer M, Lorkowski S, Wallert M (2016) Complexity of vitamin E metabolism. *World J Biol Chem* 7(1), 14–43. https://doi.org/10.4331/wjbc.v7.i1.14.

Sen CK, Khanna S, Roy S (2007) Tocotrienols in health and disease: The other half of the natural vitamin E family. *Mol Aspects Med* 28(5–6), 692–728. https://doi.org/10.1016/j.mam.2007.03.001.

Siegrist M, Stampfli N, Kastenholz H, Keller C (2008) Perceived risks and perceived benefits of different nanotechnology foods and nanotechnology food packaging. *Appetite* 51(2), 283–290.

Silva HD, Cerqueira MA, Vicente AA (2012) Nano emulsions for food applications: Development and characterization. *Food Bioprocess Technol* 5(3), 854–867.

Singh H, Singh J, Singh SK, Singh N, Paul S, Sohal HS, … Jain SK (2019) Vitamin E TPGS based palatable, oxidatively and physically stable emulsion of microalgae DHA oil for infants, children and food fortification. *J Dispers Sci Technol* 41(11), 1–16.

Tadros T, Izquierdo P, Esquena J, Solans C (2004) Formation and stability of nano-emulsions. *Adv Colloid Interface Sci* 108, 303–318.

Tan Y, McClements DJ (2021) Improving the bioavailability of oil-soluble vitamins by optimizing food matrix effects: A review. *Food Chem* 348(November 2020), 129148. https://doi.org/10.1016/j.foodchem.2021.129148.

Teleki A, Hitzfeld A, Eggersdorfer M (2013) 100 years of vitamin: The science of formulation is the key to functionality. *KONA Powder Part J* 30, 144–163.

Turck D, Bresson JL, Burlingame B, Dean T, Fairweather-Tait S, … Neuhäuser-Berthold M (2017) Dietary reference values for vitamin K. *EFSA J* 15(5), e04780.

Ulatowski L, Manor D (2013) Vitamin E trafficking in neurologic health and disease. *Annu Rev Nutr* 33, 87–103. https://doi.org/10.1146/annurev-nutr-071812-161252.

Van Hasselt PM, Janssens GEPJ, Slot TK, van der Ham M, Minderhoud TC, Talelli M, … van Nostrum CF (2009) The influence of bile acids on the oral bioavailability of vitamin K encapsulated in polymeric micelles. *J Control Release* 133(2), 161–168. https://doi.org/10.1016/j.jconrel.2008.09.089.

Verma K, Tarafdar A, Badgujar PC (2021) Microfluidics assisted tragacanth gum based sub-micron curcumin suspension and its characterization. *LWT* 135, 110269. https://doi.org/10.1016/j.lwt.2020.110269.

Walker R, Decker EA, McClements DJ (2015) Physical and oxidative stability of fish oil nanoemulsions produced by spontaneous emulsification: Effect of surfactant concentration and particle size. *J Food Eng* 164, 10–20.

Wang H, Yan Y, Feng X, Wu Z, Guo Y, Li H, Zhu Q (2020) Improved physicochemical stability of emulsions enriched in lutein by a combination of chlorogenic acid–whey protein isolate–dextran and vitamin E. *J Food Sci* 85(10), 3323–3332.

Wang D, Zhong M, Sun Y, Fang L, Sun Y, Qi B, Li Y (2021) Effects of pH on ultrasonic-modified soybean lipophilic protein nanoemulsions with encapsulated vitamin E. *LWT* 144, 111240.

Webb GP (2011) *Dietary Supplements and Functional Foods.* Wiley-Blackwell Publishing Ltd., West Sussex.

Webb AL, Villamor E (2007) Update: Effects of antioxidant and non-antioxidant vitamin supplementation on immune function. *Nutr Rev* 65(5), 181–217.

Weston BW, Monahan PE (2008) Familial deficiency of vitamin K dependent clotting factors. *Haemophilia* 14(6), 1209–1213.

Willson RL (1983) Free radical protection: Why vitamin E, not vitamin C, β-carotene or glutathione. In: *Biology of Vitamin E* (Porter R, Whelan J, eds.), pp. 19–44. Ciba Foundation Symposium 101. Pitman Books Ltd., London.

Wong RS, Radhakrishnan AK (2012) Tocotrienol research: past into present. *Nutr Rev* 70(9), 483–490. https://doi.org/10.1111/j.1753-4887.2012.00512.x.

World Health Organization. (2004) *Vitamin and mineral requirements in human nutrition.* World Health Organization.

Yang Y, McClements DJ (2013a) Vitamin E and vitamin E acetate solubilization in mixed micelles: Physicochemical basis of bioaccessibility. *J Colloid Interface Sci* 405, 312–321. https://doi.org/10.1016/j.jcis.2013.05.018.

Yang Y, McClements DJ (2013b) Vitamin E bioaccessibility: Influence of carrier oil type on digestion and release of emulsified α-tocopherol acetate. *Food Chem* 141(1), 473–481. https://doi.org/10.1016/j.foodchem.2013.03.033.

Part III

Mineral Fortification

8 Iron Fortification

Harshdeep K. Khurana, Mehvish Habib,
Sakshi Singh, Amita Shakya, and Khalid Bashir

8.1 INTRODUCTION

Iron, being one of the essential micronutrients, is responsible for several functions in the living body. Iron mainly binds to protein (hemoprotein) and exists in complex form in the human body as heme compounds (myoglobin or hemoglobin), heme enzymes, or non-heme compounds (transferrin, ferritin, and flavin-iron enzymes). Iron is essential for the production of its oxygen transport proteins (myoglobin and hemoglobin) and the creation of heme enzymes and other iron-containing enzymes involved in electron transfer and oxidation-reduction reactions. About 60% of the iron in a human body is contained in the hemoglobin present in circulating erythrocytes, 25% is in a store of easily mobilizable iron, and the other 15% is linked to myoglobin in muscle tissue and numerous enzymes involved in oxidative metabolism and different cell activities. There is no physiological system involved in excreting excess iron from the body other than blood loss through menstruation, pregnancy, or any other form of bleeding (Nadadur et al., 2008). It is recycled as well as conserved in the human body. Iron gets bound and transferred to various parts of the body with the help of transferrin. The cell surface TfR1 (transferrin receptor 1) binds to the iron-laden transferrin and results in endocytosis followed by metal cargo uptake. The iron that has been internalized is subsequently transferred to mitochondria for use in the formation of iron-sulfur clusters or heme. This is an integral part of numerous metalloproteins. Thus, excess iron is stored as well as detoxified within cytosolic ferritin (Abbaspour et al., 2014; Wang et al., 2011).

There are two forms of dietary iron: heme iron, which derives from myoglobin and hemoglobin found only in animal-based foods, and non-heme iron, which is found in animal tissues and plant foods. Consequently, heme iron contributes around 10–15% of the total iron intake in meat-eating populations. Iron is capable of contributing an overall 40% of the total absorbed iron because of its uniform and higher absorption (15–35%). In contrast, non-heme iron content, which is less efficiently absorbed than heme, enters the digestive tract's common iron pool and is absorbed to the same extent, based on the ratio of absorption inhibitors to enhancers as well as the person's iron status (Hurrell et al., 2010).

8.2 ABSORPTION AND METABOLISM OF IRON IN THE BODY

Some iron-binding proteins, like lactoferrin, transferrin, bacterioferritin, and hemoglobin, along with other functional agents present at numerous key locations, control iron transport in the human digestive system. In the stomach's acidic environment, mucins attach to iron, helping to keep it in solution form for subsequent absorption in the duodenum's alkaline circumstances and for simple transit over the mucosal cell membrane. After it enters into the cells, it gets finally transported to the basolateral side by the cytoplasmic iron-binding protein, mobilferrin, and gets exported to blood plasma. Several transport mechanisms and regulatory proteins are included in this method of heme and non-heme iron absorption in the intestinal mucosal cells. As discussed in the previous section, dietary iron is mainly expressed as two types: heme and non-heme iron. Generally, there are two valence states for non-heme (elemental) iron: reduced ferrous (Fe^{2+}) and oxidized ferric (Fe^{3+}). It is crucial that the duodenal cytochrome b reductase (duodenal cytochrome b) or other reducing agents found in the apical membrane of duodenal enterocytes first convert ferric iron into ferrous iron. Later, the DMT-1 ("Divalent Metal Iron Transporter-1") transports it to the duodenal cytoplasm. The elemental heme iron requires acidic conditions for uptake and supplies protons along with it through DMT-1 and hence aids in ferrous iron solubility. The human body utilizes absorbed ferrous iron in three ways.

- To produce heme molecules after being transferred to the mitochondria,
- To be stored within the enterocyte followed by transfer into ferritin, and
- By transporting it to different sites in the body where iron is required.

In case of genetic protein deficiency, iron oxidation gets hindered, which leads to an excess of iron in the pancreas, brain, as well as liver. This condition can result in causing diabetes and neurological dysfunction. In the case of chronic anemia, iron sequestration caused by hepcidin, a small "glycopeptide" molecule (released from the liver and helps in the circulation and absorption of iron), can be suppressed with the increase in levels of RBC production, hypoxia, and iron deficiency. During enhanced hepcidin levels, absorption of iron is reduced by retarding the transcription of DMT-1 at the enterocyte. Iron absorption is reduced with enhanced levels of hepcidin (WHO, 2006; Shubham et al., 2020).

8.3 BIOLOGICAL ROLE OF IRON

There are numerous vital functions performed by iron in the human body. Red blood cell pigment (hemoglobin) acts as a transporter for oxygen from the lungs to tissues within the body. Hemoglobin (Hb) is present in the blood of vertebrates and is known as an oxygen transport protein, as it primarily transports oxygen for efficient energy production. Additionally, for a variety of different important enzymes, Fe is considered important for the normal development of the brain in the child and fetus, for adolescent girls, for optimum immune defense, and young pregnant women. In the case of oxidative metabolism, they transfer energy in the cell, especially in mitochondria. Other vital functions include foreign substances detoxification in the liver, steroid hormones, and bile acids synthesis, and signal control in approximate neurotransmitters (like the dopamine and serotonin systems responsible for cognitive development in infants and children) in the brain. Iron is reversibly deposited in the liver as hemosiderin and ferritin and transferred across body compartments via transferrin (a protein) (Gupta, 2014).

Deficiency of iron disrupts certain processes and leads to consistent structural, behavioral, and electro-physiologic abnormalities both during the period of deficiency as well as long-term after iron repletion. These findings imply impaired cognitive and poor motor development in the case of children and infants (Hurrell, 2018). Since iron enzymes depend on the levels of hemoglobin, they can thus be impaired before their levels decrease critically. According to Hurrell (2018), it has been estimated that about 2 billion people are affected by iron deficiency, which mainly involves young women and children. Apart from this, decreased levels of iron impact the absorption of micronutrients as well as their functions. For example, people with iron deficiency are more prone to iodine deficiency, as iron is required for the effective metabolism of the thyroid due to thyroperoxidase impairment, a heme-dependent enzyme that iodizes thyroglobulin. Another important consequence of iron deficiency, as discussed above, is decreased energy production. Low levels of Hb downregulate those iron enzymes that are important for sufficient oxygen and energy metabolism. This leads to fatigue and declined work capacity in people, which affects the growth and economy of the nation. Other negative effects associated with this are a lack of oxygen due to anemia, resulting in poor pregnancy outcomes such as low birth weight, prematurity, and maternal deaths at childbirth (Hurrell, 2018).

8.4 IRON DEFICIENCY AND PREVALENCE

According to the mortality data from the World Health Organization (WHO, 2000), iron deficiency has been responsible for approximately 0.8 million deaths each year. Severe anemia is reported to be the primary cause of 20% of maternal deaths in developing countries. In other words, it can be said that in developing countries, around one-tenth of cases of maternal death and one-fifth of the cases of perinatal death are due to iron deficiency. This makes anemia caused by a serious public health concern in most of the nations (Figure 8.1 and Table 8.1) (Sadighi et al., 2009).

The pervasiveness of anemia in developing nations has always had direct as well as indirect impacts on the country's national economy. In most cases, anemia is acute, which can be developed into chronic if not treated within time or even worsen at a later stage. About 1.62 billion individuals (about 24.8% of the world's population) are impacted by anemia with 40% of pregnant women (WHO, 2011). According to data suggested by the WHO in 2000, it is estimated that around 40% of the world's population (such as higher than 2 billion people) is said to suffer anemia, namely low blood level, out of which pregnant women and non-pregnant women account for 35%; infants and children (1–2 yrs) for 50%; adolescents 30 to 55%; school children 40%; and preschool-aged children 25%. These groups are considered the mean incidence among particular population groups, with pregnant women and children being the most susceptible (Shubham et al., 2020). Though the efficacy of iron absorption is increasing, the amount of absorbed iron from food is still not fulfilled due to a diet lacking fruits, meat, fish, and vegetables, and our comforting sedentary lifestyle. Inadequacy of nutrients leads to a deficiency that impacts more than 50% of the world's population (Shubham et al., 2020). The prevalence of anemia in children in Africa is presented in Figure 8.2.

8.4.1 FACTORS CONTRIBUTING TO IRON DEFICIENCY

Over an extended period in the human body, if iron absorption is not able to keep up with the metabolic needs of iron required for replenishment of iron loss and sustenance of growth, it leads to depletion of stored iron. When iron stores are exhausted and an inadequate amount of iron is available for "erythropoiesis", hematologic signs of iron deficiency, i.e., anemia, appear because of impairment of hemoglobin synthesis in erythrocyte precursors (Figure 8.3) (Hurrell, 2021). The main factors that contribute to an iron deficiency are a low intake of bioavailable iron, increased iron requirements, restrictions on iron use due to fast development, menstruation, and pregnancy, and excessive blood loss brought on by pathological infections such as whipworm and hookworm, which cause loss of gastrointestinal blood and hinder iron absorption (Figure 8.4). The prevalence of iron deficiency increases in female teenagers due to the combination of monthly iron losses and the necessity for fast development (Oppenheimer, 2001). It results in a deficiency of nutritional iron when physiological needs could not be fulfilled by iron absorption obtained from the diet. The bioavailability of dietary iron is found to be low in a group of people who consume monotonous plant-based diets as they are rarely fortified with iron, leading to an increase in the frequency of anemia in children younger than four years by 50% (Abbaspour et al., 2014; Reddy et al., 2006).

High-income nations have anemia mostly due to its deficiency, whereas low and middle-income countries (LMICs) experience anemia primarily due to widespread infections, inflammation, and hemoglobinopathies. Hurrell (2021) estimated that iron deficiency accounts for only 25% of anemia in children as well as 37% of anemia in reproductive-age women living in LMICs. Most nations with a high incidence of anemia consume little to no bioavailable iron from animal tissue diets and only poor bioavailable iron from plant foods (Hurrell, 2021). There are cases where the cause of anemia is not a low intake of iron. For example, in the case of endemic malarial infection, there is a reduction in red cell iron recycling, an increase in the lysis of infected red cells, as well as an inflammation-related obstruction of iron absorption (Zimmermann 2020). Thus, estimating the incidence of iron deficiency based on anemia would be unreliable and imprecise. Hence, it should not only be the considerable parameter utilized to assess the requirement for iron fortification. In other words,

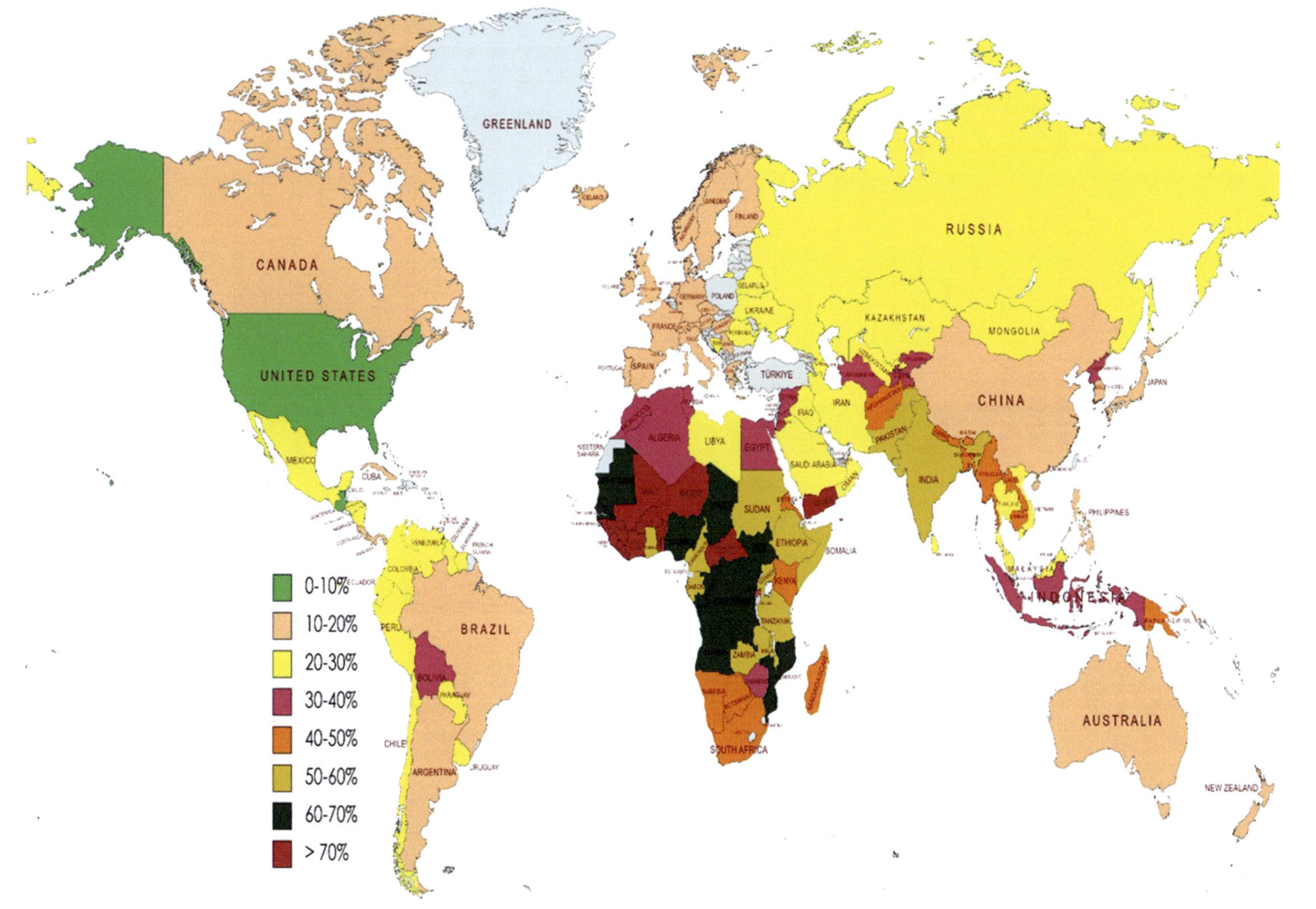

FIGURE 8.1 Global status of anemia in children (Our World in Data, 2023).

TABLE 8.1

Global Hemoglobin Concentration and Percentage Anemia

Regions	Children (Aged 6 to 59 months)		All women of reproductive age (15–49 years)	
	Mean (95% CI) blood hemoglobin conc (g/L)	Percentage (95% CI) of the population with anemia	Mean (95% CI) blood hemoglobin conc (g/L)	Percentage (95% CI) of the population with anemia
Africa	105 (103 to 106)	60.2 (57.0 to 63.1)	123 (121 to 125)	37.6 (32.4 to 43.0)
Latin America and the Caribbean	117 (114 to 120)	29.1 (22.5 to 36.9)	130 (126 to 134)	19.1 (13.1 to 29.4)
Northern America	124 (122 to125)	7.0 (4.9 to 12.3)	131 (130 to 133)	12.4 (9.3 to 17.1)
Asia	112 (109 to 115)	42.0 (34.1 to 49.9)	124 (122 to 127)	31.9 (24.6 to 40.6)
Oceania	117 (112 to 122)	26.2 (14.5 to 41.6)	128 (123 to 132)	20.0 (12.0 to 35.5)
Europe	120 (116 to 123)	19.3 (10.9 to 30.7)	129 (126 to 131)	20.1 (13.8 to 28.3)
Global	111 (110 to 113)	42.6 (37.7 to 47.4)	125 (124 to 127)	29.4 (24.5 to 35.0)

Source: WHO, 2020; Blanco-Rojo et al., 2019.

the prevalence of anemia alone cannot not be utilized to check iron fortification interventions, as the additional iron may only affect the proportion of anemia resulting from its deficiency and not from other causes (Hurrell, 2021). To improve the status of iron in the body, it is therefore always advised to inculcate sources of non-heme iron (turnip, mustard, green-leafy vegetables, cereals, and sprouted pulses, roots, tubers, and eggs) and heme iron (red meat, chicken, liver). Anemia could be inhibited by improving the bioavailability of iron which further depends on its intake along with vitamin B, folic acid, and ascorbic acid (Vitamin C). The sedentary lifestyle of modern times has contributed significantly to the increase in the incidence of Fe deficiency, and this may even lead to obesity. This alarming scenario can be seen in those socioeconomic groups who tend to consume fast foods and have low nutrient levels. Therefore, it is important to come up with ideas to combat iron deficiency keeping the population's eating habits in consideration.

8.4.2 RECOMMENDED DAILY ALLOWANCE (RDA)

Vegetarians are advised to consume about 1.8 times more iron than meat eaters because heme iron obtained from meat products is more bioavailable compared to non-heme iron. The demand for iron also increases about three times in a pregnant woman due to fetal-placental progression and enlargement of the maternal red-cell mass. Iron requirements in the early infancy stage are met by iron present in breast milk and it rises to around 0.7 to 0.9 mg/day during the remaining first part of the year. The need for iron content doubles between 1 and 6 years of age. Also, the need for iron increases during puberty both in girls and boys. On average, 0.5 mg Fe is lost by women per day during menstruation (values vary for women with higher losses). In boys, an increase in hemoglobin mass and concentration is observed during adolescence. However, there is a high iron requirement during postnatal development which can be achieved by including it in the diet. Although, it has been observed that the iron level in the human body becomes constant upon reaching adulthood. The upper intake level (UL) of iron, which is tolerable and hence recommended for adults, is 45 mg/day. RDA of iron required at various phases of life is presented in Table 8.2. The requirement of iron in a pregnant woman is dependent on the women's iron status prior to pregnancy (Schümann et

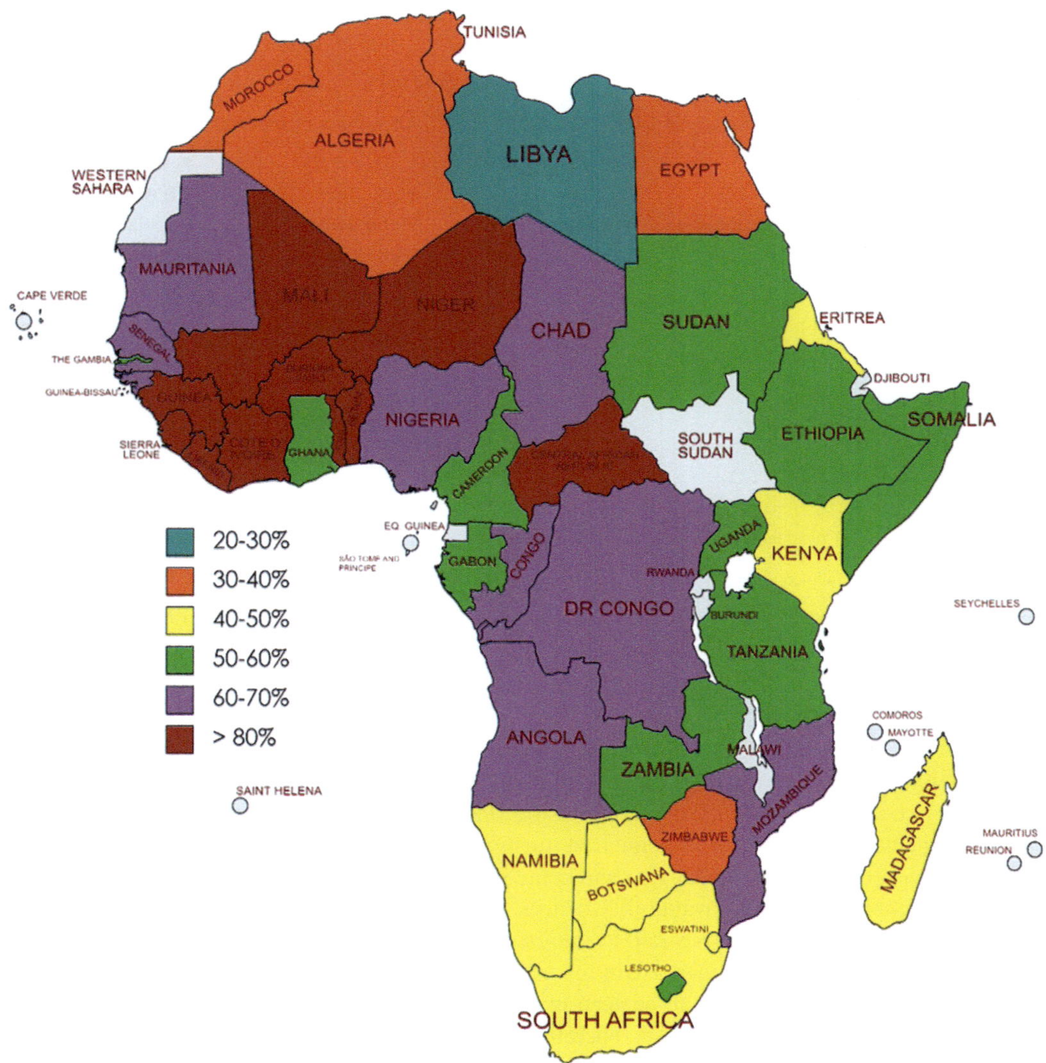

FIGURE 8.2 Prevalence of anemia in children in Africa (Our World in Data, 2023).

al., 1998; Hurrell, 2021; Abbaspour et al., 2014). On average, 1 to 3 g of iron is stored by the human body with a regularly maintained balance between dietary uptake as well as loss. Each day, approximately 1 mg of iron is lost by shedding of cells from the skin as well as mucosal surfaces (such as gastrointestinal tract lining). To replace the iron lost in urine and stools, which represent around 0.8 mg for an adult female and 0.9 mg of iron for an adult male, dietary intake of iron has significant importance (Abbaspour et al., 2014; Beard and John, 2000).

8.4.3 Iron Toxicity

An iron overload in the human system leads to increased liver storage of ferritin. It generally occurs in individuals with genetic disorders like thalassemia and hemochromatosis, which alter the functions of ferroprotein or hepcidin. This reduces the efficiency of the regulators involved in preventing excessive aggregation of iron. There is an increased level of insulin resistance and oxidized LDL ("low-density lipoprotein") in patients with increased iron overload, which results in cardiovascular

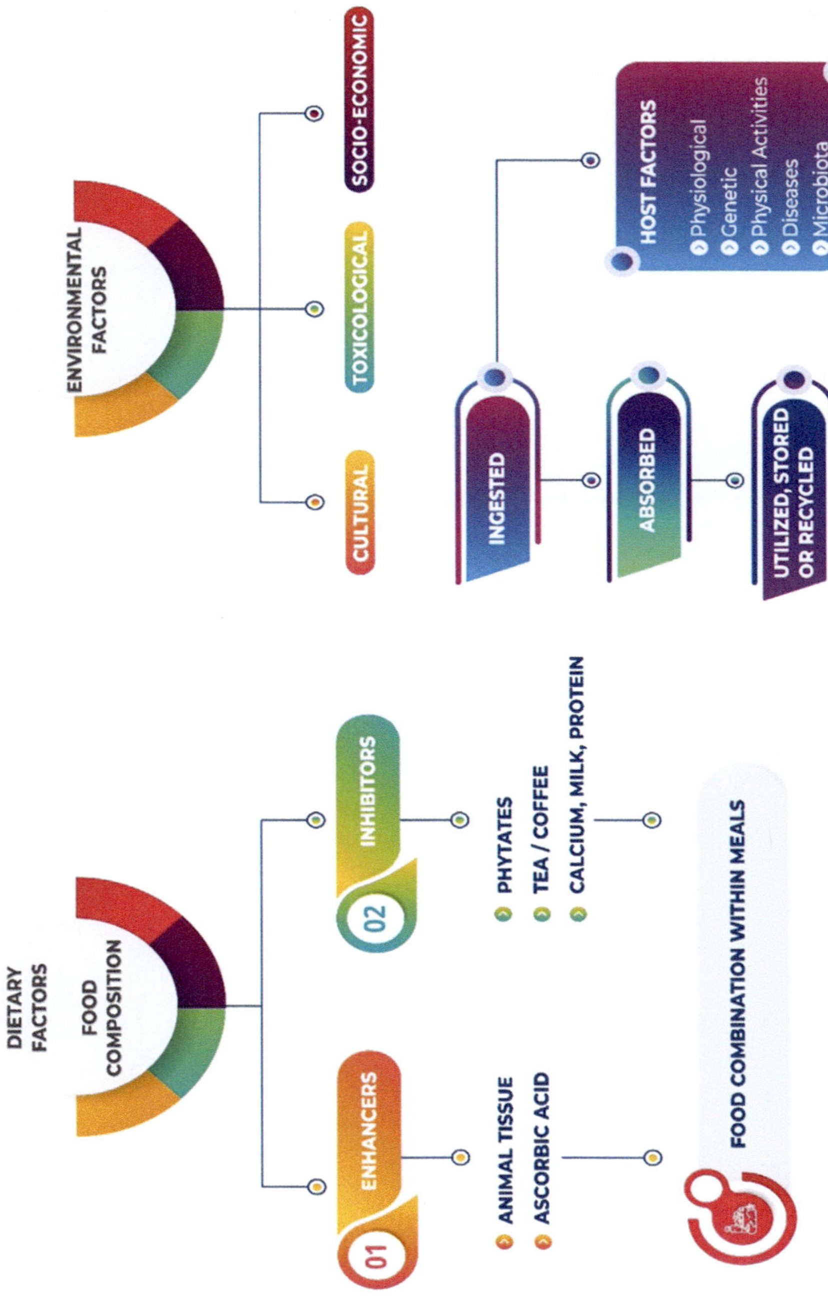

FIGURE 8.3 Factors modulating iron bioavailability (Blanco-Rojo et al., 2019).

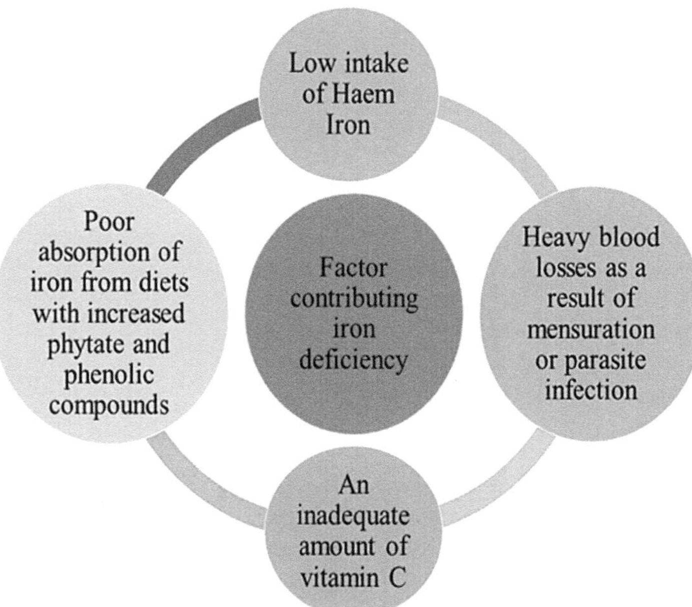

FIGURE 8.4   Risk factors of iron deficiency (Hurrell & Richard, 2021).

TABLE 8.2

Recommended Daily Allowance (RDA) of Iron in Different Groups

Category	Iron mg/day
Sedentary, moderate, and heavy workers (men)	19
Sedentary, moderate, and heavy workers (women)	29
Pregnant women, second and third trimester	40
Lactation (0 to 12 months)	23
0 to 6 months (infants)	-
7 to 12 months (infants)	3
1 to 3 years (Children)	8
4 to 6 years (Children)	11
7 to 9 years (Children)	15
Boys 10 to 12 years	16
Girls 10 to 12 years	28
Boys 13 to 15 years	22
Girls 13 to 15 years	30
Boys 16-18 years	26
Girls 16-18 years	32

disease. According to Hurrell (2018), patients with increased iron load have to deal with the fate of inordinate absorption of iron and thus demand special care, even if the diet is not iron fortified.

Conditions that are associated with the increased load of iron are as follows:

a) **Inflammatory oxidative stress response:** Iron is used as a biocatalyst in proteins and as an electron carrier in energy metabolism because of its easy redox cycling properties. It is therefore known as a reactive element. However, it is a potent pro-oxidant. Free iron can become toxic under anaerobic conditions which form reactive oxygen species, including superoxide and other free radicals. Hence, iron is present both as a nutrient and a potentially toxic element in circulation. The majority of iron is bound to the heme molecule in hemoglobin while the remaining part is stored in the form of ferritin. During transport, it becomes tightly bound to transferrin (Blanco-Rojo et al., 2019; Hurrell, 2018).

b) **Gastrointestinal problems:** According to HurrellRichard (2018) the primal considerations for the safety of iron-fortified formulas, including the ones used since birth, were based on trials at different degrees of fortification. The gastrointestinal tract of an infant is specifically susceptible to any kind of imbalance, which can change the development of the innate immune system, a function of the mucosal barrier, maturation of intraepithelial tight junctions, and intestinal permeability. The author reported the presence of several prebiotic factors, including a scale of oligosaccharides in human milk, that marked the absence of gastrointestinal signs (Hurrell, 2018). Lactoferrin is prebiotic with important functional characteristics that promote a healthy intestinal microbiome, binding as well as facilitating iron uptake by the enterocyte. It also binds iron, which contributes towards a low iron environment in the gut lumen of breastfed infants. It also has immunomodulatory activity using the induction of T-helper cells, which protect the young infant against infection.

c) **Cardiovascular disease**: There have been various pieces of evidence that suggest increased iron storage may lead to the development of coronary heart disease (CHD). Elevated levels of serum ferritin, which is an acute phase protein and not a good biomarker of iron status in the presence of any inflammation common in chronic disease.

d) **Diabetes type 2:** A positive correlation was observed between increased concentrations of serum ferritin and type 2 diabetes mellitus. However, increased serum ferritin values are generally met with the presence of obesity and the metabolic syndrome associated with the risk of development of type 2 diabetes (Blanco-Rojo et al., 2019; Hurrell, 2018).

8.5 STRATEGIES TO COMBAT IRON DEFICIENCY

8.5.1 Food Diversification

It involves modifying and adding variety in the diet, which increases the intake of iron-rich foods (mainly flesh foods), and consumption of fruits and vegetables enriched with vitamin C, which not only enhances the absorption of non-heme iron but also reduces the intake of tea and coffee, as these suppress the Fe absorption. Further methods to make the supplied iron from food more available could be applied by reducing the anti-nutrient contents. Techniques that enhance the activity of endogenous or exogenous phytase enzymes (such as fermentation and germination) by enzymatic hydrolysis of phytic acid are employed under this strategy to increase the bioavailability of iron in whole grains and legumes. Non-enzymatic techniques, such as soaking, milling, and thermal processing, have been proved to be successful in improving iron bioavailability in plant-based staples by reducing their content of phytic acid (Larocque et al., 2005).

8.5.2 Supplementation

Absorption of iron is reported to be higher in research conducted by Abbaspour et al. (2014) and Zimmermann et al. (2007) by giving iron supplements to individuals on an empty stomach. However, nausea and epigastric pain developed due to higher doses (~60 mg Fe/day). In some developing countries, women with low iron stores during pregnancy are advised to take iron supplementation. Ferrous salts such as ferrous gluconate and ferrous sulfate are recommended due to their low cost and high bioavailability for oral supplementation of iron. However, according to some evidence, iron supplementation has been seen to increase the severity in the case of malaria (Cavalli-Sforza et al., 2005; Zimmermann et al., 2007)

8.5.3 Fortification

Fortifying food with iron is more difficult than fortifying it with any other nutrient. To avoid unnecessary sensory changes, forms of iron that are less soluble yet less well absorbed are often selected for the same to elude undesirable sensory changes. Ferric pyrophosphate, ferrous sulfate, electrolytic iron powder, and ferrous fumarate are the recommended iron compounds for food fortification (De Benoist et al., 2006; Hurrell and Richard, 2007)

8.5.4 Biofortification

Biofortification is a biotechnological process concentrating on the aggregation of micro and macronutrients in plant cells. Modification of the actual genetic traits for the development or improvement of micronutrients is involved in genetic biofortification. According to Borrill et al. (2014), this technique is now being accepted as well as used worldwide to improve crop yield with increased nutritional content. The transgenic approach of biofortification comprises the transfer of the micronutrient-producing specific genes from one plant to the other plant which is devoid of it. This is practiced by inserting the desired genetic traits liable for trace element binding proteins or by over-expression of the existing storage proteins or the ones subjected to the uptake of micronutrients. Plants take up an oxidized form of iron (Fe^{3+}) from the nutrients present in the soil. But, after a certain time, the predominant inorganic iron present in the soil becomes unavailable to the plants due to processes such as adsorption, precipitation, and oxidative reactions. In such cases, iron is delivered by using iron chelators. Some microorganisms, for example, bacteria and fungi have the capability to produce compounds of low molecular weight (< 10 kD) and siderophores for sequestering iron compounds.

The roots of a plant can take up siderophores that are secreted by microorganisms that promote plant growth and enhance iron uptake. The microorganisms taking part in the production of such compounds include bacterial species such as Enterobacter, *Bacillus, Arthrobacter, Pseudomonas, Azotobacter, Serratia, Rhizobium, Azospirillum*, and fungal species including *Penicillium, Rhizopus, and Aspergillus*. Biofortified crops include rice, wheat, and maize (Borrill et al., 2014). Fortunately, fortification of food has had an impressive past success in public health and has helped in eliminating or decreasing micronutrient deficiency to some extent (Hurrell and Richard, 2021).

8.6 HISTORY OF IRON FORTIFICATION

According to the WHO and the FAO (2006), food fortification has been used in some countries for a very long time and has succeeded in controlling micro-nutrient deficiencies. During the 19th century, the importance of trace elements like iodine, zinc, and iron was established and considered important for health. Iron was proven to be an essential nutrient for animals in history as it was used to treat anemia in young women (Semba and Richard, 2012). To have integrated results of food fortification, it is assumed that the benefitting micro-nutrient must have satisfactory absorption, easy accessibility, be a part of the usual diet, ideal organoleptic characteristics, as well as be accepted by the mass population. In the 1940s, iron fortification in wheat flour was introduced nationally to target widespread anemia in the U.S. and the U.K. Now, it has become mandatory in almost 81 countries. If designed, implemented, and monitored correctly, it is predicted to enhance or maintain an adequate amount of iron. According to Hurrell (2021), iron interventions must not only be monitored by hemoglobin (Hb) alone because when limited biomarkers, such as serum ferritin, are used, studies show that regular consumption of iron-fortified foods and a significant improvement in iron status are observed. Iron fortification in foods should necessarily decrease the risk of anemia, especially in young children (Alina et al., 2019). The most common foods fortified with iron include (Pachón et al., 2015; Whiting et al., 2016; Allen et al., 2006; Cardoso et al., 2019):

a) Tea, as well as other beverages;
b) Dairy and milk products;

c) Oils and fats; and
d) Cereals and cereal products.

8.7 CHALLENGES IN IRON FORTIFICATION

Today, though, there are enormous benefits offered by fortification programs in the world, along with some abrupt consequences such as overconsumption of micronutrients by groups beyond the target population and hence not reaching the intended one. This requires examining additional intakes along with the nutritional status of fortified foods consumed. Thus, the countries which fortify foods to enhance intake and nutritional status face certain challenges, such as adopting convenient vehicles for fortification, influencing the target population, and avoiding overconsumption in groups not targeted (Dwyer et al., 2015).

Iron is undoubtedly considered the most difficult micronutrient to be added to the food matrix. Risks for cardiovascular disease, diabetes mellitus, and cancer arise due to high iron stores. It is thus important to maintain and constantly monitor the balance between iron uptake and its utilization. Adverse effects of iron overload can be declined by taking in dietary sources rich in iron chelators such as fruits (except citrus), turmeric milk, and black tea. This can be implemented and enforced by effective regulation. However, the elemental part of any fortification program should involve monitoring additional iron intake among non-targeted groups. The initiatives made by food fortification industries have estimated that fortification could hike commodity prices by 2–5% (Shubham et al., 2020). Unlike many alternate micronutrients, iron could cause inadmissible sensory modifications to foods and thus the advancement of effective iron-fortified foods was ratified to be more difficult than developing nutritionally effective foods fortified with another micronutrient (Hurrell, 2021). Based on the requirements of the target population, iron salts are tested and added to the food matrix to produce iron-fortified food. One such salt is ferrous sulfate which is economical and effectively swallowed, but it is considered to be unstable as its oxidation depends on air exposure and temperature. Due to this, it may give rise to organoleptic changes which is the challenging part of fortification (Blanco-Rojo et al., 2019).

8.8 DIFFERENT FORTIFICANTS USED AND THEIR CHEMISTRY

Fortifying iron is based on various factors, including the bioavailability of the iron compound used, the inhibitory and boosting effects of the meals' capacity to absorb iron, the quantity of addition, and the way the vehicle is consumed. Commonly used fortificants along with their relative bioavailability values compared to ferrous sulfate are presented in Table 1.3. Compounds with high bioavailability dissolve in dilute acid or water and can change in their off-flavors, fat oxidation, and color after reacting with other food elements. These changes adversely affect the storage properties of the vehicle. Less soluble forms of iron are often used to prevent such spoilage and ensure consumer approval of the fortified product. In many developing nations, where wheat flour is stored for a longer duration, dried ferrous sulfate is used in limited concentrations due to its solubility in water and high bioavailability. Therefore, less reactive elemental iron powders are widely adopted widely to prevent spoilage (Lynch and Sean, 2005)

The highest relative bioavailability (RBV), i.e. absorbability, is the major factor taken into account while selecting any appropriate iron compound as a food fortificant. Another factor considered is the cost. Ferrous sulfate is one such fortificant used which does not cause any significant change to sensory properties (texture, color, taste) of the food vehicle and hence is used often. The characteristics of good fortificants are as follows (Joshi et al., 2020):

i). Must be a GRAS (Generally Recognized as Safe) substance (as per FDA recommendations)
ii). Minimal effect on sensory and processing quality of the food matrix
(iii). No negative effects of the added fortificants on storage or production.

TABLE 8.3

Iron Compounds Used as Iron Fortificants and Their Characteristics [Fe= Iron Content; RBV= Relative Bioavailability

Compound	Fe (%)	RBV
Water soluble		
Sodium iron EDTA	13	>100
Ferric ammonium citrate	17	51
Ferrous bisglycinate	20	>100
"Ferrous lactate	19	67
Ferrous gluconate	12	89
Ferrous sulfate, dried	33	100
Ferrous sulfate .7 H20	20	100
Poorly water-soluble, soluble in dilute acid		
Ferric saccharate	10	74
Ferrous succinate	33	92
Ferrous fumarate	33	100
Water insoluble, poorly soluble in dilute acid		
Carbonyl	99	5–20
Electrolytic	97	75
CO-reduced	97	12–32
Atomized	96	24
H-reduced	96	13–148
Elemental iron	-	-
Ferric pyrophosphate	25	21–74
Ferric orthophosphate	29	25–32
Encapsulated forms		
Ferrous fumarate	16	100
Ferrous sulphate"	16	100

Source: Lynch and Sean, 2005; HurrellRichard, 2002.

Currently, there are ample varieties of iron compounds being utilized as fortificants (Table 8.3). On the basis of solubility, they can be largely segregated into three categories (Lynch and Sean, 2005) which include: soluble in water; poorly soluble in water but soluble in dilute acid, and insoluble in water as well as poorly soluble in dilute acid.

i) **Water-soluble compounds:**

a) *Ferrous sulfate* (FeSO4): This is one of the most extensively used and affordable iron compounds utilized to fortify food. In 2009, according to the WHO, low extraction products fortification such as wheat flour, dried milk powders, infant formula, pasta, and bread is carried out by ferrous sulfate. Despite the presence of phytic acid and the absence of ascorbic acid, its usage has been proved to be efficacious especially for the fortification of white wheat flour. However, it can also cause some objectionable olfactory changes which include fat oxidation in stored cereal foods; metallic taste in beverages; precipitation of peptides in sauces; and color changes in complementary foods containing fruits and vegetables, salt; cocoa products; and extruded rice (HurrellRichard, 2018). Reportedly, ferrous sulfate showed the highest iron absorption due to its stability in full cream milk. However, on the other hand, it may also cause rancidity depending on the climate, physical characteristics, and fat content of the flour to which it is added (Ahmad and Ahmad, 2019). It may however result in rancidity in the case of cereal flours stored for longer periods. Being less prooxidant in cereals than the hydrated form, it is available as heptahydrate which constitutes about 20% Fe or dried form constituting 33% Fe (Hurrell, 2002).

b) *Ferrousbisglycinate (C4H8FeN2O4):* This is an "iron–amino" acid chelate in which the iron gets bound to amino acid and glycine. This offers protection against the action of absorption inhibitors. It is ideal for adding nutrition to dairy products like liquid whole milk. It is more costly than other iron compounds and can promote rancidity by oxidizing lipids in food (cereals, cereal flours). Fortification with ferrous bis-glycinate is two to three times higher than ferrous sulfate and is protective against phytic acid and calcium (Hurrell, 2018).

c) *Ferrous Gluconate (C12H24FeO14):* Ferrous gluconate contains 12% Fe. It is utilized to fortify infant formula, milk powder, and certain beverages. Fewer times, it gets more expensive as compared to ferrous sulfate. Like ferrous sulfate, it also causes color variations and thus promotes fat oxidation within cereals. It was observed that young Mexican children consuming 5.8mg iron per day as ferrous gluconate along with ascorbic acid in milk can help decrease anemia by 41% to 12% (Hurrell, 2018).

ii) **"Iron compounds are poorly soluble in water but soluble in dilute acid":** They dissolve thoroughly but very slowly in the gastric juice dilute acid (HurrellRichard, 2002).

a) *Ferrous fumarate (C4H2FeO4):* This is a dark red-brown colored compound containing about 33% Fe. It hardly causes any sensory changes in susceptible foods; however, its poor water solubility makes it a less likely chosen compound for fortification. It is frequently added to cereal flours and chocolate drink powders, as well as other supplementary meals made from cereal. It can dissolve within the diluted acid of gastric juice, except in children with low iron status. Absorption of iron through ferrous fumarate is upregulated to a lesser extent in children with iron deficiency than ferrous sulfate. There have been many studies conducted to show an increase in iron stores and hemoglobin in infants as well as young children after consuming ferrous fumarate. Ferrous fumarate has shown success in countries such as Venezuela and Costa Rica in the case of wheat flour and maize within national fortification programs (Hurrell, 2018; Hurrell, 2002)

iii) **Iron compounds that are poorly soluble in dilute acid and insoluble in water:** These compounds have lower as well as variable bioavailability and so do not dissolve completely within gastric juice. Due to this attribute, the impact on foods' sensory properties as well as cost is minimized in comparison to other soluble compounds. According to the WHO (2006), iron absorption via water-insoluble compounds ranges from 20% to 75% approximately. However, these compounds are usually considered the last option, particularly in those scenarios in which the diet of the target population is high inside iron absorption and inhibitors. Insoluble compounds such as ferric phosphate compounds which include ferric pyrophosphate are and ferric orthophosphate are utilized to fortify chocolate-containing foods, rice, as well as some infant cereals. The ferric pyrophosphate's relative iron bioavailability is about 21 to 74%, and that of "ferric orthophosphate" is approximately 25–32%. However, the latter is not constant and affects the processing of food (WHO, 2006).

a) *Micronized ferric pyrophosphate {Fe4P2O73}:* On adding the food fortificant micronized ferric pyrophosphate, the bioavailability of insoluble iron salt in the elemental iron powders increases, thereby reducing their particle size. According to Shubham et al. (2020), a micronized form of ferric pyrophosphate is found available in dried as well as liquid forms. To make it dispersible within liquids, its particles have a coating of emulsifiers. Although this remains to be tested adequately due to being water insoluble nature. It is currently used in Japan by adding it to liquid milk and yogurt products (Blanco-Rojo et al., 2019). Since FPP causes minor sensory alterations to the food vehicle, it is utilized to fortify those foods which are receptive to color like vegetables, infant foods comprising fruits, chocolate drink powders, and bouillon cubes. Additionally, as it is the only iron compound added to extruded fortified rice grains with kernel-premix, it does not result in any undesirable modifications and is used to fortify salt and margarine (Hurrell, 2018).

b) *Encapsulated ferrous fumarate (C4H2FeO4) and ferrous sulfate (FeSO4):* The central objective of encapsulation here is to isolate the elemental iron from other components of food, and further alleviate its sensory changes. Distinct compounds of iron available commercially in this form are ferrous fumarate and ferrous sulfate. The encapsulation of iron compounds is done by using mono- and diglycerides, hydrogenated vegetable oils, maltodextrins, and ethyl cellulose. It is currently being utilized in dry infant cereals and infant formulas (especially in industrialized countries). The cost of such encapsulated compounds increases three to five times and is presented in terms of iron amounts. Iron Encapsulation was verified to avoid iodine losses and slow down the color variations (Hooda et al., 2014).

c) *Encapsulated ferrous sulfate (FeSO4):* Olfactory changes caused by ferrous sulfate can be prevented or decreased by encapsulation with hydrogenated lipids. The most frequently used material for partially hydrogenated capsules is soybean oil. It fortifies infant formula as well as infant cereal to avoid color changes. Nevertheless, its use is prohibited due to its melting point which is 65°C, and is eliminated by either processing of heat or the addition of water or hot milk (Hooda et al., 2014).

d) *Elemental iron powders:* These are mainly used for fortifying cereal flours as well as other cereal products, namely complementary foods and breakfast cereals. These powders are usually cited as "reduced iron". They are not one entity since they are produced using five separate methods: the atomization, electrolytic, carbonyl, and reduction of CO and H. Particle size, shape, porosity, purity, and surface area are the main known characteristics that influence their solubility within gastric juice when differentiated in different powder grades made by a process of single manufacturing. According to the "Food Chemical Codex", the passage of reduced iron powders should be via a 100-mesh sieve (149 μm) and 325-mesh sieve (44 μm) for electrolytic iron and carbonyl powders. Also, in industrialized countries, the reduced iron powders applied to fortify cereal foods generally use a particle size of 44 μm. However, these sieve size values are not enough to guarantee adequate absorption (Hurrell, 2002).

8.9 DIFFERENT FOODS FORTIFIED WITH IRON

The most important requirement for food fortification is the appropriate identification as well as the suitability of a food vehicle often used by the target population to obtain high micronutrient bioavailability. Understanding how food components and iron compounds interact chemically is another important thing to take into account when choosing any meal vehicle. Dietary patterns, geographical regions, food habits, and socioeconomic status are the factors studied while selecting any food vehicle for fortification. The optimal nature of any vehicle utilized for fortification involves focusing on the "target population" without making any consequential modifications after consuming it regularly in their diet; otherwise, it may result in an improved level of iron within the body.

Cereal-based products, sugar, salt, fruit juices (orange, guava, grape juice), milk, and other dairy products are generally treated as food vehicles for iron fortification (Figure 8.5 and Table 8.4). In Southern Asia, rice is a staple food, and to obtain significant amounts of iron and beta carotene from them, they are biofortified through genetic engineering. In the Western world, iron fortification of bakery products such as bread, biscuits, and cookies is a general practice. Similarly, in Indian diets, pulses are considered to be a cheaper vehicle for iron delivery as they hold an important part of the meal. Therefore, the fortification of polished and milled split dal or its flour can legitimately be justified as being a less expensive method of delivering iron. Children who drank iron-fortified water (20 mg of Fe/L) had a substantial decline in their anaemic levels throughout an 8-month intervention experiment (WHO, 2006). The food quantity of food consumed and the degree of fortification

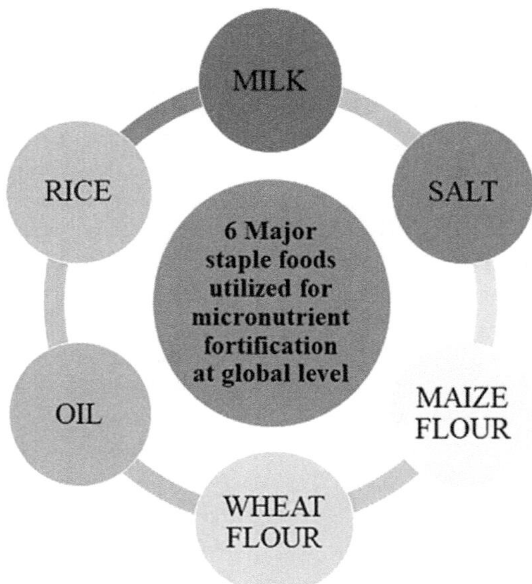

FIGURE 8.5 Commonly fortified food items.

affect how much iron is delivered by fortification. Table 8.4 provides a list of iron compounds that can be utilized to fortify various dietary items (Hurrell, 2018).

8.9.1 FORTIFICATION OF CEREAL-BASED PRODUCTS

a. *Wheat flour:*

To enhance the population's health, fortification of wheat flour began globally during the 1940s (Cardoso et al., 2019). Tripathi et al. (2011), fortified finger millet flour with two iron fortificants, namely ferric pyrophosphate and ferrous fumarate, and assessed it for approximately 60 days on the basis of different parameters and found that EDTA along with other salts provided a high amount of bioaccessible iron to cure deficiency of iron (Tripathi et al., 2011). Sadighi et al. (2008) reported that wheat flour was an ideal vehicle for a food fortification program in Iran and so determined the effectiveness of the program by targeting two groups of women and found increased iron levels. Secondly, they analyzed the iron levels in fortified bread and flour and found coverage of approximately 99.7–100% iron from the fortified samples. Because of the high consumption of wheat flour worldwide, it is regarded as among the most appropriate vehicles for "multi-micronutrient fortification" (Sadighi et al., 2008). Based on a controlled analysis carried out by Muthayya et al. (2012), whole wheat flour was found to be a highly effective vehicle for iron fortification which reduced anemia by 35%, IDA by 51%, and ID by 67% in school-going Indian children who were deficient in iron (Muthayya et al., 2012). The stats of iron deficiency in India are presented in Figure 8.6.

b. *Cereal-based complementary foods:*

These foods are those depending on dry cereals and are usually consumed by infants in the form of porridge (either with water or milk). Furthermore, they depend on cereals and legumes (eaten as porridge with water) blends. Hurrell (2018) suggest that dried ferrous sulfate can be used to fortify flours with moderate to low amounts of phytic acid. When ferrous sulfate is added, low-quality flours or those kept in unfavorable moisture or temperature conditions may occasionally get rancid or turn colored. Its substitutes include ferrous fumarate, NaFeEDTA, and electrolytic iron. Only if

TABLE 8.4

Iron Fortificants Used for Fortifying Different Food Vehicles

Food vehicle	Fortificant used
Low extraction (white) wheat flour or degermed corn flour	Dry ferrous sulfate, Ferrous fumarate, Electrolytic iron (x2 amount)
	Encapsulated ferrous sulfate
	Encapsulated ferrous fumarate
High-extraction wheat flour, corn flour, corn masa flour	Sodium iron EDTA
	Ferrous fumarate (x2 amount)
	Encapsulated ferrous sulfate (x2 amount)
	Encapsulated ferrous fumarate (x2 amount)
Pasta	Dry ferrous sulfate
Rice	Ferric pyrophosphate (x2 amount)
Dry milk (Milk powder, dried infant formulas)	Ferrous sulfate+ ascorbic acid, Ferric ammonium citrate
	Ferrous bis-glycinate
	Micronized ferric pyrophosphate
Fluid milk	Ferric ammonium citrate
	Ferrous bis-glycinate, Micronized ferric pyrophosphate
Cocoa products	Ferrous fumarate plus ascorbic acid
	Ferric pyrophosphate (×2 amount) plus ascorbic acid
Salt	Encapsulated ferrous sulfate
	Ferric pyrophosphate (×2 amount)
Sugar	Sodium iron EDTA
Soy sauce, Fish sauce	Sodium iron EDTA
	Ferrous sulfate plus citric acid
Juice, soft drinks	Ferrous bisglycinate, ferrous lactate
	Micronized ferric pyrophosphate
Bouillon cubes, Spice mixes	Micronized ferric pyrophosphate
Cereal-based complimentary foods	Ferrous sulfate
	Encapsulated ferrous sulfate
	Ferrous fumarate
	Electrolytic iron (×2 amount)
	All with ascorbic acid ($\geq$2:1 molar ratio of ascorbic acid: Fe)
Breakfast cereals	Electrolytic iron (×2 amount)
Micronutrient powders	Encapsulated ferrous fumarate, ferric pyrophosphate"

Source: Hooda et al., 2014; Hurrell, 2018.

the particle size is fine enough to pass the final wheat sifting process might encapsulated sulfate or fumarate be employed. The only iron component advised for whole grain; high phytic acid flours is NaFeEDTA (Hurrell, 2018).

c. *Rice*:

Polished rice has low phytate content compared to wheat or corn flour. Hence, its fortification to detrimental interactions including sensory effects (Dobe et al., 2018). A study conducted by Losso et al. (2017) gives an insight that fortifying rice with iron is challenging due to the easy leaching of iron from the surface of rice during cooking and washing, which decreases the iron concentration of the fortified rice (Losso et al., 2017). Germinated brown rice was used as a vehicle for the iron

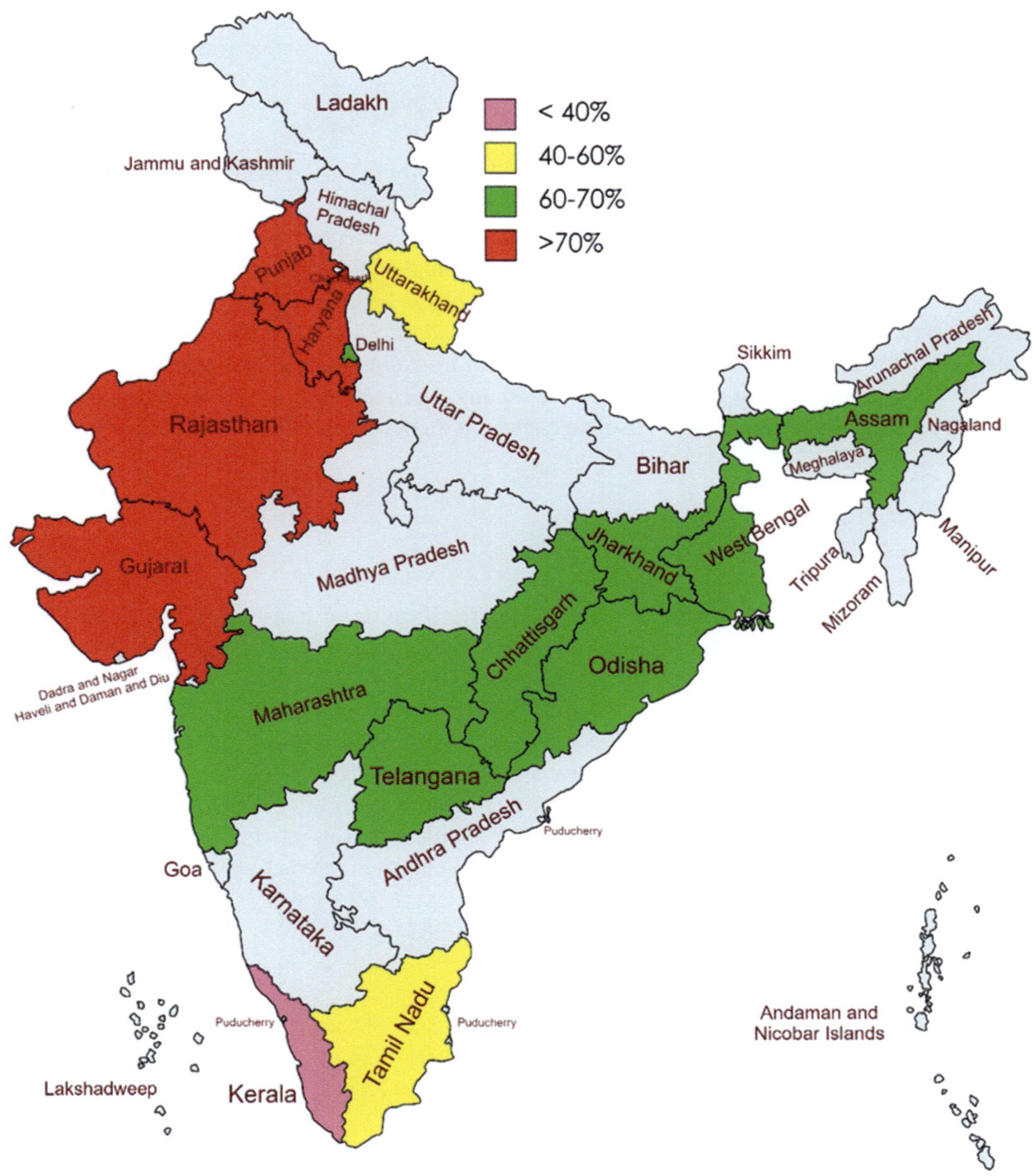

FIGURE 8.6 Anemia status in children in India (Our World in Data, 2023).

fortification program due to its palatability and nutritional value. It was concluded that germinated brown rice with the fortification solution had a higher amount of iron in the grain and bioavailability than ordinary germinated rice and thus was regarded as a cost-effective method to improve Fe intake and its bioavailability in a short time (Wei et al., 2013). It was found difficult to fortify extruded premix rice as it is combined with regular rice and has high levels of iron. FPP (ferric pyrophosphate) is generally ground to a particle size of 2.5μm and is the only compound that causes no color variations (Hurrell, 2018). A new approach was explored by sonication and modifying the rice kernel followed by the addition of a fortificant. Facilitation of lost water-soluble phosphorus was calculated, which resulted in higher iron bioavailability and improved softness in its texture (Bonto et al., 2020). Losso et al. (2017) fortified rice kernels with ferrous sulfate and combined them with unfortified rice in a ratio of 1:200, so that the iron content in the fortified rice was 18mg/100g. It was found that to retain iron from the fortified rice, cooking them with absorbing cooking water is more considerable than the method involving spilling excess water because a significant amount of iron gets lost in the latter. Further, it was observed that 1.13% of the iron content was absorbed from the rice and transformed into hemoglobin (Losso et al., 2017).

8.9.2   Dairy Products

The characteristics of milk or dairy products are significantly impacted by iron fortification. Based on the components of the added iron, all the milk compounds namely caseins, lipids, whey minerals, and proteins are modified directly or indirectly with global acceptance (Gaucheron et al., 2000). Chilled dairy products that contain fruits are notably complex to fortify with iron without causing any color changes (Hurrell, 2018). Zhang et al. (1989) fortified cheddar cheese with iron fortificants namely, ferric citrate, ferric casein complex, ferric chloride ($FeCl_3$), and ferric polyphosphate-whey protein complex for about 46–182 mg, and the level of iron along with its recovered quantities was studied. The amount of iron recovered was 52–81 mg, i.e., increased iron concentration with no effect on the overall recovered iron. Also, no change in the level of calcium in the cheese was seen (Zhang et al., 1989).

8.9.3   Other Products

a. *Bouillon cubes and spice mixes:*

FPP ("Ferric Pyrophosphate") may be used for the fortification of bouillon cubes. Since FPP is seen as a market-driven voluntary fortification, it is typically added to supply 30% to 60% of RDA instead of the necessary 15% to 30% to make up for its lower absorption. However, the consumption of bouillon cubes is low, therefore the cube could not give all the lacking iron by itself in the diet. It should hence be united with other foods in the national program. Hurrell (2018) suggest that consumption of bouillon cubes might give a minimum of 10% of the EAR (1.5 mg per day) to young women who consume a diet rich in iron with 10% bioavailability or 20% in cases where FPP is added (3 mg Fe/day) along with it. Although FPP should be taken into consideration and evaluated, polyphenols may specifically cause color development when interacting with iron compounds in spices (Hurrell, 2018).

b. *Soy sauce and fish sauce:*

Studies conducted by Fidler et al. (2003) confirmed iron absorption in individuals fed with NaFeEDTA-fortified fish or soy sauce (Fidler et al., 2003). It is recommended that fish sauce if fortified with "NaFeEDTA" must always be packaged in brown bottles as a precautionary measure to avoid EDTA sunlight degradation (HurrellRichard, 2018).

c. *Salt:*

Salt iron fortification is associated with unacceptable color changes and degradation of the added compounds of iodine due to oxidation into the gaseous form of iodine, which evaporates further (Hurrell, 2018).

d. *Beverages:*

Nowadays, consumers want healthier versions of beverages, such as soymilk, yogurt-based drinks, nutrient-rich mineral water, and similar kind. When school-age children (in either high, moderate, or low doses) drank juices enriched with many micronutrients, their hemoglobin levels increased significantly. This led to a substantial decrease in levels of anemia in the tested subjects. According to a similar research trial conducted by Ahmad and Ahmad (2019), some infants in the low socio-economic group who were given samples of wheat-milk beverages comprising ferrous fumarate or ferric pyrophosphate with extra amounts of 3 to 5 mg per day showed a convincing increase in levels of serum ferritin and hemoglobin (Ahmad and Ahmad, 2019). According to studies by Hurrell (2018), high-sugar drinks that promote obesity are not regarded as the best method for iron fortification. Either the iron combination is soluble or dispersible. Occasionally observable variations in color and metallic flavor depend on the beverage's ingredients. Ferrous gluconate, ferrous bisglycinate, ferrous sulfate, and micronized dispersible ferric pyrophosphate (FPP) can be considered ideal in the case of iron fortification of beverages (Hurrell, 2018).

e. *Horticulture produce:*

A horticulture matrix may be included in a consumer's diet without any changes in the form of freshly cut or minimally processed food, and it may be the best vehicle for fortification. The following are the main factors that can make fresh-cut horticultural products a reliable matrix for fortification that also focus on low-value crops (Joshi et al., 2020):

   i). Recommendation to be used among every age group;
  ii). An option for the economic class of consumers;
 iii). Achievement of RDI of minerals after consuming fresh-cut produce;
  iv). Inherent porosity makes fortificants quickly ingestible and adherent;
   v). Shelf-life extension along with nutrient enhancement; and
  vi). Complete absence or low levels of antinutritional factors in comparison to cereal crops, which further aid in the better absorption of essential nutrients by the human body.

8.10 BIOACCESSIBILITY AND BIOAVAILABILITY OF IRON IN DIFFERENT FORTIFIED FOODS

Iron bioavailability is described as the ratio of iron that is consumed, followed by absorbed by the intestine, and either utilized through regular metabolic pathways or stored (Blanco-Rojo et al., 2019). Thus, it can be said that the ratio between dietary ligands (which decrease or boost iron absorption within the intestine) determines its iron bioavailability (Schümann et al., 1998). The quantity of the nutrient that can be used for metabolism, typical physiological processes, and storage is known as bioavailability. This amount can be limited or improved by the presence of additional dietary ingredients and processing methods. To put it another way, the bioavailability of a nutrient may be used to characterize its availability in a biological system (Joshi et al., 2020). According to Blanco-Rojo et al. (2019), the more soluble the iron component is, the higher its absorption capacity and hence bioavailability. This path of solubility and bioavailability depends on its process of digestion simulated and transported by a semipermeable membrane. The bioavailability of iron can also be defined as the percentage of iron intake influenced both by dietary as well as host factors. Heme iron enters the body by a different route than non-heme iron. It is not affected by interactions among other food components. However, as discussed previously, non-heme is the main iron source and heme iron constitutes approximately 10–15% of the total dietary iron only (Blanco-Rojo et al., 2019). The type of compound, its oxidation state, the food to which it is added, other foods within the diet, as well as physiological conditions, are different factors that impact the bioavailability of iron in a food matrix. Ferrous salts are more effective because ferric salts have a greater potential for reduction before absorption (Park et al., 2009). The bioavailability of different iron types may also be affected by their solubility and valence state (Shubham et al., 2020).

Accordingly, the order of bioavailability of iron-based salts followed within the human body is *ferric sulfate, ferrous pyrophosphate>ferric citrate, ferrous tartrate, ferrous gluconate>ferrous citrate, ferrous glutamate, ferrous glycine sulfate, ferrous succinate, ferrous fumarate, ferrous lactate,* and *ferrous sulfate.* Iron chelation with NaFeEDTA or ferrous glycinate enhances its luminal solubility. This influences the negative effects of phytates moderately which thereby enhances the bioavailability of iron (Hunt et al., 2005). The percentage of a substance that is made available for intestinal absorption by being released from the food matrix to the gastrointestinal system is known as bioaccessibility. This comprises a series of processes that occur during digestion that turn food into a form that the body can easily absorb. However, the incidence of other dietary components interacts via some type of bonding, which may either increase or restrict the absorption of minerals. Bioaccessibility of iron in bread and biscuits was determined by calculating the fraction of a mineral released against its total content (Dipika, 2020). According to studies, it has been found that fermentation, germination, or soaking are some methods to enhance the mineral's bioavailability such as iron as these processes reduce anti-nutrient levels in pulses and cereals. Also, heat processes such as baking as well as roasting soften the food matrix and deliver protein-bound minerals like iron, which as a result facilitate its absorption (Agrahar-Murugkar and Dipika, 2020). Parboiled rice was fortified and found that the amount of Fe levels were improved along with an increased rate of absorption (Wahengbam et al., 2019). Also, in another analysis, Oluyimika et al. (2019) found that the addition of baobab fruit powder such as citric acid increased the bioaccessibility of iron (Oluyimika et al., 2019).

8.11 SYNERGISTIC/ANTAGONISTIC EFFECT OF OTHER MINERALS ON ITS BIOAVAILABILITY

Always keep in mind how each nutrient interacts with the others before focusing on one particular nutrient. For instance, when minerals like Zn, Fe, and Ca combine with certain inhibitors like phytates or oxalates, they become insoluble or nonexistent. This demonstrates that the primary crucial concept employed when fortifying any suitable variety is the natural antinutrient component. This is because the delivery of nutrients in the human body is subdued by such factors. Calcium addition reduces the bioavailability of iron in foods and beverages that are fortified with the mineral. To address these issues, the product's iron salt content is either raised or given a boost by the addition of an organic acid like ascorbic acid (Joshi et al., 2020; Ahmad and Ahmad, 2019)

8.12 NATIONAL AND INTERNATIONAL GUIDELINES FOR IRON FORTIFICATION

WHO and the "Food and Agricultural Organization" (FAO) of the UN have delineated fortification as "the purposeful practice of raising the amount of any vital micronutrient, such as vitamins, minerals, and trace elements, in a food to increase the nutritional quality of the food supply and benefit public health with the least amount of danger to health" (World Health Organization, 2006). In Codex Alimentarius, fortification of food (FF) is described as "the process of adding one or more important nutrients to food that are not ordinarily present in it to prevent or treat a population's or a particular demographic group's proven nutritional deficiency."

The fortificant added must be within secure limits so that it does not result in any kind of toxicity of iron. In addition to this, the dose should also be amended to maintain its efficient concentration along with a healthy diet. Certain tolerable ULs ("upper intake limits") have also been described by the United States Institute of Medicine's Food and Nutrition Board for long-term intakes. The hidden consequences of elevated iron intake on infection rates as well as the risk of cancer and CVD have drawn more attention. Nevertheless, most of these concerns are linked to pharmaceutical iron supplement intake and not to fortified foods. Therefore, to guarantee the required levels of nutrient concentrations, a long-term evaluation must be included as part of the fortification process. In many nations, the legal foundation for managing the regional fortification program has to be developed. Due to the additional cost of fortification and the additional processing stages, fortification typically increases the price of food products. In 2009, the WHO issued "recommendations on wheat and maize flour fortification" considering five main nutrients-folate, iron, zinc, vitamin B12, and vitamin A. It should not be assumed that countries need to add only the above-mentioned nutrients, or all of the nutrients to their product; instead, they should add those which are the most limited ones in the diet of the target population. For each nutrient, the minimum target levels on the basis of daily/capita availability of "fortifiable" flour are cleared. The target levels are based on flour as it is the most commonly used item all over the world. Furthermore, recommended fortification compounds are determined for every nutrient (Marks et al., 2018).

8.13 NOVEL AND EMERGING TECHNIQUES OF IRON FORTIFICATION

Fortification concentrates on expanding the amount of one or more "target micronutrients" in the food during processing and value addition. The following techniques are the techniques applied while fortifying food with iron:

i). *Encapsulation:* It is a method that traps the desired component inside a matrix that surrounds it, enabling the component's targeted and contained release. Encapsulated iron compounds are generally appropriate for use as dry food components as well as in baby feeds, dry beverage mixes, along with minimally processed foods. Infant formulas and encapsulated cereal flours made from maize and wheat are fortified with ferrous sulfate and ferrous fumarate. Increased bioavailability can be achieved with nanoencapsulation (Figure 8.7), which can also balance oxidation protection, controlled release, improved handling, as well as targeted delivery. Iron remains as Fe^2+ at a lower pH, whereas it gets oxidized to Fe^{3+} at a high system redox potential (Eh) (Gutiérrez et al., 2016).

ii. *Chelation and redox modulation:*

For this technique, ferrous bis-glycinate, an amino acid-chelated iron, is commonly utilized. Both pH and Eh are lessened by adding reducing agents (ascorbic acid) and organic acids (citric acid), which creates a reducing setting to prevent the expansion of off-colors along with oxidation of ferrous ion ($Fe^{2+)}$ to ferric ion (Fe^{3+}) effectively. Nevertheless, Fe(II) is stabilized by redox modulation and sets limitations to avoid the benefit of reducing the adverse chemical reactions of modifications occurring in the sensory properties as well as shelf-life. Techniques such as micronization and emulsification have already been used for improving iron bioavailability. The elemental iron

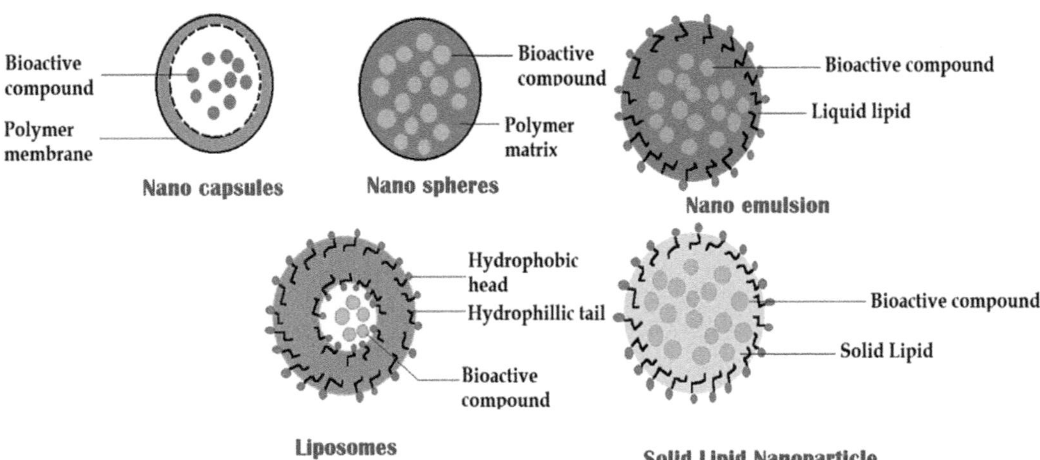

FIGURE 8.7 Nanoencapsulation technique (Shubham et al., 2020).

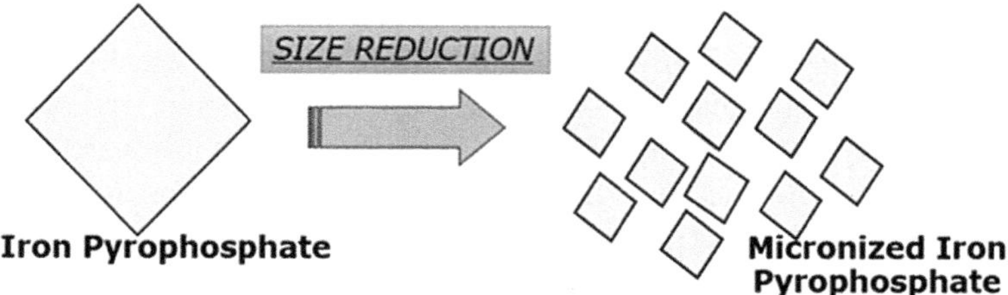

FIGURE 8.8 Micronization of iron pyrophosphate molecule for iron fortification (Shubham et al., 2020)

powder's bioavailability in its pure form at a valence state of zero is linked to the particle size, solubility, and surface area, which contradicts with its corresponding process of manufacturing (Shubham et al., 2020; Hunt, 2005)

iii. *Micronization:*

Micronization is a method that involves the breakage of "ferric pyrophosphate" into smaller parts to solubilize it in the food matrix (Figure 8.8). Though the present process has larger prospects in the future, its high cost is the major limitation. The diameter of the "micronized ferric pyrophosphate" type ranges from 0.3 to 0.5µm and was recently established for the fortification of food. The utmost benefit of this strategy is that it does not modify the food's chemical characteristics or its organoleptic properties. When compared with other techniques, no negative impacts on the sensory, color, and textural properties of foods of the micronized ferric pyrophosphate (FPP) were reported. However, this approach leads to low iron bioavailability. (Shubham et al., 2020; Haro-Vicente et al., 2006; Mashkour et al., 2018)

iv. *Fortification with bioactive compounds:*

The benefits of food fortification are attributed to the associated bioactive compounds, which include dietary fiber, phenolic compounds, essential amino acids, minerals, vitamins, and other substances. Ingestion of necessary micronutrients can be guaranteed to stay healthy by fortifying regular food items. Mashkour et al.(2018), fortified potatoes by ultrasound pre-treatment using vacuum impregnation and found enhancement in the overall iron content (Mashkour et al., 2018; Del Pino-García et al., 2018).

8.14 NATIONAL AND INTERNATIONAL PROGRAMS/ CASE STUDIES AND LESSONS LEARNED

Iron fortification is adopted in various parts of the world today. Currently, there are 77 nations that have accepted wheat flour fortification. All the obligatory countries (except Australia) fortify wheat flour along with at least iron. Iron-fortified seasonings such as fish or soy sauce are approaching as a central vehicle in various parts of Asia (Thailand, China, and Vietnam). Nearly 31% of the industrially milled wheat flour consumed worldwide is anticipated to be iron-fortified due to both required and voluntary initiatives (Mannar and Wesley, 2017). A progressive contraction in the pervasiveness of iron deficiency was experienced among the Venezuelan population during the period 1960 to 1985. In 1993, in response to this crisis, the Venezuelan government appointed a special commission for the advancement of food (CENA), which led to a program involving the fortification of both precooked corn and wheat with iron. An impressive decline was observed in the prevalence of deficiency of iron and anemia after continuous fortifying for 2 years. However, for the next three reviews (1997–1999), the incidence of anemia began to escalate and returned to 19%. In 1994, the prevalence of iron deficiency was efficiently shrunk efficiently and continued to drop for almost 3 years. The main reason behind this prevalence was the partial replacement of a highly bioavailable iron compound ("ferrous fumarate") for a less available iron type (i.e., elemental iron), along with the degraded condition of society. Overall, under the fortification program, improved stores of iron and maintained anemic prevalence were documented (García-Casal and Layrisse, 2002). According to Mannar and Gallego (2002), the Philippines' Food and Nutrition Research Institute developed a coating method for the iron fortification of rice. After six months of eating coated rice, schoolchildren in a successful study showed a rise in hemoglobin levels, despite significant iron losses caused by washing and cooking. Different iron compounds were also tested for color and cost. In 2000, a food fortification law was passed which stated that fortification of certain foods (including rice) must become mandatory. Today, efforts are being made to improve the technology and processing/distribution systems (Mannar and Gallego, 2002).

In Vietnam, where 80% of the population regularly consumes fish sauce, it is seen to be a valuable vehicle for programs of iron fortification. Before the product is bottled, fish sauce is fortified with iron with only a few small production process changes. Mannar and Gallego (2002) conducted a random, double-blind efficacy trial among anaemic, non-pregnant female factory workers aged 17 to 49 years old which showed that after an intervention for 6 months, an important enhancement in the iron status with a decrease in the occurrence of anemia and deficiency was found in those women who consumed fish sauce fortified with 100 mg iron, i.e., NaFeEDTA/100 mL fish sauce on a regular basis. Education plans on food fortification, which emphasized the fortification of iron, were also launched (Mannar and Gallego, 2002). Soy sauce has been used as the food carrier in China for research of a similar nature by Mannar and Gallego (2002) because soy sauce is consumed by 70% of the country's people. NaFeEDTA was chosen as the iron-fortifying substance due to its increased bioavailability and benefit for those following a plant-based diet. Soy sauce comprising 5/20 mg Fe, in the form of NaFeEDTA, was proven to be highly effective in treating anaemic children within only 3 months (Mannar and Gallego, 2002).

Food fortification is largely practiced in Mexico. Fortification of maize and wheat flour with iron has contributed to a relatively low incidence of anemia and iron deficiency. Children receive approximately 70% of their dietary iron through fortification sources. It has been identified and quantified that the prevalence of iron sources, along with the existence of important enhancers and iron absorption inhibitors, within the diet of children aged 3-14 years is about 3.7% and 4%, respectively. Fortified iron accounts for 72% of all iron consumption, which varies from 1.9 to 3.3 times the RDA. Despite the fact that maize flour's iron fortification has been in use for more than ten years, it has caused some children to become overloaded (Jyväkorpi et al., 2006).

In Senegal, micronutrient deficiency is a serious health issue. According to data provided by the National Food Fortification Alliance, the spread of anemia was high (76.4%) among children under 5 years old. In another survey, it was found that 54.3% of women of reproductive age (15 to 49 years old) suffered from anemia. In 2006, FRAT ("Fortification Rapid Assessment Tool") analysis was conducted in Senegal, which disclosed that wheat flour and oil are the potential vehicles for food fortification. This choice was based on their widespread use in relatively high numbers among its geographic and socioeconomic categories. This program not only strengthened quality control but also allowed higher compliance with standards set for the production of flour and oil. Thus, the Senegal food fortification program made impressive growth within a very short duration. Fortified flour and oil coverage among women of childbearing age who consumed fortified oil and flour once a week was realized to be high, at 85% and 73%, respectively **(Abdoulaye and Manus, 2018)**

Approximately, 50% of the population in the Philippines, especially in rural areas, suffered anemia. For the case study, Spiced Vinegar (which was a broadly available, inexpensive, culturally acceptable condiment in their homes) was physiologically characterized and evaluated for consumer acceptability by fortifying it with FF (Ferrous Fumarate), FeSO4 (Ferrous Sulfate), or NaFeEDTA (Sodium Iron Ethylenediaminetetraacetate) at 0.2 mg Fe per ml. In the case of ferrous sulfate and ferrous fumarate, it was not possible to store vinegar beyond 3 days as they did not allow maximum solubility. Whereas, in terms of consumer acceptability and physicochemical stability, spiced vinegar-NaFeEDTA depicted potential could be used as a vehicle for the delivery of iron and at the same time address iron deficiency anemia with 100% recovery and was the most accepted fortified vinegar among Filipinos. (Lopez et al., 2018)

The notion and application of fortification of food is not so peculiar in India. The FSSAI ("Food Safety and Standards Authority of India") made it mandatory and broadened the exemplar of food fortification. In the current governance, staple foods like salt, rice, wheat flour, and refined wheat flour were chosen as food vehicles for iron fortification. Currently, trials on double fortification are being practiced in India (Wang et al., 2011).

8.15　IMPACT OF IRON FORTIFICATION ON THE SENSORY AND TEXTURAL FOOD PROPERTIES

The micro nutrient is said to must-have ideal organoleptic, textural as well as sensory characteristics or not affect or alter the color as well as the flavor of the fortified food so that it has good acceptance by the consumers (Siddique and Park, 2019).

i. *Effect on textural properties:*

Textural changes involve changes such as green or bluish coloration in cereals, darkening of salt to red/brown or yellow, and greying of chocolate and cocoa (WHO, 2006). Bonto et al. (2020) studied that fortifying milled rice by ultrasonication led to decreased hardness, increased absorption, decreased cooking time, and increased water uptake ratio, and hence favored not only the nutritional aspect but also the consumer acceptability as well as palatability. (Bonto et al., 2020). An increase in textural softening was reported in potato when it was subjected to Fe fortification using vacuum impregnation (Mashkour et al., 2018). Similarly, the textural properties of Cheddar cheese were affected by iron fortification followed by storage (Siddique and Park, 2019). Significant differences in color and a decrease in luminosity on fortifying unhurt with nano as well as micro-sized iron were also reported by Santillán-Urquiza et al. (2017). During potato flakes fortification with Fe, some discoloration was seen after adding ferripolyphosphates to the mash and got more intense after the addition of 10 mg Fe/100g (Sapers et al., 1974).

ii. *Effect on sensory properties*:

Sensory changes are said to be variable but not predictable. Iron is considered a more challenging micronutrient to be added to beverages as well as food due to its best bioavailability, which interacts with food constituents and, as a result, generates undesirable changes. It is also known to catalyze the oxidation of lipids, which leads to rancidity along with the development of odor and off-flavor in different food items. This can be prevented by deaeration in the case of milk before adding iron into it (Gaucheron, 2000). Similarly, iron-fortified food with an elevated level of the unsaturated fatty acid system cannot be stored for a prolonged time as it can cause rancidity and subsequently produce off flavor as discussed above. This happens due to the pro-oxidant properties of iron, which promote the oxidation process in unsaturated fat-containing liquid beverages. Reportedly, ferrous and not ferric compounds were responsible for the development of an oxidized flavor in whole milk fortified with iron (Ahmad and Ahmad, 2019). Saade and Arijaje (2020), reported that a metallic taste was imparted by the reduced iron in the food, which could be alleviated by decreasing its particle size. Though ferric orthophosphate does not impart a metallic taste, it is quite expensive, and to increase the low iron content, more of it may be required to meet the same nutritional content. The shelf life of every iron-fortified product, especially cereals, should be tested prior, as iron is subjected to accelerate rancidity (Saade and Arijaje, 2020). The rancid flavor in iron-fortified food is caused due to ferric compounds, whereas ferrous compounds resulted in an oxidized flavor and not rancidity

(Park, 2009). Therefore, it can be concluded that iron fortification brings along increased rancidity due to the oxidation of unsaturated lipids and unwanted color changes.

8.16 CHALLENGES WITH CURRENTLY ADOPTED METHODS

As discussed above, numerous iron compounds are colored and cannot not be utilized to fortify foods that are light in color. The main problem with grains is iron's catalytic influence on fat oxidation during storage. Water-soluble substances like ferrous sulfate, for example, encourage the oxidation of fat, which shortens the product's shelf life and causes product discoloration (Akhtar et al., 2011). Increased iron intakes and greater iron stores have also been a source of concern, particularly in light of their possible implications on infection rates as well as the risk of cancer and cardiovascular disease. Since by far, there is no such biological mechanism to excrete excess iron from the human body which may otherwise have potential toxic impacts. Levels of serum ferritin >20 µg/l in humans are linked with poor absorption of heme as well as non-heme iron. It was presumed that the incidence of unabsorbed fortificant iron within the body, most of which reaches the colon and results in the generation of free radicals in the stool that damages the colon mucosa. The reason behind the production of free radicals in the colon is the highest insolubility of iron in the pH of the same which increases the amount of unabsorbed ferrous sulfate. This increased level of iron due to fortification brings the risk, especially in infants which could seriously affect their microbiota equilibrium and thus replace the pathogenic bacteria growth instead of useful gut flora. Surveys have also proclaimed that adding "prebiotic galactooligosaccharides" in the infant formulae could reduce the adverse impacts of iron efficiently in the gut microflora. In some cases, iron excess might result from dietary fortification in hemochromatosis patients. However, there is no evidence yet which suggests the long survival of free radicals to cause tissue damage (Hooda et al., 2014; Shubham et al., 2020). Though, iron fortification of basic foods and condiments is still a viable strategy for addressing an iron deficit. Malaria itself can be brought on by excess iron since it helps to create the environment for germs that make it a deadly pandemic. Calculating the cost ratio takes into consideration the expenses associated with dietary treatments, the incidence of nutrient insufficiency, and its effects on economic output (Shubham et al., 2020).

8.17 IRON BIOAVAILABILITY

The amount of reinforced iron's bioavailability varies depending on the iron compound utilized (Table 8.5). Heme iron (15–40%) is better absorbed compared to non-heme iron (1% to 15%), as was previously mentioned. Hunt et al. (2005) claim that cutting back on a meat diet can lower heme iron levels. While the overall quantity of non-heme iron obtained "from a plant-based diet is only 10%, the absolute amount of accessible iron in a mixed diet that includes meat is predicted to be 18%. In such a scenario, the only method to increase the iron availability in a vegetarian diet is through cereal fortification" and other plant-based foods with iron. Dietary variables, the kind and concentration of an iron supplement, a person's health, and nutritional interactions are the main factors determining iron bioavailability. The absorption pattern of non-heme iron is significantly influenced by a person's iron status, whereas heme iron appears to be less affected by it. The status and bioavailability of iron can be improved by a smart mixture of diverse meals. Additionally, prior to dietary fortification, knowledge of iron boosters and inhibitors must be prioritized to combat anemia (Shubham et al., 2020). According to the WHO, the dietary iron bioavailability and iron from added iron fortificants are significantly influenced by a population's dietary patterns. Therefore, understanding the fate of iron and its corresponding bioavailability becomes primary in developing strategic methods to combat anemia via iron fortification. There have been several ongoing techniques in practice for anemia prevention for decades, but today an effective long-term running strategic approach is the need of the hour. In this thought, food fortification with iron could be

TABLE 8.5

Classification of Regular Diets Based on Iron Bioavailability

Category	Iron bioavailability (%)	Dietary characteristics
Low	1–9	Simple, uniform diet based on the basis of cereals, tubers, or roots, with inconsequential amounts of poultry, fish, meat, or ascorbic acid-rich foods. A diet high in foods that hinder iron absorption such as beans, maize, sorghum, and whole wheat flour.
Intermediate	10–15	A diet rich in cereals, tubers, or roots, with some foods of animal origin (fish, meat, or poultry) and/or comprising some vitamin C (from vegetables and fruits).
High	>15	An expanded diet comprising larger amounts of poultry, fish, meat, and/or foods high in vitamin C.

Source: Shubham et al., 2020; Blanco-Rojo et al., 2019; Leclère et al., 2002.

proven to not only be a promising but also a cost-effective technique that targets a specific group of individuals in the long term (Shubham et al., 2020).

8.17.1 Nutritional and Anti-nutritional Factors (Inhibitors and Enhancers)

The amount of iron absorbed is a precondition for having bioavailability of iron. This illustrates that no matter how potent an enhancer or inhibitor is, its influence on iron absorption will be completely inconsequential if it is ingested in a meal when iron is missing. In this perspective, it's also critical to keep in mind that interactions in the digestive system take place a few hours after food is consumed. Thus, the idea of bioavailability as the "proportion" of the iron eaten employed or stored for bodily activities must also take into account the amount of iron after ingestion, the make-up of the meal, and the interval between them.

Another vital aspect of bioavailability is its form. Animal foods such as fish and meat contain about 40% heme iron content, of the total iron. The iron bioavailability in a mixed diet is approximately 90% with about 5–15% non-heme content. The reports of Shubham et al. (2020) and Blanco-Rojo et al. (2019) suggested that ferrous iron is generally absorbed better than ferric iron. However, if both ions reach the mucosa in a soluble state, they might be successfully absorbed. The limiting factor is solubility because ferric salts may precipitate if the stomach's pH increases to that of the duodenum. Additionally, it is claimed that dietary substances that change iron from ferric to ferrous would boost its bioavailability. Absorption of non-heme Fe can be increased by retinoids, carotenes, citric, tartaric, and malic acids, and alcohol (by improving gastric acid secretion promoting valence state), and are thus said to be enhancers, whereas minerals like Mg, P, Ca as well as other chemical compounds like oxalic, phytates, and malonaldehyde cause a negative effect on the absorption of iron and are hence known as inhibitors (Shubham et al., 2020; Blanco-Rojo et al., 2019). Table 8.6 summarized the important enhancer and inhibitors that affect Fe bioavailability.

   a. **Enhancers**:

a) **Ascorbic acid:** Ascorbic acid, also known as a pro-oxidant in the incidence of iron, increases the absorption of iron by reducing mechanisms responsible for converting ferric ions to ferrous ions. This is then effectively transported via duodenum microvilli. Usually, ascorbic acid complements enhanced non-heme iron absorption from meals, thus meat and fish must be included in the meal. It is also capable of chelating iron, aiding in the

TABLE 8.6
Iron Bioavailability's Main Enhancers and Inhibitors

	Food or food group	Components	Mechanism
Enhancers	Citrus fruits, citrus fruit juices, vegetables	Ascorbic acid	Reduction of $Fe3+$ to $Fe2+$ and the formation of "soluble iron-ascorbate complexes" that stay soluble within the intestine. Both favor absorption
	Meat, fish, poultry	Animal tissue	Iron binding to digestion products, primarily proteins, leads to soluble complexes
Inhibitors	Nuts, whole grain cereals, legumes	Phytic acid	Formation of insoluble iron phytate complexes in the gut
	Coffee, cocoa, tea	Polyphenols	Insoluble complexes formation with iron inside the gut
	Dairy products, cow's milk	Milk protein	Iron is strongly bound by whole casein and αs -casein phosphopeptides, inhibiting iron absorption. The whey protein does not inhibit.
		Calcium	Disruption of the enterocyte's blood transport system. Reduced heme and non-heme iron absorption

Source: Shubham et al., 2020; Blanco-Rojo et al., 2019.

development of a soluble complex over a wide pH 2–1 range. A study has reported that guava juice effectively increases the erythrocyte index by 3.99% (REF). According to the results of another study, ascorbic acid can decrease the absorption of iron by polyphenols (REF).

Enhancers in animal tissue: Enhancers present in animal tissues, from poultry, fish, or meat, are said to exert an enhancing impact on iron absorption. Several mechanisms explaining improved iron absorption by animal tissue were proposed (Shubham et al., 2020; Blanco-Rojo et al., 2019; Oluyimika et al., 2019; Leclère et al., 2002).

  b. **Inhibitors:**
  I. **Phytates***:* Phytates, myo-inositol phosphates, are present in whole-grain legumes and cereals and vary in the "number of phosphates connected to the inositol ring". Inositol pentaphosphate and inositol hexaphosphate are the most effective iron inhibitors with dose-dependent effects. Phytates present in soy isolates at a high level act as inhibitors. Some common food processing techniques, such as malting, soaking, and fermenting, which hydrolyze phytate from its myo-inositol hexaphosphate form into intermediate myo-inositol phosphates, can enhance the absorption of iron from phytate-rich" meals. As a result, higher derivatives of phosphorylated have a greater ability to prevent the absorption of iron than their less phosphorylated counterparts (Makowska et al., 2018; Shubham et al., 2020; Blanco-Rojo et al., 2019).
  II. **Polyphenols:** The interactions of polyphenols, a diverse collection of plant chemicals, with iron and other metals are well recognized. The composition of polyphenol determines how well iron is chelated. Black tea is the finest polyphenol inhibitor, but other examples include seeds, spices, herbs, wine, chocolate, coffee, and green tea. Additionally, polyphenols often have antioxidant properties and may provide protection against chronic illnesses including diabetes, cardiovascular disease, and certain forms of cancer (Teng and Chen, 2019; Shubham et al., 2020).

III. **Calcium and other divalent metals:** Lower iron status is caused by increased calcium or other dairy product intake. It was shown that consuming a product of milk enriched with iron that delivered 100% of the daily essential iron intake did not enhance the iron status within iron-deficient women over 4 months. As a result, its impact was due to the product's absence of iron boosters and partial presence of inhibitors, including calcium and casein. According to reports, increased amounts of zinc are observed during pregnancy and infancy (Blanco-Rojo et al., 2019).

IV. **Phytic acid:** The phytic acid's molar ratio to iron must be reduced to at least 1:1, or less than 0.5:1, to significantly boost iron absorption. Even though milling removes 90% of the phytic acid from cereal grains, the remaining 10% still has a powerful inhibitory effect. Traditional procedures including soaking, germination, and fermentation can activate naturally existing cereal phytases. According to studies conducted by Blanco-Rojo et al. (2019) and Hurrell (2002), at the level of industrial, it is completely feasible to diminish phytic acid in complementary food mixes of legumes and cereals by the addition of exogenous phytases or by adding sources naturally high in phytases such as whole wheat or whole rye. (Shubham et al., 2020; Hurrell, 2002; Blanco-Rojo et al., 2019).

8.18 CONCLUSION

Iron fortification helps to combat iron deficiency, reducing the risk of anemia and its associated health complications, thereby improving overall public health. This cost-effective strategy contributes to better iron intake and addresses nutritional gaps in populations, particularly in vulnerable groups and has the major advantage of aligning with the functioning and physiology of the human body. Thus, fortification remains the safest intervention to address the concern of iron deficiency. Most efforts to combat iron deficiency using fortification strategies have so far been concentrated on overcoming technical problems like discoloration, off-flavor development, and fatty acid oxidation, often neglecting the consideration of other practical problems such as large-scale production, marketing, and quality control, which are crucial for successful implementation.

REFERENCES

Abbaspour, Nazanin, Richard Hurrell, and Roya Kelishadi. "Review on iron and its importance for human health." *Journal of Research in Medical Sciences: The Official Journal of Isfahan University of Medical Sciences* 19, no. 2 (2014): 164.

Abdoulaye, Ka, and Caroline Manus. "Food fortification in Senegal: A case study and lessons learned." In *Food fortification in a globalized world*, pp. 327–331. Academic Press, 2018.

Agrahar-Murugkar, Dipika. "Food to food fortification of breads and biscuits with herbs, spices, millets and oilseeds on bio-accessibility of calcium, iron and zinc and impact of proteins, fat and phenolics." *LWT* 130 (2020): 109703.

Ahmad, Asif, and Zaheer Ahmed. "Fortification in beverages." In Alexandru Mihai Grumezescu and Alina Maria Holban (Eds), *Production and management of beverages*, pp. 85–122. Woodhead Publishing, 2019.

Akhtar, Saeed, Faqir M. Anjum, and M. Akbar Anjum. "Micronutrient fortification of wheat flour: Recent development and strategies." *Food Research International* 44, no. 3 (2011): 652–659.

Albretsen, Jay. "The toxicity of iron, an essential element." *Veterinary Medicine-Bonner Springs Then EDWARDSVILLE* 101, no. 2 (2006): 82.

Alina, Vaic Romina, Mureşan Crina Carmen, Muste Sevastita, Mureşan Andruţa, Muresan Vlad, Suharoschi Ramona, Petruţ Georgiana, and Mihai Mihaela. "Food fortification through innovative technologies." *Food Engineering* (2019): 1-25.

Allen, Lindsay, Bruno De Benoist, Omar Dary, and Richard Hurrell. *Guidelines on food fortification with micronutrients.* WHO/FAO, 2006.

Beard, John L. "Iron requirements in adolescent females." *The Journal of Nutrition* 130, no. 2 (2000): 440S–442S.

Blanco-Rojo, Ruth, and M. Pilar Vaquero. "Iron bioavailability from food fortification to precision nutrition: A review." *Innovative Food Science and Emerging Technologies* 51 (2019): 126–138.

Bonto, Aldrin P., Nichada Jearanaikoon, Nese Sreenivasulu, and Drexel H. Camacho. "High uptake and inward diffusion of iron fortificant in ultrasonicated milled rice." *LWT* 128 (2020): 109459.

Borrill, Philippa, James M. Connorton, Janneke Balk, Anthony J. Miller, Dale Sanders, and Cristobal Uauy. "Biofortification of wheat grain with iron and zinc: Integrating novel genomic resources and knowledge from model crops." *Frontiers in Plant Science* 5 (2014): 53.

Cardoso, Rossana V. C., Ângela Fernandes, Ana M. González-Paramás, Lillian Barros, and Isabel C. F. R. Ferreira. "Flour fortification for nutritional and health improvement: A review." *Food Research International* 125 (2019): 108576.

Cavalli-Sforza, Tommaso, Jacques Berger, Suttilak Smitasiri, and Fernando Viteri. "Weekly iron-folic acid supplementation of women of reproductive age: Impact overview, lessons learned, expansion plans, and contributions toward achievement of the millennium development goals." *Nutrition Reviews* 63, no. 12 Pt 2 (2005): S152–S158.

De Benoist, Bruno, Omar Dary, and Richard Hurrell. *Guidelines on food fortification with micronutrients*, Vol. 126, edited by Lindsay Allen. World Health Organization, 2006.

Del Pino-García, Raquel, Daniel Rico, and Ana Belén Martín-Diana. "Evaluation of bioactive properties of Vicia narbonensis L. as potential flour ingredient for gluten-free food industry." *Journal of Functional Foods* 47 (2018): 172–183.

Dobe, Madhumita, Pankaj Garg, and Gaurav Bhalla. "Fortification as an effective strategy to bridge iron gaps during complementary feeding." *Clinical Epidemiology and Global Health* 6, no. 4 (2018): 168–171.

Dwyer, Johanna T., Kathryn L. Wiemer, Omar Dary, Carl L. Keen, Janet C. King, Kevin B. Miller, Martin A. Philbert et al. "Fortification and health: Challenges and opportunities." *Advances in Nutrition* 6, no. 1 (2015): 124–131.

Fidler, Meredith C., Lena Davidsson, Thomas Walczyk, and Richard F. Hurrell. "Iron absorption from fish sauce and soy sauce fortified with sodium iron EDTA." *The American Journal of Clinical Nutrition* 78, no. 2 (2003): 274–278.

García-Casal, María Nieves, and Miguel Layrisse. "Iron fortification of flours in Venezuela." *Nutrition Reviews* 60, no. suppl_7 (2002): S26–S29.

Gaucheron, Frédéric. "Iron fortification in dairy industry." *Trends in Food Science and Technology* 11, no. 11 (2000): 403–409.

Gupta, C. P. "Role of iron (Fe) in body." *IOSR Journal of Applied Chemistry* 7, no. 11 (2014): 38–46.

Gutiérrez, G., M. Matos, P. Barrero, D. Pando, O. Iglesias, and C. Pazos. "Iron-entrapped niosomes and their potential application for yogurt fortification." *LWT* 74 (2016): 550–556.

Haro-Vicente, J. F., C. Martinez-Gracia, and G. Ros. "Optimisation of in vitro measurement of available iron from different fortificants in citric fruit juices." *Food Chemistry* 98, no. 4 (2006): 639–648.

Hooda, Jagmohan, Ajit Shah, and Li Zhang. "Heme, an essential nutrient from dietary proteins, critically impacts diverse physiological and pathological processes." *Nutrients* 6, no. 3 (2014): 1080–1102.

Hunt, Janet R. "Dietary and physiological factors that affect the absorption and bioavailability of iron." *International Journal for Vitamin and Nutrition Research* 75, no. 6 (2005): 375–384.

Huo, Junsheng, Jing Sun, Hong Miao, Yu Bo, Tao Yang, Zhaoping Liu, Chengqian Lu et al. "Therapeutic effects of NaFeEDTA-fortified soy sauce in anaemic children in China." *Asia Pacific Journal of Clinical Nutrition* 11, no. 2 (2002): 123–127.

Hurrell, Richard F. "Efficacy and safety of iron fortification." In M.G. Venkatesh Mannar and Richard F. Hurrell (eds.), *Food fortification in a globalized world*, pp. 195–212. Academic Press, 2018.

Hurrell, Richard F. "Fortification: Overcoming technical and practical barriers." *The Journal of Nutrition* 132, no. 4 (2002): 806S–812S.

Hurrell, Richard F. "Iron fortification practices and implications for iron addition to salt." *The Journal of Nutrition* 151, no. Supplement_1 (2021): 3S–14S.

Hurrell, Richard F. "Iron fortification: Its efficacy and safety in relation to infections." *Food and Nutrition Bulletin* 28, no. Supplemen_4 (2007): S585–S594.

Hurrell, Richard, and Ines Egli. "Iron bioavailability and dietary reference values." *The American Journal of Clinical Nutrition* 91, no. 5 (2010): 1461S–1467S.

Joshi, Alka, Uma Prajapati, Shruti Sethi, Bindvi Arora, and Ram Roshan Sharma. "Fortification in fresh and fresh-cut horticultural products." In Mohammed Wasim Siddiqui (ed.), *Fresh-cut fruits and vegetables*, pp. 183–204. Academic Press, 2020.

Jyväkorpi, Satu K., Homero Martínez, Alicia Pineda, Salvador Pizarro, and Joel Monárrez-Espino. "Iron nutrition in schoolchildren of western Mexico: the effect of iron fortification." *Ecology of Food and Nutrition* 45, no. 6 (2006): 431–447.

Larocque, Renee, Martin Casapia, Eduardo Gotuzzo, and Theresa W. Gyorkos. "Relationship between intensity of soil-transmitted helminth infections and anemia during pregnancy." *The American Journal of Tropical Medicine and Hygiene* 73, no. 4 (2005): 783–789.

Leclère, Juliette, Inès Birlouez-Aragon, and Michaël Meli. "Fortification of milk with iron-ascorbate promotes lysine glycation and tryptophan oxidation." *Food Chemistry* 76, no. 4 (2002): 491–499.

Lopez Barrera, Emely C., Shashank Gaur, Juan E. Andrade, Nicki J. Engeseth, Christine Nielsen, and William G. Helferich. "Iron fortification of spiced vinegar in the Philippines." *Journal of Food Science* 83, no. 10 (2018): 2602–2611.

Losso, J. N., N. Karki, J. Muyonga, Y. Wu, K. Fusilier, G. Jacob, Y. Yu, J. C. Rood, J. W. Finley, and F. L. Greenway. "Iron retention in iron-fortified rice and use of iron-fortified rice to treat women with iron deficiency: A pilot study." *BBA Clinical* 8 (2017): 78–83.

Lynch, Sean R. "The impact of iron fortification on nutritional anaemia." *Best Practice and Research: Clinical Haematology* 18, no. 2 (2005): 333–346.

Makowska, Agnieszka, Magdalena Zielińska-Dawidziak, Przemysław Niedzielski, and Michał Michalak. "Effect of extrusion conditions on iron stability and physical and textural properties of corn snacks enriched with soybean ferritin." *International Journal of Food Science and Technology* 53, no. 2 (2018): 296–303.

Mannar, M. G. Venkatesh, and Annie S. Wesley. *Food fortification*, edited by Stella R. Quah. International Encyclopedia of Public Health, 2017, 143–152.

Mannar, Venkatesh, and Erick Boy Gallego. "Iron fortification: Country level experiences and lessons learned." *The Journal of Nutrition* 132, no. 4 (2002): 856S–858S.

Marks, K. J., C. L. Luthringer, L. J. Ruth, L. A. Rowe, N. A. Khan, L. M. De-Regil, X. López, and H. Pachón. "Review of grain fortification legislation, standards, and monitoring documents." *Global Health: Science and Practice* June 27;6, no. 2 (2018): 356–371.

Mashkour, Mana, Yahya Maghsoudlou, Mahdi Kashaninejad, and Mehran Aalami. "Effect of ultrasound pretreatment on iron fortification of potato using vacuum impregnation." *Journal of Food Processing and Preservation* 42, no. 5 (2018): e13590.

Muthayya, Sumithra, Prashanth Thankachan, Siddhivinayak Hirve, Vani Amalrajan, Tinku Thomas, Himangi Lubree, Dhiraj Agarwal et al. "Iron fortification of whole wheat flour reduces iron deficiency and iron deficiency anemia and increases body iron stores in Indian school-aged children." *The Journal of Nutrition* 142, no. 11 (2012): 1997–2003.

Nadadur, S. S., K. Srirama, and A. Mudipalli. "Iron transport & homeostasis mechanisms: Their role in health & disease." *Indian Journal of Medical Research* 128, no. 4 (2008): 533.

Oluyimika, Y. Adetola, Johanita Kruger, Zelda White, and John R. N. Taylor. "Comparison between food-to-food fortification of pearl millet porridge with moringa leaves and baobab fruit and with adding ascorbic and citric acid on iron, zinc and other mineral bioaccessibility." *LWT* 106 (2019): 92–97.

Oppenheimer, Stephen J. "Iron and its relation to immunity and infectious disease." *The Journal of Nutrition* 131, no. 2 (2001): 616S–635S.

Pachón, H. "Wheat and maize flour fortification." In M. G. Venkatesh Mannar and Richard F. Hurrell (Eds), *Food fortification in a globalized world*, pp. 123–129. Academic Press, 2018.

Pachón, H., R. Spohrer, Z. Mei, and M. K. Serdula. "Evidence of the effectiveness of flour fortification programs on iron status and anemia: A systematic review." *Nutrition Reviews* 73, no. 11 (2015): 780–795.

Papanikolaou, George, and Kostas Pantopoulos. "Iron metabolism and toxicity." *Toxicology and Applied Pharmacology* 202, no. 2 (2005): 199–211.

Park, Y. W. (2009). Potential for improving health: Iron fortification of dairy products. *Bioactive Componentsin Milk and Dairy Products*, 379.

Reddy, Manju B., Richard F. Hurrell, and James D. Cook. "Meat consumption in a varied diet marginally influences nonheme iron absorption in normal individuals." *The Journal of Nutrition* 136, no. 3 (2006): 576–581.

Saade, Carol, and Emily O. Arijaje. "Fortification." In Alicia A. Perdon, Sylvia L. Schonauer and Kaisa S. Poutanen (eds.), *Breakfast cereals and how they are made*, pp. 343–359. AACC International Press, 2020.

Sadighi, J., K. Mohammad, R. Sheikholeslam, M. A. Amirkhani, P. Torabi, F. Salehi, and Z. Abdolahi. "Anaemia control: Lessons from the flour fortification programme." *Public Health* 123(12) (2009): 794–799.

Sadighi, Jila, Robabeh Sheikholeslam, Kazem Mohammad, Hamed Pouraram, Zahra Abdollahi, Koroush Samadpour, Fariba Kolahdooz, and Mohsen Naghavi. "Flour fortification with iron: A mid-term evaluation." *Public Health* 122, no. 3 (2008): 313–321.

Santillán-Urquiza, Esmeralda, Miguel Ángel Méndez-Rojas, and Jorge Fernando Vélez-Ruiz. "Fortification of yogurt with Nano and micro sized calcium, iron and zinc, effect on the physicochemical and rheological properties." *LWT* 80 (2017): 462–469.

Sapers, G. M., O. Panasiuk, S. B. Jones, E. B. Kalan, F. B. Talley, and R. L. Shaw. "Iron fortification of dehydrated mashed potatoes." *Journal of Food Science* 39, no. 3 (1974): 552–554.

Schümann, K., B. Elsenhans, and A. Mäurer. "Iron supplementation." *Journal of Trace Elements in Medicine and Biology* 12, no. 3 (1998): 129–140.

Semba, Richard D. "The historical evolution of thought regarding multiple micronutrient nutrition." *The Journal of Nutrition* 142, no. 1 (2012): 143S–156S.

Shubham, K., T. Anukiruthika, S. Dutta, A. V. Kashyap, J. A. Moses, and C. Anandharamakrishnan. "Iron deficiency anemia: A comprehensive review on iron absorption, bioavailability and emerging food fortification approaches." *Trends in Food Science & Technology* 99 (2020): 58–75.

Siddique, Aftab, and Young W. Park. "Effect of iron fortification on microstructural, textural, and sensory characteristics of caprine milk Cheddar cheeses under different storage treatments." *Journal of Dairy Science* 102, no. 4 (2019): 2890–2902.

Teng, Hui, and Lei Chen. "Polyphenols and bioavailability: An update." *Critical Reviews in Food Science and Nutrition* 59, no. 13 (2019): 2040–2051.

Tripathi, Bhumika, and Kalpana Platel. "Iron fortification of finger millet (Eleucine coracana) flour with EDTA and folic acid as co-fortificants." *Food Chemistry* 126, no. 2 (2011): 537–542.

Wahengbam, Elizabeth Devi, Arup Jyoti Das, Brian Desmond Green, Joanna Shooter, and Manuj Kumar Hazarika. "Effect of iron and folic acid fortification on in vitro bioavailability and starch hydrolysis in ready-to-eat parboiled rice." *Food Chemistry* 292 (2019): 39–46.

Wang, Jian, and Kostas Pantopoulos. "Regulation of cellular iron metabolism." *Biochemical Journal* 434, no. 3 (2011): 365–381.

Wei, Yanyan, M. J. I. Shohag, Feng Ying, Xiaoe Yang, Chunyong Wu, and Yuyan Wang. "Effect of ferrous sulfate fortification in germinated brown rice on seed iron concentration and bioavailability." *Food Chemistry* 138, no. 2–3 (2013): 1952–1958.

Whiting, S. J., W. M. Kohrt, M. P. Warren, M. I. Kraenzlin, and J.-P. Bonjour. "Food fortification for bone health in adulthood: A scoping review." *European Journal of Clinical Nutrition* 70, no. 10 (2016): 1099–1105.

WHO. https://www.who.int/data/nutrition/nlis/info/anaemia#:~:text=WHO%20defines%20anaemia%20in%20children,concentration%20%3C120%20g%2FL.

WHO (World Health Organization). Anaemia data. 2011. https://www.who.int/data/gho/data/themes/topics/anaemia_in_women_and_children

Zhang, Dejia, and Arthur W. Mahoney. "Effect of iron fortification on quality of Cheddar cheese." *Journal of Dairy Science* 72, no. 2 (1989): 322–332.

Zimmermann, M. B. "Global look at nutritional and functional iron deficiency in infancy." *Hematology 2014, The American Society of Hematology Education Program Book* December 4, no. 1 (2020): 471–477.

Zimmermann, G. R., J. Lehar, and C. T. Keith. "Multi-target therapeutics: When the whole is greater than the sum of the parts." *Drug Discovery Today* 12, no. 1–2 (2007): 34–42.

Zimmermann, Michael B., and Richard F. Hurrell. "Nutritional iron deficiency." *The Lancet* 370, no. 9586 (2007): 511–520.

9 Iodine Fortification

Sonia Morya, Chinaza Godswill Awuchi,
Arno Neumann, and Deepika Sandhu

9.1 INTRODUCTION

Iodine deficiency (ID) is the condition where there is the lack of (or insufficient) iodine, an essential (trace element) nutrient in diet. The deficiency of iodine often results in several metabolic disorders including goiter, sometimes as endemic goiter and also as cretinism because of untreated or improperly treated congenital hypothyroidism, resulting in developmental delays, swelling of the goiter, and other health issues. ID is a key health issue to mostly fertile and pregnant women, as well as infants and children. ID is a preventable cause of poor intellect and intellectual disability. The World Health Organization (WHO) defines adequate intake of iodine in adults as a median urinary iodine concentration (UIC) value ≥100 µg/L (Iodine Global Network, 2020a). The WHO Recommended Dietary Allowances (RDAs) for iodine can be found in Table 9.1. Thyroid iodine accumulation and turnover were used to calculate the estimated average requirement. The trace element is essential for neurodevelopment among toddlers and children. Thyroxin and triiodothyronine (thyroid hormones) have iodine as a major constituent. Iodine deficiency is common in regions where very little iodine is detected in their diets, usually remote localities where seafood consumption is uncommon. ID is also common in mountainous areas of the world where food cultivation is often done in soil having inadequate iodine. The prevention of iodine deficiency includes addition of small quantities of iodine compounds in the form of iodates or iodides to table salt. Additionally, iodine compounds such as potassium iodates or iodide have been added to foods, water during production, grain flours, milk and dairy products, and bread, in areas with ID (Eastman and Michael, 2018).

Iodine deficiency leading to goiter has been reported in 187 million individuals' worldwide (Vos et al., 2012). The number of people deficient in iodine has reduced recently, thanks to the wide acceptance of iodization of salts. In the United States, the iodization has reduced over overdose concern since the middle of the 20th century; at same time, bromine, perchlorate, and fluoride, which are iodine antagonists, have become more ubiquitous (Meletis, 2011).

9.2 SIGNS AND SYMPTOMS FOR IODINE DEFICIENCY

9.2.1 GOITER

Goiter is the enlargement of the front as well as the sides of the neck as a result of the inflammation of the thyroid. Low amounts of thyroxin levels in the blood because of ID results in relatively high levels of thyroid stimulation hormones (TSH) that stimulate thyroid and initiate several metabolic processes, and cell growths and proliferation that may result in goiter. In mild ID, triiodothyronine (T_3) levels could elevate in low levothyroxine levels, which will lead to the body transforming additional levothyroxine to triiodothyronine. Patients may have goiter with no elevated thyroid stimulating hormone. Inception of salt iodization as early as 1900 has helped to eliminate iodine deficiency disorders (IDD) in most affected nations; however, in the New Zealand, Australia, as well as many European countries, ID remains a public health concern (Andersson et al., 2005). The total goiter rate of a group is used to define severity using the following criteria: <5%, iodine

DOI: 10.1201/9781003160663-12

TABLE 9.1
The RDAs for Iodine

Age	Female (µg)	Male (µg)	Pregnancy (µg)	Lactation (µg)
6 months and below	110*	110*		
7 months to 1 yrs	130*	130*		
1 to 3 yrs	90	90		
4 to 8 yrs	90	90		
9 to 13 yrs	120	120		
14 to 18 yrs	150	150	220	290
19 yrs and above	150	150	220	290

Source: Institute of Medicine, Foods & Nutrition Board, 2001; World Health Organization, 2006, 2014
* Adequate Intake (AI), Yrs = years

sufficiency; 5.0%–19.9%, mild deficiency; 20.0%–29.9%, moderate deficiency; and >30%, severe deficiency (Zimmermann, 2020). ID is more commonly reported in the developing and underdeveloped nations, especially the areas which consume little or no seafood. The health initiatives to reduce the risks of cardiovascular and circulatory diseases have caused reduced discretionary usage of table salt. Goiter is considered endemic as soon as its prevalence goes higher than 5% in a population (Philip and Lawrence, 2001). If goiter is left not treated for about 5 years, thyroxin treatment or iodine supplementation may not lower the thyroid gland size and it may be impaired for life.

9.2.2 Congenital Iodine Deficiency Syndrome (CIDS) *(Cretinism)*

Congenital iodine deficiency syndrome (CIDS) is the health condition associated with ID and goiters, often characterized with deafness, mental impairment, stance and gait impairment, squint, and stunted growth resulting from hypothyroidism. Children with impaired mental capacities and stunted bodies were found to have a thyroid hormone insufficiency as a result of restricted diet, intermarriage, seclusion, and ID in their foods. In 2007, the WHO reported nearly 2 billion people had ID. It was concluded that ID is the single greatest preventable cause of intellectual disability (World Health Organization, 2014).

9.2.3 Fibrocystic Breast Changes

Preliminary evidence suggests that ID increases the sensitivities of the breast tissues to the hormone estrogen (Cann et al., 2000; Pizzorno and Michael, 2012). In a few studies, iodine supplementation was found to have beneficial effects including the reduction of the breast cyst, breast pain, and fibrous tissues plaque in the women suffering from fibrocystic breast changes (Cann et al., 2000; Patrick, 2008). As a result of iodine's antiproliferative properties in breast tissue, it has been suggested that molecular iodine supplementation can be used as adjuvant in breast cancer therapy (Aceves et al., 2005).

9.3 MAJOR SOURCES OF IODINE

a. Iodized salt

Presently, sodium and potassium iodates or iodides, which are iodine salts, are permitted to be added to table salt. The US, Canada, Nigeria, India, and many other countries have salt iodization programs (World Health Organization, 2020; UNICEF, 2020; Global Fortification Data Exchange,

2020). In the US, manufacturers have been iodizing salt ever since 1920.Voluntarily and presently iodized salt contains 45 µg iodine/g salt, which may vary but is within that range (Dasguptaet al., 2008). The US FDA approved cuprous and potassium iodides for salt iodization, whereas the WHO recommend use of potassium iodates because of its very high stability, especially in hot temperatures (World Health Organization, 2020; US Food and Drug Administration, 2009 and 2016). However, most of salt intake in developed countries comes regularly from processed food products.

b. Iodine dietary supplementations

Iodine supplements are in use nowadays. Numerous multivitamins as well as mineral supplements have iodine in the form of sodium or potassium iodide. Iodine-rich seaweed (kelp) or dietary supplements of iodine are available. Aquaron et al. (2002) studied the bioavailability of seaweed iodine in human beings. The study looked at the bioavailability of pure mineral iodine, such as potassium iodide, which was 96.40%, and pure organic iodine, such as monoiodotyrosine, which was a comparatively lower at 80%.

c. Dietary (food) iodine

Seaweed is known to be among the excellent dietary source of iodine. Kelp, nori, wakame, and kombu are among the richest sources of iodine, however it is greatly varied (Table 9.2) (Zimmermann, 2009, Teas et al., 2004). Other sources of dietary iodine include other seafoods such as oysters, snapper, tinned salmon and seaweed, grains and their products, and eggs. Dairy products contain sufficient iodine, partly because of the usage of iodophors sanitizing agents and iodine feedstock in dairy industries. It is important to note that the addition of seaweed to plant-based goods to reinforce them with calcium might enhance the iodine content, although the iodine concentration varies depending on the type of seaweed extract used. Moreover, dairy products should be regularly consumed, particularly in the countries where it is not mandatory to fortify the iodine or where the availability of iodized salt is lower. Iodine is present in infant formulas and human breast milk (Zimmermann, 2020; Krela-Ka 'zmierczak et al., 2021). Around 90% of iodine is absorbed by the stomach and duodenum and is mostly delivered from fortified salt and other sources such as fish, seafood, dairy, water, eggs, broccoli, peas, or spinach. Fruits and vegetables are also good iodine

TABLE 9.2

Selected Dietary Source of Iodine (Approximate Values)

Foods	µg per serving	Daily value (%)
Banana, one medium	3	2
Apple juice, one cup	7	5
Egg, one large	24	16
Ice cream, chocolate, half cup	30	20
Shrimp, three ounces	35	23
Bread, enriched, two slices	45	30
Fish stick, three ounces	54	36
Milk, with reduced fat, a cup of	56	37
Iodized salt, 1.5 g	71	47
Yogurt, low-fat, plain, one cup	75	50
Cod, baked, three ounces	99	66
Seaweed, whole or sheet, a gram	16–2,984	11–1,989

Source: Teas et al., 2004; World Health Organization, 2014.

sources, and the amount of which depends on fertilizer use, iodine in soil, and irrigation practices used. The concentrations of iodine in plant foods usually range from 1 mg/kg to 10 mg/kg of dry weight (Zimmermann, 2009). Approximately 100 mcg of iodine is found in one ounce of cranberries. Strawberries are very tasty and popular due to their appealing flavor. A cup of fresh strawberries contains about 13 mcg of iodine. Iodine is abundant in a variety of beans. Among vegetables, potatoes are a good source of iodine. The iodine content of one medium baked potato is 60 mcg (Choudhry and Nasrullah, 2018). There is variability in iodine levels in meat as well as other animal products because of the influence of iodine levels in animal feed (Teas et al., 2004).

9.4 TREATMENT OF IODINE DEFICIENCY

Iodine deficiency can be reverted by consuming iodine or its supplements. Mild cases of ID can be treated by drinking more milk, eating saltwater fish, consuming egg yolks, or adding iodized salt in foods consumed daily. For animal product or salt restricted diets, seaweed could be regularly included in foods as the source of iodine (Stephanie et al., 2017). The WHO stated that "the RDI (recommended daily intake) of iodine for women (adult) is 150 to 300 µg for the maintenance of the normal functioning of thyroid; 150 µg is recommended for men". However, there are insufficient data available on the iodization of foods, except salt, which has resulted in the prevention of goiter incidence or the improvement of physical and intellectual development (Santos et al., 2019).

9.5 HEALTH RISKS FROM EXCESSIVE IODINE

Excess iodine intake is rare and may occur due to overconsumption of iodine supplements or other iodine rich sources. High iodine intake has toxic impacts which can end up as hyperthyroidism and subsequently hyperthyroxinemic (high amount of thyroid hormones in blood). High intakes of iodine (Table 9.3) can cause symptoms the same as ID, such as goiter, hypothyroidism, and elevated amount of TSH. Excessive iodine in some people inhibits the synthesis of thyroid hormones, increasing the stimulation of TSH which could cause goiter (Institute of Medicine, Foods & Nutrition Board, 2001). Hyperthyroidism may also result when iodine is administered for the treatment of iodine deficiency. Studies have indicated that excess intake of iodine can cause thyroid papillary cancers and thyroiditis. Acute poisoning of iodine is rare; it is often due to doses of several grams. The symptoms of acute poisoning include stomach problems, fever, a burning in the

TABLE 9.3
Iodine Tolerable Upper Intake Level (UL)

Age of population	Male and female (in µg)	Pregnant women (in µg)	Lactating mothers (in µg)
Below 6 months	NPE*		
7 months to 1 yrs	NPE*		
1 to 3 yrs	200		
4 to 8 yrs	300		
9 to 13 yrs	600		
14 to 18 yrs	900	900	900
19 yrs and above	1,100	1,100	1,100

*NPE= Not possible to establish. Formula/diet has to be only source of iodine for any infant. Yrs = years. Source: World Health Organization, 2006 and 2014; Institute of Medicine, Foods & Nutrition Board, 2001

throat and mouth, abdominal pain, diarrhea, weak pulse, coma, nausea, and vomiting (Institute of Medicine, Foods & Nutrition Board, 2001).

The reactions to the consumption of too much iodine and the dosage usually cause an undesirable impact that differs from individual to individual. A few individuals, such as people who have iodine deficiency and autoimmune thyroid disease, could experience adverse effects even with iodine intake that is safe for the population in general (Institute of Medicine, Foods & Nutrition Board, 2001). In most individuals, consumption of dietary iodine and iodine supplements are very unlikely to go beyond the tolerable upper intake levels (ULs). Continuous intake of iodine beyond ULs notably raises the risks of the adverse health effects such as thyroiditis, hypothyroidism, hyperthyroidism, and thyroid papillary cancer (Southern and Jwayyed, 2022). However, ULs are not applicable to those receiving iodine as medicine; these people have to be placed in the care of a nutritionist or physician.

9.6　PREVALENCE

The indicators for generally assessing rate of the ID in given population are total goiter prevalence and median urinary iodine (MUI). According to the WHO and Iodine Global Network,

> iodine deficiency becomes a public health problem in the populations where the concentration of MUI is less than 100 µg/l, or in the areas where there is endemic goiter, as defined by more than 5% of children aged 6 to 12 years having goiter.

As the concentration of the MUI (Table 9.4) shows, the current iodine intake as well as relative responses for the correction of ID is often preferred to be used to monitor impacts of the interventions for regulation and control of IDD. These indicators are recommended by the WHO and relate not only to the iodine status in a population as estimated by the urinary concentrations but also

---

TABLE 9.4

Indicators Used to Assess Iodine Status to a Given Population

Indicators	Population group	Samples	Definition of "mild deficiency"	Definition of "severe deficiency"	Guiding comments
Prevalence of total goiter	Children 6 to 12 years	Clinical examination	>5%	>30%	Reflects past or recent thyroid dysfunction. Can be estimated by ultrasonography or by clinical examination. Not for use in monitoring the impacts of interventions becauseresponse of goiter to the status of iodine correction is proven to delay.
Iodine	Children 6 to 12 years	Urine	Median <100µg/l	Median <20µg/l	Recommended for monitoring andestimating iodine status at given population. As the distribution of urinary iodine is abnormal, cut-off is alwaysknownby median values.

Source: Iodine Global Network, 2017–2020; World Health Organization, 2006

covers many program indicators that estimate the sustainability of the salt iodization program itself. ID manifesting as goiter was reported in 187 million individuals worldwide in 2010 (Vos et al., 2012), while it reduced to about 2 billion individuals in 2017, among which nearly 50 million people presented with clinical and subclinical manifestations (Biban and Lichiardopol, 2017). Despite the reduced numbers of people showing clinical signs of ID in recent years, ID as IDD, goiter, and other forms, remains a global health problem. Due to unavailability of iodine and its natural deficiency, some parts of the world including South-East Asia, Africa, and the Western Pacific have severe ID, affecting about two billion population worldwide at present (Eastman et al., 2018). Countries affected by ID, such as Kazakhstan, Russia, and China, have started taking actions to overcome it.

The developing world has reported the prevalence of ID for many decades; however, recently, a decrease in salt consumption and changes in dairy processing practices, which have eliminated the use of iodine-containing decontaminants, has resulted in increasing predominance of ID in some developed countries, including Australia and New Zealand (FSANZ, 2013). The criteria for the evaluation of iodine deficiency have been mentioned in Table 9.5. In October 2009, a proposal making the use of iodized salt compulsory in many commercial bread manufacturing plants was agreed in Australia and New Zealand (FSANZ, 2013). In a UK study in 2011, nearly 70% of people were reported to have ID (Vanderpump et al., 2011). The authors of the study recommended investigation into factual recommendations for iodine supplements usage (Vanderpump et al., 2011).

ID is widely observed during pregnancy in areas of Europe that are iodine-deficient or have mild ID, despite widespread use of iodized salt, posing a risk to embryonic neurodevelopment (Trumpff et al., 2013). Research in a mild ID area found that it affected more than half of the breastfeeding women; conversely, the majority of the newly born infants by these women had excess iodine in their systems, owing to neonatal exposure to iodine-based decontaminants (Yaman et al., 2013). A 2014 meta-analytical study reported that consumption of iodine supplements improves certain maternal thyroid measures and might benefit cognitive function in school-age children, including the areas marginally deficient of iodine (World Health Organization, 2014; Taylor et al., 2014; Völzke et al., 2018).

9.6.1 Modern Day Iodine Status and Prevalence

It is over three decades ever since The 1990 World Summit for Children, in which leaders met to sign a unanimous declaration aimed at promoting a much-needed healthier population as well as the survival of the children worldwide. The ID and IDD eradication was generally adopted among "the key targets of new global strategy" at that time. National governments and international organizations all over the world have worked towards the implementation of policies and programs with regard to the global iodization of salt in order to improve intakes of iodine to avoid cognitive impairment due to iodine deficiency. Monitoring the progress via frequent surveys for iodine status nationally is a key component of the efforts against ID and IDD worldwide (Völzke et al., 2018). New data within school-age children (SAC) as well as other population groups if SAC data is absent

---

TABLE 9.5

Criteria for Evaluating Iodine Deficiency Severity

Severity of public health problems	Indicator for total goiter prevalence	Median urinary iodine indicator
Mild	5.0 to 19.9%	50 to 99 µg/l
Moderate	20 to 29.9 %	20 to 49 µg/l
Severe	>30 %	<20 µg/l

Source: WHO, 2006

---

are key for measuring the recent levels of iodine status and ID, tracking efficacy of the national/ international programs on salt iodization, as well as identifying the regions or population more prone to ID or excess of it. In the last three decades, 22 of 194 WHO nations have not declared any data relating to their population's iodine status (Iodine Global Network, 2020b).

"Global iodine status in 2020 periodic universal updates" by Iodine Global Network (IGN) was designed as a summary of the latest global progress to eliminate ID (Iodine Global Network, 2020a,b). Of the 155 WHO nations with recent available data, two countries have moderate ID (Morocco, Angola), while 18 are reported to be mildly iodine deficient (Iodine Global Network, 2020). This is similar to ones reported in 2015 to 2017 by Susanne et al. (2018), which reflected that several countries are still sustaining their optimal success in iodine nutrition via continuing implementation of "GAIN's Salt Iodization Program". In 2019, less than 8.5% of world population captured by the latest data are residing in nations with ID, however the number of WHO countries with deficient iodine has declined from 1993 to 2020 (Iodine Global Network, 2020). The trend is expected to continue if the current efforts can be sustained. Till 2020 many WHO nations have been attained and still positively sustaining their optimum dieting in iodine. The approximation of ID in 1993 was based on the total goiter rates (TGR) (WHO/UNICEF/ICCIDD, 1993) and afterwards on the urinary iodine concentrations (UIC) in SAC as well as other populations if SAC data is absent. Several nations monitor their iodine status and ID in both "SAC and women of reproductive age (WRA)" or adults. The WHO defines adequate intake of iodine in adults as a "median UIC value greater than or equal to 100 µg/L" (Iodine Global Network, 2020). However, the scientific basis for this WHO threshold has been described as weak, and the assessments based on the groups other than SAC have to be interpreted with carefulness.

9.6.2 IODINE DEFICIENT POPULATIONS OF THE WORLD

In regions of the world where little dietary iodine is common, such as remote inland terrain and semi-arid climates where no aquaticdiets are consumed, ID results in hypothyroidism and other IDD with symptoms, such as severe fatigue, mental delay, goiter, depressions, low body temperature, and weight gain (Felig and Frohman, 2001). Patrick (2008) and Monahan et al. (2015) reported that "ID is the main cause of preventable mental retardationthatoccurs primarily when small children or babies are rendered hypothyroidic by lack of iodine; Iodization of salt has largely eliminated this health problem in the developed and wealthier countries; However, ID is also a problem in someparts of Europe; In Germany it was estimated to cause 1 billion US dollars in health care

---

TABLE 9.6

Insufficient Iodine Intakes in Populations by WHO Regions (Approximate Statistics)

WHO Regions	Insufficient iodine intake (UIC less than 100 µg/l) in the general population in millions (%)	Insufficient iodine intake (UIC less than 100 µg/l) in the School-aged children in millions (%)
Southeastern part of Asia	503.6 (30.0)	73.1 (30.3)
Europe	459.7 (52.0)	38.7 (52.4)
West Pacific	374.7 (21.2)	41.6 (22.7)
Africa	316.7 (42.0)	58.5 (41.4)
East Mediterranean	259.3 (47.2)	43.3 (48.8)
Americas	98.6 (11.0)	11.6 (10.6)
Globally	2,012.6 (30.7)	266.8 (31.5)

Source: Andersson, et al., 2010; World Health Organization, 2014
UIC= urinary iodine concentration.

costs each year; a modelling estimation suggests universal iodine supplementation for the pregnant women in England can save £199 to the health services per pregnant woman and also save £4476 per pregnant woman in the societal costs."

9.7 HISTORY OF IODINE FORTIFICATION

9.7.1 Discovery and Early Research

The discovery of the iodine seems to be a French enterprise. It was identified by French scientists, including Gay-Lussac, in 1809–1810 (Abraham, 2004). An early Chinese text of around 3600 BC was the first to suggest the reduction in goiter after the consumption of seaweeds as well as burnt sea sponges (Rosenfeld, 2000). Even though iodine had not yet been discovered, these dietary treatments remained in place. Their usage continued worldwide, as was reported by Hippocrates and others in subsequent centuries (Leung et al., 2012; Rosenfeld, 2000; Abraham, 2004). Discovery of the element iodine was accidentally made in the 19th century. In 1811, Bernard Courtois, while extracting some of the sodium salts required for manufacturing gunpowder, observed unusual purplish vapor emanating from the seaweed ashes treated using sulfuric acids (Rosenfeld, 2000). The studies of this purplish vapor were continued by Joseph Louis Gay-Lussac, Sir Humphry Davy, and others. The first article presentation of this new substance, iodine (named after ioeides, a Greek word referring to its violet-color), by Gay-Lussac was read in 1813 (Rosenfeld, 2000). Shortly after this, J.F. Coindet, a physician from Switzerland, published experimental observation that iodine used as grain in distillated alcohol reduced the size of goiters in patients (Carpenter, 2005; Rosenfeld, 2000). In 1852, a chemist, Adolphe Chatin, published the first hypothesis of ID in the population connected with endemic goiters. This was confirmed by Eugen Baumann, and following his report, iodine was discovered in the thyroid in the year 1896 (Rosenfeld, 2000).

Although it was Courtois who discovered iodine in 1811, Gay-Lussac showed it was new and also named it *"iode"*. Davy anglicized *"iode"* by referring to it as *"iodin"* which was later named iodine in 1930s (Abraham, 2004). By 1813, Gay-Lussac synthesized many products from the new element, iodine, and entirely characterized it, although he reserved for Courtois the total credit for iodine discovery. In early 1850s, Chatin studied the relationship between goiter prevalence and concentrations of iodides in water, soil, as well as foods of different places (Chatin, 1853). Chatin also examined the effects of eating iodine supplements on the goiter endemic. Chatin came up with the following observations with regard to iodine supplementation is particular for goiter prevention; goiter and cretinism are infrequent in iodine-rich areas; while goiter and cretinism do occur on a regular basis in iodine-deficient areas. From 1891 to 1892, numerous publications were made in the British Medical Journal (BMJ) reporting the effectiveness of using thyroid extracts "parentally" and "orally" in the patients with "hypothyroidism" (Fox, 1892).

Bauman, in 1895 reported high iodine concentration in thyroid and postulated that the active ingredient in thyroid extractions had the new element, iodine. Before Bauman identified this, apothecary and pharmaceutical preparations having iodine as an integral part, without extracts from thyroid, were already used widely as treatment for all human illnesses. Kelley said "in the first rush of the enthusiasm for newcomer, surgeons and physicians tested iodine and also tried it for all conceivable pathological conditions" (Kelly, 1961). The number of illnesses for which iodine was used in the early decades is truly amazing. These included scrofula, hip-joint disease, lacrimal fistula, paralysis, deaf, syphilis, chorea, spine distortions, dropsy, acute inflammation, carbuncles, chilblains, gout, whitlow, lupus, croup, catarrh, gangrene, and burn. There was a remarkable acceptance of products with iodine. From over ten preparations listed in a pharmacopoeia in 1851, this grew to 1,700 approved pharmacopoeia names designated to products with iodine in 1956. This is a convincing indication of the widespread usage of iodine in traditional and orthodox medicine (Kelly, 1961).

In early 1900s, Professor Kocher, a Nobel Prize winner in 1909, joined the league and reported adverse iodophobic effects on the management of hyperthyroidism and claimed that he suffered hyperthyroidism after ingesting iodide. Kocher was against using iodine or iodide for hyperthyroidism (Abraham, 2004). However, Redisch and William reported that only iodine was extensively used by medical experts for treatment of hyperthyroidism (Redisch and William, 1940). Using only iodine in patients presenting with hyperthyroidism, Thompson, and others, in 1930, reported 88% success, and Starr et al. (1924) reported 92% success with Lugol solutions at dosage of 690 mg/day.

In the first centenary after iodine discovery, clinical uses were completely experimental, and necessary doses were reached by trial and error. In 1911, Bertrand proposed "a concept of essential trace elements" (which was never put in place), required for normal functions and growth of plants. Marine and Kimball (1917) later applied Bertrand's concept to iodine using human subjects by choosing adolescent school girls 10 to 18 years of age residing in Ohio, a city with 56% of goiter incidence at that time (Marine and Kimball, 1917). After observations of 2.5 years, 495 people (22%) from the control had enlarged thyroid size; 0.2% of the iodine-supplementation group had goiter "Iodism" was reported in 0.5%. In Switzerland with high goiter incidence (82 to 95%), Marine reported that Klinger administered 10 to 15 milligrams of iodine per week to 760 people of the age group 10 to 18 years. Their iodine intake per day was 1.4 to 2 mg. A first examination showed that 90% developed thyroid enlargement. After one and half years of this study, 28.3% still had enlarged gland; no "iodism" was reported. In response, the Swiss Goiter Commission recommended using iodine supplements in all cantons. An iodized fat in the form of tablet with 3 to 5 mg of iodine in each tablet was applied.

9.7.2 IODINE FORTIFICATION AND SUPPLEMENTATION

In 1831, JG Boussingault, a French chemist, proposed iodized table salts as the way of avoiding goiters and other IDD. The proposal/decision was first implemented in European nations as well as the US in 1920. Salt iodization gave false sense of the sufficiency of iodine and lead to public reliance on iodized salt for iodine instead of the iodine and iodide such as the Lugol solution (Abraham, 2004). Prior to salt iodization, autoimmune thyroiditis was nearly non-existent in the US, although potassium iodide and Lugol solution were extensively used in medicine in amounts higher than the daily iodine intake from iodized table salts. In 1912, Hashimoto published findings regarding four thyroid glands removed during surgery, which he branded "pathology of the thyroid strumalymphomatosa" later branded the "Hashimoto's thyroiditis". In 2004, Abraham stated that

> Hashimoto's thyroiditis is currently classified as goitrous autoimmune thyroiditis; the two conditions are chronic, degenerating with time to hypothyroidism in a large number of patients; but both conditions showed improvement following a nutritional program with emphasis on magnesium combined with iodine supplements for overall body sufficiency.
>
> (Abraham, 2004)

A proposed mechanism detailing the oxidative damage resulting from low iodide levels in combination with antithyroid drug is that an insufficient supply of iodide to the thyroid, worsened by goitrogenic particles, activates "thyroid peroxidase (TPO) system" via increased TSH, low iodinated lipid amount, and a high cytosolic free calcium level, leading to too much H_2O_2 making. The excess production of H_2O_2 is supported by known facts that the antioxidant substances used in study conducted by Bagchi et al. (1995) didn't affect the organification and oxidation of iodide and as a result neutralized the excess oxidant alone. This production of H_2O_2 is more than normal because of an insufficient feedback system resulting from high cytosolic calcium due to the deficiency of magnesium and low iodinated lipid levels (Abraham, 2004). Thyroid peroxidases in presence of organic substrate and H_2O_2 return completely to their functions as peroxidases, being the main halo peroxidases functions, resulting in oxidative damages to the molecular substances closest to the action sites the substrate thyroglobulin (Tg) and thyroid peroxidase (TPO). According to Abraham (2004),

Oxidized Tg and TPO elicit autoimmune reaction with antibodies production against the altered protein with subsequent destruction to the thyroid cells apical membrane, leading to lymphocytic infiltration and then clinical manifestations of Hashimoto's thyroiditis; based on this mechanism (magnesium deficiencies and iodine induce autoimmune thyroiditis), the implementations of total nutritional program with emphasis on magnesium in combination with iodine supplementation was very effective in the patients with thyroid pathology, with inclusive of autoimmune hyperthyroidism.

(Abraham, 2004)

9.7.3 Iodized Salt

Since 1993, when the WHO and UNICEF approved iodine use in salt, several regions globally have reported considerable positive improvement in the prevention and control of ID and IDD. In the US, "iodized salt" became public in groceries and stores on May 1, 1924 (Kimball and David, 1992). Initially, salt was iodized at 100 mg/kg, causing average estimated intake of iodine of around 500 µg/day (Zimmermann, 2008). Iodized salt intake is seasonally consistent, with relatively small costs, global food, and easily distributed (Zimmermann, 2012). At least 120 countries, including Canada, Nigeria, and Uganda, have adopted mandatory iodization of food-grade salt (Dasgupta et al., 2008). However, IDD still remain a concern in numerous nations where the salt iodization is inadequately reinforced. Conversely, a statement in Dietary Supplement Fact Sheet says "iodization of salt in the United State is not compulsory; the US FDA does not make the listing of iodine on food packaging compulsory (Dietary Supplement Fact Sheet, 2020). Currently, salt iodization is compulsory in many parts of the globe.

Salt is generally considered an appropriate for iodine fortification, due to the following

a) Iodine remains in industrially processed foods with salt as their main recipe;
b) Iodization of salt is inexpensive (costing estimated US$ 0.02–0.05 per person covered per year, may even be less);
c) Adding iodide or iodate to salt has no effect on the salt flavor and also no effect on the foods containing the iodized salt;
d) The technological requirements for iodization of salt is cost effective, well established, and easy to move from one country to another;
e) In most countries, the production of salt is handled by a few centers, making it easier for quality control and quality assurance;
f) It is consumed worldwide, with little variation in the consumption patterns;
g) The iodine (iodide or iodate) concentration in salt is easily adjusted to meet the level of salt needed to avoid the risks of cardiovascular disease. (World Health Organization, 2014)

9.8 CHALLENGES IN IODINE FORTIFICATION

Mining solid rock deposit is major source of salt in European nations, North American countries, and Australia. In other parts of the globe, such as in Asian, South American, and African countries, solar evaporation of lake, underground brines, or sea water remains the main source of salt. Following extractions, the crude salt is refined to increase its purity from 85%–95% then to 99% salt (NaCl)" (Winger et al., 2008). Iodine is often added to salt after refining and drying of the salt, by any of two techniques; wet or dry. For wet dry, potassium iodate (KIO_3) is usually sprayed or dripped at uniform rates on the salt passing through conveyor. Wet technique is cheap. The dry technique is painstaking than wet which requires total mixing after adding iodine compound to make sure there is a uniform distribution of the iodine compound. Improper mixing is the leading cause of inadequate iodization of "table salt". The iodine stability in salt rests on rate of acidities, water level, and the purity of salt in which the iodine compound is included (WHO, 2014). To prevent the "loss of iodine" in storage time, "iodized salt" should be properly dried and pure as possible and must also be packed properly. Iodine often migrates from the upper to the lower part of the container in salt with high water content. Iodine evaporates if there is high acidity. Losses also occur when packages with impervious linings are used; as the package dampens, iodides migrate from table salt to the fabrics and ending up being evaporated. Chances of occurrence of loses is less when "potassium iodate" is used instead of KIO_3 since KIO_3 resists oxidation better and is less soluble comparatively. Packaging which helps to prevent the loss of iodine include high density polyethylene (HDPE) bags lined with continuous puncture-resistant film or laminated with a low-density polyethylene (LDPE). High humidity in combination with porous packing (e.g., jute bags) will cause 30 to 80% iodine loss within 6 months (WHO, 2006).

Nevertheless, iodide is a great reducing agent, and iodates a great oxidizing agent. Its usage in food and food products could cause redox reactions and as a result influences the characteristics of foods, food stability, and shelf life of food products. Current methods used for iodine fortification have not been proven to prevent these effects, and the data is scarce regarding the reactions of iodine and iodine salts in foods. In general, many food industries use non-iodized salts in processed food products (WHO, 2006). Iodine salt might be volatile and can be lost during the rigor industrial processing. There could be 20% iodine loss via industrial processing and additional 20% via food preparations (WHO, 2014).

As previously stated, "iodide is a great reducing agent, and iodate acts as a great oxidizing agent," and its presence in iodized salts used in foods may theoretically result in reactions in foods that may potentially increase the level of oxidative reaction, thereby reducing "shelf life" of foods and result in color reactions and changes in foods. It also reduces iodine bioavailability and decrease the bioavailability of other nutrients. Iodine losses due to iodide oxidation are exacerbated by humidity, sunshine exposure, moisture, heat exposure, or contaminants in the table salt to which the iodine has been added. (WHO, 2006, 2014).

9.8.1 Safety Concerns

Generally, iodine fortification is very safe. Table salt and bread have been iodized for more than half a century without any known toxic effects (Santos et al., 2019). As the thyroid hormones are synthesized, the release is often well regulated via mechanisms which enable the adjustment of the body to a range of iodine ingestion; intake of up to 1000 µg (1 mg) a day is tolerated by many individuals (WHO, 2014). Acute, excess increases in iodine intake can increase risks of iodine toxicities in susceptible people; those who have had previous chronic ID can develop a condition called iodine-induced hyperthyroidism (IIH). IIH is most commonly reported complication of iodine prophylaxis and is attributed to the abrupt introduction of excess iodized salt in people who have been severely deficient in iodine for long periods. IIH outbreaks have been seen in nearly all programs involving the use of iodine supplements. An IIH outbreak often occurs in early phase of program implementation and mostly affects elderly people who have long-term thyroid nodules and often transitory in nature with incidence rates reverting to normal levels after 1 to 10 years of intervention. IIH outbreak has also been reported in some parts of Africa, like Zimbabwe and DR Congo (World Health Organization, 2006). If IIH outbreak occurs after introduction of iodized salt, same pattern of outbreak has been reported in iodine supplementations programs would be expected to be followed. Iodine-induced thyroiditis is also a condition which can be worsened or initiated through increasing intake of excess of iodized salt (Todd, 1998).

9.9 FORTIFICANTS USED FOR IODINE

The two forms of iodine iodates and iodides are suitable as fortificants and are often added as potassium salt, and sometimes as sodium or calcium salt (Santos et al., 2019). Potassium iodide has been used in bread and salt manufacturing for over eight decades now, while potassium iodate has been used for over five decades. KO_3 is less soluble in aqueous than KI and more stable under adverse environments. It also does not require stabilizers and is more resistant to oxidation and evaporation. KO_3 is preferred to KI, particularly in humid and hot environments (Unit et al., 1996). For salt iodization, KO_3 is preferred due to it is higher stability. However, European and North American countries prefer potassium iodide, but most nations in with hot temperatures make use of KO_3 (WHO, 2006)

The level of fortification used is between 30 to 200 ppm (WHO, 2014; Food and Agriculture Organization, 1995). Salt is the most commonly used food substance for iodine fortification. Universal salt iodization (USI) is the strategy recommended by the WHO for the control and prevention of IDD (Santos et al., 2019). The choice of the WHO strategy relies on following factors:

a) Consumption of table salt is fairly stable all through the year;
b) Salt is among the few commodities eaten by everyone;
c) Iodine addition to salt does not have any effect on its taste, color, or odor;
d) The technology for salt iodization is available at a reasonable cost throughout the developing countries and easy to implement;
e) Salt making is often limited to very few geographic areas;
f) Quality of iodized salt is easily monitored at the production, household, and retail levels.

9.10 NOVEL AND EMERGING TECHNIQUES AND METHODS OF IODINE FORTIFICATION

Suitable technologies for salt iodization exist for small and large-scale production plants. The four main technologies used for iodization to salt include dry mixing, spray mixing, submersion, and drip feed addition (Santos et al., 2019; WHO, 2014; Food and Agriculture Organization, 1995). Recent data showed that dietary iodine intake can be increased by adding slow-release resins to water, however the overall effectiveness is yet to be established (WHO, 2004).

The potassium iodate is added to salt after drying the refined salt and just before packaging. Iodine fortification can be connected with refining lines or existing production line or both. This is accomplished by addition of potassium iodate solution or dry powder to the salt. According to the Iodine Global Network

> in wet method, first, potassium iodate should be dissolved in water to prepare concentrated solution; the concentrated solution can be sprayed or dripped on the salt uniformly; In dry method, first, potassium iodate mixed with filler for example calcium carbonate or dry salts and the powder is sprinkled on the dry salt.

> (Iodine Global Network, 2020a)

In addition, "in dry and wet methods, thorough mixing of salt after adding the potassium iodate is required to ensure uniform penetration and even distribution of the potassium iodate" (Iodine Global Network, 2020).

9.10.1 DRY MIXING

In dry mixing method, KO_3 is mixed in a 1:9 ratio with calcium carbonate, magnesium carbonate, or tricalcium phosphates which are anti-caking agents, then one part of the stock is mixed with ten parts of salt, forming a premix. The premix is constantly introduced on a screw conveyor. The salt is introduced to the screw conveyor and the mixing begins as the mixture gradually goes through. The processes are suitable for even and fine-grained salt less than 2 mm grain size. It has been adopted in many nations in Central and South America including Bolivia, Guatemala, Argentina, and Peru. According to the Iodine Global Network, in China, a unique compact dry mixing machine is used. It consists of an inclined screw conveyor made of a feed hopper at the lower end where salt is fed to the machine, and within the feed hopper is a slide that controls the rate of addition of salt. Then a premix of potassium iodate (KIO_3) and salt at an approximate 1:2000 ratio is made separately and fed into the bulk salt at a controlled rate through a rotating arm in a conical feeder or through a screw in the screw feeder near the salt feed hopper over the conveyor (Iodine Global Network, 2020). The premix and salt are mixed while moving up the screw conveyor, and the desired iodine content in the salt is then achieved. The mixture is further homogenized by passing through a pin mill or a roll crusher that reduces it to a uniform size of 1 to 3 mm. The iodized salt is pushed through a second inclined screw for extra mixing before packaging (FAO, 1995; Winger et al., 2008).

9.10.2 Spray Mixing

Iodine fortification is usually integrated with preexisting salt production and refining. Usually, slurries of salt from a thickener are dewatered using a centrifuge and then dried in fluid bed or rotary driers. A sensor installed onto the thickener send signals to the pump dosing solution that sprays a solution of iodates within a rate that is proportional to the flow rate of the solids to centrifuge. In more orthodox operation without available refining vessels, establishing plants for salt iodization are needed. Salt in crystal forms are crushed, becoming a coarse powder using a roller mill and fed by hand into a feed hopper fitted with a wire mesh screen or grating at its upper part to avoid large lumps of salt falling into it. Another shaft that has four plates is attached in the feed hopper outlet and checks the flow rate onto inclined screw conveyor belt. The shafts are drove by varied speed drive system. The rotation rate is amended to the necessary throughput (Winger et al., 2008).

Sheet of salt that is releasing from belt onto the spray chamber receives the finely atomized spray containing the solution of KO_3 (or KI) from 2 nozzles, at 1.4 kg per cm^2 pressure. The spray nozzle is made in such a way to release flattened spray which entirely distributes all over the salt stream side-to-side. The solution concentration as well as the spray rates is usually amended to give the requisite dose of iodates in the salt. Solution of the iodate remains pressurized in two drums made with stainless steel, with each around 80 L. The pressure inside the drums is relatively kept constant by an air compress or using a regulator. The salt with KO_3 slides into a screw conveyor of about 20 to 25 cm width and 2.5 to 3.0 m length. Traveling via the screw ensures mixing uniformity. The screw conveyor releases into twin outlets with bags ready for filling (FAO, 1995).

9.10.3 Drip Feed Addition

This process is often preferred for the iodization of salt crystals. The salt crystals are normally fed by hand into a feed hopper which uniformly releases onto belt conveyor of around 35 to 40-centimeter width and 5.5-meter length inclined to a 20 degrees slope. The screw conveyor is fitted with a tensioning device. The feed hopper has about a 300 kg capacity. The salt flow rate onto conveyor is regulated and controlled using a slide valve. Fitted rubber curtains (very flexible) on three sides form the salts into a band and stop it from spillage. The solution of KIO_3 is then stored in two different 200 L stock tanks made of polyethylene with discharge valves situated at bottommost part to allow the filling of two different 25 L feed bottles, attached to make sure constant circulations of the solution come from main tanks to feed the bottles. The solution constantly releases at the needed rate onto salt crystals. At this point, the iodized salt empties into a hopper for final discharging and collections in bags. IGN recommends that for continuous operation the hopper should have a twin spout with a diversion valve (Iodine Global Network, 2020; FAO, 1995).

9.10.4 Methods for Iodization of Salt at Village Level

Salt iodization at village level is valuable, and the simplest method is dry mixing of NaCl and KI or KIO_3. For a 1:20,000 (50 ppm) iodization level, addition of 84 milligram of KIO_3 or KI for every kilogram of NaCl is needed. Quantities of KIO_3 included are minute in comparison to the salt volume, so even though with maintained pouring by hand and mixing back and forth, the KIO_3 distributions within the NaCl will remain fairly uneven. KIO_3 and salt could be mixed with a big wooden spoon, including "back and forth" pouring between two vessels if possible. However, the process is unreliable and not efficient, and not recommended except as the only temporary alternative measure when there is no other available means to achieve iodization. As the KIO_3 amount is too small, it has to be made available in pre-weighed packs because accurate weighing balances are not often available in villages. Other village methods and techniques exist. Overall, the village iodization technique costs almost nothing in terms of equipment. They allow villages to be in control of their own iodization of salt, to reduce ID in rural areas (FAO, 1995; Winger et al., 2008).

9.11   DIFFERENT FOODS AND BEVERAGES BEING FORTIFIED

9.11.1   DRINKING WATER

Drinking water can be a useful means for iodine fortification, as water is consumed regularly every day. The major limitation of water compared with table salt is that the sources of all drinking water are ubiquitous and too many, and as such, iodization control would be a big challenge. In addition, iodine stability in water is limited – not more than a day. Although it is technically more difficult to use drinking water for iodine fortification than using salt, there are some circumstances where iodization of water may be suitable for ID and/or IDD correction. The simplest technique to iodizing water is by adding concentrated solution of iodine as potassium iodate or iodide in drop still specific concentration in the drinking water is attained (Santos et al., 2019). This method of iodizing water has been applied in schools in northern parts of Thailand (WHO, 2006). For open wells or hand pumps, porous polymer containing iodine could be introduced to drinking water supplies. Porous vessels allow slow discharge of solution of KO_3 into drinking water supplies, although such vessels have short "shelf-life". They must be replaced every year. Water iodization has been used in some countries such as Sudan, Sicily, the Central African Republic, Mali, Thailand, and Malaysia. Another approach is by diverting pipe water via canister with iodine crystals, and reintroduce iodized water into stream of "main water." Many other methods have been used in other countries including China (FAO, 2006; Santos et al., 2019).

9.11.2   MILK

Milk enriched with iodine has been vital in preventing and controlling ID in many countries (Santos et al., 2019). Milk enriched with iodine is an essential source of iodine in some nations in northern parts of Europe, the US, and the UK (WHO, 2006). Iodizing bread was stopped in Tasmania as soon as other iodine sources, especially milk and dairy products, became available. Milk containing iodine is a main source in some countries due to iodophors usage in the dairy industries.

9.11.3   BREAD

Technically, bread is an adequate means for iodization and has shown effectiveness in ensuring continuous dietary iodine supply (Santos et al., 2019). Bread has been used in some European countries, Russia, and Tasmania. In the Netherlands, the main carrier for iodine is the baker's salt included in bread, which has been iodized since 1942. Recently, the amount of KI of the "Dutch baker's salt" has increased. please write the old and present level (Santos et al., 2019; WHO, 2006).

9.11.4   OTHER IODIZED FOODS

The feasibility of using iodized sugar been assessed in Sudan. In South-East Asia, fish sauce has also been assessed for iodization (WHO, 2006). Chicken eggs are a good source of dietary iodine.

Modifying the feed formulations of poultry feed can fortify the accumulation of iodine in the egg (Sumaiya et al., 2016). Iodated vegetable oil, such as sunflower oil, is low cost and readily available source option for dietary iodine (Sturza, 2006).

9.12   NATIONAL AND INTERNATIONAL GUIDELINES

Many national and international guidelines for iodine fortification of foods have been systematized.

The WHO developed informed and evidence-based recommendations. The steps in these recommendations include identification of priority outcomes and questions; evidence retrieval; evidence assessment and synthesis; recommendation formulation and setting research priorities; making plans for dissemination and implementation; and impact evaluation and guideline updating. Food-grade salts in food processing and household have to be iodizes for preventing/controlling IDD in populations with stable and emergency settings (WHO, 2014; Santos et al., 2019). Concentrations for iodization in food-grade salt have been mentioned in Table 9.7.

Most of the national and international guidelines are as follow

a) The recommendation acknowledges that salt iodization is compatible with salt reduction. The monitoring of intakes of sodium and iodine at regional and national levels is required for adjusting salt iodization as needed, with reference to the observed salt consumption pattern in a given population. This will ensure that people consume iodine in sufficient amount despite the reduction in salt intakes (WHO, 2014; Santos et al., 2019).
b) Nationwide salt consumption distribution must offer vital guidance, control, and regulation for the iodine concentration in salt (table salt).
c) Sufficient iodine has to be provided to most of the population, including people with lowest intakes of salt, while simultaneously preventing excess supply of iodine to individuals with constant high intake of salt.
d) Iodine concentration may require adjustment by responsible national authorities for implementing and monitoring of USI, regarding their own data on dietary salt intake.

TABLE 9.7

Suggested Concentrations for Iodization in Food-grade Salt

Average iodine to be added in salt(mg/kg)	Estimated daily consumption of salt[a], g/day
65*	3*
49*	4*
39*	5*
33	6
28	7
24	8
22	9
20	10
18	11
16	12
15	13
14	14

Source: WHO, 2014

[a] include table salt consumption and the consumption of salt from processed food; *correspond to the WHO salt reduction guidelines.

e) If ID is found in pregnant women, 220 µg (may be up to 250 µg) iodine per day is recommended. Other interventions including the intake of iodine supplements can be considered. Salt intake correlates with calorie intake; women who are pregnant often have an increase energy intakes (World Health Organization, 2014; Santos et al., 2019).

f) Iodized salt should be available to all members in a population after 12 months old. Those below 12 months are believed to be covered by breastfeeding or iodized infant formula milk. Salt addition to foods consumed by children below 12 months requires regulation to avoid excessive or inadequate sodium or iodine consumption.

g) Salt iodization policies and its reduction to less than 5 g/day can go hand-in-hand, are inexpensive, and have considerable benefits to public health. Salt iodization should not warrant promoting the intake of salt to the public.

h) Salt iodizing has to be properly regulated and controlled by governments in harmony with other country programs, including the local level programs, to make sure iodized salt is safely delivered in acceptable range of doses.

i) Monitoring the quality of food-grade salt is vital to ensure the safety and effectiveness of iodization. Monitoring of the urinary iodine concentration (UIC) is essential to spot any ID and to identify excess intakes of iodine and prevent the associated health risks, through adjusting levels of iodization accordingly, as a monitoring tool. Nations should establish "iodine loss from the iodized salt in local conditions of climate, production, packaging and also storage." Consequently, iodine losses could vary greatly and may influence the additional iodine quantities that have to be added at factories.

j) Cogent legislation has to be established for food manufacturers, including food distributors, particularly where processed food and meal outside households are the major source of table salt. In addition to proper iodization of salt, the legislation has to also cover salt contents in food products made in food industries.

k) Country programs should appropriately align with the culture of the target populations, in order to facilitate the acceptability, adoption, and sustainability of the intervention.

l) Regular monitoring and evaluation (M&E) can detect the barriers which may affect equal access to iodized salt and as a result preserve health inequity (WHO, 2014).

m) The guidelines also stated that establishing efficient system for ongoing and routine data collection, including the measures for quality control and quality assurance as well as household use of table salt and measures of program performance, is critical to make sure that iodized salt programs are efficacious and sustainable (WHO, 2014; Santos et al., 2019).

9.13 THE IMPACT OF FORTIFICATION ON FOOD TEXTURAL AND SENSORY PROPERTIES

The directly inclusion of KO_3 or KI to foods and food products has not received enough attention in the field of research. The majority of the study reported in this direction involved replacing non-iodized salt by adding iodized salt in standards for formulation of foods. It is insufficient to justify evidence of adverse reactions of iodine fortification in foods from the studies done so far. No adverse effects of iodized salt on bottled olives, canned green beans, bulk sauerkraut, canned tomato juice, and yellow sweet corn has been reported (Winger et al., 2008). Outside the change in flavor in tomatoes for high iodine mixture (200 ppm), no noticeable changes in quality, including functional and sensory properties has been reported. Iodized salt used to produce cheese in Switzerland had no noticeable impact on taste, smell, appearance, and quality of the cheese. In Germany, a study used iodized salt in many processed meat products and found no changes in the sensory properties and nitrite curing characteristics; there was no additional formation of nitrosamine in the meat products (Santos et al., 2019).

Winger et al. (2008) on the other hand, reported that "12.7 ppm iodine from potassium iodate caused proteolysis in casein during UHT milk processing." It was not reported with 6.3 ppm iodine, indicating that iodine concentration is important. This iodate content was higher than expected following iodized salt addition. The report stated that there was "interaction between iodide and cresol of lemon flavorings in a cake mixture, resulting in an undesirable flavor." To avoid this issue, use non-iodized salt (Winger et al., 2008).

9.13.1 CHALLENGES WITH CURRENT ADOPTED METHODS OF IODINE FORTIFICATION

Determination of most species of iodine in foods is very complicated and often needs significant investment in the technical equipment and daily operating expenditure. The iodine retention in industrial processed foods has not received sufficient attention. The WHO stated in 2001 that in circumstances where iodine loss from salt is 20% from production to household, another 20% is lost in cooking before consumption, and average salt intake is 10 g per person per day. Iodine concentration at the stage of manufacture should be within the range of 20 to 40 mg of iodine per kg of salt in order to provide 150 mg of iodine per person per day, and the iodine should be added as potassium (or sodium) (WHO, 2001).

Iodine losses were observed from 3% to 67% with an average of 6%– % in 50 distinct recipes in India prepared in hospital kitchens using diverse processes, while the result variability was very significant (Goindi et al., 1995). Some foods, such as carrot, "Kadhi," and spinach, had losses of up to 50%. Cooking procedures also had a significant impact on iodine loses, shallow frying caused a 27% loss, steaming caused a 20% loss, boiling caused a 37% average loss, pressure cooking caused a 20% loss, roasting caused a 6% loss, and deep frying caused a 6% loss (20% loss) of iodine from food (Winger et al., 2008).

9.14 IODINE BIOAVAILABILITY

No recent or sufficient data is available on bioavailability of dietary iodine from foods. All the methods for testing iodine bioavailability require radioisotopes. Radioisotope techniques cannot be used (ethically) for the assessment of iodine availability in humans, and therefore no accurate figures exist. However, useful guides to iodine bioavailability has been reported in early studies. Two features to iodine bioavailability include (1) iodine absorption from bloodstream by thyroid gland and (2) iodine absorption in the gastrointestinal (GI) tract (Santos et al., 2019; Winger et al., 2008). Inorganic iodide is totally and readily absorbed from GI tract. For absorption, protein-bound iodine and iodate are believed to be reduced to iodide. However, the reducing agent and reduction site both are unclear (Winger et al., 2008). While the absorption of iodine from the GI tract has been shown to be 100%, iodine absorption by the thyroid gland is around 10 to 15%. A study reported that with a 100 mg dose of iodine radioisotope, around 20% was absorbed by thyroid gland (Santos et al., 2019).

Early studies showed high absorption of iodine from many diets except water cress (Winger et al., 2008). Evidence suggests that iodine from plant sources may have lesser rate of absorption, mostly caused by reduced iodine release from plant matrix during digestion process. Excessive cooking of "hijiki", a Japanese seaweed, resulted in increased levels of iodine in urine in men by 53% of dose in contrast to 10% for uncooked material (Santos et al., 2019).

There is almost no information in the literature on the true absorption rate of iodine from the gastrointestinal tract. The present information, which suggests 100% absorption from GIT, is based on studies conducted in the first half of the last century using small sample sizes and only basic analytical methodology. There is less information on the interactions between iodine and food components with respect to their impact on bioavailability from processed foods. There are known reactions, particularly with mineral salts where bioavailability appears to be significantly reduced (Aquaron et al., 2002).

The existing literature tends to suggest that iodine and its salts when used as iodized salt to replace added non-iodized salt will have limited impact on food quality. However, these studies were focused on very few food systems and at low iodine concentrations, often not directly measured.

While the use of iodized salt to replace non-iodized salt is one method of adding iodine to foods, the levels of added iodine are very low. There is no data available to predict the outcome of the use of higher levels of iodine (for example, 25–65 mg iodine per kg food, as is required for salt). Literature evidence suggests there will be reactivity between iodine and some food components when the level of iodine is increased above 100 ppm (Aquaron et al., 2002; Meletis, 2011).

9.15 FUTURE PROSPECTS FOR IODINE FORTIFICATION

Concerns of overindulgence of salt intake and the concomitant deleterious health effects is resulting in lower salt consumption and low salt formulations for many processed foods. Lower iodized salt intake equates to lower iodine intake.

Alternative strategies to simply using iodized salt are needed to ensure proper iodine intake

a) Bio-fortification of iodine in crop plants provides an attractive opportunity to increase iodine intake in humans and to prevent and control IDD. In soil, iodine was able to be absorbed and concentrated to 89.4 and 144.0 µg/100 for potato tuber and tomato fruit (Caffagni et al., 2012).

b) Soilless cultivation of crops in hydroponics is an increasingly used technology for food production. Hydroponically grown tomatoes accumulated up to 2423 µg/100 of iodine (Caffagni et al., 2012).

c) Greater consumption of seafood and marine aquatic plants is a natural direction to nutritional adequacy. Foods of marine origin have higher iodine content because marine plants and animals concentrate iodine from the saltwater seawater (Carlsen et al., 2018).

d) Salt tolerant (halophytes) plants are increasingly being researched for food as climate change increases the sea levels in coastal areas, creates saline semi-deserts, mangrove swamps, marshes, and sloughs (Biosalinity awareness project, 2020; Centofanti and Bañuelos, 2019).

9.16 CONCLUSIONS

In conclusion, there has been significant advancement in the most recent decade in the elimination of iodine deficiency. Improved iodine nourishment is underlined by recommendations of the WHO that suggest salt iodization supplemented with iodine supplementation in remote locations where obtaining iodized salt is difficult. This mirrors the endeavors made by nations to actualize viable IDD control programs and is confirmation of the fruitful coordinated effort between all parties involved in IDD control, specifically wellbeing specialists and the salt business. Having said that, every effort should be made to guarantee that projects continue to cover populations at risk if the objective of wiping out IDD is to be reached. Current iodine measurements lack gauges dependent on UI, which would provide standardization of future worldwide assessments. The current challenge is to improve the nature of the information so as to ensure the provision of appropriate and relevant mechanisms that can allow precise and fast progress.

CONFLICT OF INTEREST

There is no conflict of interest between the authors of this manuscript.

REFERENCES

Abraham, Guy E. "The safe and effective implementation of orthoiodo supplementation in medical practice." *The Original Internist* 11, no. 1 (2004): 17–36.

Aceves, Carmen, Brenda Anguiano, and Guadalupe Delgado. "Is iodine a gatekeeper of the integrity of the mammary gland?" *Journal of Mammary Gland Biology and Neoplasia* 10, no. 2 (2005): 189–196.

Andersson, Maria, Bahi Takkouche, Ines Egli, Henrietta E. Allen, and Bruno de Benoist. "Current global iodine status and progress over the last decade towards the elimination of iodine deficiency." *Bulletin of the World Health Organization* 83 (2005): 518–525.

Andersson, Maria, Bruno de Benoist, and Lisa Rogers. "Epidemiology of iodine deficiency: Salt iodisation and iodine status." *Best Practice & Research Clinical Endocrinology & Metabolism* 24, no. 1 (2010): 1–11.

Aquaron, Robert, Francois Delange, Pascal Marchal, Vincent Lognoné, and Leon Ninane. "Bioavailability of seaweed iodine in human beings." *Cellular and Molecular Biology (Noisy-le-Grand, France)* 48, no. 5 (2002): 563–569.

Bagchi, N., T. R. Brown, and R. S. Sundick. "Thyroid cell injury is an initial event in the induction of autoimmune thyroiditis by iodine in obese strain chickens." *Endocrinology* 136, no. 11 (1995): 5054–5060.

Biban, Bianca Georgiana, and Corina Lichiardopol. "Iodine deficiency, still a global problem?" *Current Health Sciences Journal* 43, no. 2 (2017): 103.

Biosalinity Awareness Project. "Liating of Halophyte & salt-tolerant plants." 2020. http://www.biosalinity.org/salt-tolerant_plants.htm.

Caffagni, Alessandra, Nicola Pecchioni, Pierluigi Meriggi, Valerio Bucci, Emidio Sabatini, Nazareno Acciarri, Tommaso Ciriaci et al. "Iodine uptake and distribution in horticultural and fruit tree species." (2012): 229–236.

Cann, Stephen A., Johannes P. Van Netten, and Christiaan van Netten. "Hypothesis: Iodine, selenium and the development of breast cancer." *Cancer Causes & Control* 11, no. 2 (2000): 121–127.

Carlsen, Monica H., Lene F. Andersen, Lisbeth Dahl, Nina Norberg, and Anette Hjartåker. "New iodine food composition database and updated calculations of iodine intake among Norwegians." *Nutrients* 10, no. 7 (2018): 930.

Carpenter, K. J. "David Marine and the problem of goiter." *The Journal of Nutrition* 135, no. 4 (2005): 675–680.

Centofanti, Tiziana, and Gary Bañuelos. "20 practical uses of halophytic plants as sources of food and fodder." *Halophytes and Climate Change: Adaptive Mechanisms and Potential Uses* (2019): 324.

Chatin, Adolphe. "Un fait dans la question du goître et du crétinisme." Impr. W. Remquet, 1853.

Choudhry, Hani, and Md Nasrullah. "Iodine consumption and cognitive performance: Confirmation of adequate consumption." *Food Science & Nutrition* 6, no. 6 (2018): 1341–1351.

Dasgupta, Purnendu K., Yining Liu, and Jason V. Dyke. "Iodine nutrition: Iodine content of iodized salt in the United States." *Environmental Science & Technology* 42, no. 4 (2008): 1315–1323.

Dietary Supplement Fact Sheet: "Iodine." 2020. http://ods.od.nih.gov/factsheets/ Iodine-HealthProfessional/

Dold, Susanne, Michael B. Zimmermann, Tomislav Jukic, Zvonko Kusic, Qingzhen Jia, Zhongna Sang, Antonio Quirino et al. "Universal salt iodization provides sufficient dietary iodine to achieve adequate iodine nutrition during the first 1000 days: A cross-sectional multicenter study." *The Journal of Nutrition* 148, no. 4 (2018): 587–598.

Eastman, Creswell J., and Michael B. Zimmermann. "The iodine deficiency disorders." In Endotext [Internet]. MDText. com, Inc., 2018.

FAO. Guidelines for soil description. Rome: Food and Agriculture Organization of the United Nations, fourth edition, 2006, 97 p.

Felig, Philip and Frohman, Lawrence A. "Endemic Goiter." In *Endocrinology & Metabolism*. McGraw-Hill Professional, 2001.

Food and Agriculture Organization. "Food fortification: Technology and quality control." Report of an FAO Technical Meeting, Rome, Italy, 20–23 November. By Food and Agriculture Organization of the United Nations, 1995.

FSANZ, Consumers' Awareness. *Attitudes and Behaviours Towards Food Fortification in Australia and New Zealand*. Canberra: FSANZ, 2013.

Fox, E. L. "A case of myxoedema treated by taking extract of thyroid by the mouth." *British Medical Journal* 2, no. 1661 (1892): 941.

Global Fortification Data Exchange. "Fortification legislation." 2020. https://fortificationdata.org/interactive-map-fortification-legislation/.

Goindi, G., M. G. Karmarkar, U. Kapil, and J. Jagannathan. "Estimation of losses of iodine during different cooking procedures." *Asia Pacific Journal of Clinical Nutrition* 4, no. 2 (1995): 225–227.

Institute of Medicine, Foods & Nutrition Board. *Dietary Reference Intakes for Vitamin A, Vitamin K, Arsenic, Boron, Chromium, Copper, Iodine, Iron, Manganese, Molybdenum, Nickel, Silicon, Vanadium, and Zinc*. Washington, DC: National Academy Press, 2001.

Iodine Global Network. "Global scorecard of iodine nutrition in 2017 in the general population and in pregnant women (PW)." Zurich: IGN, 2017.

Iodine Global Network. "Global scorecard of iodine nutrition in 2020 in the general population based on school-age children (SAC)." *IDD Newsletter*, 2020a.

Iodine Global Network. "How is salt iodized?" IDD, 2020b. https://www.ign.org/how-salt-is-iodized---more-detail.htm. Accessed 10 July 2020.

Kelly, Francis C. "Iodine in medicine and pharmacy since its discovery—1811–1961." (1961): 831–836.

Kimball, O. P., and David Marine. "The prevention of simple goiter in man: Second paper." *Nutrition* (1992): 200–204.

Krela-Kaźmierczak, Iwona, Agata Czarnywojtek, Kinga Skoracka, Anna Maria Rychter, Alicja Ewa Ratajczak, Aleksandra Szymczak-Tomczak, Marek Ruchała, and Agnieszka Dobrowolska. "Is there an ideal diet to protect against iodine deficiency?" *Nutrients* 13, no. 2 (2021): 513.

Lee, Stephanie L., Elizabeth N. Pearce, and Sonia Ananthakrishnan. "Iodine deficiency." *Medscape*, 2017. Retrieved 06 July 2020.

Leung, A. M., L. E. Braverman, and E. N. Pearce. "History of US iodine fortification and supplementation." *Nutrients* 4, no. 11 (2012): 1740–1746.

Marine, David, and O. P. Kimball. "The prevention of simple goiter in man: A survey of the incidence and types of thyroid enlargements in the schoolgirls of Akron (Ohio), from the 5th to the 12th grades, inclusive—The plan of prevention proposed." *The Journal of Laboratory and Clinical Medicine* 3, no. 1 (1917): 40–48.

Meletis, Chris D. "Iodine: Health implications of deficiency." *Journal of Evidence-Based Complementary & Alternative Medicine* 16, no. 3 (2011): 190–194.

Monahan, Mark, Kristien Boelaert, Kate Jolly, Shiao Chan, Pelham Barton, and Tracy E. Roberts. "Costs and benefits of iodine supplementation for pregnant women in a mildly to moderately iodine-deficient population: A modelling analysis." *The Lancet Diabetes & Endocrinology* 3, no. 9 (2015): 715–722.

Patrick, Lyn. "Iodine: Deficiency and therapeutic considerations." *Alternative Medicine Review* 13, no. 2 (2008).

Pizzorno Joseph, E., and Michael T. Murray. "Textbook of natural medicine." (2012): 1371.

Redisch, Walter, and William H. Perloff. "The medical treatment of hyperthyroidism." *Endocrinology* 26, no. 2 (1940): 221–228.

Rosenfeld, Louis. "Discovery and early uses of iodine." *Journal of Chemical Education* 77, no. 8 (2000): 984.

Santos, Joseph Alvin R., Anthea Christoforou, Kathy Trieu, Briar L. McKenzie, Shauna Downs, Laurent Billot, Jacqui Webster, and Mu Li. "Iodine fortification of foods and condiments, other than salt, for preventing iodine deficiency disorders." *Cochrane Database of Systematic Reviews* 2 (2019).

Southern, A. P., and S. Jwayyed. "Iodine toxicity." 9 January 2022. In: StatPearls [Internet]. Treasure Island (FL): StatPearls Publishing. PMID: 32809605.

Starr, Paul, Henry P. Walcott, Harold N. Segall, and James H. Means. "The effect of iodin in exophthalmic goiter." *Archives of Internal Medicine* 34, no. 3 (1924): 355–364.

Sturza, Rodica. "Iodination of vegetable oil as a method for correcting iodine deficiency." *Chemistry Journal of Moldova* 1, no. 1 (2006): 65–72.

Sumaiya, S., S. Nayak, R. P. S. Baghel, A. Nayak, C. D. Malapure, and R. Kumar. "Effect of dietary iodine on production of iodine enriched eggs." *Veterinary World* 9, no. 6 (2016): 554.

Taylor, Peter N., Onyebuchi E. Okosieme, Colin M. Dayan, and John H. Lazarus. "Impact of iodine supplementation in mild-to-moderate iodine deficiency: Systematic review and meta-analysis." *Eur J Endocrinol* 170, no. 1 (2014): 1–15.

Teas, Jane, Sam Pino, Alan Critchley, and Lewis E. Braverman. "Variability of iodine content in common commercially available edible seaweeds." *Thyroid* 14, no. 10 (2004): 836–841.

Todd, Charles H. "Hyperthyroidism and other thyroid disorders: A practical handbook for recognition and management." World Health Organization, 1998.

Trumpff, Caroline, Jean De Schepper, Jean Tafforeau, Herman Van Oyen, Johan Vanderfaeillie, and Stefanie Vandevijvere. "Mild iodine deficiency in pregnancy in Europe and its consequences for cognitive and psychomotor development of children: A review." *Journal of Trace Elements in Medicine and Biology* 27, no. 3 (2013): 174–183.

UNICEF Data. "Iodized salt data." 2020. https://data.unicef.org/resources/dataset/iodized-salt-consumption/

Unit, Nutrition, and World Health Organization. "Recommended iodine levels in salt and guidelines for monitoring their adequacy and effectiveness." No. WHO/NUT/96.13. World Health Organization, 1996.

US Department of Agriculture (USDA), Agricultural Research Service. "Food data central." 2019. https://fdc.nal.usda.gov/.

US Food and Drug Administration. "Code of federal regulations." CFR 21, Sections 184.1634 and 184.1265 Revised. 2009. (http://www.accessdata.fda.gov/SCRIPTs/cdrh/cfdocs/cfCFR/CFRSearch.cfm).

US Food and Drug Administration, HHS. "Food labeling: Revision of the nutrition and supplement facts labels. Final rule." *Federal Register* 81, no. 103 (2016): 33741.

Vanderpump, Mark PJ, John H. Lazarus, Peter P. Smyth, Peter Laurberg, Roger L. Holder, KristienBoelaert, Jayne A. Franklyn, and British Thyroid Association UK Iodine Survey Group. "Iodine status of UK schoolgirls: A cross-sectional survey." *The Lancet* 377, no. 9782 (2011): 2007–2012.

Völzke, Henry. "The krakow declaration on iodine: Tasks and responsibilities for prevention programs targeting iodine deficiency disorders." *European Thyroid Journal* 7, no. 4 (2018): 201–204.

Vos, Theo, Abraham D. Flaxman, Mohsen Naghavi, Rafael Lozano, Catherine Michaud, Majid Ezzati, Kenji Shibuya et al. "Years lived with disability (YLDs) for 1160 sequelae of 289 diseases and injuries 1990–2010: A systematic analysis for the Global Burden of Disease Study 2010." *The Lancet* 380, no. 9859 (2012): 2163–2196.

WHO, UNICEF. "ICCIDD. Global prevalence of iodine deficiency disorders." Micronutrient Deficiency Information System (MDIS) Working Paper no. 1. Geneva: MDIS/WHO, 1993.

Winger, Ray J., Jürgen König, and Don A. House. "Technological issues associated with iodine fortification of foods." *Trends in Food Science & Technology* 19, no. 2 (2008): 94–101.

World Health Organization. "Assessment of iodine deficiency disorders and monitoring their elimination: A guide for programme managers." No. WHO/NHD/01.1. World Health Organization, 2001.

World Health Organization. Iodine status worldwide: WHO global database on iodine deficiency, 2004.

World Health Organization. "Guidelines on food fortification with micronutrients." Edited by Lindsay Allen, Bruno de Benoist, *Omar Dary and Richard Hurrell.* WHO Library Cataloguing-in-Publication Data. WHO/FAO, 2006.

World Health Organization. "Guideline: Fortification of food-grade salt with iodine for the prevention and control of iodine deficiency disorders." 2014.

World Health Organization. United Nations Children's Fund & International Council for the Control of Iodine Deficiency Disorders. Assessment of iodine deficiency disorders and monitoring their elimination." World Health Organization, 2020.

Yaman, ArzuKutlu, FatmaDemirel, BahriErmiş, and I. EtemPişkin. "Maternal and neonatal urinary iodine status and its effect on neonatal TSH levels in a mildly iodine-deficient area." *Journal of Clinical Research in Pediatric Endocrinology* 5, no. 2 (2013): 90.

Zimmermann, M. B. "Iodine and the iodine deficiency disorders." In *Present knowledge in nutrition* (pp. 429–441). Academic Press, 2020.

Zimmermann, M. B. "The effects of iodine deficiency in pregnancy and infancy." *Paediatric and Perinatal Epidemiology* 26 (2012): 108–117.

Zimmermann, Michael B. "Iodine deficiency." *Endocrine Reviews* 30, no. 4 (2009): 376–408.

Zimmermann, Michael B. "Research on iodine deficiency and goiter in the 19th and early 20th centuries." *The Journal of Nutrition* 138, no. 11 (2008): 2060–2063.

10 Zinc, Calcium, Fluoride, and Selenium Fortification

Sakshi Singh, Syed M. Rahman,
Kulsum Jan, Khalid Bashir, and P. Karthik

10.1 INTRODUCTION

The obligatory role of nutrients in human health is well debated in the literature. Increased awareness on health has furthered the development of nutrient-rich food commodities, keeping in view the world hunger index. The Hidden Hunger Index is an index of three deficiency prevalence estimates: *preschool* children affected by stunting, *anemia* due to iron deficiency, and *vitamin-A* deficiency. The World Health Organization estimated that more than 2 billion people are directly or indirectly affected by micronutrient deficiency across the globe. In the past two decades, a greater proportion of the population relies on processed foods due to rapid urbanization and increasing household purchasing power, which has caused micronutrient enrichment to become increasingly popular in low to middle-income countries (Lotfi et al. 1996; Shankar 2020; Baskar and Prathap 2020). Deficiency of minerals like iron, zinc, calcium, fluoride, selenium, or any other micronutrient is common and widespread in underdeveloped or low to middle-income countries (LMICs), while mandatory food fortification has been used as a strategy to prevent micronutrient deficiencies in high-income countries (HICs). These minerals are majorly classified into two groups (**Table 10.1 & Table 10.2**). A diverse diet is required to ensure adequate micronutrient intake through diet alone (by including fortified & processed foods in the diet). On one hand, fruits, vegetables, meat, dairy products, pulses, seafood, nuts, and other seeds are micronutrient-rich commodities. On the other hand, food commodities including cereals, roots, and tubers are primarily consumed as the primary source of energy, remaining deficient in micronutrients (**Figure 10.1**) (**Wessells, Singh, and Brown 2012; Polekkad et al. 2021; Vatandoust et al., 2023**). The first depiction of fortification traces back to 1920 when iodized salt was introduced to curtail the iodine deficiency in the European and North American population (Leung, Braverman, and Pearce 2012; Hollowell et al. 1998). The deficiency becomes more prevailing in the population of LMICs as a result of high consumption of starchy food crops that are poor in micronutrients and high in phytates. The underlying problem lies with the food systems that are not delivering a nutritionally balanced diet to LMICs, and that may be the reason behind food fortification being ranked as one of the most cost-effective development priorities by Copenhagen Consensus in 2008 and 2012. It's the need of the hour to look into the health issues of the upcoming generations who will shape the future of our country.

To demonstrate good health, economic, and social benefits, it's necessary not to compromise the physical and cognitive capacity of our future generations. In some countries, food fortification is the only cost-effective strategy to treat deficiencies. Literature describes three major strategies: (i) diet diversification, (ii) biofortification, and (iii) food fortification (Maurya, Bashir, and Aggarwal 2020; Maurya et al. 2023). Diet diversification deals with the inclusion of micronutrient-rich food commodities in the daily diet while biofortification deals with improving the micronutrient content in crops using crop selection, plant breeding, mineral fertilizer and genetic engineering. Biofortification is the process by which the nutritional value of food crops is improved. Biofortification projects mainly concentrate on boosting essential vitamins and minerals like iron, zinc, and provitamin

TABLE 10.1

Major Minerals and Their Functions in the Body

Minerals	Sources	Recommended daily allowances	Functions	Problems associated with deficiency
Calcium	Dairy products, dark green leafy vegetables, blackstrap molasses, nuts, brewer's yeast, and some fish	Sedentary, Moderate, Heavy M 1000mg/d Sedentary, Moderate, Heavy: F 1000mg/d Pregnant women 1000mg/d Lactation (0-6m) 1200mg/d Lactation (6–12m) 1200mg/d Children (4–6y) 550mg/d Children (7–9y) 650mg/d	Bone structure and health, as well as nerve and muscle functions, especially cardiac function	Slow growth, weak and brittle bones
Phosphorous	Meat, milk	Sedentary, moderate, heavy M – 600 mg Sedentary, moderate, heavy F – 600 mg Pregnant women – 1200 mg Lactation – 1200 mg Children (4–6y) 600 mg Children (7–9y) 600 mg	Bone formation, metabolism, ATP Production	Rare
Sodium	Table salt, milk, beets, celery, processed foods	Sedentary, moderate, heavy M – 2100mg Sedentary, moderate, heavy F – 3225 mg Children (4–6y) 1010 mg	Blood pressure, blood volume, muscle and nerve function	Rare
Chloride	Most foods, salt, vegetables, especially seaweed, tomatoes, lettuce, celery, olives	Sedentary, moderate, heavy: M – 1800–2300 mg Sedentary, moderate, heavy: M – 1800–2300 mg Pregnant women – 2300 mg Lactation – 2300 mg Children (4–9y) 1900 mg	Balance of body fluids, digestion	Loss of appetite, muscle cramps
Potassium	Meat, some fish, fruits, vegetables, legumes, dairy products	Sedentary, moderately heavy M – 3750 mg/d Sedentary, moderate heavy F – 3225 mg/d Children (4–6y) 1550 mg/d	Nerve and muscle functions; acts as an electrolyte	Hypokalemia; weakness, fatigue, muscle cramping, gastrointestinal problems, cardiac problems
Magnesium	Whole grains, nuts, green leafy vegetables	Sedentary, moderately heavy M – 385 mg/d Sedentary, moderate, heavy F – 325 mg/d Pregnant women 385mg/d Lactation (0–6m) 325 mg/d Lactation (7–12m) 325mg/d Children (4–6y) 155mg/d Children (7–9y) 215mg/day	Enzyme activation, production of energy, regulation of other nutrients	Agitation, anxiety, sleep problems, nausea and vomiting, abnormal heart rhythms, low blood pressure, muscular problems

TABLE 10.2

Trace Minerals and Their Function in the Body

Mineral	Sources	Recommended dietary allowances	Functions	Problems associated with deficiency
Iron	Meat, poultry, fish, shellfish, legumes, nuts, seeds, whole grains, dark leafy green vegetables	Sedentary, Moderate, Heavy M – 19mg/d Sedentary, Moderate, Heavy F – 29mg/d Pregnant women – 40mg/d Lactation (0–6) – 23 mg/d Lactation (7–12) – 23mg/d Children (4–6y)- 11mg/d Children (7-9y)- 15mg/d	Transport of oxygen in blood, production of ATP	Anemia, weakness, fatigue
Zinc	Meat, fish, poultry, cheese, shellfish	Sedentary, moderate, heavy M – 17 mg/d Sedentary, moderate, heavy F – 13.2 mg/d Pregnant women –- 14.5 mg/d Lactation (0–6) –- 14 mg/d Lactation (7–12) –- 14 mg/d Children (4–6y) –- 4.5 mg/d Children (7–9y) – 5.9	Immunity, reproduction, growth, blood clotting, insulin and thyroid function	Loss of appetite, poor growth, weight loss, skin problems, hair loss, vision problems, lack of taste or smell
Copper	Seafood, organ meats, nuts, legumes, chocolate, enriched breads and cereals, some fruits and vegetables	2 mg/d	Red blood cell production, nerve and immune system function, collagen formation, acts as an antioxidant	Anemia, low body temperature, bone fractures, low white blood cell concentration, irregular heartbeat, thyroid problems
Iodine	Fish, shellfish, garlic, lima beans, sesame seeds, soybeans, dark leafy green vegetables	Sedentary, moderate, meavy M – 150 µg/d Sedentary, moderate, heavy F –- 150 µg/d Pregnant women – 250 µg/d Lactation (0–6) – 280 µg/d Lactation (7–12) – 280 µg/d Children (4–6y) – 120 µg/d Children (7–9y) – 120 µg/d	Thyroid function	Hypothyroidism: fatigue, weight gain, dry skin, temperature sensitivity
Sulfur	Eggs, meat, poultry, fish, legumes	-	Component of amino acids	Protein deficiency
Fluoride	Fluoridated water	Infants <0.7 mg/d Children – 3 mg/d Adult 4 mg/d	Maintenance of bone and tooth structure	Increased cavities, weak bones and teeth
Manganese	Nuts, seeds, whole grains, legumes	4 mg/d	Formation of connective tissue and bones, blood clotting, sex hormone development, metabolism, brain and nerve function	Infertility, bone malformation, weakness, seizures

(Continued)

TABLE 10.2 (CONTINUED)

Trace Minerals and Their Function in the Body

Mineral	Sources	Recommended dietary allowances	Functions	Problems associated with deficiency
Cobalt	Fish, nuts, leafy green vegetables, whole grains	5–8 mg/d	Component of B12	None
Selenium	Brewer's yeast, wheat germ, liver, butter, fish, shellfish, whole grains	40–50 µg/d	Antioxidant, thyroid function, immune system function	Muscle pain
Chromium	Whole grains, lean meats, cheese, black pepper, thyme, brewer's yeast	Sedentary, moderate, heavy M – 30 µg/d Sedentary, moderate, heavy F – 20 µg/d Pregnant women – 30µg/d Lactation – 44–45 µg/d Children – 20–30 µg/d	Insulin function	High blood sugar, triglyceride, and cholesterol levels

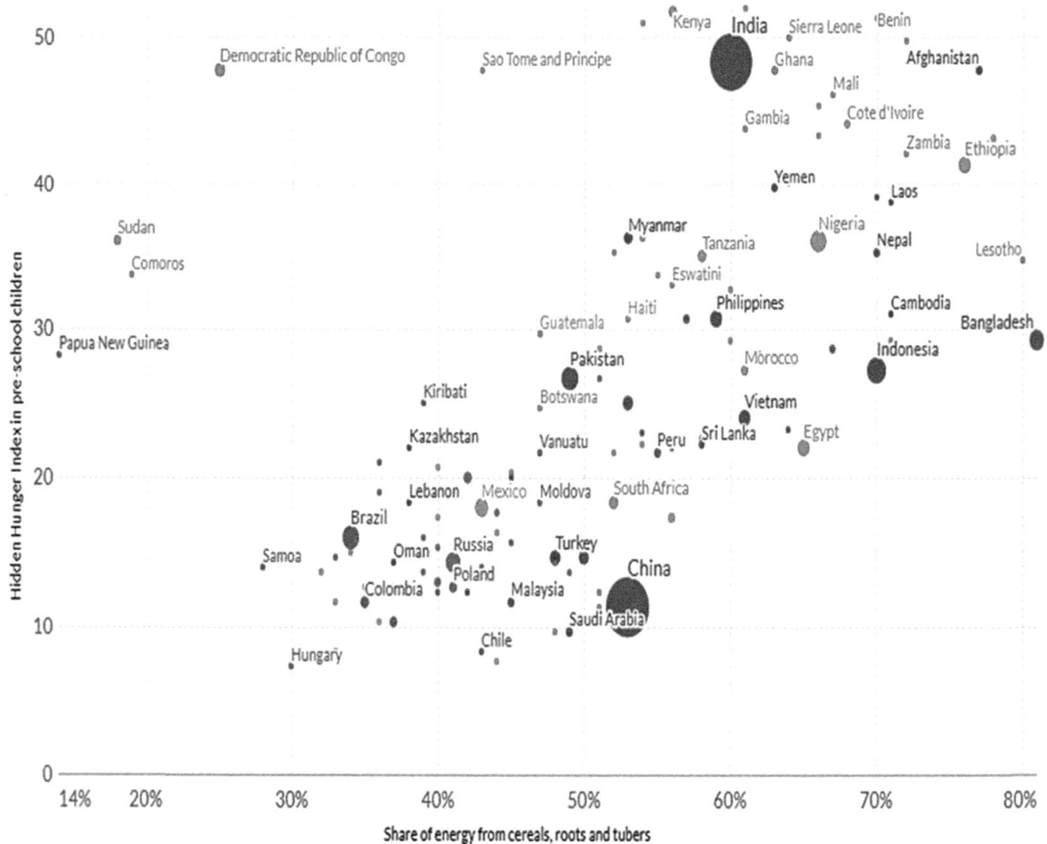

FIGURE 10.1 Country-wise Hidden Hunger Index in preschool children and share of energy from different foods (the size of dots represents the population) (Muthayya et al., 2013; Our World in Data, 2023).

A carotenoid in different food crops through plant breeding or by adding mineral fertilizer to the growing crops. Biofortification can enable food systems to deliver more nutritious foods cost-effectively, where foods are biofortified with amino acids and protein to enhance the nutrient value of food (Maurya, Bashir, and Aggarwal 2020; Buturi et al. 2021; Dhaliwal et al. 2021; Duborská et al., 2022). Examples of biofortification include iron biofortification of rice, beans, maize, and sweet potato; zinc biofortification of wheat, rice, beans, sweet potato, and corn; and Vitamin A biofortification of rice, sweet potatoes, corn, and cassava. On the contrary, in some countries, bovine hemoglobin concentrate is often used as a heme-iron fortification due to its high bioavailability (Polekkad et al. 2021; Vatandoust et al. 2023; Masuda et al. 2020).

Food fortification is the practice of deliberately adding of vitamins and minerals to foods that are consumed on a daily basis to increase their nutritional value. Fortification seems to be the most effective means to address micronutrient deficiency. On regulatory grounds, fortification falls into two major categories: (i) mandatory and (ii) voluntary fortification. Mandatory fortification mandates food manufacturers to add certain vitamins or minerals, or both, to specified foods. While in voluntary fortification, food manufacturers are free to fortify particular foods as per national food regulation or under distinct circumstances and also encouraged by the government. Depending on the population coverage, fortification is again classified into two classes (i) mass or large-scale fortification and (ii) in specific population groups. Mass fortification mandates the consumption of fortified food to the entire population irrespective of age, gender, ethnicity, geographical differences while specific fortification targets a defined section of the population such as infants, preschool children, or the elderly population (Wesley and Horton 2011; Flynn et al. 2003; Sumithra et al. 2013). To develop effective fortification methods in order to eradicate micronutrient deficiency, it is significant to review the prevalence of deficiency, its risk factorss, factor contributing to deficiency, benefits & consequence of vital minerals' fortification, potential food carrier systems, fortification methods, and the role of nanotechnology in food fortification.

10.2 ZINC

Around 17% of the total human population is deficient in Zinc with the highest prevalence in South Asia (30%) **(Figure 10.2) (Ritchie and Roser 2017; WHO 2019, 2006; Das, Khan, and Bhutta 2018)**.Inadequate consumption of critical micronutrients can lead to consequences that raise the risks of morbidity and mortality in people who are malnourished (Godswill et al. 2020). The RDA for zinc is presented in Table 10.2. Zinc (Zn) is a cofactor for metalloenzymes that are required for the proper functioning of the immunological, gastrointestinal, dermatologic, neurologic, and reproductive systems of the human body (Parisi and Vallee 1969; Mayo-Wilson et al. 2014). It can be found in a variety of foods, with the largest quantities found in meat, fish, nuts, seeds, legumes, and whole-grain cereals (Hunt 2002). The amount of zinc absorbed from each item varies depending on whether the food is of animal or plant origin, as well as whether the meal contains dietary components that influence zinc absorption (Roohani et al. 2013).

Approximately 155 million children under five years of age are stunted (height-for-age <-2 SD below the WHO Child Growth Standards median), with the vast majority living in Africa and Asia (Ohanenye et al. 2021). Various studies have also found that the deficit causes mortality from diarrhea, pneumonia, and malaria sequelae (Black et al. 2013). The phytate:zinc molar ratio, the overall amount of animal-based protein, and the amount of calcium (particularly calcium salts) in the diet were all considered in the bioavailability estimations (Morris and Ellis 1980). These dietary zinc bioavailability category estimates were then corrected to roughly 15%, 30%, and 50%, respectively (Moretti et al. 2014). Because of the low total zinc level in vegetable protein, overall absorption of zinc is lower than that of animal protein (Kristensen et al. 2006; Ball and Ackland 2000; Wessells and Hess 2007).

Zinc is important for cellular growth, cellular differentiation, and metabolism and its deficiency limits childhood growth and decreases resistance to infections. The central role of zinc in cell

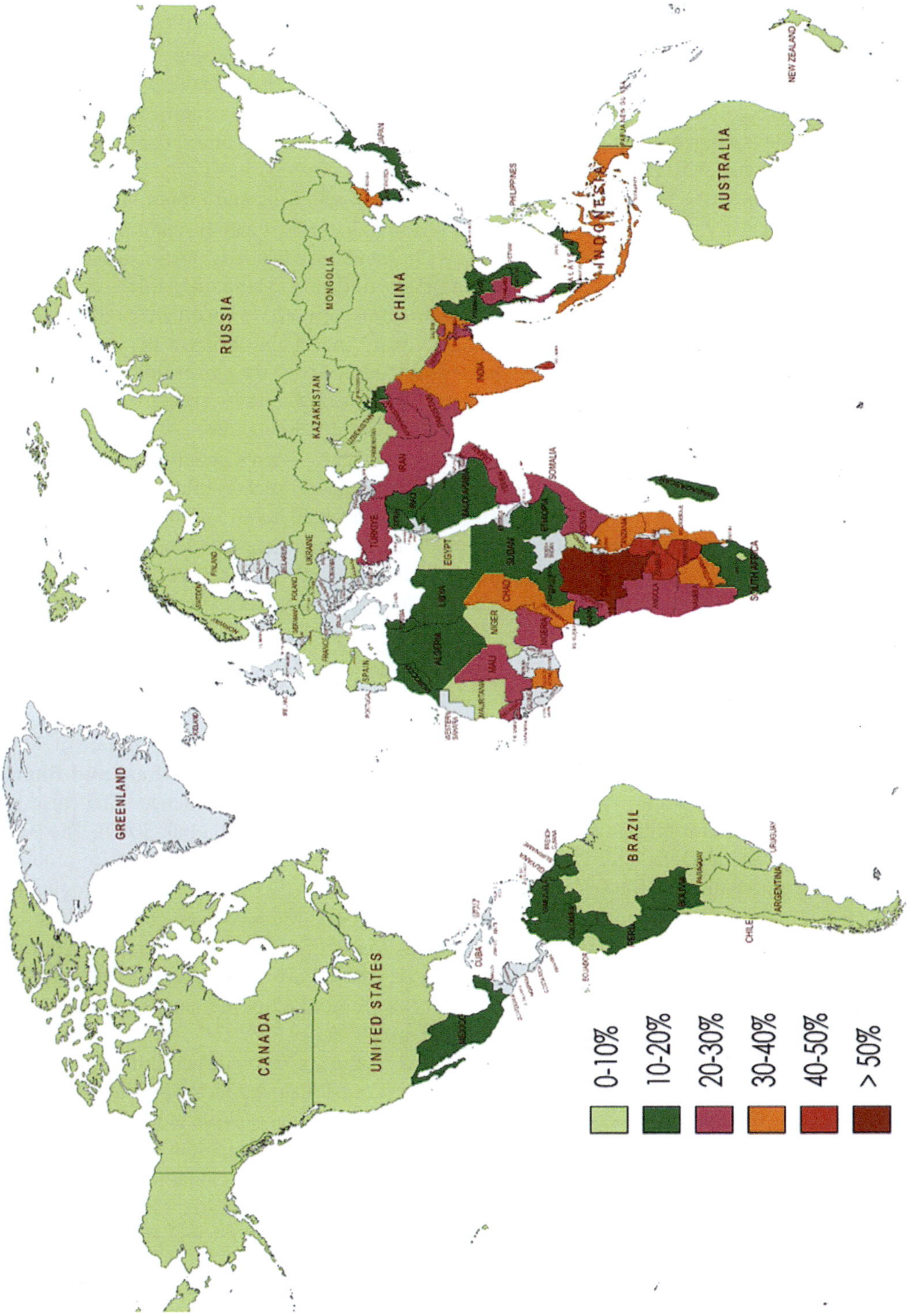

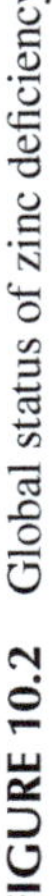

FIGURE 10.2 Global status of zinc deficiency.

division, protein synthesis, and growth means that an adequate supply of zinc is especially important for pregnant women (Lee et al. 2011). During pregnancy, zinc and other micronutrient deficiencies are common due to increased nutrient requirements of the mother and the developing fetus (Lee et al. 2011). These deficiencies can negatively impact pregnancy outcomes, including the health of the mother and newborn infant (Masters and Moir 1983; Durrani and Parveen, 2022).

10.2.1 HEALTH IMPACT OF ZINC DEFICIENCY

The lack of reliable and widely accepted indicators of zinc status of adequate sensitivity means that the global prevalence of zinc deficiency is uncertain (Hess et al. 2022). Those indicators that are available, such as zinc concentration in plasma and hair, detect changes in zinc status only in cases of severe deficiency and may fail to detect marginal deficiency. As suggested above, there are several good reasons to suspect that zinc deficiency is common, especially in infants and children (Deshpande, Joshi, and Giri 2013). Firstly, a high prevalence of low plasma zinc, which is a reasonable indicator of relatively severe depletion, has been observed in some population groups. Secondly, several randomized control trials have demonstrated that stunted children and/or those with low plasma zinc respond positively to zinc supplementation, a finding that suggests that zinc deficiency was a limiting factor in their growth (Polekkad et al., 2021; Das, Khan, and Bhutta 2018; Salgueiro et al., 2002).

Growth stunting affects about a third of children in less wealthy regions of the world and is very common in settings where diets are of poor quality. Using estimates of zinc intake and bioavailability derived from Food and Agriculture Organization of the United Nations' (FAO) food balance data, it has been calculated that about 20% of the world's population could be at risk of zinc deficiency. The geographical regions most affected are believed to be, in descending order of severity, South Asia (in particular, Bangladesh and India), Africa, and the western Pacific. It is probable that the occurrence of zinc deficiency is strongly associated with that of iron deficiency because both iron and zinc are found in the same foods (i.e. meat, poultry, and fish) and, in both cases, their absorption from foods is inhibited by the presence of phytates.

10.2.2 FACTORS CONTRIBUTING TO ZINC DEFICIENCY

The central role of zinc in cell division, protein synthesis, and growth means that an adequate supply is especially important for infants, pregnant, and lactating women. Principal risk factors for zinc deficiency include diets low in zinc or high in phytates, malabsorption disorders (including the presence of intestinal parasites and diarrhea), impaired utilization of zinc, and genetic diseases (e.g., acrodermatitis enteropathy, sickle-cell anemia) (Masters and Moir 1983). Sandström (1989) showed that the bioavailability of zinc is dependent on dietary composition, in particular, on the proportion of high-phytate foods in the diet (i.e., selected cereals and legumes) (Sandberg 2002). He studied the molar ratio of phytate: zinc in meals or diets provides a useful measure of zinc bioavailability. At high ratios (i.e., above 15:1), zinc absorption from food is low; less than 15%. The inclusion of animal proteins can improve the total zinc intake and the efficiency of zinc absorption. The extent to which the presence of phytates inhibits the absorption of zinc is not precisely known at present. It is interesting to note that several studies have shown that zinc absorption from some legume-based diets is comparable to that from a diet based on animal products, despite the relatively high phytate content of the former (Sandberg 2002; Salgueiro et al., 2002). Competitive interactions can occur between zinc and other minerals that have similar physical and chemical properties, such as iron and copper. When present in large amounts (e.g., in the form of supplements) or in aqueous solution, these minerals reduce zinc absorption. However, at the levels present in the usual diet and in fortified foods, zinc absorption is not generally affected. On the other hand, high levels of dietary calcium (i.e. >1g per day), which might be consumed by some individuals, can inhibit zinc absorption, especially in the presence of phytates (Salgueiro et al. 2002; Gupta et al. 2020). The degree of impairment varies depending on the type of diet and the source of the calcium. Unlike iron, zinc absorption is neither inhibited by phenolic compounds nor enhanced by vitamin C.

10.2.3 ZINC FORTIFICATION

The different zinc compounds approved for zinc fortification include: Zinc oxide, Zinc sulfate heptahydrate, Stearate, ascorbate, Zinc methionine, Zinc gluconate, acetate, citrate, carbonate, etc. Several countries like Indonesia, Jordan, Mexico, South Africa, and other African nations have made it mandatory to fortify different flours with Zinc. However, countries like China, Bangladesh, Azerbaijan, Palestine, Vietnam, and others are voluntarily participating in zinc fortification (Hall and King 2022; Salgueiro et al., 2002).

Shankar (2000) conducted a meta-analysis of randomized controlled supplementation trials reporting an 18% decrease in diarrhea incidence, a 25% reduction in diarrhea prevalence, and a 41% fall in the incidence of pneumonia. It also confirmed that zinc supplementation led to fewer episodes of malaria and fewer clinic visits due to complications of malaria in Papua New Guinea.

Vatandoust et al. employed extrusion technology for the triple fortification of salt with Zinc, Iodine and Iron (Vatandoust et al., 2023). They reported more stability of the salt sample fortified with Zinc fortificants. Polekkad et al. studied the stability of Zinc sulfate heptahydrate encapsulated in maltodextrin and whey protein isolate by spray drying (Polekkad et al., 2021). The encapsulated zinc was then incorporated into the milk samples and no significant changes in the organoleptic properties of the milk were reported. Tripathi & Platel studied the suitability of finger millet as a vehicle for fortification of zinc (zinc oxide and zinc stearate) (Tripathi and Platel, 2010). They reported a significant increase in the bioaccessibility of zinc upon addition of EDTA. The food products (dumplings and flat bread) prepared from the fortified samples were reported to be of good quality.

10.3 CALCIUM (CA)

Calcium is the most abundant mineral in the body. Around 99% of the body's calcium is located in the skeleton where it exists as hydroxyapatite. In addition to its role in maintaining the rigidity and strength of the skeleton, calcium is involved in a large number of metabolic processes, including blood clotting, cell adhesion, muscle contraction, hormone and neurotransmitter release, glycogen metabolism, and cell proliferation and differentiation (Weaver and Heaney 2007). The RDA of calcium is presented in Table 10.1. Osteoporosis, a disease characterized by reduced bone mass and thus increased skeletal fragility and susceptibility to fractures, is the most significant consequence of a low calcium status. Although adequacy of calcium is important during the whole lifespan, it is especially important during childhood and adolescence due to rapid skeletal growth, and for postmenopausal women and the elderly whose rate of bone loss is high (Chaiwanon et al. 2000; Gras et al. 2003).

10.3.1 PREVALENCE AND HEALTH IMPACT OF CALCIUM DEFICIENCY

Unfortunately, there are no practical population-level indicators of calcium status. Serum calcium, for example, is regulated by a complex homeostatic mechanism, which makes it an unreliable indicator of calcium status. For this reason, in most countries, the prevalence of deficiency is not known. In the absence of reliable biochemical indicators, the best indication of calcium adequacy at present, especially for developing countries, is probably provided by comparing dietary intakes with recommended nutrient intakes (RNIs), despite the variability and uncertainty in the currently recommended intakes for calcium as reported by Joint FAO/WHO (WHO 2006; Saremnezhad, Zargarchi, and Kalantari 2020)

On the basis of the fact that intakes of dairy products are low (**Figure 10.3), it is thus highly likely that low or very low calcium intakes are very common in developing countries.** Measurements

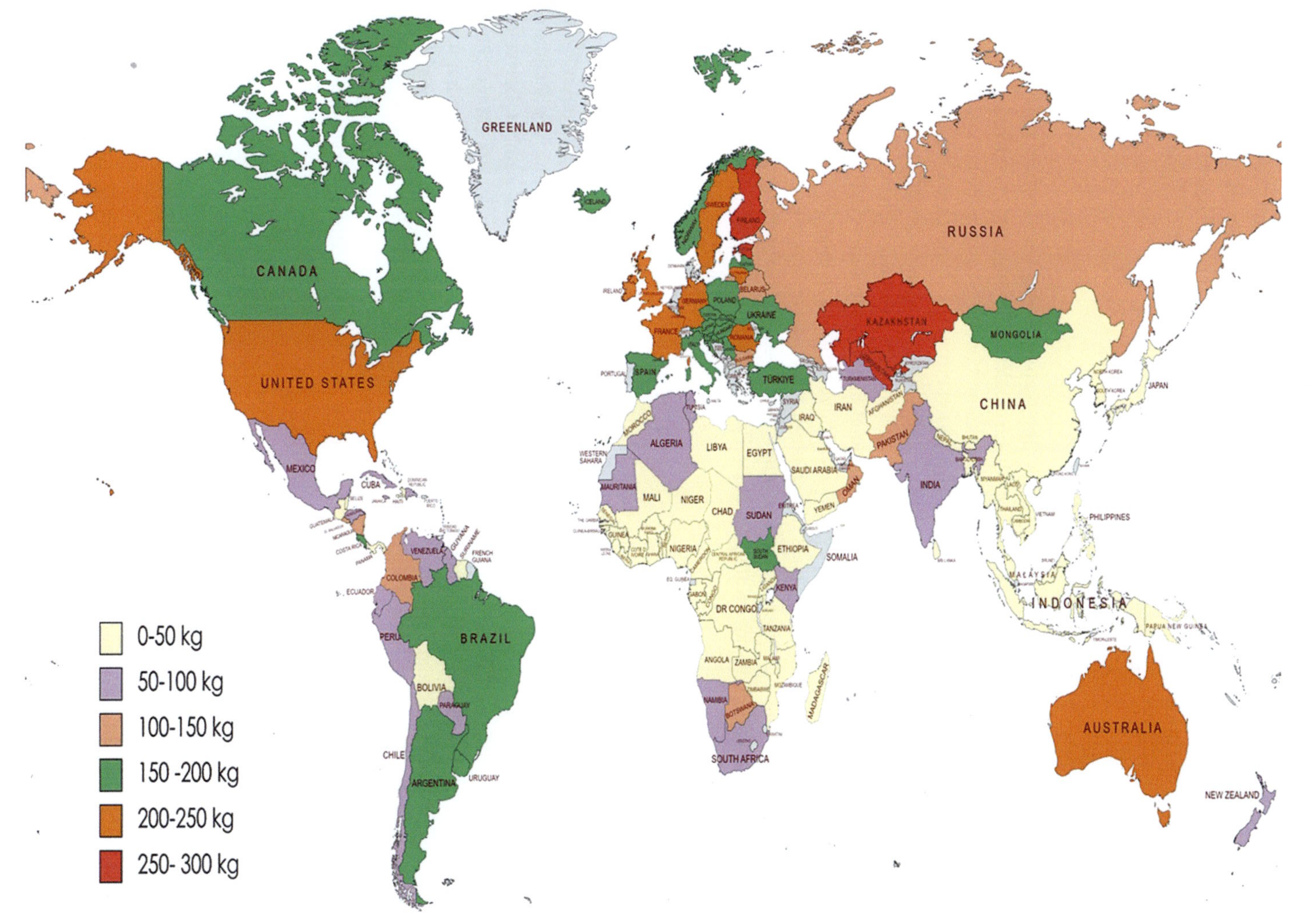

FIGURE 10.3 Global per capita consumption of fresh milk.

of bone mineral density (BMD) and bone mineral content (BMC) have provided an alternative means of assessing the likely extent of calcium deficiency in some countries. In the United States, for example, it has been estimated that 5–6 million older women and 1–2 million older men have osteoporosis. Other approaches include measuring markers of bone resorption in urine or plasma, which tend to be higher in calcium-deficient individuals. Such methods are, however, relatively expensive. All of the above measures are affected by, among many other factors, vitamin D status, level of physical activity, and hormone levels, which further complicates the assessment of calcium adequacy at the population level (Saremnezhad, Zargarchi, and Kalantari 2020; Palacios et al. 2021; Cormick et al. 2021)

10.3.2 Factors Contributing to Calcium Deficiency

Intakes of calcium will almost certainly fall below the recommended levels where dairy product intake is low. Dairy products supply 50–80% of dietary calcium in most industrialized countries, while foods of plant origin supply about 25%. The calcium content of and contribution from most other foods is usually relatively small. Calcium absorption efficiency is increased by a low calcium status and by a low dietary calcium content. Absorption is homeostatically controlled through regulation by vitamin D. The strongest known inhibitor of calcium absorption is dietary oxalate, followed by the presence of phytates. Oxalate is not an important factor in most diets (although it is high in spinach, sweet potatoes, and beans) but phytates are often consumed in large amounts, for instance, in legumes and whole-grain cereals. Thus, the countries with more dependency on cereal-based foods are more prone to calcium deficiency (**Figure 10.1**) (Aditya, Stephen, and Radhakrishnan 2021; Saremnezhad, Zargarchi, and Kalantari 2020; Palacios et al. 2021).

10.3.3 Calcium Fortification

Several interventions have been adopted to evaluate the effectiveness of calcium fortification in addressing calcium deficiency. The recommended daily allowances of calcium are very difficult to meet in the absence of milk and milk products. Globally, only the United Kingdom has adopted mandatory guidelines for calcium fortification in wheat flour. Voluntary calcium fortification of flour has reduced the risk associated with its deficiency in countries like China, Zambia, Barbados, the United Arab Emirates, Colombia, Dominica, Jamaica, Jordan, Saint Lucia, the United States of America, Qatar, Bangladesh, and others. Similar observations were registered in Finland, when the European Union distributed calcium-fortified products free of charge. The different fortificants used in calcium fortification are listed in **Table 10.3.** Further, it is also noticed that the high availability of calcium-fortified products in American's food basket also ensured sufficient consumption of calcium (Palacios et al. 2021; Saremnezhad, Zargarchi, and Kalantari 2020; Barone et al. 2021). The excessive calcium intake may also pose some health complications like renal insufficiency, milk-alkali syndrome, kidney stones, etc.

Saremnezhad et al. studied the impact of calcium fortification on the quality attributes of prebiotic ice-cream (Saremnezhad, Zargarchi, and Kalantari 2020). They reported that the calcium salts (tricalcium citrate, tricalcium phosphate & calcium chloride) resulted in a decrease in the texture (firmness) of the ice-creams, however, the taste, sweetness, and color of the fortified samples were not altered.

Meza et al. evaluated the calcium-enriched milks for their thermal stability (Meza, Zorrilla, and Olivares 2019). They reported that the milk viscosity decreased as the temperature was increased from 25 to 60 °C; however, at temperatures above 60 °C, the viscosity sharply increased. They also reported a decrease in the stability as the fortification level increased from 20–30 mmol kg^{-1}. The increase in viscosity above 60 °C can be ascribed to the denaturation of the casein protein fragments. Also, the addition of calcium results in a decrease in the pH, which causes poor electrostatic and steric repulsions, resulting in more aggregations either by sulfhydryl linkages or -Ca- bridge

TABLE 10.3

Commercially Available Calcium Fortified Foods

Product type	Fortificants used
Breads and wheat flour	Calcium sulfate; calcium stearoyl-2-lactylate; calcium peroxide
Breakfast cereals	Calcium pantothenate; calcium carbonate
Cereal bars	calcium carbonate
Infant cereals	Calcium carbonate; tricalcium phosphate
Plant-based beverages (soymilk, almond milk; and rice milk)	Calcium carbonate; calcium phosphate Tri-calcium phosphate
Orange juice, tangerine juices, and other juices	Tricalcium citrate; calcium lactate + calcium phosphate; calcium hydroxide; Calcium pantothenate
Tofu	Calcium sulfate; Calcium chloride
Infant formulas	Calcium carbonate; tricalcium phosphate; and calcium glycerophosphate
Dairy products (milk, yogurt, cheese, and malted milk)	Carbonate de calcium

formation. Several other researchers have reported an increase in the sedimentation rate as the supplementation of calcium was increased. The sedimentation has a negative impact on the processing time and also requires intensive cleaning to prevent fouling (Bansal and Chin 2006). Calcium fortified (calcium carbonate, calcium lysinate, and tricalcium phosphate) rice noodles and extrudates were analyzed for their physicochemical, functional, and calcium bioaccessibility by Janve and Singhal (Janve and Singhal 2018).They reported a significant decrease in bulk density, expansion index, and oil uptake capacity with fortification. The better organoleptic properties were reported in fortified samples. The higher bioaccessibility was reported in calcium carbonate and calcium lysinate fortified samples as compared to tricalcium phosphate. Barone et al. studied the impact of different calcium fortificants (calcium chloride and micronized calcium citrate) on the physicochemical, functional, and thermal properties of model infant milk formula (Barone et al. 2021). They reported that fortification level at 1500 ppm and at pH 6.8 demonstrated unchanged physicochemical properties.

10.4 FLUORIDE

Fluorine, an electronegative element, is the 13th most abundant in the earth. Fluorine combines with different compounds to form fluoride including sodium fluoride, stannous fluoride, and fluoride mono-fluorophosphate. Compared to other micronutrients, fluoride is required in relatively very trace amounts for preventing dental caries, reducing the ability of bacteria to cause decay, promoting the remineralization of decayed areas, and enhancing bone formation. High consumption of fluoride may cause health implications; hence, the Food and Nutrition Board of the National Academies of Sciences, Engineering, and Medicine has set the RDA for fluoride to be 3mg/day for women and 4 mg/day for men (Table 10.2) (Gotzfried 2006; Gupta, Gupta, and Gupta 2009). According to studies carried out by Phipps *et al.* the results confer that fluoride might even lower the risk of osteoporosis in older individuals (Phipps et al. 2000). Excessive fluoride intake carries a risk of enamel fluorosis, especially during the first 8 years of life. In severe cases of this condition, the enamel of the tooth becomes stained and pitted; in milder forms, the enamel acquires opaque lines or patches. Enamel fluorosis does not occur at fluoride intakes ≤ 0.10 mg/kg bodyweight per day. In adults, excessive fluoride intake can result in skeletal fluorosis, with symptoms that include bone pain, and in more severe cases, muscle calcification and crippling. Mild skeletal fluorosis only occurs at fluoride intakes that are in excess of 10 mg/day for more than 10 years. Symptoms of skeletal fluorosis are rarely seen in communities where the fluoride content of water supplies is below 20 ppm (20 mg/l).

10.4.1 Prevalence and Health Impact of Fluoride Deficiency

There are no universally agreed methods for assessing fluoride status and no generally accepted criteria with which to define deficiency. However, concentrations in urine have sometimes been used as an indicator of fluoride status. Approximately 2.4 billion population globally suffer with dental caries while 486 million children suffer with dental caries in primary teeth across the globe. Though the percentage contribution of fluoride in total dental caries is still a matter of debate as high sugar consumption also significantly participates in dental caries (Kalaycıoğlu 2023; Fordyce 2011; National Research 1993)

10.4.2 Factors Contributing to Fluoride Deficiency

Fluoride intake from most natural water supplies will be relatively low; a low fluoride content of water is thus the main risk factor for a low intake of this mineral. In Canada and the United States, for instance, water sources typically contain less than 0.4 mg/l, which compares with concentrations of 0.7–1.2 mg/l in fluoridated supplies. Moreover, the fluoride content of breast milk is low and foods contain well below 0.05 mg per 100 g, with the exception of those prepared with fluoridated water and infant formulas (Kalaycıoğlu 2023; Fordyce 2011; National Research 1993)

10.4.3 Fluoride Fortification

Addition of fluoride to most consumed food commodities seems to be the most effective strategy to curb fluoride deficiency in the masses. Water, salt, and milk become the most potential food carriers for fluoride fortification as they are consumed in significant quantities across the globe, irrespective of age, gender, geographical, and climatic variations. Water fluoridation, which is the controlled adjustment of fluoride to a public water supply, has meant to attract a good deal of interest in the past many years. Fluoridated water contains fluoride at a level that is safe to drink and has fluoride levels higher than the maximum recommended level. It is an effective procedure to reduce or prevent dental cavities in small age groups. Hexa-fluoro-silicate acid (HUSIAC) is the most used fluoride compound for large-scale water fortification. It is prepared as a concentrated aqueous solution before adding it to the water supply. The other alternative option for fluoridation is the addition of fluoride to salt and milk. The enrichment of milk with fluoride has been used in some parts of the world. Noticeably, whichever country has introduced fluoridation programs to overcome its deficiency has associated their population with a large reduction in dental decay in children, and it's not wrong to say that fortification is feasible and desirable. Wherever in the world it has been impractical or objectionable to fluoridate water or salt, the addition of fluoride to milk is an alternative approach. The level of fluoridation can be regulated appropriately to the usual volume of milk consumed by young children in a particular country. Guidelines for fluoride fortification of milk and milk products should be provided by the concerning authorities or by several government agencies of different countries (Walls 2018; Swan 2000; Buzalaf 2018; Árnadóttir et al. 2004). Fluoride fortification in water inherits several limitations and raises several concerns including unacceptability to fluoride levels suitable for young children, lack of legitimate scientific research on fluoridation, lack of freedom of choice issues, and government conspiracy theories. The allowed fluoride levels greatly vary according to the annual average air temperature in the place where the water is sold. In places with colder temperatures, the maximum fluoride level allowed is 2.4 mg/L for bottled water with no fluoride added and 1.7 mg/L for water in which fluoride is already added. However, if fluoride is added, the FDA recommends that manufacturers not go above 0.7 mg/L, which is the maximum allowed fluoride level in line with the PHS recommendation. Literature also reports a possible linkage between fluoride and cancer. Most of the concern about cancer seems to be around osteosarcoma based on the factual findings of fluoride deposits in newly growing bone tissues. One theory on how fluoridation might affect the risk of osteosarcoma is that fluoride tends to collect in growing parts known as growth plates where osteosarcomas typically develop. The theory is that fluoride might somehow cause the cells in the growth plate to grow faster, which might make them more likely to eventually become cancerous. Studies concluding research on the tolerable levels of fluoride are still on the go, but according to the US Environmental Protection Agency (EPA), the maximum amount of fluoride allowable in drinking water shall not exceed more than 4.0 mg/L. Long-term exposure to levels higher than this can cause a condition called skeletal fluorosis, which is also related to fluoride buildup in the bones. This condition can eventually result in joint stiffness, pain, and could also lead to weak bones or fractures in older adults (Fordyce 2011; Phipps et al. 2000)

10.5   SELENIUM

Selenium is one of the essential elements. However, it was believed to be a poison but later found necessary for life. The only difference is the dosage that is recommended as normal, but more or less deficient diets contain less than 0.1 ppm of selenium; normal diets contain 0.1 ppm or may be somewhat more; yet toxic diets may contain as little as 5 ppm. At higher concentrations of about 25 ppm, it is more likely to be more toxic, which is an unacceptable quantity of selenium to be used in food. Much higher concentrations may occur in plants either because the soil in which they grow contains up to 30 ppm or because certain ones (Kovács et al. 2023; Di et al. 2023).

Selenium has been shown to be an essential trace element, one kind of which is sodium selenite. Sodium selenite which is used as an ingredient in dietary supplements, particularly in multi-vitamins. However, for supplements that provide only selenium, L-selenomethionine or a selenium-enriched yeast is used to fulfill the necessity of this mineral (Ponce de León et al. 2002). There are several ways in which selenium intakes can be increased, such as the organic form of selenium, selenomethionine. Hypothetically, it poses higher risk of toxicity when absorbed as it remains longer in the body. For this reason, it is not widely used for the food fortification process.Yeast-Free Selenium Sodium Selenite (selenium) is a trace mineral that is essential for many bodily processes and is needed for optimal health. Brazil nuts, seafood, and organ meats are the richest food sources of selenium. Other sources include muscle meats, cereals and other grains, and dairy products. Selenium is toxic in high concentrations. As for sodium selenite, the chronic toxic dose for human beings was described as about 2.4 to 3 milligrams of selenium per day (Lönnerdal, Vargas-Fernández, and Whitacre 2017). Sodium selenite may damage the testes, and high exposure can cause a headache, nausea, vomiting, coated tongue, metallic taste, and a garlic odor of the breath. Repeated exposures can cause paleness, lightheadedness, anxiety, and mood changes. Sodium selenite may damage the liver and kidneys when taken above the recommended levels. Since the biological roles of selenium include the protection of tissues against oxidative stress, the maintenance of the body's defense systems against infection, and the modulation of growth and development, severe deficiency can result in Keshan or Kaschin-Beck disease, which is endemic in several world regions (Köhrle 2021)

The selenite is a white, water-soluble compound, from which absorption is about 50%. It is readily reduced to unabsorbable elemental selenium by reducing agents, such as ascorbic acid and sulfur dioxide. Sodium selenate is colorless and is less soluble in water and more stable than selenite, especially in the presence of copper and iron. It has better absorption (nearly 100% from the fortificant alone or 50–80% depending on the food vehicle to which it has been added) and increases the activity of the enzyme glutathione peroxidase more effectively. When tested in milk-based infant formulas, more selenium was absorbed from selenate (97% versus 73%), but as more selenium was excreted in the urine with selenate (36% versus 10%), the net retention of selenium appears to be similar regardless of which chemical form is used. The relative retention of selenium from other fortified foods, including salt, has not been investigated (Chomchan, Siripongvutikorn, and Puttarak 2017; Chen, Zhao, and Zhang 2021).

10.5.1 Prevalence and Health Impact of Selenium

There are several reliable indicators of selenium status, such as the concentration of selenium in plasma, urine, hair, or nails. However, the measurement of selenium in human samples presents a number of technical difficulties, a factor that limits the usefulness of such measures as indicators of status. Indeed, the lack of simple assay techniques for selenium means that currently there are no suitable biochemical indicators of selenium status that are appropriate for use in population surveys. Information regarding the prevalence of selenium deficiency is thus largely based on clinical observations and limited to the more severe forms, i.e., Keshan or Kaschin-Beck disease. Keshan disease is a cardiomyopathy associated with a low selenium intake and low levels of selenium in blood and hair. Reports of its occurrence across a wide zone of mainland China first appeared in the mainstream scientific literature in the 1930s. It has since also been observed in some areas of southern Siberia (Alsamadany et al. 2023; Lv et al. 2014).

Yang *et al.* reported that selenium deficiency is endemic in some regions of China, where Keshan disease was first described, and also in parts of Japan, Korea, Scandinavia, and Siberia. Endemic deficiency tends to occur in regions characterized by low soil selenium (Yang et al. 1984). For example, the distribution of Keshan disease and Kaschin-Beck disease in China reflects the distribution of soils from which selenium is poorly available to rice, maize, wheat, and pasture grasses. Fortification of salt and/or fertilizers with selenium is crucial in these parts of the world. Low intakes of selenium have been linked to a reduced conversion of the thyroid hormone, T4 to T3. The metabolic interrelations between selenium and iodine are such that deficiencies in one can sometimes exacerbate problems with the other. In the Democratic Republic of Congo, for instance, combined selenium and iodine deficiencies were shown to contribute to endemic myxoedematous cretinism. Administration of selenium alone appeared to aggravate this disease; by restoring selenium-dependent deiodinase activity, the synthesis and use of thyroxine (T4) and iodine is increased, thereby exacerbating the iodine deficiency (Vanderpas et al. 1993). Low selenium intakes have also been associated by some researchers with an increased incidence of cancer, in particular, oesophageal cancer, and also with cardiovascular disease (Giray et al. 2001; Baldelli et al. 2023; Khan et al. 2023).

10.5.2 Factors Contributing to Selenium Deficiency

Usual diets in most countries satisfy selenium requirements. As indicated in the previous section, deficiency occurs only where the soil, and consequently the foods produced on those soils, is low in available selenium. Worldwide, the selenium content of animal products and that of cereals and plants varies widely depending on soil selenium content (Fox et al. 2004). The selenium content of foods of plant origin ranges from less than 0.1 μg/g to more than 0.8 μg/g, while the amount in animal products ranges from 0.1 to 1.5 μg/g. Where animal feeds are enriched with selenium, such as in the United States, the selenium content of animal products may be much higher. Concentrations of less than 10 ng/g in the case of grain and less than 3 ng/g in the case of water-soluble soil selenium have been proposed as indexes to define selenium-deficient areas (Duborská et al., 2022).

10.5.3 Selenium Fortification

In areas of endemic selenium deficiency, fortification with selenium has been shown to rapidly increase plasma glutathione peroxidase levels and urinary selenium. In addition, according to the results of a large-scale survey (over 1 million people), selenium fortification of table salt has significantly reduced the prevalence of Keshan disease in China (Baldelli et al. 2023; Giacosa et al. 2014)

10.6 FORTIFICATION: A TOOL TO ADDRESS DEFICIENCY

In general, the WHO identifies food fortification as the most reasonable and operative means of curtailing micronutrient deficiencies. Fortification becomes more effective with the selection of suitable food commodities, existing food supply (Maurya et al. 2022a; Maurya, Bashir, and Aggarwal 2020). The following section of this chapter is intended to evaluate the potential of different food commodities to be an ideal carrier food system for fortification of micronutrients and also review the existing fortification method adopted for these food commodities.

10.6.1 Carriers for Food Fortification

a. **Wheat**: Wheat is an important cereal crop and, together with maize and rice, accounts for 94% of total cereal consumption worldwide. Fortification of industrially processed wheat flour, when appropriately implemented, is an effective, simple, and inexpensive strategy for supplying vitamins and minerals to the diets of large segments of the world's population. The different fortificants that have been successfully used in wheat include: Iron, Folic Acid, Vitamin B_{12}, Zinc, Vitamin A, Thiamine (Vitamin B_1), Riboflavin (Vitamin B_2), Niacin (Vitamin B_3), and Pyridoxine (Vitamin B_6) (Das, Khan, and Bhutta 2018; Maurya, Bashir, and Aggarwal 2020; Maurya et al. 2022b; Duborská et al., 2022). Wheat flour fortification should be considered when industrially produced flour is regularly consumed by large population groups in a country. Decisions about which nutrients to add and the appropriate amounts to be added should be based on the following factors:
   i).   The nutritional needs and deficiencies of the population;
   ii).   The usual consumption profile of "fortifiable" flour (i.e. the total estimated amount of flour milled by industrial roller mills, produced domestically or imported, which could in principle be fortified);
   iii). Sensory and physical effects of the added nutrients on flour and flour products;
   iv).   Fortification of other food vehicles; and
   v).   Cost.

Advantages of fortifying wheat flour

   i).   Wheat flour fortification is a safe and effective means of improving public health;
   ii).   Fortified wheat flour is an excellent vehicle for adding nutrients to the diet, as wheat flour is commonly consumed by all people around the globe;
   iii). A cost-effective method to prevent nutritional deficiencies;
   iv).   During milling of wheat, nutrient losses take place. Fortification helps in adding back these nutrients.

10.6.2 Flour Fortification Methods

The first step is designing the micronutrient premix. A premix contains a uniform mixture of the desired nutrients in the required amounts, which will help in the uniform distribution of the fortificants in the flour. The designed micronutrient premix is accurately metered through a volumetric feeder into the flour. These feeders consist of a rotating feed screw which is driven by a motor, the speed of which can be adjusted to modify the rate of addition of the premix. These feeders either make use of gravity or a pneumatic system to dispense the premix into the flour.

In order to achieve a uniform distribution of the fortificants in the flour, the feeders must be placed at a centralized location with respect to the conveyor carrying the flour. A centrally located feeder will ensure that there is sufficient time provided for the fortificants to mix before the flour is collected and sent for packaging and storage. The plant should have the right mixers, feeders, and quality control equipment to ensure that the fortified flour has effective levels of the desired fortificants present in the finished product (Das, Khan, and Bhutta 2018; Wessells and Hess 2007; Palacios et al. 2021). WHO also recommended fortifying maize flour with folic acid and iron to combat deficiencies of micronutrients. Industrial fortification of maize flour and cornmeal with at least iron has been practiced for many years in several countries in the Americas and Africa, where these ingredients are used in the preparation of many common national dishes. When appropriately implemented, fortification is an efficient, simple, and inexpensive strategy for supplying vitamins and minerals to the diets of large segments of the population (Salgueiro et al., 2002).

a) **Rice:** Rice is grown in many parts of the world because it thrives in a variety of climates. For many years, several countries around the world have practiced industrial fortification of rice with vitamins and minerals, where rice is a staple consumed regularly in the preparation of many common local dishes. Micronutrient deficiencies of public health significance are common in most rice-consuming countries; thus, rice fortification has the potential to help vulnerable populations that are currently not reached by wheat or maize flour fortification programs (Junaid-ur-Rahman et al. 2022). However, rice production is often done domestically or locally, which may make mass fortification programs difficult to reach all those in need. Rice fortification is a cost effective, culturally appropriate strategy to address micronutrient deficiency in countries with high per capita rice consumption. The cost of fortification is determined by a multitude of context specific variables such as the structure and capacity of the rice industry, the complexity of the supply chain, the policy and regulatory environment and the scale of the relevant program. Rice can be fortified by adding a micronutrient powder to the rice that adheres to the grains or spraying of the surface of ordinary rice grains in several layers with a vitamin and mineral mix to form a protective coating (Steiger et al. 2014). Rice can also be extruded and shaped into partially precooked grain-like structures resembling rice grains, which can then be blended with natural polished rice. Rice kernels can be fortified with several micronutrients, such as iron, folic acid and other B-complex vitamins, vitamin A and zinc (Hennigar and Hamaker 2022).

Fortifying rice makes it more nutritious by adding vitamins and minerals in the post-harvest phase; many of which are lost during the milling and polishing process. Rice fortification may be considered as having the highest potential to fill the gap in current staple food fortification programs as it is the staple food of many countries and reaches the most vulnerable and poorer section (Mehansho 2006a; Steiger et al. 2014)

10.6.2.1   Fortification Methods for Rice

Rice can be fortified using dusting, coating, or extrusion technology. Extrusion is the technology of choice for rice fortification given the stability of micronutrients in the rice kernels across processing, storage, washing, and cooking, also in view of cost considerations. In extrusion technology, milled broken rice is pulverized and mixed with a premix containing fortificants (Steiger et al. 2014). Fortified rice kernels (FRK) are produced from this mixture using an extruder machine. The kernels resemble rice grains. FRK is added to non-fortified rice in a ratio ranging from 1:50 to 1:200 (ideal is 1:100), resulting in fortified rice nearly identical to traditional rice in aroma, taste, and texture. It is then distributed for regular consumption (Roks 2014; Mishra, Mishra, and Srinivasa Rao 2012).

The cost of fortification is determined by a multitude of context-specific variables such as the structure and capacity of the rice industry, the complexity of the supply chain, the policy and regulatory environment, and the scale of the relevant program. A typical low-cost, 150 kgs per hour, twin-screw extruder with all ancillary equipment costs around 40 thousand USD or higher. A good quality extrusion line may cost up to 1 million USD. Utility costs like purified water plant, steam generator, air compressor, and packaging lines are not included here. The retail price increase for fortified rice ranges from an additional 1% to 10%. As rice fortification expands, production and distribution achieve economies of scale, costs are expected to reduce. Fortifying rice is cost-effective; the additional cost to the consumer inclusive of all associated costs is INR 0.45 per kg depending on the nutrients added (Mehansho 2006b; Wesley and Horton 2011).

b. **Oil:** Multiple micronutrient deficiencies are rampant in developing countries and continue to be significant public health problems, which adversely impact the health and productivity of all population groups. Since vitamin A and D are fat-soluble vitamins, fortification of edible oils and fats with vitamin A and D is a good strategy to address micronutrient malnutrition, and fortified oil is known to provide 25%–30% of the recommended dietary allowances for vitamins A & D (Pignitter et al. 2016; Fiedler and Afidra 2010; Bagriansky and Ranum, 1998).

10.6.2.2 Fortification Methods for Oil

Producers can easily implement fortification technology because it is simple. Oils can be easily and inexpensively fortified without the use of specialized equipment, and the fortificants can be distributed uniformly. It is common practice to use dosing technology to add antioxidants and other micro ingredients to oil. While many oils require a temperature of 40–50 °C to ensure uniform mixing, the temperature required to ensure a uniform liquid state for oils such as soybean oil is less than about 25 °C. Producers generally report minor changes to add vitamins A, D, and E. Personnel and technology are already in place because other additives are frequently used. The cost of fortification is 7 paise per liter of oil fortified with vitamins A and D (Pignitter et al. 2016; Fiedler and Afidra 2010; Bagriansky and Ranum 1998; Chaudhry 2018).

c. **Milk:** Milk is a rich source of high-quality protein, calcium, and of fat-soluble vitamins A and D. Vitamins A and D are lost when milk fat is removed during processing (Zahedirad et al., 2019). Many countries have a mandatory provision to add back the vitamins removed as it is easily doable. It is called replenishment as the nutrients lost during processing are added back. Fortification of milk with Vitamin A and Vitamin D is required in developing countries because of the widespread deficiencies present in the population. Since milk is consumed by all population groups, fortification of milk with certain micronutrients is a good strategy to address micronutrient malnutrition (Ocak and Rajendram 2013).

10.6.2.3 Fortification Methods for Milk

The technology to fortify milk is simple, low-cost, well-established, and readily available. Vitamin A and D premixes are widely available in the market. The cost of fortification is less than 2 paisa per liter of milk. The premix for vitamin A is in the form of Retinyl Acetate/ Retinyl Palmitate/ Retinyl Propionate and vitamin D is in the form of Ergocalciferol/ Cholecaliferol (Ocak and Rajendram 2013; Chaiwanon et al. 2000; Maurya, Bashir, and Aggarwal 2020; Maurya and Aggarwal 2019, 2019)

10.7 CONCLUSION

More than 2 billion individuals worldwide suffer from micronutrient deficiencies. Mineral deficiencies may trigger a cascade of health complications. Increased awareness about the role of micronutrients has further caused movement in the development of new nutrient enrichment methods. Diversification, biofortification, and food fortification are the most explored strategies to address micronutrient deficiency. However, the WHO suggests that food fortification is the most effective means to address nutrient deficiency. Wheat, oil, milk, and rice are the most potential food carrier systems for micronutrient fortification. Furthermore, fortification inherits several challenges including heterogeneous distribution, taste change, and low bioavailability of fortification, which are being overcome by the use of nanotechnology. However, more scientific examination is required to evaluate the technology's performance in different food commodities under realistic storage conditions.

REFERENCES

Aditya, Sanprit, Jaspin Stephen, and Mahendran Radhakrishnan. 2021. Utilization of eggshell waste in calcium-fortified foods and other industrial applications: A review. *Trends in Food Science and Technology* 115:422–432.

Alsamadany, Hameed, Hesham F. Alharby, Hassan S. Al-Zahrani, Alpaslan Kuşvuran, Sebnem Kuşvuran, and Mostafa M. Rady. 2023. Selenium fortification stimulates antioxidant-and enzyme gene expression-related defense mechanisms in response to saline stress in Cucurbita pepo. *Scientia Horticulturae* 312:111886.

Árnadóttir, Inga B., Clare E. Ketley, Cor Van Loveren, et al. 2004. A European perspective on fluoride use in seven countries. *Community Dentistry and Oral Epidemiology* 32:69–73.

Bagriansky, Jack, and Peter Ranum. 1998. *Vitamin A fortification of PL480 vegetable oil*. Sharing United States Technology to Aid in the Improvement of Nutrition.

Baldelli, Alberto, Melinda Ren, Diana Yumeng Liang, et al. 2023. Sprayed microcapsules of minerals for fortified food. *Journal of Functional Foods* 101:105401.

Ball, M. J., and M. L. Ackland. 2000. Zinc intake and status in Australian vegetarians. *British Journal of Nutrition* 83(1):27–33.

Barone, Giovanni, Saeed Rahimi Yazdi, Søren K. Lillevang, and Lilia Ahrné. 2021. Calcium: A comprehensive review on quantification, interaction with milk proteins and implications for processing of dairy products. *Comprehensive Reviews in Food Science and Food Safety* 20(6):5616–5640.

Baskar, S., and Prathap, L. 2020. Awareness of micronutrient deficiency in India. *International Journal of Pharmaceutical Research* (09752366).

Black, Robert E., Cesar G. Victora, Susan P. Walker, et al. 2013. Maternal and child undernutrition and overweight in low-income and middle-income countries. *The Lancet* 382(9890):427–451.

Buturi, Camila Vanessa, Rosario Paolo Mauro, Vincenzo Fogliano, Cherubino Leonardi, and Francesco Giuffrida. 2021. Mineral biofortification of vegetables as a tool to improve human diet. *Foods* 10(2):223.

Buzalaf, Marília Afonso Rabelo. 2018. Review of fluoride intake and appropriateness of current guidelines. *Advances in Dental Research* 29(2):157–166.

Chaiwanon, Piangjan, Prapasri Puwastien, Anadi Nitithamyong, and Prapaisri P. Sirichakwal. 2000. Calcium fortification in soybean milk and in vitro bioavailability. *Journal of Food Composition and Analysis* 13(4):319–327.

Chaudhry, S. A. 2018. "Business considerations for food fortification: Cargill India experience with oil fortification." In *Food Fortification in a Globalized World*, edited by M. G. Venkatesh Mannar and Richard F. Hurrell, pp. 357–362. Academic Press.

Chen, Nan, Changhui Zhao, and Tiehua Zhang. 2021. Selenium transformation and selenium-rich foods. *Food Bioscience* 40:100875.

Chomchan, Rattanamanee, Sunisa Siripongvutikorn, and Panupong Puttarak. 2017. Selenium bio-fortification: An alternative to improve phytochemicals and bioactivities of plant foods. *Functional Foods in Health and Disease* 7(4):263–279.

Cormick, Gabriela, Ana Pilar Betran, Iris Beatriz Romero, et al. 2021. Effect of calcium fortified foods on health outcomes: A systematic review and meta-analysis. *Nutrients* 13(2):316.

Das, Jai K., Raja S. Khan, and Zulfiqar A. Bhutta. 2018. Zinc fortification. In M.G. Venkatesh Mannar and Richard F. Hurrell (eds.), *Food fortification in a globalized world*. Elsevier.

Deshpande, Jayant D., Mohini M. Joshi, and Purushottam A. Giri. 2013. Zinc: The trace element of major importance in human nutrition and health. *International Journal of Medical Science and Public Health* 2(1):1–6.

Dhaliwal, Salwinder Singh, Vivek Sharma, Arvind Kumar Shukla, et al. 2021. Comparative efficiency of mineral, chelated and Nano forms of zinc and iron for improvement of zinc and iron in chickpea (Cicer arietinum L.) through biofortification. *Agronomy* 11(12):2436.

Di, Xuerong, Xu Qin, Lijie Zhao, et al. 2023. Selenium distribution, translocation and speciation in wheat (Triticum aestivum L.) after foliar spraying selenite and selenate. *Food Chemistry* 400:134077.

Duborská, Eva, Martin Šebesta, Michaela Matulová, Ondřej Zvěřina, and Martin Urík. 2022. Current strategies for selenium and iodine biofortification in crop plants. *Nutrients* 14(22):4717.

Durrani, A. M., and H. Parveen. 2022. Zinc deficiency and its consequences during pregnancy. In *Microbial biofertilizers and micronutrient availability: The role of zinc in agriculture and human health*: (pp. 69–82).

Fiedler, John L., and Ronald Afidra. 2010. Vitamin A fortification in Uganda: Comparing the feasibility, coverage, costs, and cost-effectiveness of fortifying vegetable oil and sugar. *Food and Nutrition Bulletin* 31(2):193–205.

Flynn, Albert, Olga Moreiras, Peter Stehle, Reginald J. Fletcher, Detlef J. G. Müller, and Valérie Rolland. 2003. Vitamins and minerals: A model for safe addition to foods. *European Journal of Nutrition* 42(2):118–130.

Fordyce, F. M. 2011. "Fluorine: Human health risks." In *Encyclopedia of Environmental Health*, edited by J. O. Nriagu, Volume 2 (pp. 776–785). Burlington: Elsevier.

Fox, T. E., E. Van den Heuvel, C. A. Atherton, et al. 2004. Bioavailability of selenium from fish, yeast and selenate: A comparative study in humans using stable isotopes. *European Journal of Clinical Nutrition* 58(2):343–349.

Giacosa, Attilio, Milena Anna Faliva, Simone Perna, Claudio Minoia, Anna Ronchi, and Mariangela Rondanelli. 2014. Selenium fortification of an Italian rice cultivar via foliar fertilization with sodium selenate and its effects on human serum selenium levels and on erythrocyte glutathione peroxidase activity. *Nutrients* 6(3):1251–1261.

Giray, Belma, Filiz Hincal, Tahsin Teziç, Ayşenur Ökten, and Yusuf Gedik. 2001. Status of selenium and antioxidant enzymes of goitrous children is lower than healthy controls and nongoitrous children with high iodine deficiency. *Biological Trace Element Research* 82(1–3):35–52.

Godswill, Awuchi Godswill, Igwe Victory Somtochukwu, Amagwula O. Ikechukwu, and Echeta Chinelo Kate. 2020. Health benefits of micronutrients (vitamins and minerals) and their associated deficiency diseases: A systematic review. *International Journal of Food Sciences* 3(1):1–32.

Gotzfried, F. 2006. Legal aspects of fluoride in salt, particularly within the EU. *Schweizer Monatsschrift für Zahnmedizin* 116(4):371.

Gras, M. L., D. Vidal, N. Betoret, A. Chiralt, and P. Fito. 2003. Calcium fortification of vegetables by vacuum impregnation: Interactions with cellular matrix. *Journal of Food Engineering* 56(2–3):279–284.

Gupta, S. K., R. C. Gupta, and A. B. Gupta. 2009. Is there a need of extra fluoride in children? *Indian Pediatrics* 46(9):755–759.

Gupta, Swarnim, Edwin Habeych, Nathalie Scheers, et al. 2020. The development of a novel ferric phytate compound for iron fortification of bouillons (part I). *Scientific Reports* 10(1):5340.

Hall, Andrew G., and Janet C. King. 2022. Zinc fortification: Current trends and strategies. *Nutrients* 14(19):3895.

Hennigar, Stephen R., and Bruce R. Hamaker. 2022. *Optimizing fortification of rice with micronutrients to improve public health*. Oxford University Press.

Hess, Sonja Y., Alexander C. McLain, Haley Lescinsky, et al. 2022. Basis for changes in the disease burden estimates related to vitamin A and zinc deficiencies in the 2017 and 2019 global burden of disease studies. *Public Health Nutrition* 25(8):2225–2231.

Hollowell, Joseph G., Norman W. Staehling, W. Harry Hannon, et al. 1998. Iodine nutrition in the United States. Trends and public health implications: Iodine excretion data from national health and nutrition examination surveys I and III (1971–1974 and 1988–1994). *The Journal of Clinical Endocrinology and Metabolism* 83(10):3401–3408.

Hunt, Janet R. 2002. Moving toward a plant-based diet: Are iron and zinc at risk? *Nutrition Reviews* 60(5):127–134.

Janve, Madhura, and Rekha S. Singhal. 2018. Fortification of puffed rice extrudates and rice noodles with different calcium salts: Physicochemical properties and calcium bioaccessibility. *LWT* 97:67–75.

Junaid-ur-Rahman, Syed, Muhammad Farhan Jahangir Chughtai, Adnan Khaliq, et al. 2022. Rice: A potential vehicle for micronutrient fortification. *Clinical Phytoscience* 8(1):1–14.

Kalaycıoğlu, Z. 2023. "Determination of fluoride content in infant food samples by sample stacking capillary electrophoresis-indirect UV detection and the estimation of daily-fluoride-intake." *Journal of Food Composition and Analysis* 105425.

Khan, Zesmin, Thorny Chanu Thounaojam, Devasish Chowdhury, and Hrishikesh Upadhyaya. 2023. The role of selenium and Nano selenium on physiological responses in plant: A review. *Plant Growth Regulation* 100(2):409–433.

Köhrle, Josef. 2021. Selenium in endocrinology—Selenoprotein-related diseases, population studies, and epidemiological evidence. *Endocrinology* 162(2):bqaa228.

Kovács, Z., Soós, Á., Kovács, B., Kaszás, L., Elhawat, N., Razem, M., ... & Domokos-Szabolcsy, É. 2023. Nutrichemical alterations in different fractions of multiple-harvest alfalfa (Medicago sativa L.) green biomass fortified with various selenium forms. *Plant and Soil*:1–23.

Kristensen, Mette Bach, Ole Hels, Catrine M. Morberg, Jens Marving, Susanne Büge, and Inge Tetens. 2006. Total zinc absorption in young women, but not fractional zinc absorption, differs between vegetarian and meat-based diets with equal phytic acid content. *British Journal of Nutrition* 95(5):963–967.

Lee, Yo A., Ji-Yun Hwang, Hyesook Kim, et al. 2011. Relationships of maternal zinc intake from animal foods with fetal growth. *British Journal of Nutrition* 106(2):237–242.

Leung, Angela M., Lewis E. Braverman, and Elizabeth N. Pearce. 2012. History of US iodine fortification and supplementation. *Nutrients* 4(11):1740–1746.

Lönnerdal, Bo, Eugenia Vargas-Fernández, and Mark Whitacre. 2017. Selenium fortification of Infant Formulas: Does selenium form matter? *Food and Function* 8(11):3856–3868.

Lotfi, Mahshid, M. G. Venkatesh Mannar, Richard J. H. Merx, and Petra Naber-van den Heuvel. 1996. *Micronutrient fortification of foods: Current practices, research, and opportunities.* IDRC, Micronutrient Initiative, Ottawa, ON, CA.

Lv, Yaoyao, Tao Yu, Zhongfang Yang, Wanfu Zhao, Meng Zhang, and Qian Wang. 2014. Constraint on selenium bioavailability caused by its geochemical behavior in typical Kaschin–Beck disease areas in Aba, Sichuan Province of China. *Science of the Total Environment* 493:737–749.

Masters, David G., and R. J. Moir. 1983. Effect of zinc deficiency on the pregnant ewe and developing foetus. *British Journal of Nutrition* 49(3):365–372.

Masuda, H., Aung, M. S., Kobayashi, T., & Nishizawa, N. K. 2020. Iron biofortification: The gateway to overcoming hidden hunger. In *The future of rice demand: Quality beyond productivity*:149–177.

Maurya, Vaibhav Kumar, and Manjeet Aggarwal. 2019. Fabrication of nano-structured lipid carrier for encapsulation of vitamin D3 for fortification of 'Lassi'; A milk based beverage. *The Journal of Steroid Biochemistry and Molecular Biology* 193:105429.

———. 2019. A phase inversion based nanoemulsion fabrication process to encapsulate vitamin D3 for food applications. *The Journal of Steroid Biochemistry and Molecular Biology* 190:88–98.

Maurya, Vaibhav Kumar, Khalid Bashir, and Manjeet Aggarwal. 2020. Vitamin D microencapsulation and fortification: Trends and technologies. *The Journal of Steroid Biochemistry and Molecular Biology* 196:105489.

Maurya, Vaibhav Kumar, Amita Shakya, Khalid Bashir, Kulsum Jan, and David Julian McClements. 2023. Fortification by design: A rational approach to designing vitamin D delivery systems for foods and beverages. *Comprehensive Reviews in Food Science and Food Safety* 22(1):135–186.

Maurya, Vaibhav Kumar, Amita Shakya, Khalid Bashir, Satish Chand Kushwaha, and David Julian McClements. 2022a. Vitamin A fortification: Recent advances in encapsulation technologies. *Comprehensive Reviews in Food Science and Food Safety* 21(3):2772–2819.

———. 2022b. Vitamin A fortification: Recent advances in encapsulation technologies. *Comprehensive Reviews in Food Science and Food Safety* 21(3):2772–2819.

Mayo-Wilson, E., Junior, J. A., Imdad, A., Dean, S., Chan, X. H. S., Chan, E. S., ... & Bhutta, Z. A. 2014. Zinc supplementation for preventing mortality, morbidity, and growth failure in children aged 6 months to 12 years of age. *Cochrane Database of Systematic Reviews* 15(5):CD009384.

Mehansho, H. 2006. "Iron fortification technology development: new approaches." *The Journal of Nutrition* 4(136): 1059–1063.

———. 2006b. Symposium: Food fortification in developing countries. *Journal of Nutrition*:1059–1063.

Meza, Bárbara Erica, Susana Elizabeth Zorrilla, and María Laura Olivares. 2019. Rheological methods to analyse the thermal aggregation of calcium enriched milks. *International Dairy Journal* 97:25–30.

Mishra, Abhinav, Hari Niwas Mishra, and Pavuluri Srinivasa Rao. 2012. Preparation of rice analogues using extrusion technology. *International Journal of Food Science and Technology* 47(9):1789–1797.

Moretti, Diego, Ralf Biebinger, Maaike J. Bruins, Birgit Hoeft, and Klaus Kraemer. 2014. Bioavailability of iron, zinc, folic acid, and vitamin A from fortified maize. *Annals of the New York Academy of Sciences* 1312(1):54–65.

Morris, Eugene R., and Rex Ellis. 1980. Effect of dietary phytate/zinc molar ratio on growth and bone zinc response of rats fed semipurified diets. *The Journal of Nutrition* 110(5):1037–1045.

Muthayya, S., J. H. Rah, J. D. Sugimoto, F. F. Roos, K. Kraemer, and R. E. Black. 2013. "The global hidden hunger indices and maps: An advocacy tool for action." *PloS One* 6(8): e67860.

National Research Council. 1993. *Health effects of ingested fluoride*. National Academies Press.

Ocak, E., and Rajendram, R. 2013. Fortification of milk with mineral elements. In *Handbook of food fortification and health: From concepts to public health applications*, Volume 1:213–224.

Ohanenye, Ikenna C., Chijioke U. Emenike, Azza Mensi, et al. 2021. Food fortification technologies: Influence on iron, zinc and vitamin A bioavailability and potential implications on micronutrient deficiency in sub-Saharan Africa. *Scientific African* 11:e00667.

Palacios, Cristina, G. Justus Hofmeyr, Gabriela Cormick, Maria Nieves Garcia-Casal, Juan Pablo Peña-Rosas, and Ana Pilar Betrán. 2021. Current calcium fortification experiences: A review. *Annals of the New York Academy of Sciences* 1484(1):55–73.

Parisi, Alfred F., and Bert L. Vallee. 1969. Zinc metalloenzymes: Characteristics and significance in biology and medicine. *The American Journal of Clinical Nutrition* 22(9):1222–1239.

Phipps, Kathy R., Eric S. Orwoll, Jill D. Mason, and Jane A. Cauley. 2000. Community water fluoridation, bone mineral density, and fractures: Prospective study of effects in older women. *BMJ* 321(7265):860–864.

Pignitter, Marc, Natalie Hernler, Mathias Zaunschirm, et al. 2016. Evaluation of palm oil as a suitable vegetable oil for vitamin A fortification programs. *Nutrients* 8(6):378.

Polekkad, Abhinash, Magdaline Eljeeva Emerald Franklin, Heartwin A. Pushpadass, Surendra Nath Battula, S. B. Nageswara Rao, and D. T. Pal. 2021. Microencapsulation of zinc by spray-drying: Characterisation and fortification. *Powder Technology* 381:1–16.

Ponce de León, C. A., M. Montes Bayón, C. Paquin, and J. A. Caruso. 2002. Selenium incorporation into Saccharomyces cerevisiae cells: A study of different incorporation methods. *Journal of Applied Microbiology* 92(4):602–610.

Ritchie, Hannah, and Max Roser. 2017. *Micronutrient deficiency*. Our World in Data.

Roks, Eveline. 2014. Review of the cost components of introducing industrially fortified rice. *Annals of the New York Academy of Sciences* 1324(1):82–91.

Roohani, Nazanin, Richard Hurrell, Roya Kelishadi, and Rainer Schulin. 2013. Zinc and its importance for human health: An integrative review. *Journal of Research in Medical Sciences: The Official Journal of Isfahan University of Medical Sciences* 18(2):144.

Salgueiro, Maria Jimena, Marcela Zubillaga, Alexis Lysionek, Ricardo Caro, Ricardo Weill, and Jose Boccio. 2002. Fortification strategies to combat zinc and iron deficiency. *Nutrition Reviews* 60(2):52–58.

Sandberg, Ann-Sofie. 2002. Bioavailability of minerals in legumes. *British Journal of Nutrition* 88(S3):281–285.

Sandström, B. "Dietary pattern and zinc supply." In *Zinc in Human Biology* (pp. 351–363). London: Springer, 1989.

Saremnezhad, Solmaz, Sina Zargarchi, and Zahra Nasrollah Kalantari. 2020. Calcium fortification of prebiotic ice-cream. *LWT* 120:108890.

Shankar, Anuraj H. 2020. Mineral deficiencies. In Ryan, E. T., Hill, D. R., Solomon, T., Aronson, N., & Endy, T. P (eds.), *Hunter's tropical medicine and emerging infectious diseases*. Elsevier.

Steiger, Georg, Nadina Müller-Fischer, Hector Cori, and Béatrice Conde-Petit. 2014. Fortification of rice: Technologies and nutrients. *Annals of the New York Academy of Sciences* 1324(1):29–39.

Swan, Euan. 2000. Dietary fluoride supplement protocol for the new millennium. *Journal-Canadian Dental Association* 66(7):362–364.

Tripathi, Bhumika, and Kalpana Platel. 2010. Finger millet (Eleucine coracana) flour as a vehicle for fortification with zinc. *Journal of Trace Elements in Medicine and Biology* 24(1):46–51.

Vanderpas, Jean B., Bernard Contempre, Ngida L. Duale, et al. 1993. Selenium deficiency mitigates hypothyroxinemia in iodine-deficient subjects. *The American Journal of Clinical Nutrition* 57(2):271S–275S.

Vatandoust, Azadeh, Kiruba Krishnaswamy, Yao Olive Li, and Levente Diosady. 2023. Triple fortification of salt with iron, iodine and zinc oxide using extrusion. *Journal of Food Engineering* 339:111258.

Walls, A. W. G. 2018. Guidelines for fluoride intake: Second discussant. *Advances in Dental Research* 29(2):179–182.

Weaver, Connie M., and Robert P. Heaney. 2007. *Calcium in human health*. Springer Science & Business Media.

Wesley, Annie S., and Sue Horton. 2011. Economics of food fortification. In Watson, R. R., Gerald, J. K., & Preedy, V. R. (eds.), *Nutrients, dietary supplements, and nutriceuticals: Cost analysis versus clinical benefits*:31–40.

Wessells, Hess, and S. Y. Hess. 2007. Zinc bioavailability from zinc-fortified foods. *International Journal for Vitamin and Nutrition Research* 77(3):174–181.

Wessells, K. Ryan, Gitanjali M. Singh, and Kenneth H. Brown. 2012. Estimating the global prevalence of inadequate zinc intake from national food balance sheets: Effects of methodological assumptions. *PLoS One* 7(11):e50565.

WHO, World Health Organization. 2006. *Guidelines on food fortification with micronutrients*. World Health Organization.

———. 2019. Meeting report: WHO technical consultation: Nutrition-related health products and the World Health Organization model list of essential medicines–practical considerations and feasibility: Geneva, Switzerland, 20–21 September 2018. World Health Organization.

Yang, Guangqi, Junshi Chen, Zhimei Wen, et al. 1984. The role of selenium in Keshan disease. *Advances in Nutritional Research*: 110:203–231.

Zahedirad, Malihe, Sepideh Asadzadeh, Bahareh Nikooyeh, et al. 2019. Fortification aspects of vitamin D in dairy products: A review study. *International Dairy Journal* 94:53–64.

Part IV

Other Food Bioactive Fortification

11 Phytochemical Fortification

Mehvish Habib, Iqra Qureshi,
Sakshi Singh, Shumaila Jan, and Khalid Bashir

11.1 INTRODUCTION

Since ancient times, medicinal plants have served as the foundation of conventional herbal medicine among rural people across the globe. Plants were undoubtedly used medicinally from the 3rd millennium BC when the Sumerian and Akkadian civilizations emerged. One of the early writers who detailed medical natural products of animal and plant origins was Hippocrates (c. 460–377 BC), who identified over 400 distinct plant species for medicinal use. Natural remedies were a crucial component of traditional medical systems such as the Ayurvedic, Egyptian, and Chinese cultures (Sarker & Nahar, 2007). They have come to play an increasingly significant role in modern civilization as a natural supply of chemotherapy and among researchers looking for new medication sources. In the poor world, almost 3.4 billion people reply on herbal remedies. This equates to around 88% of the world's population, the majority of whom depends on conventional medicine for their primary healthcare. As per WHO ("World Health Organization"), a medical plant is any plant that contains compounds that may be utilized therapeutically or that serve as precursors for the semi-synthesis of chemo-pharmaceuticals. Such a plant would have parts applied in the treatment or prevention of a disease condition, such as roots, leaves, stems, rhizomes, flowers, barks, grains, fruits, or seeds, and as a result, it includes chemical elements that are therapeutically effective. These non-nutritive plant bioactive molecules or chemical compounds, commonly recognized as phytochemicals or phytoconstituents (from the Greek "phyto," meaning "plant"), are in charge of plant protecting against insect infestations or microbial infections (Abo et al., 1991; Liu, 2004; Nweze et al., 2004; Doughari et al., 2009). On the other side, the analysis of natural compounds is known as phytochemistry. Fruits and vegetables like apples and grapes, spices like turmeric, alcoholic drinks like red wine and green tea, and many more sources have all been applied to extract and describe phytochemicals (Obidah & Doughari, 2008; Doughari et al., 2009).

Although the practice extends back to antiquity, the study of using these local or native medicinal remedies, such as plants for the ailment's treatment, is now recognized as pharmacy. The foundation of all conventional medications across the globe has been ethnopharmacology, which is now being incorporated into modern medicine. Many catalogs, such as Historia Plantarum, De Materia Medica, and Species Plantarum, were produced to offer scientific details on the therapeutic use of plants. Approximately 80% of rural residents depend on plants to cure a variety of diseases, while the varieties and application techniques differ from place to place. For example, different manuals of phytotherapy note the use of bearberry (*"Arctostaphylos uva-ursi"*) and cranberry juice (*"Vaccinium macrocarpon"*) to treat urinary tract infections, whereas plants such as tea tree (*"Melaleuca alternifolia"*), garlic (*"Allium sativum"*), and lemon balm (*"Melissa officinalis"*) are termed as broad-spectrum antimicrobial compounds (Heinrich et al., 2011). Depending on the population, a single plant may be utilized to cure a variety of illness problems. Traditional medicinal herbs have been used to cure a number of health conditions, including hypertension, esophageal cancer, constipation, asthma, and fever (Cousins and Huffman, 2002; Saganuwan, 2010). The plants are delivered in many ways, such as nasally (smoking, sniffing, or steaming), orally, through bathing, rectally (enemas), topically (creams, oils, or lotions), or in the form of poultices, formulations of various plant mixes, and infusions as teas or tinctures. To treat infectious illnesses of the biliary, gastrointestinal, urinary, and respiratory systems through treatment of the skin, many plant parts and components (stem barks, leaves, roots, essential oils, flowers, or their mixtures) have been used (Rojas et al., 2001; Recio Rıos, 2005; Adekunle Adekunle, 2009).

Even literate people in urban areas are beginning to recognize the benefits of medicinal plants, likely as a result of the growing inefficiency of several modern drugs utilized to treat illnesses like tuberculosis, gonorrhea, and typhoid fever and the rise in bacteria resistant to different antibiotics as well as the rising cost of prescription medications for maintaining one's health (Levy, 1998; Van den Bogaard, 2000; Smolinski et al., 2003). Unfortunately, the tremendous growth in the global population has made it almost impossible for contemporary healthcare facilities to keep up with demand, increasing the need for natural herbal health cures. Increased incidence of MDR ("Multiple-Drug Resistant") strains of certain harmful bacteria, including MDR, *Helicobacter pylori*, and methicillin-resistant *Staphylococcus aureus* are current issues related to the usage of antibiotics. The interest in plants with antibacterial characteristics has increased as a result of *Klebsiela pneumonia* (Voravuthikunchai & Kitpipit, 2003). Additionally, the emergence of AIDS ("Acquired Immune Deficiency Syndrome") patients and people receiving immunosuppressive chemotherapy, the rise in opportunistic infection cases, and the toxicity of various antiviral and antifungal medications have put pressure on the pharmaceutical companies as well as the scientific community to look for alternative and new sources of drugs.

11.2 SOURCES OF PHYTOCHEMICALS

11.2.1 ALKALOIDS

These comprise the majority of the secondary chemical components and are mostly formed of ammonia compounds, which are nitrogen bases generated from building blocks of amino acids with different radicals substituting one or more hydrogen atoms within the ring of peptide. The majority of these radicals contain oxygen. As a result of the compounds' basic characteristics as well as alkaline reaction, "red litmus paper" becomes blue. In actuality, one or more atoms of nitrogen, often in the form of 1°, 2°, or 3° amines, present in an alkaloid contribute to the alkaloid's basicity. The molecule's structure, the presence of functional groups, and their placement all influence how basic a compound is. (Sarker & Nahar, 2007). They don't produce any water since they react with acids to create crystalline salts (Firn, 2010). Most alkaloids, like atropine, are solids, but some are also liquids made of nitrogen, hydrogen, and carbon. Although they are only slightly soluble in water, most alkaloids are easily soluble within alcohol and are typically soluble in their salt forms. Alkaloids have very bitter solutions. Due to their powerful biological activity, these nitrogenous compounds are frequently used as medicines, stimulants, narcotics, and poisons. They play a significant role in plants' defense against herbivores and pathogens. Alkaloids are compounds that are found in abundance in nature in plant roots and seeds, commonly in association with vegetable acids. Alkaloids are used pharmacologically as CNS stimulants and anesthetics (Madziga et al., 2010). Only a few of the more than 12,000 alkaloids, which are known to occur in approximately 20% of plant species, were used as medicines. The word "alkaloid" has the prefix "in" and includes plant-derived alkaloids such as the muscle relaxant (+)-tubocurarine, analgesics morphine and codeine, anticancer drug vinblastine, antibiotics sanguinarine and berberine, pupil dilater atropine, antiarrhythmic ajmaline, and sedative scopolamine. Nicotine, cocaine, ergotamine, morphine, atropine, ephedrine, codeine, nicotine, and addictive stimulants caffeine are among other significant plant-derived alkaloids (Figure 11.1). Ornithine and lysine are often utilized as starting materials in the production of alkaloids because amino acids function as precursors. (Table 11.1 Table 11.1 provides a summary of several screening techniques for the identification of alkaloids.

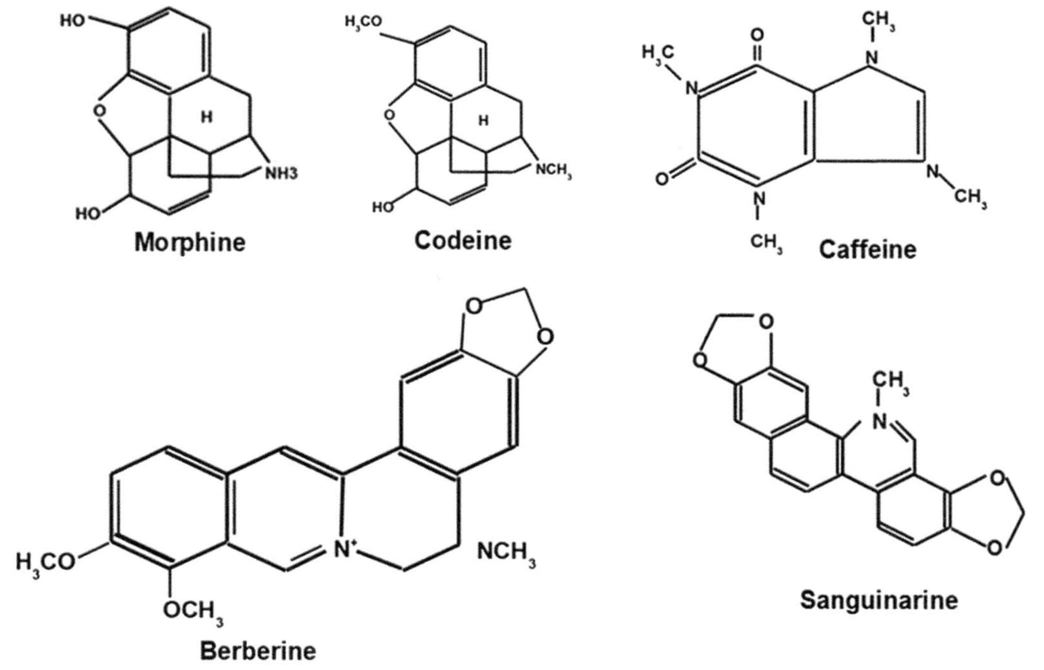

Morphine

Codeine

Caffeine

Berberine

Sanguinarine

FIGURE 11.1 Basic structures of some important plant derived alkaloids.

11.2.2 GLYCOSIDES

Glycosides are generally thought of as the end products of condensation between sugars (such as polysaccharides) and a variety of organic hydroxy (rarely thiol) compounds (always monohydrate in character), with the condition that the "hemiacetal entity" of the carbohydrate must essentially participate in the condensation. Glycosides are phytoconstituents present in cell sap; they are colorless, crystalline, and composed of oxygen and hydrogen (some also include nitrogen and sulfur). Chemically, glycosides include a portion that isn't a carbohydrate (glucose) (genin or aglycone) (Kar, 2007; Firn, 2010). Aglycones are represented by ethanol, phenol, or glycerol. Glycosides are reactively neutral and easily hydrolyze into their parts when combined with mineral or ferment acids. Based on the kind of sugar component, the chemical composition of the aglycone, or the pharmacological effect, glycosides are categorized. The suffix 'in' is often used in very outdated or trivial

TABLE 11.1
Methods for Alkaloid Detection

Reagent/test	Reagent composition	Result
"Wagner's reagent	Potassium iodide's iodine	Precipitate that is reddish-brown
Meyer's reagent	Potassiomercuric iodide solution	Cream Precipitate
Hagers reagent	A picric acid's saturated solution	Yellow Precipitation
Tannic acid	Tannic acid	Precipitation
Dragendorff's reagent	Hydrochloric acid drop containing a potassium bismuth iodide and potassium chlorate solution is evaporated to dryness, and the resultant residue is then subjected to ammonia vapor.	Reddish-brown or orange precipitation (except with a few additional alkaloids and caffeine)"

names of glycosides, and these names effectively contained the glycoside source, for instance: digitoxin from *Digitalis,* strophanthidin from *"Strophanthus",* salicin from *Salix,* barbaloin from *Aloes,* prunasin from *Prunus,* and cantharidin from *Cantharides.* Nevertheless, the systematic names are generally created by substituting the "side" suffix for the parent sugar's "ose" suffix. Typically, this class of medications is used to increase appetite and facilitate digestion. Even though they are chemically unrelated, glycosides—pure bitter fundamentals often observed in plants of the Gentianaceae family and have the same taste—intense bitterness. Saliva and gastric juice flow are enhanced as a consequence of the bitters' effect on gustatory neurons. Chemically, the bitter components comprise the lactone group, which may be either triterpenoids or diterpene lactones (such as andrographolide) (e.g., amarogentin). Owing to the existence of tannic acid, several bitter components are utilized as astringents, antiprotozoals, or to lower metabolism and thyroxine. Examples such as anthracene glycosides (purgative, and for skin diseases treatment), amarogentin, cardiac glycosides (acts on the heart), gentiopicrin, chalcone glycoside (anticancer), andrographolide, polygalin and ailanthone (Figure 11.2). According to Sarker and Nahar (2007), several medicinal products include plant extracts that comprise cyanogenic glycosides as flavoring ingredients. Amygdalin has been utilized as a cough suppressant in a variety of formulations as well as in cancer treatment (HCN released within the stomach destroys "malignant cells"). Consuming cyanogenic glycosides in excess might be deadly. If not handled appropriately, some foods that contain cyanogenic glycosides might result in poisoning (severe gastrointestinal irritations and damage) (Nahar & Sarker, 2007). The samples of the plant are hydrolyzed with HCl/H_2O to create their respective aglycones, and then an aqueous base, such as NH_4OH or NaOH solution, is added to test for "O-glycosides". The plant materials are hydrolyzed with $HCl/FeCl_3$ to produce C-glycosides, and then an aqueous base, such as NH_4OH or NaOH solution, is added. In both instances, a violet or pink color in the base layer after the addition of the "aqueous base" shows the existence of glycosides within a sample of plant.

11.2.3　Flavonoids

The plant flora contains a significant category of polyphenols called flavonoids. They are structurally formed of a variety of C15 aromatic compounds, which have more than one "benzene ring" and various studies support their usage as free radical scavengers or antioxidants (Kar, 2007). The

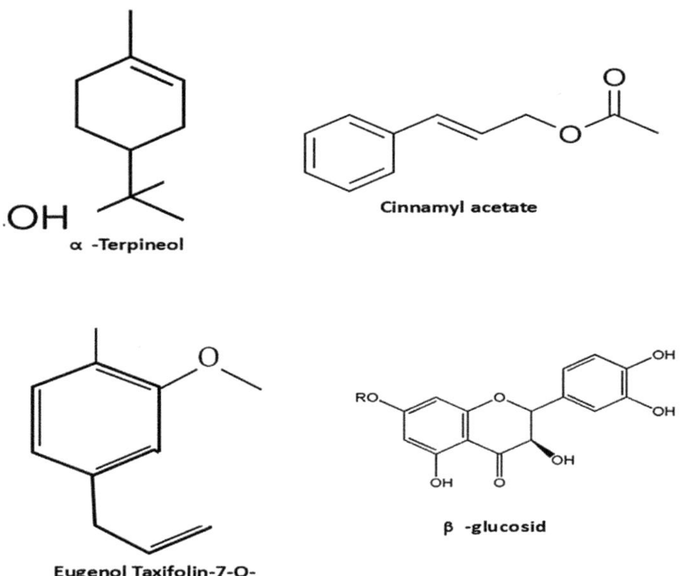

FIGURE 11.2　Basic structures of some important plant derived glycosides.

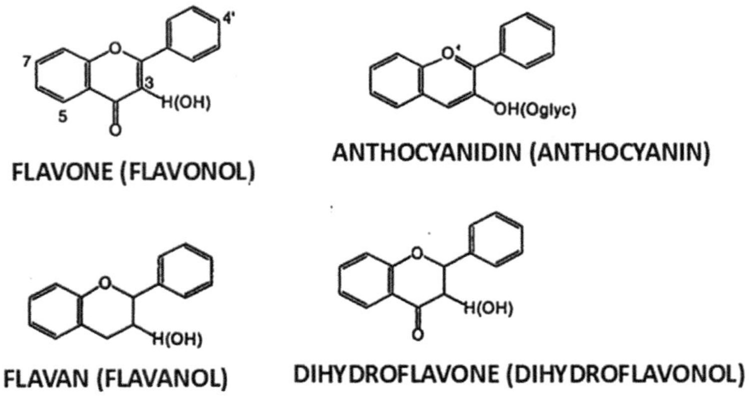

FLAVONE (FLAVONOL)

ANTHOCYANIDIN (ANTHOCYANIN)

FLAVAN (FLAVANOL)

DIHYDROFLAVONE (DIHYDROFLAVONOL)

FIGURE 11.3   Basic structures of some important plant derived flavonoids.

substances are generated from flavan-class parent substances. It is estimated that there are around 4,000 flavonoids, some of which are pigments within higher plants. Common flavonoids including quercetin, quercitrin, and kaempferol are observed in around 70% of plants. Dihydroflavons, flavones, anthocyanidins, flavonols, flavans, (Figure 11.3), proanthocyanidins, leucoanthocyanidins, catechin, and calchones are other groups of flavonoids.

11.2.4   PHENOLICS

The chemical elements known as phenols, polyphenolics, or phenols ("polyphenol extracts") are widely distributed as natural color pigments that give fruits of plants their color. The majority of phenolics within plants are produced from phenylalanine due to the activity of PAL ("Phenylalanine Ammonia Lyase"). They serve a variety of purposes and are crucial to plants. The most significant function could be in protecting plants against herbivore and pathogen predators, and as a result, they are used in the management of human pathogenic illnesses (Puupponen-Pimiä et al., 2008). They are divided into (a) flavonoid polyphenolics (xanthones, flavones, catechins, and flavonones), (b) non-flavonoid polyphenolics, and (c) phenolic acids. The most prevalent phenolic chemical found in plant flora is caffeic acid, followed by "chlorogenic acid," which is followed by "chlorogenic acid", which is referred to induce allergic dermatitis in humans (Kar, 2007). Phenolics are basically a variety of naturally occurring antioxidants that are used as nutraceuticals and are observed in foods like red wine, green tea, and apples because of their extraordinary ability to fight cancer. They are also considered to significantly decrease the risk of heart disease and occasionally act as anti-inflammatory agents. Chlorogenic, hesperidin, naringin, rutin, and flavones acid are more examples (Figure 11.4).

11.2.5   SAPONINS

The plant *Saponaria vaccaria* (also known as *Quillaja saponaria*), which is rich in saponins and was historically used as soap, is where the word 'saponin' originates. Therefore, saponins exhibit "soaplike" behavior in water, producing foam. An aglycone known as sapogenin is generated during hydrolysis. Triterpenoidal and steroid sapogenins are the two varieties. Due to the existence of a "hydroxyl group" at C-3 in the majority of sapogenins, the sugar is often connected at C-3 in saponins. The sapogenin senegin and the poisonous glycoside quillajic acid are both found in Quillaja Saponaria. Both senegin and quillajic acid are poisonous. Polygala senega contains senegin as well. When a sugar molecule is coupled with a steroid or triterpene aglycone, the result is a chemical with a high molecular weight known as a saponin. The two main categories of saponins are as follows: triterpene and steroid saponins. Saponins hydrolyze to produce aglycones, just

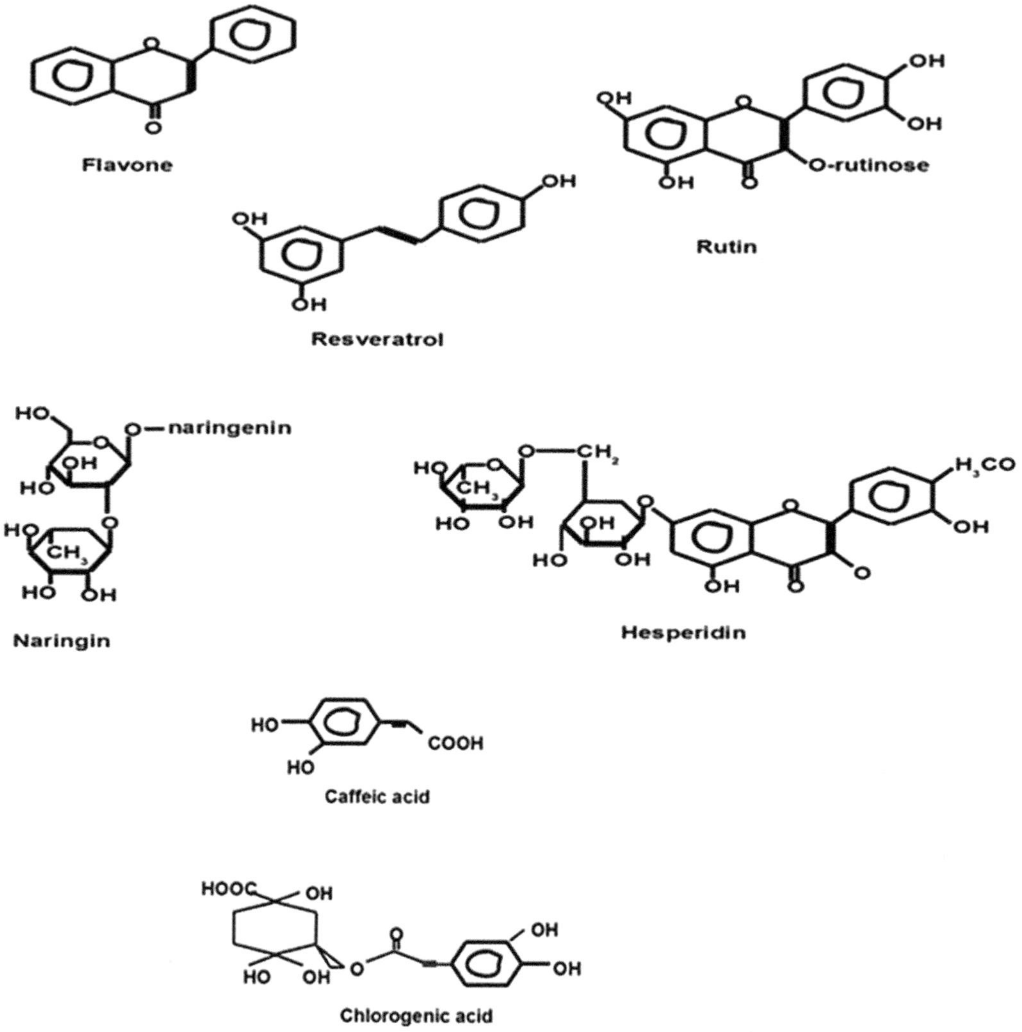

FIGURE 11.4 Basic structures of some important plant derived phenolics.

like glycosides, and are soluble within water but insoluble in ether. Because they hemolyze blood and are known to harm cattle, saponins are particularly dangerous (Kar, 2007). They taste sour and caustic, and irritate mucous membranes in addition to this. They mostly have an amorphous character and are soluble in water and alcohol but insoluble in "non-polar organic" solvents such as n-hexane and benzene.

Because they were found to have anticancer and hypolipidemic properties, saponins are also significant in therapeutics. Additionally, important for cardiac glycoside action are saponins. Diosgenin and hecogenin are the two main varieties of steroidal sapogenin. In the industrial manufacture of sex hormones for medical purposes, steroidal saponins are used. Diosgenin, for instance, is used to make progesterone. Diosgenin extracted from a species of Dioscorea, historically obtained from Mexico and currently obtained from China, is the most common precursor for the synthesis of progesterone. Hecogenin, which can be extracted from Sisal plants widely observed in East Africa, may be used as a starting material to create other steroidal hormones such as hydrocortisone and cortisone (Sarker & Nahar, 2007).

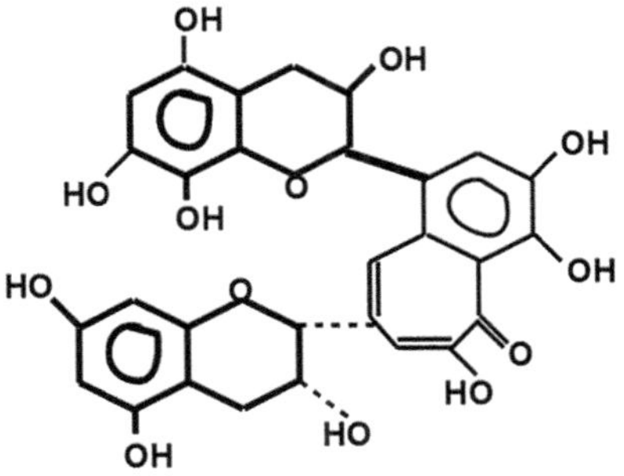

Gallic acid

Theaflavin

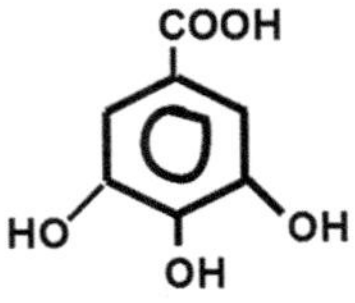

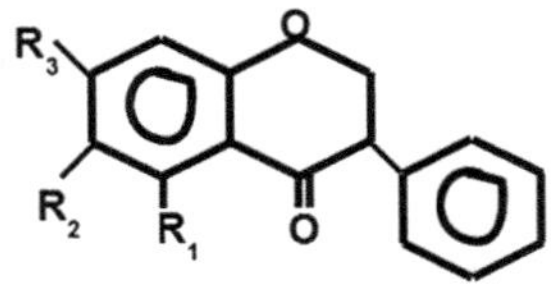

Daidzein: R$_1$=R$_2$=H; R$_3$=OH
Genistein: R$_1$=R$_3$=OH; R$_2$=H
Glycitein: R$_1$=H; R$_2$=OCH$_3$; R$_3$=OH

FIGURE 11.5   Basic structures of some important plant derived tannins.

11.2.6   TANNINS

Tannins are quite common in plant flora. High molecular weight phenolic compounds make up these substances. The stem, bark, root, as well as outer layers of plant tissue all contain tannins. They may dissolve in both alcohol and water. The ability to tan, or turn materials into leather, is a property of tannins. They undergo an acidic reaction, which is explained by the existence of carboxylic or phenolic groups (Kar, 2007). Complexes are created between them and alkaloids, gelatin, proteins, and carbohydrates. The two types of tannins are hydrolyzable and condensed tannins. Based on the kind of acid generated, hydrolyzable tannins are referred to as egallitannins or gallotannins when hydrolyzed to yield ellagic or gallic acid. The result of heating them is pyrogallic acid. The tannins' phenolic group is what gives them their antibacterial properties. Typical instances of hydrolyzable tannins are theaflavins (found in tea), glycitein, genistein, and daidezein (Fig 11.5). Tannin-rich medicinal herbs are utilized as cures for a variety of illnesses. Treatments for disorders including leucorrhea, rhinorrhea, and diarrhea have been utilized in Ayurveda which is based on tannin-rich herbs.

11.2.7   TERPENES

One of the most common and chemically varied categories of natural compounds is terpenes. They are combustible unsaturated hydrocarbons that are often present in liquid form in oleoresins, resins, as well as essential oils (Firn, 2010). According to the number of carbon atoms, terpenoids are hydrocarbons of plant origin with the general formula (C$_5$H$_8$) n and are categorized as mono-, di-, tri-, as well as sesquiterpenoids (Figure 11.6). Monoterpenes like terpinen-4-ol, menthol, eugenol, camphor, and thujone are examples of often significant monterpenes. Traditional definitions of

FIGURE 11.6   Basic structures of some important plant derived terpenes.

TABLE 11.2

Some Pharmacologically Significant Plant-derived Terpenes' Basic Structures

Forms of terpenoids	Number of carbon atoms	Number of isoprene units	Example
Monoterpene	10	2	Limonene
Sesquiterpene	15	3	Artemisinin
Diterpene	20	4	Forskolin
Triterpene	30	6	a-amyrin
Tetraterpene	40	8	b-carotene
Polymeric terpenoid	Several	Several	Rubber

diterpenes (C20) as resins include the anticancer drug taxol. The triterpenes (C30) are composed of sterols, steroids, and cardiac glycosides that have cytotoxic, sedative, anti-inflammatory, or insecticidal properties. Sesquiterpene (C15), a common triterpene like monoterpenes and one of the main components of many essential oils, includes oleanic acid, ursolic acid, and amyrins (Martinez et al., 2008). When ingested, sesquiterpenes have an effect similar to that of gastrointestinal tract irritants and function as irritants when administered topically. Many sesquiterpene lactones have been identified; in general, they function as neurotoxins and have antibacterial (especially antiprotozoal) properties. In addition to having anthelmintic properties, the "sesquiterpene lactone palasonin", which was isolated from "Butea monosperma", also slows glucose absorption and lowers Ascaridia galli's glycogen levels. According to how many isoprene units go into their creation, terpenoids are divided into several categories. Table 11.2 displays the major groups.

11.2.8 ANTHRAQUINONES

These are chemicals that are phenolic and glycosidic derivatives. Only anthracene is used in their production, which results in a variety of oxidized derivatives like anthranols and anthrones (Maurya et al., 2008; Firn, 2010). A "double hydroxylation" at positions C-1 & C-8 is a feature shared by several other compounds, including luteolin, salinosporamide, rhein, aloe-emodin, chrysophanol, and emodin (Figure 11.7). To check for free anthraquinones, powdered plant material is combined with an organic solvent, filtered, and then given a final addition of an aqueous base, such as an NH_4OH or NaOH solution. The existence of anthraquinones in the sample of the plant is indicated by a violet or pink color in the base layer (Sarker Nahar, 2007).

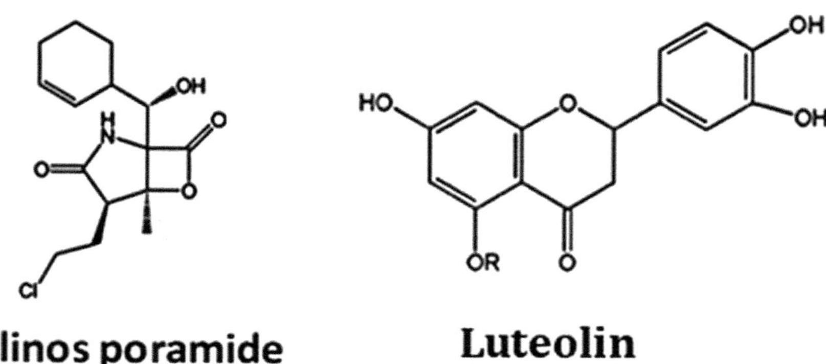

FIGURE 11.7 Basic structures of some important plant derived anthraquinones.

11.2.9 ESSENTIAL OILS

The pungent and flammable byproducts of numerous plant and animal species are known as essential oils. They are sometimes recognized as volatile or ethereal oils due to their tendency to evaporate when exposed to air, even under normal circumstances. They primarily contribute to the odoriferous components, or "essences," of aromatic plants, which are often utilized to improve the scent of various spices (Martinez et al., 2008). These oils are either released directly using the protoplasm of the plant or through the hydrolysis of certain glycosides and compounds. The following plant parts are linked to the release of essential oils: Glandular hairs (Lamiaceae, such as "Lavandula *angustifolia*"), modified parenchymal cells (Piperaceae, like *"Piper nigrum"*—Black pepper), oil tubes (or vittae) (Apiaceae, like Pimpinella anisum and Foeniculum *vulgare*—Aniseed), and lysigenous or Schizogenous passages (Rutaceae, such as *Pinus palustris*—Pine oil).

A variety of plant components, such as leaves, stems, flowers, roots, and rhizomes, have been linked to specific essential oils. More than 200 different chemical elements make up a single volatile oil; however, the bulk of these trace components are the only ones that give the oil its particular flavor and odour (Firn, 2010).

Direct steam distillation, extraction, expression, and enzymatic hydrolysis are all viable methods for producing essential oils from a wide range of plant materials. Plant materials are boiled in a distillation flask to extract their volatile oil, the steam is transferred via a water condenser, and the oil is collected in Florentine flasks. Water distillation, steam distillation, or direct distillation are all possible distillation methods based on the kind of plant source used. Volatile oils may be expressed or extracted via the sponge technique, rasping, scarification, or a mechanical process. In the sponge technique, a cleaned plant component, such as citrus fruit (such as orange, lemon, grapefruit, or bergamot), is sliced in half to thoroughly drain the juice, turned inside out by hand, as well as squeezed until the "secretary glands" burst. The sponge is used to catch the volatile oil that has leaked out and is then pressed into a container. Separation is done using the oil that is surfacing. The Ecuelle a Piquer equipment, a large bowl designed for pricking citrus fruit's skin, is used for the scarification procedure. It is a large copper funnel with a fully tinned inner layer. There are several tiny, pointed metal needles in the inner layer that are just long enough to pierce the skin. The bottom stem of the device has two functions: it receives the oil and also acts as a handle. When the cleaned lemons are put in the bowl and turned repeatedly, the oil glands are perforated (scarified) and the oil drains directly into the handle. The liquid is then transferred to a new container, where the clear oil could be stored and decanted before being filtered. A grater is skillfully used to remove the outer layer of the citrus fruit peel containing the oil gland for the rasping procedure. Now, the "raspings" are put in horsehair sacks and vigorously squeezed to release the oil that has been stored within oil glands. The liquid first seems muddy, but after standing for a while, the oil separates and could be decanted and filtered afterward. Heavy-duty centrifugal equipment is used in the mechanical process to help separate the naturally occurring oil/water emulsions. Since the invention of contemporary mechanical equipment, oil production has grown significantly. The extraction procedures can be done using either non-volatile solvents, such as lard, tallow, or olive oil, which produce perfumes, or volatile solvents, such as hexane, petroleum ether, or benzene, which produce "floral concretes," or oils with a solid consistency that is only partially soluble in 95% alcohol. Volatile oils include things like eugenol found in things like gein (Geum urbanum's volatile oil), sinigrin (black mustard's volatile oil), and amygdaline (bitter almond's volatile oil) (Figure 11.8).

11.2.10 STEROIDS

Steroid glycosides (or plant steroids) are among the most prevalent phytoconstituents identified in plants that have been used as arrow poisons or heart medicines, sometimes known as "cardiac glycosides" (Firn, 2010). The cardiac glycosides are essentially steroids that, when given intravenously to humans or animals, have the innate capacity to supply an extremely focused and potent activity, mainly on the heart muscle. Animals having osteoporosis and wasting disease have been observed

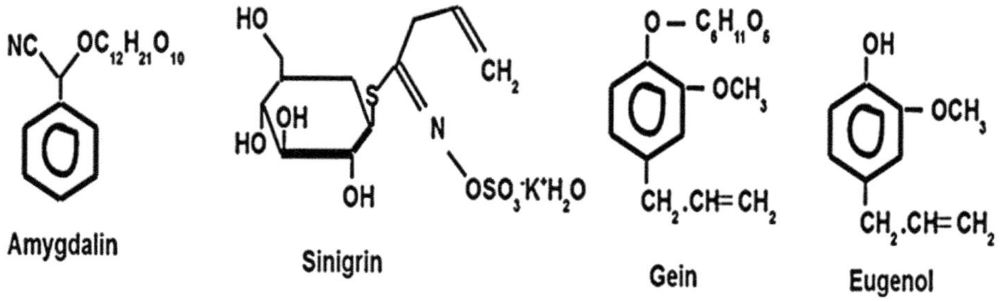

Amygdalin   **Sinigrin**   **Gein**   **Eugenol**

FIGURE 11.8   Basic structure of some important plant-derived essential oils.

to benefit from anabolic steroids' promotion of nitrogen retention (Maurya et al., 2008; Madziga et al., 2010). When employing steroidal glycosides, caution must be used since although modest doses might demonstrate the required stimulation on a sick heart, excessive dosage may even result in mortality. Plant steroids include Cevadine as well as diosgenin (from *Veratrum veride*).

11.3   EXTRACTION OF PHYTOCHEMICALS

11.3.1   Solvent Extraction

Different phytoconstituents have been extracted using various solvents. To reduce the initial high moisture content and permit a longer storage life, the plant parts are quickly dried, ideally in the shade or a controlled atmosphere at a low temperature (50–60°C). Mechanical grinders are used to grind up the dried berries, and solvent extraction is used to remove the oil. The defatted substance is subsequently extracted using a Soxhlet device, water, or alcohol (95% v/v), as well as other methods. After filtration, the resulting alcoholic extract is concentrated under vacuum or by the method of evaporation. Then, it is treated with HCl (12N) and refluxed for a minimum 6 hrs. The presence of phytoconstituents may subsequently be detected using a concentrated version of this. Since saponins often have large molecular weights, it might be difficult to isolate them in their purest form. The plant's components (roots, tubers, leaves, and stems) are cleaned, cut, and extracted for several hours in ethanol or hot water (95% v/v). The needed component is precipitated using ether after the final extract has been filtered and vacuo-concentrated.

EE ("Exhaustive extraction") is often done using several solvents when polarity is increased to remove as many of the biologically active components as feasible.

11.3.2   SFE (Supercritical Fluid Extraction)

The cutting-edge extraction method available (Patil & Shettigar, 2010). SFE involves using gases— typically CO_2—and compressing them into a thick liquid. The substance to be removed is then pushed through the cylinder holding this liquid. The liquid containing the extract is then pushed into a chamber where the extract and gas are separated and the gas is collected for later use. By adjusting the temperature as well as pressure at which one works, one may control and modify the solvent characteristics of carbon dioxide. The flexibility SFE offers in determining the components that you take out from a particular material and the fact that your completed product has very little residue of solvent are its two significant benefits (CO_2 evaporates completely). This technique has the drawback of being quite costly. Other gases and liquids may be particularly effective extraction solvents while exerting pressure, pressed (Patil & Shettigar, 2010).

(1) ***Coupled SFE-SFC***: A system that extracts a sample along with the supercritical fluid and provides the material into the input of a supercritical fluid chromatographic machine. Following that, the supercritical fluid is used to chromatograph the extract.

(2) *Coupled SFE-LC and SFE-GC:* A system that extracts a sample utilizing supercritical fluid and then depressurizes it to deposit the sample's extracted material, if necessary, in a column or the intake portion of a liquid or gas chromatographic system. SFE is distinguished by robust sample preparation, dependability, speed, and high yield. It also has the potential to be coupled with a variety of chromatographic techniques.

11.3.3 MICROWAVE-ASSISTED EXTRACTION

Microwave-Assisted Processing (MAP), as described by Patil and Shettigar (2010), is a brand-new method for solvent extraction assisted by microwaves. Applications for MAP include the extraction of valuable compounds from biomass, such as phytonutrients, components for functional and nutraceutical foods, and pharmaceutical actives. The following benefits of MAP technology may be seen in some combination when compared to traditional solvent extraction techniques: 1. Better products, more pure crude extracts, better marker compound stability, and the capacity to utilize fewer hazardous solvents; 2. Lower processing costs, more efficient extraction rates, purer and recovered marker compounds, and less energy and solvent consumption. It is possible to attain very rapid extraction rates and better solvent flexibility using extraction derived from microwaves as compared to diffusion. It is possible to adjust many factors to obtain desired product characteristics and enhance process economics, use energy-efficient technologies, such as microwave power. Excellent extracts are generated from a broad range of substrates with the process being able to be tailored to optimize for commercial and financial factors. The following are only a few examples: antioxidants derived from dried herbs, taxanes derived from taxus biomass, carotenoids derived from plant sources as well as single cells, important fatty acids derived from microalgae and oilseeds, polyphenols derived from green tea, phytosterols derived from medicinal plants, flavor constituents derived from black pepper as well as vanilla, essential oils derived from many sources, and many others (Patil & Shettigar, 2010).

11.3.4 SOLID PHASE EXTRACTION

The same rules that maintain molecules on stationary phases in chromatography also apply to the solutes adsorbed from a liquid medium on a solid adsorbent. Similar to chromatographic media, these adsorbents are available as resins or beads that may be utilized in columns or batches. They are often employed in the form of syringes, which are commercially available and contain a medium that may be gently driven through with a plunger or by suction. The amount of medium in these syringes ranges from a few hundred milligrams to a few kilos. Reverse phase, ion-exchange, and normal phase media are examples of solid phase extraction mediums. This sample purification method concentrates and separates the analyte from a mixture of isolated compounds via adsorption onto a single-use solid-phase cartridge. Usually, the analyte is cleaned, kept in the stationary phase, and then assessed with various mobile phases. When an aqueous extract is run through a packing material of a reverse-phase-filled column, anything that is relatively nonpolar would bind, but then everything polar will pass through.

11.4 BIOPOLYMER-BASED METHODS FOR ENCAPSULATION OF BIOACTIVE PHYTOCHEMICALS

Encapsulation is a cutting-edge technique that aims to increase the usefulness of materials with applications in a variety of fields, including health, pharmacy, agriculture, and the food industry (Jiang et al., 2015). Phytochemical offer several health benefits (Table 11.3) The goal of this technology is to entrap a solid, liquid, or gaseous substance, known as the fill, payload, core, active, encapsulate, or internal phase, inside another substance, known as the external phase, carrier material, coating material, wall material, shell material, or support phase, in the range of sized particles.

TABLE 11.3

Potential Health Benefits from Some Phytochemical Compounds (Dumitriu et al., 2016)

Source food	Phytochemicals	Health benefits
Carrots, tomatoes, and tomato Products	Carotenoids (Lycopene, beta-carotenes)	Neutralization of free radicals that cause cell damage
Olives, leeks, onions, garlic, scallions	Sulfides, thiols	Reduction in LDL cholesterol
Broccoli and other cruciferous vegetables such as kale, horseradish	Isothiocyanates (Sulforaphane)	Neutralization of free radicals that cause cell damage and provide protection against some cancers
Grape juice, red wine, grape extract, Cocoa	Proanthocyanidins and flavan-3-ols, Resveratrol	LDL oxidation inhibition, inhibition of proinflammatory responses, and cellular oxygenases inside arterial wall
Red wine, Strawberries, blueberries	Anthocyanins	Vision enhancement, nitric oxide production, induction of apoptosis, decreased platelet aggregation, as well as neuroprotective impacts Enhanced relaxation of calf aortas
Tofu, Soy Milk, and Soy Beans	Isoflavones (Daidzein and Genistein)	A drop in blood pressure and a rise in vessel dilation
Aromatic plants (essential Oils)	Terpene, perpenoide	Flavors, anti-inflammatory, antibacterial, antifungal, analgesic, sedative, and spasmolytic. They also exhibit anticancer, antiviral, and antidiabetic activities
Wheat bran	Sphingolipids, Phytosterols, Lignans, Phenolic acids	Protects against colon cancer

The particles between 1 nm and 100 nm are referred to as nanoparticles, while the particles between 100nm and 1000nm are referred to as microparticles (McClements, 2014).

There are two distinct kinds of micro- and nanoparticles in terms of morphology: Micro- and nanoparticle matrix kinds, also known as micro- and nanospheres, at which the primary substance is distributed in polymeric network spaces, and micro- and nanoparticle reservoir kind (or capsules), at which the active ingredient is included in a homogeneous core or cores (micro reservoirs), suspended in a protective membrane.

The most significant of the many factors requiring the encapsulation of phytochemicals is the protection of bioactive phytochemicals against environmental factors, like microorganisms, moisture, enzymes, oxygen, light, temperature, the improvement of bioactive compound solubility; the separation of bioactive molecules from other molecules of the food and drug matrix with which they could interact, as well as the controlled release of bioactive ingredients (Targeted medication delivery systems and controlled-release drug delivery systems); increasing the bio-availability of bioactive compounds because of an increase in bio accessibility; and allowing biomolecules to pass via biological barriers. Masking the flavor and smell that are offensive to some phytochemicals, like flavones, saponins, and vegetal peptides; converting liquids into freely flowing powders while reducing the bulk and volume of the biocomponent (McClements, 2014). The carrier substance affects bioactive component bioavailability, microparticle properties, and encapsulation effectiveness. Carrier materials that ensure food safety are used in the production of micro fond nanocapsules for the food and medicinal industries (GRAS-Generally Regarded as Safe). The carrier materials should be non-toxic, tasteless, odorless, and solvent-dispersible (ethyl alcohol, water). They need to be able to create elastic or thin films, have emulsifying capabilities, and be well-resistant to physicochemical agents including light, heat, ionic strength, pH, enzymes, moisture, and oxygen.

Proteins and polysaccharides are two of the major groups of natural biopolymers that are used to create biopolymer-based nanoparticles and microparticles because they are usually regarded as safe (Cerqueira et al., 2014). Proteins and polysaccharides both provide bioactive chemicals that are vulnerable to physicochemical influences, excellent protection, and enable their regulated release into the GIT ("Gastrointestinal Tract"). Pectin, for instance, is quickly damaged by the microflora in the colon that is undigested by the enzymes that are present in the stomach or small intestine. Pectin might be utilized to form polymer particles that release bioactive substances at the colon as a result (Livney et al., 2010).

Individual biopolymers or mixes of biopolymers may be used to make nano- or microparticles of biopolymers. Creating a three-dimensional network using polymer chains (gel) is the basis for the production of polymer particles from a single polymer. The pores of the gel allow for the entrapment of various colloidal systems (like liposomes, solid lipid particles, microemulsions, nanoemulsions, and conventional emulsions) that contain bioactive compounds (Dima et al., 2015).

When polymeric chains self-assemble via hydrophobic contacts or disulfide bonds, heating at a higher temperature than the temperature for thermal denaturation causes globular proteins to gel. Other proteins gel when their pH changes, their ionic strength changes, or they are in the presence of certain chemicals (genipin, glutaric aldehyde and multivalent ions) or enzymes, like transglutaminase. By increasing the pH level close to the isoelectric point (pH level is 4.6), by adding Ca^{2+} ions, or by utilizing rennet, casein gels. The following proteins are most often utilized in the production of polymer particles: whey protein (β- lactoglobulin, α-lactalbumin), gliadin, zein, gelatine, soy proteins, casein, and bovine serum albumin (Tavares et al., 2014).

Ionotropic, cold-set, and heat-set gelation are the three basic methods by which polysaccharides jellify (Wichchukit et al., 2013). Ionic groups ($-COO^-$, $-OSO3^-$, and $-NH3^+$) are grafted onto various kinds of sugar units in polysaccharides to form ionotropic gelation.

Pectin, a linear polysaccharide mostly made of (1-4)-D-galacturonic acid residues and alginate, a block-copolymer made of residues of β-Mannuronic acid as well as α–L-guluronic acid, both gel when exposed to multivalent cations (Ca^{2+}, Mg^{2+}). In particular GIT sectors, the pH-dependent behavior of alginate hydrogel particles enables the release of bioactive compounds. As a result, alginate beads shrink in the stomach due to the carboxyl groups protonation and prevent the release of biocomponents, while in the small intestine, where there are many ionized carboxyl groups, the polyanionic chains electrostatically reject one another, the alginate beads get swollen and encourage the release of biocomponents release (Fathia et al., 2014).

Another form of linear anionic polysaccharide is carrageenan, which has α-1,4 and β-1,3 linked residues of anhydrogalactose in its structure, each with a different amount of sulfate groups. There are several varieties of carrageenan according to the number of sulfate groups, but the three most significant ones are lambda (λ-), kappa (κ-), and iota (ι-). Carrageenans have a variety of topologies and gelation methods depending on how many negative charges they contain. As a result, whereas the gelation of λ-carrageenan takes place when potassium ions are present, that of κ-carrageenan takes place when calcium ions are present. Carrageenan creates strong, stiff gels when heated, while γ-carrageenan creating a soft, elastic gels. λ-carrageenan does not produce gels (Fathia et al., 2014).

N-acetylated glucosamine and glucosamine units combine to form the linear polysaccharide known as chitosan. By alkaline deacetylating chitin, a polymer with bioadhesive and antibacterial characteristics that is nontoxic, biocompatible, and biodegradable is created. The degree of deacetylation affects the substance's characteristics. It reacts with polyanions (tripolyphosphate, carrageenan, alginate) to generate hydrogels that are pH-sensitive when the amino groups are protonated and become soluble at low pH. Due to the protonated amino groups' electrostatic attraction to one another, the chitosan hydrogel expands at low pH, while chitosan precipitates at high pH. Chitosan may be used in the creation of gastric delivery systems because of its behavior (Cotarlet et al., 2014).

There are several different methods used to create biopolymeric-based micro and nanoparticles. The primary techniques for entrapping bioactive phytochemicals into polymeric particles, used in the creation of functional foods, are briefly described here.

External gelation:
Making hydrophilic or lipophilic phytochemical bioactive micro and nanospheres is a simple process. The following processes are covered in the external gelation approach for encapsulating bioactive components: a) Bioactive phytochemicals (carotenoids, essential oils, and polyphenols) are prepared in biopolymer solutions (pectin, κ-carrageenan, alginate, chitosan) that are then dropped into a gelation bath using a variety of techniques: b) concentric nozzle, coaxial air flow, electric field, atomizing disc, jet cutter, spraying nozzle, vibrating nozzle, syringe, and pipette; c) multivalent ions (Ca^{2+}, Mg^{2+}, K^{2+}, sodium tripolyphosphate) may cause the gelation of polymer droplets; d) Polymer microspheres are separated, dried, and stored. Microspheres between 10-800µm in size are produced by this process. The concentration of the gelation agents, the viscosity of the biopolymer solution, and the dropping technique all have an impact on the properties of the extruded microspheres (Dima et al., 2015).

The Emulsion-internal gelation method is used for the encapsulation of hydrophilic biocomponents. A W/O emulsion is created using the approach, and the discontinuous phase is mostly made up of droplets that include the bioactive component, the biopolymer's aqueous solution, and a source of multivalent ions that cause the biopolymers to gel (like $CaCO_3$). Acetic acid is added in the W/O emulsion to release them from their salts. A lipophilic surfactant (Span 80) may have been solubilized in vegetable oil to create the outer phase of the W/O emulsion. The process of separating the system's microspheres is the most challenging step in this approach. Sizes of the produced microspheres range from 0.5 to 100 µm (Cotirlet et al., 2013).

Complex coacervation involves entrapping bioactive substances into the complex of protein-polysaccharide or complex of an anionic-cationic polysaccharide. In certain pH and ionic strength conditions, the two biopolymers produce biopolymer complex coacervates due to electrostatic interaction between their electrical charges. Either cooling or the denaturation of the protein during heat causes the protein-polysaccharide complex to gel. Enzymatic reticulation agents (transglutaminase) or agents of chemical reticulation (epichlorohydrin, derivatives of malonic adipic acid, derivatives of glutaraldehyde acid) may be used to solidify the particles. Pimenta dioica is an important oil that was recently encapsulated in a complex of chitosan-κ- carrageenan in our lab to explore the effects of pH level and the temperature's impact on the processes of release as well as swelling (Dima et al., 2014).

Spray drying: It is among the first and most widely used techniques for encapsulating bioactive substances. The water in the small droplets quickly vaporizes when a biopolymers suspension or solution in water containing a bioactive component is atomized and sprayed in a drying chamber, creating microparticles with sizes between 10-100µm (Dima et al., 2015). The physicochemical features of the materials (water content, biopolymer concentration, viscosity, and wall material-to-core material ratio), as well as the technical factors, affect the characteristics of the microparticles (outlet temperature, inlet temperature, drying gas flow rate, and atomizing gas flow rate). This method's drawbacks include the limited usage for only hydrophilic wall materials and the usage when the temperature is high that might damage components that are thermosensitive.

11.5 GREEN TECHNOLOGIES AS FORTIFICATION ELICITORS OF FRUITS AND VEGETABLES IN BEVERAGES

Literature reports about a range of green technologies that can be utilized for phytochemical fortification (Table 11.4) and comprehensively discussed in upcoming sections.

TABLE 11.4

The Impact of Different Techniques on the Phytochemical Profile of Beverages with Quality Attributes

Technology	Beverage type and components	Treatment conditions	Optimum conditions	Shelf life	Results obtained	Reference
HPP	Apple-carrot-zucchini-pumpkin-leek smoothie	350 MPa; 10°C; 5 min	350 MPa 10°C; 5 min	28d at 4°C	High retention of vitamin C during storage	Hurtado et al. (2019)
					Retention of antioxidants (TPC and flavonoids)	
	Carrot juice	550 MPa;<38°C; 6min	550 MPa;<38°C;6min	20d at 4°C	Higher rheological properties Better Sensory attributes	Zhang et al. (2016)
	Carrot pumpkin smoothie	300–600 MPa; 23 ₒC; 5 min	400 MPa, 23°C, 5 min	7d at 5°C	Higher microbial count (approx. 6 logs lower counts) No high physicochemical changes (SSC, pH, and color)	Formica et al. (2018)
	Grape juice	500 MPa; 45°C; 5 min	500 MPa, 45°C,5 min	-	Reduction of 17-29% in aflatoxins	Pallares et al. (2021)
	Apple-orange-strawberry-banana smoothie	350–600 MPa; 10°C; 3–5 min	350MPa	48 h at 4°C	Preserves vitamin C and flavours Ensures microbial quality	Hurtado et al. (2019)"
PEF	Indian gooseberry juice	200–500 MPa; 30–60 °C; 5 min	500 MPa	-	Increase in TPC and TAC up to 50°c	Raj et al. (2019)
			30°C 5min		Less vitamin C degradation	
	Jucara-mango juice	600 MPa; 25°C; 5 min	600 MPa	-	Doesn't affect anthocyanin content	Moreira et al. (2017)
			25°C 5min		Good Sensory Properties	
	Orange juice	0–200 MPa; 25°C; 1 min (whole peeled orange) + 400 MPa; 40°C; 1 min (juice)	200 MPa 25°C 1 min + 400 MPa 40°C 1 min	-	Non-additive effect on flavonoids and vitamin C	De Ancos et al. (2020)
					A 12-fold increase in the content of colorless carotenoids	

(Continued)

TABLE 11.4 (CONTINUED)

The Impact of Different Techniques on the Phytochemical Profile of Beverages with Quality Attributes

Technology	Beverage type and components	Treatment conditions	Optimum conditions	Shelf life	Results obtained	Reference
	Grape juice	238 pulses up to 500 kJ/kg; sample: 215 mL;<75°C	238 pulses <75°C	-	Reductions by 24-84% in aflatoxins	Pallares et al. (2021)
	Grapefruit juice	20 kV/cm; 1 kHz; 80 mL/min;<45°C	20 kV/cm; 1 kHz <45°C	-	Lower non-enzymatic browning and viscosity than the untreated sample	Aadil et al. (2015)"
Ultrasound	Apple juice	2 W/cm2; 25 kHz; 70%; 30–60 min; 20°C; sample: 60 mL	2 W/cm2; 25 kHz 30 min 20°C	-	The highest phenolic and Sugar content Higher mineral and total carotenoids (60 min)	Abid et al. (2014)
	Apple-carrot-stevia juice	750W; 20 kHz; 20–80%; 15 min; sample: 100 mL	750W; 20 kHz; 60% 15 min	-	Better Phenolic profile Better radical scavenging activity	Khan et al. (2019)
	Grape-apple juice	750W; 20 kHz; 100%; 20–40 min; sample: 100 mL	750W; 20 kHz; 100% 20 min	-	Increased phenolic profile and TAC Higher organic acids	Ahmad et al. (2020)
	Grapefruit juice	720W; 28 kHz; 70%; 30–90 min; 20°C	720 W; 28 kHz; 70% 90 min 20°C	-	Improvement in sugar, carotenoid, mineral, and phenolic content Decreased spoilage microbe population	Aadil et al. (2015)

(Continued)

TABLE 11.4 (CONTINUED)

The Impact of Different Techniques on the Phytochemical Profile of Beverages with Quality Attributes

Technology	Beverage type and components	Treatment conditions	Optimum conditions	Shelf life	Results obtained	Reference
Cold Plasma	Pomegranate juice	Single-electrode atmospheric jet; 2.5 kV; 25 kHz; argon gas flow: 0.75–1.25dm3/min; time: 3–5 min; sample: 3–5 cm3	Single-electrode; 2.5 kV 0.75 dm3/min 3 min 5 cm3	-	Greater anthocyanin stability Less color changing with higher gas flow	Bursac Kovacevic et al. (2016)
	White grape juice	DBD; 80 kV; 60 Hz; time: 1–4 min; 24°C	DBD; 80 kV 1 min 24°C	-	Higher bioactive compound levels	Pankaj et al. (2017)
	Tomato juice	DBD; 10kV; 5 min; 30°C	DBD 10kV 5min 30°C	-	No effect on flavor and aroma Lower volatile Compound release	Ma et al. (2015)
	Tomato-coconut water-beetroot juice-based beverage	DBD; 60 kV; 50 Hz; time: 10 and 15 min; sample: 100 mL	DBD; 60 kV 10 min	-	Improvement of TPC	Mehta et al. (2019)"
Combined technologies	Mango juice	US-UV: 600 W; 20 kHz; pulses of 5 s on and 5 s off; 10 min; 3600 J/mL; sample: 100 mL UV: 254 nm; 8W lamp;	US-UV: 600 W; 254 nm; 3.6 kJ/mL; 10 min	30d at 4°C	Increased the bioaccessibility of ascorbic acid, TPC, and carotenoids by 102%, 114%, and 32% respectively	Wang et al. (2021)
	Orange juice	US + PEF US: 500 W; 30 kHz; 55°C; 10 min; sample: 800 mL PEF: 40 kV/cm; 15 Hz; 100 _s	US: 500 W; 10 min; 55°C + PEF: 40 kV/cm	168 d at 25°C	Lower colour differences Similar attributes to heat treatment	Walking-Ribeiro et al. (2009)

(Continued)

TABLE 11.4 (CONTINUED)

The Impact of Different Techniques on the Phytochemical Profile of Beverages with Quality Attributes

Technology	Beverage type and components	Treatment conditions	Optimum conditions	Shelf life	Results obtained	Reference
	Cranberry juice	US + HPP	US: 1.2 kW/L; 5 min; 25°C +	-	Higher anthocyanin content	Gomes et al. (2017)
		US: 600–1200 W/L; 18 kHz; <25°C; 5 min	HPP: 450 MPa; 5 min;		Good preservation of FOS	
		HPP: 450 MPa; 11.5°C; 5 min	11.5°C			
	Apple juice (cloudy)	HPP + UV	HPP: 300 MPa;	-	Increased TPC by 277.6%	Sauceda-Gálvez et
		HPP: 0–300 MPa; 32°C; sample: 13 L	32°C		Reduced PME activity	al. (2021)
		UV: 254 nm; 55 W lamp; 14.3–28.7 J/mL; 20°C;	+			
		sample: 70 mL	UV: 254 nm; 28.7 J/mL			

11.5.1 Ultraviolet

The phytochemical content of beverages may not be affected by utilizing UV light to decrease microbial loads (Table 11.4). It is important to keep in mind that many UV treatment characteristics, such as UV dosage, intensity, distance from the product, and others, should be taken into account when assessing the efficacy of these treatments (Martínez-Hernández et al., 2015). UV-C treating at 1.08 kJ m^{-2} of kale juice only reduced TPC (Total Phenolic Content) by a low (<20%) amount while achieving up to a 5-log decrease of inoculated E. coli (Pierscianowski et al., 2021). When Pseudomonas fluorescens, E. coli, and Saccharomyces cerevisiae were inoculated in juice of carrot-orange (Carrillo et al., 2017), microbial inactivation rate is high (2.5–5.9 log decreases) has also been also observed after the treatment of UV-C (10.6 kJ m^{-2}); meanwhile, a similar UV-C dosage (11.4 kJ m^{-2}) was intended to maintain the carrot juice sensory quality at the time of storage at 5°C.

TPC or TAC ("Total Antioxidant Capacity") alterations were not produced by greater UV-C dosages inside pineapple-mango or melon juice (8 or 16 kJ m^{-2}) (Fundo et al., 2019). TPC and flavonoid concentration were unaffected by treatment of UV-A (1.5 J m^{-2}), however, microbial spoilage was reduced by 1 log (Turkmen et al., 2018). UV treatments also improved the preservation of TPC, TAC, as well as physicochemical quality (viscosity, SSC, pH, and color) (Baykus et al., 2021). Interestingly, a mutual UV-C/UV-A treatment increased TAC as well as TPC in grape juice, lemon, ginger, carob, and carrot by 1.8 and 4.6 times, respectively (Baykus et al., 2021). Better preservation of phytochemical composition in such beverages may result from enhanced phytochemical extraction, impaired some phenolic molecules, broken down polyphenols into smaller components of phenolics, or/and antioxidant biosynthesis in response to free radicals generated by exposure to UV light (Fundo et al., 2019). However, the small phytochemical reductions that have been seen in certain studies might be attributed to the compounds' oxidation processes and double bond breakage, which are assisted by free radicals formed in UV exposure and subsequent photon absorption by oxygen or double bonds (Pierscianowski et al., 2021)

11.5.2 Pulsed Electric Field (PEF)

PEF treatment had no impact on the anthocyanin concentration in fruit juice. However, electroporation brought on by PEF treatments is known to improve the extractability of beneficial substances like anthocyanins, even if PEF could also encourage processes that lower the concentration of these phytochemicals. Anthocyanin content was predicted to be preserved when the previous authors added an antioxidant (Stevia) to fruit juice, and greater quantities were seen (Carbonell et al., 2016). The same scientists noted that Stevia had a similar advantageous impact to maintain the anthocyanin content in fruit juice after HPP (Carbonell-Capella et al., 2013). Additionally, they discovered that the element that affected anthocyanin content the most was the electric field, as opposed to treatment duration.

The carotenoid concentration of fruit juice after PEF was affected by both the electric field and the duration of treatment, demonstrating that lower electric fields may boost carotenoids. This is described by greater extractability after PEF and the low formation of ROS (which could promote the carotenoid chain oxidation).

However, PEF (20 to 40 kV/cm; 100 to 360 s; μs;<50°C) of fruit juice caused reductions in ascorbic acid, with stronger electric fields causing bigger reductions in this beneficial molecule regardless of the treatment duration (Carbonell et al., 2016). The authors attributed these adverse effects to the increased intracellular contents extractability during PEF owing to electroporation and the ensuing increased ascorbic acid oxidation processes.

As previously documented, PEF treatment may also lower the amount of aflatoxin in drinks. Following treatment of PEF (3 kV per cm, 238 pulses; 75°C) within grape juice, these authors obtained decreases of AFB1/AFG2, AFG1, and AFB2 of 24–25%, 84%, and 72%, respectively. PEF causes membrane holes to develop in the cell membrane, either temporarily or permanently. These membrane pores may change the structure of polysaccharides, proteins, and amino acids. The lower aflatoxin decreases following HPP evaluated with PEF discovered by Pallarés et al. (2021), may be explained by the fact that treatment of HPP has no influence on the breaking of covalent bonds and is transferred instantly. As a result, no gradients are

As shown, PEF is a superb green technology that achieves good microbial reduction, prolonging the shelf life of the product, while having no effect on phytochemical content of drinks. Therefore, following PEF treatment (25 kV per cm; 63 μs; <55°C), >5 log reductions of Staphylococcus aureus, Salmonella typhimurium, E. coli, and L. monocytogenes were obtained in apple juice, with the potential to increase product shelf life as shown in PEF-treated fruit juices & smoothies (Timmermans et al., 2016).

11.5.3 ULTRASOUND

Most phytochemical components, including phenolic compounds, flavonoids, and carotenoids, are frequently increased in drinks that have undergone ultrasonic processing. To enhance the extraction of chemicals from the food matrix, sonication is a common process in labs.

Fruit juice's TPC rose by up to 60% following treatments lasting 30–60 minutes (20–25 kHz) as evidence that phenolic contents may be boosted with US processing (Guerrouj et al., 2016). Even yet, the TPC was not enhanced by quicker (30min) beverage processing (Ahmad et al., 2020). Because of the strong antioxidant character of these molecules, such phenolic boosts were closely associated with increases in TAC (Albuquerque et al., 2021). After fruit juice's US processing (60 minute; 25 kHz), epicatechin & phlorizin raises could reach up to 76% and 130%, respectively. Chlorogenic acid was increased by 40%, caffeic acid by 20%, and catechin by 16% when researchers isolated these phenolic components to study them separately (Jabbar et al., 2014). Though there was no difference in the effects of treatments lasting 10 to 40 minutes, it is interesting to note that the concentrations of certain compounds of phenolic within nopal beverage were similarly increased following treatment. The authors' use of a high frequency (42 kHz) may help to explain their observations of phenolic improvements with brief treatments (10–20 minutes) that are comparable to processing for 40 min. Likewise, a vegetable-coconut beverage treated for 10 minutes at 37 kHz increased sinapic and gallic acid levels (Mehta et al., 2019). Because of this, each phenolic chemical responds to ultrasonic waves differently, and its concentration may either be raised or retained (Albuquerque et al., 2021). Just 15 to 30 minutes of processing at 20 to 24 kHz increased the overall flavonoid concentration of numerous fruit juices by 30 to 90%, suggesting that flavonoids are enhanced more than previously thought (Guerrouj et al., 2016).

Several factors might account for the rise in phenolic compounds in drinks after the US: (i) increased extractability as a result of the breakage of cell walls in the abrupt shift in liquid pressure caused by cavitation's shear force, which may enable the release of bound polyphenolic content; and (ii) hydroxyl radicals are attached to the phenolic substances' aromatic ring throughout sono-chemical events that take place while being processed in the US (Aadil et al., 2015). Moreover, the phenolic compounds' antioxidant activity has been demonstrated to enhance when a second hydroxyl group is added to the positions of ortho- or para-. Unchanged phenolic contents may be observed with short US processing times because sonication generates free radicals (such as hydroxy & hydrogen-free radicals) by dissociating water molecules in aqueous solutions due to the high pressure and temperature of collapsing gas bubbles related to cavitation (Guerrouj et al., 2016). U.S. processing also results in higher carotene levels in fruit juices. Apple juice's total carotenoid content rose by 27% after 60 minutes of processing at 25 kHz (Abid et al., 2014), while grapefruit juice's total carotenoid content increased by up to 40% after 90 minutes of processing at 28 kHz. Sonicated carrot juice made from carrots that had been blanched (put in a water bath at 100°C for 4 minutes) before processing of US had a content of higher carotenoid (1.7 &1.9 -fold more lutein and lycopene) than juice made from unblanched carrots (Jabbar et al., 2014). Blanching the raw product may cause the deactivation of degrading enzymes and/or additional damage of plant cells, which increases their extractability following the subsequent ultrasonic treatment, thus explaining the improved outcomes after US processing.

However, typical American processing methods tend to lower a beverage's ascorbic acid level. Ascorbic acid concentration in a vegetable-coconut drink was shown to decrease by 6–7% after being processed for 10–15 minutes at 37 kilohertz (Mehta et al., 2019). The amount of ascorbic acid lost during US processing, as shown in a nopal beverage after 20-40 minutes of treatment (42 kHz), increases as the treatment duration increases (Albuquerque et al., 2021). Ascorbic acid levels increased by 30% independent of treatment duration when the frequency of the ultrasound (US) was lower (24 kHz) and the temperature was higher (43–46°C) (Guerrouj et al., 2016). For the latter result, greater temperatures may account for more dissolved oxygen being removed from the mixture (leading to less ascorbic acid oxidation) in the sonication process. However, the water molecules sonication may produce hydrogen peroxide (H_2O_2), free radicals (O, OH), and hydrogen ions (H+) all of which can degrade ascorbic acid, which may explain the deterioration of ascorbic acid in drinks at temperatures 30°C (Mehta et al., 2019). In addition, medications available in the United States may not be enough to affect the stability of ascorbic acid's essential enzymes, the ascorbate oxidases. When compared to traditional heat treatments (80°C for 10 minutes), the reduction of ascorbic acid following US processing (40 minute; 42 kHz; <34°C) was still much lower (two-fold) (De Albuquerque et al., 2021).

11.5.4 Cold Plasma

A number of phytochemicals may also be enhanced by cold plasma processing. An ionized gas (helium, nitrogen, argon, carbon dioxide, air, or oxygen) that contains active particles like atoms, free radicals, ions, electrons, and makes up cold plasma. The covalent connections between phytochemical molecules and cell membranes may be broken by these reactive species because they possess enough electrical energy to do so. This accelerates the release of the compounds, increasing the number of free phytochemicals. The plasma source's properties, including its voltage, frequency, plasma generating system, kinds of gas, and flow rate, as well as their impact on the reactive species produced and the phytochemical contents, are determined by these factors.

In detail, a 3-minute, 2.8 mL sample volume along with 0.75 L per h gas (argon) flow were the ideal parameters to increase the sour cherry juice's TPC using single-electrode ("atmospheric plasma jet") at 2.5 kV, achieving a TPC enhancement of 15%. In the same research, a total anthocyanin increase in sour cherry juice of 34% was achieved by raising the volume of sample (3.2 mL) as well as the gas flow (1.25 L per hour) while retaining the treatment duration (3 minutes) (Garofulic et al., 2015). The existence of unidentified tiny particles or agglomerates within the juice that may get dissociated in plasma processing, according to the scientists, is what caused the larger anthocyanin levels. Associating with other anthocyanins intra- and intermolecularly or pigmenting with co-pigments like flavonoids and hydroxycinnamic acids may have also boosted the stability of the anthocyanin. Additionally, after being exposed to a plasma jet for 5 minutes at 1 L per minute argon for 3 mL of the sample, the pomegranate juice's TPC was found to increase by up to 50%, as reported by the aforementioned authors (Herceg et al., 2016). Hydroxycinnamic acids appear to be more stable than anthocyanins in chokeberry juice following treatment with a single-electrode plasma jet (Kovacevic et al., 2016). The authors claimed that since hydroxycinnamic acids are less efficient at reducing reagents, which accounts for their better stability, they react less with radical species produced using plasma. Maximum TPC improvements (7% greater) were achieved in blueberry juice employing a plasma jet (single electrode) with higher voltage (11 kV), oxygen (1%)-argon at 1 L per minute, and a longer treatment duration (6 minutes) (Hou et al., 2019). Only a 5% increase in the TPC was caused by "dielectric barrier discharge plasma" (60 kV) when applied to a 100 mL sample of coconut-vegetable juice for 10 minutes (Mehta et al., 2019). Regarding the sample's plasma orientation, dielectric barrier discharge (atmospheric direct plasma) application was more successful than indirect application in maintaining the TPC and, as a result, the TAC (Almeida et al., 2015).

TPC in cashew apple juice increased by 14–28% with the longest treatment (15 minutes) and highest levels at 50 mL/min of gas flow and greater phenolic retention. Researchers observed that whereas polyphenol retention was higher with plasma treatments of the same gas flow and duration, flavonoid retention was lower. This suggests that flavonoids may require less energy to be freed from their bindings (Rodriguez et al., 2017). Spark discharge (10.5 kV) for 4 to 5 minutes has been shown to boost the phenolic content of cloudy apple juice by the most (64% to 69% of the TPC). This significant phenolic increase may be a result of the cold plasma treatment's effective PPO inactivation (70 to 80%) (Illera et al., 2019). The scientists also noted that spark discharge results in a larger content of various radical species (NO_3 and H_2O_2), which may result in a higher enzyme inactivation, in line with earlier work. Furthermore, this cold plasma approach, which can promote cell membrane disintegration in comparison to other procedures, might be responsible for the substantial phenolic augmentation. However, further research is required to examine phenolic alterations in drinks using various cold plasma techniques. When processing beverages, vitamin C is one of the most labile phytochemicals; yet, cold plasma has minimal impact compared to traditional heat treatments. Because of this, a quick (2 minute) single-electrode plasma jet (cold plasma treatment) with various oxygen contents (up to 1%) in blueberry juice demonstrated greater vitamin C retention as compared to heat treatment (85°C, 15 minutes). Vitamin C was retained better, even after longer treatments of cold plasma, when the oxygen component of the ionized gas was 0%, the investigators found (up to 6 minutes) (Hou et al., 2019). In order to reduce vitamin C oxidation in cold therapy of plasma, orange juice was also tested. It was shown that using an air environment improved vitamin C retention compared to using an "oxygen-rich atmosphere" (65% O_2, 30% N_2, and 5% CO_2) (Xu et al., 2017). Although they may have a detrimental impact on the stability of vitamin C, reactive oxygen (nitrogen) species have been shown to initiate the most significant mechanisms responsible for the potent microbicidal action of cold plasma processing (Surowsky et al., 2014). Therefore, the levels of vitamin C within apple juice were increased by as much as 11% after being treated with cold plasma (indirect plasma field; 80 kHz) with nitrogen gas flow (Rodriguez et al., 2017). The dehydroascorbate reductase enzyme, which converts dehydro-ascorbic into ascorbic acid, is activated by some RNS, primarily nitric oxide, which is typically produced in cold plasma treatment, according to the authors, who explained why vitamin C levels increased after processing of cold plasma.

11.5.5 COMBINED TECHNOLOGIES

An efficient method to increase each processing technology's specific influence on product quality is to combine them to produce additive or synergistic benefits (Formica-Oliveira et al., 2016). Thus, high microbial inactivation (up to 5.8 log drop of S. cerevisiae) of the individual PEF and US was increased when they were used together in apple juice (30-minute US+60 s PEF) (Ferrario et al., 2015). Likewise, Alicyclobacillus acidoterrestris counts in apple juice were reduced by as much as 5 logs after being treated with US (120W for 5 minutes) and UV-C (20.2 kJ per m2) (Tremarin et al., 2017).

Additionally, using several technologies is a great way to increase the phytochemical content of drinks. Most research on combination technologies for the treatment of drinks involves the US owing to the great efficacy of ultrasound to maintain the quality of food, enhance the extractability of phytochemicals, and its relatively simple use in beverages. For instance, the combination of US (750W, 3 min) and cold plasma ("dielectric barrier discharge"; 70 kV, 3 to 4 minutes) boosted the effects of each technology individually by 22% & 34%, respectively, for TPC and entire carotenoid levels in carrot juice. In example, combined cold plasma + US treatment increased the lutein and lycopene content of carrot juice by around four and two times, respectively (Umair et al., 2019).

In prebiotic cranberry juice, US (18 kHz, 5 min) and HPP (450 MPa, 5 min) were applied; when 1200 W/L of power was applied, a rise in anthocyanins ranging from 14% to 20% was seen;

however, 600 W/L did not exhibit the same advantages (Gomes et al., 2017). Additionally, independent of the power level (600-1200 W per L), the same treatment of US-HPP enhanced the concentration of various fructo oligosaccharides (DP4-nystose as well as DP5-1-fructofuranosylnystose) (Gomes et al., 2017). The pectin methylesterase activity of turbid apple juice was likewise decreased by an HPP combination (300 MPa) along with UV (28.7 J per mL) from 190% relative activity (after a rise owing to a single treatment of HPP) to 60% (Gálvez et al., 2021). The TPC, ascorbic acid, as well as total carotenoid concentration of mango juice rose by around 35, 47, and 200%, when US (20 kHz, 10 minutes) and UV were combined. Additionally, these phytochemicals were more easily available in the treated mango juice, and the juice's bioaccessibility was increased, suggesting that it would be better preserved during cold storage if subjected to this combination treatment (Wang et al., 2021).

11.6 BEVERAGE FORTIFICATION USING NATURAL PRODUCTS

The process of fortification is used to proactively improve the bioactive, nutritional, and health-promoting elements in beverages. However, the content ("organic acids") and physicochemical characteristics (dissolved oxygen, pH) of the beverages may change the stability of the additional components (oxidations, interactions with other components). Table 11.5 lists the methods used to fortify drinks using natural ingredients. These methods are also discussed in the next section.

11.6.1 Beverages Fortification by Adding Plant Extracts as well as Other Health-Promoting Compounds

Plants are tremendous sources of phytochemicals, which have several health benefits. These substances can be utilized for technological purposes, such as antimicrobials (which prolong the shelf life of microorganisms and ensure food safety), antioxidants (which prevent degradation of quality caused by enzymatic systems), and masking of flavors which are undesirable flavors (for instance, hides the bitter taste of olive oil by adding it to beverages). Additionally, compared to edible plant parts, non-edible plant parts (often referred to as by-products in the food sector) may contain more phytochemicals. To minimize environmental effects and valorize phytochemicals for future use, their recycling becomes essential for the food industry's transition to a circular economy.

The amount of beverage phytochemicals increases practically proportionally (to the increased extract/compound content) in response to fortification with additional plant extracts. However, as previously mentioned, interactions between various substances (such as phenolic chemicals) may result in synergistic effects (McCarthy et al., 2013). However, great care must be taken when using certain extracts of plants (due to their sensory properties or when higher concentrations are being utilized), otherwise the fortified beverage's sensory acceptability may be threatened (Aderinola et al., 2018).

In general, research regarding the fortification of beverages from plant extracts has shown improvements in their antioxidant qualities, which are often linked to high phenolic compounds in the plant extracts. Therefore, adding beet leaf extract (30% of the total smoothie volume) to a fruit-vegetable smoothie resulted in a 50% rise in TPC and TAC (Fernandez et al., 2020). TAC rose by 120% when white or black brewers' waste grain extract was added to cranberry juice of cranberry at a 10% concentration (McCarthy et al., 2013). Similarly, adding clove extract (200g per mL) and pomegranate peel dry extract (2.5 mg per mL of beverage) to carrot juice of carrot and the juice of cucumber, respectively, increased the TAC, TPC, and flavonoid contents of the drinks contents which are flavonoid (Saad et al., 2021). By adding a tomato polyphenol extract (beverage's 0.5%), tomato juice's carotenoid content (lycopene and carotene) of tomato juice was fortified. Finally, adding vitamins from plant extracts to drinks before processing them is a great way to prevent further vitamin degradation, which is particularly important for extremely labile vitamins, during processing and subsequent storage of the beverages. As a result, adding mint leaf extract (8% of the beverage) enhanced the amount of vitamins C and A in a fruit smoothie (Aduloju et al., 2020)

TABLE 11.5

An Examination of the Main Phytochemical Fortification Circumstances in Beverages

Beverage type and component	Fortification conditions	Optimum conditions	Shelf life	Results	References
Apple and orange juices	Encapsulation of folic acid (synthesized) in mesoporous Silica particles	Encapsulated	-	Improved stability Controlled release after consumption by modifying vitamin bioaccessibility	Ruiz-Rico et al.,(2017)
Apple and orange juices	Free and microencapsulated L. acidophilus (10% and 30%)	Microencapsulated	63d at 4°C	Extended Survival	Da Silva et al. (2021)
Watermelon-apple-banana smoothie	Mint leaf extract (0–8%)	Mint leaf extract (8%)	-	High vitamin A, C, flavonoid, and TPC	Aduloju et al. (2020)
Watermelon juice	Citric acid, malic acid, or lemon juice (pH=3.8)	Non-centrifuged and addition of citric acid or lemon juice (pH=3.8)	20d at 4°C	Retention of sensory and functional qualities	Tarazona-Diaz et al. (2013)
Pineapple-banana-apple smoothie	Moringa Leaves (0–4.5%)	4.5% of moringa leaves	-	Increased vitamin C and E by 227% and 102% respectively Highest TPC and TAC	Aderinola et al. (2018)
Spinach-green apple-cucumber smoothie	2.5% of alga (Chlorella vulgaris and Dunaliella salina)	D.salina	28d at 5°C	Higher TPC and TAC Good Sensory attributes	Sahin et al. (2021)
Pasteurized peach juice	Lactobacillus acidophilus PTCC1643 and Lactobacillus Fermentum PTCC 1744	L.acidophilus	-	Inhibited Maillard reaction by 36.7%; 38.5% of anti-inflammatory activity Increased TAC by 74%	Mohammad Bagher et al. (2021)
Orange-celery-carrot lemon juice	Fermentation by L. plantarum strain HFC8	L.Plantarum	-	Low mesophilic aerobic bacteria and yeast and mold counts	Dogan et al. (2021)

The phytochemical pre-enhancement content of the raw materials of the plants that is utilized for the future production of the beverage is another method of beverage fortification. Several natural (chemical-free) postharvest abiotic stresses, including wounding, UVC, and hyperoxic environments, have been utilized for this. As a result, smoothie of carrot made with pre-incubated carrots and hyperoxia (80 kPaO2) treatment enhanced the beverage's TPC by 2060%. (Formica-Oliveira et al., 2017). Similar to this, after earlier blanching unpeeled carrots for 6 minutes at 80 degrees Celsius, the chlorogenic acid concentration of juice made from carrot increased by 3600% (Santana-Gálvez et al., 2019)

It has also been investigated to fortify drinks with purified compounds taken from plants and fish. The phytochemicals present in drinks may be preserved more effectively with the addition of specific substances. For instance, adding glutathione (500 mg/L) improved the anthocyanin concentration of blackberry juice for five weeks at 30°C (Stebbins et al., 2017). Melon juice that had been fortified with (-) epicatechin (beverage's 2.5 g/mL) had a 720% TPC increase (Aller et al., 2021).

In this regard, Tarazona-Daz & Aguaya investigated the impacts of pasteurizing, centrifuging, acidifying, and storing watermelon juice in the refrigerator (Table 10). According to their findings, juices that are non-centrifuged held at 4°C with a concentration of citrulline, polyphenols, and lycopene exhibited little degradation (Daz et al., 2013). In fact, it has been demonstrated that adding L-citrulline to Fashion watermelon juice (3.45 g per 500 mL) can reduce muscle pain in inexperienced male runners' insights from 24- 72hours after completing a half-marathon and maintenance of reduced plasma lactate concentrations after hard activity (Sánchez et al., 2017). Additionally, L-citrulline has been seen reduce recovery of heart rate and the soreness in muscle after 24 hours of muscle relief to the athletes (Daz et al., 2013), and this effect appears to be increased when combined with 22 mg of pomegranate ellagitannins per 200 mL for a juice of watermelon (Sánchez et al., 2017).

However, as phytochemicals have a low bioavailability in the human body, extra care must be taken to the supplemented concentration. For instance, polyphenols might have poor absorption, ranging from 0.3-43%, resulting in lower concentrations of circulating plasma (Manach et al., 2005). To regulate release after intake by changing vitamin bioavailability, mesoporous silica particles have been used to encapsulate the vitamin folic acid (to be used in fruit juices) to increase stability, lower the amount of component required, and control release after consumption (Rico et al., 2017). Similarly, when fish oil of 0.1% that had been microencapsulated using sophisticated coacervation was introduced to pomegranate juice, 16% of eicosanoid acid as well as 11% of docosahexaenoic acid were liberated (Habibi et al., 2017).

11.6.2 Fortification of Beverages by Adding Algae

Oriental societies (mostly in China, Korea, and Japan) have known for a long time about the culinary as well as health-improving benefits of algae found in marine, or seaweeds. Algae usage in medicines, food, and cosmetics (primarily as thickening agents) as well as its popularity as a gourmet condiment is both expanding daily in Western countries. According to Silva et al. (2021) marine algae are abundant sources of polysaccharides, proteins, minerals, polyphenols, and vitamins, including additional things, therefore adding them to beverage formulas may raise the phytochemical content of certain drinks (Castillejo et al., 2018). More intriguingly, according to Martínez-Hernández et al. (2018), these drinks may be enhanced with special health-promoting algal chemicals that are not present in plant-based goods, such as phlorotannins and fucoidans. However, as the fortified drinks may include undesired algal-related subtleties, careful attention has to made while number of algae is adding. As a result, kombu- and wakame-fortified drinks had the sensory ratings that are at the lowest out of 9 green smoothies made by the various marine algae, mostly because of the off-odors associated with those two algae (Castillejo et al., 2018).

The addition of microalga Dunaliella salina (beverage's 2.5%) boosted the TAC as well as TPC while maintaining the smoothie's high sensory quality (Sahin et al., 2021). Additionally, the amount of sugar in date nectar with spirulina added (at 10%) was increased, which is highly desired for covering up certain unfavorable flavors, like the bitter taste of kale drinks (Klug et al., 2019). The spirulina-fortified date nectar had greater sensory qualities, according to the authors.

11.6.3 Phytochemical Fortification of Beverages During Fermentation

The live organisms known as probiotics are consumed for their positive health effects, such as the reduction of serum cholesterol levels, the prevention of digestive illnesses, the promotion of antibacterial action, the control of lactose metabolism, the activation of the immune system, and antitumor and antimutagenic properties, among others (Dogan et al., 2021).For fermentation, which produces recognizable flavors and other sensory elements, lactic acid bacteria are often utilized in the food processing sector. In order to offer consumers drinks whose quality and increased capabilities of promoting health, the food industry makes significant use of probiotics *(such as Bifidobacteria and Lactobacillus)* for the processes of fermentation and/or fortification.

A beverage's phytochemical content may be increased by adding probiotics. For instance, after fermenting blueberry juice with Lactobacillus plantarum for 24 hours at 37°C and then for 2 hours at 4°C, the amount of phenolic and anthocyanin was raised by 43% and 15%, respectively (Zhang et al., 2021). Following L. plantarum fermentation (2 hours at 30°C continued for 28 days at 4°C), TPC and TAC in a variety of fruit juices were enhanced (by 6-8 times) (Dimitrellou et al., 2021). Similar to this, TPC and flavonoid and anthocyanin levels in fruit and vegetable juice rose following fermentation with L. plantarum (24h at 37°C followed by chilling at 4°C) (Dogan et al., 2021). Other Lactobacillus species also caused increases in other compounds such as sulforaphane, _-carotene, and riboflavin (Pedioccoccus pentosaceus-fermented broccoli juice). These include phenolic (Following fermentation with Lactobacillus paracasei, TPC increased by 49%) and TAC (Following fermentation with Lactobacillus acidophilus, there was an increase of 74%.) (Xu et al., 2021). Due to the strains employed having greater resistance to such conditions (enzymatic reactions and acidic pH), under gastrointestinal circumstances, probiotic life could be limited; probiotics may also be effectively microencapsulated to increase their survival (Dogan et al., 2021).

As shown, fermenting the drinks with the help of probiotic bacteria may result in a significant rise in the phytochemical content of certain beverages, especially phenolic compounds. The breakdown of large anthocyanin structures or macromolecular polyphenols into a smaller phenols, which also enhances their bioaccessibility via specific metabolism (such as deglycosylation) for a probiotic bacterium, is suggested to be the cause of such improvement (Zhang et al., 2021).

11.7 USE OF NANOTECHNOLOGY IN PHYTOCHEMICAL FORTIFICATION

Due to their low solubility, stability, and bioavailability, bioactive phytochemicals often have limited use in foods, supplements, and medications. These problems may be solved by phytochemical oral delivery systems (PODS), which are nanoparticles or microparticles that have been loaded with phytochemicals. PODS may be created in paste, gel, liquid, or solid form. They need to be properly developed to be cheap, durable, and sustain phytochemical bioactivity while being compatible with the product matrix. In this study, current developments in the creation of PODS, such as microgels, solid lipid nanoparticles, liposomes made of biopolymers, and microemulsions, are evaluated. An increase in the use of phytochemicals in commercial goods will result from properly built PODS.

The creation of colloidal delivery methods to encapsulate phytochemicals has attracted a lot of attention since it may enhance their chemical stability, water dispersibility, biological activity, and release profile. For usage as PODS (Phytochemical Oral Delivery Systems), a variety of various types of delivery systems are available, including emulsions, solid lipid nanoparticles, micelles, biopolymer microgels, and liposomes. For certain applications, each of these PODS offers benefits and downsides. Therefore, the maker must choose and refine the ideal colloidal delivery method for both the commercial product it will be put into and the phytochemical that has to be supplied. The pH, component interactions, ionic strength, thermal stability, and mechanical stresses of the final product matrix should be taken into consideration when choosing the composition, loading properties, size, and charge for the colloidal particles. The finished product should then have its appearance, shelf life, profile, rheology, and taste determined. Following that, PODS that work with the matrix of food may be found.

Upcoming research should concentrate on developing PODS that can be sold commercially. By utilizing GRAS (Generally Recognized as Safe) food components and widely used processing techniques, this will allow for their cost-effective production in huge volumes. Because they involve materials or processing techniques that are not economically viable, the majority of procedures explained in academic literature are not suitable for use in the industry. Additionally, many PODS developed for academic purposes are incompatible with product matrices because they have a negative impact on their taste, texture, appearance, or stability. Using animal and human feeding trials, an appropriate delivery method must be thoroughly examined to determine its effectiveness and possible toxicity. It will be vital for the delivery mechanism to be able to keep the phytochemical constant and functional during both product production and gastrointestinal tract transit. It should be shown that the encapsulated phytochemical can be absorbed in the proper part of the gastrointestinal system. It is obvious that PODS are crucial to improving the efficacy and utilization of phytochemicals. To make them economically feasible for use in commerce, further effort is needed (Zhang et al., 2021).

11.8 CONCLUSION

Green nonthermal methods may be used to fortify fruit and vegetable drinks, while high pressure and UV processing could provide increases in phytochemicals of beverages such as vitamin C, phenolic compounds, and anthocyanins, which ensure considerable antioxidant properties. Cell disruption is already a well-known use of ultrasound technology. Cold plasma and pulsed electric fields are two more environmentally friendly non-thermal technologies that may be utilized to boost the phytochemical content of fruit and vegetable drinks while also extending the shelf life of vegetable and fruit beverages. Additionally, since the majority of the literature now accessible focuses on the integration of such innovative approaches with ultrasound, the optimum combination of the technologies employed for therapy may significantly boost fortification rates. Future research may enhance our knowledge of various combinations of these technologies, where the appropriate processing conditions might change depending on the beverage's features (rheology, composition, pH). Additionally, it should be clarified if the fortification seen with these technologies is mostly the result of improved extractability after cell rupture or of biosynthetic processes, such as those that occur when substrate and enzymes come into contact.

REFERENCES

Aadil, R.M.; Zeng, X.A.; Sun, D.W.; Wang, M.S.; Liu, Z.W.; Zhang, Z.H. Combined effects of sonication and pulsed electric field on selected quality parameters of grapefruit juice. *LWT. Food Sci. Technol.* 2015, 62(1), 890–893.

Abid, M.; Jabbar, S.; Wu, T.; Hashim, M. M.; Hu, B.; Lei, S.; Zeng, X. Sonication enhances polyphenolic compounds, sugars, carotenoids and mineral elements of apple juice. *Ultrasonics sonochemistry*, 2014, 21(1), 93–97.

Abo, K.A.; Ogunleye, V.O.; Ashidi, J.S. Antimicrobial poteintial of *Spondias mombin*, *Croton zambesicus* and *Zygotritonia crocea*. *J. Pharm. Res.* 1991, 5(13), 494–497.

Action. *J. Nutr.*, 134(12), 3479S–3485S.

Adekunle, A.S.; Adekunle, O.C. Preliminary assessment of antimicrobial properties of aqueous extract of plants against infectious diseases. *Biol. Med.* 2009, 1(3), 20–24.

Aderinola, T. Nutritional, antioxidant and quality acceptability of smoothies supplemented with Moringa oleifera Leaves. *Beverages* 2018, 4(4), 104.

Aduloju, A.T.V.; Nwanja, N.M.; Ezegbe, C.C.; Okocha, K.S.; Aduloju, T.A. Phytochemicals and Vitamin Properties of Smoothie Flavoured with Mint Leaves Extract. *Int. J. Biochem. Res. Rev.* 2020, 29, 24–30.

Ahmad, K.; Imran, M.; Ahmad, T.; Ahmad, M.H.; Khan, M.K. Impact of ultrasound processing on physicochemical and bioactive attributes of grape based optimized fruit beverage. *Pak. J. Agric. Sci.* 2020, 57, 1117–1124.

Ahmad, K.; Imran, M.; Ahmad, T.; Ahmad, M.H.; Khan, M.K. Impact of ultrasound processing on physicochemical and bioactive attributes of grape based optimized fruit beverage. *Pak. J. Agric. Sci.* 2020, 57, 1117–1124.

Almeida, F.D.L.; Cavalcante, R.S.; Cullen, P.J.; Frias, J.M.; Bourke, P.; Fernandes, F.A.N.; Rodrigues, S. Effects of atmospheric cold plasma and ozone on prebiotic orange juice. *Innov. Food Sci. Emerg. Technol.* 2015, 32, 127–135.

Bayku͟s.; G.; Akgün, M.P.; Unluturk, S. Effects of ultraviolet-light emitting diodes (UV-LEDs) on microbial inactivation and quality attributes of mixed beverage made from blend of carrot, carob, ginger, grape and lemon juice. *Innov. Food Sci. Emerg. Technol.* 2021, 67, 102572.

Bursać Kovačević, D.; Gajdoš Kljusurıć, J.; Putnik, P.; Vukušić, T.; Herceg, Z.; Dragović-Uzelac, V. Stability of polyphenols in chokeberry juice treated with gas phase plasma. *Food Chem.* 2016, 212, 323–331.

Carbonell-Capella, J.M.; Barba, F.J.; Esteve, M.J.; Frígola, A. High pressure processing of fruit juice mixture sweetened with Stevia rebaudiana Bertoni: Optimal retention of physical and nutritional quality. *Innov. Food Sci. Emerg. Technol.* 2013, 18, 48–56.

Carbonell-Capella, J.M.; Buniowska, M.; Barba, F.J.; Grimi, N.; Vorobiev, E.; Esteve, M.J.; Frígola, A. Changes of Antioxidant compounds in a Fruit Juice-stevia rebaudiana Blend Processed by Pulsed Electric Technologies and ultrasound. *Food Bioprocess Technol.* 2016, 9(7), 1159–1168.

Castillejo, N.; Martínez-Hernández, G.B.; Goffi, V.; Gómez, P.A.; Aguayo, E.; Artés, F.; Artés-Hernández, F. Natural vitamin B12 and fucose supplementation of green smoothies with edible algae and related quality changes during their shelf life. *J. Sci. Food Agric.* 2018, 98, 2411–2421.

Cerqueira, M.A.; Pinheiro, A.C.; Silva, H.D.; Ramos, P.E.; Azevedo, M.A.; Flores-López, M.L.; Rivera, M.C.; Bourbon, A.I.; Ramos, O.L. *Food Eng Rev.* 2014, 6, 1–19.

Cousins, D.; Huffman, M.A. Medicinal properties in the diet of gorillas: An ethno-pharmacological evaluation. *Afr Study Monogr.* 2002, 23(2), 65–89.

Da Silva, T.M.; Pinto, V.S.; Soares, V.R.F.; Marotz, D.; Cichoski, A.J.; Zepka, L.Q.; Jacob Lopes, E.; de Bona da Silva, C.; de Menezes, C.R. Viability of microencapsulated Lactobacillus acidophilus by complex coacervation associated with enzymatic crosslinking under application in different fruit juices. *Food Res. Int.* 2021, 141, 110190.

De Albuquerque, J.G.; Escalona-Buendía, H.B.; de Magalhães Cordeiro, A.M.T.; dos Santos Lima, M.; de Souza Aquino, J.; da Silva Vasconcelos, M.A. Ultrasound treatment for improving the bioactive compounds and quality properties of a Brazilian nopal (Opuntia ficus-indica) beverage during shelf-life. *LWT. Food Sci. Technol.* 2021, 149, 111814.

De Ancos, B.; Rodrigo, M.J.; Sánchez-Moreno, C.; Pilar Cano, M.; Zacarías, L. Effect of high-pressure processing applied as pretreatment on carotenoids, flavonoids and vitamin C in juice of the sweet oranges "Navel" and the red-fleshed "cara cara". *Food Res. Int.* 2020, 132, 109105.

Dima, C.; Cotârlet, M.; Alexe, P.; Dima, S. Reprint of" Microencapsulation of essential oil of pimento [Pimenta dioica (L) Merr.] by chitosan/k-carrageenan complex coacervation method". *Innov Food Sci Emerg Technol.* 2014, 25, 97–105.

Dima, S.; Dima, C.; Gabriela Iordachescu, G. *Food Eng. Rev.* 2015. DOI 10.1007/s12393-015-9115-1.

Dimitrellou, D.; Kandylis, P.; Kokkinomagoulos, E.; Hatzikamari, M.; Bekatorou, A. Emmer-based beverage fortified with fruit juices. *Appl. Sci.* 2021, 11(7), 3116.

Dogan, K.; Akman, P.K.; Tornuk, F. Role of non-thermal treatments and fermentation with probiotic Lactobacillus plantarum on in vitro bioaccessibility of bioactives from vegetable juice. *J. Sci. Food Agric.* 2021, 101(11), 4779–4788.

Doughari, J.H.; Obidah, J.S. Antibacterial potentials of stem bark extracts of Leptadenia lancifolia against some pathogenic bacteria. *Pharmacologyonline* 2008, 3, 172–180.

Doughari, J.H.; Human, I.S.; Bennade, S.; Ndakidemi, P.A. Phytochemicals as chemotherapeutic agents and antioxidants: Possible solution to the control of antibiotic resistant verocytotoxin producing bacteria. *J. Med. Plants Res.* 2009, 3(11), 839–848.

Dumitriu, B.O.; Dima, S. Biopolymer-based techniques for encapsulation of phytochemicals bioactive in food and drug. *Rev. Mater. Plast.* 2016, 53, 126–112.

Elez Garofulıć, I.; Režek Jambrak, A.; Miloševıć, S.; Dragovıć-Uzelac, V.; Zorıć, Z.; Herceg, Z. The effect of gas phase plasma treatment on the anthocyanin and phenolic acid content of sour cherry Marasca (Prunus cerasus var. Marasca) juice. *LWT. Food Sci. Technol.* 2015, 62, 894–900.

Fathia, M.; Angel Martin, A.; Mcclements, D.J. *Trends Food Sci. Technol.* 2014, 39(1), 18–39.

Fernandez, M.V.; Bengardino, M.; Jagus, R.J.; Agüero, M.V. Enrichment and preservation of a vegetable smoothie with an antioxidant and antimicrobial extract obtained from beet by-products. *LWT* 2020, 117, 108622.

Ferrario, M.; Alzamora, S.M.; Guerrero, S. Study of the inactivation of spoilage microorganisms in apple juice by pulsed light and ultrasound. *Food Microbiol.* 2015, 46, 635–642.

Firn, R. *Nature's chemicals.* Oxford University Press, 2010, pp. 74–75.

Formica-Oliveira, A.C.; Martínez-Hernández, G.B.; Díaz-López, V.; Otón, M.; Artés, F.; Artés-Hernández, F. High hydrostatic pressure treatments for keeping quality of orange vegetables smoothies. *Acta Hortic.* 2018, 1194, 575–580.

Formica-Oliveira, A.C.; Martínez-Hernández, G.B.; Díaz-López, V.; Artés, F.; Artés-Hernández, F. Effects of UV-B and UV-C combination on phenolic compounds biosynthesis in fresh-cut carrots. *Postharvest Biol. Technol.* 2017, 127, 99–104.

Formica-Oliveira, A.C.; Martínez-Hernández, G.B.; Aguayo, E.; Gómez, P.A.; Artés, F.; Artés-Hernández, F. UV-C and hyperoxia abiotic stresses to improve healthiness of carrots: Study of combined effects. *J. Food Sci. Technol.* 2016, 53, 3465–3476.

Fundo, J.F.; Miller, F.A.; Mandro, G.F.; Tremarin, A.; Brandão, T.R.S.; Silva, C.L.M. UV-C light processing of Cantaloupe melon juice: Evaluation of the impact on microbiological, and some quality characteristics, during refrigerated storage. *LWT* 2019, 103, 247–252.

García Carrillo, M.; Ferrario, M.; Guerrero, S.; García Carrillo, M. Study of the inactivation of some microorganisms in turbid carrot-orange juice blend processed by ultraviolet light assisted by mild heat treatment. *J. Food Eng.* 2017, 212, 213–225.

Gomes,W.F.; Tiwari, B.K.; Rodriguez, Ó.; de Brito, E.S.; Fernandes, F.A.N.; Rodrigues, S. Effect of ultrasound followed by high pressure processing on prebiotic cranberry juice. *Food Chem.* 2017, 218, 261–268.

Guerrouj, K.; Sánchez-Rubio, M.; Taboada-Rodríguez, A.; Cava-Roda, R.M.; Marín-Iniesta, F. Sonication at mild temperatures enhances bioactive compounds and microbiological quality of orange juice. *Food Bioprod. Process.* 2016, 99, 20–28.

Habibi, A.; Keramat, J.; Hojjatoleslamy, M.; Tamjidi, F. Preparation of fish oil microcapsules by complex coacervation of gelatin–gum arabic and their utilization for fortification of pomegranate juice. *J. Food Process Eng.* 2017, 40(2), 1–11.

Heinrich-Salmeron, A.; Cordi, A.; Brochier-Armanet, C.; Halter, D.; Pagnout, C.; Abbaszadeh-Fard, E.; … Arsène-Ploetze, F. Unsuspected diversity of arsenite-oxidizing bacteria as revealed by widespread distribution of the aoxB gene in prokaryotes. *Appl Environ Microbiol.* 2011, 77(13), 4685–4692.

Herceg, Z. Kovačević, D.B.; Kljusurić, J.G.; Jambrak, A.R.; Zorić, Z.; Dragović-Uzelac, V. Gas phase plasma impact on phenolic compounds in pomegranate juice. *Food Chem.* 2016, 190, 665–672.

Hou, Y.; Wang, R.; Gan, Z.; Shao, T.; Zhang, X.; He, M.; Sun, A. Effect of cold plasma on blueberry juice quality. *Food Chem.* 2019, 290, 79–86.

Hurtado, A.; Guàrdia, M.D.; Picouet, P.; Jofré, A.; Bañón, S.; Ros, J.M. Shelf-life extension of multi-vegetables smoothies by high pressure processing compared with thermal treatment. Part II: Retention of selected nutrients and sensory quality. *J. Food Process. Preserv.* 2019, 43(11), e14210.

Illera, A.E.; Chaple, S.; Sanz, M.T.; Ng, S.; Lu, P.; Jones, J.; Carey, E.; Bourke, P. Effect of cold plasma on polyphenol oxidase inactivation in cloudy apple juice and on the quality parameters of the juice during storage. *Food Chem. X* 2019, 3, 100049.

Jabbar, S.; Abid, M.; Hu, B.; Wu, T.; Hashim, M.M.; Lei, S.; Zhu, X.; Zeng, X. Quality of carrot juice as influenced by blanching and sonication treatments. *LWT. Food Sci. Technol.* 2014, 55(1), 16–21.

Joye, I.J.; McClements, D.J. Biopolymer-based nanoparticles and microparticles: Fabrication, characterization, and application. *Curr. Opin. Colloid Interface Sci.* 2014, 19(5), 417–427.

Kar, A. *Pharmaocgnosy and pharmacobiotechnology (revised-Expanded Second Edition).* New Age International Limted Publishres, 2007, 332–600.

Khan, M.K.; Asif, M.N.; Ahmad, M.H.; Imran, M.; Arshad, M.S.; Hassan, S.; … Muhammad, N. Ultrasound-assisted optimal development and characterization of stevia-sweetened functional beverage. *J Food Qual.* 2019, 2019, 1–6.

Klug, T.V.; Collado, E.; Martínez-Hernández, G.B.; Artés, F.; Artés-Hernández, F. Effect of stevia supplementation of kale juice spheres on their quality changes during refrigerated shelf life. *J. Sci. Food Agric.* 2019, 99(5), 2384–2392.

Kovačević, D.B.; Kljusurić, J.G.; Putnik, P.; Vukušić, T.; Herceg, Z.; Dragović-Uzelac, V. Stability of polyphenols in chokeberry juice treated with gas phase plasma. *Food Chem.* 2016, 212, 323–331.

*Leptadenia lancifoli.*against some pathogenic bacteria. *Pharmacologyonline*, 3, 172–180.

Levy, S.B. The challenge of antibiotic resistance. *Sci. Am.* 1998, 278(3), 32–39.

Liu, R.H. Potential synergy of phytochemicals in cancer prevention: Mechanism of action. *J. Nutr.* 2004, 134(12), 3479S–3485S.

Ma, T.J.; Lan, W.S. Effects of non-thermal plasma sterilization on volatile components of tomato juice. *Int. J. Environ. Sci. Technol.* 2015, 12(12), 3767–3772.

Madziga, H.A.; Sanni, S.; Sandabe, U.K. Phytochemical and elemental analysis of *Acalypha wilkesiana* Leaf. *J. Am. Sci.* 2010, 6(11), 510–514.

Manach, C.; Williamson, G.; Morand, C.; Scalbert, A.; Rémésy, C. Bioavailability and bioefficacy of polyphenols in humans. I. Review of 97 bioavailability studies. *Am. J. Clin. Nutr.* 2005, 81, 230–242.

Martinez, M.J.A.; Lazaro, R.M.; del Olmo, L.M.B.; Benito, P.B. Anti-infectious activity in the Anthemideae tribe. In Atta-ur-Rahman (Ed.), *Studies in natural products chemistry*, Vol. 35. Elsevier, 2008, pp. 445–516.

Martínez-Hernández, G.B.; Huertas, J.-P.; Navarro-Rico, J.; Gómez, P.A.; Artés, F.; Palop, A.; Artés-Hernández, F. Inactivation kinetics of foodborne pathogens by UV-C radiation and its subsequent growth in fresh-cut kailan-hybrid broccoli. *Food Microbiol.* 2015, 46, 263–271.

Martínez-Hernández, G.B.; Castillejo, N.; Carrión-Monteagudo, M.M.; Artés, F.; Artés-Hernández, F. Nutritional and bioactive compounds of commercialized algae powders used as food supplements. *Food Sci. Technol. Int.* 2018, 24, 172–182.

Martínez-Sánchez, A.; Ramos-Campo, D.J.; Fernández-Lobato, B.; Rubio-Arias, J.A.; Alacid, F.; Aguayo, E. Biochemical, physiological, and performance response of a functional watermelon juice enriched in L-citrulline during a half-marathon race. *Food Nutr. Res.* 2017, 61(1), 1–12.

Maurya, R.; Singh, G.; Yadav, P.P. Antiosteoporotic agents from Natural sources. In Atta-ur-Rahman (Ed.), *Studies in Natural Products Chemistry*, Vol. 35. Elsevier, 2008, pp. 517–545.

McCarthy, A.L.; O'Callaghan, Y.C.; Neugart, S.; Piggott, C.O.; Connolly, A.; Jansen, M.A.K.; Krumbein, A.; Schreiner, M.; FitzGerald, R.J.; O'Brien, N.M. The hydroxycinnamic acid content of barley and brewers' spent grain (BSG) and the potential to incorporate phenolic extracts of BSG as antioxidants into fruit beverages. *Food Chem.* 2013, 141(3), 2567–2574.

McClements, D.J. Designing biopolymer microgels to encapsulate, protect and deliver bioactive components: Physicochemical aspects. *Adv Colloid Interface Sci.* 2017, 240, 31–59.

Mehta, D.; Sharma, N.; Bansal, V.; Sangwan, R.S.; Yadav, S.K. Impact of ultrasonication, ultraviolet and atmospheric cold plasma processing on quality parameters of tomato-based beverage in comparison with thermal processing. *Innov. Food Sci. Emerg. Technol.* 2019, 52, 343–349.

Mohammad Bagher Hashemi, S.; Jafarpour, D.; Jouki, M. Improving bioactive properties of peach juice using Lactobacillus strains fermentation: Antagonistic and anti-adhesion effects, anti-inflammatory and antioxidant properties, and Maillard reaction inhibition. *Food Chem.* 2021, 365, 130501.

Moreira, R.M.; Martins, M.L.; de Castro Leite Júnior, B.R.; Martins, E.M.F.; Ramos, A.M.; Cristianini, M.; da Rocha Campos, A.N.;Stringheta, P.C.; Silva, V.R.O.; Canuto, J.W.; et al Development of a juçara and Ubá mango juice mixture with added Lactobacillus rhamnosus GG processed by high pressure. *LWT. Food Sci. Technol.* 2017, 77, 259–268.

Nweze, E.L.; Okafor, J.L.; Njoku, O. Antimicrobial Activityies of methanolic extracts of *Trume guineesis* (Scchumn and Thorn) and *Morinda lucinda* used in Nigerian Herbal Medicinal practice. *J. Biol. Res. Biotechnol.* 2004, 2(1), 34–46.

Pallarés, N.; Berrada, H.; Tolosa, J.; Ferrer, E. Effect of high hydrostatic pressure (HPP) and pulsed electric field (PEF) technologies on reduction of aflatoxins in fruit juices. *LWT* 2021, 142, 111000.

Pankaj, S.K.;Wan, Z.; Colonna,W.; Keener, K.M. Effect of high voltage atmospheric cold plasma on white grape juice quality. *J. Sci. Food Agric.* 2017, 97(12), 4016–4021.

Patil, P.S.; Shettigar, R. An advancement of analytical techniques in herbal research. *J. Adv. Sci. Res.* 2010, 1(1), 08–14.

Pierscianowski, J.; Popoviⓒc, V.; Biancaniello, M.; Bissonnette, S.; Zhu, Y.; Koutchma, T. Continuous-flow UV-C processing of kale juice for the inactivation of E. coli and assessment of quality parameters. *Food Res. Int.* 2021, 140, 110085.

Puupponen-Pimiä, R.; Nohynek, L.; Ammann, S.; Oksman-Caldentey, K.M.; Buchert, J. Enzyme-assisted processing increases antimicrobial and antioxidant activity of bilberry. *J. Agric. Food Chem.* 2008, 56(3), 681–688.

Raj, A.S.; Chakraborty, S.; Rao, P.S. Thermal assisted high-pressure processing of Indian gooseberry (Embilica officinalis L.) juice Impact on colour and nutritional attributes. *LWT. Food Sci. Technol.* 2019, 99, 119–127.

Rios, J.L.; Recio, M.C. Medicinal plants and antimicrobial activity. *J Ethnopharmacol.* 2005, 100(1–2), 80–84.

Rodríguez, Ó.; Gomes, W.F.; Rodrigues, S.; Fernandes, F.A.N. Effect of indirect cold plasma treatment on cashew apple juice (Anacardium occidentale L.). *LWT. Food Sci. Technol.* 2017, 84, 457–463.

Rúa Aller, J.; González González, S.; Sanz Gómez, J.; del Valle Fernandez, M.P.; García-Armesto, M.R. Assessment of (-) epicatechin as natural additive for improving safety and functionality in fresh "Piel de Sapo" melon juice. *Food Sci. Nutr.* 2021, 9(6), 2925–2935.

Ruiz-Rico, M.; Pérez-Esteve, É.; Lerma-García, M.J.; Marcos, M.D.; Martínez-Máñez, R.; Barat, J.M. Protection of folic acid through encapsulation in mesoporous silica particles included in fruit juices. *Food Chem.* 2017, 218, 471–478.

Saad, A.M.; Mohamed, A.S.; El-Saadony, M.T.; Sitohy, M.Z. Palatable functional cucumber juices supplemented with polyphenolsrich herbal extracts. *LWT* 2021, 148, 111668.

Saganuwan, A.S. Some medicinal plants of Arabian Pennisula. *J. Med. Plants Res.* 2010, 4(9), 766–788.

Sahin, O.I.; Ozturk, B. Microalgal biomass—A bio-based additive: Evaluation of green smoothies during storage. *Int. Food Res. J.* 2021, 28(2), 309–316.

Santana-Gálvez, J.; Santacruz, A.; Cisneros-Zevallos, L.; Jacobo-Velázquez, D.A. Postharvest wounding stress in horticultural crops as a tool for designing novel functional foods and beverages with enhanced nutraceutical content: Carrot juice as a case study. *J. Food Sci.* 2019, 84, 1151–1161.

Sarker, S.D.; Nahar, L. *Chemistry for pharmacy students general, organic and natural product chemistry.* John Wiley and Sons, 2007, pp. 283–359.

Sauceda-Gálvez, J.N.; Codina-Torrella, I.; Martinez-Garcia, M.; Hernández-Herrero, M.M.; Gervilla, R.; Roig-Sagués, A.X. Combined effects of ultra-high pressure homogenization and short-wave ultraviolet radiation on the properties of cloudy apple juice. *LWT* 2021, 136, 110286.

Smolinski, M.S.; Hamburg, M.A.; Lederberg, J. Microbial threats to health. Emergence, detection and response. 2003. Institute of Medicine, National Academies Press, pp. 203–210.

Stebbins, N.B.; Howard, L.R.; Prior, R.L.; Brownmiller, C.; Mauromoustakos, A. Stabilization of anthocyanins in blackberry juice by glutathione fortification. *Food Funct.* 2017, 8(10), 3459–3468.

Surowsky, B.; Fröhling, A.; Gottschalk, N.; Schlüter, O.; Knorr, D. Impact of cold plasma on Citrobacter freundii in apple juice: Inactivation kinetics and mechanisms. *Int. J. Food Microbiol.* 2014, 174, 63–71.

Tarazona-Díaz, M.P.; Aguayo, E. Influence of acidification, pasteurization, centrifugation and storage time and temperature on watermelon juice quality. *J. Sci. Food Agric.* 2013, 93(15), 3863–3869.

Tavares, G.M.; Croguennec, T.; Carvalho, A.F.; Bouhallab, S. Milk proteins as encapsulation devices and delivery vehicles: Applications and trends. *Trends Food Sci. Technol.* 2014, 37(1), 5–20.

Timmermans, R.A.H.; Nederhoff, A.L.; Nierop Groot, M.N.; van Boekel, M.A.J.S.; Mastwijk, H.C. Effect of electrical field strength applied by PEF processing and storage temperature on the outgrowth of yeasts and moulds naturally present in a fresh fruit smoothie. *Int. J. Food Microbiol.* 2016, 230, 21–30.

Tremarin, A.; Brandão, T.R.S.; Silva, C.L.M. Application of ultraviolet radiation and ultrasound treatments for Alicyclobacillus acidoterrestris spores inactivation in apple juice. *LWT. Food Sci. Technol.* 2017, 78, 138–142.

Türkmen, F.U.; Takci, H.A.M. Ultraviolet-C and ultraviolet-B lights effect on black carrot (Daucus carota ssp. sativus) juice. *J. Food Meas. Charact.* 2018, 12(2), 1038–1046.

Umair, M.; Jabbar, S.; Senan, A.M.; Sultana, T.; Nasiru, M.M.; Shah, A.A.; Zhuang, H.; Jianhao, Z. Influence of combined effect of ultra-sonication and high-voltage cold plasma treatment on quality. *Foods* 2019, 8(11), 593.

Van den Bogaard, A.E.; Stobberingh, E.E. Epidemiology of resistance to antibiotics: Links between animals and humans. *Int. J. Antimicrob. Agents* 2000, 14(4), 327–335.

Voravuthikunchai, S.P. Activities of crude extracts of Thai medicinal plants on methicillin-resistant Staphylococcus aureus. *J. Clin. Microbiol. Infect.* 2003, 9, 236.

Walkling-Ribeiro, M.; Noci, F.; Cronin, D.A.; Lyng, J.G.; Morgan, D.J. Shelf life and sensory evaluation of orange juice after exposure to thermosonication and pulsed electric fields. *Food Bioprod. Process.* 2009, 87(2), 102–110.

Wang, J.; Xie, B.; Sun, Z. Quality parameters and bioactive compound bioaccessibility changes in probiotics fermented mango juice using ultraviolet-assisted ultrasonic pre-treatment during cold storage. *LWT* 2021, 137, 110438.

Wichchukit, S.; Oztop, M.H.; McCarthy, M.J.; McCarthy, K.L. Whey protein/alginate beads as carriers of a bioactive component. *Food Hydrocoll.* 2013, 33(1), 66–73.

Xu, L.; Garner, A.L.; Tao, B.; Keener, K.M. Microbial inactivation and quality changes in orange juice treated by high voltage atmospheric cold plasma. *Food Bioprocess Technol.* 2017, 10(10), 1778–1791.

Xu, X.; Bi, S.; Lao, F.; Chen, F.; Liao, X.; Wu, J. Induced changes in bioactive compounds of broccoli juices after fermented by animal- and plant-derived Pediococcus pentosaceus. *Food Chem.* 2021, 357, 129767.

Zhang, Y.; Liu,W.;Wei, Z.; Yin, B.; Man, C.; Jiang, Y. Enhancement of functional characteristics of blueberry juice fermented by Lactobacillus plantarum. *LWT* 2021, 139, 110590.

Zhang, Y.; Liu, X.C.; Wang, Y.; Zhao, F.; Sun, Z.; Liao, X. Quality comparison of carrot juices processed by high-pressure processing and high-temperature short-time processing. *Innov. Food Sci. Emerg. Technol.* 2016, 33, 135–144.

12 Multigrain Flour Fortification

Mehvish Habib, Iqra Qureshi, Sadaf Nazir,
Khalid Bashir, Kulsum Jan, Vaibhav K. Maurya,
and Amita Shakya

12.1 INTRODUCTION

A flour is a powder made by grinding raw grains, beans, nuts, seeds, or roots. Numerous foods are prepared from these flours. Cereal flour, specifically wheat flour, is the primary component of bread, a basic food in various cultures. Since ancient times, maize flour has played an important part in Mesoamerican cuisine and remained a staple in the Americas. Central and northern Europe uses rye flour as a component of bread. Composite flour is a combination of starches, flours, and other components that is designed to completely or partly replace wheat flour in pastry and bakery items, according to Milligan et al. (1981).

Cereal flour contains either the bran together (whole-grain flour), germ, and endosperm, or the endosperm alone (refined flour). The food is either distinguishable from flour on the basis of its somewhat coarser particle size (comminution degree) or is identical to flour; the term is used in both contexts. For instance, the term *cornmeal* commonly connotes a coarser texture while *corn flour* connotes fine powder, although there is no standardized dividing line. Raw flour must be cooked like other foods since it may contain germs such as *E. coli*.

A popular and affordable strategy to provide an adequate intake of micronutrients is to fortify flour with necessary vitamins and minerals. As part of the overall program for fortifying wheat, it is important to take into account other vitamins and minerals in addition to vitamin A. Multigrain has been a staple of the human diet for many various years. Eating whole grains is associated with several health benefits. Obesity, Heart disease, high blood pressure, diabetes, cholesterol, and CVD are all at a decreased risk. In order to maintain a healthy state of health, millets, and grains may provide additional nutrients, phytochemicals, and antioxidants. By increasing the amount of macro and micronutrients, dietary fibers, and phenolic compounds in a meal, antimutagenic, anti-glycemic, and anti-protective properties may be imparted (Agrawal et al. 2015)

12.2 MULTIGRAIN FOODS

The extension of the cereal-based industry has generated a need for the development of wholegrain-based multigrain products that can cater the need of consumers due to several health benefits. Consumption of wholegrain-based multigrain food products has been connected with the reduction of developing chronic disorders, like type 2 diabetes, cardiovascular diseases, some cancers, and other disorders. Food scientists are continuously developing new multigrain products by keeping the balance between nutrition and sensory parameters.

Multigrain food is a mixture of two or more grains such as flax, millet, corn, barley, oats, rice, and wheat among others. There are so many multigrain foods like multigrain cookies, multigrain bread, multigrain pasta, multigrain snacks, multigrain khakra and others. Micronutrients and dietary fiber are rich in multigrain food items. They are also linked with therapeutic health advantages such as bowel transition, maintaining gut health, decreasing blood sugar, and decreasing cholesterol levels

DOI: 10.1201/9781003160663-16

TABLE 12.1

Nutritional Composition of Different Flours

Flour	Composition	References
Wheat	Moisture=12.67%, Phosphorus (mg/100g)=107, Iron (mg/100g)=2.1, Calcium (mg/100g)=18, Ash=0.94%, Crude fiber=0.36%, Total Carbohydrate=74.88%, Fat=0.94%, Protein=10.55%,	Sahoo et al. (2012)
Pearl millet flour	Moisture= 10.1%, P (mg)==314, Ca (mg)=42, Ash=2.2%, Oil=4.0%, Protein=16.0%	Burton et al. (1972)
Chickpea flour	Moisture= 7.53%, Protein=20.1%, Fat content=5.7%, Crude fiber=8.3%, DPPH inhibition activity=80.64%, TPC (mgGAE/g)=3.44, Total solids=92.47%,	Hussein et al. (2020)
Soyabean flour	Moisture=8.0%, Crude fat=25.0%, Total carbohydrate=23.8%, Crude fiber=25.2%, Ash=6.0%, Crude protein= 38.2%	Akubor et al. (2003)
Oat flour	Protein=17.31g, Total starch=61.49g, β-glucan=4.68g, water absorption Index=279.24%, Water solubility Index=10.97%	Shuyi Liu et al. (2019)
Dried apple Pomace (g/100g)	Moisture=3.97 to 9.75, Fat=0.26 to 8.49, Protein= 1.2 to 6.91, Fructose= 11.5 to 49.8, Glucose==2.5 to 22.7, TDF (Total dietary fibers)=26.8 to 80.0, Pectin=3.5 to 14.32, Total polyphenolics=0.17 to 0.99, Malic acid=0.05 to 3.28, Ash=0.5 to 4.29	Antonic et al. (2020)
Raw pumpkin seed meal	Moisture=10.31%, Crude protein= 17.20%, Crude fat= 2.72%, Ash=1.85%, Carbohydrate=67.92%	Soukkary (2001)
Buckwheat Flour	Moisture=5.47%, Ash= 1.68%, Protein= 16.66%, Fat=3.42%, Fiber= 0.58%, Carbohydrate= 72.19%, Sodium (mg/100g)=20.59, Potassium (mg/100g)=360.89, Copper (mg/100g)=1.12, Zinc (mg/100g)=5.19, Iron (mg/100g)=4.95, Manganese (mg/100g)=1.23	Mohajan et al. (2019)
Flaxseed Flour	Moisture=8.66%, Ash= 3.59%, Fat=41.63%, Protein=16.07%, Dietary Fibre=9.97%, Carbohydrate= 30.13%	Masoodi et al. (2012)
Watermelon seed Flour (grams per 100g of dry weight sample)	Crude protein=35.66, Crude oil=50.10, Crude fiber=4.83, total ash=3.60, N-free extract=5.81	El-Adawy et al. (2001)
Paprika seed flour (grams per 100g of dry weight sample)	Crude protein=24.43, Crude oil=25.61, Crude fiber=34.91, total ash=4.32, N-free extract=10.73	El-Adawy et al. (2001)
Sorghum flour	Moisture=10.4%, Fat=9.5%, Protein=3.4%, Water absorption capacity (g/g)=1.40, Oil absorption capacity (g/g)==1.63, Emulsion capacity (g/g)=1.5	Singh et al. (1991)
Mango peel powder	Moisture (%)=10.5, Fat (%)=2.2, Ash (%)=3.0, Total protein (%)=3.6, Total carbohydrate (%)=80.7, Total dietary fiber (%)=51.2, Total polyphenols (mg GAE/g MPP)=96.2, Total Carotenoids (µg/g MPP)=3092, Free radical scavenging activity (µg MPP)=79.6	Ajila et al. (2010)

owing to their ideal content of dietary fiber (Angioloni and Collar 2011). The variety of fermentable soluble fibers was expanded by multigrain products.

12.2.1 MULTIGRAIN KHAKRA

Foods made from many grains are being considered more and more as a way to boost the nutritional value of goods. The crisp bread known as Indian *khakra* has a lot of potential for usage as a nutritious snack outside of local markets.

TABLE 12.2

Benefits of Adding Different Flours

Flour	Health Benefits	References
Wheat flour	Helps to control obesity in the body and helps maintain cholesterol levels. It also protects against high blood pressure, Type 2 Diabetes, colon cancer, heart attack, and breast Cancer	Randhawa et al. (2013)
Pearl millet flour	It helps in increasing HB and helps in dealing with constipation. It also has anti-cancer properties and inhibits the growth of tumors. It has a low glycemic index, which aids in the treatment of diabetes.	Malik et al. (2015)
Chickpea flour	Chickpea flour comprises agglutinins and lectins which may be anti-obesity, immunomodulatory, and anticancer in nature. It contains a high amount of protein, antioxidants, and phenolic content and solves the nutritious problems	Sahoo et al. (2012)
Soyabean flour	In addition, soybeans contain up to 45% protein, have a 91.41% digestibility rating, and are a strong source of vitamins and minerals. They also provide enough of the various amino acids needed to repair damaged body tissues. Consuming soy is linked to a decline in many diseases, such as cancer, atherosclerosis, and diabetes.	Taghdir et al. (2016).
Oat flour	Oats hold many opportunities for development as foods, industrial, and pharmaceutical products, which all add value to the oat crop. In nutraceutical production, components are isolated and purified from grain and sold in medicinal forms not usually associated with food products. Tailoring the nutritional and functional components of the oat grain will support many novel and potential uses of this crop	Sterna et al. (2016)
Dried apple pomace	It is rich in healthy bioactive compounds. The high content of dietary fibers in apple pomace is transferred to the treated final product, improving its nutritional profile. The increase in the antioxidant activity of the most fortified products has been observed because apple pomace is the main polyphenolic source of the whole fruit.	Antonic et al. (2020)
Raw pumpkin seed meal	Pumpkin seed products increase the water absorption, dough development time and softening but decrease the stability and loaf volume. Increasing the lysine, total sulfur amino acids, chemical score, total essential amino acids, crude protein, protein digestibility, crude fat, ash and mineral content qualifies pumpkin seed products as a good source of protein and nutrients for fortification of baked products.	Soukkary (2001)
Buckwheat flour	Buckwheat has many health benefits such as reducing plasma cholesterol levels, improving hypertension conditions, providing anti-inflammatory, neuro-protective, anticancer, and antidiabetic effects. These nutritional and health benefits make buckwheat a great choice for the formulation of different bakeries, crepes, pasta-noodles, cookies, cakes, breads, breakfast cereal formulations etc.	Mohajan et al. (2019)
Flaxseed flour	Flax seed has a pleasant nutty flavor with few or no significant side effects. The principal effect of flaxseed in humans and animal nutrition is the high levels of alpha linolenic acid (ALA c18; 3 n-3) in its oil and the high fiber in the seed.	Masoodi et al. (2012)
Watermelon seed flour	Watermelon seed kernels could be utilized successfully as sources of edible oils and protein for human consumption	El-Adawy et al. (2001)
Paprika seed flour	Paprika seed oil might be an acceptable substitute for highly unsaturated oils. Paprika seed flour has great potential for addition to food systems, not only as a nutrient supplement but also as a functional agent.	El-Adawy et al. (2001)
Sorghum flour	Sorghum possesses potential antioxidant, anti-inflammatory, and anticancer activities, and could improve glycemic response and insulin-related disorders, prevent dyslipidemia and cardiovascular diseases, and influence gut microbiota and promote colonic health.	Xiong et al. (2019)
Mango peel powder	Mango peel powder is a good source of phytochemicals such as polyphenols, carotenoids, and dietary fibers. Incorporation of MPP increased the polyphenol, carotenoid, and dietary fiber contents and it also exhibited improved antioxidant activity.	Ajila et al. (2010)

A large amount of arabinoxylans is offered by finger millet, sorghum, and pearl millet in khakra. Conventional unfermented flat crisp bread is called khakra. Dietary fibers in multigrain items improve the heart patient's health owing to their improved lipid-binding characteristics than single-grain products, and people who used mixed cereal products had lower lipid levels than those who did not. The starch molecules go through many physical alterations during the processing of starchy foods, resulting in the development of resistant starch based on the kind of starch and the degree of the processing conditions. Resistant starch's slow digestion has consequences for its usage in regulated glucose release applications; as a result, reduced insulin response and easier access to fat stores could be anticipated. This is essential for diabetes and has resulted in significant variations in dietary suggestions for diabetics. Whole wheat flour is an abundant source of energy 1670kJ/100g. Additionally, it is an abundant source of macronutrients such as vitamin K & B complex. The protein gluten of wheat is the primary component influencing its viscoelasticity and consequently its textural attributes.

It was reported that whole wheat khakra had a much greater moisture content (5.1%) than multigrain khakra (3.6%). Multigrain khakra had a considerably greater fat content as compared to control khakra. Again, this is explained by the fact that maize (4.5%) and pearl millet (4.8%) have greater fat concentrations than whole wheat flour (2.1%). Multigrain khakra had an ash content (3.6%) that was much greater than whole wheat khakra (2.3%). The ash level of sorghum (1.6%) and finger millet (1.8%) was shown to be substantially greater than that of whole wheat flour (1.1%), whereas the maize flour ash content (1.1%) was found to be comparable to that of wheat flour (Chauhan et al., 2017).

12.2.2 MULTIGRAIN COOKIES

Wheat flour-based cookies are key food products. These are taken by all age groups, but school-aged children like them the most. Cookies are perfect for their nutritive value, compactness, palatability, and convenience. For this purpose, wheat flour can be supplemented with other flour to make nutrient-enriched final food products, especially bread, biscuits, and cookies. Considering the health benefits of oat and barley flour in other food products, attempts have been made to explore the effect of the replacement of oat and barley with wheat flour on the Physicochemical and Sensory evaluation of multigrain cookies (Arshad et al. 2014).

In order to explore how the chemical makeup and organoleptic characteristics of multigrain cookies change after storage, Rathod et al. (2015) looked at standardizing the technique for making them. By performing preliminary testing with maida substitutions of 0, 10, 15, 20, and 25% (w/w) with defatted soy flour and oat flour, the level of all the components was standardized for the creation of cookies. The amounts of the other substances were kept constant throughout the studies. The fresh cookies had an average of 3.59% moisture content, 1.62% ash content, 22.97% crude fat content, 9.75% crude protein content, 2.55% crude fiber content, and 60.12% carbohydrate content. Defatted soy flour and oat flour were shown to have higher moisture, ash, protein, crude fiber, and crude fat contents in composite flour as compared to lower amounts of carbohydrates. To evaluate the storage stability, standardized cookies comprising 15% defatted soy flour and 15% oat flour were packaged in LDPE ("Low Density Polyethylene") and HDPE ("High Density Polyethylene") containers and kept at room temperature for 90 days. The sensory assessment of cookies was done on a regular basis, every 15 days. On the basis of the sensory quality of the cookies, the HDPE box was found to be the most suitable. According to the findings, it can be said that 15% defatted soy flour and 15% oat flour may be used to make cookies without having an impact on their general acceptance.

The use of therapeutic multigrain cookies in food items is now of attention due to consumer nutritional awareness and shifting demographics. According to Naval et al. (2018), fortification may improve the nutritional content of foods, especially baked items such as cookies. Sample cookies range in moisture content from 3.30 to 3.39%. Products ranged from 10.38–12.60, 16.32–17.10,

63.60–66.90, 1.33–1.76, and 1.30–1.55% in terms of fat, protein, ash, crude fiber, and carbohydrates (by difference). The inclusion of chickpea, wheat, maize, barley, finger millet, and pearl millet flour greatly enhanced the nutritional value in terms of carbohydrate, protein, and crude fiber content, according to sensory assessment and chemical characteristics of several formulations of multigrain cookies. Therefore, it was determined that (finger millet 6% + pearl millet 6% + maize 6% + barley 6% + chickpea 6% + wheat 70%) was the best.

The nutritional and sensory qualities of multigrain cookies were reported to be improved by the functional exploitation of mango waste by Aamir et al. (2022). An investigation was made on the functional qualities of multigrain cookies that had mango waste added as a source of bioactive chemicals. In addition to 10% barley flour and 10% oat, fortified cookies were made on purpose using kernel powder and mango peel in ratios of 2%, 4%, and 6% in plain wheat flour. The physical, proximate, antioxidant, and sensory properties of the baked cookies were examined. The findings showed that compared to mango peel powder, "mango kernel powder had higher levels of moisture (6.62±0.15%), ash (2.17±0.13%), protein (8.17±0.05%), and fat (1.67±0.02%) than mango peel powder. Additionally, mango peel" powder has greater concentrations of antioxidants and dietary fiber. The color and texture of the cookies were unaffected by the 2% mango kernel powder and 4% mango peel while replacing up to 6% led to unfavorable color changes. Due to the inclusion of mango kernel and mango peel, which improved their antioxidant activity, the quantity of phenolic and flavonoid compounds in cookies rose. The greatest sensory rating was given to cookies made with 2% peel powder and 2% kernel powder, and these cookies also generated the best findings for physicochemical and antioxidant analyses. The sensory research revealed that replacing up to 2% mango kernel and 4% mango peel powder produced cookies with a passable mango taste. The 2% mango kernel and peel were discovered to be quite important in enhancing the nutritious qualities of enriched cookies, it was established.

Effect of sugar and fat substitution on the textural, rheological, and nutritional properties of multigrain cookies was reported by Ashwath et al. (2021). In the current study, cookies with improved nutrition using whole wheat flour (WWF) and incorporation of multigrain mix (MM-oats, peas and fenugreek flours) at 0, 25, 50, 75 and 100% levels was studied. Further, fat was replaced using pumpkin seed (PS) or watermelon seed in the present study, cookies with improved nutrition using whole wheat flour (WWF) and incorporation of multigrain mix (MM-oats, peas and fenugreek flours) at 0, 25, 50, 75 and 100% levels was studied. Further, fat was replaced using pumpkin seed (PS) or watermelon seed (WS) at 25, 50 and 75% level and sugar was replaced using dry dates (DD) or raisins (RS) separately at 20, 40 and 60%. MM having protein at 15.13% and dietary fiber at 12.83% significantly decreased the water absorption (68.1–60.6%), stability (2.52–1.35 min), amylograph peak viscosity (665–821 BU), and cookie dough hardness (1737–690.5) at 100% MM. Based on the physico-sensory analysis, 75% replacement of WWF with MM was selected for replacement of fat or sugar. Addition of PS or WS increased the dough hardness (1235–4103 g), whereas the spread ratio of cookies decreased from 6.25 and 6.31 to 5.54 and 4.06 respectively. Replacement of fat with PS at 50%, sugar by DD at 40% along with a combination of sodium stearoyl lactylate (SSL) and glycerol mono stearate (GMS) showed improvement in the cookie texture. The scanning electron microscope (SEM) of cookie showed coating of starch granules and appearance of sheet-like covering of protein network. The mono and polyunsaturated fatty acid profile of cookies improved apart from a two-fold increase in protein and three-fold increase in dietary fiber.

12.2.3 MULTI-GRAIN BREAD

Bread is a vital processed ready-to-eat food. Bread has poor nutritional value due to the usage of processed wheat flour. Therefore, adding affordable staples like grains and pulses to wheat flour helps to enhance the nutritional level of wheat products (Sharma et al., 1999). To improve the flavor, fragrance, look, and nutritional value of bread, a current trend in the baking business is to include a variety of grains, legumes, and seeds rather than just one non-wheat cereal. Multigrain bread is also a better alternative since they include more important fiber, proteins, minerals, and vitamins. Multigrain breads are loaves produced from a variety of grains, nuts, and legumes. These breads are tastier, more flavorful, and more nutrient-dense than regular bread. The most popular source of vegetable proteins for strengthening baked products is soya beans. (Indrani 2010).

Indrani (2010) examined the effects of substituting 5, 10, 15, and 20% of the wheat flour with a multigrain mix (MGM) made of oats, soya bean, flaxseed, sesame seeds, and fenugreek seeds on the rheological and bread-making features of wheat flour. MGM has a negative impact on the texture and volume of bread when used in dosages ranging from 0% to 20%. It also causes protein matrix disruption, increases crumb stiffness value, and decreases volume. The dough strength and overall bread quality significantly improved with the addition of a mixture of additives to wheat flour with 15% MGM. Bread with 15% MGM had 1.5, 5.0, and 2.5 times more fat, protein, and dietary fiber than the control, respectively. The findings indicated that adding 15% MGM and a mixture of additives might create bread with better quality features and detectable multigrain flavor.

The primary objective of the invention of multigrain bread was to satisfy the rising demand for a healthy diet in light of the economy. The development of multigrain bread included substituting 5.10, 15, 20, and 30% of wheat flour with maize, barley, oat, and rice flour. To boost the bread's medicinal value, 1% of flax seeds were also added to the bread. A noticeable shift in the protein content was seen by changing the replacement levels. Similarly, ash, fiber, and fat are also affected differently by altering the flour ratios. The L*, a*, and b* values changed as a result of the color analysis. The samples' brown color intensified when additional fiber was added to them. The analysis of the texture profile (cohesiveness, chewiness, springiness, and hardness) improved by rising the proportion of composite flours in the blends. Physical properties (bread volume, dough expansion, and specific volume) increased when the number of mixes in the bread samples was decreased, and vice versa (Malik et al. 2015)

Dietary fiber (DF) with antioxidants and bioactive substances is thought to be a key component of cereal products' nutritional and health-promoting qualities. There is a growing market for wholegrain bread and seeds, thus it is important to look at how these additives affect the nutritional content of bread. The effect of wheat flour fortification on the content of PC (Polyphenols), NGOS ("Non-Glycemic Oligosaccharides"), antioxidant DF, and AA ("Antioxidant Activity") of the finished bread products was reported by Benítez et al. (2018). Depending on the nature of the flour, breadmaking had a variable impact on bioactive chemicals. While white bread had a higher fructan content, whole grain flour was the most sensitive to the breadmaking process. Both wholegrain and multigrain bread differentiated for their amounts of DF, PC, and AA. According to the findings, substituting refined flour with other grains like rye and oats, fiber like wheat bran, and seeds like flax, sesame, and sunflower would increase the potential health advantages of bread.

In an attempt to make functionally enhanced multigrain bread, different ratios of germinated kodo millet seeds (multigrain flour) and wheat were used to make the multigrain bread (30:70, 40:60, 50:50, 60:40, and 70:30). Of these, the 50:50 ratio was standardized for more research. Control and Set A are the two sets that are created. When this product's nutritional assessment was contrasted with that of commercial wheat bread, it was shown to be much higher in proteins, fiber, and antioxidants. Comparing Set, A to Control, the sensory investigation revealed the high acceptability of Set A. This demonstrates that it is a superior offering for customers who care about their health. (Sharma et al. 2018).

One of the main problems for the food business is the production of functional foods with a longer shelf life. By using OSP ("Onion Skin Powder") as a natural preservative, Sagar and Pareek (2021) intended to increase the phenolics, shelf life, and flavonoids content, and antioxidant potential of multigrain bread. OSP at 1, 2, 3, or 4% was combined with 9 multigrain flour (40%) and whole wheat flour (60%) to create 4 distinct multigrain functional bread and a control (without OSP). Total flavonoids, phenols, and AA, as well as textural and sensory qualities, were measured in the products. In order to compare the shelf-life of bread stored at "ambient 28 ± 2 °C and refrigerated 5 ± 2 °C," the total viable count was also examined. Texture examination revealed that the inclusion of OSP reduced the chewiness and hardness, while sensory analysis revealed that MGB with 3% OSP was highly favored in terms of taste, mouthfeel, color, and general acceptability. The OSP-enriched samples showed a considerable increase in total "phenols and total flavonoids. Antioxidant assays showed that 4% OSP MGB had the highest DPPH* (53.1 ± 0.3) %, ABTS*+ (39.4 ± 0.3) %, and FRAP activity (38.8 ± 0.4 µmol gallic acid equivalents/g) while the control was the lowest." According to a storage trial, ambient and refrigerated settings increased the "shelf life of OSP-enriched" MGB by 11 and 13 days, respectively. The shelf life of the items was found to be increased by refrigeration more so than by keeping them at room temperature.

12.2.4 MULTI-GRAIN COMPOSITE MIXES

Various beans, grains, millets, nuts, and sauces were combined using diverse techniques to make multigrain composite mixtures. The percentages of moisture, carbohydrates, protein, crude lipids, and ash in multigrain composite mixes were 10–12%, 56–61%, 15–20%, and 2–3%, respectively. Energy content was between ~1600 to 1700 kJ/100g (Itagi and Singh 2012).

12.2.5 MULTI-MILLET DUMPLINGS

The product designed had various millets such as jowar, bajra, kodo, proso, barnyard, and ragi. It was observed to be rich in fiber, protein, B-complex vitamins, iron, and calcium. Multi-millet dumplings were a millet-based product as millets are coarse grains, they are a repository of minerals, vitamins, fiber, and protein. In semi-arid locations, millets are an essential food and fodder crop, and their importance is rising every day as the number of malnourished people rises. More than 90 million people in Africa and Asia and around 500 million people in more than 30 nations depend on millet as a staple food. These millets are employed as nutraceuticals because they have substantially greater antioxidant content than the main cereal crops. According to reports, they help prevent heart attacks, migraines, blood pressure, diabetic heart disease, and atherosclerosis. Millets may be employed as nutraceuticals and in functional meals due to their advantages. Therefore, they are also known as "nutricereals" (Kulkarni et al., 2011).

12.2.6 MULTI-GRAIN PASTA

Traditional durum wheat pasta is a staple diet that contains 11 to 15% protein and 74 to 77% carbohydrates but is deficient in phenolic, vital amino acids, vitamins, and minerals compounds (Brennan, 2013). Currently, there is a rising public interest in the intake of nutrient-rich, low GI (Glycemic Index) foods that may slow down or limit the amount of quickly digested starch and offer protection against the dangers of cardiovascular disease, diabetes, and neurodegenerative conditions (Desai, 2018). Pasta may be a useful supplement for the addition of many nutrients, like bioactive substances ("phenolic compounds"), protein, and dietary fibre to the diet to provide low GI and other health advantages (Kamble et al., 2019). Due to their strong antioxidant activity, fortifying pasta with compounds of phenolic could be beneficial in preventing many illnesses including cardiovascular disease, cancer, and neurological illnesses that are linked to oxidative damage (Seczyk et al. 2016). High fibre, low gluten whole grain flours of barley, wheat, maize, soybean,mungbean, millets, flaxseeds, and oats were combined in different proportions as a partial substitute for wheat flour for the production of nutritionally enriched pasta by Harsimran Kaur et al. (2017). Selected samples of the prepared pasta were packaged in HDPE bags and kept in storage for 90 days. After that time, the stored samples were examined for changes in their free fatty acids, moisture content, cooking quality, water activity, and general acceptability at 30-day intervals. When using multigrain flour instead of wheat flour to make pasta, the volume expansion and water absorption rose dramatically (p<0.05), and the cooking time dropped greatly (p<0.05).When using multigrain flour, the loss of solids in the cooking water rose considerably (p<0.05). Multigrain pasta's color attributes and general acceptance ratings were on par with those of wheat-based pasta. The 90-day storage duration has no appreciable impact on the pasta's cooking quality. During the 90 days of storage, a significant (p<0.05) rise in water activity, moisture content, & free fatty acid was discovered. However, the quality criteria changed during storage, although they did so within acceptable bounds.Dinkar et al., (2019) used a central composite model to optimize the flour combination of gluten (3 to 5%), finger millet (10 to 25%), and sorghum (20 to 35%), on 5 responses, including OA ("Overall Acceptability"), firmness, WAI ("Water Absorption Index"), CT ("Cooking Time"),and CL ("Cooking Loss") of multigrain pasta. The physicochemical,antioxidant/antinutritional, IVPD ("In Vitro Protein Digestibility"), and structural features of the improved pasta were assessed. The Ideal concentrations of gluten finger millet, and sorghum, were 3.40, 13.04, and 31.96%,respectively, resulting in OA=7.30 $\pm$ 0.20, firmness=11.05 $\pm$ 2.40 N, WAI=119.54 $\pm$2.54%, CT = 6.10 $\pm$ 0.05 min and CL=9.69 $\pm$ 0.44%. The antioxidant analysis indicated greater "1,1-diphenyl-2-picryl- hydrazyl activity", decreasing power of ferric, and ("2,2-azino-bis-3-ethylbenzothiazoline-6-sulfonic acid") activity of improved pasta. The optimized pasta IVPD (76.51%) changed considerably (89.69%) than the control pasta. However, XRD and FTIR patterns showed that the secondary structure of both varieties of pasta was comparable.The cereal sector has been encouraged to develop millet-based goods due to the increased demand for eating nutritiously enhanced healthful meals. Nevertheless, little research has been done on the qualitative characteristics of pasta made from finger millet or sorghum.Sorghum and finger millet might both be utilized to manufacture pasta that has several health advantages, according to higher total phenolic content as well as antioxidant activity.The physical, chemical, technical, and sensory properties of fresh multigrain pasta were assessed by de Oliveira Filho et al., (2021). This study's objectives were to first manufacture fresh multigrain pasta with a focus on sensory qualities, and then assess how the unconventional flours (soy, oat, and rice)affected the physicochemical characteristics as well as sensory profile. The majority of the changes were caused by soy flour alone, including an increase in yellow color (b* parameter: 8.9 to 13.5), a loss in homogenous look and texture (4.4 to 1.7 N), and a reduction in cooking time (13-8.3 min). The sensory hardness (2.1–4.1) was increased by the combination of the three flours, making the pasta more al dente as would be anticipated for fresh pasta. Fresh pasta made from a blend of oat, soy, and rice flour had better nutritional value than the control, with higher protein and ash contents (2.5-2.9 g/100g) and lower carbohydrate contents (77.6 to 62.4 g/100g) than the control. Additionally, "for fresh control & enriched pasta, the lipid (1.95 & 2.0 g/100g) and caloric values" (399.6 &398.3kcal), respectively, were unchanged.

12.2.7 MULTIGRAIN NOODLES

For noodles, a traditional food in most eastern countries, are usually made with wheat flour and have been widely consumed by ancient people. But the recent increasing demand in the multigrain food market is due to an increased awareness of reducing disease risks and managing chronic diseases by intaking health-promoting food.

Through the extrusion technique, Jing et al. (2018) developed multigrain noodles using wheat flour and com as the primary components and small amounts of oat flour, soybean flour, and millet flour. The rheological features of the dough, the cooking features, and the sensory aspects of the noodles were all examined using the simplex centroid design. As a result, the following recipe for multigrain noodles was developed: 45% corn flour, 40% wheat flour, 4% oat flour, and 11% millet flour. While the multigrain noodles included less protein than wheat noodles, they did contain larger amounts of fat, dietary fiber, and total resistant starch. As wheat gluten is reduced in multigrain noodles, extrusion can be a viable method for noodle processing. Pasting and viscosities properties of multigrain flour were important for noodle processing. As shown by this study, soybean flour had negative effects on viscosity of the mixed flour and sensory qualities of the noodles, but oat flour increased the viscosity of the mixtures. The addition of millet flour could improve the sensory qualities of the multigrain noodles.

By combining soy flour (13.2%), sorghum (24.6%), and refined wheat (62.2%), Rani et al. (2020) developed multigrain soy-enriched noodles that could be utilized as supplementary food to meet the nutritional needs of primary as well as upper primary class children as suggested in the "Mid-Day Meal program by the Indian Government. In contrast to refined wheat noodles and Mid-Day Meal recipes, the nutritional makeup, antioxidant activity, and in vitro protein digestibility of multigrain noodles were assessed. In comparison to refined wheat noodles (14.82 ± 0.95%), the results showed that multigrain noodles had considerably (P≤0.05) more protein (19.10±0.63%) and dietary fiber (5.48±0.04%)." The developed noodles had increased protein digestibility, excellent total phenol content, and "1,1-diphenyl-2-picrylhydrazyl scavenging activity (85.57±1.42mg Gallic Acid Equivalent/100g, 19.64±0.20%, and 95.57±0.33" %). Multigrain noodles were shown to have lower manufacturing costs ($1.57/kg) than traditional noodles. Thus, whether supplied in 115.5g (basic level) and 179.75g (upper primary level) packaging, these protein-fiber dense noodles may significantly enhance schoolchildren's nutrition. When compared to the RDA ("Recommended Dietary Allowance"), "multigrain noodles can fulfill 20.5% of the energy and 55.3% of the protein needs at the primary level, and 25.5% of the energy and 63.2% of the protein needs at the upper primary level. Due to their high dietary fiber content, the produced noodles may be" marketed as a food high in fiber that offers a number of advantages over other foods in terms of health.

12.2.8 MULTIGRAIN SNACKS

Since so many individuals now work outside the house, they are increasingly reliant on snacks to meet some of their daily nutritional needs. Most frequently, snack foods don't offer the body enough

TABLE 12.3

Effect of Multigrain Flour Fortification on Physicochemical Properties of Different Products

Product	Key findings	References
Mango waste in multigrain cookies	Phenolic and flavonoid content increased due to the addition of mango peel and mango kernel, which boosted their antioxidant activity.	Aamir et al. 2022
Sorghum, finger millet and gluten in multigrain pasta	utilization of sorghum (31.96%) and finger millet (13.04%), along with gluten incorporation (3.40%), could be commercialized to make nutritionally enriched pasta with good quality attributes. Marketing high fiber and antioxidant-rich pasta will also be helpful for consumers suffering from diabetes, cancer, and other health complication.	Kamble et al. (2019)
Soyabean, oats, fenugreek, flaxseeds and sesame seeds in multigrain bread	Addition of increasing amounts of multigrain mix from 0 to 20% increased water absorption, decreased the strength, elasticity, and extensibility of the dough, decreased volume, increased crumb firmness value, and adversely affected the overall quality of bread above 15% addition of multigrain mix.	Indrani et al. (2010)
Refined wheat flour, and sorghum soy flour in multigrain noodles	Substitution of refined wheat flour with 24.6% sorghum and 13.2% soy flour did not alter the swelling power and water as well as oil absorption capacity. However, color and solubility varied significantly. Pasting and thermal properties of multigrain flour showed good compatibility with refined wheat flour. The composite flour-based noodles showed a higher nutritional score, antioxidant activity along with good cooking, textural, and sensory attributes.	Rani et al. (2019)
Maize, sorghum, pearl millet, and finger millet in multigrain khakra	Multigrain *khakra* was found to have functional benefits including high fiber content, good mineral content, good protein digestibility, and low GI.	Chauhan et al. (2017)

nutrients in sufficient amounts (Omueti and Morton, 1996). This can be because of their structure or the manufacturing procedure they underwent. It is important to make sure that every item that human consumes has the essential nutrients in an acceptable quantity, regardless of what is to blame for the inadequate nutritional content. Therefore, it is essential to create highly palatable, nutrient-dense snacks that may be used in nutritional programs to fight malnutrition as well as nutrient shortage. It is crucial to improve the nutritional content of a product since it is so widely eaten. Vegetable protein, like textured vegetable protein, is one way to boost the nutritional value of the diet by adding more protein (Rosa, 2003).

12.2.9 Foods with Added Plant Parts and Extracts

Due to their antibacterial and antioxidant properties, which prevent the formation of off-tastes and enhance color stability in ready-to-eat (RTE) products, plant extracts are becoming more and more significant additions to the food business. They are good replacements for synthetic compounds, which are often thought to have toxicological and cancer-causing effects because of their synthetic nature. For researchers and those involved in the food chain, the effective extraction of these antioxidant compounds from their natural sources and the assessment of their effectiveness in commercial goods have presented significant challenges (Nikmaram et al. 2018).

Emelike et al. (2015) examine the impact of drying techniques on the physicochemical as well as sensory characteristics of cookies made with wheat flour and fortified with moringa powder. Fresh Moringa leaves utilized in the formulations as well as powder obtained via several drying processes (shadow drying, sun drying, and oven drying) were used to make cookies. The cookies' physical characteristics, chemical compositions, and sensory characteristics were examined. The weight of all the cookie samples from different drying processes varied significantly, according to the data, although there was no significant variation between the fresh leaf and control cookie samples. In comparison to the other drying techniques examined, the oven-dried sample exhibited the lowest height (0.67), highest diameter (3.80), and highest spread ratio (5.79). All cookies with Moringa supplements exhibited considerably lower fiber levels than the control sample. The control sample had the greatest sensory acceptability (8.30), although there was no discernible difference in the general acceptability of the samples that were dried in the shade and the samples that were exposed to the sun (6.95 and 6.70, respectively). The results point to early possibilities for the creation of moringa-fortified cookies with potential nutritional and health advantages and functional component properties.

Kamble et al. (2021) sought to create pasta by substituting Moringa oleifera pod powder (MOPP) for durum wheat semolina (DWS) in a range of ratios (0, 5, 10, 15, and 20 g/100g; dry basis). On nutrition, cooking, textural, and consumer acceptance, the impact of MOPP addition was looked at. With rising MOPP levels (0% to 20%), the polyphenol content, nutritional composition, and "antioxidant potential of semolina pasta all increased (p ≤ .05). According to the FSSAI's guidelines, all MOPP-enriched pasta demonstrated a reduced cooking time and an acceptable level of cooking loss (≤ 8%). While the ratings for cohesion and springiness did not change substantially (p >.05), the addition of MOPP" decreased the protein hardness, digestibility, chewiness, and gumminess. In comparison to the control sample, fortified pasta had a greater intensity of functional groups, according to infrared spectroscopy. Pastas made with MOPP demonstrated strong sensory acceptance up to a 15% replacement level (with the exception of 20%) and are rated above the "slightly like" scale on the hedonic scale. According to the latest research, adding 15% MOPP to DWS could be a good way to produce pasta with increased nutritional and functional value without sacrificing sensory acceptance. M. oleifera will be a desirable component for the food sector to address the constantly expanding demand for food that promotes health since it is a rich source of bioactive chemicals.

12.3   TECHNIQUES ADOPTED FOR THE PRODUCTION OF MULTIGRAIN FOODS

(a)   Extrusion technology as a vehicle for snack production

For creating lead pipes, **Joseph Bramah** invented the first extrusion technique in 1797. For the last 250 years, extrusion technology has been utilized to create plastics, metals that are molded, and synthetic materials. The application of extrusion technology in the food business has only been seen relatively lately (since the 1970s). Extruders have promise for the food sector because of their ability to combine various components into inventive food structures, which may aid in the creation of functional meals. The composition of the feed moisture, raw materials, screw speed, barrel temperature, type of extrusion, and screw arrangement is some extrusion characteristics that might affect the quality of the final products (Miller & Mulvaney, 2000). Extrusion includes two types: cold (below 70°C) and hot (over 70°C). Instead of focusing on cold extrusion, this review focuses on how hot extrusion affects RTE quality.

We might think of extrusion as a continual cooking process, despite its apparent simplicity. There are three main kinds of screws: single-screw, intermeshing twin screws, and co-rotating twin screws (Miller & Mulvaney, 2000). During extrusion, a screw of any sort continuously turns within the barrel, moving the food material forward and generating continual pressure and shear. The product is driven through a limiting orifice known as the die at the screw's end under high pressure and temperature. The single screw type is the most often used for continuous cooking since it is less expensive and easier to operate. The ingredients are moved down the barrel in a laminar flow due to the rotating action of the screw(s), which also creates shear forces and drag. Paddles and reverse flights may be added to the extrusion screw depending on its design to improve material flow backward, which results in a minimal quantity of back mixing (Janssen, 1989). The screw arrangement often affects the product's variable residence time, which is the amount of time the product spends traveling down the barrel before exiting the extruder die, in intermeshing twin-screw extruders. This might then impact the substance as it moves down the barrel, perhaps having an impact similar to a positive displacement pumping mechanism (Janssen, 1989).

The product passes through the die assembly at the other end of the screw-barrel system as it moves down the barrel, building up pressure and heat that causes the product to melt before it exits the die. The water present evaporates as a consequence of the flash pressure drop at the die face's interaction with the environment, which causes the molten product to expand significantly at the die face. The geometry of the shaping insert, as well as the rheological and thermal characteristics of the molten material, determine the rate of expansion (Guy, 2001a). On the outside of the die, a revolving knife is often used to cut the expanding extrudate. The pieces, sometimes known as collets, may have sugar, flavoring, or colored molasses applied to them (Burns et al. (2000). The features of the final product are significantly influenced by the rheology of the extrudate pastes within the extruder (degree of expansion, bubble growth rate, and subsequent product collapse due to relaxation). By altering the process variables for screw flow dynamics, die pressure drop, or screw energy consumption, different flow rates and therefore product characteristics may be produced.

Consequently, extrusion is a technique that involves the blending, shaping, texturizing, and cooking of basic materials into a food product (Sumathi, 2007). The relative fluid product has to be flashed off moisture and its extruder pressure equalized in order to make RTE. In essence, the product's fast inflation and expansion after the extruder die create a puffy texture with plenty of inflated gas cells (Brennan, 2008a; Parada, 2011; Robin, 2011a, b; Karkle, 2012). Extrusion technique is very adaptable in the manufacture of RTEs because of the texture and cooked qualities that are so alluring to customers. For example, Mishra (2012) shows how extrusion technology may be used to produce rice analogues, which can then be used in RTE meals. The capacity to convert raw components into puffed extrudates has also been exploited in the manufacturing of grain-type analogs.

Consumers may find the production of expanded extruded items attractive due to their increased crispiness or crunchiness, but from an industrial perspective, the expansion ratio has an impact on the product's ability to break, as detailed by Robin (2011a, b).

If the product is too much enlarged, this might have major consequences in terms of loss of product intactness during the transit of commercial samples. As with everything else, there must be an ideal level that satisfies both industrial productivity standards and customer organoleptic expectations.

(b) **Roasting**

The basic dry heating food processing procedure of roasting or toasting has a variety of goals to satisfy particular needs. When exposed to high temperatures for a brief period of time, roasting causes food grains to pop, using the whole grain, including the outermost bran, husk, and germ. A low-density product (like 100kg/m3) is produced when wet grains are suddenly exposed to high temperatures, expanding them by many folds. The genetic element has a significant impact on the pericarp thickness of the grain, which is a key predictor of popping quality. The grain's moisture level, roasting duration, and temperature all have an impact on the popping quality, which includes volume expansion. Here, roasting generates a pleasing caramelized/aromatic flavor and adds the desired porous and crispy texture. The high roasting temperature creates larger, more homogeneous air cells, which results in the sufficient expansion (Manay & Shadaksharaswamy, 2008). Popped cereals include things like popcorn and rice.

(c) **3D-Printing**

Our everyday lives have been considerably enhanced by robots and software by providing us with many conveniences. A good illustration of this is 3D printing, which will usher in a new age of localized production on the basis of the digital fabrication by layer-by-layer deposition in 3 dimensions. The food industry is undergoing a transformation due to 3-D printing technology, which makes it simpler to produce individualized goods at low cost and even with precise nutrition monitoring. A three-dimensional item may be created using the 3D printing technique by depositing layers one at a time. Computer-Aided Design (CAD) software is used to develop the designs for the things, and a USB cable is used to connect and operate the 3D printer from the computer. The items may theoretically be manufactured into any form since they are generated using 3D models or other computerized data sources.

The SFF (Solid Free-Form) approach is the foundation of 3D food printing. SL ("Stereolithography"), FDM ("Fused Deposition Modeling"), and SLS ("Selective Laser Sintering") are all components of the SFF process. The primary 3D printing technique that may be utilized for food is called "Fused Deposition Modeling". FDM is applicable to the 3D printing of a variety of meals, including molten foods like chocolate, gelatin, sugars, and others that are in liquid condition. The structures of purees, gels, and doughs are either added with structuring hydrocolloids as the deposited material or are directly deposited without the use of any structuring agents. The most basic design of a food printer typically includes a syringe with one end connected to an electric motor and the other end to one or more nozzles. The nozzle is securely fixed in place by plastic clips. The components are mixed and stored in a number of containers or reservoirs, and are delivered via a syringe or nozzle. An extruder is a device that uses an electric motor to push or extrude ingredients from a syringe onto a platform. Extruders may be added to the printer in various numbers. The extruder may also be a dual-feed extruder, which modifies the mixing ratio under the direction of a color mix generator to push two separate materials with different colors out of the nozzle to produce a third color (a computer application software). This particular kind of dual-feed extrusion food printer with color mixing was created by a company named Builder in Holland (Builder Ltd., 2015). The uncooked food may be cooked on the same heating surface where the components are put. Additionally, 3D printing technology allows for the addition of specific tastes, colors, and ingredients to generate delectable meals and meet the need for individualized nutrition (Yang et al. 2016).

Healthy food intake should be enhanced to lower the risk of chronic illnesses. Islas-Rubio et al. (2014) created and analyzed a multigrain snack as a nutritious substitute for the widely used maize tortilla chips. The multigrain snack contains 153% more protein, 53% more nutritional fiber, as well as 43% less fat as compared to commercial tortilla chips. When compared to tortilla chips, its lysine, as well as isoleucine content, boosted digestibility from 83.5 to 91.8%, helping to raise the corrected-net protein consumption by 10%. Finally, a nutritionally improved "multigrain tortilla" snack was designed that contains much more dietary protein and fiber and less fat than standard corn tortilla chips while maintaining a similar look and high acceptability.

Youssif & Ghoneim (2017) reported the usage of many seeds for multigrain bread manufacturing. The effect of replacing wheat flour with varying amounts of MGM (Multigrain Mix) including flaxseeds, black, sesame, and chia seeds (5%, 10%, 15%, and 20%) on the rheological and bread-forming characteristics of wheat flour was examined. In comparison to wheat and barley flour, the findings showed that both chia, flax, sesame, and black seeds included significant quantities of crude protein, lipids, ash, and crude fibers. They also contained high levels of minerals such as zinc, iron, potassium, and calcium. Also, high amounts of amino acids especially lysine were reported. A high number of phenolic compounds was also detected.

According to Čukelj et al. (2017), flax seed and multigrain combinations were used in the creation of functional cookies. Omega-3 and lignin-rich flaxseeds are identified as essential components in market studies indicating a rise in demand for healthy biscuits. Utilizing millet flax seeds rather than whole seeds may increase lignin and omega-3 bioavailability, but it also has the unintended side effect of accelerating lipid oxidation, which lowers consumer acceptance.

12.4 DIGESTIBILITY OF MULTIGRAIN FOODS

A variety of internal and external variables affect how easily proteins may be digested. Internal factors include protein folding, amino acid content, and cross-linking. Examples of external factors include the presence of secondary molecules like emulsifiers and antinutritional components, as well as pH, temperature, and ionic strength conditions. These elements and, thus, protein digestibility is significantly impacted by food processing. Following is a discussion of the internal and external elements that contribute to variations in protein digestibility, as well as the influence of processing on protein digestibility in cereals. Additionally, growth circumstances (such as drought and heat stress) could have an impact on both internal and external elements during plant development (Impa et al. 2019).

12.4.1 DIGESTIBILITY OF MULTIGRAIN KHAKRA

The nutritional benefits of multigrain khakra made from a combination of finger millet, pearl millet, whole wheat, sorghum, and maize flour were investigated by Chauhan et al. (2017). Nutritional factors based on the glycemic index, fiber content, resistant starch content, and in vitro protein digestibility were assessed. It was discovered that multigrain *khakra* (2.4g/100g) contained considerably more total dietary fiber as compared to control whole "wheat *khakra*" (1.8g per 100g). It was observed that multigrain *khakra* had a lower glycemic index (52), greater resistant starch (1.2g per 100g), and substantially greater protein digestibility (85%) when compared with control whole "wheat *khakra*" with 70.2% protein digestibility, 55.2 glycemic indexes, and 0.6 g per 100 g resistant starch.

The khakra sample's protein digestibility was determined, and multigrain khakra (84.8%) had a protein digestibility that was substantially greater than that of the control whole wheat khakra (70.2%). It was demonstrated that maize has a greater protein digestibility than wheat, which should have been the cause of the improved protein digestibility of Khakra as a result of the integration of maize. This variation might be explained by many variables, including the technique used, potential component inclusion levels, potential antinutritional effects, or probable amino acid balances. Furthermore, it is well known that the in vitro protein digestibility of cereals is adversely linked with phytic acid levels. Whole wheat flour was reported to have a phytic acid level of up to 22mg/g. However, in millet, sorghum, and corn its average value is 10mg per g. The phytic acid level of multigrain bread has been determined to be between 1.5 and 7.5 mg/g. This might be the cause of multigrain khakra's higher protein digestibility Chauhan et al. (2017)

12.4.2 Digestibility of Multigrain Pasta

The nutritional value of traditional DWS (Durum Wheat Semolina) may be greatly improved by using millet flour because of its high dietary fiber content and therapeutic health effects. In the current study, the potential for making pasta was investigated using the physiochemical and functional properties of FMF ("Finger Millet Flour"), SF (Sorghum Flour), MF ("Multigrain Flour") and DWS prepared with a mixture of FMF (13.04%), SF (31.96%), and DWS (51.60%). Multigrain pasta that had been developed was evaluated for its anti-nutritional, antioxidant, in vitro protein, and starch digestibility along with microstructural characteristics. The pasting profile (breakdown and peak viscosity) rheological properties (departure time, development time, and water absorption), as well as transition temperatures of MF, were greater than DWS. Uncooked multigrain pasta significantly outperformed the control sample in terms of total dietary fiber, glycemic index, and antioxidant activity. In comparison to the control, the uncooked multigrain pasta showed limitations in terms of protein digestibility and structural strength, as well as greater antinutritional elements. As a result of lowered antinutritional elements, heating pasta boosted the protein as well as starch digestibility of both the control and multigrain pasta. Findings recommend that both FMF and SF may be used in the production of pasta with better nutritional content, glycemic index, and antioxidant levels (Kamble et al. (2021).

12.5 EFFECT OF OTHER FLOURS ON DOUGH RHEOLOGY AND FUNCTIONAL PROPERTIES

Many other ingredients are added to the dough mixture that might be categorized as processing aids in general and may have an indirect impact on the rheology of the dough. A helpful overview of these is provided by Stear (1990). But in this case, the chemicals that primarily affect rheology and provide information on the variables governing the reaction to the application of mechanical energy are the ones that we are most interested in.

The following is a list of the substances of interest:

- D_2O (deuterium oxide – heavy water)
- Water
- Agents affecting disulfide bonding
- Urea
- Esterifying agents for glutamine residues
- Salts
- The protein subunits present

Of course, water is necessary to make dough since it gives it its plasticity, and mixing requires careful attention to the water quantity. It establishes the proportion of loops to trains, which affects the dough's extensibility and resistance to extension (Belton, 2003). The amount of water that is really hydrated in the dough is extremely minimal; generally, 0.6 g of water is supplied for every g of flour. Since flour naturally contains 14% water, there is around 0.75g of water/g of flour. There will be around 0.75g of water/g of gluten if the water is distributed evenly among the flour's constituent parts. This implies that there will be around 5.5 water molecules per amino acid residue on a molecular level. This shows a protein system that is very concentrated. In the synthesis of HMW ("High-Molecular-Weight") subunits of gluten, results from NMR ("Nuclear Magnetic Resonance") measurements of the amount of mobile protein (Belton & Gil 1994) show that the quantity of mobile material is very sensitive to water concentration. Since small variations in water content are likely to result in big changes in the behavior of the proteins, the water content of the dough is purposefully selected in breadmaking to be in this zone. The impact of converting H_2O to D_2O is to fortify the dough (Tkachuk & Hlynka, 1968). Given that the "hydrogen bonds created by D_2O are substantially stronger than those formed by H_2O and that there don't seem to be any other major variations between the 2 isotopic forms that may have an impact, this must point to a function for hydrogen bonding in the dough." While treating glutamine residues to esterify them and remove their potential to form hydrogen bonds weakens the dough, increasing hydrogen bonds strengthens it (Beckwith, 1963; Mita and Matsumoto, 1981). The fact that the glutamine amino side chains are engaged in some hydrogen-bonding network that is crucial for regulating doug rheology is a sign of this impact. Similarly, to this, it has been suggested that the weakening effects of urea on dough rheology (Wrigley et al., 1998) are caused by the breakdown of hydrogen bonding.

Salts' molecular impacts may be rather subtle, and we still don't fully understand the mechanics behind how salts interact with proteins. Salts may alter protein solubility and hydrogen bonds, both of which have an impact on the system's cohesion. The impact of salts on the behavior of dough was examined by Eliasson and Larsson (1993). Adding sodium chloride to dough affects how much gas is retained, lengthens the time it takes for the dough to develop to its full potential, and makes the dough more stable. These outcomes might result from a number of factors that are not directly related to how the proteins interact. Although more thorough investigations have demonstrated that salt addition modifies both gluten strength and protein extractability, there may be impacts on enzymes, yeast, and other organisms.

Disulfide linkages have a crucial function in regulating the rheology of dough. The strength of the dough is dramatically diminished when disulfide linkages are lowered by a chemical, such as dithiothreitol (Wrigley et al. 1998), but it is restored upon reoxidation. Major impacts also result from the inclusion of different oxidizing and reducing chemicals that might influence how disulfide bonds are exchanged (Eliasson Larsson, 1993). It is still unclear how exactly different agents work to influence both the quantity of disulfide bonds and their interaction with one another (Weegels et al. 1994). However, they have a significant impact. In fact, due to the part that disulfide exchange plays in dough rheology, Bushuk (1998) quoted that "The importance of the disulfide interchange reaction in the development and stress relaxation of bread doughs cannot be overemphasized".

In-depth study has been done on the effect of the different protein subunits' natures in dough rheology and loaf quality. It is commonly accepted that glutenins contribute to the elastic quality of the dough and gliadins to their viscous nature. Despite making up just 1–1.7% of the dry weight of the flour or 12% of the total wheat proteins, the HMW subunits of the glutenins are the most crucial ones (Shewry et al. 2001).

12.6   NATIONAL AND INTERNATIONAL PROGRAMS

According to the FSSAI, fortification of common foods, such as wheat flour, is gaining popularity. The scientific fortification panel, professionals in the medical field, and academics all firmly support it. "One of the most practical and affordable methods to combat anemia and other micronutrient deficiencies that affect more than 50% of India's population across all population groups and geographical areas, affecting all socioeconomic classes equally, is to fortify wheat flour with iron, folic acid, and vitamin B_{12}. With an average daily intake of 200–250 gram per person and a total yearly consumption of 63.3 million tons, India has a rather high consumption of wheat flour."

To develop the capacity or skill of MF fortification and empower them to undertake the specific work in the line of processing and the value addition of various flours, a series of training were organized involving eminent scientists associated with the national agriculture innovation project

(NAIT) having the knowledge and skill in the area of processing, value addition, product development, and quality control. The training programs also included demonstrations and training on hands besides lectures.

The FDA ("Food and Drug Administration") of the US has approved rules dictating the folic acid fortification of enhanced cereal grain products. This measure was made to persuade women to consume more folate in order to lessen the possibility that a neural tube birth abnormality may harm their pregnancy (FDA 1996a).

Additionally, manufacturers may gradually add nutrients, like folic acid, to fortified goods to guarantee that the product will always have at least the quantity of nutrients listed on the label. Because label values instead of analytical values have been utilized to describe the folate composition of certain fortified products (for instance, RET cereals, and dietary supplements), the agency determined that the numbers it used were probably going to underestimate the actual folate level of such foods (Food Drug and Administration (1996b).

Among the businesses that have committed to start fortifying their wheat flour in India in accordance with the FSSAI ("Food Safety and Standards Authority of India") are General Mills, Cargill, Inc., Ashirwad, Patanjali, Hindustan Unilever, and Annapurna.

According to the FSSAI, the firms have also agreed to work together to raise consumer awareness by promoting the national logo for fortified foods on their goods.

"Fortification guidelines and a logo for fortified foods were recently issued by FSSAI, and this has given food firms motivation to use fortification on a broad scale," said Pawan Agarwal, chief executive officer of the FSSAI.

12.7 CONCLUSION

Micronutrient insufficiency is a severe issue that has negative effects and leads to economic and health catastrophes for at-risk groups across the globe. Food fortification is viewed as a great practice in the repair of nutritional deficiencies since food fortification schemes are crucial to overcoming and ensuring the right consumption of micronutrients using these populations. Flours are often a viable vehicle for fortification due to their widespread usage. However, the achievement of flour fortification depends on the accurate assessment of the occurrence of micronutrient deficiency, choice of fortifiers, political views and their application, fortification levels, the typical flour consumption level, and products obtained from that basic food, fortification of other food vehicles, cost, feasibility, and acceptability studies. Monitoring and inspection systems for fortified goods are also required.

Multigrain flour fortification is a crucial objective for worldwide businesses to reduce micronutrient deficiencies, yet it is still a problem in many nations. Therefore, each nation's national health and nutrition programs must incorporate food fortification as a method of addressing micronutrient deficiencies. These fortifications must involve the incorporation of multiple micronutrients into staple meals to address the diverse needs of millions of people throughout the globe as a low-cost, successful technique with a minimal toxicity risk. Additionally, the research that was conducted until now demonstrates the significant potential of fortifying flours as a substitute for treating micronutrient deficiencies, which will subsequently help manage and/or eradicate a number of illnesses while providing nutritional advantages and enhancing overall health.

REFERENCES

Aamir, M., Saleem, A., Rakha, A., Nadeem, M., Ateeq, H., Saeed, F., ... & Qamar, A. (2022). Functional utilization of mango waste for improving the nutritional and sensorial properties of multigrain cookies. *Journal of Food Processing and Preservation*, 46(12), e17173.

Ajila, C. M., Aalami, M., Leelavathi, K., & Rao, U. P. (2010). Mango peel powder: A potential source of antioxidant and dietary fiber in macaroni preparations. *Innovative Food Science and Emerging Technologies, 11*(1), 219–224.

Akubor, P. I., & Ukwuru, M. U. (2003). Functional properties and biscuit making potential of soybean and cassava flour blends. *Plant Foods for Human Nutrition, 58*(3), 1–12.

Angioloni, A., & Collar, C. (2011). Physicochemical and nutritional properties of reduced-caloric density high-fibre breads. *LWT-Food Science and Technology, 44*(3), 747–758.

Antonic, B., Jancikova, S., Dordevic, D., & Tremlova, B. (2020). Apple pomace as food fortification ingredient: A systematic review and meta-analysis. *Journal of Food Science, 85*(10), 2977–2985.

Arshad, U., Anjum, M., Rehman, S. U., & Sohaib, M. (2014). Development and characterization of multigrain cookies. *Pakistan Journal of Food Sciences, 24*(1), 2226–5899.

Ashwath Kumar, K., & Sudha, M. L. (2021). Effect of fat and sugar replacement on rheological, textural and nutritional characteristics of multigrain cookies. *Journal of Food Science and Technology, 58*(7), 2630–2640.

Beckwith, A. C., Wall, J. S., & Dimler, R. J. (1963). Amide groups as interaction sites in wheat gluten proteins: Effects of amide ester conversion. *Archives of Biochemistry and Biophysics, 103*, 319–330.

Belton, P. S. (2003). The molecular basis of dough rheology. In *Breadmaking Improving Quality*, pp. 273–287. Cambridge: Woodhead Publishing.

Belton, P. S., & Gil, A. M. (1994). IR and Raman spectroscopic studies of the interaction of trehalose with hen egg white lysozyme. *Biopolymers: Original Research on Biomolecules, 34*(7), 957–961.

Benitez, V., Esteban, R. M., Moniz, E., Casado, N., Aguilera, Y., & Molla, E. (2018). Breads fortified with wholegrain cereals and seeds as source of antioxidant dietary fibre and other bioactive compounds. *Journal of Cereal Science, 82*, 113–120.

Brennan, M., Derbyshire, E., B. T…. J. of F., & 2013, undefined. (2013). Ready-to-eat snack products: The role of extrusion technology in developing consumer acceptable and nutritious snacks. *Wiley Online Library, 48*(5), 893–902.

Brennan, M. A., Merts, I., Monro, J., Woolnough, J., & Brennan, C. S. (2008a). Impact of guar and wheat bran on the physical and nutritional quality of extruded breakfast cereals. Starch – St€arke, *60*, 248–256.

Burns, R. E., Caldwell, E. F., Fast, R. B., & Jones, W. H. (2000). Application of nutritional and flavoring/ sweetening coatings. In *Breakfast Cereals, and How They Are Made* (edited by R. B. Fast, & E. F. Caldwell), pp. 279–314. St. Paul, MN: American Association of Cereal Chemists.

Burton, G. W., Wallace, A. T., & Rachie, K. O. (1972). Chemical composition and nutritive value of pearl millet (Pennisetum Typhoides (Burm.) Stapf and E. C. Hubbard) Grain 1. *Crop Science, 12*(2), 187–188.

Builder Ltd. (2015). *Color mixing - Builder 3D Printers.*

Bushuk, W. (1998). Interactions in wheat doughs. In: *Hamer*, R. J. and R. C. Hoseney. *Interactions: Keys to Cereal Quality*. American Association of Cereal Chemists, Minnesota, USA. pp 10

Chauhan, S., Sonawane, S. K., & Arya, S. S. (2017). Nutritional evaluation of multigrain Khakra. *Food Bioscience, 19*, 80–84.

Čukelj, N., Novotni, D., Sarajlija, H., Drakula, S., Voučko, B., & Ćurić, D. (2017). Flaxseed and multigrain mixtures in the development of functional biscuits. *LWT, 86*, 85–92.

Desai, A., Brennan, M., Foods, C. B., & 2018, undefined (n.d.). Effect of fortification with fish (Pseudophycis bachus) powder on nutritional quality of durum wheat pasta. Mdpi.Com.

El-Adawy, T. A., & Taha, K. M. (2001). Characteristics and composition of watermelon, pumpkin, and paprika seed oils and flours. *Journal of Agricultural and Food Chemistry, 49*(3), 1253–1259.

El-Soukkary, F. A. H. (2001). Evaluation of pumpkin seed products for bread fortification. *Plant Foods for Human Nutrition, 56*(4), 365–384.

Eliasson, A. C., & K. Larsson. (1993). *Cereals in Breadmaking*. Marcel Dekker, NY, USA. pp. 261–324.

Emelike, N. J. T., Uwa, F. O., Ebere, C. O., & Kiin-Kabari, D. B. (2015). Effect of drying methods on the physico-chemical and sensory properties of cookies fortified with Moringa (Moringaolelfera) leaves. *Asian Journal of Agriculture and Food Sciences, 3*(4).

Food and Drug Administration. Food standards: Amendment of standards to identify for enriched grain products to require addition of folic acid. Final rule. *Fed. Regist.* 1996a, *61*, 8781–8797.

FDA. 1996b Food Standards: Amendment of standards of identity for enriched grain products to require addition of folic acid. *Final Rule 61 Federal Register 44* (5 March 1996) pp 8781–8797. Available online at: http://vm.cfsan.fda.gov/~lrd/fr96305b.html. Food and Drug Administration: Washington, DC.

Guy, R. (2001a). Raw materials for extrusion cooking. In *Extrusion Cooking: Technologies and Applications*(edited by R. Guy), pp. 5–28. Cambridge: CRC Press; Woodhead Publishers.

Impa, S. M., Perumal, R., Bean, S. R., John Sunoj, V. S., & Jagadish, S. V. K. (2019). Water deficit and heat stress induced alterations in grain physico-chemical characteristics and micronutrient composition in field grown grain sorghum. *Journal of Cereal Science, 86*, 124–131.

Indrani, D., Soumya, C., Rajiv, J., & Venkateswara Rao, G. (2010). Multigrain bread–its dough rheology, microstructure, quality and nutritional characteristics. *Journal of Texture Studies, 41*(3), 302–319.

Islas-Rubio, A. R., de la Barca, A. M. C., Molina-Jacott, L. E., del Carmen Granados-Nevárez, M., & Vasquez-Lara, F. (2014). Development and evaluation of a nutritionally enhanced multigrain tortilla snack. *Plant Foods for Human Nutrition, 69*, 128–133.

Itagi, H. B. N., & Singh, V. (2012). Preparation, nutritional composition, functional properties and antioxidant activities of multigrain composite mixes. *Journal of food science and technology, 49*, 74–81.

Janssen, L. (1989). Engineering aspects of food extrusion. In *Extrusion Cooking* (edited by C. Mercier, P. Linko, & J. Harper), pp. 17–38. St. Paul, MN: American Association of Cereal Chemists.

Jing, Q. I., Yingguo, L. Y. U., Yuanhui, W. A. N. G., Jie, C. H. E. N., & Panfeng, H. O. U. (2018). Formula and quality study of multigrain noodles. *Grain & Oil Science and Technology, 1*(4), 157–162.

Kamble, D. B., Singh, R., Rani, S., Kaur, B. P., Upadhyay, A., & Kumar, N. (2019). Optimization and characterization of antioxidant potential, in vitro protein digestion and structural attributes of microwave processed multigrain pasta. *Journal of Food Processing and Preservation, 43*(10), e14125.

Kamble, D. B., Singh, R., Rani, S., Upadhyay, A., Kaur, B. P., Kumar, N., & Thangalakshmi, S. (2021). Evaluation of structural, chemical and digestibility properties of multigrain pasta. *Journal of Food Science and Technology, 58*(3), 1014–1026.

Karkle, E. L., Alavi, S., & Dogan, H. (2012). Cellular architecture and its relationship with mechanical properties in expanded extrudates containing apple pomace. *Food Research International, 46*(1), 10–21.

Kaur, H., Bobade, H., Singh, A., Singh, B., & Sharma, S. (2017). Effect of formulations on functional properties and storage stability of nutritionally enriched multigrain pasta. *Chemical Science International Journal, 19*(1), 1–9.

Kulkarni, L. R., Naik, R. K., & Rokade, C. (2011). Development of nutri cereals based value added traditional products and its acceptability. NAIP: National Symposium Recapturing Nutritious Millets for Health and Management of Diseases, UAS, Dharwad, December, 16–17, p. 44.

Malik, H., Nayik, G. A., & Dar, B. N. (2015). Optimisation of process for development of nutritionally enriched multigrain bread. *Journal of Food Processing and Technology, 7*(544), 2.

Manay, N. S., & Shadaksharaswamy, M. (2008). *Foods: Facts and Principles*. 3rd edn. New Delhi: New Age International (P) Ltd, Publishers, p. 467.

Masoodi, L., & Bashir, V. (2012). Fortification of biscuit with flaxseed: Biscuit production and quality evaluation. *IOSR Journal of Environmental Science, Toxicology and Food Technology, 1*(2), 06–09.

Miller, R., & Mulvaney, S. (2000). Extrusion and extruders. In *Breakfast Cereals, and How They Are Made* (edited by R. Fast, & E. Caldwell), pp. 215–278. St. Paul, MN: American Association of Cereal Chemists.

Milligan, E. D., Amlie, J. H., Reyes, J., Garcia, A., & Meyer, B. (1981). Processing for production of edible soy flour. *Journal American Oil Chemistry Social, 58*, 331.

Mishra, A., Mishra, H. N., & Srinivasa Rao, P. (2012). Preparation of rice analogues using extrusion technology. *International Journal of Food Science & Technology, 47*(9), 1789–1797.

Mita, T., & Matsumoto, H. (1981). Flow properties of aqueous gluten and gluten methyl ester dispersions. *Cereal Chemistry, 58*, 57–61.

Mohajan, S., Munna, M. M., Orchy, T. N., Hoque, M. M., & Farzana, T. (2019). Buckwheat flour fortified bread. *Bangladesh Journal of Scientific and Industrial Research, 54*(4), 347–356.

Naval, Kishorgoliya, Mehra, M., and Goswami, P. (2018) "Nutritional quality of the developed multigrain flour and cookies", *Journal of Pharmacognosy and Phytochemistry*, 2886–2888.

Nikmaram, N., Budaraju, S., Barba, F. J., Lorenzo, J. M., Cox, R. B., Mallikarjunan, K., & Roohinejad, S. (2018). Application of plant extracts to improve the shelf-life, nutritional and health-related properties of ready-to-eat meat products. *Meat science, 145*, 245–255.

Omueti, O., & Morton, I. D. (1996). Development of extrusion of soybean snack stick: A nutritional improved soy-maize product based on the Nigeria snack (kokoro). *International Journal of Food Science Nutrition 47*, 5–13.

Parada, J., Aguilera, J. M., & Brennan, C. (2011). Effect of guar gum content on some physical and nutritional properties of extruded products. *Journal of Food Engineering, 103*(3), 324–332.

Randhawa, M., Asghar, A., Jahangir, M. A., Afzal, S., Shehzad, A., Randhawa, M. A., & Shoaib, M. (2013). Health benefits and importance of utilizing wheat and rye. *Pakistan Journal of Food Sciences, 23*(4), 212–222.

Rani, S., Singh, R., Kamble, D. B., Upadhyay, A., & Kaur, B. P. (2019). Structural and quality evaluation of soy enriched functional noodles. *Food Bioscience, 32*, 100465.

Rani, S., Singh, R., Kamble, D. B., Upadhyay, A., Yadav, S., & Kaur, B. P. (2020). Multigrain noodles: Nutritional fitness and cost effectiveness for Indian mid-day meal. *Food Security, 12*(2), 479–488.

Rathod, R. N., Kotecha, P. M., & Thorat, S. S. (2015). Standardization process for preparation of multigrain cookies. *Trends in Biosciences, 8*(20), 5627–5632.

Raw dough can contain germs that make you sick. *CDC.* 28 July 2021.

Robin, F., Dubois, C., Pineau, N., Schuchmann, H., & Palzer, S. (2011a). Expansion mechanism of extruded foams supplemented with wheat bran. *Journal of Food Engineering, 107*(1), 80–89.

Robin, F., Theoduloz, C., Gianfrancesco, A., Pineau, N., Schuhman, H., & Palzer, S. (2011b). Starch transformation in bran-enriched extruded wheat flour. *Carbohydrate Polymers, 85*(1), 65–74.

Sagar, N. A., & Pareek, S. (2021). Fortification of multigrain flour with onion skin powder as a natural preservative: Effect on quality and shelf life of the bread. *Food Bioscience,* 41, 100992.

Sahoo, A. K., Desai, A. D., Ranveer, R. C., & Sahoo, A. K. (2012). Development of nutrient rich noodles by supplementation with malted ragi flour. *International Food Research Journal, 19*(1), 309–331.

Shabir, G. A. (2003). Validation of high-performance liquid chromatography methods for pharmaceutical analysis: Understanding the differences and similarities between validation requirements of the US food and drug administration, the US Pharmacopeia and the international conference on harmonization. *Journal of Chromatography A, 987*(1–2), 57–66.

Sharma, S., Sekhon, K. S., & Nagi, H. S. (1999). Suitability of durum wheat for flat bread production. *Journal of Food Science and Technology (Mysore), 36*(1), 61–62.

Sharma, S., Sharma, N., Sharma, R., & Handa, S. (2018). Formulation of functional multigrain bread and evaluation of their health potential. *International Journal of Current Microbiology and Applied Sciences, 7*(7), 4120–4126.

Singh, U., & Singh, B. (1991). Functional properties of sorghum-peanut composite flour. *Cereal Chemistry, 68*(5), 460–463.

Stear, C. A. (1990). Measurement and control techniques for raw materials and process variables. In *Handbook of Breadmaking Technology* (pp. 306–393). Boston, MA: Springer US.

Sterna, V., Zute, S., & Brunava, L. (2016). Oat grain composition and its nutrition benefice. *Agriculture and Agricultural Science Procedia, 8*, 252–256.

Sumathi, A., Ushakumari, S., & Malleshi, N. (2007). Physico-chemical characteristics, nutritional quality and shelf-life of pearl millet based extrusion cooked supplementary foods. *International Journal of Food Sciences and Nutrition, 58*(5), 350–362.

Taghdir, M., Mazloomi, S. M., Honar, N., Sepandi, M., Ashourpour, M., & Salehi, M. (2017). Effect of soy flour on nutritional, physicochemical, and sensory characteristics of gluten-free bread. *Food Science & Nutrition, 5*(3), 439–445.

Tkachuk, R., & I. Hlynka. (1968). Some properties of dough and gluten in D2O. *Cereal Chemistry, 45*, 80–87.

Wrigley, C. W., Andrews, J. L., Bekes, F., Gras, P. W., Gupta, R. B., Macritchie, F., & Skerrit, J. H. (1998). Protein-protein interaction–essential to dough rheology. In Hamer, R. J. and R. C. Hoseney. *Interactions: Keys to Cereal Quality.* American Association of Cereal Chemists. Minnesota, USA. pp. 17–46.

Xiong, Y., Zhang, P., Warner, R. D., & Fang, Z. (2019). Sorghum grain: From genotype, nutrition, and phenolic profile to its health benefits and food applications. *Comprehensive Reviews in Food Science and Food Safety, 18*(6), 2025–2046.

Yang, Y., Chen, Y., Wei, Y., & Li, Y. (2016). 3D printing of shape memory polymer for functional part fabrication. *The International Journal of Advanced Manufacturing Technology, 84*, 2079–2095.

Yang, F., Zhang, M., & Bhandari, B. (2017). Recent development in 3D food printing. *Critical Reviews in Food Science and Nutrition, 57*(14), 3145–3153.

Youssif, M. R., & Ghoneim, G. A. (2017). Utilization of different seedss for production multigrain pan bread. *Journal of Food and Dairy Sciences, 8*(1), 9–16.

13 Prebiotics and Probiotics

Dugeshwar Karley, Kush Kumar Nayak, and Varaprasad Kolla

13.1 PREBIOTICS

Glenn Gibson and Marcel Roberfroid first used a prebiotic approach in 1995 (Gibson & Roberfroid, 1995). A prebiotic can be explained as "an indigestible food ingredient that has beneficial effects on the host by selectively stimulating the growth and activity of an individual or a group of microorganisms in the gastrointestinal tract and thereby improving host health (Davani- Davari et al., 2019)." According to this explanation, only certain ingredients can act as prebiotics, such as some oligosaccharides and polysaccharides [Fructo-oligosaccharides (FOS) and inulin], certain proteins and peptides, lactulose, Galacto-oligosaccharides (GOS), etc. The International Scientific Association for Probiotics and Prebiotics (ISAPP) has defined a "dietary prebiotic" as "a specific fermented food ingredient responsible for a notable change in microbiome composition and/or activity." These compounds can help improve the health of colonic animals, thus providing the host's health benefits (Gibson et al., 2010)." A compound can be classified as a prebiotic based on the following criteria:

1. The compound is not absorbed from the gastrointestinal tract, resists low gastric pH, resists hydrolysis, and must not react with mammalian enzymes.
2. It cannot be used as a fermentation product by the gastrointestinal microflora.
3. The growth and/or activity of the gastrointestinal microbiota must be selectively modified from prebiotic compounds to improve host health (Davani-Davari et al., 2019; Gibson et al., 2010). However, not all prebiotics are defined as carbohydrate-based prebiotics.

They can be distinguished from fiber based on the following characteristics.

1. Fiber is made from carbohydrates by polymerization (Slavin, 2013) and 2. Enzymes within the small intestine do not hydrolyze these fibers. He pointed out that the permeability and solubility of fiber are not important (Howlett et al., 2010). Several other revised definitions are also reported in scientific publications (Rastall, 2006). The ability of prebiotics or "selectivity" to stimulate a specific gastrointestinal microbiota is another essential part of the definition, although this concept has recently been questioned (Bindels et al., 2015). Prebiotics earlier, the effects of prebiotics are enhanced by cross-feeding, defined as the product of one species ingested by another (Scott et al., 2013). This suggests that the word "selective" raises doubts and ongoing debate over the definition of prebiotics (Hutkins et al., 2016).

13.2 SOURCES OF PREBIOTICS

The main sources of prebiotics are classified into several groups. Most of them are a subset of carbohydrates and are mainly oligosaccharide carbohydrates (OSCs). This chapter mainly focuses on OCS, but other research demonstrates that prebiotics are more than just carbohydrates.

DOI: 10.1201/9781003160663-17

13.2.1 GALACTO-OLIGOSACCHARIDES

Galacto-oligosaccharides are products of extended lactose forms and are classified into two main subclasses (1) galacto-oligosaccharides with abundant galactose at the C_3, C_4 or C_6 positions and (2) galacto-oligosaccharides synthesized by lactose with an enzymatic glycosylation reaction (Martins et al., 2019). In this reaction, the product is usually a mixture of trisaccharides and penta-saccharides at β (1,6), β(1,3) and β(1,4) bonds with galactose. This galacto-oligosaccharide is also known as trans-galacto-oligosaccharide (Macfarlane et al., 2008). Galacto-oligosaccharides can significantly stimulate lactobacilli and *bifidobacteria*. In neonates, *bifidobacteria* showed a greater association of GOS (O'Callaghan & van Sinderen, 2016). Other probiotics such as Firmicutes, Enterobacteria and Bacteriodetes were also stimulated by GOS but to a lesser extent Bifidobacteria (Zhang et al., 2020a). The nutritional properties of GOS are also important; They are low-calorie sweeteners, help maintain body weight, relieve constipation, have a low glycemic index, and are not involved in carcinogenic activities (Nichol et al., 2018). GOS is widely used at the industrial level to formulate dairy products, baked goods, beverages, and various sweets, all of which are transformed into functional foods (Mudgil & Barak, 2019). Furthermore, GOS is widely used in neonatal prescribing to support the formation of the neonatal microbiome (Romano, Santos et al., 2016; Romano, Schebor et al., 2016).

13.2.2 FRUCTO-OLIGOSACCHARIDES

Fructo-oligosaccharides, also sometimes called oligofructose or oligofructose, are composed of short linear fructose chains linked by a (2, 1) β-glycosidic bond and with only one D-glucosyl unit at the non-reducing end (Kelly, 2009)). Short-chain FOS produces a mixture of the smallest oligosaccharides known as 1-ketose (degree of polymerization (DP) equal to 3), nystose (DP4), and 1F-fructofuranosylnystose (DP5) (Bali et al., 2015). The production of FOS can be obtained by enzymatic methods or by hydrolysis of inulin. Inulin can be obtained from natural sources like chicory root, dahlia or agave, artichoke, and yacon, etc. (Singh et al., 2019). In enzyme synthesis for FOS production, transfructosylation is involved (Kashyap et al., 2015), in which fructosyltransferases (β-fructofuranosidase or β-D-fructosyltransferases act as biological catalysts) (Vega-Paulino & Zúniga-Hansen, 2012) In fructosylation metabolism reactions, β-2,1 glycosidic bond is broken and fructosyl radicals are transferred from carbohydrates as donors to any acceptor other than water (Plou et al., 2007) FOS is a complex process in which different reactions are catalyzed simultaneously in parallel or in series (Meyer et al., 2015) , the production of FOS proceeds through the first steps in which fructosyl acts as a donor and acceptor, leading to the formation of short-chain FOS by DP (DPn+1, DPn+2,......, DPn+) N immediately (Fuller, 2001).

13.2.3 STARCH AND GLUCOSE-DERIVED OLIGOSACCHARIDES

Oligosaccharides are derived from starch and glucose in the upper part of the digestive tract. Starches are resistant to digestion. Hence, this type of starch is called resistant starch (RS). RS can produce high levels of butyrate, which affects health benefits. This is why it has been suggested as a prebiotic (Fuentes-Zaragoza et al., 2011). Different Firmicutes species showed a maximum association with increased RS numbers (Walker et al., 2011). As previously reported, RS can also be partially digested in the presence of Ruminococcusbromii and Bifidobacteriaadolescentis, as well as minimally via rectal Eubacteria and Bacteroides thetaiotaomicron. Although during the incubation of a mixture of bacteria and feces, RS cannot degrade without Ruminococcusbromii (Ze et al., 2012). Polydextrose (PDX) is a sugar-free, glucose-derived oligosaccharide with a neutral taste. It can be used commercially as a low-calorie bulking agent in food industries such as dairy, confectionery, functional beverages, and baked goods. Studies show that PDX has been certified for use in food in about 60 countries and is recognized as a dietary fiber and prebiotic in about 20 countries (led by Carmo et al., 2016; Probert et al., 2004). PDX is a randomly linked glucose-based oligosaccharide with a mean DP of 12, ranging up to 120. PDX is a highly branched molecule with all possible combinations of α and β bonds such as glycosidic bonds 1-2, 1-3, 1-4, and 1-6 (Lahtinen et al., 2010). Due to its complex structure, mammalian digestive enzymes are unable to hydrolyze it in the small intestine, delivering it completely to the colon, where it is gradually and partially fermented by the sugar microbiota. digested, and the remaining ~60% is excreted in the feces (Holscher et al., 2015). Another study conducted by Costabile et al. have published some evidence that PDX can stimulate the growth of bifidobacteria, but this has yet to be proven (Costabile et al., 2012).

13.2.4 ANOTHER OLIGOSACCHARIDE

Some oligosaccharides are derived from an original polysaccharide called pectin. Pectin is composed of a heterogeneous polysaccharide bound to the primary cell wall and intercellular region of higher plants (Chen et al., 2013). Pectin consists of a group of acidic polymers, called rhamnogalacturonan (RG) and homogalacturonan (HG), bound to several sugars/neutral polymers such as galactans, arabinans, and arabinogalactans on the side chain (Øbro et al., 2004). These neutral and acidic polymers were extracted as pectic oligosaccharides (POS) and used in prebiotic production from agricultural waste (Lama-Muñoz et al., 2012). Pectic Oligosaccharides (POS) are considered indigestible oligosaccharides that directly enhance host health benefits by selectively stimulating the growth and/or activity of a specific bacterium or bacterial group in the gastrointestinal tract (Bifidobacteria and Lactobacilli) (Garthoff et al., 2010; Mussatto and Mancilha, 2007). As mentioned previously, pectic oligosaccharides inhibit the growth of some enteric pathogens and other pathogenic bacteria such as Enterobacteriaceae and Clostridia spp. (Clostridium difficile and Clostridium perfringens) respectively (Baldan et al., 2003; Manderson et al., 2005). The fermentation of pectin oligosaccharides in the gastrointestinal tract leads to the synthesis of short-chain fatty acids (SCFAs), which confer numerous health benefits to the host, such as pathogen inhibition, relieving constipation, and improving absorption of minerals, reducing the risk of colon cancer, reducing blood sugar and modulation of the immune system (Babbar et al., 2016).

13.3 HEALTH BENEFITS OF PREBIOTICS

Prebiotics can be used as cancer prevention agents based on observations and reported data that health-affecting bacteria such as bifidobacteria do not contain synthetic carcinogens or xenobiotic compounds, but instead synthesize SCFA, which enhances the health benefits of the host. As shown, 10 g/day of short-chain fructo-oligosaccharides in adenoma-free and adenoma-free subjects was observed to have more healthy markers than in those without adenomas (Slavin, 2013). The use of FOS together with inulin showed a significant reduction in indicators of clinical severity, a decrease in inflammatory immune responses, and calprotectin, a protein released by plasma neutrophils, and stool increases in patients with inflammatory bowel disease (Konikoff & Denson, 2006). Ingestion of GOS significantly increased the number of bifidobacteria in the stool. As a result, there was a noticeable change in stool consistency, improvement in flatulence, bloating problems, and subjective global assessment (SGA) score (Slavin, 2013). Therefore, it can be concluded based on the reported data that a prebiotic is an indigestible compound that confers health benefits to the host in various respects. In addition, it is recommended that consumption of prebiotics can:

1) Prebiotics is associated with reduced incidence and duration of infectious and antibiotic-associated diarrhea.
2) Reduces symptoms and inflammatory responses caused by the host immune system in people with inflammatory bowel disease.
3) Has been largely involved in the prevention of colon cancer.

4) Increases bioavailability and absorption of minerals such as magnesium, calcium, and possibly iron.
5) Reduce several risk factors associated with cardiovascular disease.
6) Promote satiety, lose weight, and prevent obesity.

13.4 PROBIOTICS

The term "probiotics" is used to refer to live microorganisms that confer a health benefit to the host when consumed in sufficient quantities (Hill et al., 2014). But definitions of probiotics have evolved concurrently with growing interest in using live bacterial cells as supplements and understanding their mechanisms (Kechagia et al., 2013). This term is used to describe substances synthesized by a particular microorganism that enhance the growth of other microorganisms and provide beneficial health effects by contributing to the balance of gut microbiota (Kechagia et al., 2013). The definition of probiotics currently in use has been suggested by the Food and Agriculture Organization of the United Nations and the World Health Organization (FAO/WHO), based on the fact that probiotics are considered "a live microorganism which, when used in sufficient quantities confers a health benefit on the organism", host" (Bielecka, 2006).In relation to food, this definition can be modified considering that beneficial effects are produced by microorganisms "when they are used in sufficient quantities in the food context" (FAO/WHO, 2002). Bifidobacteria strains are the bacteria most used to exhibit probiotic properties and are present in a variety of functional foods and nutritional supplements (Plaza-Diaz et al., 2019). In 2014, the International Scientific Association conference on Probiotics and Prebiotics stated that the production of by-products from bacterial metabolism, dead microorganisms, and other non-viable components is associated with potential bacteria; but not classified as probiotics (Salminen et al., 2021). However, numerous studies have shown that dead bacteria and their non-viable components exhibit biological properties (Plaza-Díaz et al., 2017, 2018). The effects of probiotics on health conditions have been published in many research articles, reviews, and systematic reviews (Fig. 13.1) (Didari et al., 2014; Hempel et al., 2012). As well documented, probiotics are involved in preventing medical complications such as antibiotic-associated diarrhea (Plaza-Diaz et al., 2019), inflammatory bowel disease (IBD), ulcerative colitis (Saez-Lara

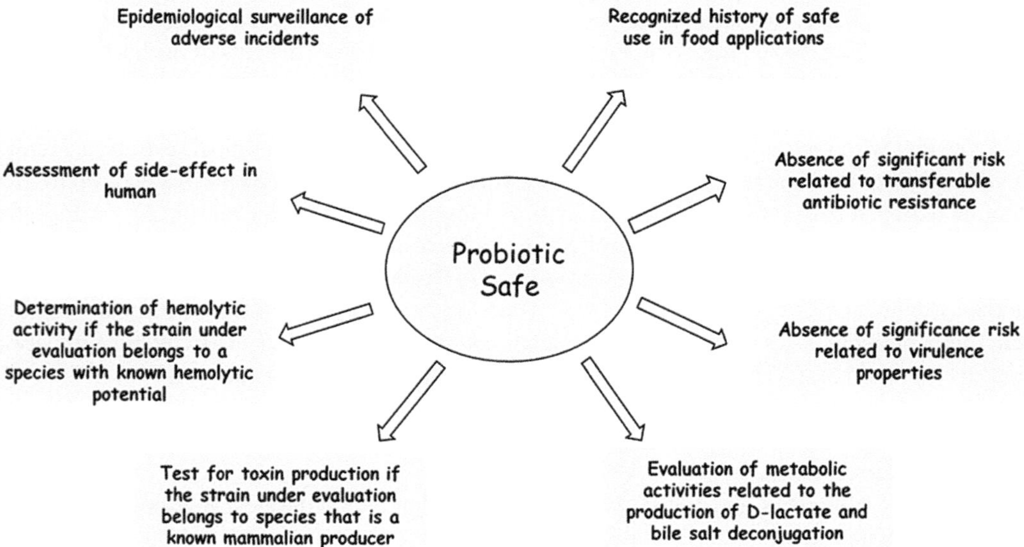

FIGURE 13.1　Guidelines for safety assessment of probiotics for human use.

et al., 2015). irritable bowel syndrome (IBS) (Moayyedi et al., 2010), eczema (Rather et al., 2016) and allergic rhinitis (Berings et al., 2017).

13.4.1 History of probiotics

The term "probiotic", which is derived from the Latin Pro and the Greek bio, which means exactly "for life", was first coined in 1953 by the German scientist Werner Kollath to refer to "compounds" active substances necessary for the existence and development of a healthy life". (Johnson et al., 2021). In 1965, the term probiotic was used by two other scientists, Lilly and Stillwell, in various references to describe "metabolic compounds produced by one organism to stimulate development of other organisms" (Markowiak & Ślizewska, 2017). Specifically, in 1992, Fuller defined a probiotic as "a live microorganism added to the diet that has a beneficial effect on the host by improving the host's balance" (McFarland, 2015). In the early 1900s, the modern era of probiotics began with the pioneering research of Nobel laureate Elie Metchnikoff and a Russian scientist at the Pasteur Institute in Paris (Mackowiak, 2013). Louis Pasteur recognized many microorganisms responsible for many different types of fermentation, while Metchnikoff continuously tried to discover the possible effects of these microorganisms on human health (Weill, n.d.). It is related to fermented dairy products regularly consumed by rural Bulgarians, such as yogurt, and it has been found to influence longevity (Veiga et al., 2020). He is trying to link this to the Bulgarian bacillus, discovered by the Bulgarian doctor Stamen Grigorov (27). It was then suggested that Lactobacillus could prevent the inactivation and aging effects of the gastrointestinal tract (Selle et al., 2014). Metchnikoff also states that bacterial putrefaction is an important source of many toxins in the gastrointestinal tract. These toxins move from there to the circulatory system and cause aging effects (Tannock, 2004). He called these bacteria inactivated bacteria, now accepted as proteolytic clostridia. Metchnikoff also clarifies that "the gut microbiota is dependent on diet, which provides an opportunity for the use of specific bacteria to modify the gut microbiota and replace harmful bacteria." by beneficial bacteria (Milner et al., 2021). This explanation is clearly expressed in terms of the "probiotic concept" (Gasbarrini et al., 2016). Metchnikoff also favored these lactobacilli as probiotics. However, the history of probiotics is as old as human civilization and is very closely linked to the use of fermented foods (Ozen & Dinleyici, 2015). However, recent research has shown that human ancestors used yeast in the production of fermented beverages long before 2000 BC (Sicard & Legras, 2011). Based on the remains of ancient Egypt, fermented dairy products such as "Leban Rayad" and "Leban Khed" are still popular in the Middle East and have been used today since 3,500 BC (Sciences et Agriculture, 2008). Along with human civilization, many other challenges take place, and all observations are carefully recorded; Different cultures have been used to produce fermented beverages (Luisa Alba-Lois, 2017).

13.5 MONGOLIAN HISTORY

It is said that at the same time as the expansion of the Mongol Empire founded by Genghis Khan, a messenger came to a village crossing the desert and asked for a bag of water. But the villagers were harmed by the Mongol invasion; The villagers filled the bag with milk and gave him water. When thirsty, the omen found that there was bubbling liquid in his pocket instead of water, but he had no choice but to drink it to quench his thirst. When Genghis Khan learned of this story, he is said to have added yogurt to his army's diet (Ozen & Dinleyici, 2015). Two Armenian scientists, Sarkis, and Rose Colombosian, who had left the United States after World War I, started selling their home-made yogurt called madzoon (yogurt in Armenian). Initially, the product was not in demand due to the unknown name, so they decided to sell it as yogurt. Originally founded in a small kitchen in Andover, Massachusetts, USA, they founded the company "Colombo and Sons Creamery" in 1929, becoming the first yogurt brand in US history (Shurtleff & Aoyagi), 2012). Yogurt was introduced in 1929, bringing the first taste to the American people, and 80 years later, the demand and popularity of this yogurt continues to grow (Hansen, 1974). As demand and popularity increased day by day, they introduced the brand "Colombo Yogurt" and then this brand started to be sold at General Mills in 1993. Meanwhile, Dr. Minoru Hirota identified the first cultured strain of Lactobacillus casei Hirota strain isolated from the human intestine by the Department of Medicine of the Microbiology Laboratory of Kyoto University, Japan in 1930. Many studies have demonstrated that the Shirata strain discovered by Hirota is resistant to the drug entering the acidic environment of the intestine, bile acids and can reach the lower intestine after oral administration. In 1935, Dr. Hirota created "Yakult", a dairy product, using this probiotic and brought it to market. The name "Yakult" originally came from yogurt in Esperanto, a universal language proposed in the 1880s. He speculated that daily consumption of Yakult (a fermented product) might be useful in promoting gut health and prolonging life by balancing the gut microbiota.

13.5.1 HEALTH BENEFITS OF PROBIOTICS

There is ample evidence supporting the beneficial effects of probiotics, including improving immune response, lowering serum cholesterol levels, improving gut health, and preventing cancer (Nazir et al., 2018). These health benefits are strain-specific and influenced by a variety of mechanisms, as previously discussed (Plaza-Diaz et al., 2019). While the mechanisms underlying various health benefits have been clearly elucidated, others need further study to confirm. There is ample evidence that probiotics can be used to prevent inflammatory conditions of the gastrointestinal tract, including antibiotic-associated diarrhea (AAD) (Milner et al., 2021), ulcerative colitis, improve lactose metabolism, Crohn's disease, infectious diarrhea, especially ileitis and infectious diarrhea (Kechagia et al., 2013; Markowiak & Ślizewska, 2017). The etiology of these infectious diseases is not fully understood, but they have been reported to be associated with chronic and existing infections or enteritis (Hendrickson et al., 2002). Some clinical studies have shown that probiotics affect the remission of ulcerative colitis, but little effect has been observed in the case of Crohn's disease (Fig. 13.2) (Geier et al., 2007).

Probiotics are believed to be involved in inhibiting certain pathogenic enzymes, which reduces the risk of colon carcinoma in animals. However, similar results were not seen in humans in clinical trials (Team et al., 2012). However, the beneficial effect on the genitourinary system in women, prevention and treatment of bacterial vaginosis and urinary tract infections (UTI), is a prime example of the positive effect of yeast use in microbiology (Falagas et al., 2006). Probiotics are also given

Probiotics		
	Normalized intestinal Microbiota	Intestinal mucosal integrity
		Colonization resistance
		Control of irritable bowel syndrome
		Control of inflammation bowel disease
	Immunomodulation	Stimulate specific immune response
		Stimulate activity macrophages
		Alleviate food allergy
		Induction of natural killer cells
	Metabolic Effect	Improve lactose tolerance
		Lower serum cholesterol
		Lower toxigenic/mutagenic reactions
		Supply of SCFA and vitamin to colon epithelium

FIGURE 13.2 Probiotics as functional foods in enhancing gut immunity

to pregnant women and infants for preventive measures against allergic diseases such as atopic dermatitis. However, this type of case study has always been controversial (Wang et al., 2019). There is some evidence that consuming probiotics (dairy products) lowers blood cholesterol, which may be helpful in overcoming diabetes, obesity, stroke, and cardiovascular disease. Simons et al. (2006). Several studies have been well documented on the effect of VSL#3 probiotic formulation and Oxalobacterformigenes on urinary oxalate elimination. As a result, the risk of urolithiasis is reduced (Lieske et al., 2005).

Several researchers have demonstrated in animals that oral administration of the Lactobacillus acidophilus strain of bacteria observed intestinal cannabinoid and µ-opioid receptor expression to modulate analgesic function in the gut, and these observations are comparable to the effects of morphine (Rousseaux et al., 2007). The effectiveness of probiotic strains has been clearly demonstrated in reducing the risk of nosocomial infections (HAIs), viral diarrhea, and non-hospital infections. Probiotics also prevent enteric viral infections by increasing levels of IgA immunoglobulins (Parvez et al., 2006) (Figure 13.2). Antibiotic-associated diarrhea (AAD) is a major drawback of most antibiotics, and Clostridium difficile disease (CDD), also caused by antibiotics, both lead to flare-ups of diarrhea and colitis due to infection. The consumption of probiotics to reduce the risk of these two related diseases remains controversial. However, some specific strains of probiotics could potentially be used as an effective treatment for these two conditions. Using meta-analyses, probiotic strains Saccharomyces boulardii, Lactobacillus rhamnosus GG, and a mixture of probiotics markedly reduced the risk of antibiotic-associated diarrhea, while Saccharomyces boulardii was able to reduce the risk of Clostridium difficile disease (McFarland, 2006). Studies conducted in Helsinki (Finland) have shown that regular consumption of the probiotic strain Lactobacillus rhamnosus GG is helpful in reducing respiratory infections (Hatakka et al., 2001). Previous reports showed that regular yogurt consumption significantly reduced gut bacteria and improved galactosidase activity in the gastrointestinal tract (Alvaro et al., 2007).

13.5.2 DESIRED PROPERTIES OF PROBIOTICS

For a potential probiotic organism to exert its beneficial effects, it needs to possess certain desirable characteristics (Fig. 13.3). The following desirable characteristics are currently identified by in vitro studies.

1. The probiotic strain must be able to tolerate acids and bile, making it suitable for oral use.
2. To successfully modulate the immune system, eliminate competitive pathogens, and prevent pathogen association and colonization, the beneficial bacterial strain must be able to adhere to mucosal and epithelial surfaces.
3. Has an antagonistic effect against cariogenic and pathogenic bacteria.
4. Produces antibacterial substances that fight disease-causing bacteria.
5. The probiotic strain must be non-pathogenic, non-toxic, and must be of human origin. BECAUSE.
6. Persistence and viability is also important characteristics of probiotic strains.
7. Has anti-cancer and anti-mutagenic properties.
8. Capable of hydrolyzing bile salts.
9. Probiotic strains must be resistant to bacteria.
10. Ensure the probiotic strain does not contain transmissible antibiotic resistance genes.

13.6 SYMBIOTIC

Symbiotics are generally defined as a combination of a prebiotic and a probiotic and thus promote the beneficial effects of both prebiotics and probiotic. The symbiosis not only improves the viability of beneficial bacteria, but also stimulates and proliferates the growth of specific native microorganisms

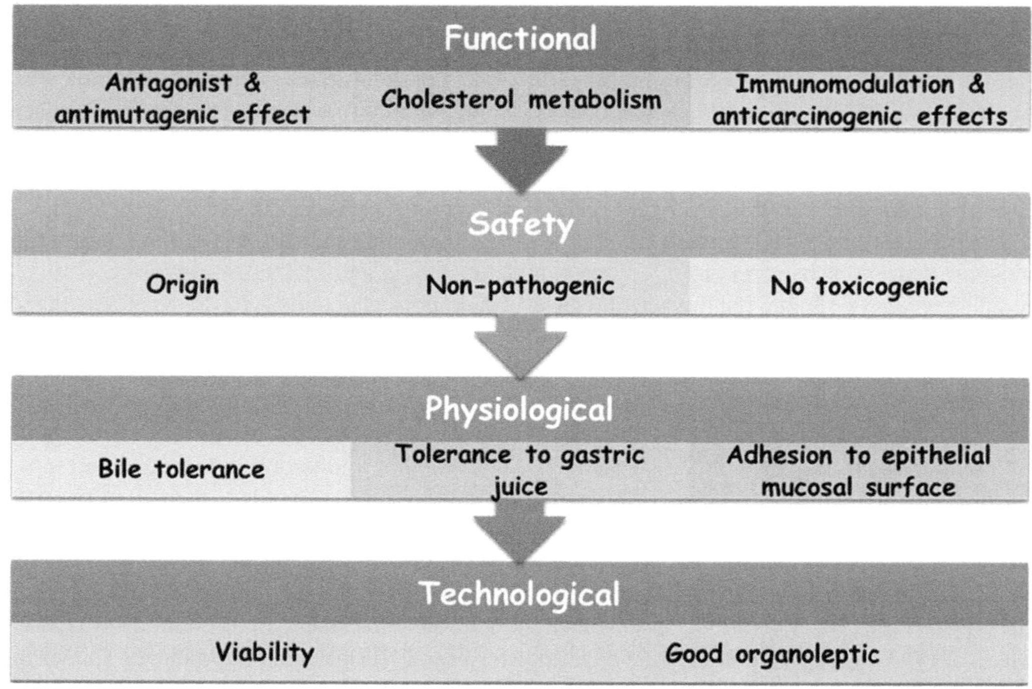

FIGURE 13.3 Characteristics of probiotics.

associated with the gastrointestinal tract (Gourbeyre et al., 2011). The health benefits of symbiotics are still poorly documented, as it appears that the health benefits of symbiotics may be related to specific combinations of prebiotics and probiotics (Guiné & Silva, 2016). Many possible symbiotic combinations hold promise for regulating the human gastrointestinal microbiome.

13.6.1 Mechanism of action of symbiotic

As mentioned earlier, the effects of probiotics are mainly observed in the small and large intestine, while the effects of prebiotics are mainly observed in the large intestine. Therefore, the combination of these two factors may have a synergistic effect (Hamasalim, 2016). Prebiotics are mainly selected based on their effectiveness to be used as a medium for beneficial bacteria, which are easy to ferment and pass through the intestinal tract. As reported in the literature, in the presence of prebiotics, probiotic bacteria gain more stability to survive in the intestinal environmental conditions including oxygen, temperature, pH, etc. (Terpou et al., 2019). However, the mechanism that works to capture the excess energy and adapt it to survive in such extreme conditions remains unclear. The combination of various ingredients ensures proper dietary supplementation and provides the right environment for the growth of beneficial bacteria strains, providing beneficial effects on host health (Nagpal et al., 2012). Two modes of action of synbiotics are explained (Wang et al., 2021).

I. By improving the viability of probiotic strains.
II. By providing a particularly positive impact on health.

Prebiotics stimulate the growth of probiotics for filtration, leading to regulation of intestinal metabolism without affecting the basic biological structure of the intestine, development of good microflora, and inhibition of microorganisms, pathogenic bacteria present in the gastrointestinal tract (Cunningham et al., 2021).

13.6.2 Benefits of symbiosis

Symbiotic has the following beneficial effects on human health (Zhang et al., 2010).

I. Balances gastrointestinal microflora with an increased number of Lactobacillus and Bifidobacteria genera.
II. Symbiosis improves liver function in patients with cirrhosis.
III. Stimulates the immune regulatory system.
IV. Reduces the risk of bacterial displacement and reduces the risk of nosocomial infections during postoperative procedures and similar operations.

13.7 ENCAPSULATION OF PROBIOTICS

Encapsulation of probiotics requires the cell to be supplied with the appropriate cell wall material to protect and prevent the discharge of the encapsulated cells (Rodrigues et al., 2020). Therefore, encapsulation of probiotic bacterial cells is a strategy that is likely to develop resistance of encapsulated cells to unfavorable conditions (Table 13.1) (Kim et al., 2017). In packaging, physicochemical or

TABLE 13.1

Main Techniques Used to Encapsulate Probiotic Bacteria

SI	Name of the technique	Encapsulated microorganisms
	Emulsion	*Akkermansiamuciniphil*
		*Lactobacillus rhamnosus*GG
		Saccharomyces boulardii
		Enterococcus faecium
		*Lactobacillus paracasei*spp. *paracasei*
		Lactobacillus acidophilus LA-5
	Extrusion	*Lactobacillus casei* 01
		*Lactobacillus casei*BGP93
		Lactobacillus acidophilus La3
		Lactobacillus lactis cremoris
		Lactobacillus acidophilus KBL409
		*Lactobacillus casei*ATCC 393
		Faecalibacteriumprausnitzii
	Spray drying	*Lactobacillus rhamnosus*GG
		Lactobacillus plantarum ATCC 8014
		*Bifidobacterium infantis*ATCC 15679
		Lactobacillus acidophilus NCDC 016
		Lactobacillus plantarum NCIM 2083
		*Kluyveromyces*VM004
		Lactobacillus fermentum K73
	Spray chilling	*Bifidobacterium animalis*spp. *lactis* BI-01
		Lactobacillus acidophilus Lac-04
		Lactobacillus acidophilus
		*Bifidobacterium animalis*spp. *lactis*
		Saccharomyces boulardii
		Lactobacillus acidophilus LA-5
		Bifidobacterium bifidum bb-12
		Lactobacillus acidophilus La3
		*Bifidobacterium animalis*spp. *lactis* BLC1

Source: Rodrigues et al., 2020.

mechanical processes are often used. During this process, the bacterial cells are encapsulated in different envelope materials that minimize the risk of injury or cell loss of the enveloped bacterial cells (Wu and Zhang, 2018). Today, many techniques are used to encapsulate probiotic cells. However, it is particularly important to analyze that when selecting a technique, it must be suitable for a particular probiotic strain, the procedure should be simple, the cell viability adequate, and the degree of viability must be adequate stable compatibility for future application purposes (Rathore et al., 2013). Several techniques are currently used for the packaging of probiotics, such as spray drying, spray cooling, extrusion, spray lyophilization, emulsification, electrospray, fluidized bed, and coagulation, etc. (Frakolaki et al., 2021). Some of these are discussed briefly in the following section:

13.7.1 Emulsion

Emulsions are widely used in the food and pharmaceutical industries to improve the stability, solubility, and physiological activity of desired compounds. The emulsion is prepared by diffusing two immiscible liquids and stabilized by the addition of a stabilizer, which has a higher affinity for the continuous phase than the dispersed phase (Alemzadeh et al., 2020). However, the dispersed phase droplets can be separated from the continuous phase by adding a solidifier (Zhang et al., 2016b). These emulsions are used to encapsulate probiotics to improve the protection and stability of the encapsulated bacterial cells. The dispersed aqueous phase is more favorable and widely used due to the hydrophilic nature of the bacterial cell wall (Wang et al., 2020). As previously reported, Lactobacillus plantarum 299v was coated using a simple emulsion and metronidazole with an aqueous phase (xanthan and guar gum) and a lipid phase (sunflower oil) (Pandey et al., 2016). This encapsulation method improves the viability of the encapsulated cells and prevents the release of encapsulated material during storage. The use of dual emulsions (Man, Rogosa and Sharp broths, grape oil) has also been reported in the Lactobacillus rhamnosus Lc705 encapsulation strategy. The protection provided by emulsifiers to encapsulated cells was well documented under osmotic pressure, equal to the satisfactory number of viable cells ($\geq$ log CFU/ml) in a hypertonic sucrose solution (Huerta-Vera et al., 2017). Emulsions have also improved the availability of probiotic cells packaged in liquid form. The encapsulation of Lactobacillus acidophilus AS 1.2686 was performed according to a two-step emulsification strategy, using a water-in-oil-in-water (w/w/w) emulsion that improved the viability of encapsulated cells in the range of 14 days compared with free cells (Hadidi et al., 2021). The first step is performed by preparing a water-in-oil emulsion with a homogenizer. In the second step, the emulsion is strongly passed through a silicon wafer containing uniform pores of micrometer size, including the outer aqueous phase. This causes the formation of water droplets; Droplets can be crosslinked. As a result, microspheres can form. These microspheres have been used to encapsulate probiotics (Sugiura et al., 2004). Hydrocolloids are also used as emulsifiers to encapsulate strains of beneficial bacteria, known as internal ionic gelling techniques. This technique uses hydrocolloid material as encapsulation with thickener. Therefore, the microbial cells are mixed in a hydrocolloid solution and a non-ionizing thickener is added to the oil, which acts as a continuous water-in-oil emulsion phase. Therefore, when the pH of the medium changes, the solid is ionized, leading to the formation of particles that can separate from the continuous phase (Zhang et al., 2020b). In the internal ion gelation technique, alginate and calcium carbonate are used as encapsulation and solidifying agents, respectively. This technique has been successfully used to encapsulate Bifidobacterial BB-12. The coated granules were effectively resistant to probiotic strains due to contact with gastric and intestinal juices, and improved cell viability and stability during storage, up to 60 days at 25°C (Holkem et al., 2016).

13.7.2 Extrusion

Extrusion is a widely used technique for encapsulating bacterial cells. It is a simple, easy to use, economically viable, and relatively gentle method that ensures high viability and stability of encapsulated bacterial cells (Krasaekoopt et al., 2003). In addition to the favorable conditions, this technique has several disadvantages, including slow speed, poor efficiency in generating microparticles smaller than 500 µm, low to medium viscosity of essential polymer solutions, and requiring a large nozzle diameter (Reis et al., 2006). Generally, in this technique, a hydrocolloid solution is used in conjunction with a microbial culture. The extrusion of the hydrocolloid through the nozzle into the crosslinking solution results in an immediate conversion of the hydrocolloid solution to a gel, which ultimately forms microspheres. The gel is very stable in acidic media, although it decomposes under alkaline conditions (Fábio J. Rodrigues et al., 2017). The size of the microsphere is affected by several factors such as the nozzle diameter, solution viscosity, flow rate of the polymer solution, distance between the droplet and the crosslinking solution, and temperature of the polymer. Brun-Graeppi et al., 2011). Furthermore, cross-linking and immediate hardening on the outer surface of the microsphere impede the movement of ions inside the inner core, affecting the instability of the microsphere (Liu et al., 2002). However, the production of microspheres is carried out on a laboratory scale; large-scale production is generally difficult due to the slow formation of microspheres (Burgain et al., 2011). Several methods have been reported to overcome the disadvantages of this simple extrusion technique, including precision granulation (PPF), coaxial extrusion, and coaxial flow method (Piazza & Roversi, 2011), multiple nozzles (Kim et al., 2012), producing electrostatically charged droplets (Zhang et al., 2016a), rotating disc atomization (Herrero et al., 2006), liquid jet cutting, and vibration or acoustic energy techniques (Whelehan & Marison, 2011). If the droplet process is controlled by pulse, jet, or vibrating nozzle, the technique is called jet or granulation. In this technique, the liquid jet is converted into a droplet by the vibration frequency generated by the vibrator (Del Gaudio et al., 2005). The size distribution of microspheres is highly dependent on the viscosity of the polymer solution and the flow rate of the solution to encapsulate the Saccharomyces boulardii bacterial sequence using jet stream disruption (Graff et al., 2008). Prepared microspheres of Saccharomyces boulardii can be coated with chitosan. However, the chitosan coating did not confer any other beneficial properties. Saccharomyces boulardii is better protected inside free microspheres than in covered microspheres (Kim et al., 2012). In coaxial flow engineering, alginate can be used in two different concentrations and injected separately into the outer surface and the inner core cavity of the coaxial nozzle, resulting in stable microsphere formation (Loftus, 2016). Therefore, in acoustic excitation engineering, the polymer droplets produced are often favorable for cross-linking in a calcium chloride solution. This technique is suitable for creating microspheres of controlled size. It is possible to prevent the escape of enveloped microbial cells from the microsphere by improving the concentrations of the outer and inner core materials (Rathore et al., 2013). Another innovative device developed by Haunge is a pneumatic micro-vibrator that can be used continuously to generate alginate microparticles in the range of 30-70 µm (Huang et al., 2010). This device offers many advantages, including flexibility, compactness, and reduced risk of damage to encapsulated cells.Spray drying technology is commonly used for microencapsulation in the food industry due to its low cost, easy operation, fast processing, and high productivity. Generally, this technique involves the atomization of the microbial cell solution in hot drying air, followed by rapid evaporation of water (Ray et al., 2016). The encapsulated cells are then separated from the carrier air in powder form. In this technique, various natural polymers can be used, including starch and gum arabic, because of their ability to form microspheres after drying. In addition, fructo oligosaccharides, inulin, gum, and alginate can also be used as encapsulation agents (Hadzieva et al., 2017). To create a good separation microscope, several parameters need to be optimized, including inlet and outlet air temperature, supply temperature, air flow rate, feed rate, etc. (Vega and Roos, 2006). Temperature regulation is of the utmost importance because low temperatures can reduce the rate of water evaporation leading to the formation of microspheres, although high temperatures can damage bacterial cell walls, directly affecting bacterial growth. viability of bacterial cells (Rathore et al., 2013) (Fig. 13.4). However, the viability of spray-drying coated probiotic bacterial cells was strongly influenced by the system outlet temperature rather than the inlet temperature (Martín et al., 2015). For example, as previously reported on the packaging of *Lactobacillus paracasei* NFBC 338

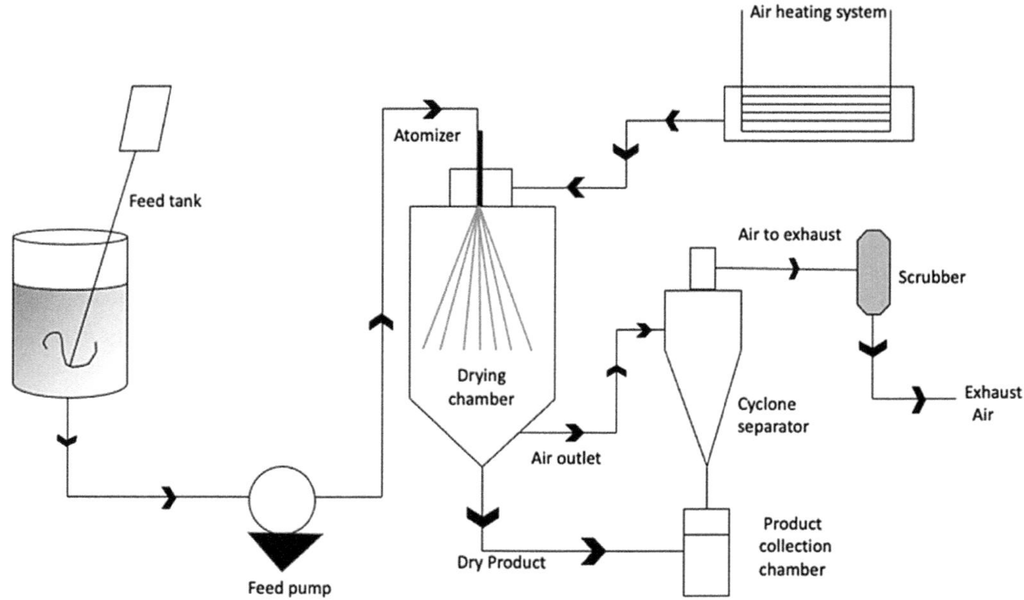

FIGURE 13.4 Schematic representation of spray-drying mechanism.

by spray drying technique using skim milk as encapsulation, a survival rate of 97% was reported at an outlet temperature of 70-75 °C and 0% at 120°C (Corcoran et al., 2006).

Lactobacillus acidophilus La-5 is currently used for packaging by spray drying, using inulin as the packaging material at inlet and outlet temperatures of 120°C and 55°C, respectively. The viability of bacterial cells measured after completion of encapsulation was 86.5% (Łopusiewicz et al., 2021). Similarly, Bifidobacteria BB-12 cells were coated by spray-drying with prebiotics using inlet and outlet temperatures of 150°C and 55°C, respectively, and bacterial cell survival rates were obtained higher than 70% (Fritzen-Freire et al., 2013). To overcome the damaging effects of high temperatures and improve the stability and viability of encapsulated cells, the addition of soluble fiber, prebiotics, gum, and mucilage with the encapsulation material has an effect used as a heat protectant and provides additional cellular resistance (Rajam & Anandharamakrishnan, 2015). Bifidobacteria bifidum BB02 cells were coated by spray drying with a combination of whey protein (concentrate), maltodextrin, and mosquito repellent gum, the symbiotic effect of which was well-documented to enhance resistance and viability of the enveloped cells to environmental conditions (Rodriguez-Huezo et al., 2007).

13.7.3 SPRAY CHILLING

The spray chilling technique is also known as spray cooling or spray freezing. The process is similar to spray drying because it produces tiny droplets. Spray refrigeration is a suitable technique for encapsulating bacterial cells and other food ingredients, as it is an economically viable, continuous, and easily scalable process (Favaro-Trindade et al., 2021) (Fig. 13.5). In addition, this technique does not require organic solvents, including alcohols or ethers, and can be used effectively for temperature-sensitive components such as omega-3 fatty acids and enzymes (Favaro-Trindade et al., 2015). However, this technique has some technological limitations, such as lower efficiency in encapsulation of microbial cells and increased ability to remove microbial cells during storage. In the cooling technique, the encapsulation material is diffused on a semi-liquid substrate. The substrate is made up of lipids and atomized in a cold-air chamber, allowing the particles to solidify (Pedroso et al., 2013). Spray cooling is not new, but it is a method that is less advantageous than

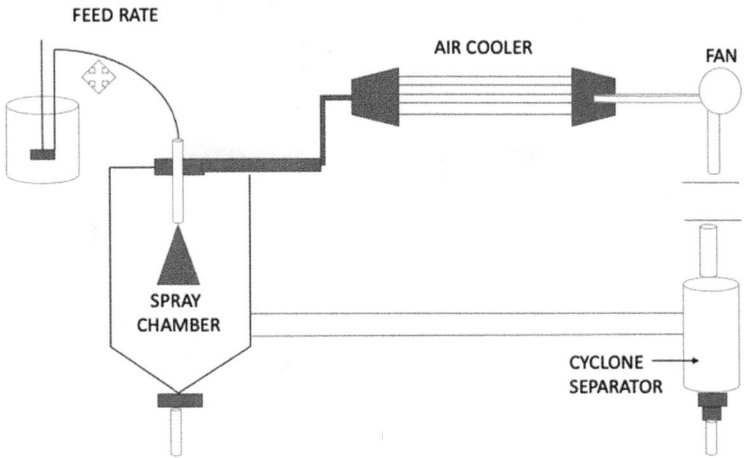

FIGURE 13.5 Encapsulation by spray-chilling.

other methods such as spray drying and ion gelation (Bertoni et al., 2019). Lipids, triglycerides, waxes, and fatty acids can be used to prepare the matrix, but hydrophilic compounds such as polysaccharides can only be used in the presence of an emulsion with an appropriate melting temperature (Queirós et al., 2020). Although spray cooling requires a specific carrier to have hydrophobic properties and a lower melting point than those that have a negative effect on the encapsulated cells, this poses the challenge of large-scale application of this technique in the food sector (Pedroso et al., 2013). Despite the limitations, Lactobacillus acidophilus and Bifidobacterial Animalis subsplectis have been functionally encapsulated by this technique using vegetable oils as encapsulation. In addition, the coated probiotic cells showed high viability during cryopreservation for about 90 days (Bampi et al., 2016). Similarly, single- and double-layer microspheres of Lactobacillus acidophilus, Saccharomyces boulardii, and Bifidobacteria bifidum obtained by the spray cooling technique are used in pastry production (Arslan-Tontul et al., 2019).

13.7.4 ENCAPSULATION MATERIAL FOR PROBIOTIC CELLS

To maintain the viability and stability of the probiotic cell, the selection of an appropriate encapsulation material is essential (Liliana & Vladimir, 2013). The packaging material must be non-toxic for microbial cells to infect it. It must provide protection against adverse environmental conditions and be sufficient to prevent leakage of the encapsulated cells (Table 13.2). During probiotic storage, relative humidity (RH) and temperature can affect cell stability and viability (Librán et al., 2017). Therefore, the material used can retain water to improve the number of viable encapsulated cells. In addition, this material does not completely release bacterial cells upon contact with gastric juice, otherwise it may not be able to protect cells (Yao et al., 2020). The production of microspheres for encapsulation using water-soluble polymers provides optimal conditions for the reactivation of microbial cells. Therefore, natural and synthetic water-soluble polymers are widely used to encapsulate microbial cells. While synthetic polymers offer better mechanical strength and chemical stability than natural polymers (ter Horst et al., 2019). The gelation of polymers can be one of several mechanisms such as thermal gelation, ionotropic gelation, thermal gelation, polymerization, and crosslinking between polymers. Hydrogen bonding, hydrophobic interactions, and/or electrostatic interactions develop intermolecular and intramolecular interactions during polymer crosslinking (Hazrati & Madadlou, 2021). The real-time in situ method was used to measure the oxygen transport in the oil/water interface of the emulsion. In this method, tris ruthenium (II) bis hexafluorophosphate dye is used to quench reversible fluorescence (Tikekar et al., 2011). The dye is

TABLE 13.2

Different Encapsulating Material Used to Encapsulate Probiotic Bacteria

Encapsulating material	Name of technique	Probiotic bacteria
Chitosan-alginate-inulin	Extrusion	*Lactobacillus rhamnosus*GG
Alginate-shellac	Fluidized bed	*Lactobacillus paracasei*BGP-1
Cellulose alginate	Fluidized bed	*Lactobacillus plantarum* IS-10506
Alginate-arabinoxylan	Extrusion	*Lactobacillus plantarum*
Alginate-goats' milk-inulin	Extrusion	*Bifidobacterium animalis spp. lactis* BB12
Alginate	Extrusion	*Lactobacillus casei*ATCC 393
Alginate-chitosan	Emulsification; internal gelation	*Bifidobacterium longum* DD98

Source: Rodrigues et al., 2020.

encapsulated in the oil phase of the emulsion by oxygen crosslinking. This technique can be used to verify the efficacy and evaluate the barrier properties of various polymers used for microencapsulation (Table 1). Therefore, suitable packaging materials can be selected using this technique (Bakry et al., 2016). In general, the polymers used to encapsulate probiotics are listed in Table 13.2.

13.8 CHALLENGES WITH CURRENTLY APPLIED METHODS

Regardless of the type of encapsulation technique used, the need to maintain aseptic conditions is common to all techniques (Chen et al., 2017). In addition, many probiotic strains are of gastrointestinal origin and therefore require anaerobic conditions for optimal growth. The presence of oxygen can significantly reduce bacterial cell viability and activity during encapsulation, especially for Bifidobacteria because this group is strictly anaerobic (Ruiz et al., 2011). In addition, antioxidants are used with different packaging materials to prevent the formation of free radicals (oxygen toxicity). Anaerobic conditions must be maintained throughout the entire encapsulation process, including packaging tools (Lobo et al., 2010). As stated earlier, no suitable deoxidation technique has been clearly identified, except for the addition of antioxidants by packaging materials. Anaerobic chambers are used to maintain anaerobic environmental conditions during packaging. However, the use of anaerobic chambers is not suitable for large-scale processing such as spray drying and other packaging applications used in industries (Chavarri et al., 2012). Although the cost of the packaging process is a major concern, including improved programming with the equipment and processes applied to the packaging, this minimizes the risk of damage to cellular viability due to oxygen toxicity during encapsulation (Singh et al., 2010).It has been reported that in vitro evaluation of encapsulated cells cannot be appropriately substituted for in vivo assessment, as the in vitro system cannot perfectly simulate physiological conditions such as those found in the gastrointestinal tract of human evolution (Costa & Ahluwalia, 2019). The existence of food ingredients in vivo may temporarily increase the pH of gastric juice. In contrast, in vitro evaluation may not account for certain biological factors such as peristaltic movement between the stomach and colon (Wojtunik-Kulesza et al., 2020). The production of mucus by epithelial tissue and its function on the gastrointestinal surface is considered a very difficult factor to mimic in vitro (Linden et al., 2008). Maintaining maximum survival during the consumption of gastrointestinal encapsulated cells is a barrier to the successful use of probiotic cells. Improving survival and shelf-life stability can be considered as another challenge in developing probiotic encapsulation techniques (Calinoiu et al., 2019). During storage, several parameters such as storage temperature, humidity, energy composition, oxygen concentration, packaging material, storage material, exposure to light, and presence of antioxidants significantly affected the survival rate of probiotics in dry powder form (Anekella,

2012). Another challenge, as discussed earlier, is the leakage of probiotics from microspheres during storage. Unnecessary leakage of probiotic cells affects the growth of undesirable bacteria on the food base, which can negatively alter organoleptic properties and perishability of food (Rad et al., 2021). Activation of microspheres to release encapsulated probiotic cells in a specific region of the gastrointestinal tract can be considered as another challenge for targeted delivery of encapsulated probiotics. Several basic approaches are mentioned here to deliver encapsulated probiotic cells to the target region (Solanki et al., 2013).

Approach 1: mechanically break down the matrix.
Approach 2: by developing the pH in the digestive tract.
Approach 3: By developing a time-dependent system.
Approach 4: use aids, biodegradable or enzymes produced by the microbiome.

While the packaging of probiotics in the form of microspheres uses different food platforms to combine, the relationship between probiotic survival rates and food background is being studied by researchers. However, the use of safety issues/challenges associated with probiotic encapsulation can be further discussed (Silva et al., 2020).

13.9 DEVELOPING PROBIOTIC FOODS.

Foods containing lactic acid occupy an important place in our usual diet today. More than 80% of the population uses the probiotic fermented food "yogurt" in their daily diet as a dietary supplement (Gómez-Gallego et al., 2018). These supplements have specific characteristics due to the addition of pure sourdough from a group of Lactic Acid Bacteria (LAB) (Fig. 13.6). By using an appropriate method, any fermented food product can obtain appropriate properties such as aroma, texture, taste, as well as other physiological and biological properties. Bintsis, 2018). These traditionally fermented LAB-containing foods have certain positive effects on human health, including reducing the risk of rotting, inhibiting the growth of pathogens, and improving metabolic activities. Sharma et al. (2020). In the composition of many fermented foods, Lactobacilli are essential. They are especially important in the production of biological foods. Several species of Lactobacillus are used as starter cultures in the production of curds, yogurts, cheeses, and other fermented dairy foods (Dimidi et al., 2019). But again, the characteristics of a strain must be applied to large-scale fermentation. Because not all strains can be grown on an industrial scale due to low reproductive rates and average or lower survival efficiency during freezing and freeze-drying (Kim et al., 2021). Therefore,

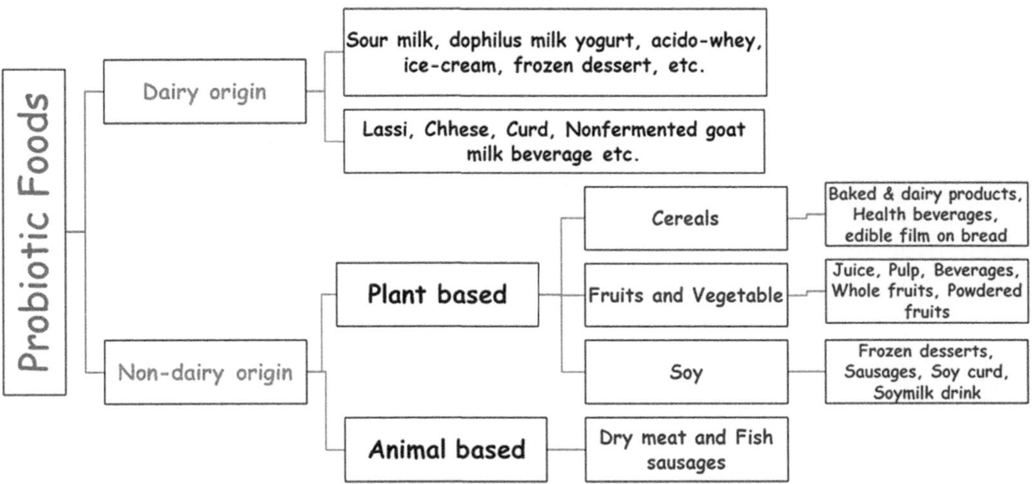

FIGURE 13.6 Classification and types of probiotic foods.

LABs used in the production of probiotics/fermented foods at the industrial level must meet certain additional requirements (table). The selection of probiotic strains for probiotic/fermented food development depends on the food microbiological safety parameters of the final product obtained using a non-pathogenic strain under appropriate hygienic conditions (Marco et al., 2021). The high number of viable cells and higher survival rates through the intestines and stomach have enabled them to confer health benefits to the host. The combination of different probiotic strains enhances the biological role of probiotic supplementation (Markowiak & Ślizewska, 2017). Fermented milk products with probiotic properties must be designed according to the probiotic development criteria. After bread, yogurt is the most used food in the world, developed from starters with a combination of probiotic strains. Lactobacillus delbrueckii subsp. Bulgaricus NBIMCC 3607 is fertile and meets all requirements as a probiotic culture medium (Widyastuti et al., 2021). The maximum number of viable cells of Lactobacillus improves the healing and preventive properties of yogurt.

13.10 VIABILITY OF PROBIOTIC CELLS

Probiotics can be in any of the FDA-regulated product categories. To date, the FDA has no specific clearinghouse or line regarding probiotics or any regulatory definition of probiotics. Probiotics are also capable of losing viability and degradation under unknown conditions (Hoffman et al., 2008). The viability of encapsulated probiotic bacterial cells refers to the ability of a cell to grow and eventually produce a colony of cells under specific habitat conditions (Wilkinson, 2018). In general, sustainability is a mandatory biological effect in relation to consumer health benefits. The researcher demonstrated that to stimulate the viability of the host's intestinal immune system, bacterial survival is required and that, during lactose digestion, the efficiency of the cells' high survival compared with nonviable probiotic cells (Pelletier et al., 2001). Recently, scientific literature has refocused on various innovative techniques used to improve the viability of probiotic cells throughout the shelf life of the product (Champagne et al., 2018). Food ingredients such as food preservatives (sugar, salt, antibacterial agents, bacteriocins or even aromatic compounds, etc.) Narvhus et al., 1993).Bacteriocins, antimicrobial compounds and/or sometimes salts have considerable difficulty in maintaining probiotic cell viability in the food matrix, especially during storage, although prebiotics are known. is to improve the viability of probiotic cells (Kumar et al., 2015). Improving probiotic cell viability, isolation, characterization, and selection of new species (e.g. *Bifidobacteriaum*, *Propionibacteria* and *Lactobacillus*) is a major area of research. The criteria for characterization and selection of a strain as a probiotic may focus on one or both or both for food and nutritional applications (Ku et al., 2016). The main aspects of these criteria can be divided into different categories: 1. Technology 2. Safety 3. Function 4. Physiological characteristics. The viability of probiotic cells during food processing and/or storage and their viability after transfer to the upper intestine, as well as the health benefits to the host, may be the main selection criterion for new probiotics (Hansen et al., 2002). As previously shown, a loss of 6 to 8 log CFU/g of probiotic cells during gastric digestion in vitro (Brinques & Ayub, 2011; Sabikhi et al., 2010) indicates there are not enough remaining probiotic cells to provide health benefits to the host (Vijayakumar et al., 2015). The guidelines and guidelines published by the Food and Agriculture Organization of the United Nations (FAO)/World Health Organization (WHO) indicate the prerequisites for the survival of probiotic strains in the gut to ensure beneficial effects on human health. To demonstrate this, for example, it was revealed that the number of viable "minimally therapeutic" probiotic cells should be at least 106 cfu/g during the shelf life of the product (Terpou et al., 2019).

13.11 CONCLUSION

Prebiotics act as an important ingredients for improving the health of individuals by stimulating the selective growth of microbes in the gastrointestinal tract. Most of the prebiotics are carbohydrates by nature, such as Galacto-oligosaccharides, Fructo-oligosaccharides, Starch, and Glucose-derived oligosaccharides, but there are other sources as well. Along with the many benefits of prebiotics, their significant role as cancer prevention agents specifically makes them the hotspot of medical and food researchers. They also help in preventing antibiotic-associated diarrhea, reducing inflammatory responses, reducing various risk factors associated with cardiovascular disease, and many more. Similarly, probiotics (microbes) are an equally important factor for the health of an individual with a documented ancient history in the food system. Like prebiotics, the benefits of probiotics are impressive but not limited to improving immune response, lowering serum cholesterol levels, improving gut health, and preventing cancer, however, the effect varies with respect to different strains. Despite having many important health benefits known for probiotics, there is a scope for further research further in health to obtain confirmations of many claims. Many observed benefits require verification through further studies. In addition to prebiotics and probiotics, a symbiotic is a combination of prebiotic and probiotic. Symbiotic is more useful as it combines the effects of prebiotics and probiotics along with minimizing the lack of using individually prebiotic or probiotic individually. The improved mechanism of symbiotics is more helpful compared to the individual (prebiotic or probiotic). This combination helps each other, such as the growth of probiotics can be improved by prebiotics. Probiotics can be used by encapsulating them with different techniques such as emulsion, extrusion, spray drying, and spray chilling. The significant use of probiotics can be achieved by developing probiotic foods.. The major concern of probiotics is their viability during food processing and storage. The applications of prebiotics and probiotics are enormous and despite of many studies and products available in the market, there are many loopholes that need to be addressed.

REFERENCES

Abou-Donia, S. A. (2008). Origin, history and manufacturing process of Egyptian dairy products: An overview. *Alexandria Journal of Food Science and Technology*, 5(1), 51–62.

Alba-Lois, L., & Segal-Kischinevzky, C. (2017). Yeast fermentation and the making of beer and yeast fermentation and the making of beer and wine. *Nature Education*, 3(January 2010), 17. https://scholar.google.com/scholar?hl=en&as_sdt=0%2C5&q=Yeast+fermentation+and+the+making+of+beer+and+wine&btnG=

Alemzadeh, I., Hajiabbas, M., Pakzad, H., Sajadi Dehkordi, S., & Vossoughi, A. (2020). Encapsulation of food components and bioactive ingredients and targeted release. *International Journal of Engineering, Transactions A: Basics*, 33(1), 1–11. https://doi.org/10.5829/ije.2020.33.01a.01

Alvaro, E., Andrieux, C., Rochet, V., Rigottier-gois, L., Lepercq, P., Galan, P., Duval, Y., & Juste, C. (2007). Composition and metabolism of the intestinal microbiota in consumers and non-consumers of yogurt, 126–133. https://doi.org/10.1017/S0007114507243065

Anekella, K. (2012). Microencapsulation of probiotics (Lactobacillus acidophilus and Lactobacillus rhamnosus) in raspberry powder by spray drying: Optimization and storage stability studies. ProQuest Dissertations and Theses, 129. https://search.proquest.com/docview/1034738787?accountid=26646%0Ahttp://link. periodicos.capes.gov.br/sfxlcl41?url_ver=Z39.88-2004&rft_val_fmt=info:ofi/fmt:kev:mtx:dissertation&genre=dissertations+%26+theses&sid=ProQ:ProQuest+Dissertations+%26+Theses+Globa

Arslan-Tontul, S., Erbas, M., & Gorgulu, A. (2019). The use of probiotic-loaded single- and double-layered microcapsules in cake production. *Probiotics and Antimicrobial Proteins*, 11(3), 840–849. https://doi.org/10.1007/s12602-018-9467-y

Babbar, N., Dejonghe, W., Gatti, M., Sforza, S., & Elst, K. (2016). Pectic oligosaccharides from agricultural by-products: Production, characterization and health benefits. *Critical Reviews in Biotechnology*, 36(4), 594–606. https://doi.org/10.3109/07388551.2014.996732

Bakry, A. M., Abbas, S., Ali, B., Majeed, H., Abouelwafa, M. Y., Mousa, A., & Liang, L. (2016). Microencapsulation of oils: A comprehensive review of benefits, techniques, and applications. *Comprehensive Reviews in Food Science and Food Safety*, 15(1), 143–182. https://doi.org/10.1111/1541-4337.12179

Baldan, B., Bertoldo, A., Navazio, L., & Mariani, P. (2003). Oligogalacturonide-induced changes in the developmental pattern of Daucus carota L. somatic embryos. *Plant Science*, 165(2), 337–348. https://doi.org/10.1016/S0168-9452(03)00193-6

Bali, V., Panesar, P. S., Bera, M. B., & Panesar, R. (2015). Fructo-oligosaccharides: Production, purification and potential applications. *Critical Reviews in Food Science and Nutrition*, 55(11), 1475–1490. https://doi.org/10.1080/10408398.2012.694084

Bampi, G. B., Backes, G. T., Cansian, R. L., de Matos, F. E., Ansolin, I. M. A., Poleto, B. C., Corezzolla, L. R., & Favaro-Trindade, C. S. (2016). Spray chilling microencapsulation of Lactobacillus acidophilus and Bifidobacterium animalis subsp. lactis and its use in the preparation of savory probiotic cereal bars. *Food and Bioprocess Technology*, 9(8), 1422–1428. https://doi.org/10.1007/s11947-016-1724-z

Berings, M., Karaaslan, C., Altunbulakli, C., Gevaert, P., Akdis, M., Bachert, C., & Akdis, C. A. (2017). Advances and highlights in allergen immunotherapy: On the way to sustained clinical and immunologic tolerance. *Journal of Allergy and Clinical Immunology*, 140(5), 1250–1267. https://doi.org/10.1016/j.jaci.2017.08.025

Bertoni, S., Albertini, B., & Passerini, N. (2019). Spray congealing: An emerging technology to prepare solid dispersions with enhanced oral bioavailability of poorly water soluble drugs. *Molecules*, 24(19), 1–21. https://doi.org/10.3390/molecules24193471

Bielecka, M. (2006). *Probiotics in Food: Chemical and Functional Properties of Food Components*, Third Edition, 413–426. https://doi.org/10.1201/9781420009613.ch16

Bindels, L. B., Delzenne, N. M., Cani, P. D., & Walter, J. (2015). Opinion. Towards a more comprehensive concept for prebiotics. *Nature Reviews. Gastroenterology and Hepatology*, 12(5), 303–310. https://doi.org/10.1038/nrgastro.2015.47

Bintsis, T. (2018). Lactic acid bacteria as starter cultures: An update in their metabolism and genetics. *AIMS Microbiology*, 4(4), 665–684. https://doi.org/10.3934/microbiol.2018.4.665

Brinques, G. B., & Ayub, M. A. Z. (2011). Effect of microencapsulation on survival of Lactobacillus plantarum in simulated gastrointestinal conditions, refrigeration, and yogurt. *Journal of Food Engineering*, 103(2), 123–128. https://doi.org/10.1016/j.jfoodeng.2010.10.006

Brun-Graeppi, A. K. A. S., Richard, C., Bessodes, M., Scherman, D., & Merten, O. W. (2011). Cell microcarriers and microcapsules of stimuli-responsive polymers. *Journal of Controlled Release*, 149(3), 209–224. https://doi.org/10.1016/j.jconrel.2010.09.023

Burgain, J., Gaiani, C., Linder, M., & Scher, J. (2011). Encapsulation of probiotic living cells: From laboratory scale to industrial applications. *Journal of Food Engineering*, 104(4), 467–483. https://doi.org/10.1016/j.jfoodeng.2010.12.031

Calinoiu, L. F., Ştefanescu, B. E., Pop, I. D., Muntean, L., & Vodnar, D. C. (2019). Chitosan coating applications in probiotic microencapsulation. *Coatings*, 9(3), 1–21. https://doi.org/10.3390/COATINGS9030194

Champagne, C. P., Gomes da Cruz, A., & Daga, M. (2018). Strategies to improve the functionality of probiotics in supplements and foods. *Current Opinion in Food Science*, 22, 160–166. https://doi.org/10.1016/j.cofs.2018.04.008

Chavarri, M., Maranon, I., & Carmen, M. (2012). Encapsulation technology to protect probiotic bacteria. *Probiotics*, 501–540. https://doi.org/10.5772/50046

Chen, J., Liang, R. H., Liu, W., Li, T., Liu, C. M., Wu, S. S., & Wang, Z. J. (2013). Pectic-oligosaccharides prepared by dynamic high-pressure microfluidization and their in vitro fermentation properties. *Carbohydrate Polymers*, 91(1), 175–182. https://doi.org/10.1016/j.carbpol.2012.08.021

Chen, J., Wang, Q., Liu, C. M., & Gong, J. (2017). Issues deserve attention in encapsulating probiotics: Critical review of existing literature. *Critical Reviews in Food Science and Nutrition*, 57(6), 1228–1238. https://doi.org/10.1080/10408398.2014.977991

Corcoran, B. M., Ross, R. P., Fitzgerald, G. F., Dockery, P., & Stanton, C. (2006). Enhanced survival of GroESL-overproducing Lactobacillus paracasei NFBC 338 under stressful conditions induced by drying. *Applied and Environmental Microbiology*, 72(7), 5104–5107. https://doi.org/10.1128/AEM.02626-05

Costa, J., & Ahluwalia, A. (2019). Advances and current challenges in intestinal in vitro model engineering: A digest. *Frontiers in Bioengineering and Biotechnology*, 7(June), 1–14. https://doi.org/10.3389/fbioe.2019.00144

Costabile, A., Fava, F., Röytiö, H., Forssten, S. D., Olli, K., Klievink, J., Rowland, I. R., Ouwehand, A. C., Rastall, R. A., Gibson, G. R., & Walton, G. E. (2012). Impact of polydextrose on the faecal microbiota: A double-blind, crossover, placebo-controlled feeding study in healthy human subjects. *British Journal of Nutrition*, 108(3), 471–481. https://doi.org/10.1017/S0007114511005782

Cunningham, M., Azcarate-Peril, M. A., Barnard, A., Benoit, V., Grimaldi, R., Guyonnet, D., Holscher, H. D., Hunter, K., Manurung, S., Obis, D., Petrova, M. I., Steinert, R. E., Swanson, K. S., van Sinderen, D., Vulevic, J., & Gibson, G. R. (2021). Shaping the future of probiotics and prebiotics. *Trends in Microbiology*, 29(8), 667–685. https://doi.org/10.1016/j.tim.2021.01.003

Davani-Davari, D., Negahdaripour, M., Karimzadeh, I., Seifan, M., Mohkam, M., Masoumi, S. J., Berenjian, A., & Ghasemi, Y. (2019). Prebiotics: Definition, types, sources, mechanisms, and clinical applications. *Foods*, 8(3), 1–27. https://doi.org/10.3390/foods8030092

Del Gaudio, P., Colombo, P., Colombo, G., Russo, P., & Sonvico, F. (2005). Mechanisms of formation and disintegration of alginate beads obtained by prilling. *International Journal of Pharmaceutics*, 302(1–2), 1–9. https://doi.org/10.1016/j.ijpharm.2005.05.041

Didari, T., Solki, S., Mozaffari, S., Nikfar, S., & Abdollahi, M. (2014). A systematic review of the safety of probiotics. *Expert Opinion on Drug Safety*, 13(2), 227–239. https://doi.org/10.1517/14740338.2014.872627

Dimidi, E., Cox, S., Rossi, M., & Whelan, K. (2019). Fermented foods : Definitions and characteristics , gastrointestinal health and disease. *Nutrients*, 11(1806), 1–26.

Do Carmo, M. M. R., Walker, J. C. L., Novello, D., Caselato, V. M., Sgarbieri, V. C., Ouwehand, A. C., Andreollo, N. A., Hiane, P. A., & dos Santos, E. F. (2016). Polydextrose: Physiological function, and effects on health. *Nutrients*, 8(9), 1–13. https://doi.org/10.3390/nu8090553

Falagas, M. E., Betsi, G. I., Tokas, T., & Athanasiou, S. (2006). Probiotics for prevention of recurrent urinary tract infections in women. *Drugs*, 66(9), 1253–1261. https://doi.org/10.2165/00003495-200666090-00007

FAO/WHO. (2002). *Guidelines for the Evaluation of Probiotics in Food*, 1–11. https://www.fao.org/3/a0512e/a0512e.pdf

Favaro-Trindade, C. S., Okuro, P., & de Matos, F. (2015). Encapsulation via spray chilling/cooling/congealing. In *Handbook of Encapsulation and Controlled Release*, November, 71–87. https://doi.org/10.1201/b19038-8

Favaro-Trindade, C. S., de Matos Junior, F. E., Okuro, P. K., Dias-Ferreira, J., Cano, A., Severino, P., Zielińska, A., & Souto, E. B. (2021). Encapsulation of active pharmaceutical ingredients in lipid micro/nanoparticles for oral administration by spray-cooling. *Pharmaceutics*, 13(8), 1–14. https://doi.org/10.3390/pharmaceutics13081186

Frakolaki, G., Giannou, V., Kekos, D., & Tzia, C. (2021). A review of the microencapsulation techniques for the incorporation of probiotic bacteria in functional foods. *Critical Reviews in Food Science and Nutrition*, 61(9), 1515–1536. https://doi.org/10.1080/10408398.2020.1761773

Fritzen-Freire, C. B., Prudêncio, E. S., Pinto, S. S., Muñoz, I. B., & Amboni, R. D. M. C. (2013). Effect of microencapsulation on survival of Bifidobacterium BB-12 exposed to simulated gastrointestinal conditions and heat treatments. *LWT - Food Science and Technology*, 50(1), 39–44. https://doi.org/10.1016/j.lwt.2012.07.037

Fuentes-Zaragoza, E., Sánchez-Zapata, E., Sendra, E., Sayas, E., Navarro, C., Fernández-Lõpez, J., & Pérez-Alvarez, J. A. (2011). Resistant starch as prebiotic: A review. *Starch/Staerke*, 63(7), 406–415. https://doi.org/10.1002/star.201000099

Fuller, R. (2001). Handbook of probiotics. *International Journal of Food Science and Technology*, 36(2). https://doi.org/10.1046/j.1365-2621.2001.00460.x

Garthoff, J. A., Heemskerk, S., Hempenius, R. A., Lina, B. A. R., Krul, C. A. M., Koeman, J. H., & Speijers, G. J. A. (2010). Safety evaluation of pectin-derived acidic oligosaccharides (pAOS): Genotoxicity and sub-chronic studies. *Regulatory Toxicology and Pharmacology*, 57(1), 31–42. https://doi.org/10.1016/j.yrtph.2009.12.004

Gasbarrini, G., Bonvicini, F., & Gramenzi, A. (2016). Probiotics history. *Journal of Clinical Gastroenterology*, 50(December), S116–S119. https://doi.org/10.1097/MCG.0000000000000697

Geier, M. S., Butler, R. N., & Howarth, G. S. (2007). Inflammatory bowel disease: Current insights into pathogenesis and new therapeutic options; probiotics, prebiotics and synbiotics. *International Journal of Food Microbiology*, 115(1), 1–11. https://doi.org/10.1016/j.ijfoodmicro.2006.10.006

Gibson, G. R., & Roberfroid, M. B. (1995). Dietary modulation of the human colonic microbiota: Introducing the concept of prebiotics. *Journal of Nutrition*, 125(6), 1401–1412. https://doi.org/10.1093/jn/125.6.1401

Gibson, G. R., Scott, K. P., Rastall, R. A., Tuohy, K. M., Hotchkiss, A., Dubert-Ferrandon, A., Gareau, M., Murphy, E. F., Saulnier, D., Loh, G., Macfarlane, S., Delzenne, N., Ringel, Y., Kozianowski, G., Dickmann, R., Lenoir-Wijnkoop, I., Walker, C., & Buddington, R. (2010). Dietary prebiotics: Current status and new definition. *Food Science & Technology Bulletin Functional Foods*, 7(1), 1–19. https://doi.org/10.1616/1476-2137.15880

Gómez-Gallego, C., Gueimonde, M., & Salminen, S. (2018). The role of yogurt in food-based dietary guidelines. *Nutrition Reviews*, 76, 29–39. https://doi.org/10.1093/nutrit/nuy059

Gourbeyre, P., Denery, S., & Bodinier, M. (2011). Probiotics, prebiotics, and Synbiotics: Impact on the gut immune system and allergic reactions. *Journal of Leukocyte Biology*, 89(5), 685–695. https://doi.org/10.1189/jlb.1109753

Graff, S., Hussain, S., Chaumeil, J. C., & Charrueau, C. (2008). Increased intestinal delivery of viable Saccharomyces boulardii by encapsulation in microspheres. *Pharmaceutical Research*, 25(6), 1290–1296. https://doi.org/10.1007/s11095-007-9528-5

Guarner, F., Khan, A. G., Garisch, J., Eliakim, R., Gangl, A., Thomson, A., ... & Kim, N. (2012). World gastroenterology organisation global guidelines: Probiotics and prebiotics october 2011. *Journal of clinical gastroenterology*, 46(6), 468–481.

Guiné, R. de P. F., & Silva, A. C. F. (2016). Probiotics, prebiotics and Synbiotics. *Functional Foods: Sources, Health Effects and Future Perspectives*, May, 143–207. https://doi.org/10.1201/b15561-2

Hadidi, M., Majidiyan, N., Jelyani, A. Z., Moreno, A., Hadian, Z., & Khanegah, A. M. (2021). Alginate/fish gelatin-encapsulated lactobacillus acidophilus: A study on viability and technological quality of bread during baking and storage. *Foods*, 10(9). https://doi.org/10.3390/foods10092215

Hadzieva, J., Mladenovska, K., Crcarevska, M. S., Dodov, M. G., Dimchevska, S., Geškovski, N., Grozdanov, A., Popovski, E., Petruševski, G., Chachorovska, M., Ivanovska, T. P., Petruševska-Tozi, L., Ugarkovic, S., & Goracinova, K. (2017). Lactobacillus casei encapsulated in soy protein isolate and alginate microparticles prepared by spray drying. *Food Technology and Biotechnology*, 55(2), 173–186. https://doi.org/10.17113/ftb.55.02.17.4991

Hamasalim, H. J. (2016). Synbiotic as feed additives relating to animal health and performance. *Advances in Microbiology*, 06(04), 288–302. https://doi.org/10.4236/aim.2016.64028

Hansen, L. T., Allan-Wojtas, P. M., Jin, Y. L., & Paulson, A. T. (2002). Survival of Ca-alginate microencapsulated Bifidobacterium spp. in milk and simulated gastrointestinal conditions. *Food Microbiology*, 19(1), 35–45. https://doi.org/10.1006/fmic.2001.0452

Hansen, R. G. (1974). Milk in human nutrition. In *Nutrition and Biochemistry of Milk/Maintenance*. https://doi.org/10.1016/b978-0-12-436703-6.50013-2

Hatakka, K., Savilahti, E., Pönkä, A., Meurman, J. H., Poussa, T., Näse, L., Saxelin, M., & Korpela, R. (2001). Effect of long term consumption of probiotic milk on infections in children attending day care centres: Double blind, randomised trial. *British Medical Journal*, 322(7298), 1327–1329. https://doi.org/10.1136/bmj.322.7298.1327

Hazrati, Z., & Madadlou, A. (2021). Gelation by bioactives: Characteristics of the cold-set whey protein gels made using gallic acid. *International Dairy Journal*, 117, 104952. https://doi.org/10.1016/j.idairyj.2020.104952

Hempel, S. J., Maher, A. R., Wang, Z., Miles, J. N. V., Shanman, R., Johnsen, B., & Shekelle, P. G. (2012). Probiotics for the prevention and treatment of antibiotic-associated diarrhea. *JAMA*, 307(18), 1959–1969.

Hendrickson, B. A., Gokhale, R., & Cho, J. H. (2002). Clinical aspects and pathophysiology of inflammatory bowel disease. *Clinical Microbiology Reviews*, 15(1), 79–94. https://doi.org/10.1128/CMR.15.1.79-94.2002

Herrero, E. P., Martín Del Valle, E. M., & Galán, M. A. (2006). Development of a new technology for the production of microcapsules based in atomization processes. *Chemical Engineering Journal*, 117(2), 137–142. https://doi.org/10.1016/j.cej.2005.12.022

Hill, C., Guarner, F., Reid, G., Gibson, G. R., Merenstein, D. J., Pot, B., Morelli, L., Canani, R. B., Flint, H. J., Salminen, S., Calder, P. C., & Sanders, M. E. (2014). Expert consensus document: The international scientific association for probiotics and prebiotics consensus statement on the scope and appropriate use of the term probiotic. *Nature Reviews. Gastroenterology and Hepatology*, 11(8), 506–514. https://doi.org/10.1038/nrgastro.2014.66

Hoffman, F. A., Heimbach, J. T., Sanders, M. E., & Hibberd, P. L. (2008). Executive summary: Scientific and regulatory challenges of development of probiotics as foods and drugs. *Clinical Infectious Diseases*, 46(Suppl. 2), 53–57. https://doi.org/10.1086/523342

Holkem, A. T., Raddatz, G. C., Nunes, G. L., Cichoski, A. J., Jacob-Lopes, E., Ferreira Grosso, C. R., & de Menezes, C. R. (2016). Development and characterization of alginate microcapsules containing Bifidobacterium BB-12 produced by emulsification/internal gelation followed by freeze drying. *LWT - Food Science and Technology*, 71, 302–308. https://doi.org/10.1016/j.lwt.2016.04.012

Holscher, H. D., Gregory Caporaso, J., Hooda, S., Brulc, J. M., Fahey, G. C., & Swanson, K. S. (2015). Fiber supplementation influences phylogenetic structure and functional capacity of the human intestinal microbiome: Follow-up of a randomized controlled trial. *American Journal of Clinical Nutrition*, 101(1), 55–64. https://doi.org/10.3945/ajcn.114.092064

Howlett, J. F., Betteridge, V. A., Champ, M., Craig, S. A. S., Meheust, A., & Jones, J. M. (2010). The definition of dietary fiber - Discussions at the Ninth Vahouny fiber symposium: Building scientific agreement. *Food and Nutrition Research*, 54, 1–5. https://doi.org/10.3402/fnr.v54i0.5750

Huang, S. B., Wu, M. H., & Lee, G. B. (2010). Microfluidic device utilizing pneumatic micro-vibrators to generate alginate microbeads for microencapsulation of cells. *Sensors and Actuators, B: Chemical*, 147(2), 755–764. https://doi.org/10.1016/j.snb.2010.04.021

Huerta-Vera, K., Flores-Andrade, E., Pérez-Sato, J. A., Morales-Ramos, V., Pascual-Pineda, L. A., & Contreras-Oliva, A. (2017). Enrichment of banana with Lactobacillus rhamnosus using double emulsion and osmotic dehydration. *Food and Bioprocess Technology*, 10(6), 1053–1062. https://doi.org/10.1007/s11947-017-1879-2

Hussein, H., Awad, S., El-Sayed, I., & Ibrahim, A. (2020). Impact of chickpea as prebiotic, antioxidant and thickener agent of stirred bio-yoghurt. *Annals of Agricultural Sciences*, 65(1), 49–58.

Hutkins, R. W., Krumbeck, J. A., Bindels, L. B., Cani, P. D., Fahey, G., Goh, Y. J., Hamaker, B., Martens, E. C., Mills, D. A., Rastal, R. A., Vaughan, E., & Sanders, M. E. (2016). Prebiotics: Why definitions matter. *Current Opinion in Biotechnology*, 37, 1–7. https://doi.org/10.1016/j.copbio.2015.09.001

Johnson, D., Thurairajasingam, S., Letchumanan, V., Chan, K. G., & Lee, L. H. (2021). Exploring the role and potential of probiotics in the field of mental health: Major depressive disorder. *Nutrients*, 13(5), 1–18. https://doi.org/10.3390/nu13051728

Kashyap, R., Palai, T., & Bhattacharya, P. K. (2015). Kinetics and model development for enzymatic synthesis of fructo-oligosaccharides using fructosyltransferase. *Bioprocess and Biosystems Engineering*, 38(12). https://doi.org/10.1007/s00449-015-1478-4

Kechagia, M., Basoulis, D., Konstantopoulou, S., Dimitriadi, D., Gyftopoulou, K., Skarmoutsou, N., & Fakiri, E. M. (2013). Health benefits of probiotics: A review. *ISRN Nutrition*, 2013, 1–7. https://doi.org/10.5402/2013/481651

Kelly, G. (2009). Inulin-type prebiotics: A review (Part 2). *Alternative Medicine Review*, 14(1), 36–55.

Kim, I. Y., Pusey, P. L., Zhao, Y., Korban, S. S., Choi, H., & Kim, K. K. (2012). Controlled release of Pantoea agglomerans E325 for biocontrol of fire blight disease of apple. *Journal of Controlled Release*, 161(1), 109–115. https://doi.org/10.1016/j.jconrel.2012.03.028

Kim, S. Il, Kim, J. W., Kim, K. T., & Kang, C. H. (2021). Survivability of collagen-peptide microencapsulated lactic acid bacteria during storage and simulated gastrointestinal conditions. *Fermentation*, 7(3). https://doi.org/10.3390/fermentation7030177

Kim, J. U., Kim, B., Shahbaz, H. M., Lee, S. H., Park, D., & Park, J. (2017). Encapsulation of probiotic Lactobacillus acidophilus by ionic gelation with electrostatic extrusion for enhancement of survival under simulated gastric conditions and during refrigerated storage. *International Journal of Food Science and Technology*, 52(2), 519–530. https://doi.org/10.1111/ijfs.13308

Konikoff, M. R., & Denson, L. A. (2006). Role of fecal calprotectin as a biomarker of intestinal inflammation in inflammatory bowel disease. *Inflammatory Bowel Diseases*, 12(6), 524–534. https://doi.org/10.1097/00054725-200606000-00013

Krasaekoopt, W., Bhandari, B., & Deeth, H. (2003). Evaluation of encapsulation techniques of probiotics for yoghurt. *International Dairy Journal*, 13(1), 3–13. https://doi.org/10.1016/S0958-6946(02)00155-3

Ku, S., Park, M. S., Ji, G. E., & You, H. J. (2016). Review on Bifidobacterium bifidum bgn4: Functionality and nutraceutical applications as a probiotic microorganism. *International Journal of Molecular Sciences*, 17(9). https://doi.org/10.3390/ijms17091544

Kumar, H., Salminen, S., Verhagen, H., Rowland, I., Heimbach, J., Bañares, S., Young, T., Nomoto, K., & Lalonde, M. (2015). Novel probiotics and prebiotics: Road to the market. *Current Opinion in Biotechnology*, 32, 99–103. https://doi.org/10.1016/j.copbio.2014.11.021

Lahtinen, S. J., Knoblock, K., Drakoularakou, A., Jacob, M., Stowell, J., Gibson, G. R., & Ouwehand, A. C. (2010). Effect of molecule branching and glycosidic linkage on the degradation of polydextrose by gut microbiota. *Bioscience, Biotechnology and Biochemistry*, 74(10), 2016–2021. https://doi.org/10.1271/bbb.100251

Lama-Muñoz, A., Rodríguez-Gutiérrez, G., Rubio-Senent, F., & Fernández-Bolaños, J. (2012). Production, characterization and isolation of neutral and pectic oligosaccharides with low molecular weights from olive by-products thermally treated. *Food Hydrocolloids*, 28(1), 92–104. https://doi.org/10.1016/j.foodhyd.2011.11.008

Librán, C. M., Castro, S., & Lagaron, J. M. (2017). Encapsulation by electrospray coating atomization of probiotic strains. *Innovative Food Science and Emerging Technologies*, 39, 216–222. https://doi.org/10.1016/j.ifset.2016.12.013

Lieske, J. C., Goldfarb, D. S., De Simone, C., & Regnier, C. (2005). Use of a probioitic to decrease enteric hyperoxaluria. *Kidney International*, 68(3), 1244–1249. https://doi.org/10.1111/j.1523-1755.2005.00520.x

Liliana, S. C., & Vladimir, V. C. (2013). Probiotic encapsulation. *African Journal of Microbiology Research*, 7(40), 4743–4753. https://doi.org/10.5897/ajmr2013.5718

Linden, S. K., Sutton, P., Karlsson, N. G., Korolik, V., & McGuckin, M. A. (2008). Mucins in the mucosal barrier to infection. *Mucosal Immunology*, 1(3), 183–197. https://doi.org/10.1038/mi.2008.5

Liu, X. D., Yu, W. Y., Zhang, Y., Xue, W. M., Yu, W. T., Xiong, Y., Ma, X. J., Chen, Y., & Yuan, Q. (2002). Characterization of structure and diffusion behaviour of Ca-alginate beads prepared with external or internal calcium sources. *Journal of Microencapsulation*, 19(6), 775–782. https://doi.org/10.1080/0265204021000022743

Lobo, V., Patil, A., Phatak, A., & Chandra, N. (2010). Free radicals, antioxidants and functional foods: Impact on human health. *Pharmacognosy Reviews*, 4(8), 118–126. https://doi.org/10.4103/0973-7847.70902

Loftus, R. (2016). Calcium alginate microbead production via an air assisted shearing process. *Honors Research Projects. 321.* http://ideaexchange.uakron.edu/honors_research_projects/321

Łopusiewicz, Ł., Bogusławska-Wąs, E., Drozłowska, E., Trocer, P., Dłubała, A., Mazurkiewicz-Zapałowicz, K., & Bartkowiak, A. (2021). The application of spray-dried and reconstituted flaxseed oil cake extract as encapsulating material and carrier for probiotic lacticaseibacillus rhamnosus gg. *Materials*, 14(18). https://doi.org/10.3390/ma14185324

Macfarlane, G. T., Steed, H., & Macfarlane, S. (2008). Bacterial metabolism and health-related effects of galacto-oligosaccharides and other prebiotics. *Journal of Applied Microbiology*, 104(2), 305–344. https://doi.org/10.1111/j.1365-2672.2007.03520.x

Mackowiak, P. A. (2013). Recycling Metchnikoff: Probiotics, the intestinal microbiome and the quest for long life. *Frontiers in Public Health*, 1(November), 1–3. https://doi.org/10.3389/fpubh.2013.00052

Manderson, K., Pinart, M., Tuohy, K. M., Grace, W. E., Hotchkiss, A. T., Widmer, W., Yadhav, M. P., Gibson, G. R., & Rastall, R. A. (2005). In vitro determination of prebiotic properties of oligosaccharides derived from an orange juice manufacturing by-product stream. *Applied and Environmental Microbiology*, 71(12), 8383–8389. https://doi.org/10.1128/AEM.71.12.8383-8389.2005

Marco, M. L., Sanders, M. E., Gänzle, M., Arrieta, M. C., Cotter, P. D., De Vuyst, L., Hill, C., Holzapfel, W., Lebeer, S., Merenstein, D., Reid, G., Wolfe, B. E., & Hutkins, R. (2021). The International Scientific Association for Probiotics and Prebiotics (ISAPP) consensus statement on fermented foods. *Nature Reviews. Gastroenterology and Hepatology*, 18(3), 196–208. https://doi.org/10.1038/s41575-020-00390-5

Markowiak, P., & Ślizewska, K. (2017). Effects of probiotics, prebiotics, and Synbiotics on human health. *Nutrients*, 9(9). https://doi.org/10.3390/nu9091021

Martín, M. J., Lara-Villoslada, F., Ruiz, M. A., & Morales, M. E. (2015). Microencapsulation of bacteria: A review of different technologies and their impact on the probiotic effects. *Innovative Food Science and Emerging Technologies*, 27, 15–25. https://doi.org/10.1016/j.ifset.2014.09.010

Martins, G. N., Ureta, M. M., Tymczyszyn, E. E., Castilho, P. C., & Gomez-zavaglia, A. (2019). Technological aspects of the production of fructo and enzymatic synthesis and hydrolysis. 6(May). https://doi.org/10.3389/fnut.2019.00078

McFarland, L. V. (2006). Meta-analysis of probiotics for the prevention of antibiotic associated diarrhea and the treatment of Clostridium difficile disease. *American Journal of Gastroenterology*, 101(4), 812–822. https://doi.org/10.1111/j.1572-0241.2006.00465.x

McFarland, L. V. (2015). From yaks to yogurt: The history, development, and current use of probiotics. *Clinical Infectious Diseases*, 60(Suppl. 2), S85–S90. https://doi.org/10.1093/cid/civ054

Meyer, T. S. M., Miguel, A. S. M., Fernández, D. E. R., & Ortiz, G. M. D. (2015). Biotechnological production of oligosaccharides — Applications in the food industry. *Food Production and Industry*. https://doi.org/10.5772/60934

Milner, E., Stevens, B., An, M., Lam, V., Ainsworth, M., Dihle, P., Stearns, J., Dombrowski, A., Rego, D., & Segars, K. (2021). Utilizing probiotics for the prevention and treatment of gastrointestinal diseases. *Frontiers in Microbiology*, 12(August), 1–13. https://doi.org/10.3389/fmicb.2021.689958

Moayyedi, P., Ford, A. C., Talley, N. J., Cremonini, F., Foxx-Orenstein, A. E., Brandt, L. J., & Quigley, E. M. M. (2010). The efficacy of probiotics in the treatment of irritable bowel syndrome: A systematic review. *Gut*, 59(3), 325–332. https://doi.org/10.1136/gut.2008.167270

Mudgil, D., & Barak, S. (2019). Dairy-based functional beverages. In *Milk-based beverages*, Vol. 9. The Science of Beverages, December 2020, 67–93. https://doi.org/10.1016/B978-0-12-815504-2.00003-7

Mussatto, S. I., & Mancilha, I. M. (2007). Non-digestible oligosaccharides: A review. *Carbohydrate Polymers*, 68(3), 587–597. https://doi.org/10.1016/j.carbpol.2006.12.011

Nagpal, R., Kumar, A., Kumar, M., Behare, P. V., Jain, S., & Yadav, H. (2012). Probiotics, their health benefits and applications for developing healthier foods: A review. *FEMS Microbiology Letters*, 334(1), 1–15. https://doi.org/10.1111/j.1574-6968.2012.02593.x

Narvhus, J. A., Axelsson, L., & Gene, N. (1993). Lactic acid bacterium starter. *Process Biochemistry*, 28(2), 126. https://doi.org/10.1016/0032-9592(93)80020-h

Nazir, Y., Hussain, S. A., Abdul Hamid, A., & Song, Y. (2018). Probiotics and their potential preventive and therapeutic role for cancer, high serum cholesterol, and allergic and HIV diseases. *BioMed Research International*, 2018. https://doi.org/10.1155/2018/3428437

Nichol, A. D., Holle, M. J., & An, R. (2018). Glycemic impact of non-nutritive sweeteners: A systematic review and meta-analysis of randomized controlled trials. *European Journal of Clinical Nutrition*, 72(6), 796–804. https://doi.org/10.1038/s41430-018-0170-6

O'Callaghan, A., & van Sinderen, D. (2016). Bifidobacteria and their role as members of the human gut microbiota. *Frontiers in Microbiology*, 7(JUN). https://doi.org/10.3389/fmicb.2016.00925

Øbro, J., Harholt, J., Scheller, H. V., & Orfila, C. (2004). Rhamnogalacturonan I in Solanum tuberosum tubers contains complex arabinogalactan structures. *Phytochemistry*, 65(10), 1429–1438. https://doi.org/10.1016/j.phytochem.2004.05.002

Ozen, M., & Dinleyici, E. C. (2015). The history of probiotics: The untold story. *Beneficial Microbes*, 6(2), 159–165. https://doi.org/10.3920/BM2014.0103

Pandey, S., Senthilguru, K., Uvanesh, K., Sagiri, S. S., Behera, B., Babu, N., Bhattacharyya, M. K., Pal, K., & Banerjee, I. (2016). Natural gum modified emulsion gel as single carrier for the oral delivery of probiotic-drug combination. *International Journal of Biological Macromolecules*, 92, 504–514. https://doi.org/10.1016/j.ijbiomac.2016.07.053

Parvez, S., Malik, K. A., Ah Kang, S., & Kim, H. Y. (2006). Probiotics and their fermented food products are beneficial for health. *Journal of Applied Microbiology*, 100(6), 1171–1185. https://doi.org/10.1111/j.1365-2672.2006.02963.x

Pedroso, D. L., Dogenski, M., Thomazini, M., Heinemann, R. J. B., & Favaro-Trindade, C. S. (2013). Microencapsulation of Bifidobacterium animalis subsp. lactis and Lactobacillus acidophilus in cocoa butter using spray chilling technology. *Brazilian Journal of Microbiology*, 44(3), 777–783. https://doi.org/10.1590/S1517-83822013000300017

Pelletier, X., Laure-Boussuge, S., & Donazzolo, Y. (2001). Hydrogen excretion upon ingestion of dairy products in lactose-intolerant male subjects: Importance of the live flora. *European Journal of Clinical Nutrition*, 55(6), 509–512. https://doi.org/10.1038/sj.ejcn.1601169

Piazza, L., & Roversi, T. (2011). Preliminary study on microbeads production by co-extrusion technology. *Procedia Food Science*, 1(Icef 11), 1374–1380. https://doi.org/10.1016/j.profoo.2011.09.204

Plaza-Díaz, J., Robles-Sánchez, C., Abadía-Molina, F., Sáez-Lara, M. J., Vilchez-Padial, L. M., Gil, Á., Gómez-Llorente, C., & Fontana, L. (2017). Gene expression profiling in the intestinal mucosa of obese rats administered probiotic bacteria. *Scientific Data*, 4, 1–10. https://doi.org/10.1038/sdata.2017.186

Plaza-Díaz, J., Ruiz-Ojeda, F. J., Gil-Campos, M., & Gil, A. (2018). Immune-mediated mechanisms of action of probiotics and Synbiotics in treating pediatric intestinal diseases. *Nutrients*, 10(1), 1–20. https://doi.org/10.3390/nu10010042

Plaza-Diaz, J., Ruiz-Ojeda, F. J., Gil-Campos, M., & Gil, A. (2019). Mechanisms of action of probiotics. *Advances in Nutrition*, 10, S49–S66. https://doi.org/10.1093/advances/nmy063

Plou, F. J., Segura, A. G. De, & Ballesteros, A. (2007). Application of glycosidases and transglycosidases in the synthesis of oligosaccharides. *Industrial Enzymes: Structure, Function and Applications*, 1, 141–157. https://doi.org/10.1007/1-4020-5377-0_9

Probert, H. M., Apajalahti, J. H. A., Rautonen, N., Stowell, J., & Gibson, G. R. (2004). Polydextrose , lactitol , and fructo-oligosaccharide fermentation by colonic bacteria in a three-stage continuous culture system, 70(8), 4505–4511. https://doi.org/10.1128/AEM.70.8.4505

Queirós, M. de S., Viriato, R. L. S., Ribeiro, A. P. B., & Gigante, M. L. (2020). Dairy-based solid lipid microparticles: A novel approach. *Food Research International*, 131(January), 109009. https://doi.org/10.1016/j.foodres.2020.109009

Rad, A. H., Aghebati-Maleki, L., Kafil, H. S., Gilani, N., Abbasi, A., & Khani, N. (2021). Postbiotics, as dynamic biomolecules, and their promising role in promoting food safety. *Biointerface Research in Applied Chemistry*, 11(6), 14529–14544. https://doi.org/10.33263/BRIAC116.1452914544

Rajam, R., & Anandharamakrishnan, C. (2015). Microencapsulation of Lactobacillus plantarum (MTCC 5422) with fructooligosaccharide as wall material by spray drying. *LWT*, 60(2), 773–780. https://doi.org/10.1016/j.lwt.2014.09.062

Rastall, B. (2006). Prebiotics. In *Chemical and Functional Properties of Food Components*, Third Edition (Issue 57186). https://doi.org/10.1201/9780849381829.ch3

Rather, I. A., Bajpai, V. K., Kumar, S., Lim, J., Paek, W. K., & Park, Y. H. (2016). Probiotics and atopic dermatitis: An overview. *Frontiers in Microbiology*, 7(April), 1–7. https://doi.org/10.3389/fmicb.2016.00507

Rathore, S., Desai, P. M., Liew, C. V., Chan, L. W., & Heng, P. W. S. (2013). Microencapsulation of microbial cells. *Journal of Food Engineering*, 116(2), 369–381. https://doi.org/10.1016/j.jfoodeng.2012.12.022

Ray, S., Raychaudhuri, U., & Chakraborty, R. (2016). An overview of encapsulation of active compounds used in food products by drying technology. *Food Bioscience*, 13, 76–83. https://doi.org/10.1016/j.fbio.2015.12.009

Reis, C. P., Neufeld, R. J., Vilela, S., Ribeiro, A. J., & Veiga, F. (2006). Review and current status of emulsion/dispersion technology using an internal gelation process for the design of alginate particles. *Journal of Microencapsulation*, 23(3), 245–257. https://doi.org/10.1080/02652040500286086

Rodrigues, F. J., Cedran, M. F., Bicas, J. L., & Sato, H. H. (2020). Encapsulated probiotic cells: Relevant techniques, natural sources as encapsulating materials and food applications – A narrative review. *Food Research International*, 137(August). https://doi.org/10.1016/j.foodres.2020.109682

Rodrigues, F. J., Omura, M. H., Cedran, M. F., Dekker, R. F. H., Barbosa-Dekker, A. M., & Garcia, S. (2017). Effect of natural polymers on the survival of Lactobacillus casei encapsulated in alginate microspheres. *Journal of Microencapsulation*, 34(5), 431–439. https://doi.org/10.1080/02652048.2017.1343872

Rodríguez-Huezo, M. E., Durán-Lugo, R., Prado-Barragán, L. A., Cruz-Sosa, F., Lobato-Calleros, C., Alvarez-Ramírez, J., & Vernon-Carter, E. J. (2007). Pre-selection of protective colloids for enhanced viability of Bifidobacterium bifidum following spray-drying and storage, and evaluation of aguamiel as thermoprotective prebiotic. *Food Research International*, 40(10), 1299–1306. https://doi.org/10.1016/j.foodres.2007.09.001

Romano, N., Santos, M., Mobili, P., Vega, R., & Gómez-Zavaglia, A. (2016). Effect of sucrose concentration on the composition of enzymatically synthesized short-chain fructo-oligosaccharides as determined by FTIR and multivariate analysis. *Food Chemistry*, 202, 467–475. https://doi.org/10.1016/j.foodchem.2016.02.002

Romano, N., Schebor, C., Mobili, P., & Gómez-Zavaglia, A. (2016). Role of mono- and oligosaccharides from FOS as stabilizing agents during freeze-drying and storage of Lactobacillus delbrueckii subsp. bulgaricus. *Food Research International*, 90, 251–258. https://doi.org/10.1016/j.foodres.2016.11.003

Rousseaux, C., Thuru, X., Gelot, A., Barnich, N., Neut, C., Dubuquoy, L., Dubuquoy, C., Merour, E., Geboes, K., Chamaillard, M., Ouwehand, A., Leyer, G., Carcano, D., Colombel, J. F., Ardid, D., & Desreumaux, P. (2007). Lactobacillus acidophilus modulates intestinal pain and induces opioid and cannabinoid receptors. *Nature Medicine*, 13(1), 35–37. https://doi.org/10.1038/nm1521

Ruiz, L., Ruas-Madiedo, P., Gueimonde, M., De Los Reyes-Gavilán, C. G., Margolles, A., & Sánchez, B. (2011). How do bifidobacteria counteract environmental challenges? Mechanisms involved and physiological consequences. *Genes and Nutrition*, 6(3), 307–318. https://doi.org/10.1007/s12263-010-0207-5

Sabikhi, L., Babu, R., Thompkinson, D. K., & Kapila, S. (2010). Resistance of microencapsulated Lactobacillus acidophilus LA1 to processing treatments and simulated gut conditions. *Food and Bioprocess Technology*, 3(4), 586–593. https://doi.org/10.1007/s11947-008-0135-1

Saez-Lara, M. J., Gomez-Llorente, C., Plaza-Diaz, J., & Gil, A. (2015). The role of probiotic lactic acid bacteria and bifidobacteria in the prevention and treatment of inflammatory bowel disease and other related diseases: A systematic review of randomized human clinical trials. *BioMed Research International*, 2015. https://doi.org/10.1155/2015/505878

Salminen, S., Collado, M. C., Endo, A., Hill, C., Lebeer, S., Quigley, E. M. M., Sanders, M. E., Shamir, R., Swann, J. R., Szajewska, H., & Vinderola, G. (2021). The International Scientific Association of Probiotics and Prebiotics (ISAPP) consensus statement on the definition and scope of postbiotics. *Nature Reviews. Gastroenterology and Hepatology*, 18(9), 649–667. https://doi.org/10.1038/s41575-021-00440-6

Scavuzzi, B. M., Henrique, F. C., Miglioranza, L. H. S., Simão, A. N. C., & Dichi, I. (2014). Impact of prebiotics, probiotics and synbiotics on components of the metabolic syndrome. *Annals of Nutritional Disorders & Therapy*, 1(2), 1–13.

Scott, K. P., Gratz, S. W., Sheridan, P. O., Flint, H. J., & Duncan, S. H. (2013). The influence of diet on the gut microbiota. *Pharmacological Research*, 69(1), 52–60. https://doi.org/10.1016/j.phrs.2012.10.020

Selle, K. M., Klaenhammer, T. R., & Russell, W. M. (2014). Lactobacillus: Lactobacillus acidophilus. In *Encyclopedia of Food Microbiology*, Second Edition, Vol. 2. Elsevier. https://doi.org/10.1016/B978-0-12-384730-0.00179-8

Sharma, R., Garg, P., Kumar, P., Bhatia, S. K., & Kulshrestha, S. (2020). Microbial fermentation and its role in quality improvement of fermented foods. *Fermentation*, 6(4), 1–20. https://doi.org/10.3390/fermentation6040106

Shurtleff, W., & Aoyagi, A. (2012). *History of Soy Yogurt, Soy Acidophilus Milk and Other Cultured Soymilks (1918–2012)*. Soyinfo Center.

Sicard, D., & Legras, J. L. (2011). Bread, beer and wine: Yeast domestication in the Saccharomyces sensu stricto complex. *Comptes Rendus - Biologies*, 334(3), 229–236. https://doi.org/10.1016/j.crvi.2010.12.016

Silva, D. R., Sardi, J. de C. O., Pitangui, N. de S., Roque, S. M., Silva, A. C. B. da, & Rosalen, P. L. (2020). Probiotics as an alternative antimicrobial therapy: Current reality and future directions. *Journal of Functional Foods*, 73(December 2019), 104080. https://doi.org/10.1016/j.jff.2020.104080

Simons, L. A., Amansec, S. G., & Conway, P. (2006). Effect of Lactobacillus fermentum on serum lipids in subjects with elevated serum cholesterol. *Nutrition, Metabolism, and Cardiovascular Diseases*, 16(8), 531–535. https://doi.org/10.1016/j.numecd.2005.10.009

Singh, M. N., Hemant, K. S. Y., Ram, M., & Shivakumar, H. G. (2010). Microencapsulation: A promising technique for controlled drug delivery. *Research in Pharmaceutical Sciences*, 5(2), 65–77.

Singh, R. S., Singh, T., & Larroche, C. (2019). Biotechnological applications of inulin-rich feedstocks. *Bioresource Technology*, 273(October 2018), 641–653. https://doi.org/10.1016/j.biortech.2018.11.031

Slavin, J. (2013). Fiber and prebiotics: Mechanisms and health benefits. *Nutrients*, 5(4), 1417–1435. https://doi.org/10.3390/nu5041417

Solanki, H. K., Pawar, D. D., Shah, D. A., Prajapati, V. D., Jani, G. K., Mulla, A. M., & Thakar, P. M. (2013). Development of microencapsulation delivery system for long-term preservation of probiotics as biotherapeutics agent. *BioMed Research International*, 2013. https://doi.org/10.1155/2013/620719

Sugiura, S., Nakajima, M., Yamamoto, K., Iwamoto, S., Oda, T., Satake, M., & Seki, M. (2004). Preparation characteristics of water-in-oil-in-water multiple emulsions using microchannel emulsification. *Journal of Colloid and Interface Science*, 270(1), 221–228. https://doi.org/10.1016/j.jcis.2003.08.021

Tannock, G. W. (2004). A special fondness for lactobacilli. *Applied and Environmental Microbiology*, 70(6), 3189–3194. https://doi.org/10.1128/AEM.70.6.3189-3194.2004

ter Horst, B., Moiemen, N. S., & Grover, L. M. (2019). Natural polymers. *Biomaterials for Skin Repair and Regeneration*, 151–192. https://doi.org/10.1016/B978-0-08-102546-8.00006-6

Terpou, A., Papadaki, A., Lappa, I. K., Kachrimanidou, V., Bosnea, L. A., & Kopsahelis, N. (2019). Probiotics in food systems: Significance and emerging strategies towards improved viability and delivery of enhanced beneficial value. *Nutrients*, 11(7). https://doi.org/10.3390/nu11071591

Tikekar, R. V., Johnson, A., & Nitin, N. (2011). Real-time measurement of oxygen transport across an oil-water emulsion interface. *Journal of Food Engineering*, 103(1), 14–20. https://doi.org/10.1016/j.jfoodeng.2010.08.030

Vega, C., & Roos, Y. H. (2006). Invited review: Spray-dried dairy and dairy-like emulsions - Compositional considerations. *Journal of Dairy Science*, 89(2), 383–401. https://doi.org/10.3168/jds.S0022-0302(06)72103-8

Vega-Paulino, R. J., & Zúniga-Hansen, M. E. (2012). Potential application of commercial enzyme preparations for industrial production of short-chain fructooligosaccharides. *Journal of Molecular Catalysis B: Enzymatic*, 76, 44–51. https://doi.org/10.1016/j.molcatb.2011.12.007

Veiga, P., Miret, S., & Jimenez, L. (2020). Danone : The gut microbiome and probiotics –100 years of shared history. *Danon Nutric Research*, 577, S26–S30.

Vijayakumar, M., Ilavenil, S., Kim, D. H., Arasu, M. V., Priya, K., & Choi, K. C. (2015). In-vitro assessment of the probiotic potential of Lactobacillus plantarum KCC-24 isolated from Italian rye-grass (Lolium multiflorum) forage. *Anaerobe*, 32(January), 90–97. https://doi.org/10.1016/j.anaerobe.2015.01.003

Walker, A. W., Ince, J., Duncan, S. H., Webster, L. M., Holtrop, G., Ze, X., Brown, D., Stares, M. D., Scott, P., Bergerat, A., Louis, P., McIntosh, F., Johnstone, A. M., Lobley, G. E., Parkhill, J., & Flint, H. J. (2011). Dominant and diet-responsive groups of bacteria within the human colonic microbiota. *ISME Journal*, 5(2), 220–230. https://doi.org/10.1038/ismej.2010.118

Wang, H., Anvari, S., & Anagnostou, K. (2019). The role of probiotics in preventing allergic disease. *Children*, 6(2), 24. https://doi.org/10.3390/children6020024

Wang, L., Song, M., Zhao, Z., Chen, X., Cai, J., Cao, Y., & Xiao, J. (2020). Lactobacillus acidophilus loaded pickering double emulsion with enhanced viability and colon-adhesion efficiency. *LWT*, 121(December 2019), 108928. https://doi.org/10.1016/j.lwt.2019.108928

Wang, X., Yang, J., Qiu, X., Wen, Q., Liu, M., Zhou, D., & Chen, Q. (2021). Probiotics, pre-biotics and Synbiotics in the treatment of pre-diabetes: A systematic review of randomized controlled trials. *Frontiers in Public Health*, 9(March). https://doi.org/10.3389/fpubh.2021.645035

Weill, R. (n.d.). Ancient food in the 21 century.

Whelehan, M., & Marison, I. W. (2011). Microencapsulation using vibrating technology. *Journal of Microencapsulation*, 28(8), 669–688. https://doi.org/10.3109/02652048.2011.586068

Widyastuti, Y., Febrisiantosa, A., & Tidona, F. (2021). Health-promoting properties of lactobacilli in fermented dairy products. *Frontiers in Microbiology*, 12(May), 1–8. https://doi.org/10.3389/fmicb.2021.673890

Wilkinson, M. G. (2018). Flow cytometry as a potential method of measuring bacterial viability in probiotic products: A review. *Trends in Food Science and Technology*, 78(May), 1–10. https://doi.org/10.1016/j.tifs.2018.05.006

Wojtunik-Kulesza, K., Oniszczuk, A., Oniszczuk, T., Combrzyński, M., Nowakowska, D., & Matwijczuk, A. (2020). Influence of in vitro digestion on composition, bioaccessibility and antioxidant activity of food polyphenols—A non-systematic review. *Nutrients*, 12(5). https://doi.org/10.3390/nu12051401

Wu, Y., & Zhang, G. (2018). Synbiotic encapsulation of probiotic Latobacillus plantarum by alginate -arabinoxylan composite microspheres. *LWT*, 93(November 2017), 135–141. https://doi.org/10.1016/j.lwt.2018.03.034

Yao, M., Xie, J., Du, H., McClements, D. J., Xiao, H., & Li, L. (2020). Progress in microencapsulation of probiotics: A review. *Comprehensive Reviews in Food Science and Food Safety*, 19(2), 857–874. https://doi.org/10.1111/1541-4337.12532

Ze, X., Duncan, S. H., Louis, P., & Flint, H. J. (2012). Ruminococcus bromii is a keystone species for the degradation of resistant starch in the human colon. *The ISME Journal*, 1535–1543. https://doi.org/10.1038/ismej.2012.4

Zhang, D., Li, S., Wang, N., Tan, H. Y., Zhang, Z., & Feng, Y. (2020). The cross-talk between gut microbiota and lungs in common lung diseases. *Frontiers in Microbiology*, 11(February), 1–14. https://doi.org/10.3389/fmicb.2020.00301

Zhang, M. M., Cheng, J. Q., Lu, Y. R., Yi, Z. H., Yang, P., & Wu, X. T. (2010). Use of pre-, pro-and Synbiotics in patients with acute pancreatitis: A meta-analysis. *World Journal of Gastroenterology*, 16(31), 3970–3978. https://doi.org/10.3748/wjg.v16.i31.3970

Zhang, R., Zhou, L., Li, J., Oliveira, H., Yang, N., Jin, W., Zhu, Z., Li, S., & He, J. (2020). Microencapsulation of anthocyanins extracted from grape skin by emulsification/internal gelation followed by spray/freeze-drying techniques: Characterization, stability and bioaccessibility. *LWT*, 123(January), 109097. https://doi.org/10.1016/j.lwt.2020.109097

Zhang, Y., Lin, J., & Zhong, Q. (2016). S/O/W emulsions prepared with sugar beet pectin to enhance the viability of probiotic Lactobacillus salivarius NRRL B-30514. *Food Hydrocolloids*, 52, 804–810. https://doi.org/10.1016/j.foodhyd.2015.08.020

Zhang, Z., Zhang, R., Zou, L., & McClements, D. J. (2016). Protein encapsulation in alginate hydrogel beads: Effect of pH on microgel stability, protein retention and protein release. *Food Hydrocolloids*, 58, 308–315. https://doi.org/10.1016/j.foodhyd.2016.03.015

14 Polyunsaturated Fatty Acid Fortification

M. Vijaykrishnaraj, P. Kiran Kumar, and P. Karthik

14.1 INTRODUCTION

In recent two decades, the food industry is facing an increasing demand for nutritive and healthy foods. There is an impact of nutrition which has a potential influence on physical function and body metabolism. Most importantly, there has been an overwhelming increase in the research examination of the relationship between polyunsaturated fatty acids and their significance to human health for wellbeing. Researchers examine the significance of consuming a healthy diet made of polyunsaturated fatty acids for effective human physiological development. At the same time, it will also be focused to improve the eschewing of problems of conditions like coronary illness, diabetes, joint pain, malignancy, weight, emotional wellness, and bone well-being (Haghighi et al., 2015; Benatti et al., 2004).

Omega-6 and omega-3 polyunsaturated fatty acids are significant underlying parts of cell layers, they influence cell layer properties with a specific functional, metabolic, and signaling role. The PUFAs in the cell membrane phospholipids exhibit expert actions on cell functions like tissue responsiveness to signals, signaling pathway and inflammatory processes (Oppedisano et al., 2020). The interest in PUFA's is expanding quickly because of dietary and way of life necessities (Timilsena et al., 2017). However, oxidative degradation, which produces a disagreeable flavor, is one of the main disadvantages of PUFA-rich oils. The greater vulnerability to oxidative deterioration is a serious problem that often leads to a decrease in quality by reducing the shelf-life, flavor and nutritional value, functionality and consumer acceptability. The oxidation of PUFA's forms a mixture of complex volatile secondary products (Contini et al., 2014; Arab-Tehrany et al., 2012). In case to minimize the impact of oxidation, to protect the bioavailability of PUFA, and storing for a long time is a great challenge. These difficulties can be tended to address by utilizing the advanced process of encapsulation technologies (Aberkane et al., 2014).

Encapsulation is a technique known for encasing substances in solid, liquid, or gaseous states in matrices, and it can release the active components under certain conditions at a controlled rate. The capsule is composed of a tiny sphere surrounded by a monolithic shell where the core material is called encapsulant and the shell material is commonly called as wall material, membrane shell, or matrix. (Nedovic et al., 2011; Anandharamakrishnan, 2014; Hundre et al., 2015). Encapsulation will have numerous significant properties such as increased handling, controlled target release, protection, and stability. Figure 14.1 depicted the schematic representation of encapsulation with their specific application. Encapsulation-based micro and nano-encapsulation technology are increasingly being used in both the food and pharma industries. Due to the high encapsulation efficiency and mild processing conditions, they are widely associated with these technologies (Karthik et al., 2016). Encapsulation is a rapidly growing technology and an extending innovation with a ton of likely applications in regions including drug and food ventures (Gouin et al., 2004). Microencapsulation (10^{-6} m) is a technique wherein bioactive (or oil) compounds are capsulated by secondary materials i.e., whey protein, starch, maltodextrin, cellulose, etc, (Karthik & Anandharamakrishnan, 2013; Rajam et al., 2012). A similar technology that is widely gaining popularity is nanoencapsulation (10^{-9} m), which has proven protection with controlled release of bioactive compounds at the targeted site with better bioavailability.

DOI: 10.1201/9781003160663-18

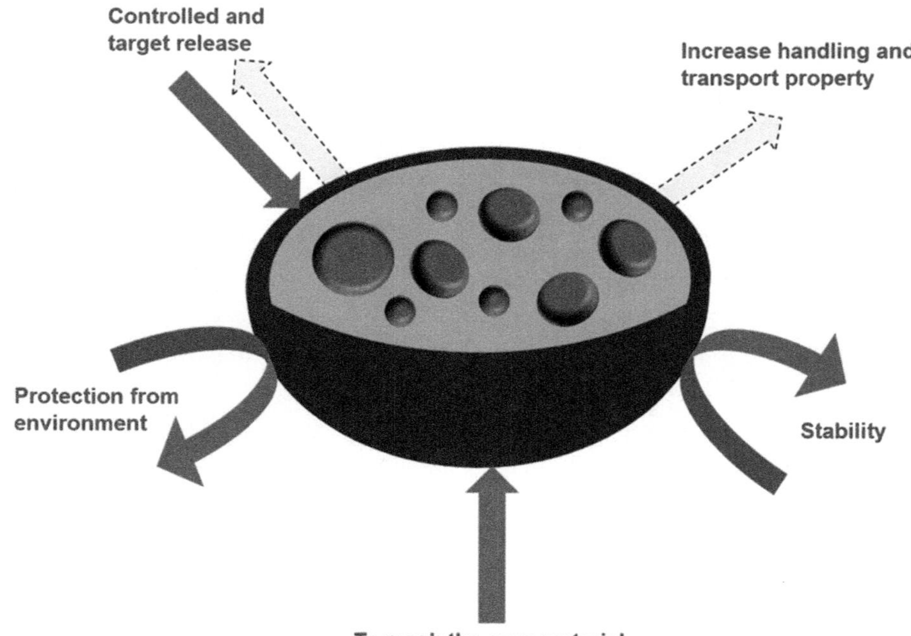

FIGURE 14.1 Schematic representation for encapsulation and their application.

Nanoencapsulation stays to be quite possibly the most encouraging innovation having the achievability to ensnare bioactive mixtures. Nanoencapsulation techniques such as emulsification, coacervation, liposome, complexation, nanoprecipitation, emulsification–solvent evaporation, and supercritical extraction are more suitable for bioactive compounds (Ezhilarasi et al., 2013). All these techniques are the highlighting factor concerning the increase of interest in food fortification which consist the value addition, cost-effectiveness, and several other advantages. Food fortification is the addition of micronutrients to processed foods, thereby incorporating a strategy that can lead to rapid improvements for nutrition deficiency at a very reasonable cost. Subsequently, it is crucial to accomplish the issues of safety, shelf life, and technological compatibility. The book chapter primarily focuses on polyunsaturated fatty acids applicable to food and nutraceutical industries. This includes various positive approaches to improve the functional properties of the food formulations for their health benefits. In this way, the section includes various techniques of micro and nano encapsulation, and fortification of food products for enveloping the bioactive ingredients.

14.2 OMEGA-6 FATTY ACIDS

14.2.1 Occurrence and sources

Omega-6 fatty acids is known as polyunsaturated fatty acids and it is also n-6 fatty acids. ω-6 fatty acids have their last double in relation to the terminal methyl end of the molecule and it is an essential fatty acid for human. Under ω-6 fatty acids classification, Linoleic acid (LA), arachidonic acid (AA), and gamma linolenic acid (GLA) (Stefanska et al., 2015). 18-carbon fatty acid with two double bonds are LA, it is the primary diet of omega-6 fatty acids. The majority of seeds and vegetable oils contain omega-6 fatty acids (Saini et al., 2014).

14.2.2 Lichens

Few lichen species (symbiosis association between fungi and algae) have the source of Arachidonic acid (AA). A trace of AA was discovered in a Cetraria pseudocomplicata (5.2 percent), Cladonia mitis (2.3 percent), and Nephroma arcticum (1.7 percent) (Yamamoto & Watanabe, 1974). Eight lichens collected from Kirghizstan's Tian Shan Mountains revealed the AA at a level of Peltigera canina contained 1.47 percent AA, 1.90 percent Xanthoria sp., 2.39 percent Acarospora gobiensis, 2.52 percent Cladonia furcate, 2.92 percent Parmelia tinctina, 3.43 percent P. comischadalis, 3.64 percent Lecanora fructulosa, and 4.17 percent Leptogium saturninum. According to Hanus et al. 2008, the level of AA content of the lichen Ramalina lacera ranged from 0.96 to 2.25% depending on the substrate. Hence, lichens are also considered to be a prevalence of PUFA and can be utilized the source in the future.

14.3 OTHERS SOURCES

In humans, majority of arachidonic acid was absorbed by ingested animal tissues and marine fish oil. However, arachidonic acid, eicosapentaenoic acid and docosahexaenoic acid are majorly present in the many aquatic creatures of marine ecosystem. In this context, aquatic shrimps, bivalves, and abalone shown a moderate level of AA. On the other side, sea cucumber, starfish, and some species of corals shown a significant level of AA (20–30 percent) (Suloma et al., 2007). Hence, it is clearly understood that fish is not actually PUFA producers; rather, fish consume PUFA by PUFA-rich aquatic creatures through the food chain (Benbrook et al., 2013). Arachidonic acid cannot produce due to the genetic absence of several of its biosynthetic enzymes in mammals including humans (Ulven et al., 2011). In spite of this, AA can be consumed thorough dietary intake of precursors (El-Sayed & Ibrahim, 2017). Crop seeds and vegetable oils such as canola, soybean, corn, and sunflower oils are significant sources of n-6 FAs in the form of LA with little n-3 FA content (ALA). At the same time, intake of n-3 FAs is typically insufficient due to their constrained sources (Moghadasian, 2008). ALA is particularly abundant in the seeds of chia (Salvia hispanica), perilla (Perilla frutescens), and flax (Linum usitatissimum). In a similar vein, green leafy vegetables have substantial levels of ALA-derived PUFAs (60–70% of total FAs).

14.4 ADVANTAGE

Tissue maintenance is one of the major roles of omega-6 fatty acids because they are significantly accumulated and generated in specific tissues based on their needs. Omega-6 FAs are also responsible for playing dual role as pro-inflammatory and anti-inflammatory eicosanoids. On the other side, omega-6 FAs participate in a wide range of physiological functions such as bone health, metabolism, regulation, and stimulating skin and hair growth. Recent reports suggest that consumption of omega-6 fatty acid reduces the risk of cardiovascular disease.

14.5 OMEGA-3 FATTY ACIDS OCCURRENCE AND SOURCES

Plant and animal kingdoms of the ecosystems shows richest diversity of polyunsaturated fatty acids. Microorganisms especially in algae, fungi & bacteria also have a richest diversity of PUFA. PUFAs are generally extractable oils and membrane lipids such lipoproteins, glycolipids, glycosphingolipids, phospholipids, esters, ethers, and glycerides. Biological synthetic pathway of saturated-fatty-acid contribute the genesis. Hence, stearate is converted to oleate and then to linoleate, which is central precursor for the ω-6 and ω-3 series. However, desaturase and elongase activities were present in algae, fungi, bacteria, insects, and some other invertebrates for the de nova production of the various PUFAs. The classification of PUFAs is linoleic acid (LA), γ linoleic acid (GLA), α linoleic acid (ALA), arachidonic acid (AA), eicosapentaenoic acid (EPA), and docosahexaenoic acid (DHA). It is generally obtained or extracted from the commercially available selected seed plants (LA, GLA and, ALA), marine fish (AA, EPA, and DHA), and certain mammals (AA) (Gill & Valivety, 1997).

14.5.1 MICROBES

One of the ubiquitous in nature is microorganisms, and it is a source of wide range of active molecules. Microbes including fungi, yeast, and few bacteria are the source of significant production of LC-PUFAs, mainly AA (Jacq et al., 1989; Ells et al., 2012). Flavobacterium strain 651 (psychrophilic bacterium) produces 1.4–2.7% AA (Nichols et al., 1997). Non-pathogenic fungi Mortierella spp. from M. Alpina 1S-4 and ATCC 32,222 produced AA up to 70% of its lipids (Dedyukhina et al., 2011).

14.5.2 ALGAE AND MICROALGAE

Blue-green algae (cyanobacteria) are unicellular, and heterocystic organisms have the ability to produce PUFAs. Numerous studies have shown that sources of omega-3 PUFA as cyanobacteria, microalgae, and algae. Unicellular red algae (Porphyridium purpureum) produce a decent amount of AA. However, stress culture conditions such as light intensity, pH and temperature, and increased salinity can increase AA production by up to 40% of total FAs. Moreover, in a favorable growth condition, PUFA is primarily represented as eicosapentaenoic acid (EPA), as many researchers have reported (Cohen, 1990; Bigogno et al., 2002; Su et al., 2016). Another algae (Euglena gracilis) contain AA synthesized from LA (C18:2). The fresh-water green alga Parietochloris incisa is the richest plant source of AA (77%) of total FAs content (Bigogno et al., 2002). Microalgae and various types of seaweeds are classified under algal divisions, such as Phaeophyceae, Rhodophyceae, Dinophyceae, and Chlorophyceae. Those algae majorly contain PUFA (Al-Hasan et al., 1991; Airanthi et al., 2011).

14.5.3 MACROALGAE

Marine macroalgae are one of the excellent sources of PUFAs with an ω-6 FA: ω-3 FA ratio of less than 10. DHA yield in the seaweed was measured in (Sargassum natans) and nine other seaweed species (4 brown, 3 red, and 2 green) (Van Ginneken et al., 2011). Red algae (Palmaria palmate) contain EPA and moderate levels (AA and LA). Another red algae species (Gracilaria Sp.,) showed 60% of the total FAs content of AA (Kumar et al., 2011). Interestingly, 17 macroalgal species from Chlorophyta, Phaeophyta, and Rhodophyta genera contain a novel dietary source of PUFAs (Pereira et al., 2012). It exerts C18 and C20 (LA, AA & EPA) PUFAs. Among these, Rhodophycean and Phaeophycean species showed higher levels of PUFAs specifically the ω-3 family. However, Ulva sp. was the only Chlorophyta genus shown a high concentration of ω-3 PUFA (ALA). Hence, macroalgae could be a potential source of essential PUFA. Consuming algae may provide essential fatty acids needed for growth and regulation.

14.6 HEALTH BENEFITS OF Ω-3 PUFAS

Growth and regulation functions are pivotal in the biological cell function, and PUFAs are precursors for many of these functions. Mainly, Omega-3 polyunsaturated fatty acids such as α-linolenic acid (ALA; 18:3 -3), stearidonic acid (SDA; 18:4 ω-3), eicosapentaenoic acid (EPA; 20:5 ω-3), docosapentaenoic acid (DPA; 22:5 ω-3), and docosahexaenoic acid (DHA; 22:6 ω-3). Omega-3 fatty acids dietary intake is very important in many biological systems. PUFAs are consumed mainly by the source of aquatic organisms, and plants are the primary source of ALA and also in some seeds, nuts, and vegetable oils. The beneficial role of omega-3 PUFAs in various disorders is emphasized below.

14.6.1 CARDIOVASCULAR DISEASE

In the modernized world, one of the deadliest diseases is cardiovascular disease (CVD). Recently, omega-3 fatty acids administration can help to resolve myocardial, coronary and arteries based vascular disorders. Regular fatty fish consumption rich in ω-3 PUFAs can ameliorate the risk of death from CVD. A study on a multicenter randomized control trial of DHA-rich canola oil improved the HDL cholesterol, triacylglycerides (TAG), and blood pressure and attenuate the risk of CVD (Jones et al., 2014). At the same time, the amelioration effects of ω-3 PUFAs may be linked with the substrate competition for cyclooxygenase (COX) enzymes that generate prostaglandins (PGs) and thromboxane (TX) (Shahidi & Miraliakbari, 2006). Moreover, anti-aggregatory effects may be due to TXA2 and PGI2 binding to TXA3 in platelets. Similarly, if it binds to PGI3 in endothelial cells resulting in an anti-inflammatory effect (Garg et al., 1988; Siscovick et al., 2017).

14.6.2 INFLAMMATORY DISEASES

These types of disorders are established by inappropriate T cell activation that destroys the host tissues of the organ. Numerous studies have demonstrated dietary changes that the use of ω-3 FAs may play role in the development of inflammatory bowel disease (IBD) (Lorente-Cebrián et al., 2015; Generoso et al., 2015). However, in IBD patients the absence of essential fatty acids and supplementing ω-3 FAs may help to inhibit natural cytotoxicity via altering AA metabolites or improving oxidative stress. Growing evidence suggests that gastrointestinal mucose responds strongly to PUFAs like ω-3. At the same time, ω-3 FA can be a choice of anti-inflammatory in the treatment of ulcerative colitis (UC) and Crohn's disease (CD) (Lorente-Cebrián et al., 2015).

14.6.3 RHEUMATOID ARTHRITIS

Rheumatoid arthritis (RA) is an auto-immune disorder, and systemic inflammatory disease is influenced by genetics and the environment. The environmental factor is a diet that has been studied in the context of other inflammatory conditions (Sperling, 1991). Modulating immune response and anti-inflammatory effects in RA were found by administering omega-3 fatty acids. The immunomodulatory activity of omega-3 fatty acids has been demonstrated in animal models. *Linum usitatissimum* fixed oil (Linseed oil), a rich source of the omega-3 FA administration, significantly reduced joint swelling and lower serum TNF alpha levels in rats with collagen-induced arthritis (CIA) (Lee et al., 1984).

14.6.4 CYSTIC FIBROSIS

Cystic fibrosis (CF) is the most common disorder in the Caucasian population, affecting approximately one in 2500 births. Lung function is majorly affected in terms of pulmonary inflammation that tend to cause morbidity and mortality in CF (Coste et al., 2007). EFA-rich food or oral supplementation may prevent the essential fatty acid deficiency (EFAD) in CF disease. At the same time, the level of PUFA ratio in the diet should be given in an adequate amount to those patients. Administration of fatty acids to CF patients may improve their nutritional status and growth rate (Steinkamp et al., 2000).

14.6.5 ASTHMA AND ALLERGIC DISEASES

Asthma is one of the common respiratory disorders, and 300 million people affect globally. Treatment or preventive medication for asthma is the inhalation of corticosteroids, and 5-10% of the patients develop resistance. It affects disease management and medication (Braman, 2006). As first-line therapy in asthma, Leukotriene receptor antagonists are widely used and also imply abnormal levels of lipid regulators to disease pathology (Bell & Busse, 2013). Fatty fish consumption during lactation leads to higher levels of EPA in breast milk and is proven to lower the risk of atopic dermatitis (Hoppu et al., 2005). However, observational studies in adults have shown that fish or fish oil consumption may reduce the risk of asthma (Barros et al., 2011; Li et al., 2013). ω-3 FAs act as protective molecules in murine asthma models that regulate the functions (Schuster et al., 2014; Yin et al., 2009). DHA inhalation during sensitization of the allergen phase reduced airway inflammation in mice. The effects were accompanied by inflammatory cell decrease in the bronchoalveolar lavage fluid (BALF (Yokoyama et al., 2000). As a result, ω-3 FAs can act as modulators in a variety of physiological functions, providing health benefits.

14.6.6 NEUROLOGICAL BENEFITS

During PUFAs presence, the neuronal membrane components may act as lipid-derived messengers to promote cellular functions. For optimal visual function and neural development, an adequate level of omega-3 PUFA intake is necessary. On the other side, there is growing evidence suggesting that higher intakes of the long-chain omega-3 PUFAs may benefit a variety of psychiatric and neurological disorders (Dyall & Michael-Titus, 2008). Interestingly, during pregnancy, DHA synthesis capacity may be increased due to gender differences in humans, especially women having higher conversion efficiency (Burdge, 2006; Bolton-Smith et al., 1997). When compared to younger women (20-48 years old), elderly women (>75 years old) had a significantly lower DHA: EPA ratio (Babin et al., 1999).

14.6.7 NEURODEGENERATIVE DISORDERS

The geriatric population is most commonly susceptible to Alzheimer's disease (AD) dementia. It's a type of neurodegenerative disorder that represent the decline of daily cognitive functions and changes in behavior (Selkoe, 2001). Amyloid-beta peptides deposition, neurofibrillary tangles, and genetic factors present in the brain diagnose the disease pathology. However, the decline in the omega-3 levels in the brain and peripheral tissues is also linked to AD. The reduced brain DHA levels are majorly found in Alzheimer's patients compared to age-related controls (Skinner et al., 1989; Soderberg et al., 1991). Similarly, elevated levels of serum DHA phosphatidylcholine are also a significant risk factor for Alzheimer's disease (Kyle et al., 1999). Hence, omega-3 PUFA administration in the diet may limit the progression of cognitive decline to boost brain functions.

14.6.8 PARKINSON'S DISEASE (PD)

Parkinson's disease is second most after Alzheimer's disease in the geriatric population. Dopaminergic neuronal loss in the substantia nigra of the mid-brain and the presence of Lewy bodies enriched with protein a-synuclein is the diagnostic hallmark of the disease. The common symptoms of the disease are resting tremors, rigidity, bradykinesia, and postural instability (de Rijk et al., 1997). MPTP (1-methyl-4-phenyl-1,2,3,6-tetrahydropyridine) treated monkeys were administered for a short duration of 100mg/kg DHA attenuates the severity of levodopa-induced dyskinesia without altering the anti-parkinsonian effects (Samadi et al., 2006). Moreover, a high-fat diet containing omega-3 FA-fed mice were treated with MPTP attenuated the severity of levodopa-induced dyskinesia (Bousquet et al., 2008). However, replacing PUFA with saturated fat was linked to an increase in the risk of Parkinson's disease in men but not women. In the second study, PUFA consumption was linked to a lower risk of Parkinson's disease (de Lau et al., 2005).

14.6.9 FATTY ACID ABSORPTION IN THE BRAIN

Brain barrier availability and functionality were studied for PUFA over the past few decades. The accumulated fatty acids were present in the brain myelin sheath; during this process, myelination of oleic acid occurs. In the final trimester of pregnancy, the brain functions the accumulation of PUFA, especially DHA. However, the DHA turnover is unknown, but the estimation must be high that could lineate the DHA demand in the developing brain. Dietary DHA may be necessary for pregnant women who accumulate the presence in breast milk and indicate the infants' neurodevelopment. In this context, post-mortem studies of infants fed with DHA-deficient formula had a lower level of brain DHA than those fed with breast milk. Likewise, infants fed with DHA-containing formula reveal healthy neuronal development scores than DHA-free formula (Kuipers et al., 2012).

14.7 CHALLENGES ASSOCIATED IN PUFA FORTIFICATION

There has been growing demand every year for nutritional foods including omega-3 and omega-6 fatty acids to increase the quality of health benefits. To provide an adequate amount of PUFA's, the incorporation of food fortification by micro and nano encapsulation is considered to be a promising technology. Despite various researches, fortification has a big challenge due to the highly susceptible oxidation nature of PUFA's, decrease in shelf life, unpleasant smell, and its influence on sensory attributes during storage and taste (Karthik, 2016). In addition, fortification requires monitoring and maintaining high-quality and there is a need for tracking to look for any adverse effects of excessive nutrient exposures (Dwyer et al., 2015). The illustration of challenges associated with PUFA fortification into food system has represented in Figure 14.2. Large-scale food fortification, especially related to improving the quality of delivery and measuring process is another challenge that should be part of efforts to control and prevent nutrient deficiencies (Osendarp et al., 2018). The encapsulated bioactive delivery system plays a major role in bioavailability and bioaccessibility. Therefore, considerable research needs to be conducted to study delivery systems and the biological actions of these fortified foods on the body (Ozturk et al., 2017). It is very crucial to understand the side effects, like cell damaging, toxicity, long term risk and abnormal biological functions (Aguilar-Pérez et al., 2021). With all these challenges it is essential to develop food fortification technologies of more adequate, cost-effective, and to maximize the stability of omega-3 fatty acids for extensive use in the food industry (Feizollahi et al., 2018). For an absolute outcome, good manufacturing practices and effective analytical techniques need to be implemented for the fortification of food products to validate health claims and enforcement of compliance with manufacturing strategies (Hernandez, 2014). Additionally, there should be a need to create and maintain consumer demand with a demonstration of positive health impact by targeting new research programs, improved regulatory monitoring, and increased transparency (Mannar, 2018). The new generation of healthy foods is now required to include the continued extension of the scientific and technical base, and to overcome the current

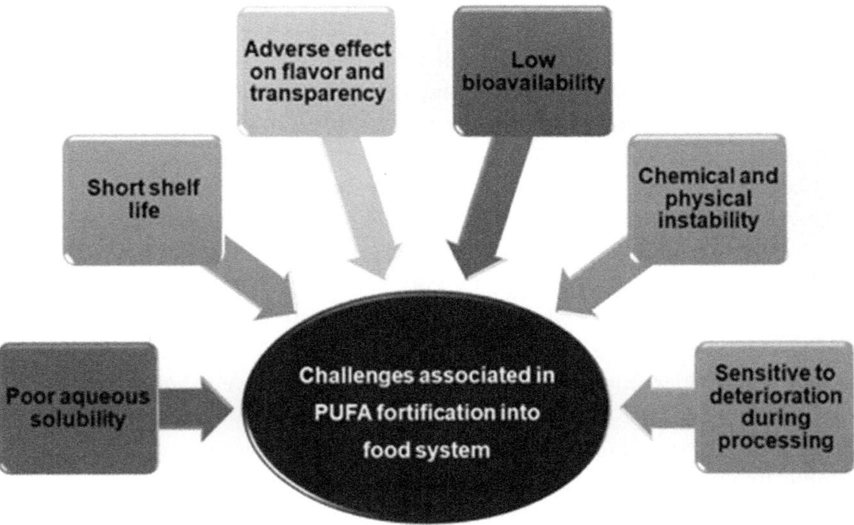

FIGURE 14.2   Illustration of challenges associated in PUFA fortification into food system.

challenges. The fortification technology needs to consider the critical issues in the design and implementation of novel food development.

14.8 MICROENCAPSULATION OF OMEGA-3 FATTY ACIDS

Microencapsulation is a promising technology where the active substances of encapsulating solid, liquid or gas are coated by secondary wall material with a diameter of 1-1000 µm. The technology is extensively used in pharmaceutical, cosmetic, medicine, food and agrochemical industries. They are often used in encapsulating flavors, oils, vitamins, minerals and other bioactive food ingredients. The advantage of microencapsulation lies in the wall material which isolates the sources from external environment without affecting the properties of core materials (Karthik, 2016). Different techniques are commonly applied for omega-3 fatty acids microencapsulation such as spray drying, spray cooling, freeze drying, spray-freeze drying, electrospinning and electrospraying are of most common (Table 14.1). Klaypradit et al. (2008) stated that the freeze-drying method for fish oil using a combination of chitosan and maltodextrin as wall materials gave the smallest particle size with good stability. Karthik and Anandharamakrishnan, (2013) microencapsulated DHA algae oil using whey protein isolate as a wall material by spray-freeze-drying showed greater oxidative stability. Besides, it was found that spray-freeze-drying had less percentage of oxidation (14%) as compared with spray drying and freeze-drying methods (Figure 14.3). Likewise, Zhang et al. (2020) encapsulated algal oil by spray drying method using buttermilk and maltodextrin, the result showed buttermilk as an emulsifier that showed good oxidation stability. Wang et al. (2011) prepared barley enriched microcapsules by spray-drying and it was observed higher capacity to prevent oxidation of fish oil. Klinkesorn et al. (2005) used lecithin and chitosan as wall material for tuna oil encapsulation by spray drying method which inhibited oxidation. Carneiro et al. (2013) used a spray drying method for encapsulating flaxseed oil using different wall materials out of which maltodextrin Hi-cap showed the best encapsulation efficiency to minimize lipid oxidation. The less encapsulation efficiency of microencapsulated fish oil was observed by Chen et al. (2013) using milk proteins (whey protein isolate and sodium caseinate) through spray drying method.

In the other study, spray drying of tuna oil using whey protein isolate and gum arabic showed better oxidation stability when compared with freeze drying (Eratte et al., 2014). Gallardo et al. (2013) prepared microcapsules made of 100% gum arabic and other combinations (using maltodextrin, methyl cellulose and whey protein isolate) for linseed oil encapsulation by spray drying method. The result exhibited the highest protection from oxidation for 100% gum arabic. Conversely, Moomand et al. (2015) used electrospinning and electrospraying methods for encapsulating fish oil using electro sprayed zein beads and zein fibers as wall materials. This study suggested the zein matrix can be used as a carrier to deliver fish oil in the gastrointestinal tract. Umesha et al. (2015) used whey protein as a wall material for encapsulating garden cress seed oil by spray drying method. The encapsulation method offered oxidative protection to ALA during backing and increased the shelf-life of biscuits over long-term storage. The spray drying method was utilized for encapsulating fish oil through maltodextrin mixed with fish gelatin and carrageenan. The microencapsulates from the maltodextrin mixed with fish gelatin showed the best encapsulation efficiency (Mehrad et al., 2015). Similar results were observed by Pourashouri et al. (2014) who encapsulated fish oil by spray drying using fish gelatin and chitosan, microbial transglutaminase with maltodextrin. All wall compositions showed increased encapsulation efficiency and higher stability of microcapsules. Haq et al. (2018) tested microencapsulation of salmon oil using polyethylene glycol by gas saturated solutions (PGSS) process. The obtained results showed significant thermogravimetric stability up to 350 °C and increased bioavailability. Thus, it is noteworthy that microencapsulation has proved itself a useful technology in a variety of commercial applications. The features and benefits of using microencapsulation include stability improvement, increasing the shelf life of the final product and bioavailability.

TABLE 14.1

Microencapsulation Techniques of Omega-3 Fatty Acids

Microencapsulation techniques	Source of omega-3 fatty acids	Encapsulating agent (wall materials)	Size	Application	References
Freeze drying	Fish oil	Chitosan, maltodextrin and whey protein isolate.	0.8 to 14.1 μm	The combination of chitosan and maltodextrin gave smallest particle size and highest stability.	Klaypradit et al.
Spray drying, freeze drying and spray-freeze-drying	DHA algae oil	Whey protein isolate	-	Spray-freeze-drying showed more oxidative stability than other drying methods.	Karthik and Anandharamakrishnan (2013)
Spray drying	Algal oil	Butter milk and maltodextrin	34.70 to 54.75 μm	The result showed, buttermilk as an emulsifier gave good oxidation stability.	Zhang et al. (2020)
Gas saturated solutions (PGSS) process	Salmon oil	Polyethylene glycol (PEG)	0.37 μm and 449 μm	The PGSS process showed significant thermogravimetric stability up to 350 °C and bioavailability.	Haq et al. (2018)
Spray drying	Fish oil	Barley protein	1 to 5 μm	Barley enriched microcapsules showed higher capacity to prevent oxidation of fish oil.	Wang et al. (2011)
Spray drying and freeze drying	Tuna oil	Whey protein, Gum Arabic	Below 5 μm	The spray dried microcapsules found to be more stable against oxidation compared to freeze dried.	Eratte et al. (2014)
Spray drying	Tuna oil	Lecithin and Chitosan		Spray drying method inhibited oxidation.	Klinkesorn et al. (2005)
Spray drying	Flaxseed oil	Maltodextrin, Gum arabic, whey protein Two types of modified starch (Hi-Cap 100TM and Capsul TA)	0.6 to 26 μm	The best encapsulation efficiency to minimize lipid oxidation was obtained for MaltoDextrin: Hi-Cap.	Carneiro et al. (2013)
Spray drying	Fish oil	Milk proteins (Whey protein isolate and sodium caseinate (4:1))		There was less efficiency for oxidation indicators.	Chen et al. (2013)

(Continued)

TABLE 14.1 (CONTINUED)

Nanoencapsulation Techniques of Omega-3 Fatty Acids

Microencapsulation techniques	Source of omega-3 fatty acids	Encapsulating agent (wall materials)	Size	Application	References
Spray drying	Linseed oil	Combinations of gum arabic, maltodextrin, methyl cellulose and whey protein isolate	10 to 50 μm	Microcapsules made of 100% gum arabic presented the highest protection from oxidation.	Gallardo et al. (2013)
Electrospinning and electrospraying	Fish oil	Electro sprayed zein beads and zein fibres	1 to 3 μm	The zein matrix was used to deliver the particles to the gut.	Moomand et al. (2015)
Spray drying	Garden cress seed oil	Whey protein	15.4 μm	ALA prevented from heat during baking process and increased the shelf-life of biscuits over long-term storage due to microencapsulation.	Umesha et al. (2015)
Spray drying	Fish oil	Maltodextrin mixed with fish gelatin, carrageenan		Maltodextrin mixed with fish gelatin showed the best encapsulation efficiency.	Mehrad et al. (2015)
Spray drying	Fish oil	Fish gelatin and chitosan Microbial transglutaminase with maltodextrin	2.41 to 3.12 μm	All wall materials showed increased encapsulation efficiency and improved stability of microcapsule powders.	Pourashouri et al. (2014)

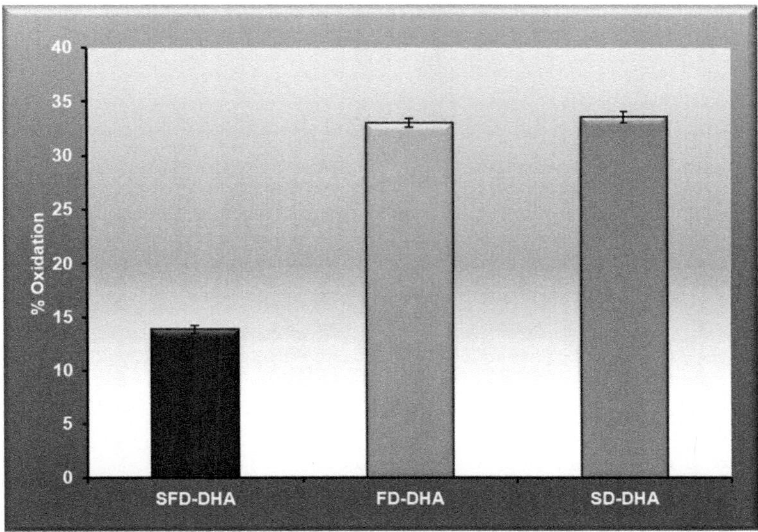

FIGURE 14.3 Influences of microencapsulation techniques on percent of oxidation of omega-3 fatty acids (Karthik and Anandharamakrishnan, 2013).

14.9 NANOENCAPSULATION OF OMEGA-3 FATTY ACIDS

Nanotechnology is one of the revolutionary technologies having applications in the food and nutraceutical industries. Nanoencapsulation of food bioactive compounds is one of the important aspects of nanotechnology for the past couple of decades. The nanoencapsulation involves the modification of materials to nano-dimension approximately below 1 μm scale (Zhou et al., 2016). The nano capsule acts as a shielding material that protects the bioactive compounds from oxidative degradation with enhanced stability. Table 14.2 shows the different techniques for the nanoencapsulation of omega-3 fatty acids. Due to the increase in surface area, the bioavailability of food ingredients is drastically increased. Nejadmansouri et al. (2016) used high-intensity ultrasound for encapsulating fish oil using zein nanofiber as wall material. The nanoencapsulation fairly increased the chemical stability of unsaturated fatty acids. According to Yang et al. (2017), the zein nanofiber mat was used as a wall material for encapsulating fish oil. The nano encapsulant showed favorable oxidative stability with an increase in bioavailability. Raeisi et al. (2019) prepared nanocapsules by electrostatic layer-by-layer deposition and used Persian gum and chitosan as a wall material the capsules were thermally stable and showed promising use in the pharmaceutical and food industries. In the other study, fish oil was encapsulated by Cetinkaya et al. (2021) using α, γ- and β Kafirin as wall material through the electro spraying method, the results showed the potential of kafirin as hydrophobic shell material. Likewise, Melgosa et al. (2019) used particle gas-saturated solutions drying process for encapsulation of fish oil using octenyl-succinic-anhydride modified starch (OSA-starch), the nano-encapsulation showed good oxidation inhibition at low temperature when compared to the other conventional methods. The highest encapsulation efficiency was achieved from nano encapsulated fish oil while using gum Arabic than whey protein as a wall material through freeze drying method (Vahidmoghadam et al., 2019). da Silva Stefani et al. (2019) encapsulated linseed oil using a spray drying method with chia seed mucilage as a wall material, the results showed the encapsulate was hydrophobic compound and provides better bioavailability to the linseed oil. Conversely, Camila de Campo et al. (2017) obtained nano capsules of chia seed oil using chia seed mucilage as wall material, the result showed good protection to the oil against lipid oxidation and also improved the stability. The spray-drying method used for encapsulating echium oil and the study suggested exhibited a good potential for fortification of daily foods (Azizi et al., 2018). Wang et al. (2011) successfully

TABLE. 14.2

Nanoencapsulation Techniques of Omega-3 Fatty Acids

Nano encapsulation techniques	Source of omega-3 fatty acids	Encapsulating agent (wall materials)	Size	Application	References
High intensity ultrasound	Fish oil		82 nm	Nano encapsulation fairly increased the chemical stability of unsaturated fatty acids.	Nejadmansouri et al. (2016)
Electrospinning	Fish oil	Zein	440 nm	The nanoencapsulant showed a favorable oxidative stability and release property.	Yang et al. (2017)
Electrostatic layer-by-layer deposition	Fish oil	Persian gum and chitosan	23.19 nm	The nano-capsules were thermally stable and had been a choice in the pharmaceutical and food industries.	Raeisi et al. (2019)
Electro spraying	Fish oil	α, γ- and β Kafirin	552 to 861nm	Kafirin showed as hydrophobic shell material for encapsulating fish oil.	Cetinkaya et al. (2021)
Particle gas-saturated solutions (PGSS)-drying	Fish oil	Octenyl-succinic-anhydride modified starch (OSA-starch)	116 nm	PGSS-drying method showed good oxidation inhibition at low temperature when compared to conventional methods.	Melgosa et al. (2019)
Freeze drying	Fish oil	Whey protein and gum arabic	50 nm	The highest encapsulation efficiency was found in the sample containing 100% gum arabic.	Vahidmoghadam et al. (2019)
Spray drying	Linseed oil	Chia seed mucilage	356 nm	Chia seed mucilage as nanoencapsulation material for hydrophobic compounds and improves bioavailability.	Stefani et al. (2019)
Spray drying	Chia seed oil	Chia seed mucilage	205 ± 4.24 nm	Protect the oil against lipid oxidation and to improve the stability	Campo et al. (2017)
Spray drying	Echium oil	Whey protein isolate	~ 200 nm	The study suggested a good potential for fortification of daily foods with whey protein stabilized echium oil.	Azizi et al. (2018)
Spray drying	Curcumin	Sodium caseinate	300 to 330 nm	Encapsulation was successfully improved antioxidant properties of solid-lipid particles	Wang et al. (2016)
Spray drying	Curcumin	Low density lipoprotein and pectin	Less than 60nm	It showed promising potential for oral delivery for lipophilic nutrients and drugs.	Zhou et al. (2016)
Nano complex formation	Docosahexaenoic acid (DHA)	β-lactoglobin and low methoxy pectin	100 nm	There was about 5-10% DHA lost during 100h at 40 °C as compared to 80% loss of unprotected DHA.	Zimet and Livney (2009)

encapsulated curcumin using sodium caseinate as wall material by a spray drying method. The result showed improved antioxidant properties of solid-lipid particles. Docosahexaenoic acid (DHA) was encapsulated by Zimet and Livney (2009), by β-lactoglobulin and low methoxy pectin as wall material through nano complex formation. The result showed a minimum loss of 5 to 10% DHA at 40 °C when compared to unprotected DHA (about 80% loss). The method serves to enhance health-promoting the properties of food and beverages against deterioration. With all this context, nanoencapsulation is a promising approach to achieve the precise release of sensitive bioactive molecules through food fortification. Nanoencapsulation can mask flavor and aroma, and also enhance thermal and oxidative stability of PUFA. Generally, the technology serves the major benefits for food and nutraceutical industries.

14.10   PUFA FORTIFICATION IN FOODS

Fatty acids from the dietary source are known to provide energy, and they are biomolecules precursors for biological membrane components. The most commonly available ω-3 PUFA is α-Linolenic acid through a dietary oil source from plants and animals. However, DHA and EPA are available through fatty fish, fish oil, and algae (Ganesan et al., 2014). Fortification of ω-3 PUFA in foods was attempted to attract a huge population in a regular consumption manner. At the same time, the fortification of PUFA should provide adequate levels of absorption to the consumers. At least 0.5 grams per day of ω-3 PUFAs, i.e., EPA and DHA, is recommended for the daily intake ratio (Harris et al., 2009). Table 3 shows a list of studies and their findings, such as omega-3 fatty acid fortification in food products and product quality characteristics. Majorly consumed food products by all age groups are milk and dairy products. Dairy products such as milk, cheese, butter, spread, yogurt & flavoured milk are widely consumed around the globe. However, due to the lack of nutrients and fortification for enrichment involved majorly in those dairy products. An overview of the PUFA and their different functionality has been depicted in Figure 14.4.

14.10.1   DAIRY PRODUCTS FORTIFICATION

Milk protein emulsion for the fortification of fish oil has been characterized in terms of its organoleptic properties. In this context, caseinate/whey protein emulsion o fish oil was found to have significant physical stability and lower oxidation levels during storage (Singh et al., 2006). Cheese is a product of high pH, a high-fat content, and a high buffering capacity. Thus, cheese can be a food-based delivery vehicle for omega-3 PUFA. It offers to protect biological activity and intestinal transit (Hayes et al., 2006). According to Ye et al. (2009), cheese samples were fortified with fish oil were processed. The study shows that fortifying PUFA in processed cheese with fish oil causes oxidation during processing and storage and enhances the "fishy" off-flavor of the cheeses (5% fish oil contained 20% soya oil). Hence, it reduces the amount of omega-3 LC PUFA fortification in the final products. Interestingly, the fat spread is also fortified with fish oil by Kolanowski et al. (2004). The author chose fat spread (Kama Foods SA) and fish oil (ROPUFA 30 n-3 Food oil) for the fortification study. Fish oil was added at levels 30, 40, and 50 g kg–1 and evaluated sensorily. Reduced-fat spread enriched with omega-3 PUFAs, EPA, and DHA, up to about 10.0g/kg by adding 30g/kg of unhydrogenated fish oil. Fat spread's textural properties were not affected by adding fish oil, but its fatty acid profile improved significantly. A daily serving of 3 g of the enriched spread contains 0.25 g of EPA and DHA. Similarly, the soft goat cheese with purified fish oil at a ratio of (0, 60, 80, and 100 g/3600 g of goat milk). The fish oil fortified soft goat cheese yielded a product that contained approximately per 28 g serving containing 127mg of EPA and DHA. It offers nearly four times the level required to meet the need of recommended allowance. Storage studies of four weeks of refrigerated storage showed no change in oxidative stability. These findings have implications for dairy product fortification with high levels of fish oil.

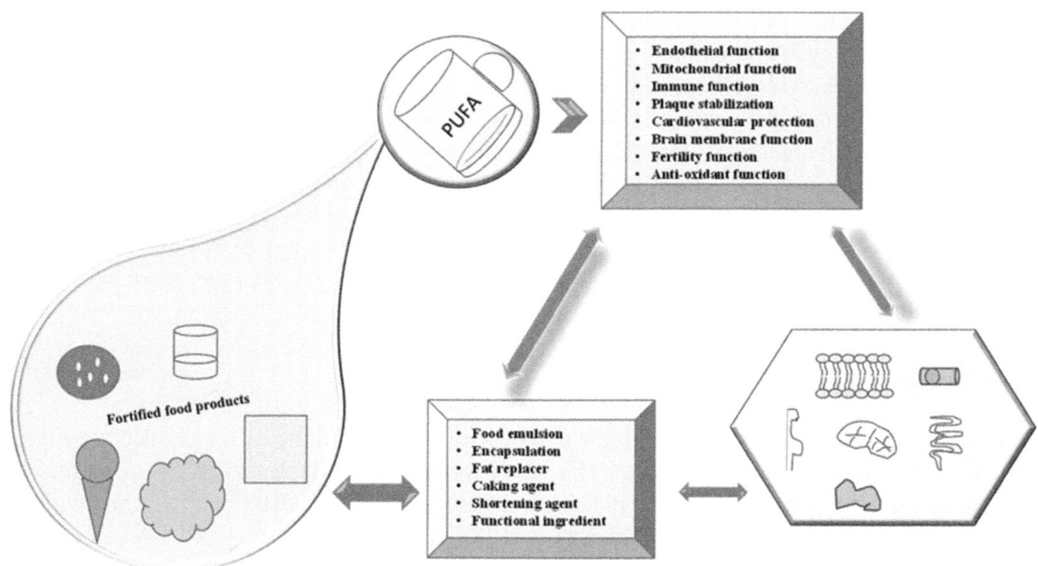

FIGURE 14.4 PUFA products and their overall functional benefits.

Enriching yogurt with n-3 FA is a quite challenging that includes consumers should receive the recommended levels needed and prevention of oxidative degradation. Consumers motivated by health trends may make different choices than other population segments. Health-conscious consumers prefer to drink yogurts with healthy ingredients rather than consumers motivated by price, convenience, mood, or familiarity (Pohjanheimo & Sandell, 2009). Flavor enriched yogurt with the addition of n-3 fatty acid by fish oil study by Rognlien et al. (2012). The study was performed with different formulations such as Chile-lime flavor and oils (butter, fish, oxidized fish) in low-fat yogurt. A single serving (170 g) of savory-flavored yogurt can provide enough heart-healthy n-3 FA to meet the recommended daily intake (145 mg of EPA + DHA). In the other study, the PUFAs are in direct contact with prooxidants during the foods are fortified with oils (unencapsulated form). When encapsulated PUFAs, the emulsifier layer may be protected because of positively charged attributes; hence, they can repel each other with the positively charged metal ions (Figure 14.5). Furthermore, the protective layer on the surface acts as a barrier between reactive oxygen species and droplets (Gumus & Gharibzahedi, 2021). The selectable emulsifiers with emulsification methods can help to eliminate unwanted odors. It can also enhance the sensory properties and improve the level of consumers' acceptability (Gharibzahedi & Jafari, 2017).

Microencapsulation is another choice for masking fishy odor or preventing biological reactions during food processing (Iafelice et al., 2008). Oxidation and chemical reactions during light, oxygen, humidity, and other environmental conditions penetration in fish oil can be prevented by specific coating emulsion materials (Kolanowski & Weibbrodt, 2008). Omega-3 from animal and vegetable sources in three cheese varieties at various stages of the cheese-making process was studied. The study also used non-thermal approaches (high hydrostatic pressure, pulsed electric fields, and ultrasound) to improve omega-3 incorporation and retention in the three types of cheese (Bermdez-Aguirre & Barbosa-Cánovas., 2012). Omega-3 from various sources can be added to cheese at a 1% level. However, the maximum amount of omega-3 can be fortified in cheese is 1.8% (FDA, 2007). Non-thermal technologies (high hydrostatic pressure, pulsed electric fields, and ultrasound) were followed in three types of cheese (Queso Fresco, Cheddar, and Mozzarella) for pasteurizing of the milk. It was determined that novel non-thermal approaches provide an intriguing way to incorporate omega-3 into the cheese matrix. High nutrient concentration with good physicochemical properties and a product with a longer storage life are achieved in this cheese processing.

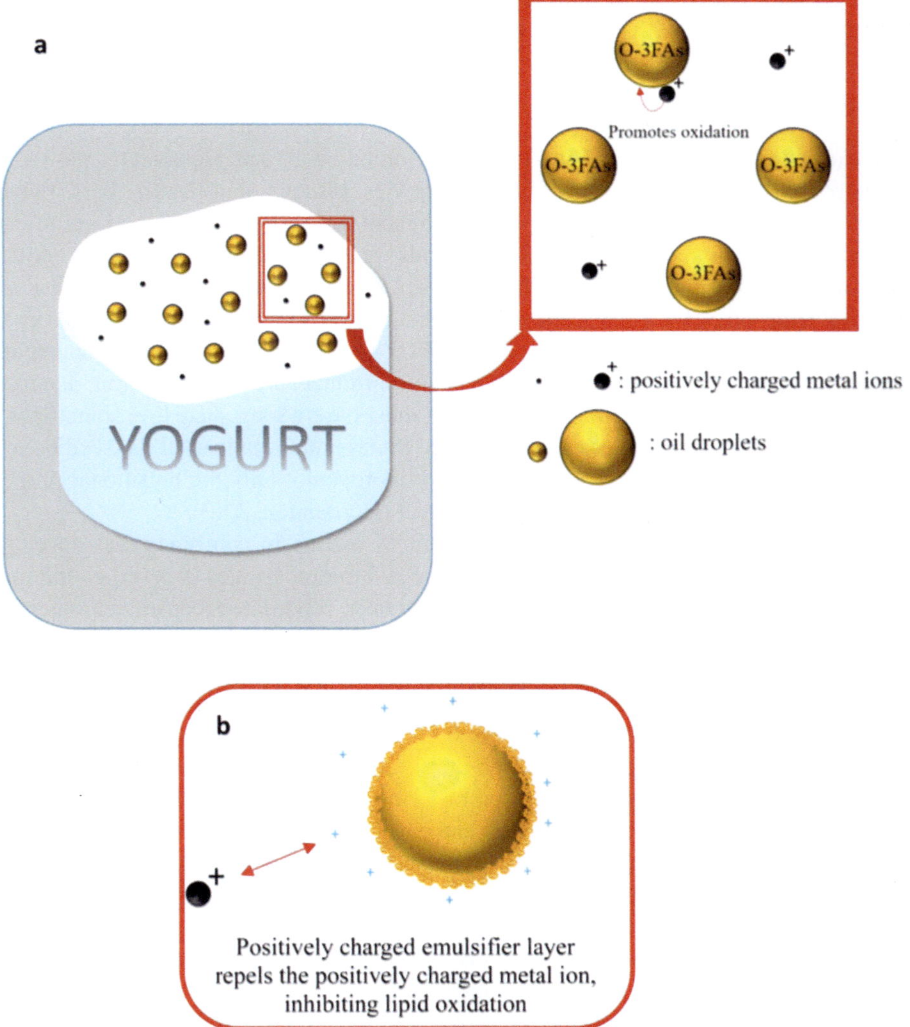

FIGURE 14.5 Schematic representation on the influence of (a) metal ions and (b) emulsifier layer on lipid oxidation in food system (Gumus and Gharibzahedi, 2021).

Dahi (Indian yogurt) was enriched with omega-3 microencapsulated flaxseed oil powder by Goyal et al. (2016). Dahi was studied for its physico-chemical, textural and oxidative stability (peroxide value) properties at a α-linolenic acid (ALA, ω-3) in a 2% fortification of microencapsulated flaxseed oil. Butter fortification using flaxseed oil (2.9-5.1%) and whey protein emulsion (4.8–8.6%) were also studied for omega-3 fatty acid fortification (Pandule et al., 2021). The addition of omega-3 fortification was suitable before maturing the cream of butter. Moreover, it is suitable for fortifying the butter with omega-3 fatty acids as maturation and subsequent aging help the consistency. Flaxseed oil and emulsion fortified at 4.1 and 6.8 percent acceptability, with overall quality comparable to control butter. Moreover, it has the technical advantage of improved spreadability. Interestingly, flaxseed oil microcapsules' fortification and processing parameters were optimized, and levels of fortification (3, 4, and 5%) and flavors (vanilla, butterscotch, and strawberry) were studied in the ice-cream mix (Gowda et al., 2018). A freshly prepared functional ice cream serving met ~ 45% of the RDA (1.4 g ALA/day). Microencapsulated flaxseed oil powder, which was used to develop omega-3 fortified ice cream, remained oxidative stable during the storage period studied.

Fortified ice cream revealed that the developed omega-3 fortified ice cream (butterscotch and strawberry flavors) was accepted with sensorial quality.

14.10.2 Bakery goods fortification

Bakery products are one of the delicious snacks consumed by all age groups. Omega-3 fatty acids can be effectively fortified in the form of bread, cookies, biscuits, and healthy bars. Fortification of PUFA in bakery products has been reported in recent years. Flour and bread are effective for nutraceutical enrichment such as omega-3 fatty acids, phenolic glucosides, lignans, phytosterols, fructooligosaccharides, inulin, and other dietary fibers (Hayta & Ozugur, 2011). Bread is also suitable for introducing omega-3 fatty acids into the diet. Bread is widely consumed and provides a stable environment for polyunsaturated fatty acids (PUFA) due to its low moisture content and short shelf life. At the same time, fortifying PUFA in bakery products is also challenging due to the heat process of product making. Moreover, baked goods must have a long shelf life, sometimes up to a year. In the late 1980s, several bakeries in the United States and Canada used flaxseed flour in their bread. However, flaxseed flour has been reported to improve not only the nutritional properties of baked goods but also the overall quality of the product (Hernandez, 2013).

Microencapsulated omega-3 fatty acid and rosemary extract in pan bread was developed (De Conto et al., 2012). Even at the highest dosage of omega-3 microcapsules (5 g/100 g total mass), the bread had good sensory acceptance (scores > 5). Moreover, 5% microencapsulated omega-3 bread and a serving of 50 g provide 0.30 g omega-3 (12.9 g/100 g EPA & DHA). It is 60% of the recommendation of the International Society for the Study of Fatty Acids and Lipids ($\geq$0.5 g/day EPA & DHA). Similarly, nano-encapsulated omega-3 fatty acid fortification was investigated to develop functional bread (Gökmen et al., 2011).

Flaxseed oil was nanoencapsulated by high amylose corn starch to form nanosized complexes with flaxseed oil, which were then spray-dried to form powder of microparticles. The addition of flaxseed oil (0-10%) during bread development and encapsulation significantly reduced lipid oxidation. The formation of acrylamide and HMF in bread was significantly reduced by increasing the number of particles in the dough. Furthermore, the particles were approximately 5 m in size and non-homogeneously distributed in the crumb. The development of biscuits to enrich the omega-3 fatty acids and calcium content was done with flaxseed grounded flour. Meanwhile, the content of carbohydrates was decreased in the biscuit substituted with flaxseed. Biscuits with 20% flaxseed had a higher content of omega-3 compared to that not containing flaxseed. The incorporation of flaxseed progressively increased the price-to-sales (P/S ratio) while the ω-6/ω-3 ratio decreased below the maximum recommended ratio. The fortification of biscuits with flaxseed progressively increased the calcium content (Elshehy et al., 2018). Garden cress seed oil (GCO), also another proven source of omega-3 fatty acid, and microencapsulated garden cress oil (MGCO) were fortified into biscuits (5% and 20%). The study reported that garden cress oil encapsulation for preventing auto-oxidation of α linolenic acid (ALA) and fortified the biscuits to enrich their functionality. MGCO or GCO supplementation improved the nutritional quality of biscuits with essential fatty acid ALA (~1.0 g/100 g) and protein in biscuits. Monoglyceride-flaxseed oil–water-gel emulsion was also utilized to develop shortened dough biscuits (Umesha et al., 2015). However, another study found that baking at high temperatures and low pressure produced biscuits with acceptable water content and color while minimizing omega-3 PUFA oxidative degradation as well as acrylamide and furan levels (Anese et al., 2016). Roasted and ground flaxseed (RGF) flours substitution at 5, 10, 15, and 20% levels on wheat flour dough properties revealed that rheological properties of resistance to extension, and extensibility values decreased as RGF substitution increased from 0 to 20%. RGF substitution in cookies above 15% had a negative impact on quality. By substituting 15% of RGF cookies were accepted with overall quality. The 15% RGF had 4.75–5.13% linolenic acid. The cookies can be stored for up to 90 days at ambient temperature (Rajiv et al., 2012).

TABLE 14.3

Omega-3 Fatty Acid Fortified Food Products and Their Functionality

Fortified foods	Material	Level of fortification	Product quality	PUFA levels	References
Yoghurt Processed fresh cheese Soft cheese Spreadable cheese Processed cheese Butter	Purified fish oil (ROPUFA)	10, 20, 30, 40, 50 & 60 g/kg	Sensorially accepted 40 & 60 g/kg of processed cheese	180, 270 and 360 mg of omega-3 LC PUFA	Kolanowski and Weißbrodt (2007)
Processed cheese	Tuna fish oil (ROPUFA '30' n-3 food oil) & Tuna fish oil emulsion	10,20,30,40,50, 60, & 75 g/kg	Fishy – off flavour cheese in all levels. Whereas, encapsulated oil fortification maintained higher quality	Not detected	Ye et al. (2009)
Soft cheese	Purified fish oil	0, 60, 80, and 100 g of fish oil per 3600 g goat milk)	Fishy- flavour aroma and 60 g/kg level accepted with control cheese quality	126 mg EPA+DHA/ 28g serving	Hughes et al. (2012)
Fat spread	ROPUFA n-3 fish oil	30, 40, & 50 g/kg	Fishy flavour and 30g/kg were accepted with sensory quality	3 g / 0.25 g of EPA+DHA	Kolanowski et al. (2004)
Flavoured yoghurt	Fish oil (DenOmega)	0.4% and 1% (w/w)	1% might be acceptable for the population in need	145 mg EPA+DHA/ 170g serving	Rognlien et al. (2012)
Queso fresco, cheddar and mozzarella cheese	Omega-3 Powder-NG, 2-DenomegaTM fish oil, flaxseed oil, & BakOmega	1% (w/w)	Non-thermal approaches to yield high quality	5.49 mg/g and 6.64 mg/g	Bermúdez-Aguirre and Barbosa-Cánovas (2012)
Indian yoghurt (Dahi)	Micro encapsulated flaxseed oil powder	1,2, & 3% level (w/w)	physico-chemical characteristics of dahi (2%) unaffected quality with control	31.00 % of the RDA of omega-3 in fortified Dahi (200 mL).	Goyal et al. (2016)
Ice cream	Microencapsulated flaxseed oil	3, 4, & 5%	4% fortified ice cream was stable	1.4g ALA/100g serving	Gowda et al. (2018)
White pan bread	Microencapsulated omega-3 (BA35 Plus)	0.73, 2.7, 4.7 & 5g/100g	Good resistance in baking parameters and 5g/100g level was accepted with overall quality	0.30 g omega-3 (12.9 g/100 g EPA +DHA) /50g of bread serve	de Conto et al. (2012)
Bread	Nanoencapsulated flax seed oil	1.0%, 2.5%, 5.0% and 10.0% (w/w)	Encapsulation significantly decreased lipid oxidation, acrylamide & HMF	Not detected	Gökmen et al. (2011)

(Continued)

TABLE 14.3 (CONTINUED)

Omega-3 Fatty Acid Fortified Food Products and Their Functionality

Fortified foods	Material	Level of fortification	Product quality	PUFA levels	References
Biscuits	Microencapsulated garden cress seed oil	20 g/100 g	Good resistance & Protection to ALA during baking parameters	ALA content was 1.02 g and 1.05 g/100 g	Umesha et al. (2015)
Biscuits	Grounded Flaxseed	10, 20, & 30%	20% fortified biscuits were acceptable with overall quality	The incorporation of flaxseed progressively increased the P/S ratio while the ω-6/ω-3 ratio decreased	Elshehy et al. (2018)
Cookies	Flaxseed oil as a shortening fat	5, 10, 20, 30, 40, & 50 % (w/w)	30 % shortening replacement with flaxseed oil as compared to the control cookies	Omega-3 Fatty acid ratio (14.4%) in 30% cookies	Rangrej et al. (2015)
Cookies	Roasted and ground flaxseed flour	5, 10, 15, & 20 % (w/w)	15% RGF cookies showed good spread ratio	15% RGF had 4.75–5.13% linolenic acid	Rajiv et al. (2012)
Cookies	Fish oil (sea cod) and microencapsulated	15%	Cookies with fish oil encapsulates by maltodextrin was accepted with good quality.	Cookies with fish oil encapsulates contained fish gelatin, maltodextrin was better oxidative stability	Jeyakumari et al. al. (2016)
Cashew apple and araca-boi pulp juice	DHA	0.1g of DHA acid in 100mL	The beverage with formulation with encapsulated omega-3 fatty acids maintained sensory acceptability and rheological parameters did not change the sample properties up to 120days storage.	-	PRADO et al. (2020)
Pomegranate juice	Fish oil	0.07% of fish oil	The pomegranate juice fortified with up to 0.07% fish oil was fair for acceptability.	100mg	
Orange juice	omega-3 and omega-6	-	The formulated juice showed significantly stable form; furthermore, there were no changes in the acceptability.	-	Marsanasco et al. (2015)
Fruit juice	DHA and EPA	-	The study indicates that nanoencapsulation found stable and useful for enrichment of non- or low-fat beverages.	56%	Ilyasoglu et al. (2014)

Replacement of shortening of dough using flax seed oil for preparing the biscuits at a 5-50 % level (Rangrej et al., 2015) was performed. At a dose dependent level, cookies' spread ratio and breaking strength were increased. Sensory score was unaffected by up to 30% shortening replacement with flaxseed oil. Increase in omega-3 fatty acid from 0 (control) to 14.14 percent in 30 percent flaxseed oil cookies were observed. Similarly, fish oil-based cookie was formulated in three forms: fish oil as such (neat), fish oil-in-water emulsion (fish oil was emulsified with tween 20 emulsifier), and fish oil encapsulate (spray dried). Six different cookie formulations were created using various formulations. Cookies fortified with fish oil or microencapsulated fish oil have a higher nutritional quality. Fish oil encapsulates contained maltodextrin cookies were comparable with control (Jeyakumari et al., 2016).

14.10.3 FORTIFICATION INTO BEVERAGES

The technology for fortifying beverages with polyunsaturated fatty acids is rapidly evolving. Nowadays, beverages are not only served to quench thirst, but they also provide a wide range of health benefits while being delicious. They are food segment beverages that allow desirable nutrients to be combined with bioactive compounds to increase nutrient values. The cashew-apple and araca-boi beverages contained omega-3 fatty acids. The encapsulated beverage formulation demonstrated good sensory acceptability and rheological properties, as well as being stable for up to 120 days in storage. The freeze-dried microencapsulated fish oil was successfully fortified with pomegranate juice. The sensory evaluation showed that up to 100 mg of omega-3 fatty acids per liter of juice can be fortified. Overall, the pomegranate juice fortified with up to 0.07% fish oil was fair for acceptability. Omega-3 and Omega-6 fatty acid was encapsulated and the constituents were added to orange juice. The study presents to enhance food nutritional value in the human diet by delivering bioactive constituents by generating functional orange juice. The obtained results suggested that the formulated juice remained significantly stable and exhibited high oxidative stability (Marsanasco et al., 2015). The nanoencapsulation of fish oil (source EPA/DHA fatty acids) by sodium caseinate and gum arabic for the enrichment of fruit juice was investigated. The finding of the study clearly showed that the nanoencapsulation was found stable and useful for enrichment of non- or low-fat beverages (Ilyasoglu et al., 2014). The increased attention to a healthy diet has given a strong input to the food market. The functional beverages encapsulated with omega-3 and omega-6 fatty acids prove an opportunity for the industrial economy and consumer health. Further, there is a need for more research to increase the shelf-life evaluation and resistance to heat treatment such as pasteurization for the encapsulated omega-3 and omega-6 fatty acids.

14.11 SUMMARY AND OUTLOOK

Omega-6 and omega-3 polyunsaturated fatty acids underlying parts of cell layers play substantially a crucial role. PUFA's stimulus cell layer properties with a specific functional, metabolic, and signaling role. These fatty acid compositions are regarded as an important building element for the body's cells, tissues, and organs, and they help to improve and sustain human health throughout life. In this atmosphere, the demand for PUFAs in everyday diets has increased, leading to the consumption of functional and practical foods. PUFAs are highly susceptible to oxidation during prolonged storage and processing; hence there is a requirement for protection through micro and nano encapsulation. These exceptional technologies can safely entrap PUFA's and it aids in the controlled release of bioactive compounds at the target place. One of the ways to prevent deficiency of omega-3 and omega-6 polyunsaturated fatty acids in the developing countries is through the food products rich in PUFA's by avoiding undesirable color, flavor, and maintaining the nutritive values in the food. Food fortification is an important nutrition intervention to fight against deficiencies with PUFA's for both developed and underdeveloped countries and to reduce the risks from various diseases. Therefore, the food fortification through microencapsulation and nanoencapsulation techniques are the added advantages together for incorporating various sources of PUFAs that are highly stable and suscep-tible to oxidation. According to the findings, the features and applications of food fortification have been demonstrated to be highly successful in incorporating PUFAs into a daily diet.

14.12 ACKNOWLEDGMENTS

Dr. Vijaykrishnaraj M (VM) is grateful to Dr. Karthik P (PK) for providing the opportunity to col-laborate and write this book chapter and also privileged to Zhejiang Gongshang University, School of Food Science and Biotechnology for providing the foreign national post-doctoral fellowship. The corresponding author PK wishes to thank Dr. C. Anandharamakrishnan, Sr. Principal Scientist, CSIR- Central Food Technological Research Institute (CFTRI) for his help and encouragement in Micro and Nanoencapsulation of bioactives studies.

REFERENCES

Aberkane, L., Roudaut, G., & Saurel, R. (2014). Encapsulation and oxidative stability of PUFA-rich oil micro-encapsulated by spray drying using pea protein and pectin. *Food and Bioprocess Technology*, 7(5), 1505–1517.

Aguilar-Pérez, K. M., Ruiz-Pulido, G., Medina, D. I., Parra-Saldivar, R., & Iqbal, H. M. (2023). Insight of nanotechnological processing for nano-fortified functional foods and nutraceutical—Opportunities, challenges, and future scope in food for better health. *Critical Reviews in Food Science and Nutrition*, 63(20), 4618–4635.

Aguilar-Pérez, K. M., Ruiz-Pulido, G., Medina, D. I., Parra-Saldivar, R., & Iqbal, H. M. (2021). Insight of nan-otechnological processing for nano-fortified functional foods and nutraceutical—Opportunities, chal-lenges, and future scope in food for better health. *Critical Reviews in Food Science and Nutrition*, 1–18.

Airanthi, M. W. A., Sasaki, N., Iwasaki, S., Baba, N., Abe, M., Hosokawa, M., & Miyashita, K. (2011). Effect of brown seaweed lipids on fatty acid composition and lipid hydroperoxide levels of mouse liver. *Journal of Agricultural and Food Chemistry*, 59(8), 4156–4163.

Al-Hasan, R. H., Hantash, F. M., & Radwan, S. S. (1991). Enriching marine macroalgae with eicosatetraenoic (arachidonic) and eicosapentaenoic acids by chilling. *Applied Microbiology and Biotechnology*, 35(4), 530–535.

Anandharamakrishnan, C. (2014). *Techniques for nanoencapsulation of food ingredients* (Vol. 8, pp. 65–67). New York: Springer.

Anese, M., Valoppi, F., Calligaris, S., Lagazio, C., Suman, M., Manzocco, L., & Nicoli, M. C. (2016). Omega-3 enriched biscuits with low levels of heat-induced toxicants: Effect of formulation and baking conditions. *Food and Bioprocess Technology*, 9(2), 232–242.

Arab-Tehrany, E., Jacquot, M., Gaiani, C., Imran, M., Desobry, S., & Linder, M. (2012). Beneficial effects and oxidative stability of omega-3 long-chain polyunsaturated fatty acids. *Trends in Food Science and Technology*, 25(1), 24–33.

Azizi, M., Kierulf, A., Lee, M. C., & Abbaspourrad, A. (2018). Improvement of physicochemical properties of encapsulated echium oil using nanostructured lipid carriers. *Food Chemistry*, 246, 448–456.

Babin, F., Abderrazik, M., Favier, F., Cristol, J. P., Leger, C. L., Papoz, L., & Descomps, B. (1999). Differences between polyunsaturated fatty acid status of non-institutionalised elderly women and younger controls: A bioconversion defect can be suspected. *European Journal of Clinical Nutrition*, 53(8), 591–596.

Barros, R., Moreira, A., Fonseca, J., Delgado, L., Castel-Branco, M. G., Haahtela, T., Lopes, C., & Moreira, P. (2011). Dietary intake of α-linolenic acid and low ratio of n-6: n-3 PUFA are associated with decreased exhaled NO and improved asthma control. *British Journal of Nutrition*, 106(3), 441–450.

Bell, M. C., & Busse, W. W. (2013). Severe asthma: An expanding and mounting clinical challenge. *The Journal of Allergy and Clinical Immunology: In Practice*, 1(2), 110–121.

Benatti, P., Peluso, G., Nicolai, R., & Calvani, M. (2004). Polyunsaturated fatty acids: Biochemical, nutritional and epigenetic properties. *Journal of the American College of Nutrition*, 23(4), 281–302.

Benbrook, C. M., Butler, G., Latif, M. A., Leifert, C., & Davis, D. R. (2013). Organic production enhances milk nutritional quality by shifting fatty acid composition: A United States–wide, 18-month study. *PloS One*, 8(12), e82429.

Bermúdez-Aguirre, D., & Barbosa-Cánovas, G. V. (2012). Fortification of queso fresco, cheddar and mozzarella cheese using selected sources of omega-3 and some nonthermal approaches. *Food Chemistry*, 133(3), 787–797.

Bigogno, C., Khozin-Goldberg, I., Boussiba, S., Vonshak, A., & Cohen, Z. (2002). Lipid and fatty acid composition of the green oleaginous alga Parietochloris incisa, the richest plant source of arachidonic acid. *Phytochemistry*, 60(5), 497–503.

Bolton-Smith, C., Woodward, M., & Tavendale, R. (1997). Evidence for age-related differences in the fatty acid composition of human adipose tissue, independent of diet. *European Journal of Clinical Nutrition*, 51(9), 619–624.

Bousquet, M., Saint-Pierre, M., Julien, C., Salem, N. Jr., Cicchetti, F., & Calon, F. (2008). Beneficial effects of dietary omega-3 polyunsaturated fatty acid on toxin-induced neuronal degeneration in an animal model of Parkinson's disease. *The FASEB Journal*, 22(4), 1213–1225.

Braman, S. S. (2006). The global burden of asthma. *Chest*, 130(1), 4S–12S.

Burdge, G. C. (2006). Metabolism of a-linolenic acid in humans. *Prostaglandins, Leukotrienes, and Essential Fatty Acids*, 75(3), 161–168.

Carneiro, H. C., Tonon, R. V., Grosso, C. R., & Hubinger, M. D. (2013). Encapsulation efficiency and oxidative stability of flaxseed oil microencapsulated by spray drying using different combinations of wall materials. *Journal of Food Engineering*, 115(4), 443–451.

Cetinkaya, T., Mendes, A. C., Jacobsen, C., Ceylan, Z., Chronakis, I. S., Bean, S. R., & García-Moreno, P. J. (2021). Development of kafirin-based nanocapsules by electrospraying for encapsulation of fish oil. *LWT*, 136, 110297.

Chen, Q., McGillivray, D., Wen, J., Zhong, F., & Quek, S. Y. (2013). Co-encapsulation of fish oil with phytosterol esters and limonene by milk proteins. *Journal of Food Engineering*, 117(4), 505–512.

Cohen, Z. (1990). The production potential of eicosapentaenoic and arachidonic acids by the red alga Porphyridium cruentum. *Journal of the American Oil Chemists' Society*, 67(12), 916–920.

Contini, C., Álvarez, R., O'sullivan, M., Dowling, D. P., Gargan, S. Ó., & Monahan, F. J. (2014). Effect of an active packaging with citrus extract on lipid oxidation and sensory quality of cooked turkey meat. *Meat Science*, 96(3), 1171–1176.

Coste, T. C., Armand, M., Lebacq, J., Lebecque, P., Wallemacq, P., & Leal, T. (2007). An overview of monitoring and supplementation of omega 3 fatty acids in cystic fibrosis. *Clinical Biochemistry*, 40(8), 511–520.

da Silva Stefani, F., de Campo, C., Paese, K., Guterres, S. S., Costa, T. M. H., & Flôres, S. H. (2019). Nanoencapsulation of linseed oil with chia mucilage as structuring material: Characterization, stability and enrichment of orange juice. *Food Research International*, 120, 872–879.

de Campo, C., Dos Santos, P. P., Costa, T. M. H., Paese, K., Guterres, S. S., de Oliveira Rios, A., & Flôres, S. H. (2017). Nanoencapsulation of chia seed oil with chia mucilage (Salvia hispanica L.) as wall material: Characterization and stability evaluation. *Food Chemistry*, 234, 1–9.

de Conto, L. C., Oliveira, R. S. P., Martin, L. G. P., Chang, Y. K., & Steel, C. J. (2012). Effects of the addition of microencapsulated omega-3 and rosemary extract on the technological and sensory quality of white pan bread. *LWT – Food Science and Technology*, 45(1), 103–109.

de Lau, L. M., Bornebroek, M., Witteman, J. C., Hofman, A., Koudstaal, P. J., & Breteler, M. M. (2005). Dietary fatty acids and the risk of Parkinson disease: The Rotterdam study. *Neurology*, 64(12), 2040–2045.

de Rijk, M. C., Rocca, W. A., Anderson, D. W., Melcon, M. O., Breteler, M. M., & Maraganore, D. M. (1997). A population perspective on diagnostic criteria for Parkinson's disease. *Neurology*, 48(5), 1277–1281.

Dedyukhina, E. G., Chistyakova, T. I., & Vainshtein, M. B. (2011). Biosynthesis of arachidonic acid by micromycetes. *Applied Biochemistry and Microbiology*, 47(2), 109–117.

Dwyer, J. T., Wiemer, K. L., Dary, O., Keen, C. L., King, J. C., Miller, K. B., … Bailey, R. L. (2015). Fortification and health: Challenges and opportunities. *Advances in Nutrition*, 6(1), 124–131.

Dyall, S. C., & Michael-Titus, A. T. (2008). Neurological benefits of omega-3 fatty acids. *NeuroMolecular Medicine*, 10(4), 219–235.

Ells, R., Kock, J. L., Albertyn, J., & Pohl, C. H. (2012). Arachidonic acid metabolites in pathogenic yeasts. *Lipids in Health and Disease*, 11(1), 1–7.

El-Sayed, E., & Ibrahim, K. (2017). Effect of the types of dietary fats and non-dietary oils on bone metabolism. *Critical Reviews in Food Science and Nutrition*, 57(4), 653–658.

Elshehy, H., Agamy, N., & Ismail, H. (2018). Effect of fortification of biscuits with flaxseed on omega 3 and calcium content of the products. *Journal of High Institute of Public Health*, 48(2), 58–66.

Eratte, D., Wang, B., Dowling, K., Barrow, C. J., & Adhikari, B. P. (2014). Complex coacervation with whey protein isolate and gum arabic for the microencapsulation of omega-3 rich tuna oil. *Food and Function*, 5(11), 2743–2750.

Ezhilarasi, P. N., Karthik, P., Chhanwal, N., & Anandharamakrishnan, C. (2013). Nanoencapsulation techniques for food bioactive components: A review. *Food and Bioprocess Technology*, 6(3), 628–647.

FDA. (2007). GRAS Notice No. GRN 000200. US Food and Drug Administration. Washington, DC: Center for Food Safety and Applied Nutrition.

Feizollahi, E., Hadian, Z., & Honarvar, Z. (2018). Food fortification with omega-3 fatty acids; microencapsulation as an addition method. *Current Nutrition and Food Science*, 14(2), 90–103.

Gallardo, G., Guida, L., Martinez, V., López, M. C., Bernhardt, D., Blasco, R., … Hermida, L. G. (2013). Microencapsulation of linseed oil by spray drying for functional food application. *Food Research International*, 52(2), 473–482.

Ganesan, B., Brothersen, C., & McMahon, D. J. (2014). Fortification of foods with omega-3 polyunsaturated fatty acids. *Critical Reviews in Food Science and Nutrition*, 54(1), 98–114.

Garg, M. L., Sebokova, E., Thomson, A. B., & Clandinin, M. T. (1988). Δ 6-desaturase activity in liver microsomes of rats fed diets enriched with cholesterol and/or ω 3 fatty acids. *Biochemical Journal*, 249(2), 351–356.

Generoso, S. D. V., Rodrigues, N. M., Trindade, L. M., Paiva, N. C., Cardoso, V. N., Carneiro, C. M., … Maioli, T. U. (2015). Dietary supplementation with omega-3 fatty acid attenuates 5-fluorouracil induced mucositis in mice. *Lipids in Health and Disease*, 14(1), 1–10.

Gharibzahedi, S. M. T., & Jafari, S. M. (2017). Nano-encapsulation of minerals. In S. M. Jafari (Ed.), *Nano-encapsulation of food bioactive ingredients: Principles and applications* (pp. 333–400). London: Elsevier.

Gill, I., & Valivety, R. (1997). Polyunsaturated fatty acids, part 1: occurrence, biological activities and applications. *Trends in Biotechnology*, 15(10), 401–409.

Gökmen, V., Mogol, B. A., Lumaga, R. B., Fogliano, V., Kaplun, Z., & Shimoni, E. (2011). Development of functional bread containing nanoencapsulated omega-3 fatty acids. *Journal of Food Engineering*, 105(4), 585–591.

Gouin, S. (2004). Microencapsulation: Industrial appraisal of existing technologies and trends. *Trends in Food Science and Technology*, 15(7–8), 330–347.

Gowda, A., Sharma, V., Goyal, A., Singh, A. K., & Arora, S. (2018). Process optimization and oxidative stability of omega-3 ice cream fortified with flaxseed oil microcapsules. *Journal of Food Science and Technology*, 55(5), 1705–1715.

Goyal, A., Sharma, V., Sihag, M. K., Singh, A. K., Arora, S., & Sabikhi, L. (2016). Fortification of dahi (Indian yoghurt) with omega-3 fatty acids using microencapsulated flaxseed oil microcapsules. *Journal of Food Science and Technology*, 53(5), 2422–2433.

Guarner, F., Khan, A. G., Garisch, J., Eliakim, R., Gangl, A., Thomson, A., … & Kim, N. (2012). World gastroenterology organisation global guidelines: Probiotics and prebiotics october 2011. *Journal of clinical gastroenterology*, 46(6), 468–481.

Gumus, C. E., & Gharibzahedi, S. M. T. (2021). Yogurts supplemented with lipid emulsions rich in omega-3 fatty acids: New insights into the fortification, microencapsulation, quality properties, and health-promoting effects. *Trends in Food Science and Technology*, 110, 267–279.

Haghighi, F., Galfalvy, H., Chen, S., Huang, Y. Y., Cooper, T. B., Burke, A. K., & Sublette, M. E. (2015). DNA methylation perturbations in genes involved in polyunsaturated Fatty Acid biosynthesis associated with depression and suicide risk. *Frontiers in Neurology*, 6, 92.

Hanus, L. O., Temina, M., & Dembitsky, V. (2008). Biodiversity of the chemical constituents in the epiphytic lichenized ascomycete Ramalina lacera grown on difference substrates Crataegus sinaicus, Pinus halepensis, and Quercus calliprinos. *Biomed Pap Med Fac Univ Palacky Olomouc Czech Repub*, 152(2), 203–208.

Haq, M., & Chun, B. S. (2018). Microencapsulation of omega-3 polyunsaturated fatty acids and astaxanthin-rich salmon oil using particles from gas saturated solutions (PGSS) process. *LWT*, 92, 523–530.

Harris, W. S., Mozaffarian, D., Lefevre, M., Toner, C. D., Colombo, J., Cunnane, S. C. et al. (2009). Towards establishing dietary reference intakes for eicosapentaenoic and docosahexaenoic acids. *Journal of Nutrition*, 139(4), 804S–819S.

Hayes, M., Coakley, M., O'Sullivan, L., Stanton, C., Hill, C., Fitzgerald, G. F., et al. (2006). Cheese as a delivery vehicle for probiotics and biogenic substances. *Australia Journal of Dairy Technology*, 61, 132–141.

Hayta, M., & Özuğur, G. (2011). Phytochemical fortification of flour and bread. In Victor R. Preedy, Ronald Ross Watson and Vinood B. Patel. UK, (ed.), *Flour and breads and their fortification in health and disease prevention* (pp. 293–300). Academic Press.

Hernandez, E. M. (2013). Enrichment of baked goods with omega-3 fatty acids. In Jacobsen, C., Nielsen, N. S., Horn, A. F., & Sørensen, A. D. M. (eds.), USA *Food enrichment with omega-3 fatty acids* (pp. 319–335). Woodhead Publishing.

Hernandez, E. M. (2014). Issues in fortification and analysis of omega-3 fatty acids in foods. *Lipid Technology*, 26(5), 103–106.

Hoppu, U., Rinne, M., Lampi, A. M., & Isolauri, E. (2005). Breast milk fatty acid composition is associated with development of atopic dermatitis in the infant. *Journal of Pediatric Gastroenterology and Nutrition*, 41(3), 335–338. https://lpi.oregonstate.edu/mic/other-nutrients/essential-fatty-acids#references.

Hughes, B. H., Brian Perkins, L., Calder, B. L., & Skonberg, D. I. (2012). Fish oil fortification of soft goat cheese. *Journal of Food Science*, 77(2), S128–S133.

Hundre, S. Y., Karthik, P., & Anandharamakrishnan, C. (2015). Effect of whey protein isolate and β-cyclodextrin wall systems on stability of microencapsulated vanillin by spray–freeze drying method. *Food Chemistry*, 174, 16–24.

Iafelice, G., Caboni, M. F., Cubadda, R., Di Criscio, T., Trivisonno, M. C., & Marconi, E. (2008). Development of functional spaghetti enriched with long chain omega-3 fatty acids. *Cereal Chemistry*, 85(2), 146–151.

Ilyasoglu, H., & El, S. N. (2014). Nanoencapsulation of EPA/DHA with sodium caseinate–gum arabic complex and its usage in the enrichment of fruit juice. *LWT – Food Science and Technology*, 56(2), 461–468.

Jacq, E., Prieur, D., Nichols, P., White, D. C., Porter, T., & Geesey, G. G. (1989). Microscopic examination and fatty acid characterization of filamentous bacteria colonizing substrata around subtidal hydrothermal vents. *Archives of Microbiology*, 152(1), 64–71.

Jeyakumari, A., Janarthanan, G., Chouksey, M. K., & Venkateshwarlu, G. (2016). Effect of fish oil encapsulates incorporation on the physico-chemical and sensory properties of cookies. *Journal of Food Science and Technology*, 53(1), 856–863.

Jones, P. J., Senanayake, V. K., Pu, S., Jenkins, D. J., Connelly, P. W., Lamarche, B., ... Kris-Etherton, P. M. (2014). DHA-enriched high–oleic acid canola oil improves lipid profile and lowers predicted cardiovascular disease risk in the canola oil multicenter randomized controlled trial. *The American Journal of Clinical Nutrition*, 100(1), 88–97.

Karthik, P., & Anandharamakrishnan, C. (2013). Microencapsulation of docosahexaenoic acid by spray-freeze-drying method and comparison of its stability with spray-drying and freeze-drying methods. *Food and Bioprocess Technology*, 6(10), 2780–2790.

Karthik, P. (2016). Micro and nanoencapsulation of omega-3 fatty acids by emulsion and spray-freeze-drying techniques (Doctoral dissertation, Academy of Scientific and Innovative Research).

Klaypradit, W., & Huang, Y. W. (2008). Fish oil encapsulation with chitosan using ultrasonic atomizer. *LWT – Food Science and Technology*, 41(6), 1133–1139.

Klinkesorn, U., Sophanodora, P., Chinachoti, P., McClements, D. J., & Decker, E. A. (2005). Stability of spray-dried tuna oil emulsions encapsulated with two-layered interfacial membranes. *Journal of Agricultural and Food Chemistry*, 53(21), 8365–8371.

Kolanowski, W., & Weisbrodt, J. (2007). Sensory quality of dairy products fortified with fish oil. *International Dairy Journal*, 17(10), 1248–1253.

Kolanowski, W., & Weibbrodt, J. (2008). Possibilities of Fisherman's friend type lozenges fortification with Omega-3 LC PUFA by addition of microencapsulated fish oil. *Journal of the American Oil Chemists' Society*, 85(4), 339–345.

Kolanowski, W., Swiderski, F., Jaworska, D., & Berger, S. (2004). Stability, sensory quality, texture properties and nutritional value of fish oil-enriched spreadable fat. *Journal of the Science of Food and Agriculture*, 84(15), 2135–2141.

Kuipers, R. S., Luxwolda, M. F., Offringa, P. J., Boersma, E. R., Dijck-Brouwer, D. J., & Muskiet, F. A. (2012). Fetal intrauterine whole body linoleic, arachidonic and docosahexaenoic acid contents and accretion rates. *Prostaglandins, Leukotrienes, and Essential Fatty Acids*, 86(1–2), 13–20.

Kumar, M., Gupta, V., Trivedi, N., Kumari, P., Bijo, A. J., Reddy, C. R. K., & Jha, B. (2011). Desiccation induced oxidative stress and its biochemical responses in intertidal red alga Gracilaria Corticata (Gracilariales, Rhodophyta). *Environmental and Experimental Botany*, 72(2), 194–201.

Kyle, D. J., Schaefer, E., Patton, G., & Beiser, A. (1999). Low serum docosahexaenoic acid is a significant risk factor for Alzheimer's dementia. *Lipids*, 34(Suppl), S245.

Lee, T. H., Mencia-Huerta, J. M., Shih, C., Corey, E. J., Lewis, R. A., & Austen, K. F. (1984). Effects of exogenous arachidonic, eicosapentaenoic, and docosahexaenoic acids on the generation of 5-lipoxygenase pathway products by ionophore-activated human neutrophils. *The Journal of Clinical Investigation*, 74(6), 1922–1933.

Li, J., Xun, P., Zamora, D., Sood, A., Liu, K., Daviglus, M., Iribarren, C., Jacobs Jr, D., Shikany, J. M., & He, K. (2013). Intakes of long-chain omega-3 (n– 3) PUFAs and fish in relation to incidence of asthma among American young adults: The CARDIA study. *The American Journal of Clinical Nutrition*, 97(1), 173–178.

Lorente-Cebrián, S., Costa, A. G., Navas-Carretero, S., Zabala, M., Laiglesia, L. M., Martínez, J. A., & Moreno-Aliaga, M. J. (2015). An update on the role of omega-3 fatty acids on inflammatory and degenerative diseases. *Journal of Physiology and Biochemistry*, 71(2), 341–349.

Mannar, M. V., Garrett, G. S., & Hurrell, R. F. (2018). Future trends and strategies in food fortification. In *Food fortification in a globalized world* (pp. 375–381). Academic Press. United Kingdom.

Marsanasco, M., Piotrkowski, B., Calabró, V., del Valle Alonso, S., & Chiaramoni, N. S. (2015). Bioactive constituents in liposomes incorporated in orange juice as new functional food: Thermal stability, rheological and organoleptic properties. *Journal of Food Science and Technology*, 52(12), 7828–7838.

Mehrad, B., Shabanpour, B., Jafari, S. M., & Pourashouri, P. (2015). Characterization of dried fish oil from menhaden encapsulated by spray drying. *Aquaculture, Aquarium, Conservation & Legislation*, 8(1), 57–69.

Melgosa, R., Benito-Román, Ó., Sanz, M. T., de Paz, E., & Beltrán, S. (2019). Omega–3 encapsulation by PGSS-drying and conventional drying methods. Particle characterization and oxidative stability. *Food Chemistry*, 270, 138–148.

Moghadasian, M. H. (2008). Advances in dietary enrichment with n-3 fatty acids. *Critical Reviews in Food Science and Nutrition*, 48(5), 402–410.

Moomand, K., & Lim, L. T. (2015). Properties of encapsulated fish oil in electrospun zein fibres under simulated in vitro conditions. *Food and Bioprocess Technology*, 8(2), 431–444.

Nedovic, V., Kalusevic, A., Manojlovic, V., Levic, S., & Bugarski, B. (2011). An overview of encapsulation technologies for food applications. *Procedia Food Science*, 1, 1806–1815.

Nejadmansouri, M., Hosseini, S. M. H., Niakosari, M., Yousefi, G. H., & Golmakani, M. T. (2016). Physicochemical properties and oxidative stability of fish oil nanoemulsions as affected by hydrophilic lipophilic balance, surfactant to oil ratio and storage temperature. *Colloids and Surfaces A: Physicochemical and Engineering Aspects*, 506, 821–832.

Nichols, D. S., Brown, J. L., Nichols, P. D., & McMeekin, T. A. (1997). Production of eicosapentaenoic and arachidonic acids by an Antarctic bacterium: Response to growth temperature. *FEMS Microbiology Letters*, 152(2), 349–354.

Oppedisano, F., Macrì, R., Gliozzi, M., Musolino, V., Carresi, C., Maiuolo, J., … Mollace, V. (2020). The anti-inflammatory and antioxidant properties of n-3 PUFAs: Their role in cardiovascular protection. *Biomedicines*, 8(9), 306.

Osendarp, S. J., Martinez, H., Garrett, G. S., Neufeld, L. M., De-Regil, L. M., Vossenaar, M., & Darnton-Hill, I. (2018). Large-scale food fortification and biofortification in low-and middle-income countries: A review of programs, trends, challenges, and evidence gaps. *Food and Nutrition Bulletin*, 39(2), 315–331.

Öztürk, B. (2017). Nanoemulsions for food fortification with lipophilic vitamins: Production challenges, stability, and bioavailability. *European Journal of Lipid Science and Technology*, 119(7), 1500539.

Pandule, V. S., Sharma, M., & HC, D. (2021). Omega-3 fatty acid-fortified butter: Preparation and characterisation of textural, sensory, thermal and physico-chemical properties. *International Journal of Dairy Technology*, 74(1), 181–191.

Pereira, H., Barreira, L., Figueiredo, F., Custódio, L., Vizetto-Duarte, C., Polo, C., … Varela, J. (2012). Polyunsaturated fatty acids of marine macroalgae: Potential for nutritional and pharmaceutical applications. *Marine Drugs*, 10(9), 1920–1935.

Pohjanheimo, T., & Sandell, M. (2009). Explaining the liking for drinking yoghurt: The role of sensory quality, food choice motives, health concern and product information. *International Dairy Journal*, 19(8), 459–466.

Pourashouri, P., Shabanpour, B., Razavi, S. H., Jafari, S. M., Shabani, A., & Aubourg, S. P. (2014). Impact of wall materials on physicochemical properties of microencapsulated fish oil by spray drying. *Food and Bioprocess Technology*, 7(8), 2354–2365.

Prado, G. M. D., Sousa, P. H. M. D., Silva, L. M. R. D., Wurlitzer, N. J., Garruti, D. D. S., & Figueiredo, R. W. D. (2022). Encapsulated omega-3 addition to a cashew apple/araçá-boi juice-effect on sensorial acceptability and rheological properties. *Food Science and Technology*, 42, e64321.

Raeisi, S., Ojagh, S. M., Quek, S. Y., Pourashouri, P., & Salaün, F. (2019). Nano-encapsulation of fish oil and garlic essential oil by a novel composition of wall material: Persian gum-chitosan. *LWT*, 116, 108494.

Rajam, R., Karthik, P., Parthasarathi, S., Joseph, G. S., & Anandharamakrishnan, C. (2012). Effect of whey protein–alginate wall systems on survival of microencapsulated Lactobacillus plantarum in simulated gastrointestinal conditions. *Journal of Functional Foods*, 4(4), 891–898.

Rajiv, J., Indrani, D., Prabhasankar, P., & Rao, G. V. (2012). Rheology, fatty acid profile and storage characteristics of cookies as influenced by flax seed (Linum usitatissimum). *Journal of Food Science and Technology*, 49(5), 587–593.

Rangrej, V., Shah, V., Patel, J., & Ganorkar, P. M. (2015). Effect of shortening replacement with flaxseed oil on physical, sensory, fatty acid and storage characteristics of cookies. *Journal of Food Science and Technology*, 52(6), 3694–3700.

Rognlien, M., Duncan, S. E., O'Keefe, S. F., & Eigel, W. N. (2012). Consumer perception and sensory effect of oxidation in savory-flavored yogurt enriched with n-3 lipids. *Journal of Dairy Science*, 95(4), 1690–1698.

Saini, R. K., Shetty, N. P., & Giridhar, P. (2014). GC-FID/MS analysis of fatty acids in Indian cultivars of Moringa oleifera: Potential sources of PUFA. *Journal of the American Oil Chemists' Society*, 91(6), 1029–1034.

Samadi, P., Gregoire, L., Rouillard, C., Bedard, P. J., Di Paolo, T., & Levesque, D. (2006). Docosahexaenoic acid reduces Levodopainduced dyskinesias in 1-methyl-4-phenyl-1, 2, 3, 6-tetrahydropyridine monkeys. *Annals of Neurology*, 59(2), 282–288.

Schuster, G. U., Bratt, J. M., Jiang, X., Pedersen, T. L., Grapov, D., Adkins, Y., Kelley, D. S., Newman, J. W., Kenyon, N. J., & Stephensen, C. B. (2014). Dietary long-chain omega-3 fatty acids do not diminish eosinophilic pulmonary inflammation in mice. *American Journal of Respiratory Cell and Molecular Biology*, 50(3), 626–636.

Selkoe, D. J. (2001). Alzheimer's disease: Genes, proteins, and therapy. *Physiological Reviews*, 81(2), 741–766.

Shahidi, F., & Miraliakbari, H. (2006). Marine oils: Compositional characteristics and health effects. In *Nutraceutical and specialty lipids and their co-products* (pp. 241–264), 5, 227. CRC Press.

Singh, H., Zhu, X. Q., & Ye, A. (2006). Lipid encapsulation. WO/2006/115420.

Siscovick, D. S., Barringer, T. A., Fretts, A. M., Wu, J. H., Lichtenstein, A. H., Costello, R. B., … Mozaffarian, D. (2017). Omega-3 polyunsaturated fatty acid (fish oil) supplementation and the prevention of clinical cardiovascular disease: A science advisory from the American Heart Association. *Circulation*, 135(15), e867–e884.

Skinner, E. R., Watt, C., Besson, J. A. O., & Best, P. V. (1989). Lipid composition of different regions of the brain in patients with Alzheimer's disease. *Biochemical Society Transactions*, 17(1), 213–214.

Soderberg, M., Edlund, C., Kristensson, K., & Dallner, G. (1991). Fatty acid composition of brain phospholipids in aging and Alzheimer's disease. *Lipids*, 26(6), 421–425.

Sperling, R. I. (1991). Dietary omega-3 fatty acids: Effects on lipid mediators of inflammation and rheumatoid arthritis. *Rheumatic Disease Clinics of North America*, 17(2), 373–389.

Stefanska, A., Bergmann, K., & Sypniewska, G. (2015). Metabolic syndrome and menopause: Pathophysiology, clinical and diagnostic significance. *Advances in Clinical Chemistry*, 72, 1–75.

Steinkamp, G., Demmelmair, H., Rühl-Bagheri, I., von der Hardt, H., & Koletzko, B. (2000). Energy supplements rich in linoleic acid improve body weight and essential fatty acid status of cystic fibrosis patients. *Journal of Pediatric Gastroenterology and Nutrition*, 31(4), 418–423.

Su, G., Jiao, K., Li, Z., Guo, X., Chang, J., Ndikubwimana, T., … Lin, L. (2016). Phosphate limitation promotes unsaturated fatty acids and arachidonic acid biosynthesis by microalgae Porphyridium purpureum. *Bioprocess and Biosystems Engineering*, 39(7), 1129–1136.

Suloma, A., Ogata, H. Y., Furuita, H., Garibay, E. S., & Chavez, D. R. (2007). Arachidonic acid distribution in seaweed, seagrass, invertebrates and dugong in coral reef areas in the Philippines. In *Sustainable production systems of aquatic animals in brackish mangrove areas* (pp. 107–111). Japan International Research Center for Agricultural Sciences.

Timilsena, Y. P., Wang, B., Adhikari, R., & Adhikari, B. (2017). Advances in microencapsulation of polyunsaturated fatty acids (PUFAs)-rich plant oils using complex coacervation: A review. *Food Hydrocolloids*, 69, 369–381.

Ulven, S. M., Kirkhus, B., Lamglait, A., Basu, S., Elind, E., Haider, T., Berge, K., Vik, H., & Pedersen, J. I. (2011). Metabolic effects of krill oil are essentially similar to those of fish oil but at lower dose of EPA and DHA, in healthy volunteers. *Lipids*, 46(1), 37–46.

Umesha, S. S., Manohar, R. S., Indiramma, A. R., Akshitha, S., & Naidu, K. A. (2015). Enrichment of biscuits with microencapsulated omega-3 fatty acid (alpha-linolenic acid) rich garden cress (Lepidium sativum) seed oil: Physical, sensory and storage quality characteristics of biscuits. *LWT – Food Science and Technology*, 62(1), 654–661.

Vahidmoghadam, F., Pourahmad, R., Mortazavi, A., Davoodi, D., & Azizinezhad, R. (2019). Characteristics of freeze-dried nanoencapsulated fish oil with whey protein concentrate and gum arabic as wall materials. *Food Science and Technology*, 39, 475–481.

Van Ginneken, V. J., Helsper, J. P., de Visser, W., van Keulen, H., & Brandenburg, W. A. (2011). Polyunsaturated fatty acids in various macroalgal species from North Atlantic and tropical seas. *Lipids in Health and Disease*, 10(1), 1–8.

Wang, R., Tian, Z., & Chen, L. (2011). A novel process for microencapsulation of fish oil with barley protein. *Food Research International*, 44(9), 2735–2741.

Yamamoto, Y., & Watanabe, A. (1974). Fatty acid composition of lichens and their phyco-and mycobionts. *The Journal of General and Applied Microbiology*, 20(2), 83–86.

Yang, H., Feng, K., Wen, P., Zong, M. H., Lou, W. Y., & Wu, H. (2017). Enhancing oxidative stability of encapsulated fish oil by incorporation of ferulic acid into electrospun zein mat. *LWT*, 84, 82–90.

Ye, A., Cui, J., Taneja, A., Zhu, X., & Singh, H. (2009). Evaluation of processed cheese fortified with fish oil emulsion. *Food Research International*, 42(8), 1093–1098.

Yin, H., Liu, W., Goleniewska, K., Porter, N. A., Morrow, J. D., & Peebles Jr, R. S. (2009). Dietary supplementation of ω-3 fatty acid-containing fish oil suppresses F2-isoprostanes but enhances inflammatory cytokine response in a mouse model of ovalbumin-induced allergic lung inflammation. *Free Radical Biology and Medicine*, 47(5), 622–628.

Yokoyama, A., Hamazaki, T., Ohshita, A., Kohno, N., Sakai, K., Zhao, G. D., Katayama, H., & Hiwada, K. (2000). Effect of aerosolized docosahexaenoic acid in a mouse model of atopic asthma. *International Archives of Allergy and Immunology*, 123(4), 327–332.

Zhang, Y., Pang, X., Zhang, S., Liu, L., Ma, C., Lu, J., & Lyu, J. (2020). Buttermilk as a wall material for microencapsulation of omega-3 oils by spray drying. *LWT*, 127, 109320.

Zhou, M., Wang, T., Hu, Q., & Luo, Y. (2016). Low density lipoprotein/pectin complex nanogels as potential oral delivery vehicles for curcumin. *Food Hydrocolloids*, 57, 20–29.

Zimet, P., & Livney, Y. D. (2009). Beta-lactoglobulin and its nanocomplexes with pectin as vehicles for ω-3 polyunsaturated fatty acids. *Food Hydrocolloids*, 23(4), 1120–1126.

15 Fiber and Bioactive Peptides for Fortification

Monjurul Hoque, Debarshi Nath,
Rahul Islam Barbhuiya, and Rahel Suchintita

15.1 INTRODUCTION

Dietary fiber (DF), universally found in grains, fruits, vegetables, and legumes, consists of non-digestible carbohydrates usually in the form of polysaccharides and few as oligosaccharides. Increased consumer preference towards processed foods in the modern diet has drastically reduced the DF intake making humans prone to several metabolic disorders and cancer. Bioactive peptides (BAP), as the name suggests are hydrolyzed versions of protein that have bioactive properties which mainly relate to physiological benefits. Fortification of both in different food systems adds nutritional value to the foods and when duly absorbed in the human body, they exhibit their health beneficial properties at the target organs. It is also imperative that they must be retained in the food after processing and should not antagonize the existing nutrients in the matrix. This chapter reviews the kinds and sources of DF and BAP, their role in human health, their metabolic utilization, fortification in different food systems, and mechanisms to protect them from getting destroyed during processing.

15.1.1 DIETARY FIBER AND ITS SOURCES

DF is a ubiquitous portion of plant-based food components that are resistant to alimentary digestive enzymes present in humans or animals. In 2000, The American Association of Cereal Chemists (AACC) defined DF as the edible parts of plants or analogous carbohydrates that are resistant to digestion and absorption in the human small intestine with complete or partial fermentation in the large intestine. DF includes polysaccharides, oligosaccharides, lignin, and associated plant substances. DFs promote beneficial physiological effects, including laxation, and/or blood cholesterol attenuation, and/or blood glucose attenuation (AACC 2000).Similarly, in 2001, Australia New Zealand Food Authority defined (ANZFA) defined DF as 'the portion of edible parts of plants or their extracts, or analogous carbohydrates (polysaccharides, oligosaccharides, and lignins), that are resistant to digestion and absorption in the human small intestine, and usually with complete or partial fermentation in the large intestine'. In 2002, a panel of National Academy of Science, United States, defined the DF complex to include DF consisting of non-digestible carbohydrates and lignin that are intrinsic and intact in plants, functional fibers consisting of isolated, non-digestible carbohydrates which have beneficial physiological effects in humans and total fiber as the sum of DF and functional fiber (Dhingra, Michael, Rajput, and Patil 2012).

As illustrated in Figure 15.1, DF can be categorized as soluble or insoluble DF based on their solubility in water(Periago, Ros, López, Martínez, and Ricon 1993, Williams, Mikkelsen, Flanagan, and Gidley 2019, Pérez-Chabela and Hernández-Alcántara 2018). Gums, mucilage, and pectin are some examples of soluble DF. Gums are water-soluble and well-fermented DF and have several applications in the food as well as the pharmaceutical sector. They are mainly obtained from leguminous seed plants such as guar or locust beans, seaweed extracts such as carrageenan and

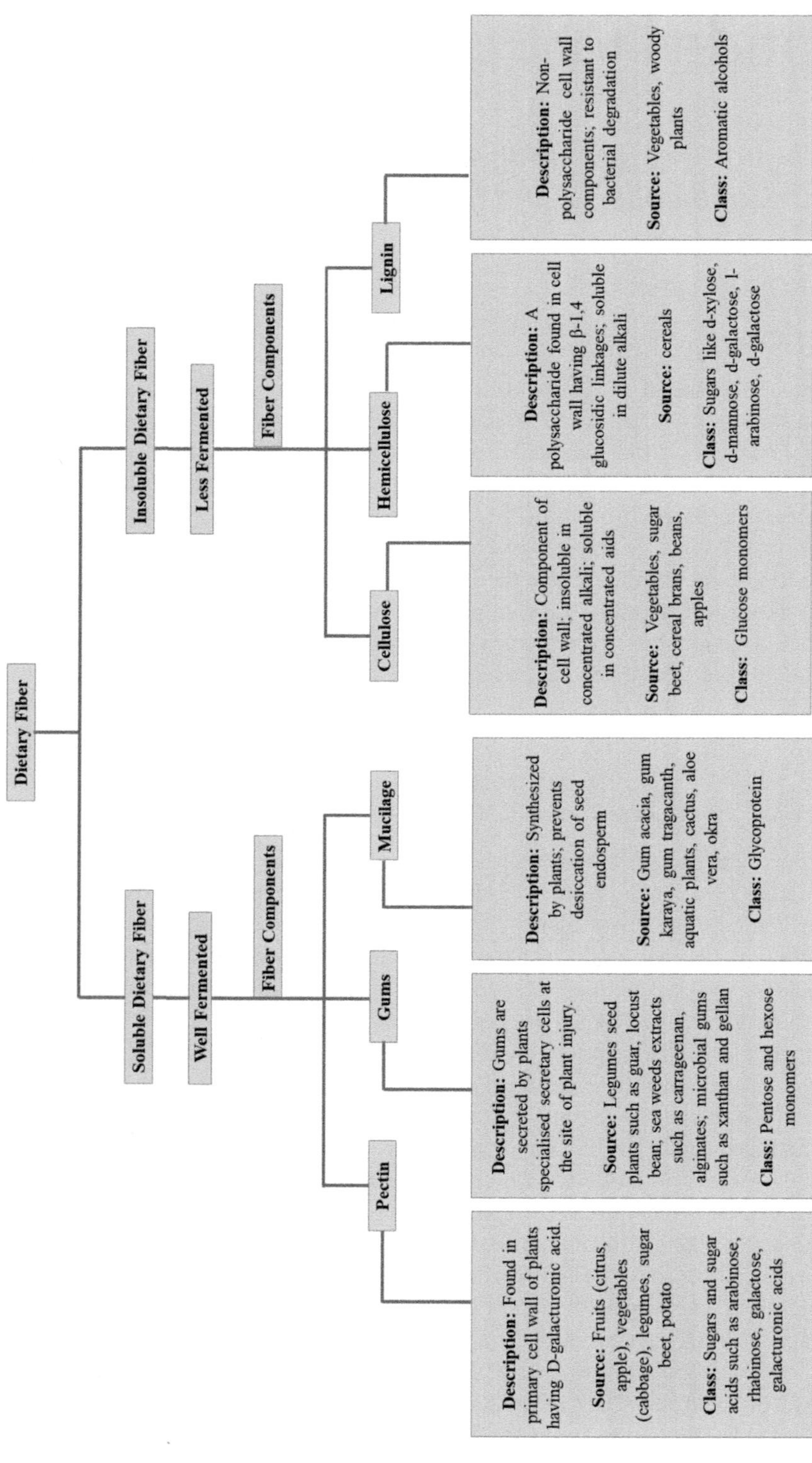

FIGURE 15.1 Classification of Dietary Fiber and their sources (Periago et al. 1993, Williams et al. 2019, Pérez-Chabela and Hernández-Alcántara 2018).

alginates, or microbial gums such as xanthan and gellan. Similarly, pectin found in the plant cell wall having D-galacturonic acid is also a water-soluble and gel-forming fiber component. Pectin is mainly obtained from fruits (citrus and apple), vegetables, legumes, potatoes, and sugar beets. Similarly, mucilages are DF obtained from plant extracts such as gum acacia, gum karaya, and gum tragacanth and are water-soluble and well fermented (Dhingra et al. 2012, Williams et al. 2019, Pérez-Chabela and Hernández-Alcántara 2018).

Lignin, cellulose, and hemicellulose are examples of insoluble DF(Dhingra et al. 2012, Williams et al. 2019). The structural component of the plant called cellulose is a fiber component insoluble in water and alkali, but soluble in concentrated acid and is primarily obtained from sugar beet and various beans. Similarly, polysaccharides in the cell wall known as hemicellulose are the fiber components insoluble in water but soluble in dilute alkali. These fibers are primarily obtained from cereal grains. Another fiber component is lignin which isa non-carbohydrate-based cell wall component, insoluble in water and less fermented, and are obtained from woody plants(Dhingra et al. 2012, Williams et al. 2019, Mišurcová, Škrovánková, Samek, Ambrožová, and Machů 2012, Xu 2010).

DF is found naturally in nuts, cereals, fruits, and vegetables but its composition and amount vary from food to food (Dhingra et al. 2012).Insoluble DFs are present in beans, nuts, wheat bran, whole-wheat flour, and in vegetables such as potatoes, green beans, cauliflower, etc. Cereals-based DF possesses substantial fecal bulking capacity because of their higher insoluble fiber content. Soluble DFs present in barley, carrots, apples, citrus fruits, oats, beans, and pea are said to play a role in in lowering blood glucose and cholesterol levels. Fibers from fruits have a lower caloric value, lower phytic acid content, better colonic fermentability, better oil holding and water holding capacity, and higher total soluble fiber content (Larrauri, Borroto, and Crespo 1997). Generally, it is recommended that fiber sources used as food ingredients preferably should possess a ratio of 1:2 (soluble DF: insoluble DF)(Jaime, Mollá, Fernández, Martín-Cabrejas, López-Andréu, and Esteban 2002).

Over the past few years, due to advanced food processing technology, there is a reduction in fiber content in human diets (Kendall, Esfahani, and Jenkins 2010).Hence, an increase in the demand for fiber-rich ingredients and products has been seen, from different sources that can be used by food industries (Chau and Huang 2003). In western countries, about 50% of the fiber intake is from cereals, 30–40% comes from vegetables, around 16% from fruits, and the remaining 3% from different minor sources(Lambo, Öste, and Nyman 2005). Various food sources of DF (per gram per 100g edible portion) are presented in Table 15.1. Additionally, fiber can also be produced from waste products such as spent brewers' grain, corn stalks, and cobs, almond and peanut skins, oat hulls, soy hulls, wheat straw, and waste portions of processed fruits and vegetables(Pérez-Chabela and Hernández-Alcántara 2018).

15.2 DIETARY FIBER IN HUMAN HEALTH

As shown in Figure 15.2, the diet containing DF has several functions, and its benefits on human health are discussed in the following section.

15.2.1 DIGESTION

The most widely known health benefit of DF is its ability to prevent constipation and promote gut mobility. It also contributes to the prevention of several diseases related to digestion by reducing the intestinal transit time, glycemic, and cholesterol levels, enhancing the fecal bulk volume and supporting the multiplication of gut microflora (Beecher 1999). Consumption of DF has varying effects on stool consistency, as different fibers show different results, such as guar gum, a fiber polysaccharide, which boosts bowel activity and reduces constipation because it is easily fermented by the intestinal microflora (Takahashi, Wako, Okubo, Ishihara, Yamanaka, and Yamamoto 1994). Overall, DFs of all types are found to escalate the rate of bowel motions thereby helping in treating constipation (Hillemeier 1995).

TABLE 15.1

Various Food Sources of Dietary Fiber (per g per 100-g edible portion)

Dietary Fiber	Soluble	Insoluble	Total
Nuts and seeds			
Flaxseed	12.18	10.15	22.33
Sesame seed	1.9	5.89	7.79
Cashew, oil roasted	-	-	6
Peanut, dry roasted	0.5	7.5	8
Coconut, raw	0.5	8.5	9
Almonds	1.1	10.1	11.2
Fruits			
Pear	1	2	3
Peach	0.9	1	1.9
Bananas	0.5	1.2	1.7
Strawberry	0.9	1.3	2.2
Plums	0.9	0.7	1.6
Oranges	1.1	0.7	1.8
Grapes	0.5	0.7	1.2
Watermelon	0.2	0.3	0.5
Pomegranate	0.11	0.49	0.6
Pineapple	0.1	1.1	1.2
Mango	0.74	1.06	1.8
Kiwi	0.8	2.61	3.39
Apple, unpeeled	0.2	1.8	2
Vegetables			
Broccoli, raw	0.29	3	3.29
Carrot, raw	0.2	2.3	2.5
Cauliflower, raw	0.7	1.1	1.8
Cucumbers, peeled	0.1	0.5	0.6
Eggplant	1.3	5.3	6.6
Green onions, raw	0	2.2	2.2
Tomato, raw	0.4	0.8	1.2
Turnips	0.5	1.5	2
Spinach, raw	0.5	2.1	2.6
Ladies finger	1.3	3	4.3
Fenugreek leaves	0.7	4.2	4.9
Beet root	2.4	5.4	7.8
Bitter gourd	3.1	13.5	16.6
Potato, no skin	0.3	1	1.3
Grains			
Wheat germ	1.1	12.9	14
Wheat (whole grain)	2.3	10.2	12.6
Rice(cooked)	0	0.7	0.7
Rice (dry)	0.3	1	1.3
Oats	3.8	6.5	10.3
Corn	-	-	13.4
Barley	-	-	17.3
Legumes & pulses			
White beans, raw	4.3	13.4	17.7
Lima beans, canned	0.4	3.8	4.2
Lentils, raw	1.1	10.3	11.4
Kidney beans, canned	1.6	4.7	6.3
Peas, green frozen	0.3	3.2	3.5
Soy	-	-	15
Green beans	0.5	1.4	1.9

Source: Habashy 2017, Khanum, Swamy, Krishna, Santhanam, and Viswanathan 2000, Spiller 2001.

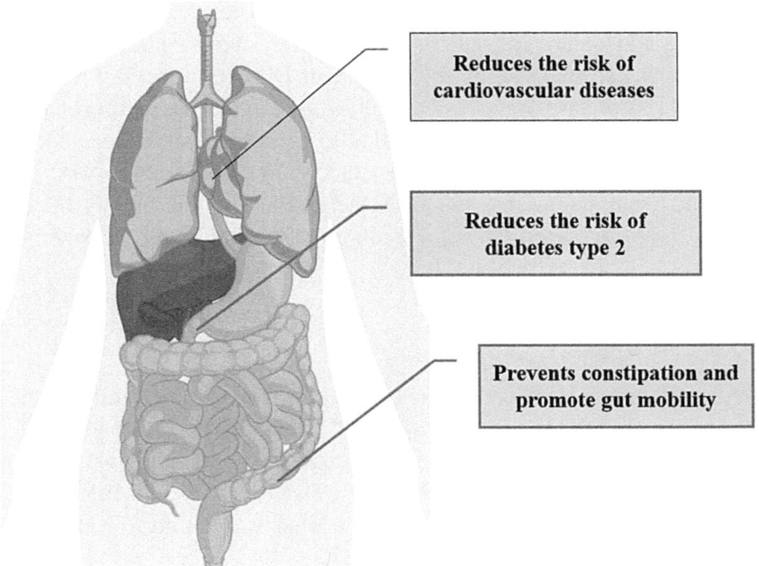

FIGURE 15.2 Role of Dietary Fiber in human health

15.2.2 Diabetes

Diabetes mellitus is a collection of metabolic ailments exhibited by hyperglycemia due to abnormalities in insulin secretion or insulin activity, or both (Association 2014, Diabetes 2006).It is classified into type 1 and type 2. Type 1 diabetes causes the pancreas' cell to secrete little or no insulin resulting in high blood sugar levels, while type 2 diabetes causes an imbalance between sugar levels and insulin levels in the blood (Chaudhury, Duvoor, Reddy Dendi, Kraleti, Chada, Ravilla, Marco, Shekhawat, Montales, and Kuriakose 2017).The primary causes of diabetes in the population include obesity, smoking, and lack of physical activity. Aside from them, nutritional factors such as a high and unregulated simple carbohydrate intake also play a significant role. It has been perceived that there exists a strong correlation between DF intake and diabetes; for instance, when DF intake is increased, the risk of diabetes in older women is considerably dropped(Schulze, Liu, Rimm, Manson, Willett, and Hu 2004).

The DFs form a viscous solution inside the small intestine and minimize the contact and binding of macronutrients with the digestive enzymes, thereby delaying glucose absorption which in turn reduces the postprandial plasma glucose and insulin levels (Bernstein, Titgemeier, Kirkpatrick, Golubic, and Roizen 2013). For example, guar gum dramatically reduces the postprandial blood glucose level thereby significantly reducing the risk of diabetes (Schulze et al. 2004). Also unlike processed grains, consumption of whole grains has displayed reduction in the glycemic index and enhanced insulin sensitivity, thereby plummeting the threat of obesity and type 2 diabetes. It has been highlighted that due to the different activities of various DF types, there is ascope for more research into DF's effect in diabetes control and to establish authentic diabetes management properties of DF.

15.2.3 Cardiovascular diseases

Cardiovascular diseases (CVDs) are a set of illnesses that deleteriously affect the heart and blood vessels. It is one of the chief reasons of death in the world accounting for 31% of all fatalities worldwide(WHO 2017). It has been stated that consumption of DF can decrease the risk of CVD. Threapleton, Greenwood, Evans, Cleghorn, Nykjaer, Woodhead, Cade, Gale, and Burley (2012)reviewed in detail the effect of DF intake on CVDs and found that ingestion of DF had an inverse relationship with the risk of CVD (0.91 risk ratio per 7g/day) and coronary heart diseases (0.91 per 7g/day). Supporting the findings of the above study, Srour, Fezeu, Kesse-Guyot, Allès, Méjean, Andrianasolo, Chazelas, Deschasaux, Hercberg, and Galan (2019), in their recent study, reported that the consumption of ultra-processed food products, especially with lower levels of DF over a median follow-up of 5.2 years, exhibited an increased risk of developing CVD, cerebrovascular disease and coronary heart disease. In a dose-dependent study, Kim and Je (2016) compiled 15 cohort studies and concluded that increasing the DF intake by 10g/day resulted in decreased risk of CVD-related death by 9%.

As comprehended from numerous human-based studies regarding DF, diverse lifestyle and health factors along with increased DF intake can contribute to apparent benefits to mitigate the risk of CVD.

15.2.4 Colo-rectal Cancer

All DFs prevent bowel cancer by lowering the bowel transit time which inhibits the formation and activity of carcinogenic cells (Takahashi et al. 1994). They also possess stool bulking properties that help to decrease the concentration of fecal carcinogens that come in contact with the gut walls, thereby preventing colon cancer(Hill 1974). Graham, Dayal, Swanson, Mittelman, and Wilkinson (1978) stated that the ingestion of vegetables rich in fiber was found to reduce the frequency of occurrence of large bowel cancer.

15.3 ABSORPTION AND METABOLISM OF DIETARY FIBER

Human beings are neither able to absorb nor digest DF as the enzymes present in the gut are unable to process the complex carbohydrates. DF instead helps in the absorption of other carbohydrates and in lipid metabolism in the digestive systems (Holscher 2017). DF is fermented by the gut micro-biome to produce short-chain fatty acids (SCFAs), carbon dioxide, hydrogen, and biomass. This regulates bowel movement while the undigested carbohydrates remain in the small intestine to provide energy through the absorption of SCFAs (Holscher 2017). A portion of by-products of this reaction is exploited by the microbes to produce carbon and energy essential for the growth of the microflora, while the other portion gets eliminated via excreta or rectal gases, although the colonic mucosa absorbs the greater part of the fermented products. Colonic fermentation helps in complete starch degradation along with alcohol-sugar, lactose, and fructans. More than 50% of the fiber intake is processed in the small intestine, while the remaining fibers are excreted (Jenkins, Taylor, Goff, Fielden, Misiewicz, Sarson, Bloom, and Alberti 1981, Cummings and Macfarlane 1991, Roediger 1995). However, purified fibers are reported to decrease the absorption of minerals and vitamins through entrapping and binding the fibers in the lumen of small intestines(Gordon 1990). Consumption of DF also helps in regulating and preventing various metabolic disorders apart from what is mentioned earlier, like glycemia, lipidaemia, and dyslipidemia, although further research is required to establish these facts (Papathanasopoulos and Camilleri 2010). Furthermore, different literature mentions the ability of DF in managing calorie intake, glucose homeostasis and lipid metabolism too (Papathanasopoulos and Camilleri 2010).

15.4 FORTIFICATION OF DIETARY FIBER IN FOOD PRODUCTS

The fortification of DF into different food products enhances their nutritional quality not only by increasing the DF content but also by enriching the food with the associated bioactive compounds such as flavonoids, carotenoids, etc. DF fortification in food matrices also helps to cut down the fat content in the formulation by adding DF as a substitute of fat, without causing any loss of quality (Byrne 1997, Martin 1999).

In a report by Tate & Lylecompiled from existing literatures and published in 2021, it was found that against the WHO's suggestion of minimum 25g fiber intake per day (Figure 15.3), much less

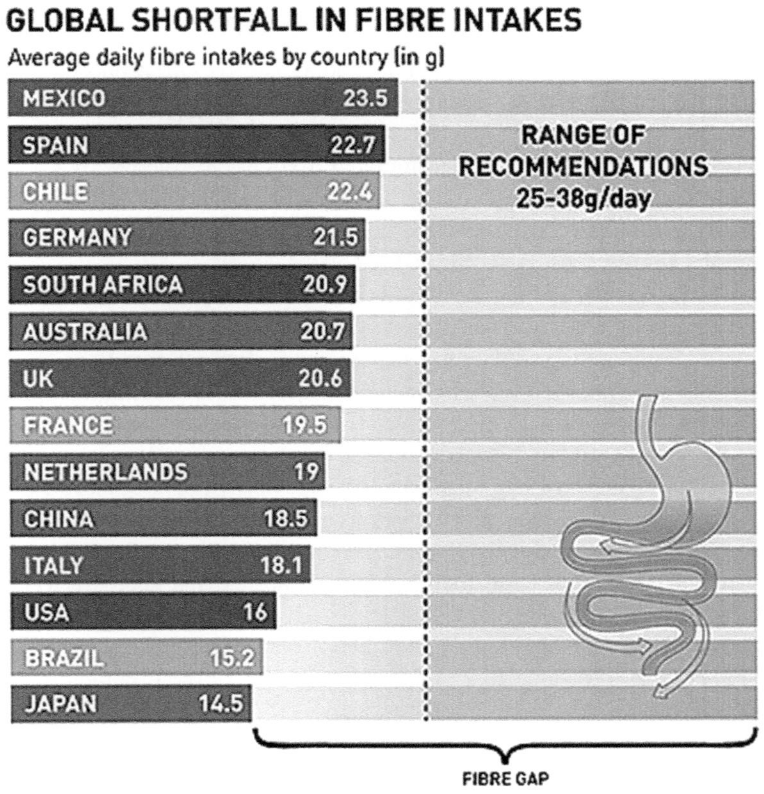

GLOBAL SHORTFALL IN FIBRE INTAKES

Average daily fibre intakes by country (in g)

FIGURE 15.3 Global shortfall in fiber intake as reported by Lyle (2021).

fiber is being consumed worldwide, which highlights the necessity of global DF fortification of commercial food products.

Toma, Orr, D'appolonia, Dlntzis, and Tabekhia (1979) investigated the incorporation of wheat bran and potato peel, both as sources of DF in bread. They found that bread incorporated with peels of potato was better in the contents of total DF and certain minerals such as silicon associated with fiber which contributes to lowering cholesterol and in binding bile acids, along with chromium which facilitates insulin attachment to its peripheral receptor sites, contributing to control of diabetes. Additionally, potato peel has a lesser quantity of starchy constituents than wheat bran and lacks phytate. Nassar, AbdEl-Hamied, and El-Naggar (2008a) reported that the incorporating orange peel and pulp in wheat flour biscuits' recipe increased the DF content of the biscuits from 2.73 to 15.31%. The authors suggested an optimum concentration of 15% orange pulp and peel in the formulation to yield highly acceptable biscuits on the basis of physical, and chemical characteristics, rheological properties, and sensory evaluation. Sharif, Butt, Anjum, and Nawaz (2009) proposed that fiber-rich rice bran supplementation considerably enhanced the DF, protein content, and mineral content of wheat flour cookies. Martinez-Saez, García, Pérez, Rebollo-Hernanz, Mesías, Morales, Martín-Cabrejas, and Del Castillo (2017) added coffee grounds to cookies resulting in increased DF-rich nutritious and flavourful cookies with prospective value in the diet of diabetic patients.

In a study conducted by Potter, Stojceska, and Plunkett (2013), different fiber-rich fruit powders derived from apple, banana, strawberry, and tangerine were incorporated into extruded snack products made of wheat flour, potato starch, corn starch, and milk powder to improve their nutritional profile. The resulting products were low in fat and sugar and were found to be a decent source of DF.

Dhingra et al. (2012) suggested that oat fiber can be incorporated into beverages such as milkshakes, instant breakfast beverages, fruit juices, vegetable based beverages, sports beverages, and cappuccinos and wine. Liquid diet beverages curated for people having special dietary needs and for either diet for weight loss or meal-replacement beverages can benefit from fortification with DF(Hegenbart 1995). Larrauri, Borroto, Perdomo, and Tabares (1995) proposed the formulation of a powdered drink named FIBRALAX, having 25% DF and 66.2% digestible carbohydrates from pineapple peel.

Staffolo, Bertola, and Martino (2004) formulated a yogurt fortified with 1.3% wheat, inulin, bamboo, and apple fibers having increased fiber content with appreciable organoleptic acceptability. Hashim, Khalil, and Afifi (2009a)evaluated the impact of fortification of fresh yogurt with 3% date fiber that produced a similar sensory profile as the control yogurt.

Dhingra et al. (2012) proposed that DFs based on cellulose, soy, pectins, wheat, maize, rice isolates, and beet fiber can be employed for enhancing the textural grain of meat products including sausages, salami, and others to formulate low-fat products, often referred to as 'Dietetic hamburgers'. Verma, Sharma, and Banerjee (2009) combined various fiber sources like pea and gram hull flour, bottle gourd, and apple pulp, in diverse blends to design low fat, low salt, and high fiber rich chicken nuggets. Garcıa, Dominguez, Galvez, Casas, and Selgas (2002) reported that adding 1.5% orange fiber in the formulation of dry fermented sausages provided a sensory profile akin to the standard high fat product. Citrus fiber, which contains associated bioactive compounds such as polyphenols, when incorporated into chicken hamburgers and bologna sausages, showed to effectively inhibit lipid oxidation, thus improving their oxidative stability and shelf life (Fernández-Ginés, Fernández-López, Sayas-Barberá, Sendra, and Pérez-Alvarez 2003, Sáyago-Ayerdi, Brenes, and Goñi 2009). Citrus fiber when added to bologna sausage was found to reduce residual nitrite levels (Fernández-Ginés et al. 2003). Table 15.2 provides a summary of representative studies on the fortification of DF rich sources in different food matrices.

TABLE 15.2

Summary of Representative Studies on Fortification of Dietary Fiber Rich Sources In Different Food Matrices

Food system	Type of DF fortification	Salient findings	Reference
Bread	Wheat bran and potato peel (hand, abrasion, steam, and lye peeled)	• After analysis of both, authors found that potato peel is better than wheat bran in terms of its water holding capacity, absence of phytate, and contents of certain minerals, such as silicon associated with the DF which contributes to lowering cholesterol and binding bile acids, along with chromium which facilitates insulin attachment to its peripheral receptor sites, contributing to control of diabetes. These dietary aspects were also found to be retained in baking quality trials. • Lye peeled potato peel incorporated bread had the best loaf volume and internal and external appearance, resembling whole wheat flour bread; steam (10%) or abrasive peeled potato peel (15%) bread had a characteristic old appearance; hand-peeled potato peel bread had the most appetizing crust color.	Toma et al. (1979)
Wheat flour biscuits	Orange peel and pulp	• DF content of biscuits increased from 2.73% to 15.31%. • At 15% incorporation, orange pulp and peel resulted in highly acceptable biscuits on the basis of physical, chemical characteristics, rheological properties, and sensory characteristics.	Nassar, Abdel-Hamied, and El-Naggar (2008b)

(Continued)

TABLE 15.2 (CONTINUED)

Summary of Representative Studies on Fortification of Dietary Fiber Rich Sources In Different Food Matrices

Food system	Type of DF fortification	Salient findings	Reference
Cookies	Wheat bran stabilized with either hot-air oven, microwave, or autoclave (SWB)	• All SWB added cookies had a darker color than the control • 30% SWB increased ash content approximately from 1.5 to 2.4%, and protein content from 18 to 19.3%, respectively. • The minimum phytic acid content (758.92 mg/100 g) was attained with autoclave-stabilized wheat bran formulated cookies. • Microwave treated 30% SWB fortification achieved superior mineral content and hot-air oven 10% SWB fortification yielded the best sensory properties.	Ertaş (2015)
Cake	Orange peel powder	• With an increase in powder concentration, DF content increased while fat and protein content decreased. • Up to 10% substitution by orange peel powder can give the best overall sensory attributes without any adverse effect.	Zaker, Sawate, Patil, Sadawarte, and Kshirsagar (2017)
Dried noodles	Passionfruit mesocarp flour (PFMF)	• Total DF content was greater than 6% over control in noodles with 9% PFMF. • PFMF inhibited gluten network formation, which impaired the cooking quality and consumer acceptance of the dried noodles.	Ning, Zhou, Wang, Zheng, Pan, Chen, Liu, Du, Cao, and Wang (2022)
Brown rice pasta	Unripe banana flour (UBF), Defatted soy flour (DSF)	• Hardness, and chewiness values decreased, and other sensory attributes improved with increasing UBF substitution, which was more distinct at more than 30% fortification. • Fiber along with protein content, is enhanced with increasing UBF replacement. • Based on, sensory, texture, color, and proximate analysis, gluten-free brown rice pasta replaced with 30% UBF and 10% DSF was considered the optimum.	Udachan, Gatade, Ranveer, Lokhande, Mote, and Sahoo (2022)
Extruded corn-snacks	Inulin (oligofructose) -type fructans (ITFS)	• Adding 13.3% oligofructose-rich inulin led to a snack having 4 g of ITFs/30 g serving size. • This had no effect on expansion but decreased cutting force and amplified fiber level seven times over the control. • ITFs-formulated snacks were organoleptically acceptable. • 13.3% ITFs enrichment either prior to or after extrusion resulted in the same reduction of glycemic index from 81 to 71 and that of glycemic load from 21 to 14, i.e. medium glycemic load.	Capriles, Conti-Silva, and Gomes Arêas (2021)
Orange juice	Citrus fibre having 29.3% soluble fiber and 41.9% insoluble fibre	• Fibre-enriched (1.4 g/100 ml) orange juice diminished postprandial serum glucose and circulating insulin levels over the placebo at 15 min after its intake. • Significant result on feelings of satiety and fullness was identified at 15 and 120 min, which were complemented with a decline in GLP1 (gut hormone- plasma glucagon-like peptide 1) response at 15 min.	Bosch-Sierra, Marqués-Cardete, Gurrea-Martínez, Valle, Talens, Alvarez-Sabatel, Bald, Morillas, Hernández-Mijares, and Bañuls (2019)

(Continued)

TABLE 15.2 (CONTINUED)

Summary of Representative Studies on Fortification of Dietary Fiber Rich Sources In Different Food Matrices

Food system	Type of DF fortification	Salient findings	Reference
Yogurt	Date fibre	• Up to 3% date fiber fortification, sweetness, sourness, smoothness, firmness, and overall acceptance values are similar to the control, which decreased at higher fortification levels by 4.5%	Hashim, Khalil, and Afifi (2009b)
Chicken nuggets	Pomegranate seed powder and tomato powder	• Both seed powder significantly amplified the fiber content of the chicken nuggets and improved their sensory quality. • Pomegranate seed powder had a significant effect on the proximate composition, pH, cooking yield, and emulsion stability of the nuggets. Tomato powder also impacted the emulsion stability, fat, and moisture content of the products, but not the cooking yield and pH	Kaur, Kumar, and Bhat (2015)

15.5 BIOACTIVE PEPTIDES (BAP)

15.5.1 BAP AND THEIR SOURCES

Peptides are short polymers of amino acids ($<$ 50 amino acid residues, and molecular weight $<$ 6000 Da) linked together by peptide bonds (Dietzen 2018, Chakrabarti, Jahandideh, and Wu 2014). One or more polypeptide subunits combine to form a protein molecule having high molecular weight. Many plant-origin or animal-based proteins are the sources of a large number of BAP(Sánchez and Vázquez 2017, Chakrabarti, Jahandideh, and Wu 2014). BAP are specific protein fragments that are functionally inactive or encrypted in the polypeptide sequence of their native protein but are released during proteolysis through enzymatic hydrolysis (*in vivo* or gastrointestinal digestion) or food processing (*ex vivo*) such as bacterial fermentation (Bhat, Kumar, and Bhat 2015, Chakrabarti, Jahandideh, and Wu 2014). Production of BAP includes extraction of protein, subsequently peptide release by hydrolysis; purification of peptides; and identification of their bioactivities(Das, Tiwari, and Garcia-Vaquero 2022).

A large number of the bioactive peptide has been isolated and identified. A database named 'Biopep' reports more than 1500 BAP (Singh, Vij, and Hati 2014). Among others, animal origin BAP from fermented dairy products (Choi, Sabikhi, Hassan, and Anand 2012), cheese(Pritchard, Phillips, and Kailasapathy 2010), and bovine milk (Mohanty, Jena, Choudhury, Pattnaik, Mohapatra, and Saini 2016) are the greatest source. However, BAP can also be obtained from other animal sources such as eggs, meat, gelatine (Lassoued, Mora, Barkia, Aristoy, Nasri, and Toldrá 2015), bovine blood (Przybylski, Firdaous, Châtaigné, Dhulster, and Nedjar 2016), and from numerous fish species such as salmon, herring, sardine, and tuna. Some vegetable sources of BAP include amaranth (Silva-Sánchez, De La Rosa, León-Galván, De Lumen, de León-Rodríguez, and De Mejía 2008), sorghum, pumpkin, mushrooms (Möller, Scholz-Ahrens, Roos, and Schrezenmeir 2008), rice (Selamassakul, Laohakunjit, Kerdchoechuen, and Ratanakhanokchai 2016), soy, maize (Singh, Vij, and Hati 2014), and wheat (ÜNAL and ŞENER 2018).

BAP are also known to be biologically active regulators and can avert fat oxidation and microbial spoilage in food (Sánchez and Vázquez 2017). They show drug-like or hormonal activities and can be characterized as antioxidative, mineral binding, immunomodulatory, opioid, anti-hypertensive, antithrombotic, and antimicrobial components. Presently, BAP are being synthesized to upgrade

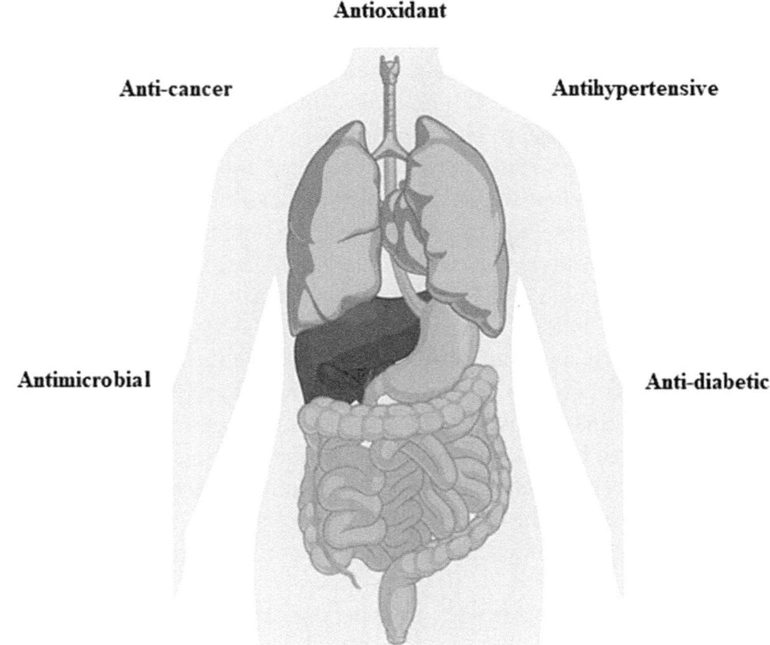

FIGURE 15.4   Role of bioactive peptides in human health.

human health by mitigating medical issues such as osteoporosis, diabetes, obesity, cancer, hypertension, stroke, and coronary heart disease (Boelsma and Kloek 2008).

15.5.2   ROLE OF BIOACTIVE PEPTIDES IN HUMAN HEALTH

BAPs are one of the most promising and healthy food additives that positively affect the nervous, cardiovascular, digestive, and immune system of the body. Furthermore, they also exhibit anticancer, anti-obesity, antioxidant, and anti-inflammatory activity (Fig. 15.4).

15.5.3   ANTI-HYPERTENSIVE

Hypertension is one of the major causes resulting in cardiovascular diseases like stroke, heart failure, myocardial infarction, and coronary heart disease. Numerous synthetic anti-hypertensive drugs available in the market have adverse effects on the body. Therefore, food-derived BAP are preferred for their anti-hypertensive activity primarily due to their tissue affinity and quick removal from the body as compared to synthetic drugs. Different anti-hypertensive BAP are studied extensively and are obtained from different sources, which help in reducing hypertension by impeding the action of angiotensin-I converting enzyme (ACE).The ACE along with renin are the primary enzymes in the renin-angiotensin system that controls blood pressure (Liu, Liu, Lu, Chen, and Zhang 2013). Other methods of lowering hypertension include the release of endothelin-1 from endothelial cells (Maes, Van Camp, Vermeirssen, Hemeryck, Ketelslegers, Schrezenmeir, Van Oostveldt, and Huyghebaert 2004), improvement of nitric oxides production derived from endothelium cells (Sipola, Finckenberg, Korpela, Vapaatalo, and Nurminen 2002) and improving the vasodilation activity to bind with the opioid receptors(Nurminen, Sipola, Kaarto, Pihlanto-Leppälä, Piilola, Korpela, Tossavainen, Korhonen, and Vapaatalo 2000).

BAPs derived from soybean are known for their anti-hypertensive action. The enzyme hydrolysates of soy proteins contain numerous ACE inhibitory peptides. Asian style fermented soybean-based food products like soy sauce(Okamoto, Hanagata, Matsumoto, Kawamura, Koizumi, and Yanagida 1995), soybean paste(Shin, Yu, Park, Chung, Ahn, Nam, Kim, and Lee 2001), and tempeh(Gibbs, Zougman, Masse, and Mulligan 2004) contain a large number of ACE inhibitory peptides(Kuba, Tanaka, Tawata, Takeda, and Yasuda 2003).*Bacillus subtilis* and *Bacillusnatto,* when used to ferment soybean, produces two ACE inhibitory peptides (YVWK and VAHINVGZK)(Singh, Vij, and Hati 2014). Additionally, fermented soybean produced more ACE inhibitory peptides like SY and GY as compared to normal soy sauce, which reduced the hypertension of rats by reducing the serum aldosterone quantity and by subduing the angiotensin-renin mechanism(Nakahara, Sano, Yamaguchi, Sugimoto, Chikata, Kinoshita, and Uchida 2010, Nakahara, Sugimoto, Sano, Yamaguchi, Katayama, and Uchida 2011). Other anti-hypertensive peptides obtained from soy proteins include YVVFK, LLF, TPRVF, PGTAVFK, IVF, IPPGVPYWT, LNF, LSW, PNNKPFQ, NWGPLV, and LEF(Singh, Vij, and Hati 2014). The variation in the ACE inhibition among hydrolysate fractions could be attributed to the amino acid composition, their sequence, and the peptide size(Paiva, Lima, Neto, and Baptista 2017). Superior anti-hypertensive action is shown by low molecular weight peptides and branched amino acids. Valine and isoleucine having branched structures exhibit ACE inhibition (Nakamura, Yamamoto, Sakai, Okubo, Yamazaki, and Takano 1995). Moreover, proline, when present in the C-terminal, displays enhanced ACE inhibition and therefore, hydrolysates containing proline in the peptide chain is useful in producing anti-hypertensive peptides (Nakamura et al. 1995).

Milk protein is used to extract anti-hypertensive peptides by fermenting milk with different lactic acid bacteria or by using their proteinases(Hayes, Ross, Fitzgerald, and Stanton 2007). Different strains of lactic acid bacteria produce different ACE inhibitory peptides, and their addition to food products improves the product's nutritional value. *L. helvelticus* is used to release ACE inhibitory enzymes like AHKAL, APLRV, IPAVF, AQSAP from milk whey proteins that show strong anti-hypertensive action; *L. brevis* is required for releasing an effective ACE inhibitory peptide known as AEKTK (Ahn, Park, Atwal, Gibbs, and Lee 2009).LQKW and LLF are two potent peptides that are extracted from β-Lactoglobulin hydrolysate, which were used to treat hypertension in hypertensive rats (Hernández-Ledesma, Miguel, Amigo, Aleixandre, and Recio 2007).BAPs like Ile-Pro-Pro, Tyr-Pro, and Val-Pro-Pro extracted from whey protein displayed ACE inhibition and reduced blood pressure in hypertensive rats (Seppo, Jauhiainen, Poussa, and Korpela 2003). A handful of microbe-fermented milk products are available in the market such as under the commercial name Evolus, DAnaten, Calpis, and Amelia for controlling blood pressure(Ricci-Cabello, Olalla Herrera, and Artacho 2012).

15.5.4 Anti-cancer

Synthetic anti-cancer drugs used for cancer treatment, available in the markets, generally cause various side effects and are found to be neurotoxic, nephrotoxic, gonadotoxic, and cardiotoxic in nature. This has resulted in the search for food-derived peptides having anti-cancer properties. Studies show that food-derived peptides, for instance, soy proteins have anti-cancer properties which diminish the risk of colon, mammary, and prostate cancers(Singh, Vij, and Hati 2014). Lunasin is a peptide obtained from the cotyledon of soybean, that exhibits anticancer activity(Singh, Vij, and Hati 2014). Lunasin peptide prevents the transformation of regular cells by oncogenes like E1A, RAS, and carcinogens(Lam, Galvez, and de Lumen 2003). Despite the cancer-preventive actions of Lunasin, it is unable to disturb the progress of normal or confirmed cancer cell lines(Lam, Galvez, and de Lumen 2003). Bow-birk inhibitor (BBI) is a low molecular weight peptide that is also used as anti-cancer peptide and has the ability to inhibit the action of chymotrypsin and trypsin (Losso 2008). BBI shows anti-cancer activity in humans and other species in the tissues of the colon, esophagus, prostate, liver, and breast; FDA approved BBI as an investigational drug in 1992(Losso 2008). Lunasin and BBI are the two primary and most effective soy peptides showing anticancer properties belonging to the 2S fraction of soybean proteins (Singh, Vij, and Hati 2014, Wang, Dia, Vasconez,

De Mejia, and Nelson 2008, Chatterjee, Gleddie, and Xiao 2018). BBI peptide also protects the Lunasin from gastrointestinal degradation when the protein is taken orally(Chatterjee, Gleddie, and Xiao 2018).Other soybean protein isolated peptides having a strong anti-proliferative effect include LSGNK, MTEEY, GLTSK, GEGSGA (Vital, De Mejía, Dia, and Loarca-Piña 2014). Soy protein peptides exhibited anti-cancer properties via several mechanisms which include enhanced differentiation of mammary glands, gene regulation of signal transducing pathways underlying tumor promotion, initiation and progression and reducing the activation of pro-carcinogens to carcinogens (Thomas, DeMarco, and Pope 2005).

Other sources of anti-cancer peptides include HVLSRAPR, extracted from the hydrolysate of *S. platensis* which displayed an enhanced inhibitory action against the escalation of cancer cells HT29 while it does not show any adverse action against liver cells(Wang and Zhang 2017). WPP a tripeptide extracted from blood clam muscle, had a strong anticancer effect against HeLa, H-1299, DU-145, and PC-3 cell lines(Chi, Hu, Wang, Li, and Ding 2015). Peptides isolated from Tuna cooking juice such as KLPPLLLAKLLMSGKLLAEPCTGR and KPEGMDPPLSEPEDRRDGAAGPK had the ability to reduce the proliferation of MCF-7 breast cancer cells (Hung, Yang, Kuo, and Hsu 2014). QPK and LANAK peptides were isolated from sepia ink protein and oyster hydrolysate, respectively. The former showed anticancer activity against PC-3, DU-145, and LNCaP cell (Huang, Yang, Yu, Wang, Li, and Ding 2012) while the latter showed action against HT-29 cells (Umayaparvathi, Meenakshi, Vimalraj, Arumugam, Sivagami, and Balasubramanian 2014). Other peptides like RQSHFANAQP isolated from chickpea(Xue, Wen, Zhai, Yu, Li, Yu, Cheng, Wang, and Kou 2015) had an anti-proliferative action against breast cancer cell; YALPAH from *Setipinnatatyca* used apoptosis in prostate PC-3 cancer cells (Song, Wei, Luo, and Yang 2014).Furthermore, peptides extracted from *Dendrobium catenatum* such as KPEEVGGAGDRWTC, RCGVNAFLPKSYLVH FGWKLLFHFD, and RHPFDGPLLPPGD are reported to reduce the proliferation of MCF-7 and HepG-2 cancel cells (Zheng, Qiu, Liu, Zhang, Cai, and Zhang 2015). Similarly, rapeseed isolated peptides also reported anti-proliferative action against liver and breast cancer cells (Xie, Wang, Zhang, Chen, Wu, and Wang 2015).

15.5.5 ANTI-DIABETIC

Diabetes mellitus is a type of metabolic disorder that causes high blood sugar levels as explained earlier. Anti-diabetic drugs currently present in the market have been reported to show a number of side effects including weight gain, pancreatitis, hypoglycemia, and gastrointestinal problems (Daliri, Oh, and Lee 2017). These side effects have led to an increased focus on the research for isolating safer alternatives, one such being the isolation of peptides possessing anti-diabetic properties. Peptides obtained from fermented food products show more anti-diabetic activity as compared to normal food products. A fermented soybean food product called 'meju' has certain peptides which are used for inducing insulin-stimulated glucose uptake by 3T3-L1 cells(Kwon, Hong, Ahn, Kim, Yang, and Park 2011). Soy proteins and isoflavones were found to be associated with a reduction in serum glucose level, fasting plasma glucose, and an increase in insulin secretion in diabetic rodents(Lu, Wang, Song, Chibbar, Wang, Wu, and Meng 2008, Nordentoft, Jeppesen, Hong, Abudula, and Hermansen 2008). Furthermore, fermented soybean products like chungkookjang and natto are also thought to prevent the inception of type 2 diabetes in human and mouse models (Chatterjee, Gleddie, and Xiao 2018).

Reports have shown that peptides such as IAVPGEVA, IAVPTGA, and LPYP derived from soy glycinin hydrolysis, demonstrated health benefits in the area of prevention of hypocholesterolaemia (Lammi, Zanoni, and Arnoldi 2015). IAVPGEVA efficiently inhibited the dipeptidyl peptidase IV (DPP-IV), which is an essential enzyme required for glucose homeostasis. YVVNPDNNEN and YVVNPDNEN like IAVPTGA are also useful in controlling blood sugar but cannot inhibit DPP-IV due to the longer peptide chain and absence of P in the fourth terminal position (Lammi, Zanoni, Arnoldi, and Vistoli 2016).

Amino acids like Leu, Thr, Val, Phe, Ile, and some peptides assist in insulin secretion and have a similar hypoglycaemic activity to that of some milk proteins. It is also observed that hydrolyzed milk proteins have higher anti-diabetic activity than unhydrolyzed milk proteins(Nongonierma and FitzGerald 2015).Enzymatic hydrolysis of camel milk peptides inhibition was found against DPP-IV, porcine pancreatic lipase (PPL), and porcine pancreatic alpha-amylase (PPA). Similarly, peptides like MPSKPPL and KDLWDDFKGL foundin B_9 and A_9 respectively were found to be most effective against PPA inhibition, while DNLMPQFM and WNWGWLLWQL found in A_9inhibited DPP-IV as they possess maximum DPP-IV binding sites (Mudgil, Kamal, Yuen, and Maqsood 2018).

Hydrolysates from black bean protein possess peptides(LSVSVL, FEELN, ATNPLF, AKSPLF) that showed potential anti-diabetic properties by blocking glucose transporter 2 and sodium-dependent glucose cotransporter 1(Mojica, de Mejia, Granados-Silvestre, and Menjivar 2017). Also, peptides from salmon fish enhanced the glucose absorption in L6 skeletal muscle cells, and thereby could control the blood sugar levels preventing the risk of type 2 diabetes(Roblet, Akhtar, Mikhaylin, Pilon, Gill, Marette, and Bazinet 2016).In another study, hydrolysate generated from salmon co-products (skin and trimmings) with Alcalase 2.4L and Flavourzyme 500Lexhibited high insulin and GLP-1 secretion, and DPP-IV inhibition(Harnedy, Parthsarathy, McLaughlin, O'Keeffe, Allsopp, McSorley, O'Harte, and FitzGerald 2018). Microbes have also been used to synthesize anti-diabetic peptides. For example, *Aspergillus awamori* is a microbe used to prepare fermented food products like shochu and awamori containing serine, thryosine, valine, and threonine amino acids, whose peptide sequence inhibited the activity of alpha glycosidase thereby controlling blood sugar level(Singh and Kaur 2016).

15.5.6 ANTIMICROBIAL

Peptides are reported to exhibit antimicrobial activity against different bacteria, virus, and yeasts and can be used as an alternative to chemical antimicrobial agents. Antimicrobial peptides (AMP) generally exhibit strong antibacterial efficacy and are preferred over standard antibiotics because of the lack of resistance developed by the pathogenic microbe against the AMP. This activity is credited to the occurrence of hydrophobic and hydrophilic amino acids at the terminal ends of the peptides which is considered the primary reason for the action of AMP(Pane, Durante, Crescenzi, Cafaro, Pizzo, Varcamonti, Zanfardino, Izzo, Di Donato, and Notomista 2017). The mechanism of microbial inactivation by AMP is contributed to the pores' development on the cell membrane which leads to the leakage of intracellular components or by the interaction between the microbial cell organelles and the AMP (Pane et al. 2017, Liu, Sameen, Ahmed, Dai, and Qin 2021). Some examples of naturally produced AMP include nisin, pediocin LA-1, lacticin 3147,etc. These AMPs show strong antimicrobial activity and thus used as preservatives (Santos, Sousa, Otoni, Moraes, Souza, Medeiros, Espitia, Pires, Coimbra, and Soares 2018).The AMPs can also penetrate the microbial cells and interact with their RNA/DNA and obstruct the metabolism of the microbe, thereby leading to cell death (Liu et al. 2021).

Peptides hydrolyzed from different protein sources also exhibit microbial inactivation. For example, *Scorpaenanotata* viscera protein, when hydrolyzed, yielded FPIGMGHGSRPA peptide, which showed remarkable antimicrobial properties against *B. cereus, B. subtitlis, Salmonella,* and *E. coli* (Tang, Zhang, Wang, Qian, and Qi 2015). AMPs are also present in bovine blood; VNFKLLSHSLLVTLASHL, a peptide extracted from bovine hemoglobin, reduced the growth of *C. albicans, E. coli, S. aureus*(Aissaoui, Chobert, Haertlé, Marzouki, and Abidi 2017). Milk proteins are also rich in antimicrobial peptides, which can help with the problem of the development of drug resistance in bacteria. Lactoferrin, a whey protein, is able to exhibit antimicrobial action by membrane disruption, penetration into dendritic cells, and prebiotic activities, eventually inhibiting bacterial growth(Morris and FitzGerald 2008, Hartmann and Meisel 2007, Brouwer, Rahman, and Welling 2011).Fermented milk, when consumed, is reported to inhibit *H. pylori* growth in humans, although other factors like the presence and action of probiotic organisms are also responsible for the inhibition (Sachdeva, Rawat, and Nagpal 2014). This inhibition is partially attributed to the cell wall disruption of bacteria by lactoferrin(Sachdeva, Rawat, and Nagpal 2014).Different peptides have different mechanisms for microbial inactivation; however, the general mechanism associated with differential behaviors of peptides regarding microbial inhibition is due to their difference in net charge, amino acid sequence, and charge distribution(Brandelli, Daroit, and Corrêa 2015).

15.5.7 ANTIOXIDANT

Antioxidants play an important role to protect the damage of a target molecule caused by oxidative stress. BAP also exhibits antioxidant activities and prevents oxidative stress inhuman bodies. Reactive oxygen species (ROS) along with reactive nitrogen species found in the human body are highly active molecules. A greater quantity of such molecules can be detrimental to the human body as they cause cell death and high oxidative stress(Pihlanto 2006). The role of antioxidant activity is mainly based on inhibiting lipid peroxidation, radical scavenging activity, and metal ion chelation by the peptides(Qian, Jung, and Kim 2008, Moure, Domínguez, and Parajó 2006).

Peptides obtained from different protein sources like soy and milk can be used to synthesize antioxidant peptides. Soy protein isolate was used to prepare antioxidative peptides by hydrolyzing with food-grade pancreatic trypsin/chymotrypsin having antioxidant activity of 113mg TEAC(Trolox Equivalent Antioxidant Capacity)/g (Beermann, Euler, Herzberg, and Stahl 2009). Soy proteins with/without isoflavones are reported to decrease the oxidants present in the body by hindering the nuclear factor-kappa B in an oxidative stress-inducible rat model, hyperlipidaemic rat model, a human suffering from final stage renal disease, and an old women above 70 years(Draganidis, Karagounis, Athanailidis, Chatzinikolaou, Jamurtas, and Fatouros 2016).Additionally, lunasin, an anti-cancer peptide, is reported to inhibit ROS production in mouse RAW 264.7 macrophages(Hernández-Ledesma, Hsieh, and Ben 2009). In a study, 50 g of soy protein when consumed by healthy men facilitated an increase in lunasin absorption by 4.5% (Cam and de Mejia 2012).This increased uptake of lunasin can prevent oxidative stress and the formation of ROS in human bodies.

Consumption of milk proteins is also expected to reduce oxidative stress in humans due to the presence of antioxidant peptides(Pihlanto 2006). According to Peng, Kong, Xia, and Liu (2010), whey protein isolate, when hydrolyzed by Alcalase enhanced the antioxidant capacity of liposomal systems. They also reported that whey protein hydrolysates (WPH) prepared by Corolase showed improved antioxidative property compared to WPH prepared using Alcalase or Flavourzyme (Mann, Kumari, Kumar, Sharma, Prajapati, Mahboob, and Athira 2015).Hernández-Ledesma, Miralles, Amigo, Ramos, and Recio (2005)in a study reported that hydrolysates obtained from α-Lactaglobulin and β-Lactaglobulin proteins exhibited good radical scavenging properties. Another study by Akan (2020) reported that donkey milk derived-peptides released from the gastrointestinal casein showed great antioxidant activity according to 2,2'-azino-bis(3-ethylbenzothi azoline-6-sulfonic acid) (ABTS), Cupric Reducing Antioxidant Capacity (CUPRAC) assays.

15.6 ABSORPTION AND METABOLISM OF BIOACTIVE PEPTIDES

Protein digestion takes place generally in the stomach where pepsinogen A and pepsinogen C gets activated as pepsin and gastricin by HCl. These enzymes break down proteins into small peptides chains. Additionally, small-sized proteins get digested by elastases released with pancreatic fluids(Sauer and Merchant 2018). The presence of different proteases in the stomach has a vital role in peptide absorption (Wang, Liu, Xie, and Tan 2017). The maximum quantity of absorption occurs in the jejunum (small intestine), where 90% of digested products including peptides are absorbed into the lymphatic circulation or bloodstream. The mucous membrane, which lines the lumen of the digestive tract acts as an important bio-chemical and physiological barrier to absorption of peptide in the digestive system and is credited to most of the absorption and secretion in the gastrointestinal tract(Wang et al. 2017).The mechanism of absorption is either through active transport or diffusion(Segura-Campos, Chel-Guerrero, Betancur-Ancona, and Hernandez-Escalante 2011). Peptide transporters (PeT1) located in the intestinal epithelium play a noteworthy role in peptide absorption. The rate and movement of peptide absorption by transporter PeT1 rely on the membrane potential and protein gradient(Adibi 1997).Moreover, PeT1 are responsible for the absorption of smaller peptides while the oligopeptides are absorbed through tight junction and membrane epithelia through a passive transport system (Udenigwe and Aluko 2012).

Miner-Williams, Stevens, and Moughan (2014), in their study, reported numerous mechanisms of absorption of peptides in the intestine. This includes i) passive transport through the intracellular junction or paracellular pathways; ii) passive diffusion through the enterocytes cell with the ability to transport peptide load or transcellular pathways, and iii) carrier-mediated transport pathways. Paracellular pathways transport the molecules through channels/pores filled with water. Hydrophilic peptides with low molecular weight prefer this method of transport. Furthermore, peptides having a molecular weight greater than 700 Da display poor bioavailability, although polypeptides could easily penetrate the tight junction due to high conformational flexibility(Wang, Xie, and Li 2019). In transcellular pathways, peptide diffusion takes place through basolateral and apical membranes. Lipophilic substances generally prefer this transport method as the cell membrane comprises a lipid bilayer that suits the absorption of lipophilic peptides over hydrophilic peptides (Renukuntla, Vadlapudi, Patel, Boddu, and Mitra 2013). In the transcellular passive absorption pathway, it is necessary to remove the peptide-water H-bondso as to facilitate the peptide absorption(Conradi, Hilgers, Ho, and Burton 1991). Carrier mediated pathway is a type of active transport where transporters are responsible for the movement of the molecules(Miner-Williams, Stevens, and Moughan 2014).

The metabolism of BAP starts in the stomach and carries on till the luminal phase in the small intestine. The digestion of BAP releases oligopeptides and proteins of which the former constitute a major part. Peptide metabolism can also take place in the intestinal brush border, which is particularly rich in amino acid activity. This is because of the activity of carboxypeptidase and aminopeptidase, which produces a mixture consisting mainly of free amino acids along with dipeptides and tripeptides(Yamada, Alpers, Kalloo, Kaplowitz, Owyang, and Powell 2011). BAP also helps in assisting the metabolic process in humans. Peptides help in carbohydrate metabolism owing to the incidence of a huge number of amino acids in their structure. It further helps in inhibiting enzymes resulting in metabolic syndrome (Jakubczyk, Karaś, Rybczyńska-Tkaczyk, Zielińska, and Zieliński 2020). It helps in the management of the manifestation of different metabolic disorders associated with lipid metabolism, glucose metabolism alongside obesity, and high blood pressure, as mentioned earlier.

15.7　FORTIFICATION OF BIOACTIVE PEPTIDES IN FOOD PRODUCTS

The various physiological functions demonstrated by the BAP make them suitable as functional food ingredients. The impact of different food matrices and storing conditions at 4 °C for 100 days on the antioxidant and antihypertensive capacity of hydrolysate from Alcalase hydrolyzed sardine proteins was evaluated after addition to commercial-grade apple juice and fish soup(Rivero-Pino, Espejo-Carpio, and Guadix 2020).ACE inhibitory and DPPH radical scavenging activity left after simulated gastrointestinal digestion were studied to substantiate its prospective for oral administration. The presence of reducing sugar in the food matrix led to the production of Maillard browning reaction end products, increasing the antioxidant and the antihypertensive activity up to 80 and 65% respectively in the juice of apple. After 100 days storage, it was depicted that enhanced bioactivity in both apple juice and fish soup was conserved. Simulated digestion processes were not found to negatively affect the bioactivity of these hydrolysates (Rivero-Pino, Espejo-Carpio, and Guadix 2020).Blue whiting fish hydrolysate powders produced using six different enzymes – Alcalase, Protamex, Savinase, Papain, Flavourzyme, Neutrase were incorporated in a vitamin-tea beverage matrix. The powders showed high>80% solubility over a wide range of pH in water, and their solubility enhanced(>85%) in the vitamin-tea beverage (Egerton, Culloty, Whooley, Stanton, and Ross 2018).Galactose-gelatin hydrolysate prepared from the skin of unicorn leatherjacket was utilized for instant coffee brew fortification, enhancing the antioxidant content of the brew without affecting the sensory characteristics (Karnjanapratum and Benjakul 2017). Chuaychan, Benjakul, and Sae-Leaw (2017) have studied the impact of different temperatures employed in spray drying on the characteristics of gelatin hydrolysate powder produced from spotted golden goatfish scales in apple juice fortification.   Lim, Lee, Park, Yoon, and Paik (2011) reported that a commercial yogurt prepared by fortification with various whey hydrolysates showed ACE-inhibitory activity. Its pH, titratable acidity, color and other sensory characteristics, and viable cell count were quite similar to the control yogurt samples during storage. The influence of pulsed electric field (15–40 kV/cm; 0–700 µs) along with that of thermal treatment (84 °C and 95 °C, 15–120 s) were evaluated on a milk mixed beverage and orange juice fortified with both water-soluble vitamins and angiotensin-I-converting enzyme (ACE) inhibitory peptides. The results depicted that peptides and vitamins remained stable after the PEF treatment and throughout storing period at 4 °C for 81 days(Rivas, Rodrigo, Company, Sampedro, and Rodrigo 2007).

Gilani, Xiao, and Lee (2008) reported that hydrolysates of casein comprising the dodecapeptide, FFVAPFPEVFGK, have been manufactured by a Dutch company, DMV, Netherlands, with the name C12 peption (Hartmann and Meisel 2007). A whey protein hydrolysate formulation constituting hypotensive peptides has also been commercially developed and patented in the United States (Henle, Deussen, and Martin 2014). Several fermented milk beverages containing ACE-inhibitory peptides (VPP, TTMPLW, RY) have been commercialized in Spain (Hernández-Ledesma et al. 2005).

15.8 CHALLENGES IN DIETARY FIBER AND PEPTIDE FORTIFICATION

Increasing demand in the market for functional food products has been prompted because of their several health benefits. Diverse functional foods with desired health benefits are being manufactured that can be used alternatively or complementarily to dietary supplements. Fortification is one of the techniques used to develop value added functional foods. Fortification is the process of the addition of nutrients to the food regardless of whether or not the nutrients were originally present in the food. To formulate any fortified food product, physicochemical and biological properties and their effects on the organoleptic properties are considered crucial. Fortification can impact the composition, texture, and sensory properties of the original product, thereby negatively affecting the shelf life of the final food product(Shegelman, Vasilev, Shtykov, Sukhanov, Galaktionov, and Kuznetsov 2019). BAPs/DF have constraints in being incorporated directly into the food system as they can react with different components in the product and are prone to degradation, causing a loss of bioactivity of both peptides and DF(Augustin and Sanguansri 2015). Therefore, for better fortification of food product using bioactive peptide/DF, proper delivery systems are required that is categorically developed for a target food product that will preserve the bioactivity of the functional agents(Augustin and Sanguansri 2012). Furthermore, the interaction between the food material, environment, and the fortification agents should be kept minimum to avoid undesirable reactions.

Fortification of food products with DF or BAP imparts several other challenges, for instance, the selection of an effective dosage required specific health benefits without altering the taste, flavor, or aroma of the original product. Another important issue is the preservation of bioactive compounds during food processing as it degrades at high temperatures. Scale-up manufacturing of BAP is hindered due to downstream bioprocessing challenges relating to bio-molecular purification(Lemes, Sala, Ores, Braga, Egea, and Fernandes 2016). Additionally, due to the incorporation of BAP/DF into the food matrix, the production cost increases, making them difficult to be made available to poor people who are most susceptible to nutrient deficiency (Spence 2006, Dwyer, Wiemer, Dary, Keen, King, Miller, Philbert, Tarasuk, Taylor, and Gaine 2015, Osendarp, Martinez, Garrett, Neufeld, De-Regil, Vossenaar, and Darnton-Hill 2018).

Other challenges include the stabilization of the fortification agents during the extended storage period, validation of proposed health benefits after consumption, and finding out the exact amount of nutrients present in the fortified product as excess consumption of nutrients might result in toxicity. Maintaining safe and effective levels of BAP/DF and also tracking its efficacy while in storage is also another vital challenge in food fortification. It is of utmost necessity to develop precise standardized approaches for characterizing and quantifying the BAP/DF and to implement dose-response studies in animals and clinical trials to evaluate the safety, possible intolerances, and potential allergenicity and efficacy of these bioactive BAP/DF. Data collected from all these dose-response investigations can be utilized to determine the safe levels of consumption of bioactive peptides/DF in different matrices and form the foundation for making potential health claims. All these are required by regulatory agencies for creating suitable policies and regulations on the fortification of BAP/DF into different food matrices and for scientifically substantiating associated health claims (Calbet and MacLean 2002). Incorporation of BAP or DF into food products at different manufacturing plants of the same company may lead to different levels of peptide/DF for the same product, if not efficiently regulated and monitored. Food Safety and Standards Authority of India (FSSAI) has laid down the regulation for designing nutritional claims for commercial products, relating to the terminologies associated with DF in those products. It can be called a 'source' of DF if the product contains at least 3 g /100 g or1.5 g/ 100kcal or 100 ml; 'high' or 'rich' in DF if the product contains minimum 6 g /100 g or 3 g /100 kcal or 100 ml. BAPs being a more recently explored area are still facing challenges in their commercialization in formulated products and have yet not been regulated by FSSAI. Regulations will help industries to follow a safe standard while fortifying these in their products and help consumers to make an informed choice during their purchase.

15.9 ENCAPSULATION OF BIOACTIVE PEPTIDES

Recently, the use of natural bioactives including omega-fatty acids, carotenoids, polyphenols, peptides and hydrolysates of protein has been carried out extensively for the formulation of functional foods(Lordan, Ross, and Stanton 2011, Pérez-Gregorio, Soares, Mateus, and de Freitas 2020). Among others, BAP has added advantages as nutraceutical food additives in functional food formulation. However, some of the shortcomings of BAP like physicochemical instability, low bioavailability, hygroscopicity, and bitterness impede their direct application in food formulation and nutraceuticals(Zhang, Liu, Wu, and Chen 2009b, Sarabandi, Gharehbeglou, and Jafari 2020). Therefore, encapsulation is one of the effective techniques to overcome these limitations. Encapsulation of BAP has been carried out by spray drying, concertation, emulsification, and film hydration.

One of the widely used encapsulation techniques for BAP is spray drying. It is rapid, simple, and has a comparatively low processing cost. The principle of spray drying involves atomizing the liquid feed containing bioactive components through an atomizer nozzle and directed into the drying chamber to generate very minute droplets, simultaneously supplying heated air inside a long drying chamber leading to instant drying of the atomized particles to form microcapsules inside the drying chamber, and consequent separation of these microcapsules through a cyclone separator(Shishir and Chen 2017). The physical properties of the encapsulated compounds depend on process conditions. Also, at high drying temperatures, some heat-sensitive or volatile compounds get degraded. Therefore, process optimization of the spray drying technique is required (Ogrodowska, Tańska, and Brandt 2017). Different peptides have been encapsulated using the spray drying technique such as whey protein hydrolysate (Rukluarh, Kanjanapongkul, Panchan, and Niumnuy 2019), casein hydrolysates(Sarabandi, Mahoonak, Hamishekar, Ghorbani, and Jafari 2018), flaxseed hydrolysates (Akbarbaglu, Jafari, Sarabandi, Mohammadi, Heshmati, and Pezeshki 2019) and chicken meat protein hydrolysate (Kurozawa, Park, and Hubinger 2009b). Freeze drying has also been explored for encapsulation of BAP but to a limited extent given the comparative advantages of spray drying over freeze-drying.

Another important encapsulation technique is coacervation. It involves three main steps, (i) developing immiscible phases while mixing continuous liquid phase, coating material, and core material, (ii) developing encapsulating layer around the active component by controlling various parameters, such as temperature and pH of the solution, the concentration of the coating materials, ionic strength(iii) solidifying the capsules through crosslinking, desolvation, or heating (Bakry, Abbas, Ali, Majeed, Abouelwafa, Mousa, and Liang 2016). The coacervation technique is divided into two categories based on the number of biopolymers involved, namely simple and complex coacervation techniques. Simple coacervation involves the use of one biopolymer, while complex coacervation constitutes two or more biopolymers. However, complex coacervation has shown to have superior functionalities compared to simple coacervation, making the former, a prime choice in the pharmaceutical and food industries (Yuan, Kong, Sun, Zeng, and Yang 2017). For example, encapsulation of oyster peptides into oyster peptides -alginate-chitosan-starch microcapsule (Zhang et al. 2009b), encapsulation of casein hydrolysate with soybean isolate-pectin (Mendanha, Molina Ortiz, Favaro-Trindade, Mauri, Monterrey-Quintero, and Thomazini 2009), microencapsulation of *Chlorella pyrenoidosa* polypeptide CPAP(Wang and Zhang 2013).

Ionotropic or ionic gelation leads to the formation of hydrogel beads that can encapsulate bioactive peptides. They are shaped by ionic polymerization or polyelectrolyte ionic bounding. An aqueous polymer solution having low molecular weight ions network with oppositely charged polyelectrolytes that react and develop an insoluble gel. The peptides encapsulated are released through diffusion via changes in gel phase, responding to external stimuli such as changes in pH, osmotic forces, enzyme and mechanical motion. Ionic gelation can be carried out by extrusion, co-extrusion, atomization, or electrostatic deposition (Kurozawa and Hubinger 2017).

Emulsification involves combining two immiscible liquid phases into a single dispersion (emulsion). It involves two chief methods, (i) single emulsion (water in oil and oil in water) (ii) double emulsion (oil in water in oil and water in oil in water). Emulsification is used mostly for the encapsulating hydrophobic and hydrophilic compounds (Jia, Dumont, and Orsat 2016, Gumus, Decker, and McClements 2017). Numerous possible applications of the emulsification technique have been executed to enhance the encapsulation process and the characteristics of the final encapsulated product, one such being emulsification-ionic gelation (Hosseini, Zandi, Rezaei, and Farahmandghavi 2013). For instance, Ying, Gao, Lu, Ma, Lv, Adhikari, and Wang (2021) applied a two-step emulsification procedure to encapsulate a soy peptide solution (40 %, w/w) in a water-in-oil-in-water emulsion.

The use of lipid-based nano-carriers is an encapsulation method in which phospholipids self-assemble thus entrapping the aqueous core loaded with bioactive compounds. This technique is broadly used in the encapsulation of BAPs. For example, encapsulation of peptides from whey protein isolates into liposomes of soybean lecithin (Mohan, McClements, and Udenigwe 2016), encapsulation of peptides from sea bream scale collagen into nanoliposomes(Mosquera, Giménez, da Silva, Boelter, Montero, Gómez-Guillén, and Brandelli 2014). A representative summary of these encapsulating methods with their advantages, disadvantages, and applications explored in literature is illustrated in Table 15.3.

15.10 BIO-ACCESSIBILITY AND BIOAVAILABILITY OF BIOACTIVE PEPTIDESIN DIFFERENT FORTIFIED FOODS

Various *in vitro* biochemical assays and *ex vivo* cell and animal prototypes have been designed and used for evaluating the bioactivity and bioaccessibility of the peptides. Bioavailability and bioaccessibility of BAP revolve around the fact that peptides should be able to repel the action of digestive enzymes throughout their gastrointestinal passage and get absorbed into the blood by crossing the intestinal epithelial barrier while maintaining their structural and functional integrity to reach undamaged to the target organs where they can exercise their beneficial properties (FitzGerald, Murray, and Walsh 2004).

TABLE 15.3

Representative Summary of Encapsulating Methods Used for Bio-Active Peptides

Encapsulation technique	Salient features	Advantages and disadvantages	Source of bioactive peptides/ carrier material	References
Spray drying	Atomizing a liquid feed (containing bioactive peptides) directed inside a drying chamber with simultaneous supply of heated air leading to instant drying of the atomized particles to form microcapsules and their separation through a cyclone separator for collection	Advantage(s): • Cost-effective, • Short process time, • Decreased product weight/volume Disadvantage(s): • Can induce slight peptide denaturation and aggregation by the increase of temperature during atomization	Flaxseed / Maltodextrin; Chicken meat / Maltodextrin - Gum Arabic; Casein / Maltodextrin	Akbarbaglu, Jafari, Sarabandi, Mohammadi, Khakbaz Heshmati, and Pezeshki (2019); Kurozawa, Park, and Hubinger (2009a); (Sarabandi, Sadeghi Mahoonak, Hamishekar, Ghorbani, and Jafari 2018); Aguilar-Toalá, Quintanar-Guerrero, Liceaga, and Zambrano-Zaragoza (2022)
Freeze-drying	Drying by sublimation of a completely frozen sample	Advantage(s):Suitable for heat-labile substances including protein/ peptides Disadvantage(s): • Time - consuming procedure, • High energy consumption, • Expensive process • Freezing stresses, ice crystal damage, freeze concentration, pH changes, and drying stress can lead to structural instability in the protein/peptides	Casein / Maltodextrin–gum arabic; Whey protein concentrate / sodium alginate	Rao, Bajaj, Mann, Arora, and Tomar (2016); Ma, Mao, Wang, Yang, Zhang, Chen, and Li (2014); Aguilar-Toalá et al. (2022)

(Continued)

TABLE 15.3 (CONTINUED)

Representative Summary of Encapsulating Methods Used for Bio-Active Peptides

Encapsulation technique	Salient features	Advantages and disadvantages	Source of bioactive peptides/ carrier material	References
Ionic gelation	On the basis of the ability of polyelectrolytes' to crosslink in the existence of counter ions. 1. External gelation (EG): Sodium alginate solution having the peptide is extruded into a solution containing calcium chloride, capsule is formed when Ca²⁺ ions diffuse into the peptide-polymer solution; 2. Internal gelation (IG): Alginate, solution hosting calcium chloride ions and the target peptide is extruded into an acidified oil medium. Ca²⁺ ions get dissolved in this acidified medium, and after reacting with alginate, and form hydrogels that have the peptides entrapped inside them	Advantage(s): • Not very expensive • No requirement of sophisticated equipment, or use of organic solvent and high thermal conditions Disadvantage(s): • More suitable for hydrophobic compounds and challenges lies in increasing encapsulation efficiency along with the enhancement of the controlled release properties of hydrophilic compounds	Chicken- or fish-derived tripeptide ACE inhibitory peptides ((leucine-lysine-proline (LKP) loaded nanoparticle (NP)) / chitosan; Jujube seed / sodium alginate Goby fish/ chitosan and tripolyphosphate	Danish, Vozza, Byrne, Frias, and Ryan (2017) Kanbargi, Sonawane, and Arya (2017) Nasri, Hamdi, Touir, Li, Karra-Chaâbouni, and Nasri (2021)
Emulsification–gelation method	Process of initial stabilization of an emulsion either containing oil droplets dispersed in an aqueous protein/ peptide solution i.e. O/W emulsion, or having an aqueous solution of protein/ peptide droplets dispersed in an oil phase i.e. W/O emulsion, and subsequent gelation of protein/peptide, prompted by heating, chemical or crosslinking with enzymes	Advantage(s): • Higher entrapment efficiency of bioactives than in case of only ionic gelation, • High-quality microspheres having small diameters Disadvantage(s): • Costly process	Collagenous by-products of smooth hounds / alginate–whey protein isolate; Insulin / alginate–chitosan	Lajmi, Gómez-Estaca, Hammami, and Martínez-Alvarez (2019); Ozkan, Franco, De Marco, Xiao, and Capanoglu (2019); Zhang, Wei, Lv, Wang, and Ma (2011)

(Continued)

TABLE 15.3 (CONTINUED)

Representative Summary of Encapsulating Methods Used for Bio-Active Peptides

Encapsulation technique	Salient features	Advantages and disadvantages	Source of bioactive peptides/ carrier material	References
Lipid-based nanocarriers (nano -liposomes)	Liposomes are prepared by aggregation of bilayers of phospholipid. These when put in an aqueous environment, they generate spherical vesicles that include aqueous sections wherein the peptides get encapsulated	Advantage (s): • Amphipathic • Bio-compatible • Non-immunogenic and non-toxic Disadvantage (s): • Physicochemical instabilities and the release of loaded compounds during the storage	Atlantic salmon (*Salmo salar*) / phospholipids derived from chitosan and milk fat globule membrane Tilapia viscera / soy-rapeseed lecithin-trehalose Rainbow trout (*Oncorhynchus mykiss*) gelatin / chitosan-coated DPPC	Li, Paulson, and Gill (2015); Sepúlveda, Alemán, Zapata, Montero, and Gómez-Guillén (2021); Ramezanzade, Hosseini, and Nikkhah (2017); Aguilar-Toalá et al. (2022); Sun, Okagu, and Udenigwe (2021)
Coacervation	Formation of microcapsules and microspheres by liquid-liquid phase separation; Peptides are distributed in the wall-forming solution of the polymer, then phases are separated by altering environmental conditions, such as pH, ionic strength, and temperature, and then solidifying the capsules through crosslinking desolvation, or heating. Simple coacervation: Through salting out leaing to incompatibilities in between the polymers; Complex coacervation: insoluble complex formation by electrostatic interaction occurring between polyelectrolyte polymers that concurrently encapsulates the target peptides	Advantage(s): • Considerable encapsulation efficiency • Reproducible • Scalable • Ambient temperature use Disadvantage (s): • High cost • Complex process • Potential concerns from the chemical cross-linkers used	Casein/soy protein isolate and pectin; Oyster / alginate-chitosan-starch; Polypeptide CPAP from *Chlorella pyrenoidosa*/ chitosan	Mendanha et al. (2009); Mukurumbira, Shellie, Keast, Palombo, and Jadhav (2022); Zhang, Liu, Wu, and Chen (2009a); Aguilar-Toalá et al. (2022); Sun, Okagu, and Udenigwe (2021)

The human and animal models which use content of digestion aspired from the stomach (Sullivan, Kehoe, Barry, Buckley, Shanahan, Mok, and Brodkorb 2014), small intestine (Boutrou, Gaudichon, Dupont, Jardin, Airinei, Marsset-Baglieri, Benamouzig, Tome, and Leonil 2013), or both (Devle, Ulleberg, Naess-Andresen, Rukke, Vegarud, and Ekeberg 2014), provide physiologically significant information on the digestion process of protein or peptides (Somaratne, Ferrua, Ye, Nau, Floury, Dupont, and Singh 2020) employing imaging techniques (Mackie, Rafiee, Malcolm, Salt, and van Aken 2013) and wireless telemetric systems (Koziolek, Schneider, Grimm, Modeß, Seekamp, Roustom, Siegmund, and Weitschies 2015, Boutrou et al. 2013). However, till now, among the few studies evaluating the *in vivo* digestion of BAPs; dairy proteins are the ones studied the most (Boutrou, Henry, and Sanchez-Rivera 2015, Bohn, Carriere, Day, Deglaire, Egger, Freitas, Golding, Le Feunteun, Macierzanka, and Ménard 2018). These studies have either used static models which consist of a sequence of bioreactors that imitate the enzymatic and physicochemical conditions of every digestive compartment involved including a standardized method finalized within the European COST Action Infogest (Asledottir, Le, Poulsen, Devold, Larsen, and Vegarud 2018, Asledottir, Le, Petrat-Melin, Devold, Larsen, and Vegarud 2017, Cattaneo, Stuknytė, Ferraretto, and De Noni 2017); or dynamic systems which are either mono-compartmental, that simulate single compartment of the gastrointestinal tract, such as the Dynamic Gastric Model (DGM) and the Human Gastric Simulator (HGS), or multi-compartmental (several compartments) such as 'TIMagc' (Bellmann, Lelieveld, Gorissen, Minekus, and Havenaar 2016), DIDGI® (Sánchez-Rivera, Ménard, Recio, and Dupont 2015), SIMGI® mimicking enzymatic and mechanical transformations happening while gastrointestinal digestion (Amigo and Hernández-Ledesma 2020). Commercially accessible human intestinal cell lines such as Caco-2 cells and HT-29 cells which combine *in vitro* gastrointestinal digestion and intestinal digestion models have been functional for assessment of peptide absorption. Absorption of whole peptide sequences across monolayers of Caco-2 cells was reported for antioxidant peptides from soy protein (Zhang, Tong, Qi, Wang, Li, Sui, and Jiang 2018), milk proteins (Pepe, Sommella, Ventre, Scala, Adesso, Ostacolo, Marzocco, Novellino, and Campiglia 2016, Wang, Xie, and Li 2016, Wang and Li 2017), antihypertensive peptides from lactoferrin(Fernández-Musoles, Salom, Castelló-Ruiz, del Mar Contreras, Recio, and Manzanares 2013), dry-cured ham (Xing, Liu, Tang, Pereira, Zhou, and Zhang 2018), corn gluten (Ding, Wang, Zhang, Yu, and Liu 2018), ovotransferrin (Ding, Wang, Zhang, and Liu 2015) and ovalbumin (Grootaert, Jacobs, Matthijs, Pitart, Baggerman, Possemiers, Van der Saag, Smagghe, Van Camp, and Voorspoels 2017).

Apart from *in vitro* studies, *in silico* analyses are also being increasingly used to identify the bioactivity of peptides sequences through peptides databases, such as BIOPEP-UWM™(Minkiewicz, Iwaniak, and Darewicz 2019), SpirPep (Anekthanakul, Hongsthong, Senachak, and Ruengjitchatchawalya 2018), StraPep (Wang, Yin, Xiao, He, Xue, Jiang, and Wang 2018), etc. These tools have been used to predict the biological activities of several peptides, for instance, as detected for barley hordein C, based on mapping of the sequences of the cereal storage proteins in BIOPEP-UWM™, which demonstrated several potential sequences of bioactive peptides, including ACE inhibitory, antioxidant, among others(Cavazos and Gonzalez de Mejia 2013).

Among the *in vivo* studies, plant peptides such as the ACE-inhibitory peptide (RGQVIYVL) obtained from quinoa bran albumin through Alcalase and trypsin hydrolysis, after oral administration was found to resist simulated gastrointestinal digestion (SGDI) and demonstrate antihypertensive activity in spontaneously hypertensive rats (SHR) (Sonklin, Alashi, Laohakunjit, Kerdchoechuen, and Aluko 2020). A peptide (DIKTNKPVIF), identified in Alcalase hydrolysate of potato protein, presented both antidiabetic activity in SHR and anti-hepatosteatosis activity in aging mice (Dumeus, Shibu, Lin, Wang, Lai, Shen, Lin, Viswanadha, Kuo, and Huang 2018, Marthandam Asokan, Wang, Su, and Lin 2019). Several BAP shave been obtained from animal sources as well; for e.g., the tryptic peptide α-casozepine (YLGYLEQLLR, fromα $_s$1-casein) from milk, was found to induce anxiolytic effects post intraperitoneal injection in Swiss mice(Benoit, Chaumontet, Schwarz, Cakir-Kiefer, Boulier, Tomé, and Miclo 2020).An oyster hydrolysate peptide, IVVPK, was found to be able to recover effects of photo-aging induced by UVB radiation in SKH-1 hairless mice(Bang, Jin, and Choung 2020). The peptide LSGYGP obtained from tilapia skin gelatin hydrolysates was found to exert a reduction in blood pressure in SHR (Chen, Ryu, Zhang, Liang, Li, Zhou, Yang, Hong, and Qian 2020). However, there is a lack of association of the *in vitro* bioactivities of peptides with their *in vivo* functions mainly because of their low bioavailability post oral administration (Wang, Ding, Du, and Liu 2020).

As suggested by Amigo and Hernández-Ledesma (2020), strategies to increase the bioavailability of peptides and retain their *in vivo* bioactivities, target accomplishing the following objectives: (i) decline in the negative impacts of various food processing on their bioactivities; (ii) encouraging favorable interactions between peptides and other components in the food matrix, decreasing the undesirable ones (iii) safeguarding the BAP from the harsh gastrointestinal conditions and enzyme activity (iv) allowing the sustained release of the peptides in their target organs; and (v) enhancement of the transport of the peptide across the intestinal epithelium and into the target cell. These different strategies include various modifications in the peptide structure to protect them from the action of digestive enzymes, increase their intestinal permeability, and/or retain their bioactivity. Modification in N- and C-amino acid terminals through acetylation/ amidation have been shown to shield the peptide from the action of amino- and carboxypeptidases, respectively (Wallace 1992, Arnesen 2011). Phosphorylation of serine's hydroxyl groups in casein-phosphopeptides prevents them from being hydrolyzed by the digestive enzymes, improves their absorption, and retains their mineral-binding capacity (Boutrou, Jardin, Blais, Tomé, and Léonil 2008). The co-administration of peptides with protease/peptidase inhibitors can avert peptides from getting degraded in the digestive tract, thus enabling optimal intestinal absorption (Mahato, Narang, Thoma, and Miller 2003, Kumar Malik, Baboota, Ahuja, Hasan, and Ali 2007, Werle, Makhlof, and Takeuchi 2009).

Food-grade permeation enhancers, including fatty acids, citric acid, and chitosan are being used in the formulation of bioactive peptides-based functional foods and nutraceuticals. Novel non-thermal food processing techniques such as ultrasound, ultrahigh hydrostatic pressure, microwave, pulsed electric field, and irradiation can be applied in processing foods formulated with peptides (Pereira and Vicente 2010). Further studies need to be carried out to gather substantial data on the prospective impacts of non-thermal processing technologies on the bioavailability and bioactivity of BAP(Piccolomini, Iskandar, Lands, and Kubow 2012). Another strategies to protect these BAP which include light, oxygen, and temperature-sensitive, or are prone to interact with other food components, and are vulnerable to the gastrointestinal tract conditions, include the encapsulation of these peptides into protective carriers as described in the previous section to retain their stability and bioavailability (Joye, Davidov-Pardo, and McClements 2014).

As pointed out by Karaś (2019), owing to the several factors affecting peptide absorption, there is still a persistent requisite to conduct further assays, especially *in vivo* assays to investigate their oral bioavailability, preservation in the digestive tract, transfer through the intestinal membranes into the bloodstream and their effect on the target site, as well as the effect of multi-component food systems on their bioavailability and the effectiveness of different strategies to protect their bioactivity.

15.11 ANTAGONISTIC EFFECT OF DIETARY FIBER ON OTHER FOOD NUTRIENTS

DF has been reported to decrease the bioavailability of certain minerals and trace elements in the human diet (Luccia and Kunkel 2002). This undesirable effect on nutrient absorption is attributed mostly to gel formation, increased viscosity, or binding and entrapment by the DF(Riedl, Linseisen, Hoffmann, and Wolfram 1999, Bohn 2008, 2014).

Idouraine, Hassani, Claye, and Weber (1995) also have reported that increased consumption of DF may negatively affect mineral absorption in the human body. This may be associated with the undesirable impact on mineral bioavailability, especially in high-risk individuals(Bosscher, Van Caillie-Bertrand, and Deelstra 2001). *In vitro* studies have demonstrated the mineral-binding ability of semi-purified insoluble and soluble fibers. (Debon and Tester 2001, Bosscher, Van Caillie-Bertrand, and Deelstra 2001, Miyada, Nakajima, and Ebihara 2011). Ion exchange interactions, possibly relating hydroxyl and carboxyl groups of DF are somewhat accountable for mineral binding (Debon and Tester 2001; Miyada, Nakajima, and Ebihara 2011). Fiber's role in antagonizing zinc, as most reported, is owing to the high phytate content of most fiber-containing foods (Barbro, Brittmarie, and Cederblad 1985). Claye, Idouraine, and Weber (1998) compared the *in vitro* binding capacity of total DF with that of individual wheat fiber fractions. The individual fiber constituents including lignin, lignocelluloses, and hemicelluloses showed selective binding on zinc, whereas a strong binding on copper was recorded with total DF. Treated fibers including the dephytinized insoluble fiber (DF fraction devoid of phytate) have exhibited mineral binding capacity at various levels (Gualberto, Bergman, and Weber 1997). Animal studies have also shown that including comparatively increased levels of DF in the diets of rats, led to reduced absorption of certain vital minerals including iron, zinc, calcium, and magnesium (Donangelo and Eggum 1986, Ward and Reichert 1986).In a study by Miyada, Nakajima, and Ebihara (2011)on the impact of pectin on absorption of iron in ileorectomized, caectomized, and normal rats, it was observed that iron absorption was decreased in the small intestine while it improved in the colon. After reviewing several literatures, Dégen, Halas, and Babinszky (2007) concluded that on increasing 1 g Neutral detergent fiber (NDF)/kg in the feed diet, there is an average reduction of ileal apparent protein digestibility by 0.03 - 0.08% while, the same results in the decrease of the digestibility energy coefficient approximately by 0.1% in growing and fattening pigs.

Wong and Cheung (2005a) and Wong and Cheung (2005b)have associated the influence of DF on mineral absorption with the three following mechanisms as seen in Figure 15.5.

The dietary fiber helps in: 1. reduced transit time of passage of DF through the small intestine, thus decreasing the time required for the absorption of minerals; 2. Directly diminishing mineral transportation when DF passes along the intestinal mucosal cells; and 3. chemical binding of DF on minerals, at any stage during food production, cooking processes, resulting in the formation of mineral-fiber complexes that are unable to be digested and absorbed by the body. Many factors including an individual's nutritional status, the pattern of meals intake, cooking/serving habits and the source of DF, its composition, particle size, as well as other physico-chemical properties when present or incorporated into different food matrices, affect the formation of these mineral-fiber complexes (van der Aar, Fahey, Ricke, Allen, and Berger 1983, Idouraine et al. 1995, Claye, Idouraine, and Weber 1998, Guillon and Champ 2000).However, soluble fibers like fructo-oligosaccharides, pectin, and inulin were alongside stated to enhance absorption of minerals(Sakai, Ohta, Shiga, Takasaki, Tokunaga, and Hara 2000).

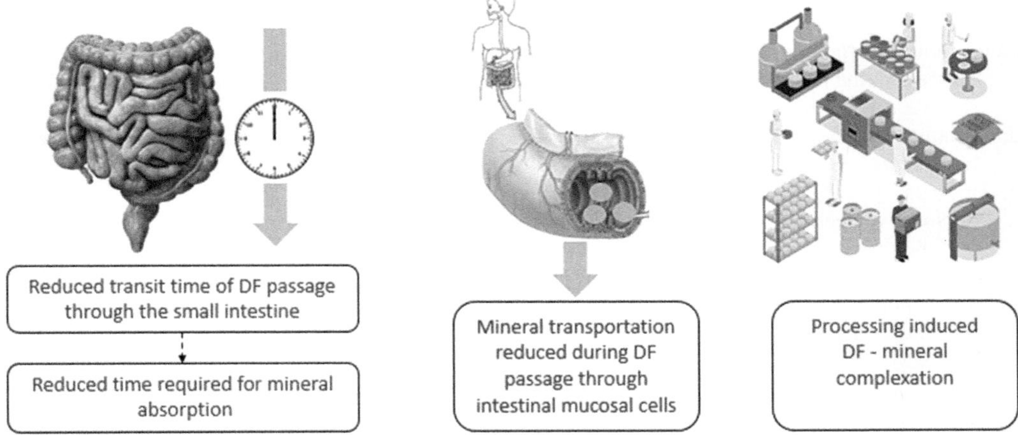

FIGURE 15.5 Mechanisms associated with the influence of dietary fiber on mineral absorption.

DF is also suspected to affect the absorption of carotenoids, mostly α-tocopherol and other poly-phenolic compounds(Palafox-Carlos, Ayala-Zavala, and González-Aguilar 2011). Pectin dimin-ished the bioavailability of β-carotene (Rock and Swendseid 1992). The incorporation of pectin was identified to drop bioavailability of vitamin E, and absorption of vitamin E was reduced by konjac mannan supplementation in diabetic and non-diseased subjects (Jenkins 1993). The leading factors assessed as causes of the influence of DF on antioxidant digestion are (i)physical entrap-ping of antioxidants inside structured assemblies as in fruit tissues, (ii) increased viscosity of gas-tric fluids limiting the peristaltic process of mixing that stimulates enzymes' transport to their respective substrates, bile salts to unmicellized fat and soluble antioxidants in to the gastrointestinal wall(Montagne, Pluske, and Hampson 2003).

Adams, Sello, Qin, Che, and Han (2018) opined that the physicochemical functionality of DF, such as bulking ability, fermentation, viscosity, gel formation, solubility, and water-absorption capacity impact absorption of nutrient along with the methods of consumption of DF.

However, it has been observed that the antagonistic effect of DF on different nutrients has not been widely studied in recent times. Hence, there lies a scope for conducting impactful investiga-tions related to this that can be referred to while fortifying DF in multi-nutrient matrix formulations.

15.12   SUMMARY AND OUTLOOK

Currently, processed foods are being increasingly consumed and these processed foods mostly get depleted of certain functional constituents leading to increased consumer demand for products for-tified with bioactive compounds such as DF and BAPs Thus, fortification of these functional food ingredients in different food products enhances their nutritional quality and facilitates their role in various physiological functions when included in the human diet. However, the formulation of functional foods with BAP impedes their direct application due to their physicochemical instability, low bioavailability, and many such associated factors as discussed above. Therefore, encapsulation of BAP has been carried out to preserve its functionality. On the other hand, DF can reduce the bio-availability of some minerals and trace elements on increased consumption in the daily diet. More research into studies relating to the incorporation of BAP and DF in food products is suggested to enable large-scale production and enhanced commercialization of these fortified foods monitored through proper guidelines specified by the food regulatory bodies.

REFERENCES

AACC. (2000). *"Approved methods of the American Association of Cereal Chemists."* American Association *of Cereal Chemists. 30.*Adams, Seidu, Cornelius Tlotlisohttps://www.cerealsgrains.org/resources/defi-nitions/Documents/DietaryFiber/Fiberpr.pdf

Adams, Seidu, Cornelius Tlotliso Sello, Gui-Xin Qin, Dongsheng Che, and Rui Han. 2018. "Does dietary fiber affect the levels of nutritional components after feed formulation?." *Fibers* 6(2):29. https://doi.org/10.3390/fib6020029

Adibi, Siamak A. 1997. "The oligopeptide transporter (Pept-1) in human intestine: Biology and function." *Gastroenterology* 113(1):332–340. https://doi.org/10.1016/S0016-5085(97)70112-4

Aguilar-Toalá, JE, D Quintanar-Guerrero, AM Liceaga, and ML Zambrano-Zaragoza. 2022. "Encapsulation of bioactive peptides: A strategy to improve the stability, protect the nutraceutical bioactivity and sup-port their food applications." *RSC Advances* 12(11):6449–6458. https://doi.org/10.1039/D1RA08590E

Ahn, JE, SY Park, A Atwal, BF Gibbs, and BH Lee. 2009. "Angiotensin I-converting enzyme (ACE) inhibi-tory peptides from whey fermented by Lactobacillus species." *Journal of Food Biochemistry* 33(4):587–602. https://doi.org/10.1111/j.1745-4514.2009.00239.x

Aissaoui, Neyssene, Jean-Marc Chobert, Thomas Haertlé, M Nejib Marzouki, and Ferid Abidi. 2017. "Purification and biochemical characterization of a neutral serine protease from Trichoderma harzia-num. Use in antibacterial peptide production from a fish by-product hydrolysate." *Applied Biochemistry and Biotechnology* 182(2):831–845. https://doi.org/10.1007/s12010-016-2365-4. PMid:27987188

Akan, Ecem. 2020. "An evaluation of the in vitro antioxidant and antidiabetic potentials of camel and donkey milk peptides released from casein and whey proteins." *Journal of Food Science and Technology*:1–9. https://doi.org/10.1007/s13197-020-04832-5. PMid:34471298

Akbarbaglu, Zahra, Seid Mahdi Jafari, Khashayar Sarabandi, Maryam Mohammadi, Maryam Khakbaz Heshmati, and Akram Pezeshki. 2019. "Influence of spray drying encapsulation on the retention of antioxidant properties and microstructure of flaxseed protein hydrolysates." *Colloids and Surfaces B: Biointerfaces* 178:421–429. https://doi.org/10.1016/j.colsurfb.2019.03.038. PMid:30908998

Amigo, Lourdes, and Blanca Hernández-Ledesma. 2020. "Current evidence on the bioavailability of food bioactive peptides." *Molecules* 25(19):4479. https://doi.org/10.3390/molecules25194479. PMid:33003506. PMCid:PMC7582556

Anekthanakul, Krittima, Apiradee Hongsthong, Jittisak Senachak, and Marasri Ruengjitchatchawalya. 2018. "SpirPep: An in silico digestion-based platform to assist bioactive peptides discovery from a genome-wide database." *BMC Bioinformatics* 19(1):1–10. https://doi.org/10.1186/s12859-018-2143-0. PMid:29678128. PMCid:PMC5910554

Arnesen, Thomas. 2011. "Towards a functional understanding of protein N-terminal acetylation." *PLOS Biology* 9(5):e1001074. https://doi.org/10.1371/journal.pbio.1001074. PMid:21655309. PMCid:PMC3104970

Asledottir, Tora, Thao T Le, Bjørn Petrat-Melin, Tove G Devold, Lotte B Larsen, and Gerd E Vegarud. 2017. "Identification of bioactive peptides and quantification of β-casomorphin-7 from bovine β-casein A1, A2 and I after ex vivo gastrointestinal digestion." *International Dairy Journal* 71:98–106. https://doi.org/10.1016/j.idairyj.2017.03.008

Asledottir, Tora, Thao T Le, Nina A Poulsen, Tove G Devold, Lotte B Larsen, and Gerd E Vegarud. 2018. "Release of β-casomorphin-7 from bovine milk of different β-casein variants after ex vivo gastrointestinal digestion." *International Dairy Journal* 81:8–11. https://doi.org/10.1016/j.idairyj.2017.12.014

Association, American Diabetes. 2014. "Diagnosis and classification of diabetes mellitus." *Diabetes Care* 37(Supplement 1):S81–S90. https://doi.org/10.2337/dc14-S081. PMid:24357215

Augustin, Mary Ann, and Luz Sanguansri. 2012. "Challenges in developing delivery systems for food additives, nutraceuticals and dietary supplements." In *Encapsulation technologies and delivery systems for food ingredients and nutraceuticals*, 19–48. Elsevier. https://doi.org/10.1533/9780857095909.1.19

Augustin, Mary Ann, and Luz Sanguansri. 2015. "Challenges and solutions to incorporation of nutraceuticals in foods." *Annual Review of Food Science and Technology* 6:463–477. https://doi.org/10.1146/annurev-food-022814-015507. PMid:25422878

Bakry, Amr M, Shabbar Abbas, Barkat Ali, Hamid Majeed, Mohamed Y Abouelwafa, Ahmed Mousa, and Li Liang. 2016. "Microencapsulation of oils: A comprehensive review of benefits, techniques, and applications." *Comprehensive Reviews in Food Science and Food Safety* 15(1):143–182. https://doi.org/10.1111/1541-4337.12179. PMid:33371581

Bang, Joon Sok, Yu Jung Jin, and Se-Young Choung. 2020. "Low molecular polypeptide from oyster hydrolysate recovers photoaging in SKH-1 hairless mice." *Toxicology and Applied Pharmacology* 386:114844. https://doi.org/10.1016/j.taap.2019.114844. PMid:31785243

Barbro, Nävert, Sandström Brittmarie, and ÅKE Cederblad. 1985. "Reduction of the phytate content of bran by leavening in bread and its effect on zinc absorption in man." *British Journal of Nutrition* 53(1):47–53. https://doi.org/10.1079/BJN19850009. PMid:2998440

Beecher, Gary R. 1999. "Phytonutrients' role in metabolism: Effects on resistance to degenerative processes." *Nutrition Reviews* 57(9):3–6. https://doi.org/10.1111/j.1753-4887.1999.tb01800.x. PMid:10568344

Beermann, Christopher, Marco Euler, Jochen Herzberg, and Bernd Stahl. 2009. "Anti-oxidative capacity of enzymatically released peptides from soybean protein isolate." *European Food Research and Technology* 229(4):637–644. https://doi.org/10.1007/s00217-009-1093-1

Bellmann, Susann, Jan Lelieveld, Tom Gorissen, Mans Minekus, and Robert Havenaar. 2016. "Development of an advanced in vitro model of the stomach and its evaluation versus human gastric physiology." *Food Research International* 88:191–198. https://doi.org/10.1016/j.foodres.2016.01.030

Benoit, Simon, Catherine Chaumontet, Jessica Schwarz, Céline Cakir-Kiefer, Audrey Boulier, Daniel Tomé, and Laurent Miclo. 2020. "Anxiolytic activity and brain modulation pattern of the α-Casozepine-derived pentapeptide YLGYL in mice." *Nutrients* 12(5):1497. https://doi.org/10.3390/nu12051497. PMid:32455588. PMCid:PMC7285003

Bernstein, Adam M, Brigid Titgemeier, Kristin Kirkpatrick, Mladen Golubic, and Michael F Roizen. 2013. "Major cereal grain fibers and psyllium in relation to cardiovascular health." *Nutrients* 5(5):1471–1487. https://doi.org/10.3390/nu5051471. PMid:23628720. PMCid:PMC3708330

Bhat, ZF, Sunil Kumar, and Hina Fayaz Bhat. 2015. "Bioactive peptides from egg: A review." *Nutrition and Food Science*. https://doi.org/10.1108/NFS-10-2014-0088

Boelsma, Esther, and Joris Kloek. 2008. "Lactotripeptides and antihypertensive effects: A critical review." *British Journal of Nutrition* 101(6):776–786. https://doi.org/10.1017/S0007114508137722. PMid:19061526

Bohn, Torsten. 2008. "Bioavailability of non-provitamin A carotenoids." *Current Nutrition and Food Science* 4(4):240–258. https://doi.org/10.2174/157340108786263685

Bohn, Torsten. 2014. "Dietary factors affecting polyphenol bioavailability." *Nutrition Reviews* 72(7):429–452. https://doi.org/10.1111/nure.12114. PMid:24828476

Bohn, Torsten, Frèderic Carriere, L Day, Amélie Deglaire, Lotti Egger, Daniela Freitas, Matt Golding, Steven Le Feunteun, Adam Macierzanka, and Olivia Ménard. 2018. "Correlation between in vitro and in vivo data on food digestion. What can we predict with static in vitro digestion models?." *Critical Reviews in Food Science and Nutrition* 58(13):2239–2261. https://doi.org/10.1080/10408398.2017.1315362. PMid:28613945

Bosch-Sierra, Neus, Roger Marqués-Cardete, Aránzazu Gurrea-Martínez, Grau-Del Valle, Clara Talens, Saioa Alvarez-Sabatel, Carlos Bald, Carlos Morillas, Antonio Hernández-Mijares, and Celia Bañuls. 2019. "Effect of fibre-enriched orange juice on postprandial glycaemic response and satiety in healthy individuals: An acute, randomised, placebo-controlled, double-blind, crossover study." *Nutrients* 11(12):3014. https://doi.org/10.3390/nu11123014. PMid:31835476. PMCid:PMC6950290

Bosscher, Douwina, Micheline Van Caillie-Bertrand, and Hendrik Deelstra. 2001. "Effect of thickening agents, based on soluble dietary fiber, on the availability of calcium, iron, and zinc from infant formulas." *Nutrition* 17(7–8):614–618. https://doi.org/10.1016/S0899-9007(01)00541-X

Boutrou, Rachel, Claire Gaudichon, Didier Dupont, Julien Jardin, Gheorghe Airinei, Agnès Marsset-Baglieri, Robert Benamouzig, Daniel Tome, and Joelle Leonil. 2013. "Sequential release of milk protein-derived bioactive peptides in the jejunum in healthy humans." *The American Journal of Clinical Nutrition* 97(6):1314–1323. https://doi.org/10.3945/ajcn.112.055202. PMid:23576048

Boutrou, Rachel, Gwénaële Henry, and Laura Sanchez-Rivera. 2015. "On the trail of milk bioactive peptides in human and animal intestinal tracts during digestion: A review." *Dairy Science and Technology* 95(6):815–829. https://doi.org/10.1007/s13594-015-0210-0

Boutrou, Rachel, Julien Jardin, Anne Blais, Daniel Tomé, and Joëlle Léonil. 2008. "Glycosylations of κ-casein-derived caseinomacropeptide reduce its accessibility to endo-but not exointestinal brush border membrane peptidases." *Journal of Agricultural and Food Chemistry* 56(17):8166–8173. https://doi.org/10.1021/jf801140d. PMid:18698795

Brandelli, Adriano, Daniel Joner Daroit, and Ana Paula Folmer Corrêa. 2015. "Whey as a source of peptides with remarkable biological activities." *Food Research International* 73:149–161. https://doi.org/10.1016/j.foodres.2015.01.016

Brouwer, Carlo PJM, Mahfuzur Rahman, and Mick M Welling. 2011. "Discovery and development of a synthetic peptide derived from lactoferrin for clinical use." *Peptides* 32(9):1953–1963. https://doi.org/10.1016/j.peptides.2011.07.017. PMid:21827807

Byrne, M. (1997). Low-fat with taste. *Food Engineering International*, 22(6), 36–41.

Calbet, Jose AL, and Dave A MacLean. 2002. "Plasma glucagon and insulin responses depend on the rate of appearance of amino acids after ingestion of different protein solutions in humans." *The Journal of Nutrition* 132(8):2174–2182. https://doi.org/10.1093/jn/132.8.2174. PMid:12163658

Cam, Anthony, and Elvira Gonzalez de Mejia. 2012. "RGD-peptide lunasin inhibits Akt-mediated NF-κB activation in human macrophages through interaction with the αVβ3 integrin." *Molecular Nutrition and Food Research* 56(10):1569–1581. https://doi.org/10.1002/mnfr.201200301. PMid:22945510

Capriles, VD, AC Conti-Silva, and JA Gomes Arêas. 2021. "Effects of oligofructose-enriched inulin addition before and after the extrusion process on the quality and postprandial glycemic response of cornsnacks." *Food Bioscience* 43. https://doi.org/10.1016/j.fbio.2021.101263. https://doi.org/10.1016/j.fbio.2021.101263

Cattaneo, Stefano, Milda Stuknytė, Anita Ferraretto, and Ivano De Noni. 2017. "Impact of the in vitro gastrointestinal digestion protocol on casein phosphopeptide profile of Grana Padano cheese digestates." *LWT* 77:356–361. https://doi.org/10.1016/j.lwt.2016.11.069

Cavazos, Ariel, and Elvira Gonzalez de Mejia. 2013. "Identification of bioactive peptides from cereal storage proteins and their potential role in prevention of chronic diseases." *Comprehensive Reviews in Food Science and Food Safety* 12(4):364–380. https://doi.org/10.1111/1541-4337.12017. PMid:33412684

Chakrabarti, Subhadeep, Forough Jahandideh, and Jianping Wu. 2014. "Food-derived bioactive peptides on inflammation and oxidative stress." *BioMed Research International* 2014. https://doi.org/10.1155/2014/608979. PMid:24527452. PMCid:PMC3914560

Chatterjee, Cynthia, Stephen Gleddie, and Chao-Wu Xiao. 2018. "Soybean bioactive peptides and their functional properties." *Nutrients* 10(9):1211. https://doi.org/10.3390/nu10091211. PMid:30200502. PMCid:PMC6164536

Chau, Chi-Fai, and Ya-Ling Huang. 2003. "Comparison of the chemical composition and physicochemical properties of different fibers prepared from the peel of Citrus sinensis L. Cv. Liucheng." *Journal of Agricultural and Food Chemistry* 51(9):2615–2618. https://doi.org/10.1021/jf025919b. PMid:12696946

Chaudhury, Arun, Chitharanjan Duvoor, Vijaya Sena Reddy Dendi, Shashank Kraleti, Aditya Chada, Rahul Ravilla, Asween Marco, Nawal Singh Shekhawat, Maria Theresa Montales, and Kevin Kuriakose. 2017. "Clinical review of antidiabetic drugs: Implications for type 2 diabetes mellitus management." *Frontiers in Endocrinology* 8:6. https://doi.org/10.3389/fendo.2017.00006. PMid:28167928. PMCid:PMC5256065

Chen, Jiali, Bomi Ryu, YuanYuan Zhang, Peng Liang, Chengyong Li, Chunxia Zhou, Ping Yang, Pengzhi Hong, and Zhong-Ji Qian. 2020. "Comparison of an angiotensin-I-converting enzyme inhibitory peptide from tilapia (Oreochromis niloticus) with captopril: Inhibition kinetics, in vivo effect, simulated gastrointestinal digestion and a molecular docking study." *Journal of the Science of Food and Agriculture* 100(1):315–324. https://doi.org/10.1002/jsfa.10041. PMid:31525262

Chi, Chang-Feng, Fa-Yuan Hu, Bin Wang, Tao Li, and Guo-Fang Ding. 2015. "Antioxidant and anticancer peptides from the protein hydrolysate of blood clam (Tegillarca granosa) muscle." *Journal of Functional Foods* 15:301–313. https://doi.org/10.1016/j.jff.2015.03.045

Choi, Jongwoo, Latha Sabikhi, Ashraf Hassan, and Sanjeev Anand. 2012. "Bioactive peptides in dairy products." *International Journal of Dairy Technology* 65(1):1–12. https://doi.org/10.1111/j.1471-0307.2011.00725.x

Chuaychan, Sira, Soottawat Benjakul, and Thanasak Sae-Leaw. 2017. "Gelatin hydrolysate powder from the scales of spotted golden goatfish: Effect of drying conditions and juice fortification." *Drying Technology* 35(10):1195–1203. https://doi.org/10.1080/07373937.2016.1236129

Claye, SS, A Idouraine, and CW Weber. 1998. "In-vitro mineral binding capacity of five fiber sources and their insoluble components for magnesium and calcium1." *Food Chemistry* 61(3):333–338. https://doi.org/10.1016/S0308-8146(97)00059-9

Conradi, Robert A, Allen R Hilgers, Norman FH Ho, and Philip S Burton. 1991. "The influence of peptide structure on transport across Caco-2 cells." *Pharmaceutical Research* 8(12):1453–1460. https://doi.org/10.1023/A:1015825912542. PMid:1808606

Cummings, JH, and GT Macfarlane. 1991. "The control and consequences of bacterial fermentation in the human colon." *Journal of Applied Bacteriology* 70(6):443–459. https://doi.org/10.1111/j.1365-2672.1991.tb02739.x. PMid:1938669

Daliri, Eric Banan-Mwine, Deog H Oh, and Byong H Lee. 2017. "Bioactive peptides." *Foods* 6(5):32. https://doi.org/10.3390/foods6050032. PMid:28445415. PMCid:PMC5447908

Danish, Minna K, Giuliana Vozza, Hugh J Byrne, Jesus M Frias, and Sinéad M Ryan. 2017. "Formulation, characterization and stability assessment of a food-derived tripeptide, leucine-lysine-proline loaded chitosan nanoparticles." *Journal of Food Science* 82(9):2094–2104. https://doi.org/10.1111/1750-3841.13824. PMid:28796309

Das, Rahel Suchintita, Brijesh K Tiwari, and Marco Garcia-Vaquero. 2022. "Bioactive peptides from algae." In *Bioactive peptides from food: Sources, analysis, and functions*, 31–54. CRC Press. https://doi.org/10.1201/9781003106524-4. PMid:34982240

Debon, Stéphane JJ, and Richard F Tester. 2001. "In vitro binding of calcium, iron and zinc by non-starch polysaccharides." *Food Chemistry* 73(4):401–410. https://doi.org/10.1016/S0308-8146(00)00312-5

Dégen, L, V Halas, and L Babinszky. 2007. "Effect of dietary fibre on protein and fat digestibility and its consequences on diet formulation for growing and fattening pigs: A review." *Acta Agriculturae Scandinavica, Section A - Animal Science* 57(1):1–9. https://doi.org/10.1080/09064700701372038. https://doi.org/10.1080/09064700701372038

Devle, Hanne, Ellen Kathrine Ulleberg, Carl Fredrik Naess-Andresen, Elling-Olav Rukke, Gerd Vegarud, and Dag Ekeberg. 2014. "Reciprocal interacting effects of proteins and lipids during ex vivo digestion of bovine milk." *International Dairy Journal* 36(1):6–13. https://doi.org/10.1016/j.idairyj.2013.11.008

Dhingra, Devinder, Mona Michael, Hradesh Rajput, and RT Patil. 2012. "Dietary fibre in foods: A review." *Journal of Food Science and Technology* 49(3):255–266. https://doi.org/10.1007/s13197-011-0365-5. PMid:23729846. PMCid:PMC3614039

Diabetes, UK. 2006. "What is diabetes." Accessed March 21.

Dietzen, Dennis J. 2018. "13 - Amino acids, peptides, and proteins." In *Principles and applications of molecular diagnostics*, edited by Nader Rifai, Andrea Rita Horvath, and Carl T Wittwer, 345–380. Elsevier. https://doi.org/10.1016/B978-0-12-816061-9.00013-8

Ding, Long, Liying Wang, Yan Zhang, and Jingbo Liu. 2015. "Transport of antihypertensive peptide RVPSL, ovotransferrin 328–332, in human intestinal Caco-2 cell monolayers." *Journal of Agricultural and Food Chemistry* 63(37):8143–8150. https://doi.org/10.1021/acs.jafc.5b01824. PMid:26335384

Ding, Long, Liying Wang, Ting Zhang, Zhipeng Yu, and Jingbo Liu. 2018. "Hydrolysis and transepithelial transport of two corn gluten derived bioactive peptides in human Caco-2 cell monolayers." *Food Research International* 106:475–480. https://doi.org/10.1016/j.foodres.2017.12.080. PMid:29579950

Donangelo, Carmen M, and BO Eggum. 1986. "Comparative effects of wheat bran and barley husk on nutrient utilization in rats: 2. Zinc, calcium and phosphorus." *British Journal of Nutrition* 56(1):269–280. https://doi.org/10.1079/BJN19860106. PMid:2823870

Draganidis, Dimitrios, Leonidas G Karagounis, Ioannis Athanailidis, Athanasios Chatzinikolaou, Athanasios Z Jamurtas, and Ioannis G Fatouros. 2016. "Inflammaging and skeletal muscle: Can protein intake make a difference?" *The Journal of Nutrition* 146(10):1940–1952. https://doi.org/10.3945/jn.116.230912. PMid:27581584

Dumeus, Stanley, Marthandam Asokan Shibu, Wan-Teng Lin, Ming-Fu Wang, Chao-Hung Lai, Chia-Yao Shen, Yueh-Min Lin, Vijaya Padma Viswanadha, Wei-Wen Kuo, and Chih-Yang Huang. 2018. "Bioactive peptide improves diet-induced hepatic fat deposition and hepatocyte proinflammatory response in SAMP8 ageing mice." *Cellular Physiology and Biochemistry* 48(5):1942–1952. https://doi.org/10.1159/000492518. PMid:30092591

Dwyer, Johanna T, Kathryn L Wiemer, Omar Dary, Carl L Keen, Janet C King, Kevin B Miller, Martin A Philbert, Valerie Tarasuk, Christine L Taylor, and P Courtney Gaine. 2015. "Fortification and health: Challenges and opportunities." *Advances in Nutrition* 6(1):124–131. https://doi.org/10.3945/an.114.007443. PMid:25593151. PMCid:PMC4288271

Egerton, Sian, Sarah Culloty, Jason Whooley, Catherine Stanton, and R Paul Ross. 2018. "Characterization of protein hydrolysates from blue whiting (Micromesistius poutassou) and their application in beverage fortification." *Food Chemistry* 245:698–706. https://doi.org/10.1016/j.foodchem.2017.10.107. PMid:29287428

Ertaş, Nilgün. 2015. "Effect of wheat bran stabilization methods on nutritional and physico-mechanical characteristics of cookies." *Journal of Food Quality* 38(3):184–191. https://doi.org/10.1111/jfq.12130

Fernández-Ginés, JM, J Fernández-López, E Sayas-Barberá, E Sendra, and JA Pérez-Alvarez. 2003. "Effect of storage conditions on quality characteristics of bologna sausages made with citrus fiber." *Journal of Food Science* 68(2):710–714. https://doi.org/10.1111/j.1365-2621.2003.tb05737.x

Fernández-Musoles, Ricardo, Juan B Salom, María Castelló-Ruiz, María del Mar Contreras, Isidra Recio, and Paloma Manzanares. 2013. "Bioavailability of antihypertensive lactoferricin B-derived peptides: Transepithelial transport and resistance to intestinal and plasma peptidases." *International Dairy Journal* 32(2):169–174. https://doi.org/10.1016/j.idairyj.2013.05.009

FitzGerald, Richard J, Brian A Murray, and Daniel J Walsh. 2004. "Hypotensive peptides from milk proteins." *The Journal of Nutrition* 134(4):980S–988S. https://doi.org/10.1093/jn/134.4.980S. PMid:15051858

Garcia, ML, R Dominguez, MD Galvez, C Casas, and MD Selgas. 2002. "Utilization of cereal and fruit fibres in low fat dry fermented sausages." *Meat Science* 60(3):227–236. https://doi.org/10.1016/S0309-1740(01)00125-5

Gibbs, Bernard F, Alexandre Zougman, Robert Masse, and Catherine Mulligan. 2004. "Production and characterization of bioactive peptides from soy hydrolysate and soy-fermented food." *Food Research International* 37(2):123–131. https://doi.org/10.1016/j.foodres.2003.09.010

Gilani, G Sarwar, Chaowu Xiao, and Nora Lee. 2008. "Need for accurate and standardized determination of amino acids and bioactive peptides for evaluating protein quality and potential health effects of foods and dietary supplements." *Journal of AOAC International* 91(4):894–900. https://doi.org/10.1093/jaoac/91.4.894. PMid:18727551

Gordon, Dennis T. 1990. "Total dietary fiber and mineral absorption." In *Dietary fiber*, 105–128. Springer. https://doi.org/10.1007/978-1-4613-0519-4_7

Graham, Saxon, Hari Dayal, Mya Swanson, Arnold Mittelman, and Gregg Wilkinson. 1978. "Diet in the epidemiology of cancer of the colon and rectum." *Journal of the National Cancer Institute* 61(3):709–714.

Grootaert, Charlotte, Griet Jacobs, Bea Matthijs, Judit Pitart, Geert Baggerman, Sam Possemiers, Hans Van der Saag, Guy Smagghe, John Van Camp, and Stefan Voorspoels. 2017. "Quantification of egg ovalbumin hydrolysate-derived anti-hypertensive peptides in an in vitro model combining luminal digestion with intestinal Caco-2 cell transport." *Food Research International* 99(1):531–541. https://doi.org/10.1016/j.foodres.2017.06.002. PMid:28784514

Gualberto, DG, CJ Bergman, and CW Weber. 1997. "Mineral binding capacity of dephytinized insoluble fiber from extruded wheat, oat and rice brans." *Plant Foods for Human Nutrition* 51(4):295–310. https://doi.org/10.1023/A:1007972205452. PMid:9650723

Guillon, Fabienne, and Martine Champ. 2000. "Structural and physical properties of dietary fibres, and consequences of processing on human physiology." *Food Research International* 33(3–4):233–245. https://doi.org/10.1016/S0963-9969(00)00038-7

Gumus, Cansu Ekin, Eric Andrew Decker, and David Julian McClements. 2017. "Gastrointestinal fate of emulsion-based ω-3 oil delivery systems stabilized by plant proteins: Lentil, pea, and faba bean proteins." *Journal of Food Engineering* 207:90–98. https://doi.org/10.1016/j.jfoodeng.2017.03.019

Habashy, Magida M El. 2017. "Health benefits, extraction and utilization of dietary fibers: A review." *Menoufia Journal of Food and Dairy Sciences* 2(2):37–60. https://doi.org/10.21608/mjfds.2017.125653. https://doi.org/10.21608/mjfds.2017.176054

Harnedy, Pádraigín A, Vadivel Parthsarathy, Chris M McLaughlin, Martina B O'Keeffe, Philip J Allsopp, Emeir M McSorley, Finbarr PM O'Harte, and Richard J FitzGerald. 2018. "Atlantic salmon (Salmo salar) co-product-derived protein hydrolysates: A source of antidiabetic peptides." *Food Research International* 106:598–606. https://doi.org/10.1016/j.foodres.2018.01.025. PMid:29579965

Hartmann, Rainer, and Hans Meisel. 2007. "Food-derived peptides with biological activity: from research to food applications." *Current Opinion in Biotechnology* 18(2):163–169. https://doi.org/10.1016/j.copbio.2007.01.013. PMid:17292602

Hashim, IB, AH Khalil, and HS Afifi. 2009a. "Quality characteristics and consumer acceptance of yogurt fortified with date fiber." *Journal of Dairy Science* 92(11):5403–5407. https://doi.org/10.3168/jds.2009-2234. PMid:19841201

Hashim, Isam, AH Khalil, and Hanan Afifi. 2009b. "Quality characteristics and consumer acceptance of yogurt fortified with date fiber." *Journal of Dairy Science* 92(11):5403–5407. https://doi.org/10.3168/jds.2009-2234. https://doi.org/10.3168/jds.2009-2234. PMid:19841201

Hayes, Maria, R Paul Ross, Gerald F Fitzgerald, and Catherine Stanton. 2007. "Putting microbes to work: Dairy fermentation, cell factories and bioactive peptides. Part I: overview." *Biotechnology Journal: Healthcare Nutrition Technology* 2(4):426–434. https://doi.org/10.1002/biot.200600246. PMid:17407210

Hegenbart, S. (1995). Using fibres in beverages. *Food Product Design, 5*(3), 68–78.

Henle, T., A. Deussen, and M. Martin, Inventors; Technische Universitaet Dresden, Assignee. 2014. "Whey protein hydrolysate containing tryptophan peptide consisting of alpha lactalbumin and the use thereof." United States patent application US 14/192,898. June 26.

Hernández-Ledesma, Blanca, Chia-Chien Hsieh, and O Ben. 2009. "Antioxidant and anti-inflammatory properties of cancer preventive peptide lunasin in RAW 264.7 macrophages." *Biochemical and Biophysical Research Communications* 390(3):803–808. https://doi.org/10.1016/j.bbrc.2009.10.053. PMid:19836349

Hernández-Ledesma, Blanca, Marta Miguel, Lourdes Amigo, Maria Amaya Aleixandre, and Isidra Recio. 2007. "Effect of simulated gastrointestinal digestion on the antihypertensive properties of synthetic [beta]-lactoglobulin peptide sequences." *The Journal of Dairy Research* 74(3):336. https://doi.org/10.1017/S0022029907002609. PMid:17466121

Hernández-Ledesma, Blanca, Beatriz Miralles, Lourdes Amigo, Mercedes Ramos, and Isidra Recio. 2005. "Identification of antioxidant and ACE-inhibitory peptides in fermented milk." *Journal of the Science of Food and Agriculture* 85(6):1041–1048. https://doi.org/10.1002/jsfa.2063

Hill, MJ. 1974. "Colon cancer: A disease of fibre depletion or of dietary excess?." *Digestion* 11(3–4):289–306. https://doi.org/10.1159/000197593. PMid:4141314

Hillemeier, C. 1995. "An overview of the effects of dietary fiber on gastrointestinal transit." *Pediatrics* 96(5 Pt 2):997–999. https://doi.org/10.1542/peds.96.5.997. PMid:7494680

Holscher, Hannah D. 2017. "Dietary fiber and prebiotics and the gastrointestinal microbiota." *Gut Microbes* 8(2):172–184. https://doi.org/10.1080/19490976.2017.1290756. PMid:28165863. PMCid:PMC5390821

Hosseini, Seyed Fakhreddin, Mojgan Zandi, Masoud Rezaei, and Farhid Farahmandghavi. 2013. "Two-step method for encapsulation of oregano essential oil in chitosan nanoparticles: Preparation, characterization and in vitro release study." *Carbohydrate Polymers* 95(1):50–56. https://doi.org/10.1016/j.carbpol.2013.02.031. PMid:23618238

Huang, Fangfang, Zuisu Yang, Di Yu, Jiabin Wang, Rong Li, and Guofang Ding. 2012. "Sepia ink oligopeptide induces apoptosis in prostate cancer cell lines via caspase-3 activation and elevation of Bax/Bcl-2 ratio." *Marine Drugs* 10(10):2153–2165. https://doi.org/10.3390/md10102153. PMid:23170075. PMCid:PMC3497014

Hung, Chuan-Chuan, Yu-Hsuan Yang, Pei-Feng Kuo, and Kuo-Chiang Hsu. 2014. "Protein hydrolysates from tuna cooking juice inhibit cell growth and induce apoptosis of human breast cancer cell line MCF-7." *Journal of Functional Foods* 11:563–570. https://doi.org/10.1016/j.jff.2014.08.015

Idouraine, Ahmed, Bibi Z Hassani, Saffiatu S Claye, and Charles W Weber. 1995. "In vitro binding capacity of various fiber sources for magnesium, zinc, and copper." *Journal of Agricultural and Food Chemistry* 43(6):1580–1584. https://doi.org/10.1021/jf00054a031

Jaime, Laura, Esperanza Mollá, Almudena Fernández, María A Martín-Cabrejas, Francisco J López-Andréu, and Rosa M Esteban. 2002. "Structural carbohydrate differences and potential source of dietary fiber of onion (Allium cepa L.) tissues." *Journal of Agricultural and Food Chemistry* 50(1):122–128. https://doi.org/10.1021/jf010797t. PMid:11754555

Jakubczyk, Anna, Monika Karaś, Kamila Rybczyńska-Tkaczyk, Ewelina Zielińska, and Damian Zieliński. 2020. "Current trends of bioactive peptides-new sources and therapeutic effect." *Foods* 9(7):846. https://doi.org/10.3390/foods9070846. PMid:32610520. PMCid:PMC7404774

Jenkins, A. L., Vuksan, V., & Jenkins, D. J. (2001). Fiber in the treatment of hyperlipidemia. CRC Handbook of dietary fiber in human nutrition, *3*, 401–412.

Jenkins, D. 1993. "Fiber in the treatment of hyperlipidemia." In *CRC handbook of dietary fiber in human nutrition.*

Jenkins, DJA, RH Taylor, DV Goff, H Fielden, JJ Misiewicz, DL Sarson, St R Bloom, and KGMM Alberti. 1981. "Scope and specificity of acarbose in slowing carbohydrate absorption in man." *Diabetes* 30(11):951–954. https://doi.org/10.2337/diab.30.11.951. https://doi.org/10.2337/diabetes.30.11.951. PMid:7028548

Jia, Zhanghu, Marie-Josée Dumont, and Valérie Orsat. 2016. "Encapsulation of phenolic compounds present in plants using protein matrices." *Food Bioscience* 15:87–104. https://doi.org/10.1016/j.fbio.2016.05.007

Joye, Iris J, Gabriel Davidov-Pardo, and David Julian McClements. 2014. "Nanotechnology for increased micronutrient bioavailability." *Trends in Food Science and Technology* 40(2):168–182. https://doi.org/10.1016/j.tifs.2014.08.006

Kanbargi, Ketaki D, Sachin K Sonawane, and Shalini S Arya. 2017. "Encapsulation characteristics of protein hydrolysate extracted from Ziziphus jujube seed." *International Journal of Food Properties* 20(12):3215–3224. https://doi.org/10.1080/10942912.2017.1282516. https://doi.org/10.1080/10942912.2017.1282516

Karaś, Monika. 2019. "Influence of physiological and chemical factors on the absorption of bioactive peptides." *International Journal of Food Science and Technology* 54(5):1486–1496. https://doi.org/10.1111/ijfs.14054

Karnjanapratum, S., & Benjakul, S. (2017). Antioxidative and sensory properties of instant coffee fortified with galactose-fish skin gelatin hydrolysate maillard reaction products. *Carpathian Journal of Food Science & Technology*, *9*(1), 90–99.

Kaur, S, S Kumar, and ZF Bhat. 2015. "Utilization of pomegranate seed powder and tomato powder in the development of fiber-enriched chicken nuggets." *Nutrition and Food Science* 45(5):793–807. https://doi.org/10.1108/NFS-05-2015-0066. https://doi.org/10.1108/NFS-05-2015-0066

Kendall, Cyril WC, Amin Esfahani, and David JA Jenkins. 2010. "The link between dietary fibre and human health." *Food Hydrocolloids* 24(1):42–48. https://doi.org/10.1016/j.foodhyd.2009.08.002

Khanum, Farhath, M Siddalinga Swamy, KR Sudarshana Krishna, K Santhanam, and KR Viswanathan. 2000. "Dietary fiber content of commonly fresh and cooked vegetables consumed in India." *Plant Foods for Human Nutrition* 55(3):207–218. https://doi.org/10.1023/A:1008155732404. PMid:11030475

Kim, Youngyo, and Youjin Je. 2016. "Dietary fibre intake and mortality from cardiovascular disease and all cancers: A meta-analysis of prospective cohort studies." *Archives of Cardiovascular Diseases* 109(1):39–54. https://doi.org/10.1016/j.acvd.2015.09.005. PMid:26711548

Koziolek, M, F Schneider, M Grimm, Chr Modeβ, A Seekamp, T Roustom, W Siegmund, and W Weitschies. 2015. "Intragastric pH and pressure profiles after intake of the high-caloric, high-fat meal as used for food effect studies." *Journal of Controlled Release* 220(A):71–78. https://doi.org/10.1016/j.jconrel.2015.10.022. PMid:26476174

Kuba, Megumi, Kumi Tanaka, Shinkichi Tawata, Yasuhito Takeda, and Masaaki Yasuda. 2003. "Angiotensin I-converting enzyme inhibitory peptides isolated from tofuyo fermented soybean food." *Bioscience, Biotechnology, and Biochemistry* 67(6):1278–1283. https://doi.org/10.1271/bbb.67.1278. PMid:12843654

Kumar Malik, Sanjula Dhirendra, Sanjula Baboota, Alka Ahuja, Sohail Hasan, and Javed Ali. 2007. "Recent advances in protein and peptide drug delivery systems." *Current Drug Delivery* 4(2):141–151. https://doi.org/10.2174/156720107780362339. PMid:17456033

Kurozawa, Louise Emy, and Miriam Dupas Hubinger. 2017. "Hydrophilic food compounds encapsulation by ionic gelation." *Current Opinion in Food Science* 15:50–55. https://doi.org/10.1016/j.cofs.2017.06.004. https://doi.org/10.1016/j.cofs.2017.06.004

Kurozawa, Louise Emy, Kil Jin Park, and Miriam Dupas Hubinger. 2009a. "Effect of carrier agents on the physicochemical properties of a spray dried chicken meat protein hydrolysate." *Journal of Food Engineering* 94(3):326–333. https://doi.org/10.1016/j.jfoodeng.2009.03.025. https://doi.org/10.1016/j.jfoodeng.2009.03.025

Kurozawa, Louise Emy, Kil Jin Park, and Miriam Dupas Hubinger. 2009b. "Effect of carrier agents on the physicochemical properties of a spray dried chicken meat protein hydrolysate." *Journal of Food Engineering* 94(3–4):326–333. https://doi.org/10.1016/j.jfoodeng.2009.03.025

Kwon, Dae Young, Sang Mee Hong, Il Sung Ahn, Min Jung Kim, Hye Jeong Yang, and Sunmin Park. 2011. "Isoflavonoids and peptides from meju, long-term fermented soybeans, increase insulin sensitivity and exert insulinotropic effects in vitro." *Nutrition* 27(2):244–252. https://doi.org/10.1016/j.nut.2010.02.004. PMid:20541368

Lajmi, Khouloud, Joaquín Gómez-Estaca, Mohamed Hammami, and Oscar Martínez-Alvarez. 2019. "Upgrading collagenous smooth hound by-products: Effect of hydrolysis conditions, in vitro gastrointestinal digestion and encapsulation on bioactive properties." *Food Bioscience* 28:99–108. https://doi.org/10.1016/j.fbio.2019.01.014. https://doi.org/10.1016/j.fbio.2019.01.014

Lam, Yi, Alfredo Galvez, and Ben O de Lumen. 2003. "Lunasin™ suppresses E1A-mediated transformation of mammalian cells but does not inhibit growth of immortalized and established cancer cell lines." *Nutrition and Cancer* 47(1):88–94. https://doi.org/10.1207/s15327914nc4701_11. PMid:14769542

Lambo, Adele M, Rickard Öste, and Margareta EG-L Nyman. 2005. "Dietary fibre in fermented oat and barley β-glucan rich concentrates." *Food Chemistry* 89(2):283–293. https://doi.org/10.1016/j.foodchem.2004.02.035

Lammi, Carmen, Chiara Zanoni, and Anna Arnoldi. 2015. "IAVPGEVA, IAVPTGVA, and LPYP, three peptides from soy glycinin, modulate cholesterol metabolism in HepG2 cells through the activation of the LDLR-SREBP2 pathway." *Journal of Functional Foods* 14:469–478. https://doi.org/10.1016/j.jff.2015.02.021

Lammi, Carmen, Chiara Zanoni, Anna Arnoldi, and Giulio Vistoli. 2016. "Peptides derived from soy and lupin protein as dipeptidyl-peptidase IV inhibitors: In vitro biochemical screening and in silico molecular modeling study." *Journal of Agricultural and Food Chemistry* 64(51):9601–9606. https://doi.org/10.1021/acs.jafc.6b04041. PMid:27983830

Larrauri, JA, B Borroto, U Perdomo, and Y Tabares. 1995. "Manufacture of a powdered drink containing dietary fibre: FIBRALAX." *Alimentaria* 260:23–25

Larrauri, José A, Bárbara Borroto, and Arelys R Crespo. 1997. "Water recycling in processing orange peel to a high dietary fibre powder." *International Journal of Food Science and Technology* 32(1):73–76. https://doi.org/10.1046/j.1365-2621.1997.00383.x

Lassoued, Imen, Leticia Mora, Ahmed Barkia, M-Concepción Aristoy, Moncef Nasri, and Fidel Toldrá. 2015. "Bioactive peptides identified in thornback ray skin's gelatin hydrolysates by proteases from Bacillus subtilis and Bacillus amyloliquefaciens." *Journal of Proteomics* 128:8–17. https://doi.org/10.1016/j.jprot.2015.06.016. PMid:26149667

Lemes, Ailton Cesar, Luisa Sala, Joana da Costa Ores, Anna Rafaela Cavalcante Braga, Mariana Buranelo Egea, and Kátia Flávia Fernandes. 2016. "A review of the latest advances in encrypted bioactive peptides from protein-rich waste." *International Journal of Molecular Sciences* 17(6):950. https://doi.org/10.3390/ijms17060950. https://doi.org/10.3390/ijms17060950. PMid:27322241. PMCid:PMC4926483

Li, Zhiyu, Allan T Paulson, and Tom A Gill. 2015. "Encapsulation of bioactive salmon protein hydrolysates with chitosan-coated liposomes." *Journal of Functional Foods* 19:733–743. https://doi.org/10.1016/j.jff.2015.09.058. https://doi.org/10.1016/j.jff.2015.09.058

Lim, Sung-Min, Na-Kyoung Lee, Keun-Kyu Park, Yoh-Chang Yoon, and Hyun-Dong Paik. 2011. "ACE-inhibitory effect and physicochemical characteristics of yogurt beverage fortified with whey protein hydrolysates." *Food Science of Animal Resources* 31(6):886–892. https://doi.org/10.5851/kosfa.2011.31.6.886

Liu, Lianliang, Lingyi Liu, Baiyi Lu, Meiqin Chen, and Ying Zhang. 2013. "Evaluation of bamboo shoot peptide preparation with angiotensin converting enzyme inhibitory and antioxidant abilities from byproducts of canned bamboo shoots." *Journal of Agricultural and Food Chemistry* 61(23):5526–5533. https://doi.org/10.1021/jf305064h. PMid:23647018

Liu, Yaowen, Dur E Sameen, Saeed Ahmed, Jianwu Dai, and Wen Qin. 2021. "Antimicrobial peptides and their application in food packaging." *Trends in Food Science and Technology.* https://doi.org/10.1016/j.tifs.2021.04.019

Lordan, Sinéad, R Paul Ross, and Catherine Stanton. 2011. "Marine bioactives as functional food ingredients: Potential to reduce the incidence of chronic diseases." *Marine Drugs* 9(6):1056–1100. https://doi.org/10.3390/md9061056. PMid:21747748. PMCid:PMC3131561

Losso, Jack N. 2008. "The biochemical and functional food properties of the Bowman-Birk inhibitor." *Critical Reviews in Food Science and Nutrition* 48(1):94–118. https://doi.org/10.1080/10408390601177589. PMid:18274967

Lu, Mei-Ping, Rui Wang, Xiuyuan Song, Rajni Chibbar, Xiaoxia Wang, Lingyun Wu, and Qing H Meng. 2008. "Dietary soy isoflavones increase insulin secretion and prevent the development of diabetic cataracts in streptozotocin-induced diabetic rats." *Nutrition Research* 28(7):464–471. https://doi.org/10.1016/j.nutres.2008.03.009. PMid:19083447

Luccia, Barbara HD, and Mary E Kunkel. 2002. "In vitro availability of calcium from sources of cellulose, methylcellulose, and psyllium." *Food Chemistry* 77(2):139–146. https://doi.org/10.1016/S0308-8146(01)00168-6

Ma, Jing-Jiao, Xue-Ying Mao, Qian Wang, Shu Yang, Dan Zhang, Shang-Wu Chen, and Ying-Hui Li. 2014. "Effect of spray drying and freeze drying on the immunomodulatory activity, bitter taste and hygroscopicity of hydrolysate derived from whey protein concentrate." *LWT - Food Science and Technology* 56(2):296–302. https://doi.org/10.1016/j.lwt.2013.12.019. https://doi.org/10.1016/j.lwt.2013.12.019

Mackie, Alan R, Hameed Rafiee, Paul Malcolm, Louise Salt, and George van Aken. 2013. "Specific food structures supress appetite through reduced gastric emptying rate." *American Journal of Physiology-Gastrointestinal and Liver Physiology* 304(11):G1038–G1043. https://doi.org/10.1152/ajpgi.00060.2013. PMid:23578786. PMCid:PMC3680687

Maes, Wim, John Van Camp, Vanessa Vermeirssen, Mattias Hemeryck, Jean Marie Ketelslegers, Jürgen Schrezenmeir, Patrick Van Oostveldt, and André Huyghebaert. 2004. "Influence of the lactokinin Ala-Leu-Pro-Met-His-Ile-Arg (ALPMHIR) on the release of endothelin-1 by endothelial cells." *Regulatory Peptides* 118(1–2):105–109. https://doi.org/10.1016/j.regpep.2003.11.005. PMid:14759563

Mahato, Ram I, Ajit S Narang, Laura Thoma, and Duane D Miller. 2003. "Emerging trends in oral delivery of peptide and protein drugs." *Critical Reviews™ in Therapeutic Drug Carrier Systems* 20(2&3). https://doi.org/10.1615/CritRevTherDrugCarrierSyst.v20.i23.30. PMid:14584523

Mann, Bimlesh, Anuradha Kumari, Rajesh Kumar, Rajan Sharma, Kishore Prajapati, Shaik Mahboob, and S Athira. 2015. "Antioxidant activity of whey protein hydrolysates in milk beverage system." *Journal of Food Science and Technology* 52(6):3235–3241. https://doi.org/10.1007/s13197-014-1361-3

Marthandam Asokan, Shibu, Ting Wang, Wei-Ting Su, and Wan-Teng Lin. 2019. "Antidiabetic effects of a short peptide of potato protein hydrolysate in STZ-induced diabetic mice." *Nutrients* 11(4):779. https://doi.org/10.3390/nu11040779. PMid:30987324. PMCid:PMC6520812

Martin, K. 1999. "Replacing fat, retaining taste." *Food Eng Int* 24(3):57–58.

Martinez-Saez, Nuria, Alba Tamargo García, Inés Domínguez Pérez, Miguel Rebollo-Hernanz, Marta Mesías, Francisco J Morales, María A Martín-Cabrejas, and Maria Dolores Del Castillo. 2017. "Use of spent coffee grounds as food ingredient in bakery products." *Food Chemistry* 216:114–122. https://doi.org/10.1016/j.foodchem.2016.07.173. PMid:27596399

Mendanha, Debora V, Sara E Molina Ortiz, Carmen S Favaro-Trindade, Adriana Mauri, Ednelí S Monterrey-Quintero, and Marcelo Thomazini. 2009. "Microencapsulation of casein hydrolysate by complex coacervation with SPI/pectin." *Food Research International* 42(8):1099–1104. https://doi.org/10.1016/j.foodres.2009.05.007. https://doi.org/10.1016/j.foodres.2009.05.007

Miner-Williams, Warren M, Bruce R Stevens, and Paul J Moughan. 2014. "Are intact peptides absorbed from the healthy gut in the adult human?." *Nutrition Research Reviews* 27(2):308–329. https://doi.org/10.1017/S0954422414000225. PMid:25623084

Minkiewicz, Piotr, Anna Iwaniak, and Małgorzata Darewicz. 2019. "BIOPEP-UWM database of bioactive peptides: Current opportunities." *International Journal of Molecular Sciences* 20(23):5978. https://doi.org/10.3390/ijms20235978. PMid:31783634. PMCid:PMC6928608

Mišurcová, Ladislava, Soňa Škrovánková, Dušan Samek, Jarmila Ambrožová, and Ludmila Machů. 2012. "Health benefits of algal polysaccharides in human nutrition." *Advances in Food and Nutrition Research* 66:75–145. https://doi.org/10.1016/B978-0-12-394597-6.00003-3. PMid:22909979

Miyada, Tomihiro, Akira Nakajima, and Kiyoshi Ebihara. 2011. "Iron bound to pectin is utilised by rats." *British Journal of Nutrition* 106(1):73–78. https://doi.org/10.1017/S0007114510005842. PMid:21521538

Mohan, Aishwarya, David Julian McClements, and Chibuike C Udenigwe. 2016. "Encapsulation of bioactive whey peptides in soy lecithin-derived nanoliposomes: Influence of peptide molecular weight." *Food Chemistry* 213:143–148. https://doi.org/10.1016/j.foodchem.2016.06.075. PMid:27451165

Mohanty, Debapriya, Rajashree Jena, Prasanta Kumar Choudhury, Ritesh Pattnaik, Swati Mohapatra, and Manish Ranjan Saini. 2016. "Milk derived antimicrobial bioactive peptides: A review." *International Journal of Food Properties* 19(4):837–846. https://doi.org/10.1080/10942912.2015.1048356

Mojica, Luis, Elvira Gonzalez de Mejia, María Ángeles Granados-Silvestre, and Marta Menjivar. 2017. "Evaluation of the hypoglycemic potential of a black bean hydrolyzed protein isolate and its pure peptides using in silico, in vitro and in vivo approaches." *Journal of Functional Foods* 31:274–286. https://doi.org/10.1016/j.jff.2017.02.006

Möller, Niels Peter, Katharina Elisabeth Scholz-Ahrens, Nils Roos, and Jürgen Schrezenmeir. 2008. "Bioactive peptides and proteins from foods: Indication for health effects." *European Journal of Nutrition* 47(4):171–182. https://doi.org/10.1007/s00394-008-0710-2. PMid:18506385

Montagne, Lucile, JR Pluske, and DJ Hampson. 2003. "A review of interactions between dietary fibre and the intestinal mucosa, and their consequences on digestive health in young non-ruminant animals." *Animal Feed Science and Technology* 108(1–4):95–117. https://doi.org/10.1016/S0377-8401(03)00163-9

Morris, PE, and RJ FitzGerald. 2008. *Whey proteins and peptides in human health.* Wiley Online Library.

Mosquera, Mauricio, Begoña Giménez, Indjara Mallmann da Silva, Juliana Ferreira Boelter, Pilar Montero, M Carmen Gómez-Guillén, and Adriano Brandelli. 2014. "Nanoencapsulation of an active peptidic fraction from sea bream scales collagen." *Food Chemistry* 156:144–150. https://doi.org/10.1016/j.foodchem.2014.02.011. PMid:24629950

Moure, Andrés, Herminia Domínguez, and Juan Carlos Parajó. 2006. "Antioxidant properties of ultrafiltration-recovered soy protein fractions from industrial effluents and their hydrolysates." *Process Biochemistry* 41(2):447–456. https://doi.org/10.1016/j.procbio.2005.07.014

Mudgil, Priti, Hina Kamal, Gan Chee Yuen, and Sajid Maqsood. 2018. "Characterization and identification of novel antidiabetic and anti-obesity peptides from camel milk protein hydrolysates." *Food Chemistry* 259:46–54. https://doi.org/10.1016/j.foodchem.2018.03.082. PMid:29680061

Mukurumbira, AR, RA Shellie, R Keast, EA Palombo, and SR Jadhav. 2022. "Encapsulation of essential oils and their application in antimicrobial active packaging." *Food Control* 136:108883. https://doi.org/10.1016/j.foodcont.2022.108883. https://doi.org/10.1016/j.foodcont.2022.108883

Nakahara, Takeharu, Atsushi Sano, Hitomi Yamaguchi, Katsutoshi Sugimoto, Hiroyuki Chikata, Emiko Kinoshita, and Riichiro Uchida. 2010. "Antihypertensive effect of peptide-enriched soy sauce-like seasoning and identification of its angiotensin I-converting enzyme inhibitory substances." *Journal of Agricultural and Food Chemistry* 58(2):821–827. https://doi.org/10.1021/jf903261h. PMid:19994857

Nakahara, Takeharu, Katsutoshi Sugimoto, Atsushi Sano, Hitomi Yamaguchi, Hiroshi Katayama, and Riichiro Uchida. 2011. "Antihypertensive mechanism of a peptide-enriched soy sauce-like seasoning: The active constituents and its suppressive effect on renin-angiotensin-aldosterone system." *Journal of Food Science* 76(8):H201–H206. https://doi.org/10.1111/j.1750-3841.2011.02362.x. PMid:22417592

Nakamura, Yasunori, Naoyuki Yamamoto, Kumi Sakai, Akira Okubo, Sunao Yamazaki, and Toshiaki Takano. 1995. "Purification and characterization of angiotensin I-converting enzyme inhibitors from sour milk." *Journal of Dairy Science* 78(4):777–783. https://doi.org/10.3168/jds.S0022-0302(95)76689-9

Nasri, Rim, Marwa Hamdi, Sana Touir, Suming Li, Maha Karra-Chaâbouni, and Moncef Nasri. 2021. "Development of delivery system based on marine chitosan: Encapsulationand release kinetic study of antioxidant peptides from chitosan microparticle." *International Journal of Biological Macromolecules* 167:1445–1451. https://doi.org/10.1016/j.ijbiomac.2020.11.098. https://doi.org/10.1016/j.ijbiomac.2020.11.098. PMid:33212105

Nassar, AG, AA AbdEl-Hamied, and EA El-Naggar. 2008a. "Effect of citrus by-products flour incorporation on chemical, rheological and organolepic characteristics of biscuits." *World Journal of Agricultural Sciences* 4(5):612–616.

Nassar, A. G., AbdEl-Hamied, A. A., & El-Naggar, E. A. (2008). Effect of citrus by-products flour incorporation on chemical, rheological and organolepic characteristics of biscuits. *World Journal of Agricultural Sciences, 4*(5), 612–616.

Ning, X, Y Zhou, Z Wang, X Zheng, X Pan, Z Chen, Q Liu, W Du, X Cao, and L Wang. 2022. "Evaluation of passion fruit mesocarp flour on the paste, dough, and quality characteristics of dried noodles." *Food Science and Nutrition.* https://doi.org/10.1002/fsn3.2788. https://doi.org/10.1002/fsn3.2788

Nongonierma, Alice B, and Richard J FitzGerald. 2015. "Bioactive properties of milk proteins in humans: A review." *Peptides* 73:20–34. https://doi.org/10.1016/j.peptides.2015.08.009. PMid:26297879

Nordentoft, I, Per Bendix Jeppesen, J Hong, R Abudula, and K Hermansen. 2008. "Increased insulin sensitivity and changes in the expression profile of key insulin regulatory genes and beta cell transcription factors in diabetic KKAy-mice after feeding with a soy bean protein rich diet high in isoflavone content." *Journal of Agricultural and Food Chemistry* 56(12):4377–4385. https://doi.org/10.1021/jf800504r. PMid:18522411

Nurminen, Marja-Leena, Marika Sipola, Hanna Kaarto, Anne Pihlanto-Leppälä, Kati Piilola, Riitta Korpela, Olli Tossavainen, Hannu Korhonen, and Heikki Vapaatalo. 2000. "α-Lactorphin lowers blood pressure measured by radiotelemetry in normotensive and spontaneously hypertensive rats." *Life Sciences* 66(16):1535–1543. https://doi.org/10.1016/S0024-3205(00)00471-9

Ogrodowska, Dorota, Małgorzata Tańska, and Waldemar Brandt. 2017. "The influence of drying process conditions on the physical properties, bioactive compounds and stability of encapsulated pumpkin seed oil." *Food and Bioprocess Technology* 10(7):1265–1280. https://doi.org/10.1007/s11947-017-1898-z

Okamoto, Akiko, Hiroshi Hanagata, Eiko Matsumoto, Yukio Kawamura, Yukimichi Koizumi, and Fujiharu Yanagida. 1995. "Angiotensin I converting enzyme inhibitory activities of various fermented foods." *Bioscience, Biotechnology, and Biochemistry* 59(6):1147–1149. https://doi.org/10.1271/bbb.59.1147. PMid:7613003

Osendarp, Saskia JM, Homero Martinez, Greg S Garrett, Lynnette M Neufeld, Luz Maria De-Regil, Marieke Vossenaar, and Ian Darnton-Hill. 2018. "Large-scale food fortification and biofortification in low- and middle-income countries: A review of programs, trends, challenges, and evidence gaps." *Food and Nutrition Bulletin* 39(2):315–331. https://doi.org/10.1177/0379572118774229. PMid:29793357. PMCid:PMC7473077

Ozkan, Gulay, Paola Franco, Iolanda De Marco, Jianbo Xiao, and Esra Capanoglu. 2019. "A review of microencapsulation methods for food antioxidants: Principles, advantages, drawbacks and applications." *Food Chemistry* 272:494–506. https://doi.org/10.1016/j.foodchem.2018.07.205. https://doi.org/10.1016/j.foodchem.2018.07.205. PMid:30309574

Paiva, Lisete, Elisabete Lima, Ana Isabel Neto, and José Baptista. 2017. "Angiotensin I-converting enzyme (ACE) inhibitory activity, antioxidant properties, phenolic content and amino acid profiles of Fucus spiralis L. protein hydrolysate fractions." *Marine Drugs* 15(10):311. https://doi.org/10.3390/md15100311. PMid:29027934. PMCid:PMC5666419

Palafox-Carlos, Hugo, Jesús Fernando Ayala-Zavala, and Gustavo A González-Aguilar. 2011. "The role of dietary fiber in the bioaccessibility and bioavailability of fruit and vegetable antioxidants." *Journal of Food Science* 76(1):R6–R15. https://doi.org/10.1111/j.1750-3841.2010.01957.x. PMid:21535705. PMCid:PMC3052441

Pane, Katia, Lorenzo Durante, Orlando Crescenzi, Valeria Cafaro, Elio Pizzo, Mario Varcamonti, Anna Zanfardino, Viviana Izzo, Alberto Di Donato, and Eugenio Notomista. 2017. "Antimicrobial potency of cationic antimicrobial peptides can be predicted from their amino acid composition: Application to the detection of "cryptic" antimicrobial peptides." *Journal of Theoretical Biology* 419:254–265. https://doi.org/10.1016/j.jtbi.2017.02.012. PMid:28216428

Papathanasopoulos, Athanasios, and Michael Camilleri. 2010. "Dietary fiber supplements: Effects in obesity and metabolic syndrome and relationship to gastrointestinal functions." *Gastroenterology* 138(1):65-72.e2. https://doi.org/10.1053/j.gastro.2009.11.045. https://doi.org/10.1053/j.gastro.2009.11.045. PMid:19931537. PMCid:PMC2903728

Peng, Xinyan, Baohua Kong, Xiufang Xia, and Qian Liu. 2010. "Reducing and radical-scavenging activities of whey protein hydrolysates prepared with alcalase." *International Dairy Journal* 20(5):360–365. https://doi.org/10.1016/j.idairyj.2009.11.019. https://doi.org/10.1016/j.idairyj.2009.11.019

Pepe, Giacomo, Eduardo Sommella, Giovanni Ventre, Maria Carmina Scala, Simona Adesso, Carmine Ostacolo, Stefania Marzocco, Ettore Novellino, and Pietro Campiglia. 2016. "Antioxidant peptides released from gastrointestinal digestion of "Stracchino" soft cheese: Characterization, in vitro intestinal protection and bioavailability." *Journal of Functional Foods* 26:494–505. https://doi.org/10.1016/j.jff.2016.08.021

Pereira, RN, and AA Vicente. 2010. "Environmental impact of novel thermal and non-thermal technologies in food processing." *Food Research International* 43(7):1936–1943. https://doi.org/10.1016/j.foodres.2009.09.013

Pérez-Chabela, Maria Lourdes, and Annel M Hernández-Alcántara. 2018. "Agroindustrial coproducts as sources of novel functional ingredients." In *Food processing for increased quality and consumption*, 219–250. Elsevier. https://doi.org/10.1016/B978-0-12-811447-6.00008-4

Pérez-Gregorio, Rosa, Susana Soares, Nuno Mateus, and Victor de Freitas. 2020. "Bioactive peptides and dietary polyphenols: Two sides of the same coin." *Molecules* 25(15):3443. https://doi.org/10.3390/molecules25153443. PMid:32751126. PMCid:PMC7435807

Periago, M. J., Ros, G., López, G., Martínez, M. C., & Ricon, F. (1993). Dietary fiber components and their physiological effects. *Revista Espanola de Ciencia y Tecnologia de Alimentos (Espana), 33*, 229–246.

Piccolomini, André F, Michèle M Iskandar, Larry C Lands, and Stan Kubow. "High hydrostatic pressure pre-treatment of whey proteins enhances whey protein hydrolysate inhibition of oxidative stress and IL-8 secretion in intestinal epithelial cells." *Food and Nutrition Research* 56(1):17549. https://doi.org/10.3402/fnr.v56i0.17549. PMid:22723766. PMCid:PMC3380274

Pihlanto, Anne. 2006. "Antioxidative peptides derived from milk proteins." *International Dairy Journal* 16(11):1306–1314. https://doi.org/10.1016/j.idairyj.2006.06.005. https://doi.org/10.1016/j.idairyj.2006.06.005

Potter, Ruth, Valentina Stojceska, and Andrew Plunkett. 2013. "The use of fruit powders in extruded snacks suitable for children's diets." *LWT – Food Science and Technology* 51(2):537–544. https://doi.org/10.1016/j.lwt.2012.11.015

Pritchard, Stephanie Rae, Michael Phillips, and Kasipathy Kailasapathy. 2010. "Identification of bioactive peptides in commercial Cheddar cheese." *Food Research International* 43(5):1545–1548. https://doi.org/10.1016/j.foodres.2010.03.007

Przybylski, Rémi, Loubna Firdaous, Gabrielle Châtaigné, Pascal Dhulster, and Naïma Nedjar. 2016. "Production of an antimicrobial peptide derived from slaughterhouse by-product and its potential application on meat as preservative." *Food Chemistry* 211:306–313. https://doi.org/10.1016/j.foodchem.2016.05.074. PMid:27283637

Qian, Zhong-Ji, Won-Kyo Jung, and Se-Kwon Kim. 2008. "Free radical scavenging activity of a novel antioxidative peptide purified from hydrolysate of bullfrog skin, Rana catesbeiana Shaw." *Bioresource Technology* 99(6):1690–1698. https://doi.org/10.1016/j.biortech.2007.04.005. PMid:17512726

Ramezanzade, Leila, Seyed Fakhreddin Hosseini, and Maryam Nikkhah. 2017. "Biopolymer-coated nanoliposomes as carriers of rainbow trout skin-derived antioxidant peptides." *Food Chemistry* 234:220–229. https://doi.org/10.1016/j.foodchem.2017.04.177. https://doi.org/10.1016/j.foodchem.2017.04.177. PMid:28551229

Rao, Priyanka Singh, Rajesh Kumar Bajaj, Bimlesh Mann, Sumit Arora, and SK Tomar. 2016. "Encapsulation of antioxidant peptide enriched casein hydrolysate using maltodextrin-gum arabic blend." *Journal of Food Science and Technology* 53(10):3834–3843. https://doi.org/10.1007/s13197-016-2376-8. https://doi.org/10.1007/s13197-016-2376-8. PMid:28017999. PMCid:PMC5147710

Renukuntla, Jwala, Aswani Dutt Vadlapudi, Ashaben Patel, Sai HS Boddu, and Ashim K Mitra. 2013. "Approaches for enhancing oral bioavailability of peptides and proteins." *International Journal of Pharmaceutics* 447(1–2):75–93. https://doi.org/10.1016/j.ijpharm.2013.02.030. PMid:23428883. PMCid:PMC3680128

Ricci-Cabello, Ignacio, Manuel Olalla Herrera, and Reyes Artacho. 2012. "Possible role of milk-derived bioactive peptides in the treatment and prevention of metabolic syndrome." *Nutrition Reviews* 70(4):241–255. https://doi.org/10.1111/j.1753-4887.2011.00448.x. PMid:22458697

Riedl, Judith, Jakob Linseisen, Jürgen Hoffmann, and Günther Wolfram. 1999. "Some dietary fibers reduce the absorption of carotenoids in women." *The Journal of Nutrition* 129(12):2170–2176. https://doi.org/10.1093/jn/129.12.2170. PMid:10573545

Rivas, A, D Rodrigo, B Company, F Sampedro, and M Rodrigo. 2007. "Effects of pulsed electric fields on water-soluble vitamins and ACE inhibitory peptides added to a mixed orange juice and milk beverage." *Food Chemistry* 104(4):1550–1559. https://doi.org/10.1016/j.foodchem.2007.02.034

Rivero-Pino, Fernando, F Javier Espejo-Carpio, and Emilia M Guadix. 2020. "Evaluation of the bioactive potential of foods fortified with fish protein hydrolysates." *Food Research International* 137:109572. https://doi.org/10.1016/j.foodres.2020.109572. PMid:33233184

Roblet, Cyril, Muhammad Javeed Akhtar, Sergey Mikhaylin, Geneviève Pilon, Tom Gill, André Marette, and Laurent Bazinet. 2016. "Enhancement of glucose uptake in muscular cell by peptide fractions separated by electrodialysis with filtration membrane from salmon frame protein hydrolysate." *Journal of Functional Foods* 22:337–346. https://doi.org/10.1016/j.jff.2016.01.003

Rock, Cheryl L, and Marian E Swendseid. 1992. "Plasma β-carotene response in humans after meals supplemented with dietary pectin." *The American Journal of Clinical Nutrition* 55(1):96–99. https://doi.org/10.1093/ajcn/55.1.96. PMid:1309477

Roediger, W. E. (1995). The place of short-chain fatty acids in colonocyte metabolism in health and ulcerative colitis: The impaired colonocyte barrier. Physiological and Clinical Aspects of Short-chain fatty acids. Cambridge University Press. https://digital.library.adelaide.edu.au/dspace/handle/2440/31015

Rukluarh, Sureerat, Kobsak Kanjanapongkul, Noppadol Panchan, and Chalida Niumnuy. 2019. "Effect of inclusion conditions on characteristics of spray dried whey protein hydrolysate/γ-cyclodextrin complexes." *Journal of Food Science and Agricultural Technology (JFAT)* 5:5–12.

Sachdeva, Aarti, Swapnil Rawat, and Jitender Nagpal. 2014. "Efficacy of fermented milk and whey proteins in Helicobacter pylori eradication: A review." *World Journal of Gastroenterology: WJG* 20(3):724. https://doi.org/10.3748/wjg.v20.i3.724. PMid:24574746. PMCid:PMC3921482

Sakai, Kensuke, Atsutane Ohta, Kazuki Shiga, Misao Takasaki, Takahisa Tokunaga, and Hiroshi Hara. 2000. "The cecum and dietary short-chain fructooligosaccharides are involved in preventing postgastrectomy anemia in rats." *The Journal of Nutrition* 130(6):1608–1612. https://doi.org/10.1093/jn/130.6.1608. PMid:10827217

Sánchez-Rivera, Laura, Olivia Ménard, Isidra Recio, and Didier Dupont. 2015. "Peptide mapping during dynamic gastric digestion of heated and unheated skimmed milk powder." *Food Research International* 77:132–139. https://doi.org/10.1016/j.foodres.2015.08.001

Sánchez, Adrián, and Alfredo Vázquez. 2017. "Bioactive peptides: A review." *Food Quality and Safety* 1(1):29–46. https://doi.org/10.1093/fqsafe/fyx006. https://doi.org/10.1093/fqsafe/fyx006

Santos, Johnson CP, Rita CS Sousa, Caio G Otoni, Allan RF Moraes, Victor GL Souza, Eber AA Medeiros, Paula JP Espitia, Ana CS Pires, Jane SR Coimbra, and Nilda FF Soares. 2018. "Nisin and other antimicrobial peptides: Production, mechanisms of action, and application in active food packaging." *Innovative Food Science and Emerging Technologies* 48:179–194. https://doi.org/10.1016/j.ifset.2018.06.008. https://doi.org/10.1016/j.ifset.2018.06.008

Sarabandi, Khashayar, Pouria Gharehbeglou, and Seid Mahdi Jafari. 2020. "Spray-drying encapsulation of protein hydrolysates and bioactive peptides: Opportunities and challenges." *Drying Technology* 38(5–6):577–595. https://doi.org/10.1080/07373937.2019.1689399. https://doi.org/10.1080/07373937.2019.1689399

Sarabandi, Khashayar, Alireaza Sadeghi Mahoonak, Hamed Hamishekar, Mohammad Ghorbani, and Seid Mahdi Jafari. 2018. "Microencapsulation of casein hydrolysates: Physicochemical, antioxidant and microstructure properties." *Journal of Food Engineering* 237:86–95. https://doi.org/10.1016/j.jfoodeng.2018.05.036

Sarabandi, Khashayar, Alireaza Sadeghi Mahoonak, Hamed Hamishekar, Mohammad Ghorbani, and Seid Mahdi Jafari. 2018. "Microencapsulation of casein hydrolysates: Physicochemical, antioxidant and microstructure properties." *Journal of Food Engineering* 237:86–95. https://doi.org/10.1016/j.jfoodeng.2018.05.036. https://doi.org/10.1016/j.jfoodeng.2018.05.036

Sauer, JM, and Hamid Merchant. 2018. "Physiology of the gastrointestinal system." In *Comprehensive toxicology*, 16–44. Elsevier. https://doi.org/10.1016/B978-0-12-801238-3.99195-5

Sáyago-Ayerdi, SG, Agustín Brenes, and Isabel Goñi. 2009. "Effect of grape antioxidant dietary fiber on the lipid oxidation of raw and cooked chicken hamburgers." *LWT – Food Science and Technology* 42(5):971–976. https://doi.org/10.1016/j.lwt.2008.12.006

Schulze, Matthias B, Simin Liu, Eric B Rimm, JoAnn E Manson, Walter C Willett, and Frank B Hu. 2004. "Glycemic index, glycemic load, and dietary fiber intake and incidence of type 2 diabetes in younger and middle-aged women." *The American Journal of Clinical Nutrition* 80(2):348–356. https://doi.org/10.1093/ajcn/80.2.348. PMid:15277155

Segura-Campos, Maira, Luis Chel-Guerrero, David Betancur-Ancona, and Victor M Hernandez-Escalante. 2011. "Bioavailability of bioactive peptides." *Food Reviews International* 27(3):213–226. https://doi.org/10.1080/87559129.2011.563395

Selamassakul, Orrapun, Natta Laohakunjit, Orapin Kerdchoechuen, and Khanok Ratanakhanokchai. 2016. "A novel multi-biofunctional protein from brown rice hydrolysed by endo/endo-exoproteases." *Food and Function* 7(6):2635–2644. https://doi.org/10.1039/C5FO01344E. PMid:27186602

Seppo, Leena, Tiina Jauhiainen, Tuija Poussa, and Riitta Korpela. 2003. "A fermented milk high in bioactive peptides has a blood pressure-lowering effect in hypertensive subjects." *The American Journal of Clinical Nutrition* 77(2):326–330. https://doi.org/10.1093/ajcn/77.2.326. PMid:12540390

Sepúlveda, Cindy T, Ailén Alemán, José E Zapata, M Pilar Montero, and M Carmen Gómez-Guillén. 2021. "Characterization and storage stability of spray dried soy-rapeseed lecithin/trehalose liposomes loaded with a tilapia viscera hydrolysate." *Innovative Food Science and Emerging Technologies* 71:102708. https://doi.org/10.1016/j.ifset.2021.102708. https://doi.org/10.1016/j.ifset.2021.102708

Sharif, Mian K, Masood S Butt, Faqir M Anjum, and Haq Nawaz. 2009. "Preparation of fiber and mineral enriched defatted rice bran supplemented cookies." *Pakistan Journal of Nutrition* 8(5):571–577. https://doi.org/10.3923/pjn.2009.571.577

Shegelman, I. R., Vasilev, A. S., Shtykov, A. S., Sukhanov, Y. V., Galaktionov, O. N., & Kuznetsov, A. V. (2019). Food fortification-problems and solutions. *Eurasian Journal of Biosciences, 13*(2), 1089–1100.

Shin, Zae-Ik, Rina Yu, Soo-Ah Park, Dae Kyun Chung, Chang-Won Ahn, Hee-Sop Nam, Kil-Soo Kim, and Hyong Joo Lee. 2001. "His-His-Leu, an angiotensin I converting enzyme inhibitory peptide derived from Korean soybean paste, exerts antihypertensive activity in vivo." *Journal of Agricultural and Food Chemistry* 49(6):3004–3009. https://doi.org/10.1021/jf001135r. PMid:11410001

Shishir, Mohammad Rezaul Islam, and Wei Chen. 2017. "Trends of spray drying: A critical review on drying of fruit and vegetable juices." *Trends in Food Science and Technology* 65:49–67. https://doi.org/10.1016/j.tifs.2017.05.006

Silva-Sánchez, C, AP Barba De La Rosa, MF León-Galván, BO De Lumen, A de León-Rodríguez, and E González De Mejía. 2008. "Bioactive peptides in amaranth (Amaranthus hypochondriacus) seed." *Journal of Agricultural and Food Chemistry* 56(4):1233–1240. https://doi.org/10.1021/jf072911z. PMid:18211015

Singh, Bahaderjeet, and A Kaur. 2016. "Antidiabetic potential of a peptide isolated from an endophytic Aspergillus awamori." *Journal of Applied Microbiology* 120(2):301–311. https://doi.org/10.1111/jam.12998. PMid:26544796

Singh, Brij Pal, Shilpa Vij, and Subrota Hati. 2014. "Functional significance of bioactive peptides derived from soybean." *Peptides* 54:171–179. https://doi.org/10.1016/j.peptides.2014.01.022. PMid:24508378

Sipola, Marika, Piet Finckenberg, Riitta Korpela, Heikki Vapaatalo, and Marja-Leena Nurminen. 2002. "Effect of long-term intake of milk products on blood pressure in hypertensive rats." *Journal of Dairy Research* 69(1):103–111. https://doi.org/10.1017/S002202990100526X. PMid:12047101

Somaratne, Geeshani, Maria J Ferrua, Aiqian Ye, Francoise Nau, Juliane Floury, Didier Dupont, and Jaspreet Singh. 2020. "Food material properties as determining factors in nutrient release during human gastric digestion: A review." *Critical Reviews in Food Science and Nutrition* 60(22):3753–3769. https://doi.org/10.1080/10408398.2019.1707770. PMid:31957483

Song, Ru, Rong-bian Wei, Hong-yu Luo, and Zui-su Yang. 2014. "Isolation and identification of an antiproliferative peptide derived from heated products of peptic hydrolysates of half-fin anchovy (Setipinna taty)." *Journal of Functional Foods* 10:104–111. https://doi.org/10.1016/j.jff.2014.06.010

Sonklin, Chanikan, Monisola A Alashi, Natta Laohakunjit, Orapin Kerdchoechuen, and Rotimi E Aluko. 2020. "Identification of antihypertensive peptides from mung bean protein hydrolysate and their effects in spontaneously hypertensive rats." *Journal of Functional Foods* 64:103635. https://doi.org/10.1016/j.jff.2019.103635

Spence, Joseph T. 2006. "Challenges related to the composition of functional foods." *Journal of Food Composition and Analysis* 19:S4–S6. https://doi.org/10.1016/j.jfca.2005.11.007

Spiller, Gene A. 2001. *CRC handbook of dietary fiber in human nutrition.* CRC Press. https://doi.org/10.1201/9781420038514

Srour, Bernard, Léopold K Fezeu, Emmanuelle Kesse-Guyot, Benjamin Allès, Caroline Méjean, Roland M Andrianasolo, Eloi Chazelas, Mélanie Deschasaux, Serge Hercberg, and Pilar Galan. 2019. "Ultraprocessed food intake and risk of cardiovascular disease: Prospective cohort study (NutriNet-Santé)." *BMJ* 365. https://doi.org/10.1136/bmj.l1451. PMid:31142457. PMCid:PMC6538975

Staffolo, M Dello, N Bertola, M Martino. 2004. "Influence of dietary fiber addition on sensory and rheological properties of yogurt." *International Dairy Journal* 14(3):263–268. https://doi.org/10.1016/j.idairyj.2003.08.004

Sullivan, Louise M, Joseph J Kehoe, Lillian Barry, Martin JM Buckley, Fergus Shanahan, KH Mok, and André Brodkorb. 2014. "Gastric digestion of α-lactalbumin in adult human subjects using capsule endoscopy and nasogastric tube sampling." *British Journal of Nutrition* 112(4):638–646. https://doi.org/10.1017/S0007114514001196. PMid:24967992

Sun, Xiaohong, Ogadimma D Okagu, and Chibuike C Udenigwe. 2021. "Chapter 15 - Encapsulation technology for protection and delivery of bioactive peptides." In *Biologically active peptides*, edited by Fidel Toldrá and Jianping Wu, 331–356. Academic Press. https://doi.org/10.1016/B978-0-12-821389-6.00028-5

Takahashi, Hidehisa, Nobuhisa Wako, Tsutomu Okubo, Noriyuki Ishihara, Junzo Yamanaka, and Takehiko Yamamoto. 1994. "Influence of partially hydrolyzed guar gum on constipation in women." *Journal of Nutritional Science and Vitaminology* 40(3):251–259. https://doi.org/10.3177/jnsv.40.251. PMid:7965214

Tang, Wenting, Hui Zhang, Li Wang, Haifeng Qian, and Xiguang Qi. 2015. "Targeted separation of antibacterial peptide from protein hydrolysate of anchovy cooking wastewater by equilibrium dialysis." *Food Chemistry* 168:115–123. https://doi.org/10.1016/j.foodchem.2014.07.027. PMid:25172690

Thomas, John C, Romano T DeMarco, and John C Pope. 2005. "Molecular biology of ureteral bud and trigonal development." *Current Urology Reports* 6(2):146–151. https://doi.org/10.1007/s11934-005-0084-4. PMid:15717974

Threapleton, Diane, Darren Greenwood, Charlotte Evans, Christine Cleghorn, Camilla Nykjaer, Charlotte Woodhead, Janet Cade, Chris Gale, and Victoria Burley. 2012. "Dietary fibre intake and cardiovascular disease: A systematic review and meta-analysis of prospective studies." *Proceedings of the Nutrition Society* 71(OCE3). https://doi.org/10.1017/S002966511200314X. https://doi.org/10.1017/S002966511200314X

Toma, RB, PH Orr, B D'appolonia, FR Dlntzis, and MM Tabekhia. 1979. "Physical and chemical properties of potato peel as a source of dietary fiber in bread." *Journal of Food Science* 44(5):1403–1407. https://doi.org/10.1111/j.1365-2621.1979.tb06448.x

Udachan, Iranna, Abhijit Gatade, Rahul Ranveer, Siddharth Lokhande, Gurunath Mote, and Akshaya Kumar Sahoo. 2022. "Quality evaluation of gluten-free brown rice pasta formulated with green matured banana flour and defatted soy flour." *Journal of Food Processing and Preservation*. https://doi.org/10.1111/jfpp.16448

Udenigwe, Chibuike C, and Rotimi E Aluko. 2012. "Food protein-derived bioactive peptides: Production, processing, and potential health benefits." *Journal of Food Science* 77(1):R11–R24. https://doi.org/10.1111/j.1750-3841.2011.02455.x. PMid:22260122

Umayaparvathi, S, S Meenakshi, V Vimalraj, M Arumugam, G Sivagami, and T Balasubramanian. 2014. "Antioxidant activity and anticancer effect of bioactive peptide from enzymatic hydrolysate of oyster (Saccostrea cucullata)." *Biomedicine and Preventive Nutrition* 4(3):343–353. https://doi.org/10.1016/j.bionut.2014.04.006

Ünal, Mümit, and Aysun Şener. 2018. "Bioactive peptides and health effects." *Annals of the University of Craiova-Agriculture, Montanology, Cadastre Series* 48(1):208–214. https://doi.org/10.15237/gida.GD18048

van der Aar, Piet J, George C Fahey Jr, Steven C Ricke, Susan E Allen, and Larry L Berger. 1983. "Effects of dietary fibers on mineral status of chicks." *The Journal of Nutrition* 113(3):653–661. https://doi.org/10.1093/jn/113.3.653. PMid:6298389

Verma, AK, BD Sharma, and R Banerjee. 2009. "Quality characteristics and storage stability of low fat functional chicken nuggets." *Fleischwirtsch Int* 24:52–57.

Vital, Diego A Luna, Elvira González De Mejía. Vermont P Dia, and Guadalupe Loarca-Piña. 2014. "Peptides in common bean fractions inhibit human colorectal cancer cells." *Food Chemistry* 157:347–355. https://doi.org/10.1016/j.foodchem.2014.02.050. PMid:24679790

Wallace, R John. 1992. "Acetylation of peptides inhibits their degradation by rumen micro-organisms." *British Journal of Nutrition* 68(2):365–372. https://doi.org/10.1079/BJN19920095. PMid:1445818

Wang, Bo, and Bo Li. 2017. "Effect of molecular weight on the transepithelial transport and peptidase degradation of casein-derived peptides by using Caco-2 cell model." *Food Chemistry* 218:1–8. https://doi.org/10.1016/j.foodchem.2016.08.106. PMid:27719884

Wang, Bo, Ningning Xie, and Bo Li. 2016. "Charge properties of peptides derived from casein affect their bioavailability and cytoprotection against H2O2-induced oxidative stress." *Journal of Dairy Science* 99(4):2468–2479. https://doi.org/10.3168/jds.2015-10029. PMid:26851854

Wang, Bo, Ningning Xie, and Bo Li. 2019. "Influence of peptide characteristics on their stability, intestinal transport, and in vitro bioavailability: A review." *Journal of Food Biochemistry* 43(1):e12571. https://doi.org/10.1111/jfbc.12571. PMid:31353489

Wang, Chun-Yang, Shu Liu, Xiao-Nv Xie, and Zhi-Rong Tan. 2017. "Regulation profile of the intestinal peptide transporter 1 (PepT1)." *Drug Design, Development and Therapy* 11:3511. https://doi.org/10.2147/DDDT.S151725. PMid:29263649. PMCid:PMC5726373

Wang, Jian, Tailang Yin, Xuwen Xiao, Dan He, Zhidong Xue, Xinnong Jiang, and Yan Wang. 2018. "StraPep: A structure database of bioactive peptides." *Database* 2018. https://doi.org/10.1093/database/bay038

Wang, Liying, Long Ding, Zhiyang Du, and Jingbo Liu. 2020. "Effects of hydrophobicity and molecular weight on the transport permeability of oligopeptides across Caco-2 cell monolayers." *Journal of Food Biochemistry* 44(5):e13188. https://doi.org/10.1111/jfbc.13188

Wang, Wenyi, Vermont P Dia, Miguel Vasconez, Elvira Gonzalez De Mejia, and Randall L Nelson. 2008. "Analysis of soybean protein-derived peptides and the effect of cultivar, environmental conditions, and processing on lunasin concentration in soybean and soy products." *Journal of AOAC International* 91(4):936–946. https://doi.org/10.1093/jaoac/91.4.936. PMid:18727556

Wang, Xiaoqin, and Xuewu Zhang. 2013. "Separation, antitumor activities, and encapsulation of polypeptide from Chlorella pyrenoidosa." *Biotechnology Progress* 29(3):681–687. https://doi.org/10.1002/btpr.1725. PMid:23606619

Wang, Zhujun, and Xuewu Zhang. 2017. "Isolation and identification of anti-proliferative peptides from Spirulina platensis using three-step hydrolysis." *Journal of the Science of Food and Agriculture* 97(3):918–922. https://doi.org/10.1002/jsfa.7815. PMid:27218227

Ward, A Thomas, and Robert D Reichert. 1986. "Comparison of the effect of cell wall and hull fiber from canola and soybean on the bioavailability for rats of minerals, protein and lipid." *The Journal of Nutrition* 116(2):233–241. https://doi.org/10.1093/jn/116.2.233. PMid:3003294

Werle, Martin, Abdallah Makhlof, and Hirofumi Takeuchi. 2009. "Oral protein delivery: A patent review of academic and industrial approaches." *Recent Patents on Drug Delivery and Formulation* 3(2):94–104. https://doi.org/10.2174/187221109788452221. PMid:19519570

WHO. 2017. "Cardiovascular diseases (CVDs)." https://www.who.int/news-room/fact-sheets/detail/cardiovascular-diseases-(cvds)

Williams, Barbara A, Deirdre Mikkelsen, Bernadine M Flanagan, and Michael J Gidley. 2019. "'Dietary fibre': Moving beyond the 'soluble/insoluble' classification for monogastric nutrition, with an emphasis on humans and pigs." *Journal of Animal Science and Biotechnology* 10(1):1–12. https://doi.org/10.1186/s40104-019-0350-9. PMid:31149336. PMCid:PMC6537190

Wong, Ka-Hing, and Peter CK Cheung. 2005a. "Dietary fibers from mushroom sclerotia: 1. Preparation and physicochemical and functional properties." *Journal of Agricultural and Food Chemistry* 53(24):9395–9400. https://doi.org/10.1021/jf0510788. PMid:16302753

Wong, Ka-Hing, and Peter CK Cheung. 2005b. "Dietary fibers from mushroom sclerotia: 2. In vitro mineral binding capacity under sequential simulated physiological conditions of the human gastrointestinal tract." *Journal of Agricultural and Food Chemistry* 53(24):9401–9406. https://doi.org/10.1021/jf0510790. PMid:16302754

Xie, Huihui, Yumei Wang, Jing Zhang, Jingyi Chen, Ding Wu, and Lifeng Wang. 2015. "Study of the fermentation conditions and the antiproliferative activity of rapeseed peptides by bacterial and enzymatic cooperation." *International Journal of Food Science and Technology* 50(3):619–625. https://doi.org/10.1111/ijfs.12682

Xing, Lujuan, Rui Liu, Changbo Tang, Jailson Pereira, Guanghong Zhou, and Wangang Zhang. 2018. "The antioxidant activity and transcellular pathway of Asp-Leu-Glu-Glu in a Caco2 cell monolayer." *International Journal of Food Science and Technology* 53(10):2405–2414. https://doi.org/10.1111/ijfs.13771

Xu, Feng. 2010. "Structure, ultrastructure, and chemical composition." In *Cereal Straw as a Resource for Sustainable Biomaterials and Biofuels*. Amsterdam: Elsevier, 9–47. https://doi.org/10.1016/B978-0-444-53234-3.00002-X. PMCid:PMC2876244

Xue, Zhaohui, Haichao Wen, Lijuan Zhai, Yanqing Yu, Yanni Li, Wancong Yu, Aiqing Cheng, Cen Wang, and Xiaohong Kou. 2015. "Antioxidant activity and anti-proliferative effect of a bioactive peptide from chickpea (Cicer arietinum L.)." *Food Research International* 77:75–81. https://doi.org/10.1016/j.foodres.2015.09.027

Yamada, Tadataka, David H Alpers, Anthony N Kalloo, Neil Kaplowitz, Chung Owyang, and Don W Powell. 2011. *Textbook of gastroenterology*. John Wiley & Sons.

Ying, Xin, Jiaxing Gao, Jing Lu, Changlu Ma, Jiaping Lv, Benu Adhikari, and Bo Wang. 2021. "Preparation and drying of water-in-oil-in-water (W/O/W) double emulsion to encapsulate soy peptides." *Food Research International* 141:110148. https://doi.org/10.1016/j.foodres.2021.110148. PMid:33642014

Yuan, Yang, Zhi-Yan Kong, Ying-En Sun, Qing-Zhu Zeng, and Xiao-Quan Yang. 2017. "Complex coacervation of soy protein with chitosan: Constructing antioxidant microcapsule for algal oil delivery." *LWT* 75:171–179. https://doi.org/10.1016/j.lwt.2016.08.045

Zaker, MA, AR Sawate, BM Patil, SK Sadawarte, and RB Kshirsagar. 2017. "Utilization of orange (Citrus sinesis) peel powder as a source of dietary fibre and its effect on the cake quality attributes." *International Journal of Agricultural Sciences* 13(1):56–61. https://doi.org/10.15740/HAS/IJAS/13.1/56-61

Zhang, Li, Yezhou Liu, Zhongchen Wu, and Haixu Chen. 2009a. "Preparation and characterization of coacervate microcapsules for the delivery of antimicrobial oyster peptides." *Drug Development and Industrial Pharmacy* 35(3):369–378. https://doi.org/10.1080/03639040802369255. https://doi.org/10.1080/03639040802369255. PMid:18941970

Zhang, Li, Yezhou Liu, Zhongchen Wu, and Haixu Chen. 2009b. "Preparation and characterization of coacervate microcapsules for the delivery of antimicrobial oyster peptides." *Drug Development and Industrial Pharmacy* 35(3):369–378. https://doi.org/10.1080/03639040802369255. https://doi.org/10.1080/03639040802369255. PMid:18941970

Zhang, Qiaozhi, Xiaohong Tong, Baokun Qi, Zhongjiang Wang, Yang Li, Xiaonan Sui, and Lianzhou Jiang. 2018. "Changes in antioxidant activity of alcalase-hydrolyzed soybean hydrolysate under simulated gastrointestinal digestion and transepithelial transport." *Journal of Functional Foods* 42:298–305. https://doi.org/10.1016/j.jff.2018.01.017

Zhang, Yueling, Wei Wei, Piping Lv, Lianyan Wang, and Guanghui Ma. 2011. "Preparation and evaluation of alginate-chitosan microspheres for oral delivery of insulin." *European Journal of Pharmaceutics and Biopharmaceutics* 77(1):11–19. https://doi.org/10.1016/j.ejpb.2010.09.016. https://doi.org/10.1016/j.ejpb.2010.09.016. PMid:20933083

Zheng, Qiuping, Daoshou Qiu, Xiaojin Liu, Lei Zhang, Shike Cai, and Xuewu Zhang. 2015. "Antiproliferative effect of Dendrobium catenatum Lindley polypeptides against human liver, gastric and breast cancer cell lines." *Food and Function* 6(5):1489–1495. https://doi.org/10.1039/C5FO00060B. PMid:25811957

Index

Taylor & Francis eBooks

www.taylorfrancis.com

A single destination for eBooks from Taylor & Francis with increased functionality and an improved user experience to meet the needs of our customers.

90,000+ eBooks of award-winning academic content in Humanities, Social Science, Science, Technology, Engineering, and Medical written by a global network of editors and authors.

TAYLOR & FRANCIS EBOOKS OFFERS:

A streamlined experience for our library customers

A single point of discovery for all of our eBook content

Improved search and discovery of content at both book and chapter level

REQUEST A FREE TRIAL
support@taylorfrancis.com